Advanced Engineering Mathematics

Second Edition

James A. Cochran
Washington State University

H. Clare Wiser
Washington State University

Bernard J. Rice
University of Dayton

Brooks/Cole Publishing Company
Monterey, California

Brooks/Cole Publishing Company
A Division of Wadsworth, Inc.

Printed in the United States of America

10 9 8 7 6 5 4 3 2 1

Library of Congress Cataloging-in-Publication Data
Cochran, James Alan.
Advanced engineering mathematics.
Includes bibliographies and index.
Rev. ed. of: Applied analysis for physicists and Engineers / Bernard J. Rice. 1972.
1. Mathematical analysis. I. Wiser, H. Clare, [date]. II. Rice, Bernard J. III. Rice, Bernard J. Applied analysis for physicists and engineers. IV. Title.
QA300.C58 1986 515 86-21536
ISBN 0-534-07008-6

Sponsoring Editor: *John Kimmel*
Editorial Assistant: *Maria Rosillo Alsadi*
Production Editor: *Penelope Sky*
Production Assistant: *Barbara Kimmel*
Manuscript Editor: *Linda L. Thompson*
Interior Design: *Jamie Sue Brooks*
Cover Design: *Albert Burkhardt*
Art Coordinator: *Lisa Torri*
Interior Illustration: *Perfect Plot*
Typesetting: *Asco Trade Typesetting Ltd., Hong Kong*
Printing and Binding: *Arcata Graphics/Fairfield, Fairfield, Pennsylvania*

Preface

This textbook began as a revision of *Applied Analysis for Physicists and Engineers* by Bernard J. Rice, published in 1972. An efficient and direct style gave the book considerable appeal for classroom use, and we have retained it. We brought to the project a number of special ideas and points-of-view engendered by our own experiences as educators and applied mathematicians. These ideas and an extensive review process led to a thorough rewriting of the entire original text, some substantive deletions and rearrangements, and the addition of a large amount of new material. One of the more important alterations is the array of new practical applications that have been incorporated both as worked examples and as exercises for the student. (In this sense the present work can be viewed as a more elementary companion to *Applied Mathematics: Principles, Techniques, and Applications* by J. A. Cochran.) A new introduction at the beginning of each chapter places the material that follows in appropriate perspective, and special reference lists are now included at the end of each chapter. These lists and the accompanying remarks provide instructors and more ambitious students with information regarding important theory and techniques that relate to and generalize the material, but are outside the scope of the text.

We have endeavored to produce a modern text for undergraduates that meets the need for a practical, brief presentation of applied analysis, one that naturally builds on an elementary calculus background. (A previous brief exposure to linear algebra and differential equations is useful, but not necessary, to those using this material.) We have sought the middle ground between a book in modeling and a formal theoretical treatise: our worked applications show how the theory may be *applied* to analyze the mathematical models that others have developed. Our personal experience, and feedback from scores of former students who have become professional engineers and scientists in their own right, convinces us that not only the techniques, but also the nature of the ideas and the style of reasoning in fundamental fields such as linear analysis and analytic function theory are of critical importance time and again in real-world problem solving.

We have leaned away from superficial treatment of a large number of areas toward significant coverage of a few basic ones. For example, complex analysis is discussed in considerable detail, while the vast area of numerical analysis is mentioned only as it naturally arises in relation to the fields we consider basic. We have tried to show the underlying unity in the intricate interplay between mathematics and science by stressing linearity, series representation, and, to a lesser extent, line and surface integrals. Specific recurring physical models provide other rich cohesive patterns. On the other hand, we are aware of the considerable variety that exists in modern applied analysis courses and have created flexibility by keeping many parts of the text as self-contained as possible. This has resulted in some repetition, considerable cross-referencing, and a built-in expectation that students can and will review material (particularly in Chapters 1 and 2) that may be skipped in individual courses. The following outline of chapter interdependence may be helpful as a rough guide to using the text in various educational settings.

A. Basic concepts of linear analysis (Chapter 1)
B. Ordinary differential equations (Chapters 2–6)
C. Partial differential equations, Fourier series, and special functions (Chapters 7–9)
D. Vector analysis (Chapters 10, 11)
E. Complex analysis (Chapters 12–16)

Unit A underlies most of the rest of the text.

Unit B branches after Chapter 3 into three independent chapters.

Unit C relies on the fundamental ideas of Chapters 2 and 3, while Chapter 9 is a natural extension of Chapter 6.

Unit D relies only on Unit A.

In Unit E, Chapters 14 and 15 rely on Chapter 11, while Chapter 16 refers to Chapters 5 and Chapter 8.

What we have, then, is a book of reasonable size (more than sufficient for a year-long course) on the essence of much of applied analysis: ordinary and partial differential equations, linear analysis, and complex variables. We have not overemphasized the theoretical aspects of the subject, but what we have said we have endeavored to say precisely. The approach is practical; the style is casual. We hope you will appreciate most of what we have done, and offer suggestions for improvement of the rest.

Acknowledgments

In addition to the splendid cooperation of the Brooks/Cole team, we would especially like to recognize Harvey Greenwald of California Polytechnic Institute, San Luis Obispo, who provided exceptionally thoughtful, in-depth critiques of a number of draft manuscripts. We thank reviewers: Ed Adams, University of North Dakota; Alfred Aeppli, University of Minnesota; J. Bustoz, Arizona State University; W.E. Conway, University of Arizona; C.L. Frenzen, Southern Methodist University; J.D. Moore, University of California at Santa Barbara;

Michael O'Flynn, San Jose State University; J.T. Scheick, Ohio State University; and Janice B. Walker, Xavier University. Shing-min Wang for his diligent reworking of all the exercises, Katherine Cochran for her cheerful help with the indexing, and Sharon Wiser for her insightful typing assistance have our grateful appreciation. Finally, special thanks go to John Kimmel of Brooks/Cole for his steadfast encouragement throughout the entire project.

J.A. Cochran
H.C. Wiser

Contents

Chapter 1

Vectors and Matrices

1.1 Introduction

In this introductory chapter we describe what is meant by mathematical objects called vectors and matrices. Many of the topics in this text could be considered in terms of vectors and matrices, even though they are not continuously stressed. For many of you, this will be familiar material. We include it since the basic concepts of vector or linear spaces and linearity are so fundamental to applied analysis. Indeed, these notions often provide important threads of unity to seemingly disconnected or unrelated topics. You should be cautioned, however; this is not a thorough discussion of linear algebra. We are content with a consideration of only those concepts and techniques needed in the remainder of the book. Initially we give the general setting, admittedly abstract, in which you should understand the concept of a vector. The specialized material appearing later softens this abstraction.

1.2 Geometric Vectors and Vector Spaces

You have, no doubt, already encountered a vector in the guise of an *arrow* in two- or three-dimensional space. We call this a *geometric vector* to distinguish it from the generalization. A geometric vector is defined as an entity having direction and length. Since "direction" and "length" are left as undefined terms, this elementary definition has its shortcomings. However, if you accept these terms in the sense of everyday Euclidean geometry, the properties of direction and length are self-evident, and consequently arrows or geometric vectors have intuitive meaning sufficient for many applications.

In what follows a geometric vector is usually represented by an arrow. You should realize that the vector is not this arrow but is the collection of *all* arrows which have the same direction and length as the representative arrow. The fact that any arrow may be replaced by a parallel arrow and still represent the same

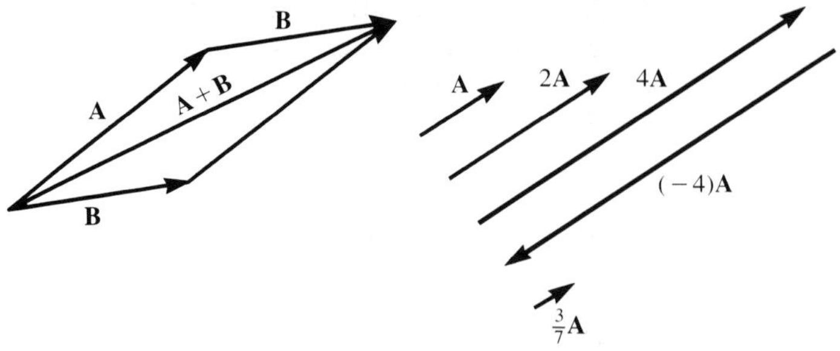

Figure 1.1 Addition and scalar multiplication of arrows

vector is expressed by saying that an arrow is "free," that is, free to move without changing direction or length.

Initially, two elementary operations are defined on the set of geometric vectors: addition and scalar multiplication. It is precisely these two operations and certain selected properties of these operations that allow a more precise definition of the idea of a vector. This more precise definition also yields an immediate generalization of the concept for use in more diverse applications.

Addition of two geometric vectors is defined in terms of representative arrows by placing the tail of the second arrow at the head of the first one. Then the sum of the two vectors is the vector represented by the arrow drawn from the tail of the first arrow to the head of the second arrow. Another way to look at this operation is to define $\mathbf{A} + \mathbf{B}$ as the diagonal of the parallelogram in Figure 1.1. This approach is often called the parallelogram law. Geometric vector addition has the following fundamental properties, which you should confirm by examining representative arrows:

1. The addition of two geometric vectors yields another geometric vector. Technically, we say that the set of geometric vectors is *closed* with respect to addition.
2. For any two geometric vectors $\mathbf{A}$ and $\mathbf{B}$, $\mathbf{A} + \mathbf{B} = \mathbf{B} + \mathbf{A}$ (commutative property).
3. For any three geometric vectors $\mathbf{A}$, $\mathbf{B}$, and $\mathbf{C}$, $(\mathbf{A} + \mathbf{B}) + \mathbf{C} = \mathbf{A} + (\mathbf{B} + \mathbf{C})$ (associative property).
4. We define a special geometric vector denoted by $\mathbf{0}$ with the property that for any other geometric vector $\mathbf{A}$, $\mathbf{A} + \mathbf{0} = \mathbf{A}$. In elementary work we think of $\mathbf{0}$ as an arrow with zero length. We leave the direction undefined.
5. Corresponding to each geometric vector $\mathbf{A}$, there is another geometric vector $\mathbf{B}$ such that $\mathbf{A} + \mathbf{B} = \mathbf{0}$. $\mathbf{B}$ is called the *additive inverse* of $\mathbf{A}$ and is denoted by $(-\mathbf{A})$. $(-\mathbf{A})$ has the same length as $\mathbf{A}$ but points in the opposite direction.

Multiplication of a geometric vector by a scalar is defined in terms of an operation by a real number on an arrow. For any real number c, $c\mathbf{A}$ is represented by an arrow $|c|$ times as long as that representing $\mathbf{A}$, pointing in the same direction if c is positive and in the opposite direction if c is negative. The following properties of scalar multiplication are fundamental:

6. Scalar multiplication is defined for each real scalar and each geometric vector and yields a geometric vector.
7. For any two real numbers b and c, $b(c\mathbf{A}) = (bc)\mathbf{A}$.
8. Scalar multiplication is distributive over vector addition:

$$a(\mathbf{A} + \mathbf{B}) = a\mathbf{A} + a\mathbf{B}$$

9. $1\mathbf{A} = \mathbf{A}$; $0\mathbf{A} = \mathbf{0}$; $(-1)\mathbf{A} = -\mathbf{A}$.

The set, or space, of geometric vectors is often called a "Euclidean vector space" and is a special case of the more general concept of an *abstract vector space*, or linear space. In general, any set of objects may be called a vector space if it has two operations, addition and scalar multiplication, which satisfy properties 1 through 9, or more precisely the more abstract statement of these properties obtained by leaving out the word "geometric" wherever it occurs. In their abstract form, properties 1 through 9 are called the *vector space axioms*. As we shall see later, a vector space may bear little or no resemblance to the set of geometric vectors other than satisfying these same basic properties.

1.3 Ordered n-tuples

By the set of ordered triples of real numbers is meant the set $\{(a_1, a_2, a_3),\ a_i$ a real number$\}$, where i indicates that this expression applies to each of the subscripts 1, 2, and 3. This set of ordered triples becomes a vector space when an appropriate addition and scalar multiplication are specified.

Addition of ordered triples is defined by component addition, that is,

$$(a_1, a_2, a_3) + (b_1, b_2, b_3) = (a_1 + b_1, a_2 + b_2, a_3 + b_3)$$

Likewise, multiplication by a real number is defined by

$$c(a_1, a_2, a_3) = (ca_1, ca_2, ca_3)$$

It is an easy task to show that the nine essential properties are satisfied, and hence the elements of the space are entitled to be called vectors.

The zero vector in this space is the ordered triple (0, 0, 0). An ordered triple of numbers has no natural length or direction, although in a geometric setting it is desirable and convenient to associate with each vector (a_1, a_2, a_3) a *length* defined by $(a_1^2 + a_2^2 + a_3^2)^{1/2}$.

One of the best-known applications of the vector space of ordered triples is its use in describing three-dimensional Euclidean space analytically. We give geometric meaning to an ordered triple (a_1, a_2, a_3) by allowing it to represent the point with Cartesian coordinates a_1, a_2, and a_3. The arrow from the origin to this point can be represented by the same ordered triple as the one representing the point (see Figure 1.2).

This identification of arrows with ordered triples is not a perfect match—only arrows with tails at the origin are involved. Also, the particular coordinate system used influences this analytic description of geometric vectors. For the Cartesian coordinate system and elementary applications encountered in cal-

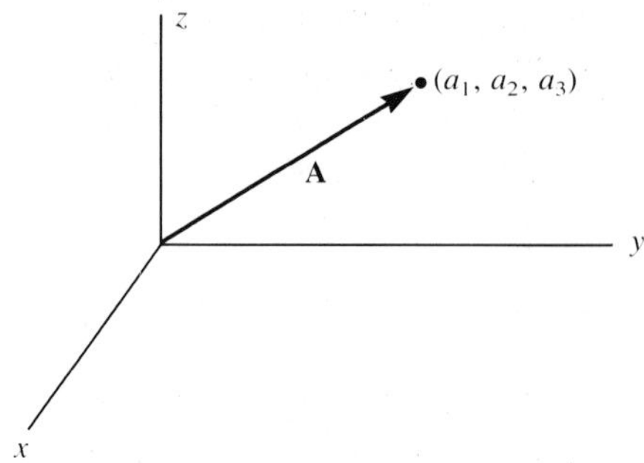

Figure 1.2 Representation of arrows with ordered triples

culus, these problems create no real difficulty, so arrows and ordered triples may be used almost interchangeably. Careful reexamination of the process is required when other coordinate systems are used or for certain applications, such as those involving couples of vectors in mechanics.

It is very common to abbreviate $(1, 0, 0)$ as $\mathbf{i}$, $(0, 1, 0)$ as $\mathbf{j}$ and $(0, 0, 1)$ as $\mathbf{k}$. Using this notation, the vector $\mathbf{A} = (a_1, a_2, a_3)$ is represented in the form $a_1\mathbf{i} + a_2\mathbf{j} + a_3\mathbf{k}$. The vectors $\mathbf{i}$, $\mathbf{j}$, and $\mathbf{k}$ are each *unit vectors* (that is, each has length one), and $\{\mathbf{i}, \mathbf{j}, \mathbf{k}\}$ forms a right-handed coordinate system (if the fingers of your right hand point from the head of $\mathbf{i}$ to the head of $\mathbf{j}$, your thumb will point in the direction of $\mathbf{k}$).

We can describe an ordered 4-tuple, 5-tuple, or n-tuple exactly as we have described ordered triples. For example, an ordered 5-tuple looks like $(a_1, a_2, a_3, a_4, a_5)$ with vector operations

$$(a_1, a_2, a_3, a_4, a_5) + (b_1, b_2, b_3, b_4, b_5) = (a_1 + b_1, a_2 + b_2, a_3 + b_3, a_4 + b_4, a_5 + b_5)$$

$$c(a_1, a_2, a_3, a_4, a_5) = (ca_1, ca_2, ca_3, ca_4, ca_5)$$

Similarly, the set of ordered n-tuples has two operations defined as natural extensions of the operations on ordered triples. As in the case of ordered triples, the operations satisfy the vector space axioms, and thus the n-tuple space is justifiably called a vector space.

Exercises for Sections 1.1–1.3

1. For the vectors $\mathbf{A} = (2, -1, 3)$ and $\mathbf{B} = (-1, 1, 4)$, determine the following:
(a) $\mathbf{A} + \mathbf{B}$ **(b)** $\mathbf{B} - \mathbf{A}$
(c) $3\mathbf{A}$ **(d)** the length of the vector $\mathbf{B}$

2. For the vectors $\mathbf{A} = (-1, 2, -3)$ and $\mathbf{B} = (0, -3, 2)$, find the following:
(a) $2\mathbf{A}$ **(b)** $3\mathbf{B} - 2\mathbf{A}$
(c) $(\mathbf{A} + \mathbf{B})/2$ **(d)** the length of the vector $\mathbf{A}$

3. Define two vectors $\mathbf{A}$ and $\mathbf{B}$ by $\mathbf{A} = (-2, -1, 3)$ and $\mathbf{B} = (3, 0, 1)$. Represent the following vectors in the form $a\mathbf{i} + b\mathbf{j} + c\mathbf{k}$:
(a) $2\mathbf{B} - \mathbf{A}$ **(b)** $3\mathbf{A} + \mathbf{B}$ **(c)** $-2\mathbf{A} - 3\mathbf{B}$

4. Define two vectors **A** and **B** by $\mathbf{A} = (3, 1, -2)$ and $\mathbf{B} = (1, -4, 1)$. Represent the following vectors in the form $a\mathbf{i} + b\mathbf{j} + c\mathbf{k}$:
   (a) $\mathbf{A} - 3\mathbf{B}$ (b) $\mathbf{B} + 4\mathbf{A}$ (c) $3\mathbf{A} + 6\mathbf{B}$
5. In three-dimensional Euclidean space, find the length of the vector **A** from the point $(1, -1, 3)$ to the midpoint of the line segment that joins the origin to the point $(6, -6, 4)$.
6. Give conditions on three vectors **A**, **B**, and **C** that ensure they can be brought together to form a triangle.
7. Show that the vectors $\mathbf{A} = \mathbf{i} - \mathbf{j}$, $\mathbf{B} = -\mathbf{i} + \mathbf{k}$, and $\mathbf{C} = -2\mathbf{i} + \mathbf{j} + \mathbf{k}$ can be brought together to form the sides of a triangle.
8. Show that the vectors $\mathbf{A} = 3\mathbf{i} - 4\mathbf{j} - 4\mathbf{k}$, $\mathbf{B} = 2\mathbf{i} - \mathbf{j} + \mathbf{k}$, and $\mathbf{C} = \mathbf{i} - 3\mathbf{j} - 5\mathbf{k}$ can be brought together to form the sides of a right triangle.
9. Graphically and algebraically, demonstrate how to find two vectors **A** and **B** if their sum **S** and difference **D** are known.
10. Assume that two nonzero vectors **A** and **B** issue from a common point. Find a vector **C** that bisects the angle formed by **A** and **B**.
11. Assume that the geometric vectors **A** and **B** extend from the origin in three-space to the points P and Q, respectively. Determine the vector **C**, that extends from the origin to the point R dividing the line segment from P to Q in the ratio m to n.
12. Show that the set of ordered triples satisfies the nine essential properties and, thus, this set is indeed a vector space.

1.4 Piecewise Continuous Functions

We now describe two important spaces in which the elements can properly be considered to be vectors even though they do not resemble geometric vectors.

Denote by $C(a, b)$ the set of continuous functions on the interval (a, b) and by $PC(a, b)$ the set of piecewise continuous functions on that interval.* Examples of graphs of members of $C(a, b)$ are shown in Figure 1.3, and some piecewise continuous functions are shown in Figure 1.4. Of course, all those shown in Figure 1.3 are also piecewise continuous, since $C(a, b)$ is contained in $PC(a, b)$.

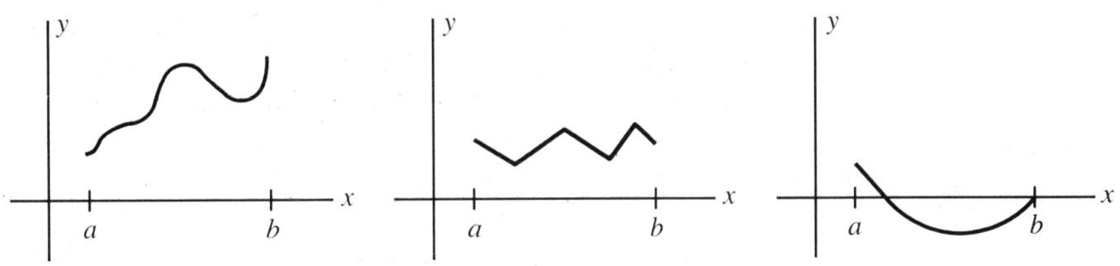

Figure 1.3 *Members of C(a, b)*

*A function f is *piecewise continuous* on the bounded interval (a, b) if it is defined at every point on the interval and the interval can be subdivided into finitely many subintervals, inside each of which the function is continuous and such that the limits from the right and from the left exist and are finite at the endpoints of each subinterval. A function on an unbounded interval (a, ∞), $(-\infty, b)$, or $(-\infty, \infty)$ is called piecewise continuous if it is piecewise continuous, by the previous definition, on each bounded subinterval of the given interval.

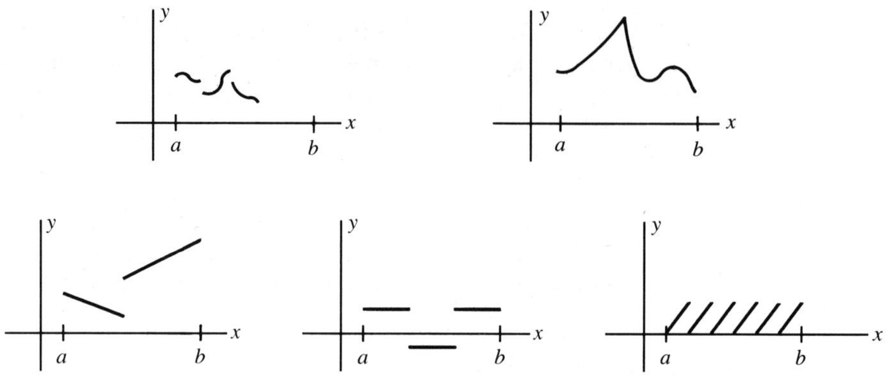

Figure 1.4 Members of $PC(a, b)$

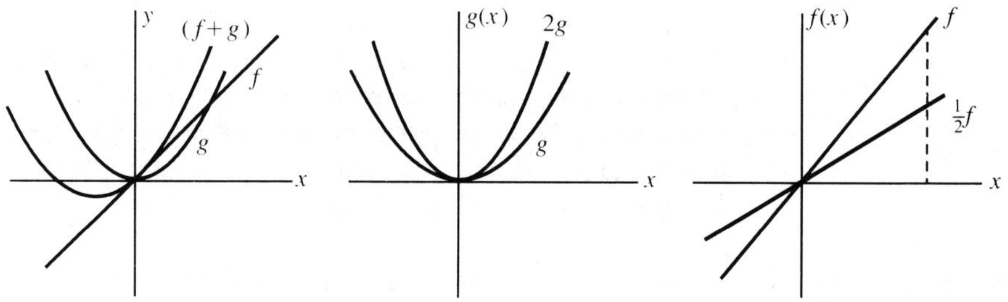

Figure 1.5

On $C(a, b)$ and $PC(a, b)$ we define operations of addition and multiplication by a scalar:

1. By $f + g$ is meant the function whose value at x is given by $f(x) + g(x)$.
2. By cf is meant the function whose value at x is given by $cf(x)$.

Example 1.1 Let $f(x) = x$ and $g(x) = x^2$. Find and sketch the functions $f + g$, $\frac{1}{2}f$, and $2g$.

Solution From the definitions,

$$(f + g)(x) = f(x) + g(x) = x + x^2$$

$$(\tfrac{1}{2}f)(x) = \tfrac{1}{2}f(x) = \frac{x}{2}$$

$$(2g)(x) = 2g(x) = 2x^2$$

The graphs are sketched in Figure 1.5. □

The fact that the nine fundamental properties are satisfied for the sets $C(a, b)$ and $PC(a, b)$, with the operations of addition and scalar multiplication as defined in the preceding paragraph, can be readily proved. We therefore say that $C(a, b)$ and $PC(a, b)$, together with these operations, form vector spaces. In Chapter 7 you will see why it is convenient to think of $PC(a, b)$ as a vector space. At present, you should at least realize that the term *vector* is often used in a much more general sense than geometric vectors in three-dimensional space. In passing we note that function spaces such as $C[a, b]$ and $PC(a, b]$ may be defined over closed intervals $[a, b]$ and half-open intervals $[a, b)$ and $(a, b]$ in exactly the same manner as we have defined $C(a, b)$ and $PC(a, b)$.

Some vector spaces consist of elements to which we do not care to associate the notion of length and direction. Vector spaces in which a length can be assigned to each element are called *normed*. We will consider only those normed vector spaces in which there is an *inner product* defined.

1.5 Inner Product Spaces

In the geometric vector space with which you are most familiar from elementary calculus, the four most common operations are (1) addition, (2) multiplication by a scalar, (3) dot product, and (4) cross product. Up to now we have generalized the first two operations by an abstraction, thus obtaining the notion of an abstract vector space. Now we shall consider those vector spaces on which a generalization of the dot product is defined, obtaining what is called an *inner product* space. (The notion of cross product does not generalize easily; we will consider it only in three-dimensional Euclidean space.)

Definition 1.1 An *inner product* is a rule that associates with each two vectors $\mathbf{v}_1$, $\mathbf{v}_2$ a real number, written $\mathbf{v}_1 \cdot \mathbf{v}_2$ or $(\mathbf{v}_1, \mathbf{v}_2)$ subject to the following axioms:

1. For all $\mathbf{v}_1$, $\mathbf{v}_2$, $\mathbf{v}_3$ in the vector space and for all real numbers a and b,

$$(a\mathbf{v}_1 + b\mathbf{v}_2)\cdot(\mathbf{v}_3) = a(\mathbf{v}_1 \cdot \mathbf{v}_3) + b(\mathbf{v}_2 \cdot \mathbf{v}_3)$$

This is known as the linearity property.

2. For all $\mathbf{v}_1$, $\mathbf{v}_2$ in the vector space,

$$\mathbf{v}_1 \cdot \mathbf{v}_2 = \mathbf{v}_2 \cdot \mathbf{v}_1$$

This property is variously called the symmetric property or commutative property.

3. For each vector $\mathbf{v}$,

$$\mathbf{v} \cdot \mathbf{v} \geqq 0$$

and $\mathbf{v} \cdot \mathbf{v} = 0$ if and only if $\mathbf{v}$ is the zero vector.
This is called the positive definiteness property.

Example 1.2 In the space of Euclidean geometric vectors, the *dot product* $\mathbf{A} \cdot \mathbf{B}$ is defined by

$$\mathbf{A} \cdot \mathbf{B} = |\mathbf{A}||\mathbf{B}| \cos\theta$$

where $|\mathbf{A}|$ and $|\mathbf{B}|$ are the lengths of $\mathbf{A}$ and $\mathbf{B}$, respectively, and θ is the smallest angle between two arrows representing $\mathbf{A}$ and $\mathbf{B}$ that are chosen with common initial points. Show that in the space of two-dimensional geometric vectors, the dot product is an inner product.

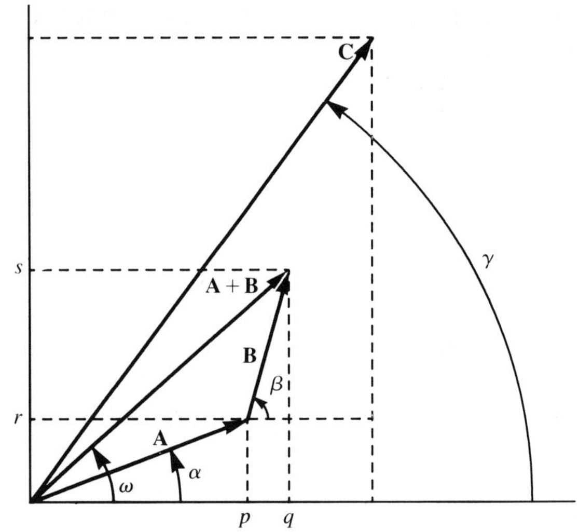

Figure 1.6

Solution

1. Referring to Figure 1.6 we see that

$$|\mathbf{A}+\mathbf{B}|\cos\omega = q = p + (q-p) = |\mathbf{A}|\cos\alpha + |\mathbf{B}|\cos\beta$$
$$|\mathbf{A}+\mathbf{B}|\sin\omega = s = r + (s-r) = |\mathbf{A}|\sin\alpha + |\mathbf{B}|\sin\beta$$

Using these relationships, the definition of dot product, and a simple trigonometric identity, we can verify the distributive property

$$(\mathbf{A}+\mathbf{B})\cdot\mathbf{C} = \mathbf{A}\cdot\mathbf{C} + \mathbf{B}\cdot\mathbf{C}$$

as follows:

$$\begin{aligned}
&(\mathbf{A}+\mathbf{B})\cdot\mathbf{C}\\
&\quad = |\mathbf{A}+\mathbf{B}||\mathbf{C}|\cos(\gamma-\omega) = |\mathbf{A}+\mathbf{B}||\mathbf{C}|(\cos\gamma\cos\omega + \sin\gamma\sin\omega)\\
&\quad = (|\mathbf{A}|\cos\alpha + |\mathbf{B}|\cos\beta)|\mathbf{C}|\cos\gamma + (|\mathbf{A}|\sin\alpha + |\mathbf{B}|\sin\beta)|\mathbf{C}|\sin\gamma\\
&\quad = |\mathbf{A}||\mathbf{C}|(\cos\gamma\cos\alpha + \sin\gamma\sin\alpha) + |\mathbf{B}||\mathbf{C}|(\cos\gamma\cos\beta + \sin\gamma\sin\beta)\\
&\quad = |\mathbf{A}||\mathbf{C}|\cos(\gamma-\alpha) + |\mathbf{B}||\mathbf{C}|\cos(\gamma-\beta) = \mathbf{A}\cdot\mathbf{C} + \mathbf{B}\cdot\mathbf{C}
\end{aligned}$$

We also have

$$(c\mathbf{A})\cdot\mathbf{B} = c|\mathbf{A}||\mathbf{B}|\cos\theta = c(\mathbf{A}\cdot\mathbf{B}) \qquad \text{for } c > 0$$

and

$$(c\mathbf{A})\cdot\mathbf{B} = |c||\mathbf{A}||\mathbf{B}|\cos(\pi-\theta) = c(\mathbf{A}\cdot\mathbf{B}) \qquad \text{for } c < 0$$

which, together with the distributive property, prove the linearity property.

2. $\mathbf{A}\cdot\mathbf{B} = |\mathbf{A}||\mathbf{B}|\cos\theta = |\mathbf{B}||\mathbf{A}|\cos\theta = \mathbf{B}\cdot\mathbf{A}$
3. $\mathbf{A}\cdot\mathbf{A} = |\mathbf{A}|^2$, which is obviously nonzero unless $\mathbf{A} = \mathbf{0}$. □

Example 1.3 Show how to define an inner product on the space of ordered n-tuples.

Solution Let $(a_1, a_2, \ldots, a_n)$ and $(b_1, b_2, \ldots, b_n)$ be any two elements of the space. Then the product defined by

$$(a_1, a_2, \ldots, a_n) \cdot (b_1, b_2, \ldots, b_n) = a_1 b_1 + a_2 b_2 + \cdots + a_n b_n$$

can be verified to be an inner product in a straightforward manner. (See Exercise 20 at the end of this section.) □

This inner product for ordered triples and the standard dot product for three-dimensional geometric vectors (defined in terms of lengths and angles) can be shown to be equivalent when the usual identification of arrows with triples is made. The vector space of piecewise continuous functions is also an inner product space.

Definition 1.2 Let f and g be any two elements of $PC[a, b]$. The *integral inner product*, denoted by (f, g), is defined to be

$$(f, g) = \int_a^b f(x)g(x)\,dx$$

The properties of the inner product are easy to verify.

Example 1.4 Calculate (f, g) in $PC[0, 1]$ and then in $PC[0, 2]$ if $f(x) = x$ and $g(x) = x^2$.

Solution In $PC[0, 1]$,

$$(f, g) = \int_0^1 (x)(x^2)\,dx = \frac{1}{4}$$

In $PC[0, 2]$,

$$(f, g) = \int_0^2 (x)(x^2)\,dx = 4$$

□

Once an inner product is defined on a vector space, a generalized length, the norm, is defined automatically.

Definition 1.3 The *norm* of a vector $\mathbf{v}$ is denoted by $\|\mathbf{v}\|$ and is defined by

$$\|\mathbf{v}\| = \sqrt{\mathbf{v} \cdot \mathbf{v}}$$

In a geometric vector space, $\|\mathbf{A}\| = \sqrt{\mathbf{A} \cdot \mathbf{A}} = |\mathbf{A}|$, which is just the length of $\mathbf{A}$. In the vector space of 3-tuples,

$$\|(a_1, a_2, a_3)\| = \sqrt{a_1^2 + a_2^2 + a_3^2}$$

which can be interpreted as the distance from the origin to the point with coordinates (a_1, a_2, a_3). In the function space $PC[a, b]$,

$$\|f\| = \left\{ \int_a^b [f(x)]^2\, dx \right\}^{1/2}$$

In the case where no confusion with absolute value arises, we may write $|\mathbf{v}|$ instead of $\|\mathbf{v}\|$. Thus for three-dimensional geometric vectors, we write $|\mathbf{A}|$ for its length, but for a function f, we write $\|f\|$ to distinguish it from $|f(x)|$.

If an inner product is defined, we not only can define the norm (which is the obvious generalization of the concept of length) but we can also make the important definition of orthogonality (which is a generalization of the concept of perpendicularity).

Definition 1.4 Two vectors $\mathbf{v}_1$ and $\mathbf{v}_2$ are said to be *orthogonal* if and only if $\mathbf{v}_1 \cdot \mathbf{v}_2 = 0$. A set of vectors is an *orthogonal set* if each two distinct vectors of the set are orthogonal.

In the case of the space of geometric vectors, if the dot product $\mathbf{A} \cdot \mathbf{B}$ of two nonzero arrows is zero, then $|\mathbf{A}||\mathbf{B}|\cos\theta = 0$, which would imply that $\theta = 90°$. Hence, in this case, orthogonality would reduce to the usual geometric concept of perpendicularity.

In the space of ordered triples, if the dot product of (a_1, a_2, a_3) and (b_1, b_2, b_3) is zero, then $a_1 b_1 + a_2 b_2 + a_3 b_3 = 0$. When ordered triples are used to describe geometric vectors analytically, this is a very convenient check for perpendicularity.

In three-dimensional analytic geometry, a plane may be defined as a surface passing through a given point such that the direction of its normal does not vary. Here, a normal to a surface at a given point is a vector which is orthogonal to each vector tangent to the surface at this point. The use of the dot product to describe this is a nice application. Let (a_1, a_2, a_3) be a fixed point through which the plane passes and let the vector $\mathbf{N} = n_1\mathbf{i} + n_2\mathbf{j} + n_3\mathbf{k}$ be normal to the plane (see Figure 1.7). Let $\mathbf{A}$ be a "position vector" to (a_1, a_2, a_3), that is, a vector from $(0, 0, 0)$ to (a_1, a_2, a_3), and let $\mathbf{R}$ be a position vector to any other point (x, y, z) on the plane. Then, the vector $\mathbf{R} - \mathbf{A}$ is in the plane, and hence the angle between $\mathbf{R} - \mathbf{A}$ and $\mathbf{N}$ is 90°. Therefore, $(\mathbf{R} - \mathbf{A}) \cdot \mathbf{N} = 0$ is the equation of the plane.

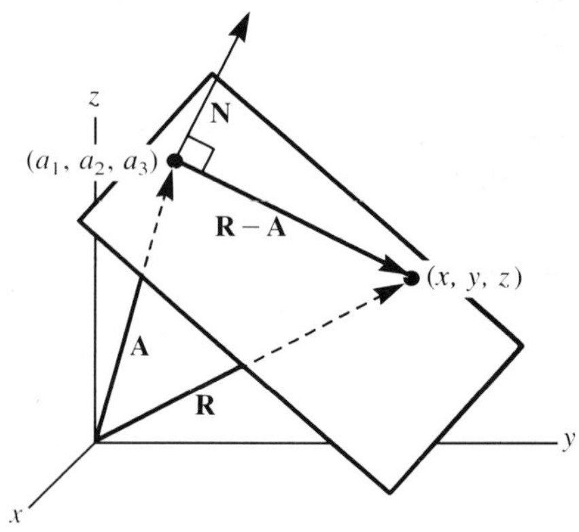

Figure 1.7

Example 1.5 Write the equation of the plane passing through $(3, -1, 4)$ to which the vector $-\mathbf{i} + \mathbf{j} + 2\mathbf{k}$ is normal.

Solution The vector $\mathbf{R}$ is given by $x\mathbf{i} + y\mathbf{j} + z\mathbf{k}$, and hence the equation of the plane is $[(x-3)\mathbf{i} + (y+1)\mathbf{j} + (z-4)\mathbf{k}]\cdot(-\mathbf{i} + \mathbf{j} + 2\mathbf{k}) = 0$. Simplifying,

$$-(x-3) + (y+1) + 2(z-4) = 0$$

or

$$-x + y + 2z = 4$$

□

In function spaces, you must be careful to specify the interval of orthogonality. It does happen that a pair of functions may be orthogonal on one interval and not on another.

Example 1.6 Show that the functions $\sin x$ and $\cos x$ are orthogonal on $[-\pi, \pi]$ but not on $[0, \pi/4]$.

Solution On $[-\pi, \pi]$,

$$(\sin x, \cos x) = \int_{-\pi}^{\pi} \sin x \cos x \, dx = 0$$

On $[0, \pi/4]$,

$$(\sin x, \cos x) = \int_{0}^{\pi/4} \sin x \cos x \, dx = \frac{1}{4} \neq 0$$

□

Do not read anything "physical" into the fact that $\sin x$ and $\cos x$ are orthogonal on $[-\pi, \pi]$. This does not mean, for example, that the graphs of the two functions intersect at right angles. Indeed, it is easy to show that they do not.

It is often necessary to divide a nonzero geometric vector $\mathbf{A}$ by its length to obtain a so-called unit vector $\mathbf{A}/|\mathbf{A}|$. In a more general setting, we want to divide a vector $\mathbf{v}$ by its norm to obtain $\mathbf{v}/\|\mathbf{v}\|$. Such a vector is said to be *normalized.* If we normalize each vector of an orthogonal set we call the new set *orthonormal.* An example of such a set is $\{\mathbf{i}, \mathbf{j}, \mathbf{k}\}$ in the space of three-dimensional geometric vectors. Why this should be a desirable quality of a set of vectors will become clearer in the next section.

Example 1.7 Show that the set of functions $\{\cos mx\}_{m=0}^{\infty}$ is an orthogonal set on $[0, \pi]$. Find the corresponding orthonormal set.

Solution If $m \neq n$,

$$\begin{aligned}(\cos mx, \cos nx) &= \int_0^{\pi} \cos mx \cos nx \, dx \\ &= \frac{1}{2}\int_0^{\pi} [\cos(m+n)x + \cos(m-n)x] \, dx \\ &= \left[\frac{\sin(m+n)x}{2(m+n)} + \frac{\sin(m-n)x}{2(m-n)}\right]\Bigg|_0^{\pi} = 0\end{aligned}$$

which proves the orthogonality.

For $m = 0$, $\cos 0x = 1$ and

$$\|1\| = \left[\int_0^\pi (1)^2\,dx\right]^{1/2} = \sqrt{\pi}$$

For $m \neq 0$,

$$\|\cos mx\| = \left[\int_0^\pi \cos^2 mx\,dx\right]^{1/2} = \sqrt{\pi/2}$$

Hence, the corresponding orthonormal set is

$$\left\{\frac{1}{\sqrt{\pi}}, \frac{\sqrt{2}}{\sqrt{\pi}}\cos x, \frac{\sqrt{2}}{\sqrt{\pi}}\cos 2x, \ldots, \frac{\sqrt{2}}{\sqrt{\pi}}\cos mx, \ldots\right\}$$ □

The norm satisfies the important *triangle inequality*

$$\|\mathbf{v}_1 + \mathbf{v}_2\| \leqq \|\mathbf{v}_1\| + \|\mathbf{v}_2\|$$

and therefore the distance between two vectors satisfies

$$\|\mathbf{v}_1 - \mathbf{v}_2\| \leqq \|\mathbf{v}_1 - \mathbf{v}_3\| + \|\mathbf{v}_3 - \mathbf{v}_2\|$$

It is this inequality that tells us that the norm acts in much the same manner as our intuitive notion of length.

Example 1.8 Find the distance between x and x^2 first in $PC[0, 1]$ and then in $PC[0, 2]$.

Solution In $PC[0, 1]$,

$$\|x - x^2\| = \left[\int_0^1 (x - x^2)^2\,dx\right]^{1/2} = \sqrt{\frac{1}{30}}$$

In $PC[0, 2]$,

$$\|x - x^2\| = \left[\int_0^2 (x - x^2)^2\,dx\right]^{1/2} = \frac{4}{\sqrt{15}}$$ □

In the first five sections of this chapter we have attempted to show how the specific concrete notions of arrows and related concepts are generalized. The following list is intended to summarize the net effect of this generalization.

Specific	*Generalization*
Arrow	Vector
Dot product	Inner product
Length	Norm
Perpendicularity	Orthogonality
Unit arrow	Normalized vector
Set of perpendicular arrows	Orthogonal set
Set of unit perpendicular arrows	Orthonormal set

You should note carefully in each case the abstraction required to implement the generalization.

Exercises for Sections 1.4 and 1.5

In Exercises 1–2, let $\mathbf{A} = (-1, 3, 1)$ and $\mathbf{B} = (2, 4, -3)$. Determine the specified quantities.

1. **(a)** $|\mathbf{B}|$ **(b)** $\mathbf{A} \cdot \mathbf{B}$ **(c)** the angle between **A** and **B**
(d) the unit vector in the direction of **A**

2. **(a)** $|\mathbf{B} - \mathbf{A}|$. What does this quantity represent geometrically?
(b) The set of all vectors orthogonal to **A**. What does this set represent geometrically in three-space?
(c) The set of all vectors orthogonal to both **A** and **B**.

3. Which of the following sets are orthogonal? For each set that is orthogonal, find the corresponding orthonormal set.
(a) $\{(1, 3), (6, -2)\}$
(b) $\{(1, 2, 2), (-2, 2, -1), (2, 1, -2)\}$
(c) $\{\mathbf{i} - 2\mathbf{j} + 3\mathbf{k}, 2\mathbf{i} + \frac{1}{2}\mathbf{j} - \frac{1}{3}\mathbf{k}, 3\mathbf{i} + 3\mathbf{j} + \mathbf{k}\}$

4. In elementary physics we learn that the *work* done by a constant force **F** acting on a particle while it moves in a straight line a distance and direction indicated by the vector **X** is given by $\mathbf{F} \cdot \mathbf{X}$. Using ergs, dynes, and centimeters as units, find each of the following:
(a) The work done by a force $(2, 3, -7)$ moving a particle along the path from the origin to the point $(1, -2, -1)$.
(b) The work done by a force of magnitude 10 dyn acting at an angle of 30° from the direction of a particle which moves 6 cm.
(c) The direction(s) of action in the xy-plane of a force of magnitude 5 dyn, which does 12 ergs of work while acting upon a particle which moves along a path represented by $(3, 3)$.

5. Determine each of the following:
(a) The work done by a constant force $\mathbf{F} = \mathbf{i} + \mathbf{j}$ when a particle moves from the point $(0, 0)$ to $(1, 0)$.
(b) The work done by a constant force $\mathbf{F} = 3\mathbf{i} - 2\mathbf{j} + \mathbf{k}$ in moving a particle from the point $(-1, 0, 2)$ to $(3, -2, 1)$.

We saw earlier that in three-space, the equation of the plane that passes through the point (a_1, a_2, a_3) and is perpendicular to the vector **N** is given by $(\mathbf{R} - \mathbf{A}) \cdot \mathbf{N} = 0$. Here **A** is the position vector to (a_1, a_2, a_3) and **R** is the position vector to any other point (x, y, z) in the plane. In Exercises 6–8, describe the geometric figures represented by the given equations:

6. $(\mathbf{R} - \mathbf{A}) \cdot \mathbf{A} = 0$

7. $(\mathbf{R} - \mathbf{A}) \cdot \mathbf{R} = 0$

8. $(\mathbf{R} - \mathbf{A}) \cdot (\mathbf{R} - \mathbf{A}) = 0$

9. Find the equation of the plane passing through the point $(5, -1, 3)$ if a vector normal to the plane is $2\mathbf{i} + \mathbf{j} - \mathbf{k}$.

10. Show that taking the dot product of two geometric vectors or of two ordered triples is a commutative operation.

11. Assume that **A**, **B**, **C** are vectors in three-space. Show that

$$|\mathbf{A} - \mathbf{B}| \leq |\mathbf{A} - \mathbf{C}| + |\mathbf{C} - \mathbf{B}|$$

(*Hint*: Apply the triangle inequality to $\mathbf{A} - \mathbf{C} + \mathbf{C} - \mathbf{B}$. This can be used to verify that the shortest distance between two points is indeed a straight line.)

12. If **A** and **B** are vectors in n-space, verify that

$$|\mathbf{A} + \mathbf{B}|^2 = |\mathbf{A}|^2 + 2\mathbf{A}\cdot\mathbf{B} + |\mathbf{B}|^2$$

13. Let **A** and **B** be nontrivial vectors in n-space and define $P(\lambda) \equiv |\mathbf{A} + \lambda\mathbf{B}|^2$.
(a) Verify that $P(\lambda)$ is a second-degree polynomial that is nonnegative for all real λ.
(b) Deduce that $|\mathbf{A}\cdot\mathbf{B}| \leq |\mathbf{A}||\mathbf{B}|$. This useful result is called the *Cauchy-Schwarz inequality*.

14. Repeat Exercise 13 for the inner product space $PC[a,b]$. In other words, if f and g are two functions in $PC[a,b]$, show that

$$|(f,g)| \leq \|f\|\,\|g\|$$

In this integrated form, the result is often called *Schwarz's inequality*.

15. Find the norm of the following functions (with respect to the integral inner product) when considered as elements of $PC[0,1]$:
(a) $f(x) = x$ **(b)** $f(x) = x^2$ **(c)** $f(x) = \sin \pi x$
(d) $f(x) = 1$ **(e)** $f(x) = \cos \pi x$

16. Find the norm of the functions of Exercise 15 when considered as elements of $PC[0,2]$.

17. The first six Legendre polynomials (see Section 9.6) are

$$P_0(x) = 1, \qquad P_1(x) = x, \qquad P_2(x) = \frac{3x^2 - 1}{2}, \qquad P_3(x) = \frac{5x^3 - 3x}{2},$$

$$P_4(x) = \frac{35x^4 - 30x^2 + 3}{8}, \qquad P_5(x) = \frac{63x^5 - 70x^3 + 15x}{8}.$$

Show that this is an orthogonal set in $PC[-1,1]$. Is the set orthonormal?

18. Show that the trigonometric set

$$\{\cos mx\}_{m=0}^{\infty} \cup \{\sin nx\}_{n=1}^{\infty}$$

is orthogonal on the interval $[-\pi, \pi]$.

19. Under what conditions is the norm of a vector equal to zero?

20. Complete Example 1.3.

21. Find constants $a_1, b_1, b_2, c_1, c_2, c_3$ such that

$$\{a_1, b_1 x + b_2, c_1 x^2 + c_2 x + c_3\}$$

is an orthonormal set in the space $PC[-1,1]$. Compare the members of the set with the first three Legendre polynomials given in Exercise 17.

In Exercises 22–24, find the "distance," in the sense of the norm, between the given functions:

22. x and x^2 considered as members of $PC[-1,0]$.

23. x and $\sin x$ considered as members of $PC[0,\pi]$.

24. x and $\sin x + \sin 3x$ considered as members of $PC[0,\pi]$.

25. Let $f(x) = 1$ on $(-1, 1)$ and $g(x) = -1$ on $(-1, 1)$. Compute:
(a) $\|f - g\|$ **(b)** $\|f\| - \|g\|$

26. Verify that $PC[-1, 1]$ possesses the nine properties necessary for it to be a vector space.

1.6 Cross Products and Triple Products

We return now to the notion of geometric vectors to discuss the operation of the *cross product* of two vectors. This operation is not as basic as that of the dot product, but it will be useful to us in later work. Our application of this operation is restricted to three-space; it does not have simple generalizations to other dimensions or to general linear spaces.

Definition 1.5 **1.** In the space of Euclidean geometric vectors, $\mathbf{A} \times \mathbf{B}$ is a vector with magnitude $|\mathbf{A}||\mathbf{B}| \sin\theta$, where θ is the angle from $\mathbf{A}$ to $\mathbf{B}$, and with the direction that a right-hand screw advances in turning through the angle θ. The cross product is $\mathbf{0}$ and thus has no unique direction in case $\mathbf{A} = \mathbf{0}$, $\mathbf{B} = \mathbf{0}$, or $\theta = 0$ or π (see Figure 1.8).

2. In the space of Cartesian ordered triples,

$$(a_1, a_2, a_3) \times (b_1, b_2, b_3) = (a_2 b_3 - a_3 b_2, a_3 b_1 - a_1 b_3, a_1 b_2 - a_2 b_1)$$

The cross product is not an associative operation (consider $\mathbf{i} \times (\mathbf{i} \times \mathbf{j})$ and $(\mathbf{i} \times \mathbf{i}) \times \mathbf{j}$) and is anticommutative, that is, $\mathbf{A} \times \mathbf{B} = -\mathbf{B} \times \mathbf{A}$. However, like the dot product, the cross product is distributive over addition.

Example 1.9 Express $|\mathbf{A} \times \mathbf{B}|$ in terms of dot products.

Solution

$$\begin{aligned} |\mathbf{A} \times \mathbf{B}|^2 &= |\mathbf{A}|^2 |\mathbf{B}|^2 \sin^2\theta \\ &= |\mathbf{A}|^2 |\mathbf{B}|^2 (1 - \cos^2\theta) \\ &= |\mathbf{A}|^2 |\mathbf{B}|^2 - |\mathbf{A}|^2 |\mathbf{B}|^2 \cos^2\theta \\ &= (\mathbf{A} \cdot \mathbf{A})(\mathbf{B} \cdot \mathbf{B}) - (\mathbf{A} \cdot \mathbf{B})^2 \end{aligned}$$

so $|\mathbf{A} \times \mathbf{B}| = \sqrt{(\mathbf{A} \cdot \mathbf{A})(\mathbf{B} \cdot \mathbf{B}) - (\mathbf{A} \cdot \mathbf{B})^2}$ □

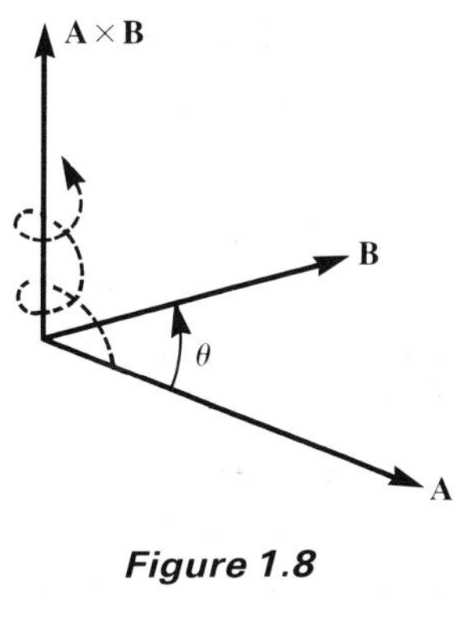

Figure 1.8

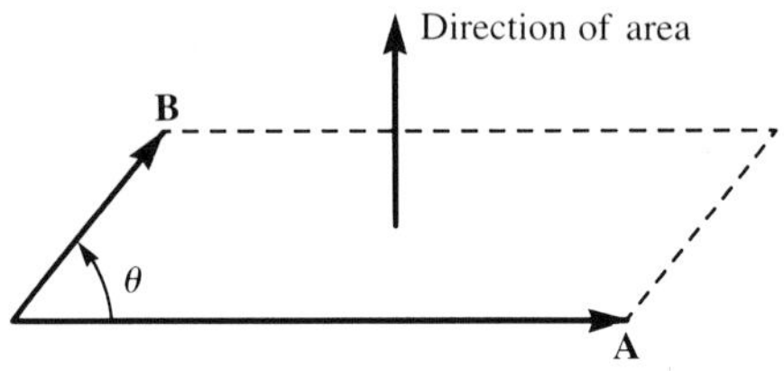

Figure 1.9 Directed area

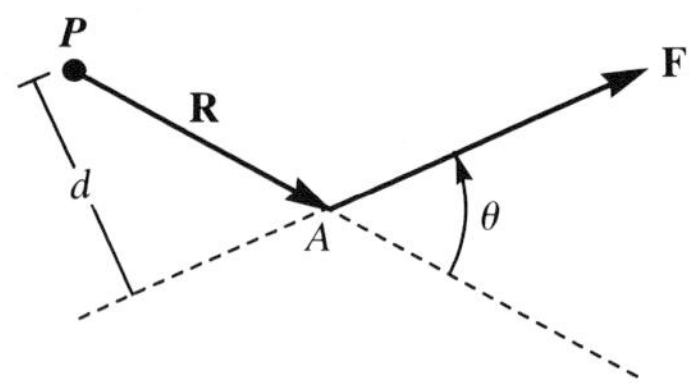

Figure 1.10 Vector moment

As with the dot product, the cross product may be used in some elementary physical definitions. For example, it sometimes is convenient to speak of a *directed area*, that is, an area represented by a vector. The area of a parallelogram with sides **A** and **B** is equal to $|\mathbf{A}||\mathbf{B}| \sin \theta$, which is $|\mathbf{A} \times \mathbf{B}|$. Hence, the directed area of the parallelogram is defined to be $\mathbf{A} \times \mathbf{B}$. (See Figure 1.9.)

In mechanics, the moment of a force **F** about a point P is defined to be $m = |\mathbf{F}| d$, where d is the distance of P from the line of action of the force. If A is any point on the line of force and **R** is a vector from P to A (see Figure 1.10), then $d = |\mathbf{R}| \sin \theta$ and therefore

$$m = |\mathbf{F}||\mathbf{R}| \sin \theta$$

or in terms of the cross product,

$$m = |\mathbf{F} \times \mathbf{R}| = |\mathbf{R} \times \mathbf{F}|$$

The *vector moment* is thus defined to be

$$\mathbf{M} = \mathbf{R} \times \mathbf{F}$$

Also, in mechanics we can analyze a rigid body rotating in three-space about an axis l with angular velocity $\mathbf{\Omega}$, where $\mathbf{\Omega}$ describes both the magnitude and, using the right-hand rule, the direction of rotation. It then turns out that the velocity **V** of a point P in the body, with position described by any vector **R** from the axis to the particle, is given (see Figure 1.11) by

$$\mathbf{V} = \mathbf{\Omega} \times \mathbf{R}$$

The dot and cross products may be combined in essentially two ways to give other meaningful quantities. (An expression such as $(\mathbf{A} \cdot \mathbf{B}) \times \mathbf{C}$ has no meaning since $\mathbf{A} \cdot \mathbf{B}$ is a real number and the cross product is defined only for two vectors.) Products such as $\mathbf{A} \cdot (\mathbf{B} \times \mathbf{C})$ and $\mathbf{A} \times (\mathbf{B} \times \mathbf{C})$ are called *triple*

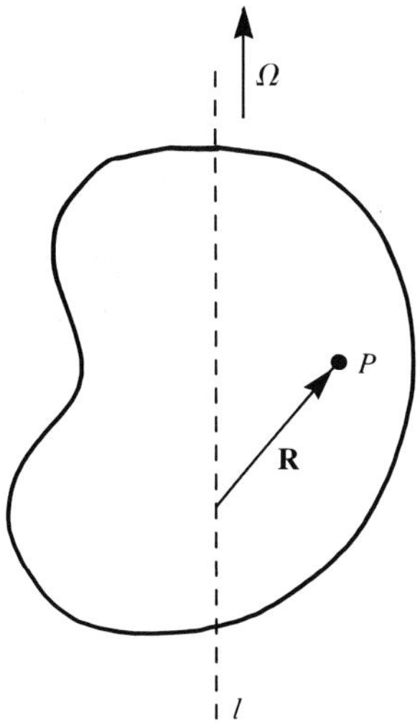

Figure 1.11

products. The first is called a *scalar triple product,* since a real number is obtained from it, and the second is called a *vector triple product,* since it yields a vector quantity.

Using the direction of the cross products involved, it is easy to see that the vector triple product $\mathbf{A} \times (\mathbf{B} \times \mathbf{C})$ is a vector that lies in the plane of $\mathbf{B}$ and $\mathbf{C}$. (All initial points are assumed to be at the origin.) Hence, if $\mathbf{B}$ and $\mathbf{C}$ do not point in the same direction, the quantity $\mathbf{A} \times (\mathbf{B} \times \mathbf{C})$ can be expressed in terms of the vectors $\mathbf{B}$ and $\mathbf{C}$ by the following formula:

$$\mathbf{A} \times (\mathbf{B} \times \mathbf{C}) = (\mathbf{A} \cdot \mathbf{C})\mathbf{B} - (\mathbf{A} \cdot \mathbf{B})\mathbf{C}$$

Exercises for Section 1.6

1. For $\mathbf{A} = (1, 2, -3)$, $\mathbf{B} = (2, -1, 1)$, and $\mathbf{C} = (4, 2, 2)$, calculate $\mathbf{A} \times \mathbf{B}$, $\mathbf{C} \times (\mathbf{A} \times \mathbf{B})$ and $\mathbf{C} \cdot (\mathbf{A} \times \mathbf{B})$.
2. Find the equation of the plane that contains the point $(1, 0, 1)$ and is parallel to the plane determined by the vectors $\mathbf{i} + \mathbf{j} - \mathbf{k}$ and $2\mathbf{i} + 3\mathbf{j} + 2\mathbf{k}$.
3. Find the area of the parallelogram with vertices having coordinates $(0, 0)$, $(3, 0)$, $(1, 2)$, and $(4, 2)$.
4. **(a)** Determine a unit vector perpendicular to the plane of the vectors $\mathbf{A} = \mathbf{i} + 3\mathbf{j} - \mathbf{k}$ and $\mathbf{B} = 2\mathbf{i} + \mathbf{j} + \mathbf{k}$.
   (b) Find the area of the triangle having these vectors as two of its sides.
5. Find the vector moment about the origin of the force $\mathbf{F} = \mathbf{i} + 3\mathbf{j} + \mathbf{k}$ applied at the point $(2, -1, 1)$.
6. Determine the scalar moment about the point $(3, -5, 4)$ of the force $\mathbf{F} = \mathbf{i} - 3\mathbf{j} + 2\mathbf{k}$ acting at the point $(1, 2, 1)$.

7. A rigid body rotates about the line $x = y = z$ with angular velocity $\mathbf{\Omega}$. Find the velocity of a particle at the point $(1, 2, 2)$.
8. Find $\mathbf{A} \times \mathbf{B}$ if $\mathbf{A} = 2\mathbf{e}_1 + 5\mathbf{e}_2 - \mathbf{e}_3$ and $\mathbf{B} = \mathbf{e}_1 - 2\mathbf{e}_2 - 4\mathbf{e}_3$, where $\mathbf{e}_1 \times \mathbf{e}_2 = \mathbf{i} - \mathbf{j}$, $\mathbf{e}_1 \times \mathbf{e}_3 = \mathbf{j} + \mathbf{k}$, and $\mathbf{e}_2 \times \mathbf{e}_3 = \mathbf{i} + \mathbf{k}$.
9. Find $\mathbf{A} \cdot \mathbf{B}$ if $\mathbf{A} = 3\mathbf{e}_1 + 5\mathbf{e}_2 - \mathbf{e}_3$ and $\mathbf{B} = \mathbf{e}_1 - \mathbf{e}_2 + \mathbf{e}_3$ if $\mathbf{e}_1 \cdot \mathbf{e}_1 = 2$, $\mathbf{e}_2 \cdot \mathbf{e}_2 = 3$, $\mathbf{e}_3 \cdot \mathbf{e}_3 = 5$, $\mathbf{e}_1 \cdot \mathbf{e}_2 = -1$, $\mathbf{e}_1 \cdot \mathbf{e}_3 = -3$, and $\mathbf{e}_2 \cdot \mathbf{e}_3 = 2$.
10. What is the scalar triple product if two of the vectors are equal?
11. Show that the scalar triple product may be interpreted as the volume of a parallelepiped with **A**, **B**, and **C** as adjacent edges.
12. Show that $\mathbf{A} \cdot (\mathbf{B} \times \mathbf{C}) = (\mathbf{A} \times \mathbf{B}) \cdot \mathbf{C}$. (This, together with the fact that $(\mathbf{A} \cdot \mathbf{B}) \times \mathbf{C}$ is meaningless, has led to the frequent use of the notation $[\mathbf{ABC}]$ for a scalar triple product.)
13. How are $\mathbf{A} \cdot (\mathbf{B} \times \mathbf{C})$ and $\mathbf{A} \cdot (\mathbf{C} \times \mathbf{B})$ related? How are $\mathbf{A} \cdot (\mathbf{B} \times \mathbf{C})$ and $\mathbf{B} \cdot (\mathbf{C} \times \mathbf{A})$ related ? Generalize.

1.7 Matrices and Determinants

As an example of how sets may be viewed as vector spaces, we now consider rectangular arrays of numbers, or, as they are technically called, *matrices*. The elements that are arranged horizontally in the array are called *rows* and those arranged vertically are called *columns*. For example, a three-row by four-column matrix is given by

$$\begin{pmatrix} a_{11} & a_{12} & a_{13} & a_{14} \\ a_{21} & a_{22} & a_{23} & a_{24} \\ a_{31} & a_{32} & a_{33} & a_{34} \end{pmatrix}$$

where each *entry* a_{ij} is a real number. The first number of the double subscript identifies the row of the matrix and the second number gives the column.

In general, a matrix with m rows and n columns is called an $m \times n$ matrix. If we wish to refer to the matrix itself as an entity, we write either **A** or (a_{ij}). Note the difference between the matrix (a_{ij}) and the typical numerical entry a_{ij}.

There are many applications of matrices ranging from use as a filing system to that of solving difficult systems of equations. Indeed, matrices naturally occur in almost every technical field of human endeavor ranging from astronomy to zoology. In some applications, there is little need of knowing more than the indexing scheme, whereas other uses of matrices depend upon being able to operate with them in an algebraic manner. It is in the latter case that the perspective of vector spaces becomes particularly important.

Definition 1.6 Two matrices (a_{ij}) and (b_{ij}) are said to be *equal* if they are of the same size—that is, they have the same number of rows and the same number of columns—and if $a_{ij} = b_{ij}$ for each i and j.

Definition 1.7 By the *sum* of two matrices of the same size, (a_{ij}) and (b_{ij}), is meant the matrix $(a_{ij} + b_{ij})$.

Example 1.10 Let

$$\mathbf{A} = \begin{pmatrix} 7 & 3 & 4 \\ -1 & 2 & 0 \\ 2 & -4 & 4 \\ 3 & 0 & -7 \end{pmatrix} \quad \text{and} \quad \mathbf{B} = \begin{pmatrix} 4 & 2 & -1 \\ 3 & -1 & 7 \\ 2 & 0 & 8 \\ 6 & 3 & -4 \end{pmatrix}$$

Find $\mathbf{A} + \mathbf{B}$.

Solution

$$\mathbf{A} + \mathbf{B} = \begin{pmatrix} 7+4 & 3+2 & 4-1 \\ -1+3 & 2-1 & 0+7 \\ 2+2 & -4+0 & 4+8 \\ 3+6 & 0+3 & -7-4 \end{pmatrix} = \begin{pmatrix} 11 & 5 & 3 \\ 2 & 1 & 7 \\ 4 & -4 & 12 \\ 9 & 3 & -11 \end{pmatrix} \qquad \square$$

It is very easy to show that addition of $m \times n$ matrices is closed, commutative, and associative. The additive identity is the $m \times n$ matrix $\mathbf{0}$ with each entry zero. The additive inverse of the matrix (a_{ij}) is the matrix $(-a_{ij})$.

Definition 1.8 By *scalar multiplication* of a matrix (a_{ij}) by the real number r is meant the matrix (ra_{ij}).

Example 1.11 Let $\mathbf{A} = \begin{pmatrix} 2 & 5 & 3 \\ -1 & 4 & 2 \end{pmatrix}$. Find $3\mathbf{A}$.

Solution

$$3\mathbf{A} = \begin{pmatrix} 3 \times 2 & 3 \times 5 & 3 \times 3 \\ -3 \times 1 & 3 \times 4 & 3 \times 2 \end{pmatrix} = \begin{pmatrix} 6 & 15 & 9 \\ -3 & 12 & 6 \end{pmatrix} \qquad \square$$

Properties 6–9 of Section 1.2 are easily verified for this operation. Consequently the set of $m \times n$ matrices forms a vector space with the two operations as given in Definitions 1.7 and 1.8. Note that there is a different vector space for each distinct pair of positive integers m and n. However, matrices are seldom called vectors except in the case of $m \times 1$ matrices or $1 \times n$ matrices, in which case the matrices are called *column vectors* and *row vectors*, respectively.

In addition to the vector space operations of addition and multiplication by a scalar, it is possible to *multiply* certain pairs of matrices. This operation is defined in such a manner that when we later consider matrices as representing linear transformations, matrix multiplication will correspond to composition of transformations. (See Exercise 6 in the next set of exercises.)

Definition 1.9 Let $\mathbf{A} = (a_{ik})$ and $\mathbf{B} = (b_{kj})$, where $\mathbf{A}$ is $m \times n$ and $\mathbf{B}$ is $n \times r$. By the *product* $\mathbf{AB}$ is meant the $m \times r$ matrix

$$\mathbf{C} = (c_{ij}) = \left(\sum_{k=1}^{n} a_{ik} b_{kj} \right)$$

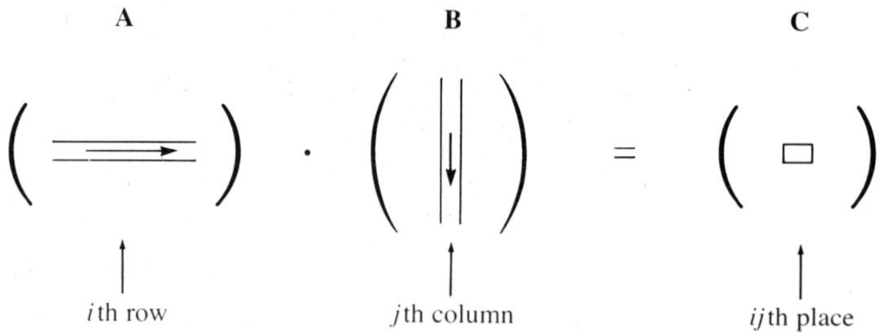

Figure 1.12 *Multiplication of matrices*

In words, the entry in the ijth place of the product matrix **C** is obtained by multiplying corresponding entries from the ith row of **A** and the jth column of **B** or, equivalently, by taking the dot product of the ith row vector of **A** and the jth column vector of **B** (see Figure 1.12).

Example 1.12 Find the product **AB** of the matrices

$$\mathbf{A} = \begin{pmatrix} 1 & 2 & -3 \\ 2 & 7 & 1 \end{pmatrix} \quad \text{and} \quad \mathbf{B} = \begin{pmatrix} 5 & 2 \\ 6 & 3 \\ -2 & 0 \end{pmatrix}$$

Solution The product matrix is 2×2 with entries

$$\begin{aligned} c_{11} &= (1)(5) + (2)(6) + (-3)(-2) \\ c_{12} &= (1)(2) + (2)(3) + (-3)(0) \\ c_{21} &= (2)(5) + (7)(6) + (1)(-2) \\ c_{22} &= (2)(2) + (7)(3) + (1)(0) \end{aligned}$$

Hence,

$$\mathbf{AB} = \begin{pmatrix} 23 & 8 \\ 50 & 25 \end{pmatrix}$$

Note that the product in reverse order is

$$\mathbf{BA} = \begin{pmatrix} 9 & 24 & -13 \\ 12 & 33 & -15 \\ -2 & -4 & 6 \end{pmatrix}$$ □

From the previous example we see that the product of matrices is *not* a commutative operation. In fact, the two products may be completely different types of matrices. Further, one of the products might not even exist since a product matrix is undefined if the number of columns of the first matrix is not equal to the number of rows of the second.

A *square* matrix has the same number of rows and columns. This number is called the *order* of the matrix. A familiar operator on square matrices, which is of great utility in the concise statement of mathematical ideas, is the *determinant*. The *determinant* of a square matrix **A**, denoted by det **A**, det(a_{ij}), or

$$\det\begin{pmatrix} a_{11}\dots a_{1n} \\ a_{21}\dots a_{2n} \\ \vdots \quad\quad \vdots \\ a_{n1}\dots a_{nn}\end{pmatrix}$$

is defined here in a recursive manner.

Definition 1.10 Let $\mathbf{A} = (a_{ij})$ be a square matrix of order n.

(a) If $n = 1$, then $\det \mathbf{A} = a_{11}$.
(b) If $n = 2$, then $\det \mathbf{A} = a_{11}a_{22} - a_{12}a_{21}$.
(c) For $n \geq 2$, let M_{ij} denote the $(n-1) \times (n-1)$ matrix obtained by deleting the ith row and jth column from $\mathbf{A}$. Furthermore, define the *cofactor* of the entry a_{ij} in $\mathbf{A}$ as the number

$$A_{ij} = (-1)^{i+j}\det(M_{ij})$$

Then

$$\det \mathbf{A} = \sum_{j=1}^{n} a_{ij}A_{ij}$$

This evaluation of $\det \mathbf{A}$ is called an expansion by elements of the ith row; a similar expression for $\det \mathbf{A}$,

$$\left(\sum_{i=1}^{n} a_{ij}A_{ij}\right)$$

is called an expansion by elements of the jth column.

The determinant has a number of useful algebraic properties:

1. If a square matrix $\mathbf{B}$ is formed from $\mathbf{A}$ by interchanging two rows (columns), then $\det \mathbf{B} = -\det \mathbf{A}$.
2. If $b_{ij} = ca_{ij}$ for fixed i (or fixed j) and constant c, then $\det \mathbf{B} = c \det \mathbf{A}$;
3. If either $b_{ij} = a_{ij} + ca_{kj}$ for fixed i, k and constant c or $b_{ij} = a_{ij} + ca_{ik}$ for fixed j, k, then $\det \mathbf{B} = \det \mathbf{A}$.
4. If $a_{ij} = 0$ for fixed i (fixed j) then $\det \mathbf{A} = 0$.
5. If two rows (columns) of $\mathbf{A}$ are proportional, then $\det \mathbf{A} = 0$.
6. $\det \mathbf{AB} = (\det \mathbf{A})(\det \mathbf{B})$.
7. $(\det \mathbf{A})\delta_{ik} = \sum_{j=1}^{n} a_{ij}A_{kj} = \sum_{j=1}^{n} a_{ji}A_{jk}$, where δ_{ik} is the familiar Kronecker delta (that is, $\delta_{ik} = 0$ if $i \neq k$, and $\delta_{ik} = 1$ if $i = k$).

Notice that, in view of this last property, each term in the linear expression for $\det \mathbf{A}$ consists of a product of n entries from $\mathbf{A}$, exactly one from each row and one from each column.

Example 1.13 Find the determinant of each of the following matrices:

(a) $\begin{pmatrix} 2 & 3 \\ 4 & 7 \end{pmatrix}$ (b) $\begin{pmatrix} 2 & 3 & 1 \\ -4 & 2 & 3 \\ 1 & 0 & -7 \end{pmatrix}$

Solution (a) $\det\begin{pmatrix}2 & 3\\ 4 & 7\end{pmatrix} = (2)(7) - (3)(4) = 14 - 12 = 2$

(b) Using the elements of the first row:

$$\det\begin{pmatrix}2 & 3 & 1\\ -4 & 2 & 3\\ 1 & 0 & -7\end{pmatrix} = (-1)^{1+1}(2)\det\begin{pmatrix}2 & 3\\ 0 & -7\end{pmatrix} + (-1)^{1+2}(3)\det\begin{pmatrix}-4 & 3\\ 1 & -7\end{pmatrix}$$

$$+ (-1)^{1+3}(1)\det\begin{pmatrix}-4 & 2\\ 1 & 0\end{pmatrix}$$

$$= (2)(-14) - 3(25) + (1)(-2) = -105$$

It would be easier to use the elements of the third row because of the zero entry:

$$\det\begin{pmatrix}2 & 3 & 1\\ -4 & 2 & 3\\ 1 & 0 & -7\end{pmatrix} = +1\det\begin{pmatrix}3 & 1\\ 2 & 3\end{pmatrix} - 7\det\begin{pmatrix}2 & 3\\ -4 & 2\end{pmatrix} = 7 - 7(16) = -105$$

□

The evaluation of 3 × 3 determinants is often done by using the following diagrammatically indicated scheme, where plus signs are attached to each product of three terms on upper-left to lower-right diagonals and minus signs are attached to products on the upper-right to lower-left diagonals:

$$\mathbf{A} = \begin{pmatrix}a & b & c\\ d & e & f\\ g & h & i\end{pmatrix} \to \begin{pmatrix}a & b & c & a & b\\ d & e & f & d & e\\ g & h & i & g & h\end{pmatrix}$$

$$\to + aei + bfg + cdh - ceg - afh - bdi = \det \mathbf{A}$$

Caution: This scheme is appropriate *only* for 3 × 3 determinants.

Example 1.14 Evaluate

$$\det\begin{pmatrix}1 & 2 & -1 & 4\\ 3 & -1 & 5 & 0\\ 2 & -2 & 1 & 1\\ -4 & 1 & 2 & 1\end{pmatrix}$$

Solution Using the diagonal attack on 3 × 3 determinants and taking advantage of the zero in row 2, we have

$$\det\begin{pmatrix}1 & 2 & -1 & 4\\ 3 & -1 & 5 & 0\\ 2 & -2 & 1 & 1\\ -4 & 1 & 2 & 1\end{pmatrix} = -3\det\begin{pmatrix}2 & -1 & 4\\ -2 & 1 & 1\\ 1 & 2 & 1\end{pmatrix} - \det\begin{pmatrix}1 & -1 & 4\\ 2 & 1 & 1\\ -4 & 2 & 1\end{pmatrix}$$

$$- 5\det\begin{pmatrix}1 & 2 & 4\\ 2 & -2 & 1\\ -4 & 1 & 1\end{pmatrix}$$

$$\begin{aligned} &= -3(2 - 1 - 16 - 4 - 4 - 2) \\ &\quad - (1 + 4 + 16 + 16 - 2 + 2) \\ &\quad - 5(-2 - 8 + 8 - 32 - 1 - 4) \\ &= 75 - 37 + 195 = 233 \end{aligned}$$

□

Some of the concepts we have considered previously can be conveniently packaged using determinants:

1. The formula for the cross product is often written in "determinant" form as

$$\mathbf{A} \times \mathbf{B} = \det \begin{pmatrix} \mathbf{i} & \mathbf{j} & \mathbf{k} \\ a_1 & a_2 & a_3 \\ b_1 & b_2 & b_3 \end{pmatrix}$$

This is not a determinant in the ordinary sense because of the vector entries in the first row. However, if we apply the rule for evaluating a third-order determinant, the result is the same as that obtained in Section 1.6. The determinant form for writing a cross product is strictly a device to assist you in remembering the formula for the cross product.

2. It is merely a matter of verifying the computations to note that if (a_1, a_2, a_3), (b_1, b_2, b_3), and (c_1, c_2, c_3) are ordered triples, then their scalar triple product is given by

$$\mathbf{A} \cdot (\mathbf{B} \times \mathbf{C}) = \det \begin{pmatrix} a_1 & a_2 & a_3 \\ b_1 & b_2 & b_3 \\ c_1 & c_2 & c_3 \end{pmatrix}$$

The absolute value of this number represents the volume of the parallelepiped determined by the geometric vectors **A**, **B**, and **C** (Exercise 11, p. 18). If the number is positive (negative), (**A**, **B**, **C**) forms a right-handed (left-handed) system.

Example 1.15 Use determinants to calculate $\mathbf{A} \times \mathbf{B}$ and $\mathbf{A} \cdot (\mathbf{B} \times \mathbf{C})$, where $\mathbf{A} = 2\mathbf{i} + 3\mathbf{j} - 4\mathbf{k}$, $\mathbf{B} = -\mathbf{i} + 2\mathbf{j} + 3\mathbf{k}$, and $\mathbf{C} = 5\mathbf{i} - \mathbf{j} - \mathbf{k}$.

Solution

$$\mathbf{A} \times \mathbf{B} = \det \begin{pmatrix} \mathbf{i} & \mathbf{j} & \mathbf{k} \\ 2 & 3 & -4 \\ -1 & 2 & 3 \end{pmatrix} = 17\mathbf{i} - 2\mathbf{j} + 7\mathbf{k}$$

(Expanding in terms of elements of the first row.)

$$\mathbf{A} \cdot (\mathbf{B} \times \mathbf{C}) = \det \begin{pmatrix} 2 & 3 & -4 \\ -1 & 2 & 3 \\ 5 & -1 & -1 \end{pmatrix} = -4 + 45 - 4 + 40 + 6 - 3 = 80$$

(Using the diagonal scheme discussed earlier.) □

Exercises for Section 1.7

Let $\mathbf{A} = \begin{pmatrix} 2 & -3 \\ 4 & 2 \end{pmatrix}$ and $\mathbf{B} = \begin{pmatrix} -1 & 2 \\ 3 & 0 \end{pmatrix}$.

1. Find $\mathbf{A} + \mathbf{B}$.
2. Find $2\mathbf{A} - 3\mathbf{B}$.
3. Find $\mathbf{AB}$ and $\mathbf{BA}$.
4. Let $\mathbf{A} = (5 \quad -1 \quad 3)$ and $\mathbf{B} = \begin{pmatrix} 2 \\ 1 \\ -1 \end{pmatrix}$. Find $\mathbf{AB}$ and $\mathbf{BA}$.
5. Calculate the determinant of each of the following matrices. In part **(c)**, use both methods discussed in Section 1.7.
   (a) (-3) **(b)** $\begin{pmatrix} 2 & -5 \\ 3 & 2 \end{pmatrix}$
   (c) $\begin{pmatrix} 2 & -1 & -1 \\ 3 & 4 & -2 \\ 2 & 1 & -1 \end{pmatrix}$ **(d)** $\begin{pmatrix} 2 & 0 & 1 & 4 \\ 1 & -1 & 2 & 1 \\ 3 & 1 & -1 & 1 \\ -2 & 0 & -1 & 1 \end{pmatrix}$
6. Consider the successive changes of variable $u = 2x - y$, $v = x + 4y$ and $r = -u + 3v$, $s = 3u + 5v$. Show that this may be written in the form
$$\begin{pmatrix} u \\ v \end{pmatrix} = \begin{pmatrix} 2 & -1 \\ 1 & 4 \end{pmatrix}\begin{pmatrix} x \\ y \end{pmatrix} = \mathbf{A}\begin{pmatrix} x \\ y \end{pmatrix}, \qquad \begin{pmatrix} r \\ s \end{pmatrix} = \begin{pmatrix} -1 & 3 \\ 3 & 5 \end{pmatrix}\begin{pmatrix} u \\ v \end{pmatrix} = \mathbf{B}\begin{pmatrix} u \\ v \end{pmatrix}$$
and that $\begin{pmatrix} r \\ s \end{pmatrix} = \mathbf{BA}\begin{pmatrix} x \\ y \end{pmatrix}$. This idea lies at the heart of the "strange" definition of matrix multiplication.
7. If $\mathbf{C}$ is a square matrix, we usually let $\mathbf{C}^2$ denote the product of $\mathbf{C}$ by itself. If $\mathbf{A}$ and $\mathbf{B}$ are square matrices of the same size, show that, in general, $(\mathbf{A} + \mathbf{B})^2 \neq \mathbf{A}^2 + 2\mathbf{AB} + \mathbf{B}^2$.
8. When is $(\mathbf{AB})^2 = \mathbf{A}^2\mathbf{B}^2$?
9. Let $\mathbf{A} = \begin{pmatrix} 1 & 2 \\ 5 & 10 \end{pmatrix}$.
   (a) Find a nonzero 2×2 matrix $\mathbf{B}$ such that $\mathbf{AB} = \mathbf{0}$.
   (b) Does $\mathbf{BA} = \mathbf{0}$?
   (c) If $\mathbf{A} = \begin{pmatrix} 1 & 2 \\ 1 & 1 \end{pmatrix}$, is part **(a)** possible?
10. Let $\mathbf{A} = \begin{pmatrix} 1 & 1 \\ 0 & 0 \end{pmatrix}$. Show that $\mathbf{A}^n = \mathbf{A}$ for any integer $n \geq 1$. A matrix with the property that $\mathbf{A}^2 = \mathbf{A}$ is called an *idempotent* matrix.
11. Find all idempotent matrices of the form $\begin{pmatrix} 1 & b \\ c & d \end{pmatrix}$.
12. Consider a communication system consisting of n stations. Form an $n \times n$ matrix $\mathbf{A}$, where $a_{ij} = 1$ if station number i can communicate directly with station j and $a_{ij} = 0$ otherwise. (Assume that $a_{ii} = 0$ for all i, that is, that a station does not communicate directly with itself.) Explore the following propositions:

(a) If $\mathbf{B} = \mathbf{A}^2$, then station i can communicate with station j through an intermediary (relay) station if and only if $b_{ij} \neq 0$.
(b) b_{ij} gives the number of one-relay links between stations i and j.
(c) If $\mathbf{C} = \mathbf{A}^3$, c_{ij} gives the number of two-relay channels of communication between station i and station j.
(d) If $\mathbf{D} = \mathbf{A} + \mathbf{A}^2 + \cdots + \mathbf{A}^{k+1}$, then d_{ij} gives the total number of channels of communication between station i and station j using k or fewer relays.

13. Is multiplication closed on the set of 2×3 matrices? Is it closed on the set of 2×2 matrices?

14. Let $\mathbf{A} = (1, 2, -3)$, $\mathbf{B} = (2, -2, 4)$, and $\mathbf{C} = (-1, 3, 2)$. Use determinants to calculate $\mathbf{A} \times \mathbf{B}$, $\mathbf{A} \times \mathbf{C}$ and $\mathbf{A} \cdot (\mathbf{B} \times \mathbf{C})$.

A matrix $\mathbf{A} = (a_{ij})$ is a *diagonal* matrix if it is square and if $a_{ij} = 0$ for $i \neq j$. The special diagonal matrix $\mathbf{I}$ with $a_{ii} = 1$ is called the *identity* matrix.

15. Show that the product of two diagonal matrices is diagonal.

16. Let $\mathbf{E}_{ij}$ be the square matrix with 1 in the ij place and 0 in all other places. For example, in 3×3 matrices

$$\mathbf{E}_{21} = \begin{pmatrix} 0 & 0 & 0 \\ 1 & 0 & 0 \\ 0 & 0 & 0 \end{pmatrix}$$

Then any square matrix $\mathbf{A}$ may be expressed as $\mathbf{A} = \sum a_{ij}\mathbf{E}_{ij}$, where the sum is taken over all values of i and j between 1 and n. In the vector space of square matrices of order n:

(a) Show that the $n \times n$ identity matrix $\mathbf{I}$ is the only matrix with the property

$$\mathbf{E}_{ij}\mathbf{I} = \mathbf{E}_{ij} = \mathbf{I}\mathbf{E}_{ij}$$

for each $\mathbf{E}_{ij}$.

(b) Deduce that $\mathbf{I}$ is the only matrix with the property $\mathbf{AI} = \mathbf{A} = \mathbf{IA}$ for an arbitrary $n \times n$ matrix $\mathbf{A}$.

The *transpose* of a matrix $\mathbf{A}$ is denoted by $\mathbf{A}^T$ and is the matrix obtained from $\mathbf{A}$ by interchanging the rows and columns.

17. If $\mathbf{A} = \begin{pmatrix} 2 & 1 & 3 \\ 3 & 0 & 1 \end{pmatrix}$, find $\mathbf{A}^T$.

18. Let $\mathbf{A}$ be as in Exercises 1–3. Find $\mathbf{A} + \mathbf{A}^T$, $\mathbf{A} - \mathbf{A}^T$, and $\mathbf{AA}^T$.

19. Demonstrate that if $\mathbf{A}$ is square, $\det \mathbf{A}^T = \det \mathbf{A}$.

20. Let $\mathbf{A}$ be as in Exercises 1–3. Find a matrix $\mathbf{B}$ such that $\mathbf{AB} = \mathbf{I} = \begin{pmatrix} 1 & 0 \\ 0 & 1 \end{pmatrix}$. Then find $\mathbf{BA}$. The matrix $\mathbf{B}$ is called the multiplicative inverse of $\mathbf{A}$.

1.8 Linear Independence

In a vector space, the fundamental operations are addition and scalar multiplication. Addition is an operation between *two* vectors, but because of the associativity of the operation, the technique of adding n vectors is well defined. If we begin with a set of vectors and add together scalar multiples of members of the set, the resulting sum is called a *linear combination.*

Definition 1.11 Let V be a vector space. A *linear combination* of vectors in V is a vector expressed in the form

$$\sum_{i=1}^{n} c_i \mathbf{v}_i$$

where the c_i are real numbers and the $\mathbf{v}_i$ are vectors in V.

Note that every linear combination of vectors from V is another vector in V. For example, some linear combinations of the set of functions $\{e^x, e^{-x}, \sinh x\}$ are:

(a) $e^x + e^{-x} - \sinh x$ (b) $e^x - e^{-x} - 2\sinh x$
(c) $0e^x + 0e^{-x} + 0\sinh x$ (d) $3e^{-x}$

Sometimes a linear combination of the vectors of a set is equal to the zero vector. In fact, if each $c_i = 0$, then

$$\sum_{i=1}^{n} c_i \mathbf{v}_i$$

is equal to the zero vector. Since that particular combination is so obvious, it is called the *trivial linear combination* of the vectors. In the examples of linear combinations given in the preceding paragraph, note that combination (b) is a nontrivial linear combination equal to the zero vector. This discussion is a prelude to the following definition and should help to motivate it.

Definition 1.12 A set of vectors $\{\mathbf{v}_i\}_{i=1}^n$ is said to be *linearly dependent* if and only if *some* nontrivial linear combination of the vectors in the set is equal to the zero vector; otherwise the set is *linearly independent.*

Hence the set of functions $\{e^x, e^{-x}, \sinh x\}$ is a linearly dependent set, since there does exist a nontrivial linear combination, specifically $e^x - e^{-x} - 2\sinh x$,which is equal to the zero function. Note that the definition of linear dependence requires that *some* nontrivial linear combination be zero, not *all* such combinations.

The definition of linear independence is often stated more positively than in Definition 1.12. We say the set

$$\{\mathbf{v}_i\}_{i=1}^n$$

is linearly independent if whenever

$$\sum_{i=1}^{n} c_i \mathbf{v}_i = \mathbf{0}$$

then $c_i = 0$ for all i.

Example 1.16 Show that the set $\{\mathbf{i}, \mathbf{j}, \mathbf{k}\}$ is linearly independent.

Solution If a linear combination of these vectors were equal to zero,

$$c_1(1,0,0) + c_2(0,1,0) + c_3(0,0,1) = (0,0,0)$$

and we would have $c_1 = 0$, $c_2 = 0$, and $c_3 = 0$; that is, the set $\{\mathbf{i}, \mathbf{j}, \mathbf{k}\}$ is linearly independent. □

Example 1.17 Show that the set $\{3\mathbf{i} + 2\mathbf{j}, \mathbf{i} + 5\mathbf{k}, 6\mathbf{i} + \mathbf{j}\}$ is linearly independent.

Solution A general linear combination of vectors from the given set is:

$$c_1(3\mathbf{i} + 2\mathbf{j}) + c_2(\mathbf{i} + 5\mathbf{k}) + c_3(6\mathbf{i} + \mathbf{j})$$

Simplifying and equating this linear combination to the zero vector, we obtain:

$$(3c_1 + c_2 + 6c_3)\mathbf{i} + (2c_1 + c_3)\mathbf{j} + 5c_2\mathbf{k} = \mathbf{0}$$

Since the set $\{\mathbf{i}, \mathbf{j}, \mathbf{k}\}$ is linearly independent, the scalar multipliers of $\mathbf{i}$, $\mathbf{j}$, and $\mathbf{k}$ are identically zero:

$$\begin{aligned} 3c_1 + c_2 + 6c_3 &= 0 \\ 2c_1 \qquad + c_3 &= 0 \\ 5c_2 \qquad &= 0 \end{aligned}$$

The only solution for this system is $c_1 = c_2 = c_3 = 0$, and hence the set is linearly independent. □

Example 1.18 Determine the linear independence or linear dependence of the set of three-dimensional vectors expressed in terms of $\mathbf{i}$, $\mathbf{j}$, and $\mathbf{k}$ as $\{\mathbf{i} + \mathbf{j} + 3\mathbf{k}, \mathbf{i} + 2\mathbf{k}, \mathbf{i} - 2\mathbf{j}\}$.

Solution We take an arbitrary linear combination of the three given vectors and equate it to the zero vector:

$$c_1(\mathbf{i} + \mathbf{j} + 3\mathbf{k}) + c_2(\mathbf{i} + 2\mathbf{k}) + c_3(\mathbf{i} - 2\mathbf{j}) = \mathbf{0}$$

from which we obtain the three equations

$$c_1 + c_2 + c_3 = 0 \qquad c_1 - 2c_3 = 0 \qquad 3c_1 + 2c_2 = 0$$

One nontrivial solution is $c_1 = 2$, $c_2 = -3$, $c_3 = 1$, which shows that the given set is linearly dependent. □

A better intuitive understanding of linear dependence may be obtained by noting (see Exercise 3 at the end of this section) the following:

1. Two nonzero two-dimensional geometric vectors are linearly dependent *if and only if* one is a constant multiple of the other or, equivalently, they have the same or opposite directions.
2. Three nonzero three-dimensional geometric vectors are linearly dependent *if and only if* one is a linear combination of the other two or, equivalently,

some plane can be found that contains arrow representations of all three vectors.

The scalar triple product can be used to put (2) in an important computational form, as in the following theorem.

Theorem 1.1 Three geometric vectors are linearly independent if and only if their scalar triple product is nonzero.

Proof In Exercise 11, p. 18, you were asked to show that the scalar triple product may be interpreted as the volume of a parallelepiped with **A**, **B**, and **C** as adjacent edges. Hence $\mathbf{A}\cdot(\mathbf{B}\times\mathbf{C})$ is zero if and only if **A**, **B**, and **C** are coplanar (see Figure 1.13). From (2), this is true if and only if **A**, **B**, and **C** are linearly dependent, and the theorem follows. ■

Example 1.19 Show that the vectors $\mathbf{A} = \mathbf{i} + 2\mathbf{j} - \mathbf{k}$, $\mathbf{B} = 2\mathbf{i} - \mathbf{j} - \mathbf{k}$, and $\mathbf{C} = -\mathbf{i} + 3\mathbf{j} + 4\mathbf{k}$ are linearly independent.

Solution Calculating the scalar triple product [**ABC**]:

$$\det\begin{pmatrix} 1 & 2 & -1 \\ 2 & -1 & -1 \\ -1 & 3 & 4 \end{pmatrix} = -20 \neq 0$$

which proves, by Theorem 1.1, that $\{\mathbf{A}, \mathbf{B}, \mathbf{C}\}$ is a linearly independent set. □

One of the distinguishing features of vector spaces is size. By this we do not mean the number of vectors in the space but rather the number of elements in a special subset of the space called a *basis*, from which the entire space can be generated.

Definition 1.13 A set of vectors $\{\mathbf{v}_i\}_{i=1}^n$ is said to be a *basis* for a vector space V if

(a) it is linearly independent, and
(b) every vector in V can be expressed as a linear combination of vectors from the set. We say that the basis *generates* the vector space, V.

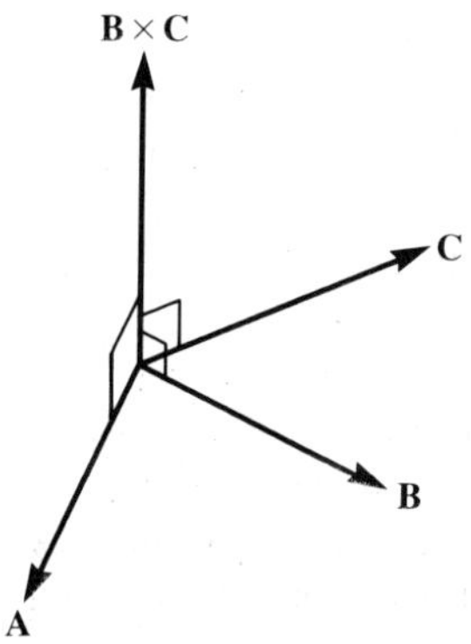

Figure 1.13

Clearly not every linearly independent set of vectors for a vector space is a basis for the space. Some such sets are too small to satisfy condition (b). The set $\{\mathbf{i}, \mathbf{j}\}$ in three-dimensional space is such a set.

Even though a set of basis elements is not unique, it is a theorem (and a very important one) that if the number of elements in a basis set is finite, then every other basis will have precisely the *same* number of elements. We can therefore make the following definition.

Definition 1.14 The number of elements in any basis of a vector space is called the *dimension* of the space.

You may have noticed that we have unashamedly used the word "dimension" in a rather informal way up to now. Indeed, without saying so, we have assumed and will continue to assume the following important theorem.

Theorem 1.2 In a vector space of dimension n, every set of n linearly independent vectors is a basis. ■

Example 1.20 Determine a basis for the set of 3×3 square matrices. What is the dimension of this vector space?

Solution Recall that $\mathbf{E}_{ij}$ is the matrix with 1 in the ij place and 0 in the other places. (See Exercise 16, p. 25.) Then, any 3×3 matrix $\mathbf{A} = (a_{ij})$ may be written

$$\mathbf{A} = a_{11}\mathbf{E}_{11} + a_{12}\mathbf{E}_{12} + a_{13}\mathbf{E}_{13} + a_{21}\mathbf{E}_{21} + \cdots + a_{33}\mathbf{E}_{33}$$

Furthermore, the set of matrices $\mathbf{E}_{ij}$ is linearly independent, and hence the set is a basis. Since there are nine matrices in the set, the dimension of the vector space of 3×3 square matrices is 9. □

If the set $\{\mathbf{v}_i\}_{i=1}^{n}$ is a basis, the symbol $V\{\mathbf{v}_1, \mathbf{v}_2, \ldots, \mathbf{v}_n\}$ is often used to denote the vector space generated by that basis.

Example 1.21 Show that $V\{1, x\} = V\{3, x - 2\}$.

Solution We show only that $V\{1, x\} \subset V\{3, x - 2\}$. The reverse inclusion is done in a similar manner. Let $\mathbf{v}$ be any function in $V\{1, x\}$. Then $\mathbf{v} = a + bx$. To show $\mathbf{v}$ is an element in $V\{3, x - 2\}$, let $a + bx = c(3) + d(x - 2) = 3c - 2d + dx$. Then $d = b$ and $c = (a + 2b)/3$ and hence $\mathbf{v}$ can be written in terms of 3 and $x - 2$. □

Whereas Theorem 1.2 presupposes knowledge of the dimension of a vector space, it is nonetheless important, since it leads to a great latitude in choosing a basis. In fact, although the orthonormal set $\{\mathbf{i}, \mathbf{j}, \mathbf{k}\}$ is a convenient basis for the set of three-dimensional geometric vectors, the linearly independent set of Example 1.17 would serve as a basis too, although far more awkwardly.

The choice of a basis for a given vector space can be very important; in effect, the choice of a basis for a space is equivalent to choosing the coordinates for the space.

When a vector is written as a linear combination of the basis vectors, the scalar multipliers are called the *components* of the vector with respect to that basis. Of course, the components change if the basis is changed. For example, in Example 1.21 we say that $\mathbf{v}$ has components a, b with respect to the basis $\{1, x\}$ and $(a + 2b)/3$, b with respect to the basis $\{3, x - 2\}$.

Example 1.22 Find the components of the vector $\mathbf{A}$ with respect to the basis $\{3\mathbf{i} + 2\mathbf{j}, \mathbf{i} + 5\mathbf{k}, 6\mathbf{i} + \mathbf{j}\}$ if the components of $\mathbf{A}$ with respect to the $\{\mathbf{i}, \mathbf{j}, \mathbf{k}\}$ basis are 1, 2, and -1.

Solution The components of $\mathbf{A}$ with respect to the basis are numbers a_1, a_2, a_3 such that

$$\mathbf{i} + 2\mathbf{j} - \mathbf{k} = a_1(3\mathbf{i} + 2\mathbf{j}) + a_2(\mathbf{i} + 5\mathbf{k}) + a_3(6\mathbf{i} + \mathbf{j})$$

Using the linear independence of the set $\{\mathbf{i}, \mathbf{j}, \mathbf{k}\}$, this vector equation yields the following three scalar equations:

$$\begin{aligned} 3a_1 + a_2 + 6a_3 &= 1 \\ 2a_1 \qquad + a_3 &= 2 \\ 5a_2 \qquad &= -1 \end{aligned}$$

from which we obtain the solution $a_1 = \frac{6}{5}$, $a_2 = -\frac{1}{5}$, $a_3 = -\frac{2}{5}$. □

In passing we note that Definitions 1.11–1.14 apply to finite linear combinations and to finite dimensional spaces. In Chapter 7 we will need extensions of these concepts to infinite dimensional spaces when we discuss the space $PC[-L, L]$ in more detail.

It is particularly advantageous to choose an orthonormal set (or at least one that is orthogonal) as a basis. All the advantages are not immediately obvious but one of the reasons the basis $\{\mathbf{i}, \mathbf{j}, \mathbf{k}\}$ is used is that resulting formulas are simpler. For example suppose you wish to find the components of a vector $\mathbf{A}$ with respect to an orthonormal basis $\{\mathbf{e}_1, \mathbf{e}_2, \mathbf{e}_3\}$. Then as in the previous example you must find a_1, a_2, a_3 such that

$$\mathbf{A} = a_1\mathbf{e}_1 + a_2\mathbf{e}_2 + a_3\mathbf{e}_3$$

To find a_1 we "dot" both sides with $\mathbf{e}_1$:

$$\mathbf{e}_1 \cdot \mathbf{A} = a_1\mathbf{e}_1 \cdot \mathbf{e}_1 + a_2\mathbf{e}_2 \cdot \mathbf{e}_1 + a_3\mathbf{e}_3 \cdot \mathbf{e}_1$$

Since the basis is orthogonal, $\mathbf{e}_2 \cdot \mathbf{e}_1 = \mathbf{e}_3 \cdot \mathbf{e}_1 = 0$, and since the basis elements are normalized, $\mathbf{e}_1 \cdot \mathbf{e}_1 = 1$. Therefore

$$a_1 = \mathbf{e}_1 \cdot \mathbf{A}$$

Similarly, $a_2 = \mathbf{e}_2 \cdot \mathbf{A}$ and $a_3 = \mathbf{e}_3 \cdot \mathbf{A}$. These are easy formulas to use and certainly far easier than the general procedure that we must employ in the case of a nonorthogonal basis, as in Example 1.22.

This technique of finding the components of a vector with respect to an orthonormal basis generalizes to inner product spaces. The method is often called the *Fourier method*, since it is even used for infinite-dimensional spaces, where Fourier series play an important role. (See Chapter 7.)

Exercises for Section 1.8

1. Show that the following sets are basis sets for the set of three-dimensional geometric vectors:
(a) $\mathbf{i} + \mathbf{j}, \mathbf{k}, \mathbf{i} + \mathbf{k}$ **(b)** $3\mathbf{i} + 2\mathbf{j} + \mathbf{k}, \mathbf{i} - 5\mathbf{j} + 2\mathbf{k}, \mathbf{j} + \mathbf{k}$

2. Express the vector $5\mathbf{i} + 3\mathbf{j} - 2\mathbf{k}$ in terms of each of the bases of Exercise 1.

3. Prove Assertions (1) and (2) on p. 27.

4. **(a)** Is it possible for a set of four three-dimensional geometric vectors to be linearly independent?
(b) Is it possible for a set of two three-dimensional geometric vectors to be linearly dependent?
Justify your answers.

5. Show that *any* set of vectors which contains the zero vector of the space is linearly dependent.

6. Show that the following bases are orthogonal:
(a) $\mathbf{i} + \mathbf{j}, \mathbf{i} - \mathbf{j}, \mathbf{k}$ **(b)** $3\mathbf{i} + 2\mathbf{j} + \mathbf{k}, \mathbf{i} + \mathbf{j} - 5\mathbf{k}, 11\mathbf{i} - 16\mathbf{j} - \mathbf{k}$

7. Find the corresponding orthonormal basis for each of the sets of Exercise 6.

8. Express the vector $6\mathbf{i} + 2\mathbf{j} + 3\mathbf{k}$ in terms of the orthonormal bases of Exercise 7.

9. Show that the set of real numbers can be called a vector space. What is its dimension?

10. Show how the Fourier method is used in a general finite dimensional vector space to obtain the components of a vector with respect to an orthonormal basis.

11. Show that $V\{1, x, x^2\} = V\{x - 1, x^2 + 1, x - 3\}$.

12. Show that the functions $3, x - 2, x + 3, x^2 + 1, 3x^2 - 2$ are linearly dependent. Find a basis for the vector space of all linear combinations of these functions. What is the dimension of the space?

13. Show that $V\{e^x, e^{-x}\} = V\{\cosh x, \sinh x\}$, where the hyperbolic functions $\cosh x$ and $\sinh x$ are defined by $\cosh x \equiv \dfrac{e^x + e^{-x}}{2}$, $\sinh x \equiv \dfrac{e^x - e^{-x}}{2}$.

14. In Chapter 12 you will learn that

$$e^{(a \pm bi)x} = e^{ax} \cos bx \pm ie^{ax} \sin bx$$

Show that

$$V\{e^{(a+bi)x}, e^{(a-bi)x}\} = V\{e^{ax} \cos bx, e^{ax} \sin bx\}.$$

This vector space will be very important to us in Chapter 3.

The set U of all functions having at least two derivatives on an interval (a, b) can be considered to be a vector space. If y_1, y_2, y_3 are in U, the function W given by the determinant

$$W(x) = \det \begin{pmatrix} y_1(x) & y_2(x) & y_3(x) \\ y_1'(x) & y_2'(x) & y_3'(x) \\ y_1''(x) & y_2''(x) & y_3''(x) \end{pmatrix}$$

is called the *Wronskian* of the functions y_1, y_2, y_3.

15. Show that if y_1, y_2, y_3 are linearly dependent functions from U, then $W(x) \equiv 0$ on (a, b). Extend this result to n functions.

16. Show that the converse of Exercise 15 is false by considering the following functions:

$$y_1(x) = \begin{cases} 1 + x^3 & , \quad x \leqq 0 \\ 1 & , \quad x \geqq 0 \end{cases}$$

$$y_2(x) = \begin{cases} 1 & , \quad x \leqq 0 \\ 1 + x^3 & , \quad x \geqq 0 \end{cases}$$

$$y_3(x) = 3 + x^3 \qquad \text{for all } x$$

Show that the three functions are linearly independent even though their Wronskian vanishes for all x.

17. Show that x and $|x|$ are linearly dependent in $PC[0, 1]$ but linearly independent in $PC[-1, 1]$.

In Exercises 18–19, show that the given functions are linearly independent on any interval (a, b).

18. **(a)** x and x^2 **(b)** x, x^2, and x^3

19. **(a)** $\sin x$ and $\cos x$ **(b)** e^x and e^{-x}

20. Show that 1, $\sin^2 x$, and $\cos^2 x$ are linearly dependent.

21. Determine if the vectors $\mathbf{i} - \mathbf{j} + \mathbf{k}$, $\mathbf{i} + \mathbf{j} - \mathbf{k}$, and $\mathbf{k}$ are linearly dependent or independent by using Theorem 1.1.

22. Determine the dimension of the vector space of 3×3 diagonal matrices.

23. Are the following two matrices linearly independent?

$$\mathbf{A} = \begin{pmatrix} 2 & -3 \\ 4 & 2 \end{pmatrix} \qquad \mathbf{B} = \begin{pmatrix} -1 & 2 \\ 3 & 0 \end{pmatrix}$$

24. Determine a basis for the set of 2×3 matrices. What is the dimension of the vector space of 2×3 matrices? Generalize to the space of $m \times n$ matrices.

1.9 Some Important Techniques

Matrices frequently appear in the role of *linear transformations* or *linear operators*. Through matrix multiplication, a relation such as $\mathbf{y} = \mathbf{Ax}$ assigns to each vector $\mathbf{x}$ in Euclidean n-space a unique vector $\mathbf{y}$ in Euclidean m-space. Indeed, matrix notation is often used to simplify and unify mathematical presentation of the operation of diverse other transformations. The essential feature of matrices in these cases is that their action obeys the two linearity properties

$$\mathbf{A}(\mathbf{x}_1 + \mathbf{x}_2) = \mathbf{A}\mathbf{x}_1 + \mathbf{A}\mathbf{x}_2$$
$$\mathbf{A}(c\mathbf{x}) = c\mathbf{A}\mathbf{x}$$

We will encounter numerous other examples of linear operators in subsequent chapters.

Perhaps the most ubiquitous of elementary mathematical chores is the solving of a set of m linear equations in n unknowns or, expressed in matrix notation, finding the n-vector $\mathbf{x}$ such that

$$\mathbf{A}\mathbf{x} = \mathbf{b}$$

given the m-vector $\mathbf{b}$ and the $m \times n$ matrix $\mathbf{A}$. The theory underlying this task makes use of the notion of matrix *rank*.

Definition 1.15 The *rank* $r(\mathbf{A})$ of an arbitrary rectangular matrix $\mathbf{A}$ is the order of the largest square submatrix with nonzero determinant.

Theorem 1.3 Let $\mathbf{A}$ be the $m \times n$ coefficient matrix of the system of linear algebraic equations $\mathbf{A}\mathbf{x} = \mathbf{b}$ and denote by $[\mathbf{A} : \mathbf{b}]$ the augmented $m \times (n + 1)$ matrix formed from $\mathbf{A}$ by appending the column vector $\mathbf{b}$ as a new last column. The solvability of the matrix equation $\mathbf{A}\mathbf{x} = \mathbf{b}$ is as follows:

(a) If $r(\mathbf{A}) = r([\mathbf{A} : \mathbf{b}]) = n$, there is only one solution.
(b) If $r(\mathbf{A}) = r([\mathbf{A} : \mathbf{b}]) < n$, there are infinitely many solutions.
(c) If $r(\mathbf{A}) < r([\mathbf{A} : \mathbf{b}])$, no solution exists. ■

The theory notwithstanding, for practical purposes it is important to recognize that effective procedures for actually solving systems of equations, independent of the relationship of the number of equations and unknowns, can be based upon the elementary operations of interchanging two equations, multiplying an equation by a given nonzero scalar, and replacing one equation by the sum of that equation plus some multiple of another. In the following simple examples, we employ these reduction operations in their most basic form, using the first term in the initial equation to eliminate the first variable in succeeding equations, the (new) first term in the second equation to eliminate the corresponding variable in all successive equations, and so on.

Example 1.23 Solve the system of equations

$$\begin{aligned} x + 2y - 3z &= 4 \\ 2x - y + z &= 1 \\ 3x + 2y - z &= 5 \end{aligned}$$

Solution Subtracting two times the first equation from the second and three times the first from the third yields

$$\begin{aligned} x + 2y - 3z &= 4 \\ -5y + 7z &= -7 \\ -4y + 8z &= -7 \end{aligned}$$

Next, subtract four times the second equation from five times the third:

$$\begin{aligned} x + 2y - 3z &= 4 \\ -5y + 7z &= -7 \\ 12z &= -7 \end{aligned}$$

Finally, we start with the last equation and work backward, finding each variable as we go:

$$z = -\frac{7}{12}$$

$$-5y = -7 - 7z = -\frac{35}{12}, \quad y = \frac{7}{12}$$

$$x = -2y + 3z + 4 = \frac{13}{12}$$

□

Example 1.24 Determine the solution of

$$\begin{aligned} 2x + y - z &= 3 \\ 4x - y + z &= 2 \end{aligned}$$

Solution Elimination of x from the second equation gives rise to

$$\begin{aligned} 2x + \ y - \ z &= 3 \\ -3y + 3z &= -4 \end{aligned}$$

Thus z can be taken to be arbitrary; $-3y = -4 - 3z$, $y = z + \frac{4}{3}$; $2x = -y + z + 3 = \frac{5}{3}$, $x = \frac{5}{6}$. □

Example 1.25 Solve:

$$\begin{aligned} x + \ y &= 4 \\ 2x - \ y &= 2 \\ 3x + 2y &= 3 \end{aligned}$$

Solution Elimination of x from the second and third equations leads to

$$\begin{aligned} x + y &= 4 \\ -3y &= -6 \\ -y &= -9 \end{aligned}$$

The last two equations yield a contradiction, so the system has no solution, or is *inconsistent*. □

For bookkeeping purposes, the steps involved in solving these systems may be streamlined by merely applying elementary operations to the rows of the augmented matrix $[\mathbf{A} : \mathbf{b}]$. (This is what is done in actual computational algorithms.) The first example would then appear as

$$\begin{pmatrix} 1 & 2 & -3:4 \\ 2 & -1 & 1:1 \\ 3 & 2 & -1:5 \end{pmatrix} \rightarrow \begin{pmatrix} 1 & 2 & -3: & 4 \\ 0 & -5 & 7: & -7 \\ 0 & -4 & 8: & -7 \end{pmatrix} \rightarrow \begin{pmatrix} 1 & 2 & -3: & 4 \\ 0 & -5 & 7: & -7 \\ 0 & 0 & 12: & -7 \end{pmatrix}$$

Example 1.26 Balance the following chemical equation:

$$I_2 + HNO_3 \to HIO_3 + NO_2 + H_2O$$

Solution Let the coefficients of the terms in the equation be a, b, c, d, and e, so that the balanced equation becomes

$$aI_2 + bHNO_3 \to cHIO_3 + dNO_2 + eH_2O$$

The number of atoms of iodine on each side must be the same, so that

$$2a = c$$

In the same way, by considering hydrogen, nitrogen, and oxygen, respectively, we have

$$b = c + 2e$$
$$b = d$$
$$3b = 3c + 2d + e$$

Using matrices we solve the system in one step as follows:

$$\begin{pmatrix} 2 & 0 & -1 & 0 & 0:0 \\ 0 & 1 & -1 & 0 & -2:0 \\ 0 & 1 & 0 & -1 & 0:0 \\ 0 & 3 & -3 & -2 & -1:0 \end{pmatrix} \to \begin{pmatrix} 2 & 0 & -1 & 0 & 0:0 \\ 0 & 1 & -1 & 0 & -2:0 \\ 0 & 0 & 1 & -1 & 2:0 \\ 0 & 0 & 0 & -2 & 5:0 \end{pmatrix}$$

If we let $e = 1$ (for the time being), then $d = \frac{5}{2}$, $c = d - 2e = \frac{1}{2}$, $b = c + 2e = \frac{1}{2} + 2 = \frac{5}{2}$, $2a = c$, and $a = \frac{1}{4}$. Since e was arbitrary, we may reselect it to be 4 and eliminate fractions to obtain the final equation

$$I_2 + 10\,HNO_3 \to 2\,HIO_3 + 10\,NO_2 + 4\,H_2O$$ □

For $n \times n$ systems of equations $\mathbf{Ax} = \mathbf{b}$, in which $\det \mathbf{A} \neq 0$, another solution approach is based upon the *inverse* matrix.

Definition 1.16 A square matrix $\mathbf{A}$ is called *nonsingular* if $\det \mathbf{A} \neq 0$; if $\det \mathbf{A} = 0$, $\mathbf{A}$ is termed *singular*.

Definition 1.17 Let $\mathbf{A}$ be a nonsingular square matrix and denote by $\text{Adj}(\mathbf{A}) \equiv (A_{ji})$ = the transpose of the matrix of cofactors of $\mathbf{A}$ (recall Definition 1.10). The *inverse* $\mathbf{A}^{-1}$ of $\mathbf{A}$ is then defined by

$$\mathbf{A}^{-1} \equiv \frac{\text{Adj}(\mathbf{A})}{\det \mathbf{A}}$$

In view of Property 7 of determinants, $\mathbf{A}^{-1}$ has the fundamental feature that

$$\mathbf{A}^{-1}\mathbf{A} = \mathbf{I} = \mathbf{A}\mathbf{A}^{-1}$$

where $\mathbf{I}$ is the identity matrix of order n. Thus if $\mathbf{A}$ is a nonsingular matrix of order n, the set of algebraic equations $\mathbf{Ax} = \mathbf{b}$ has the unique solution $\mathbf{x} = \mathbf{A}^{-1}\mathbf{b}$, that is, $\mathbf{x} = (x_1, x_2, \ldots, x_n)$ with

$$x_i = \frac{1}{\det \mathbf{A}} \sum_{j=1}^{n} A_{ji} b_j \qquad i = 1, 2, \ldots, n$$
$$= \frac{\det \mathbf{A}_i}{\det \mathbf{A}}$$

where $\mathbf{A}_i$ is the $n \times n$ matrix obtained from $\mathbf{A}$ by replacing the ith column of $\mathbf{A}$ with the column vector $\mathbf{b}$. This last expression is commonly designated as *Cramer's rule.*

Unfortunately, although this solution approach is of considerable theoretical and notational utility, because of the numerous operations needed to determine $\det \mathbf{A}$ and the $\det \mathbf{A}_i$, it does not provide an efficient method to solve a given system of algebraic equations unless n is small.

If $\mathbf{b} = \mathbf{0}$ in the system $\mathbf{Ax} = \mathbf{b}$, we have a homogeneous set of equations. A theorem concerning such systems, which has some claims to being the fundamental theorem of applied mathematics, is given next.

Theorem 1.4 The system $\mathbf{Ax} = \mathbf{0}$, where $\mathbf{A}$ is $n \times n$ and $\mathbf{x}$ is $n \times 1$, has a nontrivial solution *if and only if* $\mathbf{A}$ is singular. ■

In solving a homogeneous system using elementary row operations on the augmented matrix, we note that the last column at any stage consists of zeros. We usually dispense with this extra baggage as can be seen in Example 1.27.

In many applications to come, we will be asked to find *eigenvalues* λ of a square matrix $\mathbf{A}$ (that is, values of λ such that $\det(\mathbf{A} - \lambda\mathbf{I}) = 0$) and the corresponding nontrivial *eigenvectors* $\mathbf{v}$ such that $(\mathbf{A} - \lambda\mathbf{I})\mathbf{v} = \mathbf{0}$ for that particular eigenvalue. The equation $\det(\mathbf{A} - \lambda\mathbf{I}) = 0$ is called the *characteristic equation.*

Example 1.27 Calculate eigenvalues and eigenvectors for each of the following matrices:

(a) $\begin{pmatrix} 2 & -1 & -1 \\ 2 & 1 & -1 \\ 0 & -1 & 1 \end{pmatrix}$ (b) $\begin{pmatrix} 1 & -1 \\ 1 & 3 \end{pmatrix}$ (c) $\begin{pmatrix} 3 & -2 & -1 \\ 3 & -4 & -3 \\ 2 & -4 & 0 \end{pmatrix}$

Solution (a) $\det(\mathbf{A} - \lambda\mathbf{I}) = \det\begin{pmatrix} 2-\lambda & -1 & -1 \\ 2 & 1-\lambda & -1 \\ 0 & -1 & 1-\lambda \end{pmatrix}$

$$= -\lambda^3 + 4\lambda^2 - 6\lambda + 4$$
$$= -(\lambda - 2)(\lambda^2 - 2\lambda + 2)$$
$$= -(\lambda - 2)[\lambda - (1 + i)][\lambda - (1 - i)]$$

The eigenvalues are thus $\lambda = 2$, $\lambda = 1 + i$, and $\lambda = 1 - i$. To find eigenvectors corresponding to $\lambda = 2$, we must solve the homogeneous system with matrix

$$\mathbf{A} - 2\mathbf{I} = \begin{pmatrix} 0 & -1 & -1 \\ 2 & -1 & -1 \\ 0 & -1 & -1 \end{pmatrix}$$

which reduces to

$$\begin{pmatrix} 2 & -1 & -1 \\ 0 & -1 & -1 \\ 0 & 0 & 0 \end{pmatrix}$$

with corresponding solutions $(0, -c, c)$. The eigenvectors are thus nonzero multiples of the vector $\begin{pmatrix} 0 \\ -1 \\ 1 \end{pmatrix}$.

For the eigenvalue $1 + i$, the reduction is

$$\begin{pmatrix} 1-i & -1 & -1 \\ 2 & -i & -1 \\ 0 & -1 & -i \end{pmatrix} \to \begin{pmatrix} 1-i & -1 & -1 \\ 0 & 1 & i \\ 0 & -1 & -i \end{pmatrix} \to \begin{pmatrix} 1-i & -1 & -1 \\ 0 & 1 & i \\ 0 & 0 & 0 \end{pmatrix}$$

which yields an eigenvector $\begin{pmatrix} 1 \\ -i \\ 1 \end{pmatrix}$. The eigenvalue $1 - i$ yields an eigenvector $\begin{pmatrix} 1 \\ i \\ 1 \end{pmatrix}$. Note that this is the conjugate of the eigenvector we obtained from the eigenvalue $1 + i$. This is not an accident!

(b) $$\det(\mathbf{A} - \lambda\mathbf{I}) = \det\begin{pmatrix} 1-\lambda & -1 \\ 1 & 3-\lambda \end{pmatrix} = \lambda^2 - 4\lambda + 4 = (\lambda - 2)^2$$

To find eigenvectors corresponding to $\lambda = 2$, we must solve the homogeneous system with matrix

$$\begin{pmatrix} -1 & -1 \\ 1 & 1 \end{pmatrix} \to \begin{pmatrix} -1 & -1 \\ 0 & 0 \end{pmatrix}$$

Such an eigenvector is $\begin{pmatrix} 1 \\ -1 \end{pmatrix}$. Note that there is only *one* linearly independent eigenvector associated with the double eigenvalue $\lambda = 2$.

(c) Straightforward calculations yield the characteristic equation $\det(\mathbf{A} - \lambda\mathbf{I}) = -\lambda^3 - \lambda^2 + 16\lambda - 20 = -(\lambda + 5)(\lambda - 2)^2 = 0$ and, corresponding to the eigenvalue $\lambda = -5$, an eigenvector

$$\begin{pmatrix} 1 \\ 3 \\ 2 \end{pmatrix}$$

In part (b), the double eigenvalue $\lambda = 2$ gave rise to only one independent eigenvector. In this case, however, not only do we have the unlikely occurrence of a double root but an equally unlikely reduced matrix with two rows of zeros, resulting in two arbitrary constants and two independent eigenvectors:

$$\begin{pmatrix} 1 & -2 & -1 \\ 3 & -6 & -3 \\ 2 & -4 & -2 \end{pmatrix} \to \begin{pmatrix} 1 & -2 & -1 \\ 0 & 0 & 0 \\ 0 & 0 & 0 \end{pmatrix}$$

$$\to \text{solutions} \begin{pmatrix} 2c_2 + c_1 \\ c_2 \\ c_1 \end{pmatrix} = c_1 \begin{pmatrix} 1 \\ 0 \\ 1 \end{pmatrix} + c_2 \begin{pmatrix} 2 \\ 1 \\ 0 \end{pmatrix} \qquad \square$$

Exercises for Section 1.9

In Exercises 1–3, use row reduction *without* resorting to matrix representation.

1. Solve the system of equations.

$$\begin{aligned} 2x + 5y + z &= 5 \\ x + 4y + 2z &= 1 \\ 4x + 10y - z &= 1 \end{aligned}$$

2. Solve the system of equations.

$$\begin{aligned} x + 3y + 5z &= 7 \\ 3x - 2y + 4z &= -1 \\ 4x + y + 9z &= 6 \end{aligned}$$

3. Solve the system of equations.

$$\begin{aligned} x + y &= 1 \\ x + z &= 1 \\ 2x + y + z &= 0 \end{aligned}$$

In Exercises 4–9, begin by representing the problem in matrix form and then use the row-reduction technique.

4. Solve the system of equations.

$$\begin{aligned} 2x_1 + 3x_2 - 4x_3 + x_4 &= 9 \\ x_1 + 2x_2 - x_3 + x_4 &= 2 \\ 3x_1 + 9x_2 + 2x_3 + 4x_4 &= 0 \end{aligned}$$

5. Solve the system of equations.

$$\begin{aligned} x_1 + 2x_2 - 6x_3 &= 2 \\ x_1 + 4x_2 + 4x_3 &= 1 \\ 3x_1 + 10x_2 + 2x_3 &= -1 \end{aligned}$$

6. Balance the chemical equation.

$$Ca(OH)_2 + H_3PO_4 \to Ca_3(PO_4)_2 + H_2O$$

7. Solve the system of equations.

$$\begin{aligned} x_1 \quad\quad + 2x_3 - x_4 &= 3 \\ x_2 + x_3 \quad\quad &= 5 \\ 3x_1 + 2x_2 \quad\quad - 2x_4 &= -1 \\ -x_3 + 4x_4 &= 13 \\ 2x_1 \quad\quad - x_3 + 3x_4 &= 11 \end{aligned}$$

8. Solve the system of equations.

$$\begin{aligned} 3x_1 + x_2 - 5x_3 - x_4 &= 0 \\ 2x_1 + x_2 - 3x_3 - 2x_4 &= 0 \\ x_1 + x_2 - x_3 - 3x_4 &= 0 \end{aligned}$$

9. Solve the system of equations.

$$\begin{aligned} x_1 + 4x_2 + 2x_3 &= 0 \\ 2x_1 + 5x_2 - x_3 &= 0 \\ x_1 + x_2 - x_3 &= 0 \end{aligned}$$

10. Employ Cramer's rule to solve the system of equations given in Exercise 1.

11. Use Cramer's rule to establish that when $\mathbf{A}$ is square, a nontrivial solution to the system $\mathbf{Ax} = \mathbf{0}$ is possible only if $\det \mathbf{A} = 0$.

As we shall see in subsequent chapters, the notions of eigenvalues and eigenvectors are particularly important and useful. Eigenvectors are very special vectors, as their name implies—in the matrix case the transformed vector $\mathbf{Ax}$ is collinear with the original vector $\mathbf{x}$. The two vectors differ only in their lengths.

The following exercises explore several fundamental applications.

12. Consider the quadratic form

$$Q(\mathbf{x}) = x_1^2 + 4x_1x_2 + x_2^2$$

Show that the change of variables

$$x_1 = \frac{1}{\sqrt{2}}(y_1 + y_2)$$

$$x_2 = \frac{1}{\sqrt{2}}(y_1 - y_2)$$

transforms $Q(\mathbf{x})$ into a new quadratic form without any crossterms.

13. **(a)** Verify that the eigenvalues of the matrix

$$\mathbf{A} = \begin{pmatrix} 1 & 2 \\ 2 & 1 \end{pmatrix}$$

are $\lambda = 3, \ -1$ and the corresponding normalized eigenvectors are $\begin{pmatrix} \frac{1}{2} \\ \frac{1}{2} \end{pmatrix}$ and $\begin{pmatrix} \frac{1}{2} \\ -\frac{1}{2} \end{pmatrix}$, respectively.

(b) Demonstrate that $\mathbf{P}^T\mathbf{A}\mathbf{P}$, where $\mathbf{P} = \begin{pmatrix} 1/\sqrt{2} & 1/\sqrt{2} \\ 1/\sqrt{2} & -1/\sqrt{2} \end{pmatrix}$, yields a diagonal matrix with the eigenvalues 3 and -1 as the diagonal entries.

14. Prove that the results of Exercises 12 and 13 are related by showing that if (1) $\mathbf{P}$ is a matrix whose columns are linearly independent normalized eigenvectors of a given real symmetric matrix $\mathbf{A} = \begin{pmatrix} a_{11} & a_{12} \\ a_{12} & a_{22} \end{pmatrix}$ and (2) $\mathbf{P}^T\mathbf{P} = \mathbf{I}$, then the change of variables $\mathbf{x} = \mathbf{P}\mathbf{y}$ diagonalizes $Q(\mathbf{x}) = a_{11}x_1^2 + 2a_{12}x_1x_2 + a_{22}x_2^2$, that is, transforms $Q(\mathbf{x})$ into a quadratic form in $\mathbf{y}$ without any crossterms.

15. The results of Exercises 12–14 are closely related to the process of finding the normal form of a conic section in the plane by way of rotation of axes. Recall that the xy term in the expression $ax^2 + bxy + cy^2 + dx + ey + f = 0$ may be eliminated by rotating the axes through an angle determined by $\tan 2\theta = b/(a - c)$ $(\theta = \pi/4$ if $a = c)$ using the transformations $x = \bar{x}\cos\theta - \bar{y}\sin\theta$, $y = \bar{x}\sin\theta + \bar{y}\cos\theta$.

Carry out this process for $x^2 + 4xy + y^2 = \mathbf{x}^T\begin{pmatrix} 1 & 2 \\ 2 & 1 \end{pmatrix}\mathbf{x} = 3 \left(\text{where } \mathbf{x} = \begin{pmatrix} x \\ y \end{pmatrix}\right)$ and compare the lengths and directions of the major and minor axes to the eigenvalues and eigenvectors of the matrix $\begin{pmatrix} 1 & 2 \\ 2 & 1 \end{pmatrix}$.

16. Generalize the result of Exercise 14 for n-dimensional quadratic forms.

In various mechanical and electrical applications, the eigenvalues of a given matrix turn out to be related to the *square* of the fundamental frequencies of vibration of the system of interest. In such situations all the eigenvalues must be nonnegative. In other applied problems, the eigenvalues must all be negative or, if complex-valued, all have negative real parts. In still other applications, whether or not the eigenvalues have absolute values smaller than 1 is of importance. For the remaining exercises, find the eigenvalues and corresponding eigenvectors of each given matrix and comment on the nature of the eigenvalues as they pertain to these practical considerations.

17. $\mathbf{A} = \begin{pmatrix} 1 & -\frac{1}{6} \\ 1 & \frac{1}{6} \end{pmatrix}$

18. $\mathbf{A} = \begin{pmatrix} 11 & -4 & -7 \\ 7 & -2 & -5 \\ 10 & -4 & -6 \end{pmatrix}$

19. $\mathbf{A} = \begin{pmatrix} -4 & 5 & 5 \\ -4 & 6 & 5 \\ -5 & 5 & 6 \end{pmatrix}$

20. $\mathbf{A} = \begin{pmatrix} 1 & -2 \\ 1 & -1 \end{pmatrix}$

21. $\mathbf{A} = \begin{pmatrix} 2 & 1 & 1 \\ 1 & 2 & 1 \\ -2 & -2 & -1 \end{pmatrix}$

1.10 Looking Ahead

Although we have laid some groundwork for important extensions to the infinite-dimensional case, the principal considerations of this chapter have been techniques and applications in vector and inner product spaces of finite dimension. Probably still the best reference for a more advanced discussion of this topic is [11]. In the appendix of [11] and in [12], moreover, the fundamental generalization of Hilbert space is explored. This extension provides theoretical support for practical application of the Fourier method, which we will discuss in Chapter 7 (also see the classical reference [7], for example).

As we have suggested previously, virtually all this introductory chapter has relevance to subsequent material. In addition to the close alliance to Fourier series representations already mentioned, the notions of linearity, linear independence, and eigenvalue and eigenvector play especially significant roles in the considerations of the next three chapters on ordinary differential equations, for instance. Discussion of nonconstant vectors depending on parameters and their differential and integral calculus will be taken up in Chapters 10 and 11.

A generalization of vectors and matrices valuable in rigid body mechanics, fluid dynamics and relativity is the construct called a tensor (see [6] and [15]). Matrix manipulations, particularly in the case of large matrices, are fundamental to mathematical programming and other areas of optimization theory and techniques (see [13] as well as parts of [1] and [3]).

There now exist computational packages prepared by experts designed to numerically perform various standard operations with, and on, matrices solving sets of algebraic equations (LINPACK, [8]) and finding eigenvalues/eigenvectors (EISPACK, [17] and [9]). References such as [10], [14], [18], or [19] provide ample additional information for ambitious readers interested in numerical linear algebra.

Basic References

[1] Agnew, J., and R. C. Knapp. 1983. *Linear Algebra with Applications*. 2d ed. Monterey, Calif.: Brooks/Cole.

[2] Anton, H. 1984. *Elementary Linear Algebra*. 4th ed. New York: John Wiley.

[3] Kolman, B. 1984. *Introductory Linear Algebra with Applications*. 3d ed. New York: Macmillan.

[4] Rorres, C., and H. Anton. 1984. *Applications of Linear Algebra*. 3d ed. New York: John Wiley.

[5] Venit, S., and W. Bishop. 1985. *Elementary Linear Algebra*. 2d ed. Boston: Prindle, Weber, and Schmidt.

More-Advanced References

[6] Borisenko, A. I., and I. E. Tarapov. 1979. *Vector and Tensor Analysis and Applications*. New York: Dover.

[7] Churchill, R. V., and J. W. Brown. 1978. *Fourier Series and Boundary Value Problems*. 3d ed. New York: McGraw-Hill.

[8] Dongarra, J. J. et al. 1979. *LINPACK Users' Guide*. Philadelphia: Society for Industrial and Applied Mathematics.

[9] Garbow, B. S., et al. 1977. "Matrix Eigensystems Routines—EISPACK Guide Extension." *Lecture Notes in Computer Science*, vol. 51. New York: Springer-Verlag.

[10] Golub, G. H., and C. F. Van Loan. 1983. *Matrix Computations*. Baltimore, Md.: Johns Hopkins University Press.

[11] Halmos, P. R. 1974. *Finite-Dimensional Vector Spaces*. 2d ed. New York: Springer-Verlag.

[12] ———. 1951. *Introduction to Hilbert Space*. 2d ed. New York: Chelsea.

[13] Hillier, F. S., and G. J. Lieberman. 1980. *Operations Research*. 3d ed. San Francisco: Holden-Day.

[14] Johnson, L. W., and R. D. Riess. 1982. *Numerical Analysis*. Reading, Mass.: Addison-Wesley.

[15] Lass, H. 1950. *Vector and Tensor Analysis*. New York: McGraw-Hill.

[16] Noble, B., and J. W. Daniel. 1977. *Applied Linear Algebra*. 2d ed. Englewood Cliffs, N. J.: Prentice-Hall.

[17] Smith, B. T., et al. 1976. "Matrix Eigensystem Routines—EISPACK Guide." *Lecture Notes in Computer Science*, vol. 6. 2d ed. New York: Springer-Verlag.

[18] Steinberg, D. I. 1974. *Computational Matrix Algebra*. New York: McGraw-Hill.

[19] Stewart, G. W. 1973. *Introduction to Matrix Computations*. New York: Academic Press.

Chapter 2

First-Order Ordinary Differential Equations

2.1 Introduction

A surprising number of natural "laws," including many from classical physics, chemistry, and engineering, can be stated in the form of a differential equation or a system of such equations. A few examples are Newton's second law of motion, Kirchhoff's laws for passive electrical circuits, Fourier's law of heat conduction, Lambert's law of light absorption, the equation of continuity in fluid mechanics, the wave equation, equations governing the vibration of beams, and chemical reaction equations. For this reason differential equations appear with great regularity in the mathematical models associated with a diverse range of practical applications.

A *differential equation* is an equation involving ordinary or partial derivatives of an unknown function. Much of the elementary theory of differential equations, which we will consider in the next few chapters, focuses on exact methods for finding this unknown function through the use of integration and linear algebra. Many important equations can be solved in this manner, and modeling real-world phenomena with these equations is a profitable pursuit. In advanced studies of differential equations, the emphasis shifts to numerical and qualitative approaches applied to more difficult equations and to sophisticated techniques for determining approximate solutions (for example, phase-plane analysis and perturbation theory).

2.2 Some Elementary Terminology and Examples

If the unknown function in a differential equation is a function of one independent variable, the equation is called *ordinary*. Some examples of ordinary differential equations are

(a) $y' = x^2$ (b) $y' = 2x + y$ (c) $xy'' + y = 0$ (d) $(y')^2 = y$

(e) $yy' = 1 + x$ (f) $xy'' = \sin y - x^3$, where $y' \equiv \dfrac{dy}{dx}$ and $y'' \equiv \dfrac{d^2y}{dx^2}$

The *order* of a differential equation is the order of the highest derivative occurring in the equation. Equations (a), (b), (d), and (e) are of the first order, and the others are of the second order. Since many of the elementary physical laws mentioned earlier can be formulated in terms of first- and second-order differential equations, we usually limit ourselves to these two orders.

A differential equation is said to be *linear* when it is of the form

$$a_n(x)y^{(n)} + a_{n-1}(x)y^{(n-1)} + \cdots + a_1(x)y' + a_0(x)y = f(x)$$

where $y^{(n)} \equiv d^n y/dx^n$. Equations (a)–(c) are linear and (d)–(f) are nonlinear. Linear equations have a particularly elegant theory based on linear algebra and, at least in the case where the coefficients a_k are constant, are easy to solve, as we shall see in the next chapter. On the other hand, when the order of the differential equation is 1, the presence of linearity is far less important. In this chapter we concentrate on the various techniques which have been developed for solving (generally nonlinear) first-order equations.

A *solution* to a differential equation is a function that when substituted into the equation reduces it to an identity. A solution is not always described by an equation or rule in the explicit form $y = f(x)$. Sometimes we will agree that a solution is defined implicitly by an equation such as $xy^2 - e^{xy} = 1$, by a graph, a table of values, or a convergent power series. We are usually interested in a solution over some interval (finite or infinite).

To emphasize that an equation is an identity, that is, the equation holds for all values of the independent variable, we often use the notation $\equiv$. For example $y_1 = e^{2x} - 2x + 1$ is a solution of $y'' - y' - 2y = 4x$ because $y_1'' - y_1' - 2y_1 = 4e^{2x} - 2e^{2x} + 2 - 2e^{2x} + 4x - 2 \equiv 4x$.

Example 2.1 Show that e^x, e^{x+c}, and ce^x are solutions to $y' = y$. Show also that e^{2x} and $e^x + c$ are not solutions (if, of course, $c \neq 0$).

Solution The fact that e^x, e^{x+c}, and ce^x are solutions follows readily since the functions and their derivatives are identically equal. Similarly, since $(d/dx)e^{2x} = 2e^{2x} \not\equiv e^{2x}$ and $(d/dx)(e^x + c) = e^x \not\equiv e^x + c$, neither of these functions satisfies $y' = y$. □

Example 2.2 Show that $y' = 4(y/x) + x^4$ has solutions $y_1 = x^5 + x^4$ and $y_2 = x^5$.

Solution Since $y_1' = 5x^4 + 4x^3$ and $4(y_1/x) + x^4 = [(4x^5 + 4x^4)/x] + x^4 \equiv 5x^4 + 4x^3$, it follows that $y_1(x)$ is a solution. Since $y_2' = 5x^4$ and $4(y_2/x) + x^4 \equiv 5x^4$, $y_2(x)$ is also a solution. □

We will usually not be content with finding one function which satisfies a differential equation but, rather, will find an infinite family of solutions. Ideally we will obtain *all* functions that satisfy a given differential equation and an expression, the *complete solution*, which describes all of these solutions. The term *general solution* is often applied to an expression representing an infinite family of solutions, but, due to past vagueness as to whether the general solution is complete or not, we do not use this terminology.

The imposition of initial or boundary conditions often determines one solution, called a *particular solution*, from a family of solutions. For example, the complete solution set for the differential equation $y' = y$ is $\{y = ce^x | c \text{ a real constant}\}$. If we specify that when $x = 0$, $y = 1$, then the particular solution $y = e^x$ is determined. The complete solution is simply $y = ce^x$.

If a (second-order) differential equation is considered over an interval $[a, b]$, there are two fundamentally different types of auxiliary conditions:

1. If all the associated supplementary conditions relate to the single value $x = a$, the problem is called an *initial value problem.*
2. If at least one of the auxiliary conditions relates to *both* of the endpoints $x = a, b$ or one condition relates to $x = a$ and another to $x = b$, the problem is generally called a *boundary value problem.*

We concentrate on initial value problems in our treatment of ordinary differential equations.

Example 2.3 A rock is thrown straight down from the top of a 1000-ft cliff with initial velocity 48 ft/s. How long does it take for the rock to reach the ground?

Solution Neglecting air resistance and assuming the acceleration of gravity to be a constant 32 ft/sec^2, we have $y'' = -32$, where $y(t)$ is the height of the rock at any time t. Antidifferentiation yields $y' = -32t + c_1$. Since $y'(0) = -48$, we must have $y' = -32t - 48$. After another antidifferentiation, we have $y = -16t^2 - 48t + c_2$, where $c_2 = 0$ if we choose our coordinate system so that $y(0) = 0$. Now $y(t) = -16t^2 - 48t$, and we must find t_1 such that $y(t_1) = -1000$, that is, $-16t_1^2 - 48t_1 + 1000 = 0$. The quadratic formula yields $t_1 = (-3 + \sqrt{259})/2 \text{ sec} \doteq 6.5$ sec.* In this example we have used the initial conditions $y(0) = 0$ and $y'(0) = -48$ to find the particular solution $y = -16t^2 - 48t$ from the complete solution $y = -16t^2 + c_1 t + c_2$. □

2.3 Separable First-Order Equations

First-order differential equations are usually written in the form

$$y' = f(x, y)$$

or in a form using differentials that does not identify which of the variables, x or y, is independent:

$$M(x, y)\,dx + N(x, y)\,dy = 0$$

We will use these forms for a first-order equation interchangeably. For example,

---

* The symbol "$\doteq$" should be read "is numerically approximated by."

the equations

$$y' = \frac{-y}{x} \quad \text{and} \quad x\,dy + y\,dx = 0$$

may be treated as equivalent if we exclude $x = 0$ from consideration.

From here to the end of Section 2.6, you will notice that all our methods for solving first-order equations amount to "tricks," which reduce each problem to one of simple integration.

Perhaps the easiest differential equation of first order to solve is the kind you considered in elementary calculus, where $f(x, y)$ is independent of y. Then the differential equation has the form

$$y' = f(x)$$

and the solution set is, therefore, the family of functions

$$y = \int f(x)\,dx + c$$

where c is an arbitrary constant.

Example 2.4 Solve the differential equation $x^2\,dx - dy = 0$.

Solution This equation may be put into the form

$$y' = x^2$$

and hence the complete solution, obtained by antidifferentiation, is

$$y = \frac{x^3}{3} + c$$

□

Example 2.5 Solve the initial value problem $y' = e^{x^2}$, $y(0) = 3$.

Solution In this case antidifferentiation does not allow us to use the information given by the initial condition. This is because the antiderivative of e^{x^2} is not an elementary function. The initial condition can be applied, however, if the definite integral is taken. Indeed,

$$\int_0^x y'(t)\,dt = \int_0^x e^{t^2}\,dt$$

so

$$y(x) - y(0) = \int_0^x e^{t^2}\,dt \quad \text{and} \quad y(x) = 3 + \int_0^x e^{t^2}\,dt$$

This last expression makes perfectly good sense because, for any particular x, $\int_0^x e^{t^2}\,dt$ can be estimated numerically to any desired accuracy. □

Equations that can be put in the form

$$y' = \frac{f(x)}{g(y)}$$

are only slightly more difficult to solve. Here the variables are said to be *separated*, and the original differential equation is said to be *separable*. In such an equation we are looking for a solution $y(x)$ satisfying

$$y'(x)g(y(x)) = f(x)$$

Thus we take the antiderivative of both sides of this equation with respect to x

$$\int y'(x)g(y(x))\,dx = \int f(x)\,dx$$

By the substitution rule of calculus, this is equivalent to

$$\int g(y)\,dy = \int f(x)\,dx$$

This reasoning is usually short-circuited by going directly from the original equation to this last expression and formally antidifferentiating both sides.

Example 2.6 Solve the differential equation

$$y' = \frac{xy}{y+1}$$

Solution The equation can be written

$$\frac{y+1}{y}\,dy = x\,dx$$

Antidifferentiating both sides, we get

$$y + \ln|y| = \frac{x^2}{2} + c$$

which defines the solution set implicitly. We could have written

$$2(y + \ln|y| + \ln c_1) = x^2$$

instead, from which

$$2y + \ln(c_1 y)^2 = x^2$$

□

Note that in the last example $y(x) \equiv 0$ describes a solution of the original equation that is not included in the expression of the solution we obtained. We lost this solution, and perhaps others, when we divided both sides of our equation by y. This is a common occurrence when dealing with separable equations. We will therefore seek only a formally obtained family of solutions for separable equations now, postponing consideration of complete solutions until Section 2.7.

One of the more important separable differential equations is one for which the derivative of a function is directly proportional to the function. Using k as the constant of proportionality, we may write

$$y' = ky$$

The method of separation of variables gives the family of solutions

$$y = ce^{kx}$$

which can be shown to be the complete solution. The constant k is fixed for the entire solution set and the constant c is the parameter. In a well-posed problem, auxiliary conditions should allow the determination of both constants to obtain a particular solution.

Example 2.7 It has been established empirically that radioactive elements decompose at approximately a fixed rate, called the *decay rate*. That is, if $y(t)$ is the mass of the substance present at any time, then the rate of change per unit mass, $(dy/dt)/y$, is a constant. The *half-life* of a substance is the amount of time it takes for it to decay to one-half of its initial amount. *Question*: Find the decay rate of a substance whose half-life is 20 years.

Solution We use the differential equation

$$\frac{dy}{dt} = -ky$$

so that the decay constant will be positive. The complete solution is

$$y = ce^{-kt}$$

At $t = 0$, $y = y_0$ and at $t = 20$, $y = y_0/2$. The first condition gives $c = y_0$ and the second one gives the equation

$$\tfrac{1}{2} = e^{-k20}$$

from which

$$k = \frac{\ln 2}{20} \doteq \frac{0.693}{20} \doteq 0.035$$ □

Other physical laws besides those involving decay can be expressed in terms of separable differential equations. The situation is similar to many areas of applied mathematics: The actual mathematics is easy; the difficulty is in establishing a useful mathematical model or abstract description of the physical situation. It is this part of the problem that is often more demanding in terms of analytical maturity.

The next example requires use of *Newton's law of cooling*. Experiments show that if an object is immersed in a medium and if the medium is kept at a constant temperature, then the temperature of the object will change at a rate proportional to the difference in temperature between the object and the medium.

Example 2.8 A steel ball is heated to a temperature of 100° and at time $t = 0$ is placed in water maintained at 10°. At $t = 4$ min, the temperature of the ball is 60°. When is the temperature of the ball reduced to 20°?

Solution The mathematical model of the physical situation may be formulated with the aid of Newton's law of cooling as

$$\frac{dT}{dt} = -k(T - 10), \qquad T(0) = 100, \qquad T(4) = 60$$

where $T(t)$ is the temperature of the ball at any time t and k is the constant of proportionality. Separating variables and antidifferentiating,

$$T(t) = ce^{-kt} + 10$$

The constants c and k may be obtained by applying the conditions at 0 and 4. We obtain $c = 90$ and

$$k = \tfrac{1}{4}\ln\tfrac{9}{5} \doteq \tfrac{1}{4}(2.1972 - 1.6094) \doteq 0.147$$

Therefore

$$T(t) \doteq 90e^{-0.147t} + 10$$

When $T = 20°$,

$$20 \doteq 90e^{-0.147t_0} + 10$$

from which

$$t_0 \doteq \frac{\ln 9}{0.147} \doteq \frac{2.1972}{0.147} \doteq 14.9 \text{ min}$$

□

Exercises for Sections 2.1–2.3

1. Verify that e^x, $3e^x$, and $-e^x$ are solutions to $y' - y = 0$ but that e^{-x} or e^{3x} is not a solution.
2. Verify that $\cos x$, $\sin x$, and any linear combination of $\cos x$ and $\sin x$ are solutions to $y'' + y = 0$. What can you say about the vector space of functions generated by $\sin x$ and $\cos x$?
3. Verify that e^x, e^{-x}, $\sinh x$, and $\cosh x$ are solutions to $y'' - y = 0$. Find a linearly independent subset of these functions. What can you say about the function space generated by this linearly independent set?

In Exercises 4–13, find a family of solutions to the given differential equation.

4. $y' + y = 0$
5. $(\sin x)y' + (\cos x)y = 0$
6. $y' = \dfrac{1 - y^2}{1 - x^2}$
7. $y' = e^{x+y}$
8. $x\cos^2 y + y'\tan y = 0$
9. $x\sqrt{x^2 - a^2}\,y' = a^2$
10. $(4 + e^{2x})y' - ye^{2x} = 0$
11. $(1 - x)y' = y^2$

12. $y' = \sin^2 y \cos^3 x$

13. The equation for the current in a passive LR electrical circuit, $Ldi/dt + Ri = 0$, L and R equal to constants.

In Exercises 14–21, find a particular solution having the specified initial condition.

14. $y' = y; \quad y(0) = 3$

15. $y' = x^2; \quad y(0) = 2$

16. $y' - 3y = 0; \quad y(1) = 7$

17. $y' + y \sin x = 0; \quad y(\pi) = 3$

18. $xyy' - y^2 = 1; \quad y(2) = 1$

19. $y' = xe^{-y-x^2}; \quad y(0) = 0$

20. $xy^2 + e^x y' = 0; \quad y(0) = 1$

21. $y' = \dfrac{x(y^2 - 1)}{(x-1)y^3}; \quad y(2) = 1$

22. A steel ball is heated to a temperature of 80° and at time $t = 0$ is placed in water, maintained at 30°. At $t = 3$ minutes, the temperature of the ball is 55°. When is the temperature of the ball reduced to 40°?

23. At time $t = 0$, a body of temperature 70° is placed outdoors, where the temperature is 20°. After 15 minutes the temperature of the body has fallen to 50°.
(a) What is the temperature of the body after $\frac{1}{2}$ hour?
(b) How long will it take the body to cool to 32°?
Assume Newton's law of cooling is applicable.

24. What is the half-life of a radioactive substance of which 20% disappears in 100 years?

25. After 20 years a quantity of radium has decayed to 60 grams and at the end of 50 years to 40 grams. How many grams were there in the first place?

26. Students sometimes confuse the expression $(y')^2$ with $(y^2)'$. They are not equal in general, of course, but for what functions are the two expressions the same?

27. Find all nontrivial functions such that the derivative is the cube of the function.

28. Find all functions with derivative two more than the square of the function.

29. Find all functions whose derivative is equal to the reciprocal of the function.

30. Experimental data suggest that a number of chemical and biochemical reactions are "second order," that is, the reaction is governed by the initial value problem

$$y'(t) = -ky^2, \qquad y(0) = C$$

where $y(t)$ is the concentration of the given reacting chemical and k is a positive constant.
(a) Determine a solution to this initial value problem.
(b) Describe in detail the graph of $1/y$ as a function of t.

31. A chemical dissociates in a second-order reaction into two products. During the dissociation the following two measurements of concentration $y(t)$ are made:

$$y(5) = 30 \text{ mol/L}; \qquad y(20) = 15 \text{ mol/L}.$$

Determine both the reaction rate k and the initial concentration $y(0)$ of the reactant.

32. The velocity of a body falling through a resisting medium can often be modeled by the equation

$$v'(t) = g - \frac{k}{m}v^2$$

where g is the acceleration due to gravity, m is the mass of the body, and k is a positive constant.

(a) Find a solution to this equation assuming that $v(0) = v_0$.
(b) Show that as $t \to \infty$, the solution approaches a terminal velocity that is independent of the initial velocity.

33. Repeat Exercise 32 under the assumption that

$$v'(t) = g - \frac{k}{m}v$$

34. The equation in Exercise 33 is a reasonable model during the free-fall portion of skydiving. Find $v(t)$ for a 160-lb skydiver if $k = \frac{1}{2}$. Assume $v(0) = 0$, $w = mg$, and $g = 32$.

35. The equation in Exercise 32 is accurate for parachute-assisted descent. Determine $v(t)$ under the same assumptions as in Exercise 34.

36. The "logistic equation"

$$\frac{dP}{dt} = aP - bP^2$$

accurately describes the growth of many populations.
(a) If $P(0) = P_0$, solve the resulting initial value problem.
(b) Find the behavior of $P(t)$ as $t \to \infty$.

37. The population $P(t)$ in a suburb of a large city is modeled by the logistic equation of Exercise 36, in which $a = 10^{-1}$, $b = 10^{-7}$, and t is measured in months.
(a) If $P(0) = 5000$, what is the limiting value of this population?
(b) When will the population reach one-half of this limiting value?

38. In the continuous compounding of interest on investments, the amount of money (capital plus interest) satisfies the differential equation

$$y'(t) = ky$$

where k is the annual interest rate and time is measured in years.
(a) If \$10,000 is invested at the annual rate of 7% compounded continuously, how much will the investment be worth in 10 years?
(b) Determine how long it will take a person to double his or her initial investment if the interest rate is 10% per year compounded continuously.

39. **(a)** Show that the normal distribution in statistics is a solution of the differential equation

$$y' = \frac{a - x}{b}y$$

(b) What is the role of the parameters a and b in the shape of the graph of this distribution?

2.4 Substitutions

Often a substitution can be made that reduces a differential equation to a form that you already know how to solve. In this section you should be more concerned with learning *how* to make a substitution than in memorizing particular types. It is the technique that you will have to recall; the actual substitutions used will often be suggested or self-evident.

If a differential equation can be put into the form

$$y' = f\left(\frac{y}{x}\right)$$

then the substitution $v = y/x$ will reduce the equation to separable form. Since $y = vx$, then $y' = v + v'x$; substituting into the equation,

$$v + v'x = f(v)$$

Separating variables,

$$\frac{v'}{f(v) - v} = \frac{1}{x}$$

which can be explicitly solved by antidifferentiating if $1/[f(v) - v]$ is relatively elementary. It is not difficult to determine when this substitution can be made profitably. Almost any book on elementary differential equations will have more details on how to make this judgment. For us, a bit of trial and error will suffice.

Example 2.9 Solve the differential equation

$$y' = \frac{x - y}{x + y}$$

Solution Note that this equation is not separable as it stands. By dividing numerator and denominator of the right-hand side by x, however, the differential equation takes the form

$$y' = \frac{1 - (y/x)}{1 + (y/x)} = f\left(\frac{y}{x}\right)$$

Using the substitution $y = vx$, we obtain

$$v + v'x = \frac{1 - v}{1 + v}$$

Simplifying,

$$v'x = \frac{1 - v}{1 + v} - v = \frac{1 - v - v - v^2}{1 + v}$$

or

$$\frac{1 + v}{1 - 2v - v^2}v' = \frac{1}{x}$$

An antiderivative of $\dfrac{v + 1}{v^2 + 2v - 1}$ is

$$\int \frac{v + 1}{v^2 + 2v - 1}\,dv = \int \frac{v + 1}{(v + 1)^2 - 2}\,dv = \frac{1}{2}\ln|v^2 + 2v - 1|$$

Using this in the differential equation,

$$\ln |v^2 + 2v - 1|^{-1/2} = \ln |cx|$$

with c an arbitrary constant, and hence

$$(v^2 + 2v - 1)x^2 = c_1$$

where $c_1 = 1/c^2$ and is therefore also "arbitrary." Since $v = y/x$, the solution, in implicit form, is

$$y^2 + 2xy - x^2 = c_1 \qquad \square$$

Example 2.10 Solve the differential equation

$$(x^2 + y^2)\,dx - xy\,dy = 0$$

Solution We may write this equation in the form

$$y' = \frac{x^2 + y^2}{xy} = \frac{1 + (y/x)^2}{y/x}$$

The substitution $y = vx$ gives

$$v'x + v = \frac{1 + v^2}{v}$$

and hence

$$v'x = \frac{1}{v}$$

Therefore

$$v\,dv = \frac{dx}{x}$$

whose solution is

$$\frac{v^2}{2} = \ln |cx|$$

In terms of y,

$$y^2 = 2x^2 \ln |cx| \qquad \square$$

Sometimes, two successive substitutions are required to reduce the equation to separable form. For example, there are occasions when we express x and y linearly in terms of two new variables X and Y followed by $v = Y/X$.

Example 2.11 Solve the differential equation

$$y' = \frac{x - y + 1}{x + y - 1}$$

Solution We let $X = x - h$ and $Y = y - k$, and assume that $Y(X) = y(x) = y(X + h)$. Then

$$\frac{dY}{dX} = \frac{dY}{dx}\frac{dx}{dX} = \frac{dy}{dx}$$

Hence, after substitution, the differential equation is

$$\frac{dY}{dX} = \frac{X + h - Y - k + 1}{X + h + Y + k - 1}$$

The given substitution will be completely specified when the values of h and k are fixed. We now choose h and k to eliminate the constant terms from the numerator and denominator by simultaneously satisfying the two equations

$$h - k + 1 = 0$$
$$h + k - 1 = 0$$

This system is satisfied by $h = 0$, $k = 1$. The differential equation then has the form

$$\frac{dY}{dX} = \frac{X - Y}{X + Y}$$

The solution to this was obtained in Example 2.9:

$$Y^2 + 2XY - X^2 = c$$

Therefore the solution to the given differential equation is

$$(y - 1)^2 + 2(y - 1)x - x^2 = c$$ □

Some substitutions are suggested by the appearance of a group of terms in the differential equation.

Example 2.12 Solve the differential equation

$$y' = \frac{x + y - 1}{x + y + 1}$$

Solution The occurrence of the expression $x + y$ in both the numerator and denominator suggests that we let $u = x + y$. From $y = u - x$, we get $y' = u' - 1$, and substituting into the differential equation,

$$u' - 1 = \frac{u - 1}{u + 1}$$

$$u' = \frac{u - 1}{u + 1} + 1 = \frac{u - 1 + u + 1}{u + 1} = \frac{2u}{u + 1}$$

$$\frac{u + 1}{u}\,du = 2\,dx$$

Antidifferentiating both sides,

$$u + \ln|u| = 2x + c$$

The solution, in implicit form, is therefore

$$y - x + \ln|x + y| = c \qquad \square$$

2.5 Exact Differential Equations

In this section we consider equations which are not separable but whose solution is found by two antidifferentiations.

First recall that the *total differential* of a function of two variables $u(x, y)$ is given by

$$du = u_x\,dx + u_y\,dy$$

Thus if we know the two partial derivatives of a function, its total differential is determined. Conversely, as the following example will show, a knowledge of the total differential is sufficient to determine the function within an arbitrary constant.

Example 2.13 The total differential of a function is given by

$$du = 3x(xy - 2)\,dx + (x^3 + 2y)\,dy$$

Find $u(x, y)$.

Solution The coefficients of dx and dy are u_x and u_y. Therefore

$$u_x = 3x^2y - 6x \quad \text{and} \quad u_y = x^3 + 2y$$

Partially antidifferentiating the first equation gives

$$u(x, y) = x^3y - 3x^2 + C(y)$$

where $C(y)$ stands for an arbitrary function of y only, since y is constant for the partial antidifferentiation. From this function, u_y is found to be

$$u_y = x^3 + C'(y)$$

Equating this to the given expression for u_y,

$$x^3 + C'(y) = x^3 + 2y$$

This means that $C(y) = y^2 + c$, where c is an arbitrary constant, and therefore

$$u(x, y) = x^3y - 3x^2 + y^2 + c \qquad \square$$

The following definition should establish the connection between the concept of a total differential and the subject of first-order ordinary differential equations.

Definition 2.1 The first-order differential equation

$$M(x, y)\,dx + N(x, y)\,dy = 0$$

is said to be *exact* if the left-hand side of the equation is the total differential of some function $u(x, y)$ so the equation may be written as $du = 0$.

The differential du may be interpreted as the change of the "height" of the plane tangent to the surface described by $z = u(x, y)$ when the independent variables x and y are given increments dx and dy. This being the case, "$du = 0$" describes a surface, which at each point has a tangent plane parallel to the xy-plane, that is, $u(x, y) \equiv c$.

Note that the crux of the problem in finding the solution to an exact equation is in finding the function $u(x, y)$. This function itself is *not* the solution but rather, when equated to a constant c, it defines *implicitly* a solution function y in terms of x.

Example 2.14 Solve the differential equation

$$3x(xy - 2)\,dx + (x^3 + 2y)\,dy = 0$$

Solution In the previous example we found a function u such that du is the left-hand side of this equation. Hence the implicit solution is

$$x^3y - 3x^2 + y^2 = c \qquad \square$$

Example 2.15 Solve the differential equation

$$(y\cos x - \sin y)\,dx + (\sin x - x\cos y)\,dy = 0$$

under the assumption that the left-hand side is of the form $du = 0$.

Solution Since the left-hand side is the differential of some unknown function, the coefficient of dx must equal u_x and the coefficient of dy must equal u_y:

$$u_x = y\cos x - \sin y$$
$$u_y = \sin x - x\cos y$$

Antidifferentiating the first of these equations,

$$u(x, y) = y\sin x - x\sin y + C(y)$$

The partial derivative of this function is

$$u_y = \sin x - x\cos y + C'(y)$$

Equating this expression for u_y to that previously given, we obtain $C'(y) = 0$ and hence $C(y)$ is a constant. Therefore the solution in implicit form is

$$y\sin x - x\sin y = c \qquad \square$$

The previous example could have been solved by grouping terms. Note that the sum $-\sin y\,dx - x\cos y\,dy$ is the differential of $-x\sin y$; the sum

$y \cos x \, dx + \sin x \, dy$ is the total differential of $y \sin x$. Hence the left-hand side is the differential of the sum $y \sin x - x \sin y$.

Now let us see what would happen if we tried to apply the technique for exact equations to one that was not of the proper form.

Example 2.16 Attempt to solve the equation $(x + y)\,dx + (y - x)\,dy = 0$ as if it were an exact equation.

Solution (Note that this differential equation *can* be solved by use of the substitution $v = y/x$.) If we assume (erroneously) that the differential equation is exact, we obtain the two requirements on the function $u(x, y)$,

$$u_x = x + y, \qquad u_y = y - x$$

Antidifferentiating the first of these, we get

$$u(x, y) = \frac{x^2}{2} + xy + C(y)$$

where $C(y)$ is assumed to be a function of y only. Finding u_y from this expression and equating it to the given one,

$$x + C'(y) = y - x$$

or

$$C'(y) = y - 2x$$

and this contradicts the assumed condition that $C(y)$ is independent of x. □

The following theorem provides us with a conclusive test of the exactness of a differential equation.

Theorem 2.1 Let $M(x, y)$ and $N(x, y)$ be functions with continuous first partial derivatives over some domain $D = \{(x, y) | a < x < b, c < y < d\}$. Then the differential equation

$$M(x, y)\,dx + N(x, y)\,dy = 0$$

is exact *if and only if* $M_y = N_x$.

Proof If the differential equation is exact, then there exists a function $u(x, y)$ such that the left-hand side of the equation is of the form $u_x\,dx + u_y\,dy$. Thus,

$$M = u_x \quad \text{and} \quad N = u_y$$

Partially differentiating the first of these with respect to y and the second with respect to x,

$$M_y = u_{xy} \quad \text{and} \quad N_x = u_{yx}$$

Since M_y and N_x are assumed to be continuous, so are u_{xy} and u_{yx}. Hence, from a theorem of elementary calculus, $u_{xy} = u_{yx}$; that is, $M_y = N_x$. The converse of the theorem is a bit more difficult for it requires the construction of a function $u(x, y)$ from $M(x, y)$ and $N(x, y)$ given that $N_x = M_y$. The procedure parallels the examples. Define $u(x, y)$ by the equation

$$u(x, y) = \int_x M(x, y)\,dx + C(y)$$

where $C(y)$ is to be determined later. (The x under the integral sign means to antidifferentiate with respect to x, holding y constant.) From this expression, we compute u_y,

$$u_y = \frac{\partial}{\partial y}\int_x M(x, y)\,dx + C'(y)$$

and this is equal to $N(x, y)$. Therefore

$$C'(y) = N(x, y) - \frac{\partial}{\partial y}\int_x M(x, y)\,dx$$

There will be a solution if $C'(y)$ is independent of x, and this will be the case if the derivative of the right-hand side with respect to x is zero. Carrying out this differentiation,

$$\frac{\partial}{\partial x}\left[N(x, y) - \frac{\partial}{\partial y}\int_x M(x, y)\,dx\right] = \frac{\partial N}{\partial x} - \frac{\partial}{\partial x}\frac{\partial}{\partial y}\int_x M(x, y)\,dx = \frac{\partial N}{\partial x} - \frac{\partial M}{\partial y}$$

under the assumption that the partial differentiations with respect to x and y may be interchanged. The last expression is zero because of the hypothesis. ■

Example 2.17 Determine the solution of the initial value problem

$$(2xy + 3y)\,dx + (4y^3 + x^2 + 3x + 4)\,dy = 0, \qquad y(0) = 1$$

Solution Since $M_y = 2x + 3 = N_x$, the equation is exact. Therefore the partial derivatives of the desired function $u(x, y)$ are given by

$$u_x = 2xy + 3y \quad \text{and} \quad u_y = 4y^3 + x^2 + 3x + 4$$

Antidifferentiating the first of these,

$$u(x, y) = x^2 y + 3xy + C(y)$$

From this expression we find u_y and equate it to the known expression for u_y,

$$x^2 + 3x + C'(y) = 4y^3 + x^2 + 3x + 4$$

Therefore

$$C'(y) = 4y^3 + 4$$

and

$$C(y) = y^4 + 4y$$

Hence, the implicit solution to the differential equation is

$$x^2y + 3xy + y^4 + 4y = c$$

The initial condition is satisfied by setting $c = 5$. □

Exercises for Sections 2.4 and 2.5

In Exercises 1–9 solve the differential equations by first making the substitution $y = vx$.

1. $y' = \dfrac{xy}{x^2 + y^2}$

2. $y' = \dfrac{2x + y}{x - 2y}$

3. $y' = \dfrac{x^2 - y^2}{xy}$

4. $y' = \dfrac{x^3 + xy^2}{yx^2 - y^3}$

5. $y' = \dfrac{xy}{(x + 2y)^2}$

6. $(x^2 - y^2e^{-x/y})y' = xy$

7. $y' = \dfrac{y - x\csc(y/x)}{x}$

8. $y' = -\dfrac{x^2 + y^2}{xy}$

9. $y' = \dfrac{x^2 + 2xy - 4y^2}{x^2 - 8xy - 4y^2}$

In Exercises 10–12, solve by first making linear substitutions.

10. $y' = \dfrac{x + 2y - 1}{x + 2y + 7}$

11. $y' = \dfrac{x - y + 8}{y - 3x + 2}$

12. $y' = \dfrac{2x + 3y - 4}{x - y}$

13. $y' = \dfrac{y + x}{y - x}$; solve by using a substitution.

14. Solve Exercise 13 by another method.

In Exercises 15–24, determine which of the equations are exact, and solve those which are.

15. $(x^2 + y^2)\,dy + 2xy\,dx = 0$

16. $(y^2 - x^2)\,dx + 2xy\,dy = 0$

17. $(2x + y)\,dx + (y - x)\,dy = 0$

18. $(2x + y)\,dx + (x - y)\,dy = 0$

19. $(ye^x - 2x)\,dx + e^x\,dy = 0$

20. $(ye^{xy} + 2xy)\,dx + (xe^{xy} + x^2)\,dy = 0$

21. $(6x + y^2)\,dx + y(2x - 3y)\,dy = 0$

22. $(\cos 2y - 3x^2y^2)\,dx + (\cos 2y - 2x\sin 2y - 2x^3y)\,dy = 0$

23. $(1 + y^2)\,dx + (x^2y + y)\,dy = 0$

24. $(15x^2y^2 - y^4)\,dx + (10x^3y - 4xy^3 + 5y^4)\,dy = 0$

25. Show that an equation of the form $f(x)\,dx + g(y)\,dy = 0$ is exact.

26. Under what conditions is the differential equation $(x + ay)\,dx + (y + bx)\,dy = 0$ exact?

27. Under what conditions is the differential equation

$$(x^2 + axy + by^2)\,dx + (y^2 + cxy + dx^2)\,dy = 0$$

exact?

28. Find the most general function $M(x, y)$ such that the differential equation $M(x, y)\,dx + [\sin(x^2) - x^2]\,dy = 0$ is exact.

29. Find the most general function $N(x, y)$ such that the differential equation $(ye^{xy} - 2y^3)\,dx + N(x, y)\,dy = 0$ is exact.

In Exercises 30–32, solve the given differential equation by *two* methods.

30. $(x + 3y)\,dx + (3x + y)\,dy = 0$

31. $(x^2 - 4xy)\,dx - (2x^2 - y^2)\,dy = 0$

32. $(x^2 + y^2)\,dx + (y^2 + 2xy)\,dy = 0$

Determine the solutions of the initial value problems in Exercises 33–38.

33. $(y - \sqrt{x^2 + y^2})\,dx - x\,dy = 0,\ y(\sqrt{3}) = 1$

34. $(y^3 - x^3)\,dx - xy^2\,dy = 0,\ y(1) = 2$

35. $(\cos x \sin x - xy^2)\,dx - y(x^2 - 1)\,dy = 0,\ y(0) = 2$

36. $x^2\,dx + ye^y\,dy = 0,\ y(0) = 1$

37. $[2x + y\cos(xy)]\,dx + x\cos(xy)\,dy = 0,\ y(2) = -\pi/4$

38. $(1 + y^2)\,dx + (x^2 y + y)\,dy = 0,\ y(1) = 0$

39. The solutions of the differential equation

$$y' = \frac{ax - y}{x + by}$$

are conic sections.

(a) Determine values of the parameters a and b that yield circles.

(b) When can straight lines result?

If a one-parameter family of curves is given by the equation

$$F(x, y) = c$$

the *orthogonal trajectories* of the family consist of curves that everywhere cross the original curves at right angles.

40. Show that the orthogonal trajectories of the family of curves $F(x, y) = c$ satisfy the differential equation

$$\frac{dy}{dx} = \frac{F_y}{F_x}$$

41. Find the orthogonal trajectories of the one-parameter family of circles $x^2 + y^2 = 2cx$.

42. Determine the orthogonal trajectories of the one-parameter family of curves

$$x^2 + 2xy - y^2 + 4x - 4y = c$$

2.6 Integrating Factors; Linear Equations of the First Order

If a differential equation is not exact, we can sometimes "make it exact" by multiplying the equation by a function called an *integrating factor*. Finding an integrating factor for a first-order equation is equivalent to solving it, for as was shown in the previous section, an exact equation may be solved by finding two partial antiderivatives. Generally, we are not concerned with the uniqueness of a given integrating factor and, in fact, they usually are not unique. For example, the functions x^{-2} and y are both integrating factors of the differential equation

$$\frac{x^2}{y}dy + 2x\,dx = 0$$

There actually exists an elegant theorem guaranteeing that, under very general conditions on M and N, there is always an integrating factor for equations of the form

$$M(x, y)\,dx + N(x, y)\,dy = 0$$

It would seem, then, that the whole study of first-order equations should be reduced to the determination of integrating factors. The trouble is that finding these integrating factors is more art than science. We restrict our study of integrating factors, therefore, to the simple but very important case of linear equations.

Recall that all first-order linear equations may be written in the form

$$a_1(x)y' + a_0(x)y = r(x)$$

or

$$y' + p(x)y = f(x)$$

This latter form is called "standard."

We will now show that an integrating factor can always be found for this kind of differential equation and hence the entire class of first-order linear equations is solvable. To show this, assume that an integrating factor exists which is a function of x only. If the integrating factor is $v(x)$,

$$v(x)\,dy + [v(x)p(x)y - v(x)f(x)]\,dx = 0$$

Applying the condition for exactness,

$$\frac{\partial}{\partial x}[v(x)] = \frac{\partial}{\partial y}[v(x)p(x)y - v(x)f(x)]$$

or

$$v' = p(x)v$$

This separable equation can be solved for $v(x)$ to give

$$v(x) = e^{\int p(x)\,dx}$$

where we have ignored the constant of integration since we only need *one* integrating factor.

Multiplying the first-order equation in standard form by this integrating factor we obtain

$$y'e^{\int p(x)\,dx} + p(x)ye^{\int p(x)\,dx} = f(x)e^{\int p(x)\,dx}$$

The left-hand side of this equation is the derivative of the product $ye^{\int p(x)\,dx}$ and hence

$$ye^{\int p(x)\,dx} = \int f(x)e^{\int p(x)\,dx}\,dx + c$$

Thus, if the indicated antidifferentiations can be performed, the solution can be found explicitly as

$$y = e^{-\int p(x)\,dx}\int f(x)e^{\int p(x)\,dx}\,dx + ce^{-\int p(x)\,dx}$$

Substitution into this expression is *not* the recommended procedure. It is better to recall that an integrating factor is $e^{\int p(x)\,dx}$, multiply both sides of the equation by this function, and find the antiderivative of both sides, as indicated in Example 2.18. If some of the antidifferentiations cannot be performed, definite integrals should be used.

Example 2.18 Solve the differential equation

$$xy' + 2y = x^2$$

Solution First recognize that this equation is linear and of the first order. To put the equation in standard form, divide both sides of the equation by x,

$$y' + \frac{2}{x}y = x$$

The integrating factor is

$$e^{\int (2/x)\,dx} = e^{\ln x^2} = x^2$$

Multiplying the differential equation (in standard form) by this factor,

$$x^2y' + 2xy = x^3$$

The left-hand side *must* be the derivative of y times the integrating factor. Thus

$$\frac{d(yx^2)}{dx} = x^3$$

and

$$yx^2 = \frac{x^4}{4} + c$$

In explicit form, this is

$$y = \frac{x^2}{4} + \frac{c}{x^2}$$ □

Linear first-order equations arise naturally in some recognizably important applications. For example, the electric circuit with negligible capacitance but nonnegligible resistance R and inductance L (see Figure 2.1) is described by the first-order linear equation

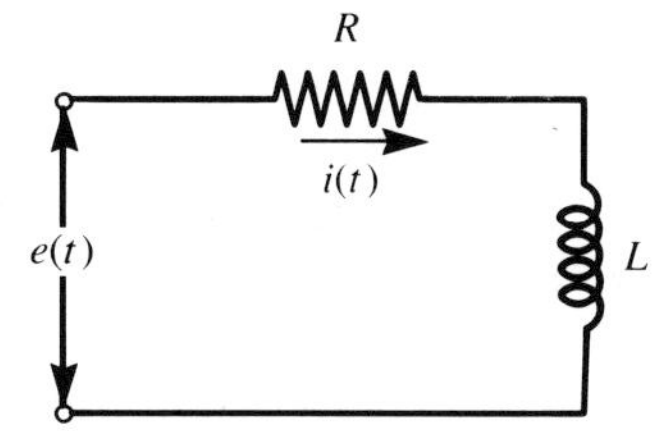

Figure 2.1 RL circuit

$$L\frac{di}{dt} + Ri = e(t)$$

where $e(t)$ is the driving voltage and $i(t)$ is the current in the circuit. Often L and R are considered constants, but generally there would be no actual requirement for this to be so. The solution function, which is the current in the circuit as a function of time, is sometimes called the *response.*

Example 2.19 Find the current in an LR circuit, where L and R are constants, $e(t)$ is a d.c. (constant) voltage, and the initial current is I_0.

Solution Let E be the input voltage. Then the differential equation of the circuit is

$$L\frac{di}{dt} + Ri = E$$

In standard form, the equation is

$$\frac{di}{dt} + \frac{R}{L}i = \frac{E}{L}$$

Hence, an integrating factor is $e^{Rt/L}$, which when used on the equation yields

$$d(ie^{Rt/L}) = \frac{E}{L}e^{Rt/L}\,dt$$

Antidifferentiating and solving for i,

$$i = \frac{E}{R} + ce^{-Rt/L}$$

The condition $i(0) = I_0$ yields $c = I_0 - (E/R)$, and therefore the current in the circuit is given by

$$i = \frac{E}{R} + \left(I_0 - \frac{E}{R}\right)e^{-Rt/L}$$

See Figure 2.2. □

Example 2.20 Find the response of an RL circuit to a unit square wave. Assume the current is initially zero. See Figure 2.3(a).

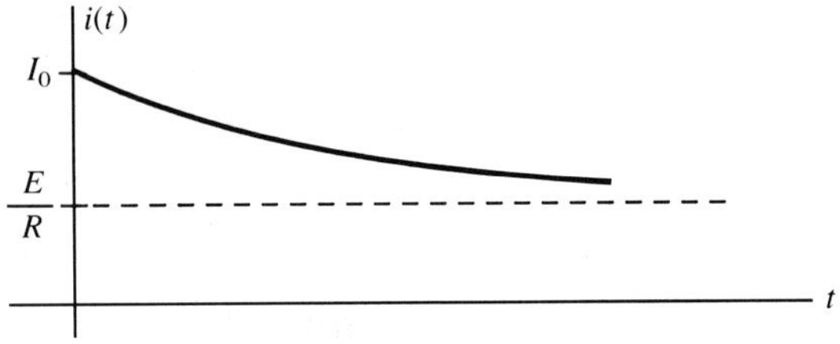

Figure 2.2

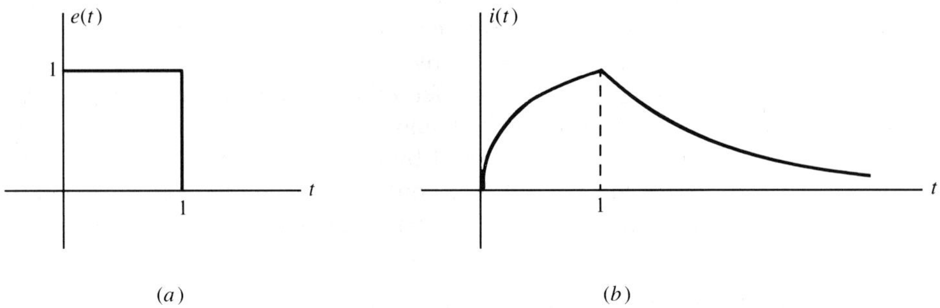

Figure 2.3

Solution For the interval $0 \leqq t \leqq 1$, the current is the same as in the previous example, with $E = 1$ and $I_0 = 0$. Substituting these values,

$$i(t) = \frac{1}{R}(1 - e^{-Rt/L}), \qquad 0 \le t \le 1$$

Since on physical grounds we expect the current in this circuit to be a continuous function of time, the mathematical model of the circuit for $t \ge 1$ is given by

$$L\frac{di}{dt} + Ri = 0$$

$$i(1) = \frac{1}{R}(1 - e^{-R/L})$$

The differential equation is easily solved by the method of separation of variables:

$$i(t) = c_1 e^{-Rt/L}$$

Applying the continuity condition at $t = 1$,

$$i(1) = c_1 e^{-R/L} = \frac{1}{R}(1 - e^{-R/L})$$

from which

$$c_1 = \frac{1}{R}(e^{R/L} - 1)$$

Therefore the current for $t \geqq 1$ (see Figure 2.3(b)) is:

$$i(t) = \frac{e^{R/L} - 1}{R} e^{-Rt/L}$$

The expression

$$i(t) = \begin{cases} \dfrac{1}{R}(1 - e^{-Rt/L}), & 0 \leq t \leq 1 \\ \dfrac{e^{-Rt/L}}{R}(e^{R/L} - 1), & t \geq 1 \end{cases}$$

whose graph is indicated in Figure 2.3(b), is patched together from solutions over two intervals to give a *quasi-solution* over $[0, \infty)$. We have assured the continuity of this expression by proper choice of the initial condition at $t = 1$, but the expression is *not* differentiable and thus is not a bona fide solution of our problem over $[0, \infty)$. This should not cause too much concern, since the nondifferentiability at $t = 1$ was caused by the discontinuity in the driving voltage $e(t)$. If $e(t)$ had been modeled by a continuous and thus more realistic function, we would have obtained a legitimate differentiable solution for the entire domain $t \geq 0$. □

Example 2.21 Find the current in an RL circuit that has zero initial current, a resistance of $2\ \Omega$, an applied voltage of 4 V, and an inductance which varies as

$$L(t) = \begin{cases} 5 - t, & 0 \leqq t \leqq 5 \\ 0, & 5 \leqq t \end{cases}$$

Solution The mathematical model for the circuit for $0 \leqq t \leqq 5$ is

$$(5 - t)\frac{di}{dt} + 2i = 4, \qquad i(0) = 0$$

The solution for this time interval is obtained by first dividing by $5 - t$ to put the equation into standard form. (*Note*: Do not make the error of applying the result of Example 2.19 with $L = 5 - t$. That result applies only to the case in which L is a constant.)

$$\frac{di}{dt} + \frac{2}{5 - t} i = \frac{4}{5 - t}$$

An integrating factor is

$$e^{\int [2/(5-t)]\, dt} = e^{-2 \ln (5-t)} = (5 - t)^{-2}$$

Applying this integrating factor to the differential equation,

$$\frac{d(i(5 - t)^{-2})}{dt} = \frac{4}{(5 - t)^3}$$

Therefore

$$i(5 - t)^{-2} = 2(5 - t)^{-2} + c$$

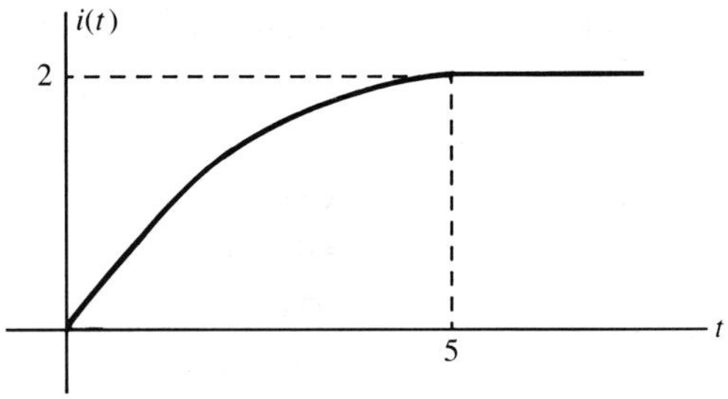

Figure 2.4

or

$$i = 2 + c(5 - t)^2$$

The value of the constant c is determined from the fact that $i = 0$ when $t = 0$,

$$0 = 2 + c(5)^2$$
$$c = -\tfrac{2}{25}$$

Hence,

$$i(t) = 2 - \frac{2(5 - t)^2}{25}$$
$$= \frac{2t}{5}\left(2 - \frac{t}{5}\right)$$

For $t \geqq 5$, the differential equation becomes

$$0\frac{di}{dt} + 2i = 4$$

whose solution is $i = 2$ (See Figure 2.4.)

Again, we have only a quasi-solution occasioned by the discontinuity at $t = 5$ in the coefficients of the equation in standard form. □

Exercises for Section 2.6

In Exercises 1–10, solve by using the standard integrating factor for linear equations of the first order.

1. $3y' + 2y + x = 0$
2. $y' \cos x - y \sin x + e^x = 0$
3. $y' - 4y = x - x^2$
4. $y' + y = \cos x$
5. $y' + 2y = e^{-2x}$
6. $y' + 2xy = 2x$
7. $xy' + (x - 2)y = 3x^3 e^{-x}$
8. $(x^2 + 9)y' + xy = 0$
9. $xy' + y = x \sin x$
10. $xy' + (1 + x)y = e^{-x} \sin 2x$

Solve the initial value problems in Exercises 11–20.

11. $xy' + y = 2x, \quad y(2) = 2$ **12.** $y' + 2xy = x^3, \quad y(1) = 1$

13. $y' + (\cos x)y = 0, \quad y(0) = 2$

14. $(x^2 + 1)y' - 2xy = x^2 + 1, \quad y(1) = \pi$

15. $(x \ln x)y' - y = 0, \quad y(e) = -1$

16. $(x - 1)y' + xy = x(x - 1)^2, \quad y(-1) = 14$

17. $y' + (\tan x)y = \cos^2 x, \quad y(0) = -1$

18. $y' + y = f(x), \quad y(0) = 0 \quad \text{where } f(x) = \begin{cases} 1 \ , & 0 \le x < 1 \\ 0 \ , & x \ge 1 \end{cases}$

19. $y' + 2xy = f(x), \quad y(0) = 2 \quad \text{where } f(x) = \begin{cases} x, & 0 \le x < 2 \\ 0, & x \ge 2 \end{cases}$

20. $y' + y = f(x), \quad y(0) = 1 \quad \text{where } f(x) = \begin{cases} e^{-x}, & 0 \le x < a \\ e^{-a}, & x \ge a \end{cases}$

21. Find the current in an RL circuit, initially quiescent (zero initial current), with input voltage as given for each of the following cases. R and L are assumed to be constant.
(a) e^{-t} **(b)** $\sin t$ **(c)** t

22. Find the current in an RL circuit with $R = 1\ \Omega$, $L = 1$ H, and input voltage as shown in Figure 2.5. Assume that the initial current is zero.

23. Consider an RL circuit with $R = 10\ \Omega$, input voltage 20 V, and

$$L = \begin{cases} 5 - t \ , & 0 \leqq t \leqq 5 \\ 0 \ , & 5 \leqq t \end{cases}$$

Find the current in the circuit if the initial current is zero.

24. Repeat Exercise 23 with the assumption that the initial current is 2 A.

25. Find the current in the RL circuit of Exercise 22 if the input voltage is as shown in Figure 2.6. Assume that the initial current is 1 A.

26. If the input voltage of the previous exercise is extended periodically, what should the initial current be to assure a periodic output?

27. Consider the RLcircuit of Exercise 22 with the periodic square wave input voltage shown in Figure 2.7. What should the initial current be to assure a periodic output?

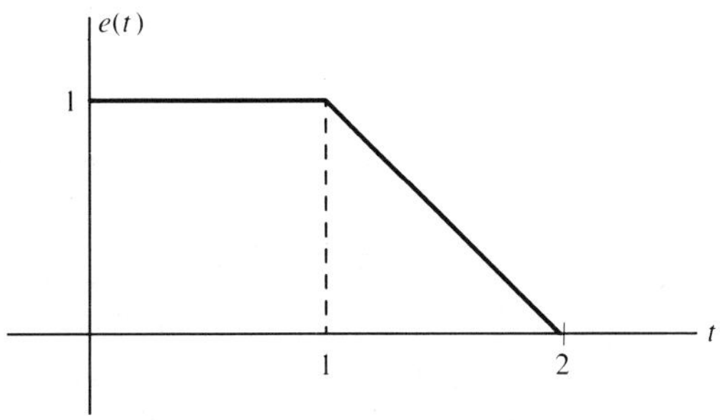

Figure 2.5

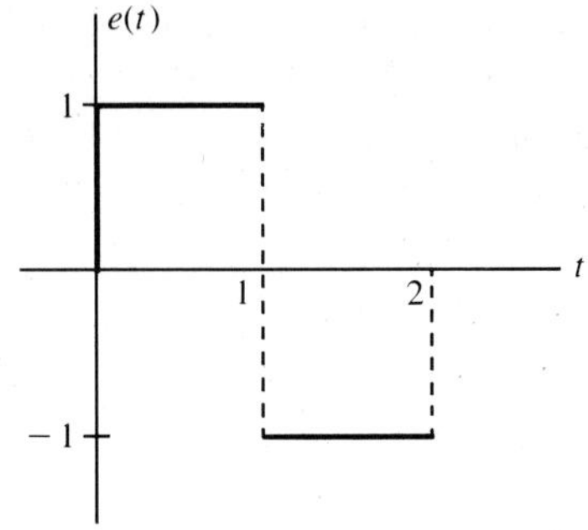

Figure 2.6

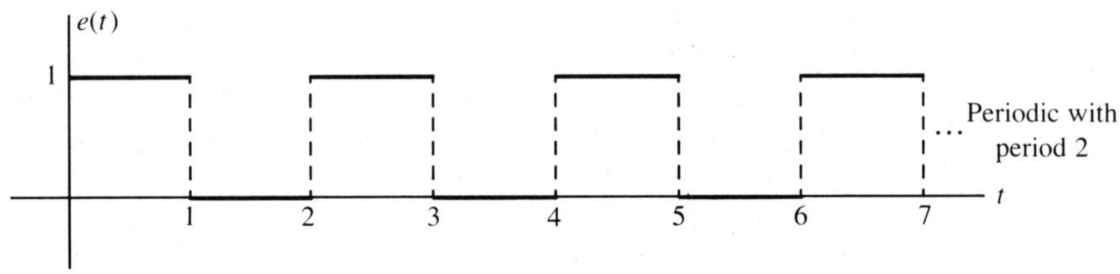

Figure 2.7

28. In Example 2.8 on p. 49, we solved the differential equation

$$\frac{dT}{dt} = -k(T - T_0)$$

associated with Newton's law of cooling by separating variables. Redetermine the solution of this equation by treating it as a first-order linear equation and finding an integrating factor.

29. A simple mathematical model of learning is given by the differential equation

$$y' = A(M - y)$$

Here $y(t)$ is the amount of material learned at time t, M is the total amount of material to be learned, and A is a positive constant that incorporates an individual's natural learning ability.

(a) Solve the initial value problem that results when you assume no material has been learned at time $t = 0$.

(b) Show that $y(t)$ approaches M asymptotically as $t \to \infty$.

30. In economics the rate of change of the price P of a commodity is approximately proportional to the difference between the demand D for that commodity and the supply S, that is,

$$\frac{dP}{dt} = \alpha(D - S)$$

Determine $P(t)$ in the case that both D and S are linear functions of P.

Interchanging the roles of the dependent and independent variables allows us easily to solve those first-order equations that are *linear in x*.

Apply this method to determine the solutions of the given differential equations in Exercises 31–33.

31. $(x + y^2)\dfrac{dy}{dx} = 1$

32. $x\dfrac{dy}{dx} = y - y^2$

33. $(xy + 2x - ye^y)\dfrac{dy}{dx} + y = 0$

The homogeneous equation

$$y' + a(x)y + b(x)y^n = 0$$

in which n is a constant is called *Bernoulli's differential equation.* When $n = 0$ or 1, it is a linear equation. When $n \neq 0$ or 1 it can be reduced to linear form by the substitution $u = y^{1-n}$.

In Exercises 34–36, solve the given Bernoulli equations.

34. $y' + \dfrac{1}{x}y + x^3y^2 = 0$

35. $y' - y + xe^{-2x}y^3 = 0$

36. $xyy' = y^2 - x^2$

2.7 Approximation by Straight Lines

In many practical situations only the numerical values of the solution for one or several values of the independent variable are of interest. The description of the solution in terms of elementary functions and algebraic operations may be relatively unimportant or, from a practical standpoint, very difficult or impossible to obtain. Thus it may be preferable to describe the solution by a table of values such as

x	2	2.5	3	3.5	4	4.5	5
y	3.72	4.62	5.19	6.94	9.11	10.01	11.23

or by way of a graph rather than by an analytic formula.

Approximate methods for solving differential equations graphically or numerically are much more powerful (that is, they apply to a much larger collection of types of differential equations) than analytical methods. In practice, the tolerance imposed on the admissible error determines the appropriate method to be used.

In this section we consider first-order equations written in the form $y' = f(x, y)$. This differential equation essentially specifies the value of the slope of any solution at every point in the plane. We therefore attach to each point (x_0, y_0) a third number, which is the value of the slope at the point. This slope is computed by evaluating $f(x_0, y_0)$. To display this graphically we sketch short line segments at a "reasonable" number of points in the plane, each line segment having slope as determined by $f(x, y)$. The complete set of line elements is known as the *direction field* of the differential equation.

The direction field specifies the direction of solution curves. Particular solutions may be sketched by beginning at a given point (a, b) and continuously modifying the slope of the curve in accordance with the direction of the line elements.

Example 2.22 Sketch the direction field associated with the differential equation $y' = x + y$ and draw the particular solution curve that passes through the point $(0, 1)$.

Solution Figure 2.8 shows both the sketch of the direction field and the particular solution. In sketching the set of line elements, it is helpful to consider first the curves along which the direction of the line segments is the same. Such curves are called *curves of constant slope* of the direction field, and are obtained by setting $f(x, y)$ equal to a constant. In this case we obtain the lines $x + y = c$. These lines are not solution curves but are meant to assist in the drawing of the direction field. The curve of constant slope zero is particularly useful. □

The value of such a graphical approach should not be underestimated. Although the degree of accuracy is largely dependent upon the completeness of the sketch of the direction field, a good qualitative view of particular solutions can be obtained from even a very crude direction field.

Even in cases where analytical methods are being applied, the direction field may be a useful auxiliary tool. In Section 2.3 we took the rather casual approach of finding only a family of solutions to a separable equation. In the following example we use the direction field to help us look for a complete solution in a touchy case.

Example 2.23 Find the complete solution of $y' = \sqrt{1 - y^2}$.

Solution Formal, and careless, separation of variables and antidifferentiation leads to

$$\int \frac{dy}{\sqrt{1 - y^2}} = \int dx, \qquad \sin^{-1} y = x + c, \qquad y = \sin(x + c)$$

Even a glance back at the differential equation alerts us to trouble. Since y' is never negative, only parts of the proposed sine curves could possibly correspond

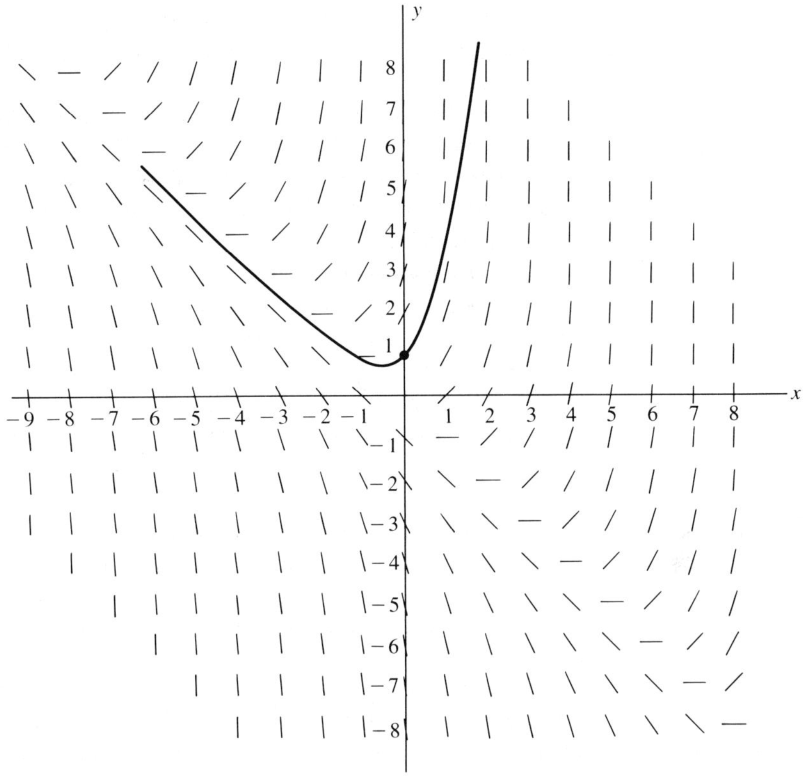

Figure 2.8

to solutions. Let us look at the direction field and sample approximate solution curves in Figure 2.9. We note from this field that $y \equiv 1$ and $y \equiv -1$ must be solutions. This is readily verified from the differential equation. We also see that the parts of sine curves with positive slope join smoothly to the two horizontal lines $y = 1$ and $y = -1$.

We can now piece together a complete solution:

$$y = 1 \quad \text{or} \quad y = -1$$

or

$$y = \begin{cases} -1, & x + c \le -\dfrac{\pi}{2} \\ \sin(x + c), & -\dfrac{\pi}{2} \le x + c \le \dfrac{\pi}{2} \\ +1, & x + c \ge \dfrac{\pi}{2} \end{cases}$$

This solution could also have been obtained by *very* careful attention to the range and domain of $\sin^{-1} y = x + c$ and the implications of having $\sqrt{1 - y^2}$ in the denominator in our analytical approach. □

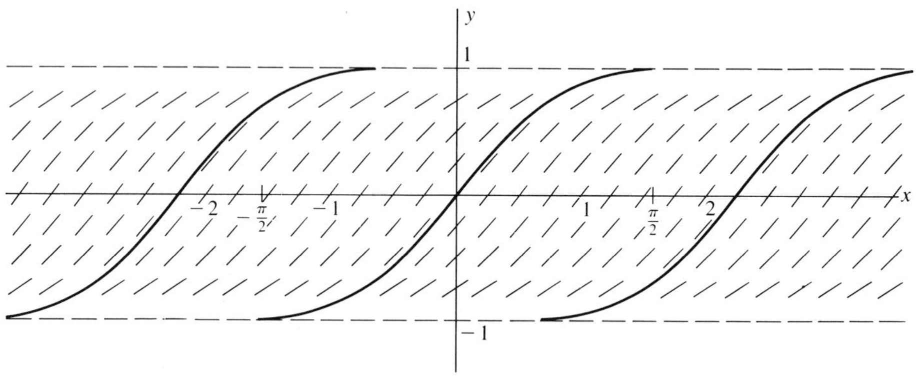

Figure 2.9

The foregoing geometric approach may be modified to give a numerical solution. The graphical method yields a solution curve whose tangent is continuously changing. Unless other information is given, it is not possible actually to write the equation of the curve directly from its graph. However, if the changes to the slope of the solution curve are made in discrete incremental steps rather than continuously, then the approximation curve is a polygonal line whose analytic description can be written on each interval.

The approximating line through (a, b) will have slope $f(a, b)$, and thus its equation is

$$\frac{y - b}{x - a} = f(a, b)$$

This equation, when written in the form

$$y = f(a, b)(x - a) + b$$

expresses y as a function that we are willing to accept as a reasonable approximation to the actual solution if x is restricted to a small interval $a \leqq x \leqq a + h$. In particular, this approximation gives the point $x_1 = a + h$, $y_1 = f(a, b)h + b$. The point (x_1, y_1) and the slope $f(x_1, y_1)$ determine a new linear approximation

$$y = f(x_1, y_1)(x - x_1) + y_1$$

that we are willing to accept over $a + h \leqq x \leqq a + 2h$. The process is continued in this manner until the desired value of x is reached. The value of the function used to describe the line segment at that value of x is considered to be the approximate value of the solution.

Example 2.24 Approximate the value of the solution at $x = 3$ to the initial value problem $y' = y + x^2$, $y(0) = 1$ by using increments of length 1 along the x-axis.

Solution The first line passes through $(0, 1)$ with slope 1. Its equation is $y = 1 + x$, from which $y = 2$ when $x = 1$. From the differential equation, the slope of the direction field at $(1, 2)$ is 3. Thus the equation of the second line is $y = 3x - 1$. Similarly, the third line passes through $(2, 5)$ with slope 9 and its equation is

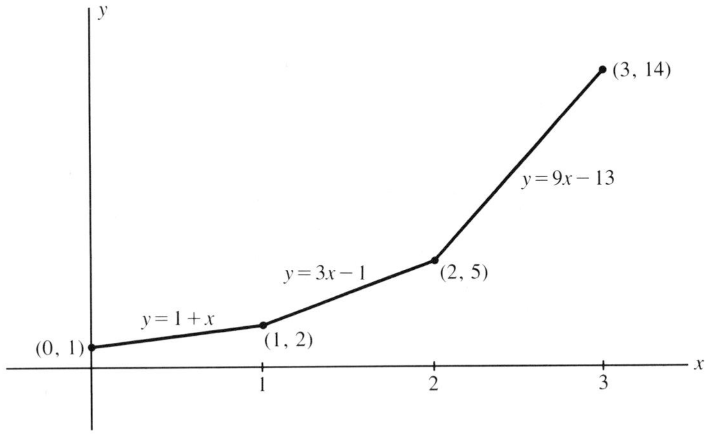

Figure 2.10

$y = 9x - 13$. From this last equation we obtain the value of y when $x = 3$ to be 14. The graph is sketched in Figure 2.10. □

It might be noted that the initial value problem of this example can be solved explicitly:

$$y(x) = 3e^x - x^2 - 2x - 2$$

The exact value of y at $x = 3$ is thus $y(3) = 3e^3 - 17 \doteq 43.3$, and the error incurred with $h = 1$ is seen to be rather large. The method just described yields more accurate results, however, if you are willing to use smaller values of h. Actually, the error can be made arbitrarily small by making h sufficiently small. Mathematically this is expressed by

$$(\text{Actual solution}) = \lim_{h \to 0} (\text{approximation functions})$$

The calculations become quite lengthy if we use a small value of h, but they are of a type that can easily be adapted to a computer. We do not actually use the equation of the nth line over the whole interval $x_{n-1} \leqq x \leqq x_n$ but rather to calculate the ordinate y_n at the one abscissa x_n.

For the initial value system $y' = f(x, y)$, $y(a) = b$, the successive approximations are given by the formulas

$$\begin{aligned}
x_0 &= a & y_0 &= b \\
x_1 &= a + h & y_1 &= y_0 + f(x_0, y_0)h \\
x_2 &= a + 2h & y_2 &= y_1 + f(x_1, y_1)h \\
&\vdots & &\vdots \\
x_n &= a + nh & y_n &= y_{n-1} + f(x_{n-1}, y_{n-1})h
\end{aligned}$$

The points (x_n, y_n) are the vertices of the polygonal approximation to the solution curve. To "start" the computational process in obtaining these points, the point (x_0, y_0), the function $f(x, y)$, and the increment h must be known.

Exercises for Section 2.7

In Exercises 1–6 sketch the direction field associated with the given differential equation. Draw particular solution curves which pass through the points (1, 0), (0, 1), and (1, 1).

1. $y' = y$ **2.** $y' = x$

3. $y' - xy + 1 = 0$ **4.** $x + 2yy' = 0$

5. $y + xy' = 0$ **6.** $y' = x^2 + y$

In Exercises 7–10, use the method of straight-line approximations to estimate the value of the solution to the given initial value problems at the indicated points. In each case, write the equations of the straight lines over the different intervals.

7. $y' = y$, $y(0) = -2$. Find $y(4)$, using increments of length 2.

8. $y' = x + y$, $y(0) = 0$. Find $y(3)$, using increments of length 1.

9. $y' = x^2$, $y(1) = 2$. Find $y(4)$, using increments of length 1.

10. $y' = 3$, $y(0) = -4$. Find $y(5)$, using increments of length $\frac{1}{2}$.

For Exercises 11–20 find the explicit particular solution of the corresponding Exercises 1–10, and compare this solution with the approximate solution determined above.

2.8 Additional Applications

In many practical applications, the natural choice for the independent variable is time. In such cases the first derivative of a quantity with respect to this independent variable then becomes the *rate of change* of the quantity.

We earlier encountered a number of "rate" problems relating change rates to other problem parameters and variables—radioactive decay, solid body cooling, population growth, *RL* circuit response, and commodity price behavior, to name a few. In the next two examples we consider another important class of rate problems involving mixtures.

Example 2.25 A tank initially contains 20 gal. of pure water. Starting at time $t = 0$, a brine containing 2 lb of salt per gallon flows into the tank at the rate of 4 gal./min. The mixture is kept uniform by stirring, and the well-stirred mixture flows out of the tank at the same rate as the inflow. How much salt is in the tank at any time $t > 0$? How much salt is in the tank after a very long time?

Solution Letting $y(t)$ denote the amount of salt present at any time t, then $y'(t)$ is the rate of change of this amount as a function of t. The basic equation which models this and related mixture problems is

$$y'(t) = \text{rate in} - \text{rate out}$$

where "rate in" denotes the rate at which the salt enters the mixture and "rate out" denotes the rate at which it leaves. In this example

$$\text{Rate in} = (2\ \text{lb/gal.})(4\ \text{gal./min}) = 8\ \text{lb/min}$$

In order to determine the rate out, we first compute the *concentration* of salt at time t, that is,

$$\text{Concentration} = \frac{\text{pounds of salt in the tank at time } t}{\text{gallons of brine in the tank at time } t}$$

$$= \frac{y(t)}{20} \text{ lb/gal.}$$

(Recall that the inflow and outflow solution rates are identical.) Hence

$$\text{Rate out} = \left(\frac{y}{20} \text{ lb/gal.}\right)(4 \text{ gal./min}) = \frac{y}{5} \text{ lb/min}$$

The differential equation appropriate for this mixture problem is, therefore,

$$y' = 8 - \frac{y}{5}$$

We also have the initial condition $y(0) = 0$ expressing the lack of any salt in the tank initially.

The above equation is both linear and separable, and the particular solution is easily found to be

$$y(t) = 40(1 - e^{-t/5})$$

This represents the amount of salt in the tank at time t. To answer the second question, we merely observe that as $t \to \infty$, $y(t) \to 40$. □

Frequently in mixture problems the inflow and outflow solution rates are different and/or the final answer sought is not merely the amount of substance present.

Example 2.26 A 400 L tank initially is half full of a fluid in which there is dissolved 50 gm of a potent chemical weed killer. Fluid containing 2 gm/L of the dissolved chemical is being added to the tank at a rate of 10 L/min, while the well-stirred mixture is being drawn off simultaneously at the rate of 4 L/min. Will the amount of weed killer in the tank reach 600 gm before the tank is filled?

Solution As before, we let $y(t)$ denote the amount of substance (in this case chemical weed killer) present at time t. Then

$$\text{Rate in} = (2 \text{ gm/L})(10 \text{ L/min}) = 20 \text{ gm/min}$$

$$\text{Rate out} = \left(\frac{y(t)}{200 + t(10 - 4)} \text{ gm/L}\right)(4 \text{ L/min})$$

$$= \frac{2y}{100 + 3t} \text{ gm/min}$$

and thus

$$y' = 20 - \frac{2y}{100 + 3t}$$

with $y(0) = 50$.

This is a linear equation with an integrating factor $(100 + 3t)^{2/3}$. The desired particular solution is readily determined to be

$$y(t) = 400 + 12t - 350\left(\frac{100}{100 + 3t}\right)^{2/3}$$

Since the *net* solution flow rate is 6 L/min, it will take $\frac{200}{6}$ min to fill the remaining half of the tank. When this is accomplished, the amount of weed killer will be

$$\begin{aligned} y(\tfrac{200}{6}) &= 400 + 400 - 350\left(\frac{100}{100 + 100}\right)^{2/3} \\ &\doteq 579.5 \end{aligned}$$

somewhat less than the desired 600 gm. □

In Chapter 3, we shall encounter applications that can be effectively modeled as rate problems in which the second derivative, or acceleration, also appears. If the independent variable time does not explicitly appear in the second-order differential equation, the problem can often be analyzed by successively solving two first-order equations. In the following examples this useful and very general technique is applied in two classic situations.

Example 2.27 In an elementary mechanical system, a spring is suspended vertically from a fixed support and attached to a mass m. Under the assumption that a vertical displacement y of the mass induces a compensating restoring force in the spring proportional to the displacement (Hooke's law), Newton's second law suggests the following applicable differential equation, in the absence of other forces:

$$m\frac{d^2y}{dt^2} + ky = 0$$

where k is a positive constant. (See Section 3.8 for the more general situation.) Find $y(t)$ as a function of the problem parameters.

Solution To solve this equation using first-order techniques, we let $v = dy/dt$ and assume that v can be regarded as a function of y so that

$$\frac{d^2y}{dt^2} = \frac{dv}{dt} = \frac{dv}{dy}\frac{dy}{dt} = v\frac{dv}{dy}$$

The above equation then becomes the first-order equation

$$mv\frac{dv}{dy} + ky = 0$$

which has the solution

$$mv^2 + ky^2 = kA^2$$

where A is a constant. Back substitution for v now gives rise to another first-order equation

$$\sqrt{\frac{m}{k}}\frac{dy}{dt} = \pm\sqrt{A^2 - y^2}$$

Without difficulty we obtain the desired final solution

$$y = A\cos\left(\sqrt{\frac{k}{m}}t + B\right)$$

where B is an additional constant. □

Although the above example can also be readily solved using techniques that we will introduce in Chapter 3, some interesting applications lead to nonlinear equations, to which alternative methods do not apply.

Example 2.28 A rocket is launched vertically upward from the surface of the earth with initial velocity v_0. If we assume that the earth is a sphere of radius R and denote by y the directed distance from the center of the earth to the center (of mass) of the rocket, then the "inverse square law" of Newtonian mechanics can be used to model the flight of the rocket. In the absence of air resistance and other forces, the applicable differential equation after fuel burnout is

$$\frac{d^2y}{dt^2} = -\frac{gR^2}{y^2}$$

where g is the acceleration due to gravity (assumed constant). Analyze this model under the initial conditions $y(0) = R$, $y'(0) = v_0$.

Solution Since the independent variable t does not appear explicitly, we again let $v = dy/dt$ and replace d^2y/dt^2 with $v(dv/dy)$. Thence

$$v\frac{dv}{dy} = -\frac{gR^2}{y^2}$$

which we can readily solve as

$$v^2 = \frac{2gR^2}{y} + \text{constant}$$

The value of the constant of integration can be fixed using the initial conditions, and thus

$$v^2 = v_0^2 - 2gR + 2g\frac{R^2}{y}$$

Note that the right-hand side of this expression will always be positive if $v_0 \geq \sqrt{2gR}$. This value is commonly called the "escape velocity" for the rocket.

Back substitution for v obviously leads to the first-order equation

$$\frac{dy}{dt} = \left(v_0^2 - 2gR + 2g\frac{R^2}{y}\right)^{1/2}$$

(The positive square root must be selected since dy/dt is positive initially.) This equation has the implicit solution

$$[by(a + by)]^{1/2} - a\ln[(by)^{1/2} + (a + by)^{1/2}] = b^{3/2}t + \text{constant}$$

where $a = 2gR^2$ and $b = v_0^2 - 2gR$ (assumed nonnegative). For large values of t, $y(t)$ is approximately $(v_0^2 - 2gR)^{1/2}t$. □

Exercises for Section 2.8

In Exercises 1–6, solve the given "rate" problem by any of the methods introduced in this chapter.

1. The *Malthusian law* of population growth suggests that populations P obey the differential equation

$$\frac{dP}{dt} = kP$$

If the population of a certain city is 50,000 in 1985 and time t is measured in years, what value of the constant k in this model will imply a doubling of the population in the year 2000?

2. The spread of a flu virus on a collegiate campus is modeled by the differential equation of *logistic* type

$$\frac{dx}{dt} = kx(10{,}000 - x)$$

where $x(t)$ is the number of students infected with the virus at time t, measured in days. Determine $x(t)$ under the assumptions that $x(0) = 1$ and $x(7) = 5000$.

3. The coupled first-order equations

$$\frac{dx}{dt} = (ay - b)x, \qquad \frac{dy}{dt} = (c - dx)y$$

in which a, b, c, and d are positive constants, were proposed by Lotka and Volterra to describe two-species populations. Divide the two equations and find y as a function of x and the other problem parameters.

4. A general "second-order" chemical reaction is governed by the initial value problem

$$\frac{dy}{dt} = k(\alpha - y)(\beta - y)$$

$$y(0) = 0$$

where k, α and β are positive constants with $\beta \geq \alpha$. Determine an explicit solution to this initial value problem in the two cases $\beta > \alpha$ and $\beta = \alpha$.

5. A 300-gal. tank initially contains 100 gal. of a brine in which 10 lb of salt is dissolved. Brine containing 1 lb of salt per gallon is then pumped into the tank at the rate of 4 gal./min; the well-stirred mixture is drawn out of the tank at this same rate. Find the number of pounds of salt in the tank after 1 hour.

6. Redo Exercise 5 under the assumption that the *outflow* rate is r and that one of the following applies.

(a) $0 < r < 4$ gal./min

(b) $r > 4$ gal./min

In part **(a)**, how small can r be? In part **(b)**, how large can r be?

In Exercises 7–11, apply the solution procedure suggested in Examples 2.27–2.28. Be sure to find *all* functions that satisfy the given differential equation.

7. $\dfrac{d^2y}{dt^2} + y^{-3} = 0$

8. $\dfrac{d^2y}{dt^2} + e^{-y}\dfrac{dy}{dt} = 0$

9. $y\dfrac{d^2y}{dt^2} - \left(\dfrac{dy}{dt}\right)^2 = 0$

10. $\dfrac{d^2y}{dt^2} - 3\dfrac{dy}{dt} + 2y = 0$

11. $\dfrac{d^2y}{dt^2} = a\sqrt{1 + \left(\dfrac{dy}{dt}\right)^2}$

Successive application of first-order methods can also be applied in the case of higher-order equations in which the *dependent* variable is absent.

In Exercises 12–16, solve the given differential equation by first making the substitution $v = dy/dt$.

12. $\dfrac{d^2y}{dt^2} - \dfrac{dy}{dt} = 2t$

13. $\dfrac{d^2y}{dt^2} + 2t\left(\dfrac{dy}{dt}\right)^2 = 0$

14. $\dfrac{d^2y}{dt^2} - \left(\dfrac{dy}{dt}\right)^2 = 1$

15. $\dfrac{d^2y}{dt^2}\exp\left(\dfrac{dy}{dt}\right) = 1$

16. The differential equation

$$\frac{d^2y}{dt^2} = -g - k\frac{dy}{dt}$$

where g is the constant of gravitational acceleration and k is a positive constant, is a reasonable model for certain falling bodies slowed by air resistance. Here y is distance measured vertically from the ground. Determine the particular solution of this differential equation that satisfies the initial conditions $y(0) = y_0$, $y'(0) = 0$. Comment on the notion of a "limiting velocity."

17. The differential equation

$$\frac{d^2y}{dx^2} = \frac{dy}{dx}\left[1 + \left(\frac{dy}{dx}\right)^2\right]$$

occurs in the analysis of satellite orbits. Find all solutions of this equation by first making the substitution $v = dy/dx$.

18. The shape of the classic "hanging chain" of uniform density is governed by the differential equation

$$\frac{d^2y}{dx^2} = \frac{d}{T}\sqrt{1 + \left(\frac{dy}{dx}\right)^2}$$

where d is the constant density and T is the tension in the chain at its lowest point. Solve this differential equation by first making the substitution $v = dy/dx$. (Compare Exercise 11.) The resulting shape involves a hyperbolic cosine whose graph is called a *catenary*.

2.9 Looking Ahead

The considerations of this chapter have been principally ordinary differential equations of order one. Higher-order ordinary differential equations and systems of equations are the subjects of Chapters 3 and 4; partial differential equations are discussed in Chapter 8.

Much more could be said about the occurrence and application of "integrating factors" in the solution of nonlinear first-order equations (see [5] and [6], for example). If explicit solutions cannot be obtained, however, qualitative considerations become much more important. How do we know a solution exists for the given differential equation? If a solution does exist, can we specify auxiliary conditions so that it is unique? What is the behavior of solutions for large values of the independent variable? The classic advanced references for these (and other) topics are [8] and [11] (see also [12]).

If a unique solution exists for a given first-order equation but there are not any known procedures that are applicable to find this solution, approximation and/or numerical methods may be of great value. The "approximation by straight lines" approach of Section 2.7 is also referred to as the Euler method. This procedure and the basic Runge-Kutta numerical algorithm are at least briefly mentioned in most modern texts on differential equations. References such as [9] and [10] provide more complete discussions of useful techniques for the numerical solution of ordinary differential equations.

Basic References

[1] Burghes, D. N., and M. S. Borrie. 1981. *Modelling with Differential Equations*. Chichester, England: Ellis Horwood.

[2] Dettman, J. W. 1974. *Introduction to Linear Algebra and Differential Equations*. New York: McGraw-Hill.

[3] Finizio, N., and G. Ladas. 1982. *Ordinary Differential Equations with Modern Applications*. 2d ed. Belmont, Calif.: Wadsworth.

[4] Rabenstein, A. L. 1982. *Elementary Differential Equations with Linear Algebra*. 3d ed. New York: Academic Press.

[5] Ross, S. L. 1980. *Introduction to Ordinary Differential Equations*. 3d ed. New York: John Wiley.

[6] Spiegel, M. R. 1981. *Applied Differential Equations*. 3d ed. Englewood Cliffs, N. J.: Prentice-Hall.

More-Advanced References

[7] Carrier, G. F., and C. E. Pearson. 1968. *Ordinary Differential Equations*. New York: John Wiley.

[8] Coddington, E. A., and N. Levinson. 1955. *Theory of Ordinary Differential Equations*. New York: McGraw-Hill.

[9] Conte, S., and C. W. DeBoor. 1980. *Elementary Numerical Analysis: An Algorithmic Approach*. 3d ed. New York: McGraw-Hill.

[10] Daniel, J. W., and R. E. Moore. 1970. *Computation and Theory in Ordinary Differential Equations*. San Francisco: W. H. Freeman.

[11] Ince, E. L. 1953. *Ordinary Differential Equations*. New York: Dover.

[12] Sanchez, D. A. 1979. *Ordinary Differential Equations and Stability Theory: An Introduction*. New York: Dover.

Chapter 3

Linear Differential Equations

3.1 Introduction

In the preceding chapter, our discussion centered on differential equations of order one. For such equations, the question of whether a given equation was linear or not played a relatively minor role. The situation is substantially different, however, for higher-order equations. In order to have any hope of a coherent theory that generally allows quantitative considerations, we have to be far more specific in the type of equations we include in our investigation—at the very least they should be linear. It is our good fortune, then, that practical applications that can either be formulated in terms of linear differential equations or are well approximated by such equations include many of great interest and importance.

In this chapter we discuss the linear differential equation of order $n \geq 2$

$$a_n(x)y^{(n)} + a_{n-1}(x)y^{(n-1)} + \cdots + a_1(x)y' + a_0(x)y = f(x)$$

which will usually be written in the *standard form*

$$y^{(n)} + a_{n-1}(x)y^{(n-1)} + \cdots + a_1(x)y' + a_0(x)y = f(x)$$

by dividing through by $a_n(x)$ if necessary. The $a_i(x)$ are called the *coefficient functions* and $f(x)$ is called the *driving*, *forcing*, or *input function*. If $f(x) \equiv 0$, the equation is called *homogeneous*; otherwise it is called a *nonhomogeneous* equation.

The left member of an nth-order linear differential equation is often represented by $L(y)$, where L is called a differential operator of order n. We can think of L as being a combination of coefficient functions and derivatives which operate on a function y to yield $f(x)$. The linear differential equation is thus written as

$$L(y) = f(x)$$

Corresponding to each nonhomogeneous equation $L(y) = f(x)$ is the homogeneous equation $L(y) = 0$, which we call the *reduced equation*. For example the reduced equation corresponding to the linear second-order equation $xy'' + y' + 3y = \sin x$ is $xy'' + y' + 3y = 0$.

Note that there may be distinct limitations on the kind of functions for which $L(y)$ has meaning. Functions for which $L(y)$ *has* meaning are called *admissible* for that operator. Usually the criteria for admissibility will be obvious. For example, only twice differentiable functions are admissible for the operator

$$\left(\frac{d^2}{dx^2} + 1\right)$$

For linear differential equations, the differential operator can easily be shown to have the two properties

$$L(y_1 + y_2) \equiv L(y_1) + L(y_2)$$
$$L(cy) \equiv cL(y)$$

where y, y_1, and y_2 are all admissible for the operator L. As we recall from Chapter 1, these two properties are called the *linearity properties*, and any operation satisfying these two properties is called a *linear operator*.

The direct connection between linear operators and the extensive and convenient methods of linear algebra has made these operators the tools most often used in applied mathematics.

3.2 The Solution Set of a Linear Differential Equation

The nth-order homogeneous linear differential equation, which we represent by $L(y) = 0$, has a simple but elegant theory associated with it. We first note some rather general vector space properties of equations of this type.

Theorem 3.1 The set of solutions of an nth-order homogeneous linear differential equation is a vector space.

Proof Since solutions must first of all be admissible functions, they form a subset of the vector space of piecewise continuous functions. Hence the only vector space properties to be verified are the so-called closure properties; that is, we must show that any linear combination of solutions is also a solution. Let y_1 and y_2 be solutions to the given differential equation. Then $L(y_1) \equiv 0$ and $L(y_2) \equiv 0$. Let $y_3 = c_1y_1 + c_2y_2$. Then, using the linearity properties of L,

$$\begin{aligned} L(y_3) &\equiv L(c_1y_1 + c_2y_2) \\ &\equiv L(c_1y_1) + L(c_2y_2) \\ &\equiv c_1L(y_1) + c_2L(y_2) \\ &\equiv c_10 + c_20 \\ &\equiv 0 \end{aligned}$$

■

Vector space terminology allows a characterization of the set of solutions to an nth-order homogeneous linear differential equation in a very concise statement. The following theorem and its corollary are important both theoretically and practically.

Theorem 3.2 Let $L(y) = 0$ be an nth-order homogeneous linear differential equation in standard form with continuous coefficients on the interval I. The dimension of the space of solutions on I is n. ■

Corollary A basis for the solution space of $L(y) = 0$ consists of n linearly independent functions, each of which satisfies the differential equation.

Since the solution space is completely determined if a basis is known, we will know all solutions if we can find n linearly independent ones. If the set

$$\{y_i\}_{i=1}^{n}$$

is a basis, the complete, or general, solution is

$$y(x) = \sum_{i=1}^{n} c_i y_i(x)$$

It is therefore important to know when n solutions to an nth-order homogeneous linear equation are linearly independent. For the case $n = 2$, we can show linear dependence on an interval by showing that one of the functions is a constant multiple of the other, over that interval. For the general case of n arbitrary functions, direct appeal to the definition of linear independence is necessary; that is, to show that the set of functions

$$\{y_i\}_{i=1}^{n}$$

is a linearly independent set on some interval, we must show that if

$$\sum_{i=1}^{n} c_i y_i(x) \equiv 0$$

on the interval, then $c_i = 0$ for each i.

There is a rather simple formal test for deciding the linear independence of a set of solutions to a homogeneous linear differential equation. This procedure is based on the Wronskian determinant being nonzero. In the exercises for Section 1.8 the Wronskian determinant of three functions y_1, y_2, y_3 was defined to be

$$\det \begin{pmatrix} y_1 & y_2 & y_3 \\ y_1' & y_2' & y_3' \\ y_1'' & y_2'' & y_3'' \end{pmatrix}$$

In general,

$$W(y_1, \ldots, y_n) = \det \begin{pmatrix} y_1 & y_2 & \cdots & y_n \\ y_1' & y_2' & \cdots & y_n' \\ \vdots & \vdots & & \vdots \\ y_1^{(n-1)} & y_2^{(n-1)} & \cdots & y_n^{(n-1)} \end{pmatrix}$$

Theorem 3.3 sets up the test for linear independence, which is then stated explicitly in Theorem 3.4.

Theorem 3.3 Consider $L(y) = 0$, where L is nth order, is in standard form, and has coefficients continuous on an interval I. If $y_1, \dots, y_n$ are solutions of this equation, then either $W(y_1, \dots, y_n) \equiv 0$ on I or there is *no* $x \in I$ such that $W(y_1, \dots, y_n)(x) = 0$. ■

Theorem 3.4 If $y_1, \dots, y_n$ are solutions of $L(y) = 0$, where L is nth order, is in standard form, and has continuous coefficients on an interval I, then $\{y_1, \dots, y_n\}$ is linearly independent on I *if and only if* $W(y_1, \dots, y_n)(x) \neq 0$ for some $x \in I$. ■

Under the conditions of Theorem 3.4, if $W(y_1, \dots, y_n)(x) \neq 0$ at some $x \in I$ then $W(y_1, \dots, y_n)(x)$ is nonzero for each x in I.

Example 3.1 Solve the differential equation $y''' = 0$.

Solution The solution space is three-dimensional, and obviously $y_1 = 1$, $y_2 = x$, and $y_3 = x^2$ are three solutions. To test for linear independence, consider

$$W(1, x, x^2) = \det \begin{pmatrix} 1 & x & x^2 \\ 0 & 1 & 2x \\ 0 & 0 & 2 \end{pmatrix} = 2$$

Since the Wronskian is never zero, the solutions are linearly independent and the complete solution is

$$y = c_1 + c_2 x + c_3 x^2$$ □

Example 3.2 Given that there are solutions of $L(y) = y'' + 3y' + 2y = 0$ of the form $y = e^{mx}$, find a basis for the solution space.

Solution Since $L(e^{mx}) = m^2 e^{mx} + 3me^{mx} + 2e^{mx} = e^{mx}(m^2 + 3m + 2) = e^{mx}(m + 2)(m + 1)$, we will have $L(e^{mx}) \equiv 0$ if $m = -1$ or $m = -2$. Further, $W(e^{-x}, e^{-2x}) = -e^{-2x} \neq 0$, so $\{e^{-x}, e^{-2x}\}$ is a basis for the solution space. □

Example 3.3 For $L(y) = x^2 y'' + xy' - y = 0$, it can be verified that $L(x) \equiv 0$ and $L(1/x) \equiv 0$. What can be said about the solution space of L?

Solution Putting $L(y) = 0$ in standard form, $y'' + (1/x)y' - (1/x^2)y = 0$, we see that the coefficients are continuous over the interval $x > 0$ and

$$W\left(x, \frac{1}{x}\right) = \det \begin{pmatrix} x & \dfrac{1}{x} \\ 1 & \dfrac{-1}{x^2} \end{pmatrix} = \frac{-2}{x} \neq 0 \qquad \text{for } x > 0$$

Thus $y = c_1 x + c_2/x$ is the complete solution for $x > 0$. Similarly, the same expression yields the complete solution over the interval $x < 0$. □

The general technique for solving a linear nth-order differential equation with *nonzero* driving function is based upon the following theorem.

Theorem 3.5 Let y_p be *any* particular solution of the nth-order linear differential equation represented by

$$L(y) = f(x)$$

and let y_c be the complete solution of the corresponding reduced equation, $L(y) = 0$. Then $y(x) = y_c(x) + y_p(x)$ is the complete solution of the given nonhomogeneous equation $L(y) = f(x)$.

Proof Since $L(y_p) \equiv f(x)$ and $L(y_c) \equiv 0$, $L(y_c + y_p) \equiv L(y_c) + L(y_p) \equiv f(x)$, which shows that any function of the form $y_c + y_p$ is a solution of the nonhomogeneous equation. Conversely, let y_{p_1} be any particular solution to the nth-order linear differential equation, $L(y) = f(x)$. Then

$$L(y_p - y_{p_1}) \equiv L(y_p) - L(y_{p_1}) \equiv f(x) - f(x) \equiv 0$$

which shows that $y_p - y_{p_1}$ is one of the solutions of the reduced equation. Thus, any particular solution can be written as the sum of y_p and one of the solutions to the reduced equation, which is the same as saying that the complete solution is

$$y(x) = y_c(x) + y_p(x)$$

■

Any two particular solutions $y_{p_1}(x)$ and $y_{p_2}(x)$ differ by a function that is a solution to the reduced equation. This accounts for the often dissimilar appearance of particular solutions to nonhomogeneous equations.

Example 3.4 Reconcile the fact that $(1 + \frac{1}{2}\tan x)(\sin x)$ and $\frac{1}{2}\sin x \tan x$ are both particular solutions to the differential equation $y'' + y = \sec^3 x$.

Solution We simply note that the difference of the two solutions is $\sin x$, which is a solution to the corresponding reduced equation $y'' + y = 0$. □

Theorem 3.5 introduces the notation y_c for the complete solution of the reduced equation. In some texts this is called the *homogeneous solution* for obvious reasons. Other authors call it the *complementary solution.*

Theorem 3.5 says that if *any* particular solution to the differential equation is known, then finding the complete solution is a matter of solving the corresponding reduced equation.

Example 3.5 Show that $x - 1$ is a solution of $y'' + 3y' + 2y = 2x + 1$ and then find the complete solution.

Solution Verification of the fact that $x - 1$ is a solution of the differential equation is a matter of substituting $y = x - 1$, $y' = 1$, and $y'' = 0$ into the equation. From Example 3.2 the complete solution of the reduced equation is

$$y_c(x) = c_1 e^{-x} + c_2 e^{-2x}$$

Hence, the complete solution of the given equation is

$$y(x) = c_1 e^{-x} + c_2 e^{-2x} + x - 1 \qquad \square$$

In certain applications, the driving function may be written as the sum of two functions, say $f(x) = f_1(x) + f_2(x)$. It is then convenient to consider the two equations $L(y) = f_1(x)$ and $L(y) = f_2(x)$ separately.

Theorem 3.6 Let y_{p_1} be a solution to $L(y) = f_1(x)$ and y_{p_2} be a solution to $L(y) = f_2(x)$. Then $y_{p_1} + y_{p_2}$ is a solution to

$$L(y) = f_1(x) + f_2(x)$$

Proof See Exercise 22. ■

This last result is sometimes referred to as the *principle of superposition.* As a consequence of this principle, finding the complete solution of a nonhomogeneous linear equation such as

$$y'' - 5y' + 2y = \sin x + x^2$$

is a matter of finding

1. the complete solution, y_c, to $y'' - 5y' + 2y = 0$,
2. a particular solution, y_{p_1}, to $y'' - 5y' + 2y = \sin x$,
3. a particular solution, y_{p_2}, to $y'' - 5y' + 2y = x^2$.

Then, $y_c + y_{p_1} + y_{p_2}$ is the complete solution of the given differential equation.

Exercises for Section 3.2

The differential equations in Exercises 1–5 have solutions of the form e^{mx}. Find the complete solution of each by finding n linearly independent solutions. Show that the functions are linearly independent by showing that their Wronskians are nonzero.

1. $y''' + y'' - 10y' + 8y = 0$
2. $y'' + 4y' + 3y = 0$
3. $y'' - 4y = 0$
4. $y'' - y' = 0$
5. $y''' - 8y'' + 7y' = 0$

The differential equations in Exercises 6–9 have solutions of the form $\cos mx$ and $\sin mx$. Write the complete solution for each.

6. $y'' + 4y = 0$
7. $9y'' + y = 0$
8. $y^{(4)} + 4y'' + 3y = 0$
9. $4y^{(4)} + 37y'' + 9y = 0$

The differential equations in Exercises 10–13 have solutions of the form $\cos mx$, $\sin mx$, and/or e^{mx}. Determine the complete solution of each by finding n linearly independent solutions.

10. $y''' + 25y' = 0$
11. $y''' + y'' + 4y' + 4y = 0$

12. $4y^{(4)} - 3y'' - y = 0$

13. $y^{(4)} - 6y'' + y = 0$

The differential equations in Exercises 14–17 have solutions of the form x^m. Write the complete solution for each.

14. $x^2y'' + xy' - 4y = 0$

15. $x^2y'' - xy' - 3y = 0$

16. $x^3y''' + 4x^2y'' - 8xy' + 8y = 0$

17. $xy''' + 3y'' = 0$

18. Show that $\{e^x, xe^x, x^2e^x\}$ is linearly independent by showing that the Wronskian of the three functions does not vanish.

19. Show that $\{e^{mx}, e^{nx}, e^{kx}\}$ is linearly independent, where m, n, and k are distinct real numbers.

20. Show that $\{\cos x, \sin x, x\sin x, x\cos x\}$ is linearly independent.

21. Using the definition of linear independence, show that the set of functions $\{e^{rx}, xe^{rx}, x^2e^{rx}, \dots, x^{k-1}e^{rx}\}$ is linearly independent.

22. Prove Theorem 3.6.

23. The functions $x + 1$, $x + e^x$, and $x - 1 - e^x$ are three solutions to a certain non-homogeneous linear equation of the second order. Find all solutions.

24. Given that e^x is a solution to $y'' + y = 2e^x$, find the complete solution.

25. Given that $e^{-x} - 3x + \frac{3}{2} - 2xe^{-x}$ is a solution to $y'' - y' - 2y = 6x + 6e^{-x}$, find the complete solution.

26. Given that $e^{2x} + e^{-2x} - xe^{-x} + x + \frac{1}{2}$ is a solution to $y''' + y'' - 4y' - 4y = 3e^{-x} - 4x - 6$, find the complete solution.

27. Show that if y_1 is a solution to $L(y) = f(x)$, then ay_1 is a solution to $L(y) = af(x)$.

28. Find the complete solution to $y''' + y'' = x + 1$.

29. The undamped oscillatory motion of a mass-spring combination of mass m and spring constant k can be described by the differential equation

$$m\frac{d^2y}{dt^2} + ky = 0$$

(a) Show that $\left\{\cos\sqrt{\frac{k}{m}}t, \sin\sqrt{\frac{k}{m}}t\right\}$ constitutes a basis for the solution space of this equation.

(b) Find the unique solution that satisfies the initial conditions $y(0) = y_0$, $\dfrac{dy(0)}{dt} = v_0$.

30. The steady-state temperature T in a region bounded by two long coaxial cylinders satisfies

$$r\frac{d^2T}{dr^2} + \frac{dT}{dr} = 0$$

where r is the distance from the common axis. Verify that

$$T(r) = \frac{T_1\ln(b/r) + T_2\ln(r/a)}{\ln(b/a)}$$

is the solution of the boundary value problem in which the cylinders are of radii a and b ($a < b$) and they are kept at constant temperatures T_1 and T_2 respectively.

3.3 Reduction of Order

It is usually more difficult, and sometimes impossible, to find the complete solution of a linear differential equation if the coefficients are not constant. The following method allows you to simplify such an equation if you have found, by some alternative means (or have been given), one (or more) nontrivial solution(s) of the corresponding homogeneous equation. The simplification takes the form of reducing the order of the equation by one for each known solution.

In order to simplify the exposition, let us consider a second-order linear differential equation

$$L(y) = y'' + a_1(x)y' + a_0(x)y = f(x)$$

and the corresponding homogeneous equation

$$y'' + a_1(x)y' + a_0(x)y = 0$$

which has a nontrivial solution y_1. Make the substitution $y = uy_1$ to obtain

$$(u''y_1 + 2u'y_1' + uy_1'') + a_1(uy_1' + u'y_1) + a_0uy_1 = f$$

or

$$y_1u'' + (2y_1' + a_1y_1)u' + (y_1'' + a_1y_1' + a_0y_1)u = f$$

Notice that the coefficient of u is zero since $L(y_1) \equiv 0$, so that by letting $u' = w$ we obtain

$$y_1w' + (2y_1' + a_1y_1)w = f$$

a first-order linear equation that can be solved by the specialized methods considered in Section 2.6.

Example 3.6 Solve $y'' - y = xe^x$ by noting that $y_1 = e^x$ is a solution of $y'' - y = 0$.

Solution Let $y = ue^x$. Then $y' = u'e^x + ue^x$ and $y'' = u''e^x + 2u'e^x + ue^x$. Substituting into the differential equation,

$$u''e^x + 2u'e^x = xe^x$$

or

$$u'' + 2u' = x$$

Letting $u' = w$,

$$w' + 2w = x$$

from which

$$u' = w = c_1e^{-2x} + \frac{x}{2} - \frac{1}{4}$$

$$u = -\left(\frac{c_1}{2}\right)e^{-2x} + \frac{x^2}{4} - \frac{x}{4} + c_2$$

and, since $y = ue^x$,

$$y = c_1\left(\frac{-e^{-x}}{2}\right) + c_2e^x + \left(\frac{x^2}{4} - \frac{x}{4}\right)e^x$$

Of course, the $-\frac{1}{2}$ can be "absorbed" into the c_1 to express the complete solution as

$$y = c_1e^{-x} + c_2e^x + \left(\frac{x^2}{4} - \frac{x}{4}\right)e^x \qquad \square$$

The disappearance of the terms involving u after the substitution $y = uy_1$ is a feature for a linear equation of any order. However, if we start with an equation of order greater than two, a single application of this method will only lower the order by one—not produce a complete solution.

Example 3.7 Noting that $L(x^2) \equiv 0$, reduce the order of

$$L(y) = y''' + 2x^2y'' - 3xy' + 2y = 0$$

Solution Let $y = ux^2$. Then $y' = x^2u' + 2ux$, $y'' = x^2u'' + 4xu' + 2u$ and $y''' = x^2u''' + 6xu'' + 6u'$. Substituting into the differential equation

$$L(y) = L(ux^2) = x^2u''' + (2x^4 + 6x)u'' + (5x^3 + 6)u' = 0$$

Letting $u' = w$, we have the second-order equation

$$x^2w'' + (2x^4 + 6x)w' + (5x^3 + 6)w = 0 \qquad \square$$

In this example, although we lowered the order, we did not seem to be a great deal closer to solving the original equation. To determine the complete solution, we would need to find a solution of our new equation so the order could be reduced one more time or, perhaps, pursue an entirely different approach to the problem.

Exercises for Section 3.3

In Exercises 1–15, use reduction of order with the given solution of the associated homogeneous equation and determine the complete solution of each differential equation.

1. $y'' - 3y' + 2y = 0;\quad y_1 = e^{2x}$
2. $y'' - y = 0;\quad y_1 = e^x + e^{-x}$
3. $x^2y'' - 3xy' + 4y = 0;\quad y_1 = x^2$
4. $x^2y'' + xy' + y = 0;\quad y_1 = \cos(\ln x)$
5. $x^2y'' + xy' + (x^2 - \frac{1}{4})y = 0;\quad y_1 = \dfrac{\sin x}{\sqrt{x}}$
6. $y'' - y = e^{-x};\quad y_1 = e^{-x}$
7. $y'' + y = \sec x;\quad y_1 = \cos x$
8. $y'' + y = \csc x;\quad y_1 = \sin x$

9. $y'' + y = \csc^3 x; \quad y_1 = \cos x$
10. $x^2 y'' + xy' - y = 8x^3; \quad y_1 = x$
11. $(x + 2)y'' - (2x + 5)y' + 2y = (x + 1)e^x; \quad y_1 = e^{2x}$
12. $x^2 y'' - (x^2 + 2x)y' + (x + 2)y = x^3 e^x; \quad y_1 = x$
13. $x(x \cos x - 2 \sin x)y'' + (x^2 + 2)\sin x\, y' - 2(x \sin x + \cos x)y = 0; \quad y_1 = x^2$
14. $(1 - x^2)y'' - 2xy' + 2y = 6(1 - x^2); \quad y_1 = x$
15. $y'' - 2y' \tan x + 3y = \sec x; \quad y_1 = \sin x$

16. The differential equation

$$(x^3 - 2x^2)y'' - (x^3 + 2x^2 - 6x)y' + (3x^2 - 6)y = 0$$

has one solution of the form $y_1 = x^n$.
(a) Find this solution.
(b) Use reduction of order to determine the complete solution.

17. If reduction of order with the known solution y_1 is applied to the differential equation

$$y'' + p(x)y' + q(x)y = 0$$

a second solution of the form

$$y_2 = y_1(x) \int \frac{e^{-\int p(x)\,dx}}{y_1^2(x)} dx$$

is obtained.
(a) By direct substitution verify that y_2 is indeed a solution of the given equation.
(b) Show that $\{y_1, y_2\}$ is linearly independent.

18. The two functions $y_1 = 1$ and $y_2 = e^{-2x}$ are known to be solutions of the differential equation

$$y''' + 4y'' + 4y' = 0$$

Employ reduction of order twice to find the complete solution.

19. Repeat Exercise 18 for

$$x^3 y''' - 3x^2 y'' + 6xy' - 6y = 0$$

with $y_1 = x$ and $y_2 = x^2$.

20. Solve the forced vibrations initial value problem

$$\frac{d^2 y}{dt^2} + y = \cos t; \qquad y(0) = 0, \qquad \frac{dy(0)}{dt} = 1$$

3.4 Solution to Homogeneous Equations with Constant Coefficients

The exponential function e^{ax} is the only function that yields a constant multiple of itself when differentiated or, more precisely, the complete solution of $y' = ay$ is $y = ce^{ax}$. When confronted by a homogeneous linear equation

$$L(y) = a_n y^{(n)} + a_{n-1} y^{(n-1)} + \cdots + a_1 y' + a_0 y = 0$$

where each a_k is constant, we might well guess a solution $y = e^{mx}$. After all, we are looking for a function such that a sum of constant multiples of the function and its derivatives will be zero. Since

$$L(e^{mx}) = a_n m^n e^{mx} + a_{n-1} m^{n-1} e^{mx} + \cdots + a_1 m e^{mx} + a_0 e^{mx}$$

e^{mx} will be a solution *if and only if*

$$P(m) = a_n m^n + a_{n-1} m^{n-1} + \cdots + a_1 m + a_0 = 0$$

The polynomial P is called the *characteristic polynomial*, and $P(m) = 0$ is called the *characteristic equation.* By this substitution, the analysis problem $L(y) = 0$ has apparently been reduced to the purely algebraic problem of finding the roots of a polynomial.

Example 3.8 Solve the constant coefficient differential equation $y'' + 3y' + 2y = 0$.

Solution The characteristic equation $m^2 + 3m + 2 = 0$ has roots $m = -1$ and $m = -2$, so e^{-x} and e^{-2x} are solutions. Since these two functions are linearly independent, we may apply the corollary of Theorem 3.2 to express the complete solution as

$$y = c_1 e^{-x} + c_2 e^{-2x}$$ □

Example 3.9 Determine the complete solution of

$$y''' - 2y'' - y' + 2y = 0$$

Solution The characteristic equation $m^3 - 2m^2 - m + 2 = 0$ leads to the roots $m = \pm 1$, 2 and linearly independent solutions e^{-x}, e^x, and e^{2x}. By virtue of the corollary of Theorem 3.2, the complete solution of the given constant coefficient differential equation is

$$y = c_1 e^{-x} + c_2 e^x + c_3 e^{2x}$$ □

In general, as suggested by the above examples, we are seeking n independent solutions of $L(y) = 0$, and $P(m)$, of course, must have n roots. However, some of these roots may be multiple.

Example 3.10 Solve the differential equation $y'' - 2y' + y = 0$.

Solution The characteristic equation $m^2 - 2m + 1$ has the double root 1, which yields only one independent solution, $y_1 = e^x$. Let us try reduction of order to find a second solution. Using the formula from Exercise 17, p. 90,

$$y_2 = e^x \int \frac{e^{2x}}{e^{2x}} dx = xe^x$$

Hence, the complete solution is

$$y = c_1 e^x + c_2 x e^x$$ □

The result of Example 3.10 may be generalized to yield the following theorem.

Theorem 3.7 If r is a root of multiplicity k of the characteristic equation of $L(y) = 0$, a homogeneous linear differential equation with constant coefficients, then $x^n e^{rx}$ is a solution of $L(y) = 0$ for $n = 0, 1, \ldots, k - 1$. The set of all such functions corresponding to the roots of $P(m) = 0$ forms a basis for the solution space. ■

Example 3.11 The roots of the characteristic equation associated with a given fourth-order homogeneous linear differential equation with constant coefficients are $m = -3, -3, -3, -3$. Determine the complete solution.

Solution As an immediate application of Theorem 3.7, we know that e^{-3x}, xe^{-3x}, x^2e^{-3x}, and x^3e^{-3x} are linearly independent solutions of the given equation $L(y) = 0$. Hence

$$y = c_1e^{-3x} + c_2xe^{-3x} + c_3x^2e^{-3x} + c_4x^3e^{-3x}$$

represents the complete solution. □

Example 3.12 Solve the differential equation $y^{(4)} - 5y''' + 6y'' + 4y' - 8y = 0$.

Solution Assuming that textbook problems are sufficiently unrealistic to have simple solutions, we check the possible rational roots $\pm 8, \pm 4, \pm 2, \pm 1$ of the characteristic equation

$$p(m) = m^4 - 5m^3 + 6m^2 + 4m - 8 = 0$$

to find

$$p(m) = (m + 1)(m - 2)^3$$

The complete solution is

$$y = c_1e^{-x} + c_2e^{2x} + c_3xe^{2x} + c_4x^2e^{2x}$$

□

Complex characteristic roots do not present an exception to the solution procedure suggested by Theorem 3.7. However, it may sometimes be preferable to express solutions corresponding to a complex root in terms of sines and cosines rather than as complex exponentials.

Here, and in much that follows, we make use of some elementary properties of the conjugate of a complex number. For a complex number $z = a + bi$, $\bar{z} = a - bi$ is called the *conjugate* of z, a is the *real part* of z (Re z), and b is the *imaginary part* of z (Im z). For all complex numbers z and w, $\overline{z + w} = \bar{z} + \bar{w}$, $\overline{zw} = \bar{z}\bar{w}$, $\operatorname{Re} z = (z + \bar{z})/2$, $\operatorname{Im} z = (z - \bar{z})/2i$, and $\bar{z} = z$ if and only if z is real.

Let us suppose that $L(y)$ has *real* coefficients. Then if $a + bi$ is a characteristic root, so is its conjugate $a - bi$. Thus

$$y_1 = e^{(a+bi)x} \quad \text{and} \quad y_2 = e^{(a-bi)x}$$

are solutions. Using Euler's formula $e^{i\theta} = \cos\theta + i\sin\theta$ (see Chapter 12, p. 481) and the linearity of the solution space, y_3 and y_4 are also solutions, where

$$y_3 = \frac{y_1 + y_2}{2} = \frac{e^{ax}}{2}(\cos bx + i \sin bx) + \frac{e^{ax}}{2}(\cos bx - i \sin bx) = e^{ax} \cos bx$$

and

$$y_4 = \frac{y_1 - y_2}{2i} = \frac{e^{ax}}{2i}(\cos bx + i \sin bx) - \frac{e^{ax}}{2i}(\cos bx - i \sin bx) = e^{ax} \sin bx$$

(Recall Exercise 14, p. 31.) Notice that $y_3 = \operatorname{Re} y_1$ and $y_4 = \operatorname{Im} y_1$. The linear independence of $\{y_3, y_4\}$ can easily be checked. A complex root of multiplicity k of a real characteristic equation will, of course, yield solutions

$$e^{ax} \cos bx,\ e^{ax} \sin bx,\ xe^{ax} \cos bx,\ xe^{ax} \sin bx,\ \ldots,\ x^{k-1}e^{ax} \cos bx,\ x^{k-1}e^{ax} \sin bx$$

Example 3.13 Solve the differential equation $y'' + 2y' + 4y = 0$.

Solution The characteristic equation is $m^2 + 2m + 4 = 0$. The roots of this equation are

$$a \pm ib = \frac{-2 \pm \sqrt{4 - 16}}{2} = -1 \pm i\sqrt{3}$$

and hence a basis for the solution space is $\{e^{-x} \cos \sqrt{3}x, e^{-x} \sin \sqrt{3}x\}$. The complete solution is

$$y = e^{-x}(c_1 \cos \sqrt{3}x + c_2 \sin \sqrt{3}x)$$

Another basis would be $\{e^{(-1+i\sqrt{3})x}, e^{(-1-i\sqrt{3})x}\}$ with corresponding complete solution

$$y = c_1 e^{(-1+i\sqrt{3})x} + c_2 e^{(-1-i\sqrt{3})x}$$

□

Example 3.14 The function $\cos x$ is a solution to the differential equation $y^{(4)} + 2y'' + y = 0$. Find three more real functions which together with $\cos x$ form a linearly independent set of solutions of the given equation.

Solution The characteristic equation $m^4 + 2m^2 + 1 = 0$ has the roots $m = \pm i$, $\pm i$. The desired real solutions are thus $\cos x$, $\sin x$, $x \cos x$, and $x \sin x$. Since

$$W(\cos x, \sin x, x \cos x, x \sin x) \equiv 4$$

they are indeed linearly independent.

□

It should be noted from the discussion of this section that solutions of homogeneous linear differential equations with constant coefficients may always be written as a sum of terms of the type $x^k e^{ax}$, $x^k e^{ax} \cos bx$ and $x^k e^{ax} \sin bx$, where a and b are real numbers. It is perhaps startling that this very important class of differential equations, found in so many applications, should have a solution space that is so easy to describe.

Exercises for Section 3.4

In the following exercises, find a basis for the solution space.

1. $y'' + 9y = 0$
2. $y'' - 9y = 0$
3. $y'' + 5y' + 4y = 0$
4. $y'' + y = 0$
5. $y''' + y = 0$
6. $4y^{(4)} - 8y''' - 7y'' + 11y' + 6y = 0$
7. $y^{(4)} + 8y''' + 16y'' = 0$
8. $y^{(5)} - y''' = 0$
9. $y''' - 3y' - 2y = 0$
10. $y'' - 2y' + 6y = 0$
11. $y''' + y'' + 4y' + 4y = 0$
12. $y''' + 2y'' + 5y' = 0$
13. $y^{(6)} + 9y^{(4)} + 24y'' + 16y = 0$
14. $y^{(5)} + 18y''' + 81y' = 0$

In Exercises 15–18, determine the complete solution of the given equation.

15. The fourth-order homogeneous linear differential equation with constant coefficients whose characteristic equation has the roots $m = -2, 1, 1, 1$.
16. The sixth-order homogeneous linear differential equation with constant coefficients whose characteristic equation has the roots $m = 0, 0, -1 \pm 2i, -1 \pm 2i$.
17. $y^{(6)} - 64y = 0$, for which we know $m = 1 - \sqrt{3}i$ is one root of the characteristic equation.
18. $y^{(6)} + 12y^{(4)} + 48y'' + 64y = 0$, for which we know $\sin 2x$ is one solution.

Solve the initial value problems in Exercises 19–28.

19. $y'' - y = 0, \quad y(0) = 1, \quad y'(0) = 1$
20. $y'' - 6y' + 9y = 0, \quad y(0) = 1, \quad y'(0) = 2$
21. $9y'' + y = 0, \quad y(0) = 0, \quad y'(0) = 3$
22. $y''' + 2y'' + y' + 2y = 0, \quad y(0) = 0, \quad y'(0) = 2, \quad y''(0) = -5$
23. $y'' - 4y' + 7y = 0, \quad y(0) = 1, \quad y'(0) = -1$
24. $y''' + 7y'' + 19y' + 13y = 0, \quad y(0) = 0, \quad y'(0) = 2, \quad y''(0) = -12$
25. $y^{(4)} - 4y''' + 6y'' - 8y' + 8y = 0, \quad y(0) = 2, \quad y'(0) = -1, \quad y''(0) = -8,$ $y'''(0) = -12$
26. $y^{(4)} + 2y''' + 10y'' = 0, \quad y(0) = 5, \quad y'(0) = -3, \quad y''(0) = 0, \quad y'''(0) = 0$
27. $y''' - 6y'' + 12y' - 8y = 0, \quad y(0) = 0, \quad y'(0) = -2, \quad y''(0) = -10$
28. $y^{(4)} + 18y'' + 81y = 0, \quad y(0) = 0, \quad y'(0) = 1, \quad y''(0) = 0, \quad y'''(0) = -27$
29. Solve the system

$$y'' - (4 + \varepsilon)y' + (4 + 2\varepsilon)y = 0, \qquad y(0) = 1, \qquad y'(0) = 1$$

first when ε is zero and then when ε is a nonzero constant. Show that the solution for $\varepsilon = 0$ is the limit of the solution for $\varepsilon \neq 0$ as $\varepsilon \to 0$.

30. Solve the system

$$y'' - 4y' + ay = 0, \qquad y(0) = 0, \qquad y'(0) = 1$$

where a is a constant. Consider the three cases $a = 4$, $a = 4 - \varepsilon^2$, and $a = 4 + \varepsilon^2$, and study the limit as $\varepsilon \to 0$.

31. Find a particular solution to

$$y'' - y' - 2y = 0, \qquad y(0) = 2$$

if $y \to 0$ as $x \to \infty$.

32. Using basic properties of conjugates, show that if y_1 is any solution of a homogeneous linear equation with real constant coefficients, then $\bar{y}_1$ is also a solution.

33. The differential equation

$$\frac{d^4u}{dx^4} - k^4u = 0$$

occurs in the analysis of elastic beam vibrations. Determine the complete solution of this equation.

34. When the oscillations are small, the motion of a pendulum of length L under the sole influence of gravity can be described by the differential equation

$$\frac{d^2\theta}{dt^2} + \frac{g}{L}\theta = 0$$

Here $\theta(t)$ is the angular deviation of the pendulum from the vertical at time t and g is the constant acceleration due to gravity.

(a) Show that the complete solution of this equation can be written in the form

$$\theta = A\cos\sqrt{\frac{g}{L}}(t + B)$$

where A and B are constants.

(b) Determine the values of A and B that are appropriate for the solution of the initial value problem arising when $\theta(0) = 1$ and $\theta'(0) = 2$ with $L = 4$ ft and $g = 32$ ft/sec^2.

35. One reasonable model of supply and demand in economics uses the assumption that, when inflation is constant, the price of individual commodities satisfies the differential equation

$$\frac{d^2P}{dt^2} + k(P - P_0) = 0$$

where P_0 is constant. Show that the price oscillates periodically about the constant level P_0, and determine the period of this oscillation.

3.5 Method of Undetermined Coefficients

In the preceding section we learned how to solve the equation $L(y) = 0$ in the constant coefficient case. We now couple this with a method for finding a particular solution of $L(y) = f(x)$ and thus derive the complete solution of the nonhomogeneous equation. As we describe this method, it is *strictly* limited to the constant coefficient case and applies only if f is very special, namely, a solution itself of some linear homogeneous equation with constant coefficients.

We continue, in the spirit of Section 3.4, to expand an educated guess into a formal solution procedure.

Example 3.15 Find a particular solution of the differential equation $L(y) = y'' - y' - 2y = 2e^{3x}$.

Solution A reasonable guess would be $y_p = Ae^{3x}$. Since

$$L(Ae^{3x}) = 9Ae^{3x} - 3Ae^{3x} - 2Ae^{3x} = 4Ae^{3x}$$

if our guess is to be a solution, we must have $4Ae^{3x} \equiv 2e^{3x}$, or $A = \frac{1}{2}$. We then have $y_p = \frac{1}{2}e^{3x}$. Since

$$p(m) = m^2 - m - 2 = (m - 2)(m + 1)$$

we also have $y_c = c_1 e^{2x} + c_2 e^{-x}$, and the complete solution is thus

$$y = y_c + y_p = c_1 e^{2x} + c_2 e^{-x} + \tfrac{1}{2}e^{3x}$$ □

Example 3.16 Solve the differential equation $y'' - y' - 2y = \sin x$.

Solution Let us start with the guess $y_p = A \sin x$. We would like

$$L(A \sin x) \equiv \sin x \quad \text{but} \quad L(A \sin x) = -A \sin x - A \cos x - 2A \sin x$$

In order to get rid of the cosine term, we must have $A = 0$, which eliminates the sine term as well. Undaunted we try another guess, $y_p = A \sin x + B \cos x$, hoping that A and B together may be chosen both to eliminate the cosine term and to salvage the sine term. In this case,

$$\begin{aligned} L(A \sin x + B \cos x) &= -A \sin x - B \cos x - A \cos x \\ &\quad + B \sin x - 2A \sin x - 2B \cos x \\ &= (-3A + B) \sin x + (-A - 3B) \cos x \stackrel{?}{\equiv} \sin x \end{aligned}$$

Solving the system

$$\begin{aligned} -3A + B &= 1 \\ -A - 3B &= 0 \end{aligned}$$

we obtain $A = -\frac{3}{10}$ and $B = \frac{1}{10}$, so $y_p = -\frac{3}{10} \sin x + \frac{1}{10} \cos x$ and

$$y = c_1 e^{2x} + c_2 e^{-x} - \tfrac{3}{10} \sin x + \tfrac{1}{10} \cos x$$ □

Example 3.17 Solve the differential equation $y'' - y' - 2y = \tan x$.

Solution (attempt) Trying $y_p = A \tan x$:

$$L(A \tan x) = 2A \sec^2 x \tan x - A \sec^2 x - 2A \tan x \stackrel{?}{\equiv} \tan x$$

This will not work, since we cannot get rid of the $\sec^2 x$ or $\sec^2 x \tan x$ terms, so we try

$$y_p = A \tan x + B \sec^2 x + C \sec^2 x \tan x$$

However, $L(y_p)$ now has terms involving $\sec^4 x \tan x$ and $\sec^2 x \tan^3 x$ (arising from taking two derivatives of $\sec^2 x \tan x$), which cannot be eliminated. Making further, more elaborate, guesses for y_p only leads us deeper and deeper into trouble. □

Example 3.17 suggests that our "guess method" will work only for an f that produces finitely many linearly independent functions when repeatedly differentiated. This is indeed the case. Such functions can be characterized as solutions of some homogeneous linear equation with constant coefficients, that is, functions that are linear combinations of terms of the type $x^k e^{ax} \cos bx$ and $x^k e^{ax} \sin bx$.

Our trial y_p should be a linear combination of all linearly independent forms that arise when f is differentiated repeatedly, as indicated in Table 3.1.

Table 3.1 *Trial Solutions for the Method of Undetermined Coefficients*

$$L(y) = a_n y^{(n)} + a_{n-1} y^{(n-1)} + \cdots + a_1 y' + a_0 y = f(x)$$

$a_n, a_{n-1}, a_{n-2}, \ldots, a_1, a_0$ are *constant*

$f(x)$	"First Guess" for y_p such that $L(y_p) = f(x)$
c	A
cx^n	$A_n x^n + A_{n-1} x^{n-1} + \cdots + A_1 x + A_0$
$c_n x^n + c_{n-1} x^{n-1} + \cdots + c_1 x + c_0$	$A_n x^n + A_{n-1} x^{n-1} + \cdots + A_1 x + A_0$
ce^{rx}	Ae^{rx}
$c \cos kx$ $c \sin kx$	$A \cos kx + B \sin kx$
$(c_n x^n + \cdots + c_1 x + c_0) \sin kx$ $(c_n x^n + \cdots + c_1 x + c_0) \cos kx$	$(A_n x^n + \cdots + A_1 x + A_0) \sin kx +$ $(B_n x^n + \cdots + B_1 x + B_0) \cos kx$
$(c_n x^n + \cdots + c_1 x + c_0) e^{rx}$	$(A_n x^n + \cdots + A_1 x + A_0) e^{rx}$
$ce^{rx} \sin kx$ $ce^{rx} \cos kx$	$Ae^{rx} \sin kx + Be^{rx} \cos kx$
$(c_n x^n + \cdots + c_1 x + c_0) e^{rx} \sin kx$ $(c_n x^n + \cdots + c_1 x + c_0) e^{rx} \cos kx$	$(A_n x^n + \cdots + A_1 x + A_0) e^{rx} \sin kx +$ $(B_n x^n + \cdots + B_1 x + B_0) e^{rx} \cos kx$

1. Note that the last entry includes all the others.
2. By Theorem 3.6, when $f(x)$ consists of a sum of several terms of the type in the left column, the trial y_p should be the sum of the corresponding terms from the right column.

Example 3.18 Solve the differential equation $y'' + 3y' + 2y = x^2 - xe^x$.

Solution The solution to the homogeneous equation is $y_c = c_1 e^{-x} + c_2 e^{-2x}$. According to Table 3.1, our trial y_p should be of the form

$$y_p = Ax^2 + Bx + C + (Dx + E)e^x$$

We must choose the constants so that $L(y_p) \equiv x^2 - xe^x$. Since

$$L(y_p) = 2Ax^2 + (6A + 2B)x + (2A + 3B + 2C) + 6Dxe^x + (5D + 6E)e^x$$

we are led to the following system of linear equations:

$$\begin{aligned} 2A \qquad\qquad\qquad\qquad &= 1 \\ 6A + 2B \qquad\qquad\quad &= 0 \\ 2A + 3B + 2C \qquad\; &= 0 \\ 6D \quad &= -1 \\ 5D + 6E &= 0 \end{aligned}$$

which has the solution $A = \frac{1}{2}$, $B = -\frac{3}{2}$, $C = \frac{7}{4}$, $D = -\frac{1}{6}$, $E = \frac{5}{36}$. The complete solution of the differential equation is then

$$y = y_c + y_p = c_1e^{-x} + c_2e^{-2x} + \tfrac{1}{2}x^2 - \tfrac{3}{2}x + \tfrac{7}{4} - \tfrac{1}{6}xe^x + \tfrac{5}{36}e^x \qquad \square$$

One last modification is needed to assure that our method of *undetermined coefficients* will always work (for the type of f about which we have been talking and, of course, assuming constant coefficients).

Example 3.19 Solve the differential equation $y'' - y' - 2y = e^{-x}$.

Solution If we try $y_p = Ae^{-x}$, we find

$$L(e^{-x}) = Ae^{-x} + Ae^{-x} - 2Ae^{-x} \equiv 0 \not\equiv e^{-x}$$

This should not be a surprise, since we know that e^{-x} is a solution of the reduced equation. For reasons directly related to the method of handling repeated roots of a characteristic equation, we try multiplying by x. That is, we try $y_p = Axe^{-x}$, yielding

$$L(Axe^{-x}) = Axe^{-x} - 2Ae^{-x} + Axe^{-x} - Ae^{-x} - 2Axe^{-x} = -3Ae^{-x} \stackrel{?}{\equiv} e^{-x}$$

Thus $y_p = -\frac{1}{3}xe^{-x}$ and $y = c_1e^{-x} + c_2e^{2x} - \frac{1}{3}xe^{-x}$. $\square$

The modification suggested by this example is:

When a term in a trial expression from Table 3.1 duplicates a term in the complementary function y_c, all the terms in that expression must be multiplied by the lowest positive integral power of x that eliminates any such duplication. After each expression from the table is modified in this fashion, the resulting new trial function *will* produce a valid y_p.

Example 3.20 Solve the equation $y'' - 2y' + y = 2e^x + 3$.

Solution The first guess corresponding to $f_1(x) = e^x$ is $y_{p_1} = Ae^x$, but since $y_c = c_1e^x + c_2xe^x$, we must multiply y_{p_1} by x^2 to eliminate duplication. There is no trouble with $y_{p_2} = B$ corresponding to $f_2(x) = 3$, so our actual trial solution is $y_p = Ax^2e^x + B$. Our now-familiar technique produces $L(y_p) = 2Ae^x + B$, so

$A = 1$ and $B = 3$, and the complete solution is

$$y = y_c + y_p = c_1 e^x + c_2 x e^x + x^2 e^x + 3 \qquad \square$$

Example 3.21 Solve $y'' + 2y' = e^x - x^2$.

Solution The first guess for y_p would be

$$y_p = \underbrace{Ae^x}_{y_{p_1}} + \underbrace{Bx^2 + Cx + D}_{y_{p_2}}$$

Since $y_c = c_1 e^{-2x} + c_2$, the D of y_{p_2} duplicates a term of y_c, and *all* y_{p_2} must be multiplied by x to find the trial y_p:

$$y_p = Ae^x + Bx^3 + Cx^2 + Dx$$

The solution is then readily found to be

$$y = c_1 e^{-2x} + c_2 + \tfrac{1}{3}e^x - \tfrac{1}{6}x^3 + \tfrac{1}{4}x^2 - \tfrac{1}{4}x \qquad \square$$

In order to apply this method of undetermined coefficients, the $f(x)$ must be a sum of terms of the form of some entry in the left-hand column of Table 3.1. There are, of course, various functions that, although apparently not appropriate, can be modified to be of the right form; for example,

$$\cos^2 x = \tfrac{1}{2} + \tfrac{1}{2}\cos 2x, \qquad \sinh x = \tfrac{1}{2}e^x - \tfrac{1}{2}e^{-x}, \quad \text{and} \quad 2^x = e^{(\ln 2)x}$$

Exercises for Section 3.5

Solve the equations in Exercises 1–25.

1. $y'' - 4y' + 4y = e^x$

2. $y'' - 4y' + 4y = e^x + 1$

3. $y'' - 4y' + 4y = e^{2x}$

4. $y'' - 4y' + 4y = \sin x$

5. $y'' - 4y' + 4y = xe^{2x} + e^{2x}$

6. $y'' - 4y' + 4y = xe^{2x}$

7. $y'' + y = \sin 2x$

8. $y'' + 4y = \sin 2x$

9. $y'' + 4y' = \sin 2x$

10. $y'' - 2y' + 5y = \sin 2x$

11. $y'' - 2y' + 5y = e^x \cos 2x$

12. $y'' - y = \cosh 2x$

13. $y'' - y = \cosh x$

14. $y'' + y = x \cos x$

15. $y'' - 3y' + 2y = e^x + e^{2x} + e^{-x}$

16. $y'' - y' = x^2$

17. $y'' - y' = 3$

18. $y'' + 3y' = x + 3$

19. $y'' + 4y = 3$

20. $y'' + 4y = 3x$

21. $y''' - 3y' - 2y = \sin 2x$

22. $y^{(4)} - y = e^{-x}$

23. $y''' + 4y'' + 9y' + 10y = -e^x$

24. $y''' + y'' = 1$

25. $y''' + y'' - 2y = x^2 + 10\cos 2x$

Determine the solutions of the initial value problems in Exercises 26–35.

26. $y'' + y = 10e^{2x}, \quad y(0) = 0, \quad y'(0) = 0$

27. $y'' + 3y' = -18x, \quad y(0) = 0, \quad y'(0) = 5$

28. $y'' + 4y' + 5y = 10e^{-3x}$, $y(0) = 4$, $y'(0) = 0$

29. $y'' + 9y = -3\sin 3x$, $y(0) = 0$, $y'(0) = -1$

30. $y''' - y'' - y' + y = 50\cos 2x$, $y(0) = 2$, $y'(0) = -10$, $y''(0) = -12$

31. $y''' - 4y'' + 4y' = 4$, $y(0) = -3$, $y'(0) = 1$, $y''(0) = 0$

32. $y'' + 4y' + 3y = 6x^2e^{-x}$, $y(0) = 1$, $y'(0) = -\frac{3}{2}$

33. $y'' + 2y' + 5y = 4e^{-x}\cos 2x$, $y(0) = 3$, $y'(0) = -3$

34. $y'' + y = 24\cos 2x - 12\sin x$, $y(\pi/2) = -3$, $y'(\pi/2) = 0$

35. $y''' - 2y'' + y' = xe^x + 5$, $y(0) = 1$, $y'(0) = 5$, $y''(0) = -1$

36. Consider the differential equation $y''' + 2y'' + y' = f(x)$. Determine the trial y_p to be used if $f(x)$ equals each of the following:
(a) x **(b)** $x + 2$ **(c)** $\sin x + x$
(d) e^{-x} **(e)** xe^{-x} **(f)** $\sinh x$
(g) $\sinh x + \cosh x$ **(h)** $x(1 + e^{-x})$

For each of the differential equations in Exercises 37–40, *set up* the expression in terms of undetermined coefficients that you would use in attempting to find a particular solution y_p. (Do not actually evaluate the coefficients.)

37. $y'' - 2y' - 3y = 2x\sin x - x^3e^{3x}$

38. $y''' - 3y'' + 2y' = 3x(1 - e^x) + 5x^2e^{2x}$

39. $y^{(4)} + y'' = \cos x + x^4$

40. $y^{(4)} + 3y'' - 4y = \cosh x - 2\cos^2 x$

41. The differential equation

$$EIu^{(4)}(x) = f(x)$$

arises in the analysis of the static deflection of a "rigid" thin homogeneous beam of uniform cross section. In this expression, E is the modulus of elasticity of the beam material and I is the moment of inertia of the beam cross section about the so-called neutral axis. Find the deflection $u(x)$ for the following transverse loads (determine only the particular solution):
(a) $f(x) = \sin x$ **(b)** $f(x) = x^2 - x$

3.6 Method of Variation of Parameters

In the previous section we found a particular solution of the nth-order linear equation with constant coefficients, $L(y) = f(x)$, by the method of undetermined coefficients. We saw that the method is limited to those driving functions that are solutions to some homogeneous linear equation with constant coefficients.

In this section we consider a method that, like reduction of order, has no such restriction. In fact, the method is applicable to all linear equations, including those with variable coefficients. The reduction-of-order method used the substitution $y = uy_1$, where y_1 was a nontrivial solution of the reduced equation, to solve completely second-order equations and at least simplify higher-order equations. If we have *all* the solutions of the reduced equation, we should be able to do even better. For purposes of discussion, we study the application of the method to the second-order equation in standard form,

$$y'' + a_1 y' + a_0 y = f(x)$$

Let y_1 and y_2 be any two linearly independent solutions to the corresponding reduced equation $y'' + a_1 y' + a_0 y = 0$. Assume that a solution to the nonhomogeneous equation may be written in the form

$$y = uy_1 + vy_2$$

where u and v are unknown functions (or *parameters*) to be determined.

Since two functions are to be determined, we need to impose two auxiliary conditions. The first condition is that $uy_1 + vy_2$ must satisfy the given differential equation, whereas the second condition, to be imposed for the convenience of simplifying the computations, is that $u'y_1 + v'y_2 \equiv 0$.

To impose the first condition, we find y' and y'' and substitute into the differential equation. Thus

$$y' = u'y_1 + v'y_2 + uy_1' + vy_2'$$

and, since $u'y_1 + v'y_2 \equiv 0$, this becomes

$$y' = uy_1' + vy_2'$$

Computing y'',

$$y'' = u'y_1' + v'y_2' + uy_1'' + vy_2''$$

Substituting into the differential equation,

$$(u'y_1' + v'y_2' + uy_1'' + vy_2'') + a_1(uy_1' + vy_2') + a_0(uy_1 + vy_2) = f(x)$$
$$(u'y_1' + v'y_2') + u(y_1'' + a_1 y_1' + a_0 y_1) + v(y_2'' + a_1 y_2' + a_0 y_2) = f(x)$$

The coefficient of u is zero, since y_1 is a solution to the corresponding reduced equation. Similarly the coefficient of v is zero, so that u' and v' must satisfy the system of equations

$$u'y_1 + v'y_2 = 0$$
$$u'y_1' + v'y_2' = f(x)$$

This system of equations may be solved for u' and v' (by Cramer's rule):

$$u' = \frac{-y_2 f(x)}{\det\begin{pmatrix} y_1 & y_2 \\ y_1' & y_2' \end{pmatrix}}; \qquad v' = \frac{y_1 f(x)}{\det\begin{pmatrix} y_1 & y_2 \\ y_1' & y_2' \end{pmatrix}}$$

Note that the denominator is the Wronskian of the two functions y_1 and y_2, and since y_1 and y_2 are linearly independent solutions, by Theorems 3.3 and 3.4 this Wronskian is nonzero. Hence u' and v' are completely determined and we see the wisdom of the artificial-appearing condition $u'y_1 + v'y_2 = 0$, which we chose earlier.

Example 3.22 Solve the differential equation $y'' + y = \sec x$.

Solution Since two linearly independent solutions to the corresponding reduced equation are $\cos x$ and $\sin x$, the complete solution to the reduced equation is

$$y_c = c_1 \cos x + c_2 \sin x$$

and we must find $y = u \cos x + v \sin x$, where u' and v' are given by

$$u' = \frac{-\sin x \sec x}{\det\begin{pmatrix} \cos x & \sin x \\ -\sin x & \cos x \end{pmatrix}}; \qquad v' = \frac{\cos x \sec x}{\det\begin{pmatrix} \cos x & \sin x \\ -\sin x & \cos x \end{pmatrix}}$$

Therefore $u' = -\tan x$ and $v' = 1$, from which $u = \ln \cos x + c_1, v = x + c_2$, and

$$y = c_1 \cos x + c_2 \sin x + (\ln \cos x) \cos x + x \sin x \qquad \square$$

As this example illustrates, by including arbitrary constants in our calculations of u and v, we not only find a particular solution of the nonhomogeneous equation, but also recapture the solution of the reduced equation as well. In practice, it is perhaps simpler to leave off these constants of integration and substitute the resulting $y_p = (\ln \cos x) \cos x + x \sin x$ into the expression $y = c_1 \cos x + c_2 \sin x + y_p$ in order to obtain the complete solution of the nonhomogeneous equation.

Example 3.23 Solve the equation $x^2 y'' + xy' - y = x^4$, $x > 0$.

Solution Recall from Example 3.3 that the complete solution of the reduced equation is $y_c = c_1 x + c_2 x^{-1}$. An essential first step is to put the equation in standard form:

$$y'' + x^{-1} y' - x^{-2} y = x^2$$

We then find $y_p = ux + vx^{-1}$, where u' and v' satisfy

$$u' = \frac{-x}{\det\begin{pmatrix} x & x^{-1} \\ 1 & -x^{-2} \end{pmatrix}}; \qquad v' = \frac{x^3}{\det\begin{pmatrix} x & x^{-1} \\ 1 & -x^{-2} \end{pmatrix}}$$

Therefore $u' = x^2/2$ and $v' = -x^4/2$, from which $u = x^3/6$, $v = -x^5/10$, and $y_p = x^4/6 - x^4/10 = x^4/15$. The complete solution is

$$y = c_1 x + c_2 x^{-1} + \frac{x^4}{15} \qquad \square$$

Notice that the undetermined-coefficients method described earlier could not have been used in Example 3.22 (because $f(x)$ is not of an appropriate form) nor could it have been employed in Example 3.23 (because the coefficients of the differential equation are not constant). In general, however, one should use the undetermined-coefficients method whenever it applies and the more cumbersome variation-of-parameters method only when necessary.

To apply variation of parameters to an nth-order linear equation in *standard form*,

$$y^{(n)} + a_{n-1}(x) y^{(n-1)} + \cdots + a_1(x) y' + a_0(x) y = f(x)$$

replace the constants (or parameters) in the complete solution

$$c_1 y_1 + c_2 y_2 + \cdots + c_n y_n$$

of the reduced equation by variables $u_1, \ldots, u_n$ to obtain a particular solution

$$y_p = u_1 y_1 + u_2 y_2 + \cdots + u_n y_n$$

where $u_1', \ldots, u_n'$ satisfy the system

$$\begin{aligned} u_1' y_1 + u_2' y_2 \quad &+ \cdots + u_n' y_n &= 0 \\ u_1' y_1' + u_2' y_2' \quad &+ \cdots + u_n' y_n' &= 0 \\ \vdots \qquad\qquad & &\vdots \\ u_1' y_1^{(n-1)} + u_2' y_2^{(n-1)} &+ \cdots + u_n' y_n^{(n-1)} &= f(x) \end{aligned}$$

This system is easy to remember, since its determinant is the Wronskian $W(y_1, y_2, \ldots, y_n)$, but actual computation of y_p is likely to be very tedious even for $n = 3$ or 4.

Example 3.24 Solve $y''' - y' = \dfrac{1}{1 + e^x}$.

Solution The standard form of the equation is

$$y''' - y' = \frac{1}{1 + e^x}$$

Since the solution to the reduced equation is

$$y_c = c_1 + c_2 e^x + c_3 e^{-x}$$

we will be looking for a y_p of the form $u_1 + u_2 e^x + u_3 e^{-x}$ where

$$\begin{aligned} u_1' + u_2' e^x + u_3' e^{-x} &= 0 \\ u_2' e^x - u_3' e^{-x} &= 0 \\ u_2' e^x + u_3' e^{-x} &= \frac{1}{1 + e^x} \end{aligned}$$

The determinant of this system is

$$\det \begin{pmatrix} 1 & e^x & e^{-x} \\ 0 & e^x & -e^{-x} \\ 0 & e^x & e^{-x} \end{pmatrix} = 2$$

so Cramer's rule yields

$$u_1' = \frac{1}{2} \det \begin{pmatrix} 0 & e^x & e^{-x} \\ 0 & e^x & -e^{-x} \\ \dfrac{1}{1 + e^x} & e^x & e^{-x} \end{pmatrix} = \frac{-1}{1 + e^x}$$

$$u_2' = \frac{1}{2} \det \begin{pmatrix} 1 & 0 & e^{-x} \\ 0 & 0 & -e^{-x} \\ 0 & \dfrac{1}{1 + e^x} & e^{-x} \end{pmatrix} = \frac{\frac{1}{2} e^{-x}}{1 + e^x}$$

and

$$u_3' = \frac{1}{2}\det\begin{pmatrix} 1 & e^x & 0 \\ 0 & e^x & 0 \\ 0 & e^x & \dfrac{1}{1+e^x} \end{pmatrix} = \frac{\frac{1}{2}e^x}{1+e^x}$$

We need only to find the appropriate antiderivatives:

$$u_1 = \int \frac{-dx}{1+e^x} = -\int \frac{e^{-x}\,dx}{e^{-x}+1} = \ln(e^{-x}+1) + c_1$$

$$u_2 = \int \frac{\frac{1}{2}e^{-x}}{1+e^x}dx = \frac{1}{2}\int\left(e^{-x} - \frac{1}{1+e^x}\right)dx = -\frac{1}{2}e^{-x} + \frac{1}{2}\ln(e^{-x}+1) + c_2$$

$$u_3 = \frac{1}{2}\int \frac{e^x\,dx}{1+e^x} = \frac{1}{2}\ln(1+e^x) + c_3$$

Putting the pieces together:

$$\begin{aligned} y &= u_1 + u_2e^x + u_3e^{-x} \\ &= c_1 + c_2e^x + c_3e^{-x} - \frac{1}{2} + \ln(e^{-x}+1) + \frac{1}{2}e^x\ln(e^{-x}+1) + \frac{1}{2}e^{-x}\ln(1+e^x) \end{aligned}$$

It is interesting (but not necessary!) to find simpler forms for this final result.

□

Exercises for Section 3.6

Solve Exercises 1–10 by using the method of variation of parameters.

1. $y'' - y' - 2y = e^{2x}$ **2.** $y'' + 2y' + y = e^{-x}/x$

3. $y'' + 4y = \tan 2x$ **4.** $y'' + 4y = \tan^2 2x$

5. $y'' + 4y = \sin^2 2x$ **6.** $y'' - 3y' + 2y = \cos(e^{-x})$

7. $y'' - 4y' + 4y = e^{2x}/x$ **8.** $y'' + 6y' + 9y = e^{-3x}/(x^2+1)$

9. $y'' + 2y' + y = e^{-x}\ln x$ **10.** $y'' + 2y' + y = e^{-x}/x^3$

11. By using the method of variation of parameters, show that the solution to $y'' + y = f(x)$ can be written in the form

$$y = c_1\cos x + c_2\sin x + \int_a^x f(t)\sin(x-t)\,dt$$

The second-order equation $a_2x^2y'' + a_1xy' + a_0y = f(x)$, where the a_i are constants, is called the *Cauchy-Euler equation* of order two. This important equation is one of the simplest second-order equations with variable coefficients to solve, as suggested by the next two exercises.

12. Show that the substitution for the independent variable $x = e^v$ (or equivalently $v = \ln x$) reduces the Cauchy-Euler equation to

$$a_2\frac{d^2y}{dv^2} + (a_1 - a_2)\frac{dy}{dv} + a_0y = f(e^v)$$

which is a second-order equation with constant coefficients and can be solved by the methods of the previous sections.

13. Alternatively, verify that solutions to the reduced Cauchy-Euler equation have the following forms:
(a) $y_1 = x^{m_1}$, $y_2 = x^{m_2}$ for $m_1 \neq m_2$; m_1, m_2 real
(b) $y_1 = x^{m_1}$, $y_2 = x^{m_1} \ln x$ for $m_1 = m_2$ real
(c) $y_1 = x^a \cos(b \ln x)$, $y_2 = x^a \sin(b \ln x)$ for $m_1 = \bar{m}_2 = a + ib$
Here, m_1, m_2 satisfy the quadratic equation

$$a_2 m^2 + (a_1 - a_2)m + a_0 = 0$$

which is obtained when the trial solution $y = x^m$ is inserted in the original equation.

Solve the Cauchy-Euler equations in Exercises 14–18.

14. $x^2 y'' + 4xy' + 2y = 0$
15. $x^2 y'' - xy' + y = 0$
16. $x^2 y'' + xy' + 9y = 0$
17. $x^2 y'' - 2xy' + 2y = 4x + \sin(\ln x)$
18. $x^2 y'' - xy' - 3y = x^2 \ln x$

Determine the solutions of the initial value problems in Exercises 19–23.

19. $x^2 y'' - 3xy' + 4y = 0, \quad y(1) = 2, \quad y'(1) = 1$
20. $x^2 y'' + xy' + 4y = 0, \quad y(1) = 1, \quad y'(1) = 4$
21. $x^2 y'' + 2xy' - 2y = 6x, \quad y(1) = 3, \quad y'(1) = -7$
22. $x^2 y'' - 3xy' + 13y = 9 + 13 \ln x, \quad y(1) = 0, \quad y'(1) = -1$
23. $x^2 y'' + xy' + y = 4 \sin(\ln x), \quad y(1) = 0, \quad y'(1) = -2$

Employ the variation-of-parameters procedure to determine the complete solution in the following exercises.

24. $y''' - y'' - y' + y = 2xe^x$
25. $y''' + y' = \tan x$
26. $y''' - 3y' - 2y = e^x + xe^{2x}$

In Exercises 27–30, find the complete solution, given the two linearly independent solutions of the corresponding reduced equation.

27. $(x^2 + 2x)y'' - 2(x + 1)y' + 2y = x^2 + 4x + 4; \quad y_1 = x + 1, y_2 = x^2$
28. $xy'' - 2(x + 1)y' + (x + 2)y = (x - 2)e^{2x}; \quad y_1 = e^x, y_2 = x^3 e^x$
29. $x^2(x + 2)y'' + 2xy' - 2y = (x + 2)^2; \quad y_1 = x, y_2 = 1 + 1/x$
30. $(\sin x)y'' - 2(\cos x)y' + (2 \csc x - \sin x)y = \sin^2 x; \quad y_1 = \sin x, y_2 = x \sin x$

3.7 Application to Electric Circuits

One of the most immediate applications of second-order linear differential equations with constant coefficients is to electric circuits. In analyzing circuits, it is convenient to consider influences that behave to impede the flow of current, the

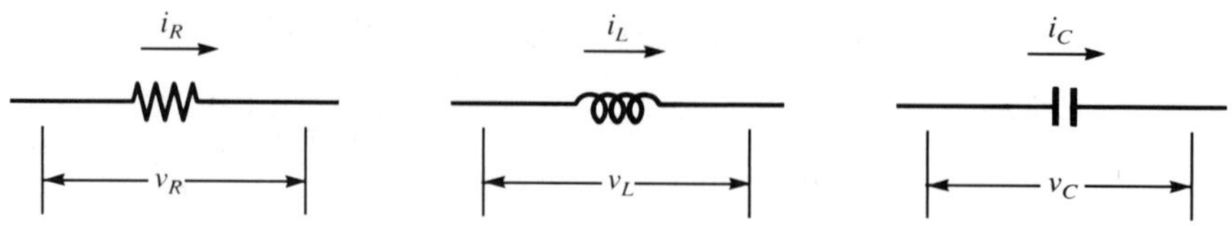

Figure 3.1

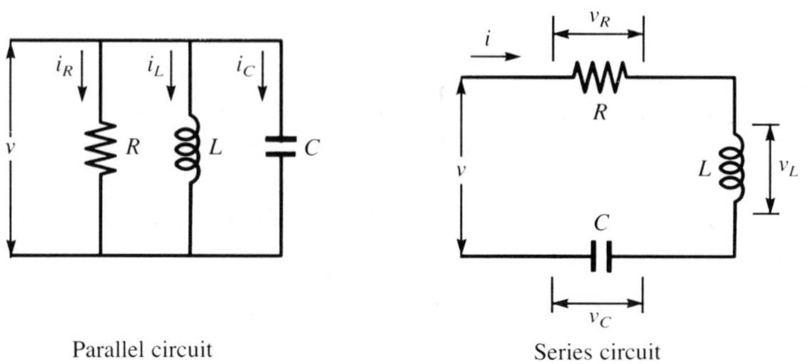

Figure 3.2

change in current, or the change in voltage. In the most elementary cases, these influences—called resistance, R, inductance, L, and capacitance, C—are considered as being constant and "lumped" in specific components. These three constants are essentially defined by the following three equations (see Figure 3.1):

$$Ri_R = v_R; \qquad L\frac{di_L}{dt} = v_L; \qquad C\frac{dv_C}{dt} = i_C$$

where the i's and v's refer to the currents through and the voltages across the lumped components. In a so-called series circuit, $i_R = i_L = i_C$, and in a parallel circuit, $v_R = v_L = v_C$ (Figure 3.2). Many electric circuits can be considered as combinations of these two arrangements.

The governing physical law for an RLC series circuit is given by Kirchhoff's law, which says that the driving voltage of the circuit is equal to the sum of the voltage drops. In differential equation form,

$$L\frac{di}{dt} + Ri + \frac{1}{C}\int i(t)\,dt = v$$

where v is the driving voltage. This equation relates the response, that is, the current, to the input voltage, or excitation.

It is often easier to handle the equation if both sides are differentiated with respect to t,

$$L\frac{d^2i}{dt^2} + R\frac{di}{dt} + \frac{i}{C} = \frac{dv}{dt}$$

Alternately, we can focus our attention on the instantaneous charge, q, on the capacitor and express the current as dq/dt. The voltage drop due to the capacitance C is then q/C, and the differential equation modeling the circuit is

$$L\frac{d^2q}{dt^2} + R\frac{dq}{dt} + \frac{1}{C}q = v$$

Notice that this equation and the earlier equation describing the current, i, have exactly the same characteristic equation

$$Lm^2 + Rm + \frac{1}{C} = 0$$

but have a different driving function.

A very common type of input to an RLC circuit is some combination of sine or cosine functions. For the remainder of this section we discuss the RLC circuit with a cosinusoidal input voltage, $E\cos\omega t$, where E is the *amplitude* of the input and $\omega \neq 0$ is the angular frequency. Since $dv/dt = -E\omega\sin\omega t$, the differential equation of the circuit is

$$L\frac{d^2i}{dt^2} + R\frac{di}{dt} + \frac{i}{C} = -E\omega\sin\omega t$$

The corresponding reduced equation is

$$L\frac{d^2i}{dt^2} + R\frac{di}{dt} + \frac{i}{C} = 0$$

If $R \neq 0$, the solution to the reduced equation is called the *transient current*, since—regardless of the specific values of R, L, and C or the initial conditions—the current approaches zero as time increases. The characteristic polynomial equation is

$$Lm^2 + Rm + \frac{1}{C} = 0$$

whose solution is

$$m = \frac{-R \pm \sqrt{R^2 - 4L/C}}{2L} = \begin{cases} -a \pm b & \text{if } R^2 - \dfrac{4L}{C} \geq 0 \\ -a \pm i\omega_0 & \text{if } R^2 - \dfrac{4L}{C} < 0 \end{cases}$$

where

$$a = \frac{R}{2L}$$

$$b = \frac{\sqrt{R^2 - 4L/C}}{2L} \qquad \text{(note that } a > b\text{)} \quad \text{if } R^2 - \frac{4L}{C} \geq 0$$

$$\omega_0 = \frac{\sqrt{4L/C - R^2}}{2L} \qquad \text{if } R^2 - \frac{4L}{C} < 0$$

The quantity a is usually called the *damping factor*, and ω_0 is called the *internal*, or *natural angular*, *frequency*.

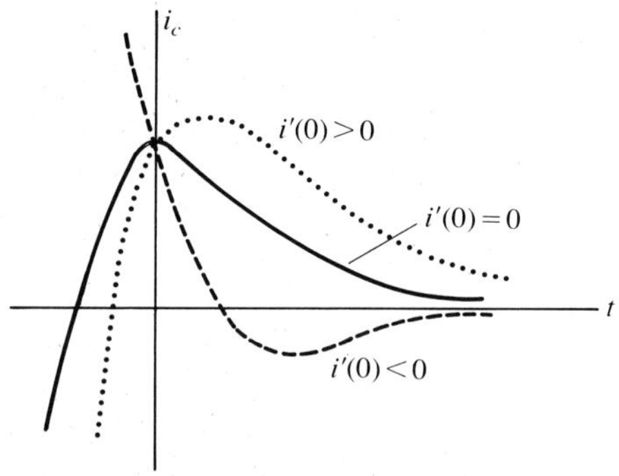

Figure 3.3 *Overdamped or critically damped*

Case 1: Overdamped If $R^2 > 4L/C$, then both roots are real and negative. The transient current has the form

$$i_c = c_1 e^{-(a+b)t} + c_2 e^{-(a-b)t}$$

Figure 3.3 shows the current for some typical initial values of $i' \equiv di/dt$.

Case 2: Critically damped If $R^2 = 4L/C$, then the characteristic equation has a double root $m = -a$. Hence the solution is

$$i_c = e^{-at}(c_1 + c_2 t)$$

The transient current is of the same general form as for the overdamped case.

Case 3: Underdamped If $R^2 < 4L/C$, then the roots to the characteristic equation are $m = -a \pm i\omega_0$ and the solution is

$$i_c = e^{-at}(c_1 \cos \omega_0 t + c_2 \sin \omega_0 t)$$

This is sometimes put into the form

$$i_c = e^{-at} c_3 \cos(\omega_0 t - c_4)$$

where

$$c_3 = \sqrt{c_1^2 + c_2^2} \quad \text{and} \quad c_4 = \tan^{-1}\left(\frac{c_2}{c_1}\right)$$

Figure 3.4 shows a typical underdamped case for the initial conditions $i(0) = 1$, $i'(0) = 0$.

The graphs of the overdamped and critically damped transients cut the t-axis at most once (see Exercise 5, p. 115). The underdamped transient is zero an infinite number of times but is *not* periodic because of the presence of the real exponential factor. Thus the description "natural angular frequency" given earlier for ω_0 refers only to the sine and cosine terms present, but does relate nicely to the equally spaced zeros of the damped oscillation.

When $R \neq 0$ the particular solution to the nonhomogeneous equation corresponding to $i_c = 0$ is called the *steady-state* current, i_{ss}. It is easily obtained

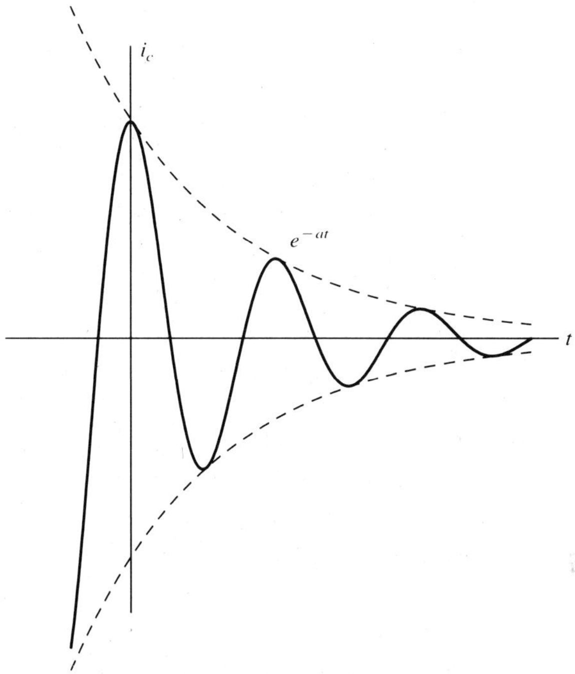

Figure 3.4 *Underdamped*

by an application of the method of undetermined coefficients. Let

$$i_{ss} = A\cos\omega t + B\sin\omega t$$

from which

$$\frac{d}{dt}(i_{ss}) = -\omega A\sin\omega t + \omega B\cos\omega t$$

$$\frac{d^2}{dt^2}(i_{ss}) = -\omega^2 A\cos\omega t - \omega^2 B\sin\omega t$$

Substituting into the differential equation, we obtain

$$-\omega^2 AL\cos\omega t - \omega^2 BL\sin\omega t - \omega AR\sin\omega t + \omega RB\cos\omega t + \frac{A}{C}\cos\omega t + \frac{B}{C}\sin\omega t = -E\omega\sin\omega t$$

Equating coefficients of $\cos\omega t$ and $\sin\omega t$, we have the following two equations:

$$\left(-\omega^2 L + \frac{1}{C}\right)A + \omega RB = 0$$

$$-\omega RA + \left(-\omega^2 L + \frac{1}{C}\right)B = -E\omega$$

Solving for A and B, we obtain

$$B = \frac{E\omega^2(\omega L - 1/\omega C)}{\omega^2(\omega L - 1/\omega C)^2 + \omega^2 R^2}, \qquad A = \frac{E\omega^2 R}{\omega^2(\omega L - 1/\omega C)^2 + \omega^2 R^2}$$

The quantity $(\omega L - 1/\omega C)$ is called the *reactance* and is denoted by X. The quantity $\sqrt{X^2 + R^2}$ is called the *impedance* and is denoted by Z. Using these notations,

$$A = \frac{ER}{Z^2} \quad \text{and} \quad B = \frac{EX}{Z^2}$$

Since the input *voltage* is $E \cos \omega t$, it is customary to express the steady-state current in the form $\alpha \cos(\omega t - \beta)$ which displays the *amplitude*, α, of the output oscillation, and the *phase shift*, β, instead of the constants A and B. Thus

$$\alpha = \sqrt{A^2 + B^2} = \sqrt{\frac{E^2X^2 + E^2R^2}{Z^4}} = \frac{E}{Z}; \qquad \beta = \text{Tan}^{-1}\frac{X}{R}$$

In summary, the total output current is given by:

$$\text{Output current} = \text{transient current} + \text{steady-state current}$$

$$i = c_1 e^{-(a+b)t} + c_2 e^{-(a-b)t} + \frac{E}{Z}\cos\left(\omega t - \text{Tan}^{-1}\frac{X}{R}\right) \qquad \textit{Overdamped}$$

$$i = e^{-at}(c_1 + c_2 t) + \frac{E}{Z}\cos\left(\omega t - \text{Tan}^{-1}\frac{X}{R}\right) \qquad \textit{Critically damped}$$

$$i = e^{-at}c_3\cos(\omega_0 t - c_4) + \frac{E}{Z}\cos\left(\omega t - \text{Tan}^{-1}\frac{X}{R}\right) \qquad \textit{Underdamped}$$

If the input voltage had been $E \sin \omega t$, the steady-state current would have been

$$\frac{E}{Z}\sin\left(\omega t - \text{Tan}^{-1}\frac{X}{R}\right)$$

in each of the three cases.

One special situation that is interesting occurs when $R = 0$. In this case, $a = 0$, $\omega_0 = \sqrt{1/(LC)}$, and, if $\omega \neq \omega_0$,

$$i = c_3\cos(\omega_0 t - c_4) + \frac{E}{X}\cos(\omega t - \pi/2) \qquad \textit{No damping}$$

If we choose initial conditions such that $c_4 = \pi/2$ and $c_3 = -(E/X)$, then

$$i = \frac{E}{X}(\sin \omega t - \sin \omega_0 t)$$

By use of the trigonometric identity $\sin u - \sin v = 2\sin\frac{u - v}{2}\cos\frac{u + v}{2}$, this becomes

$$i = 2\frac{E}{X}\left[\sin\left(\frac{\omega - \omega_0}{2}\right)t\cos\left(\frac{\omega + \omega_0}{2}\right)t\right]$$

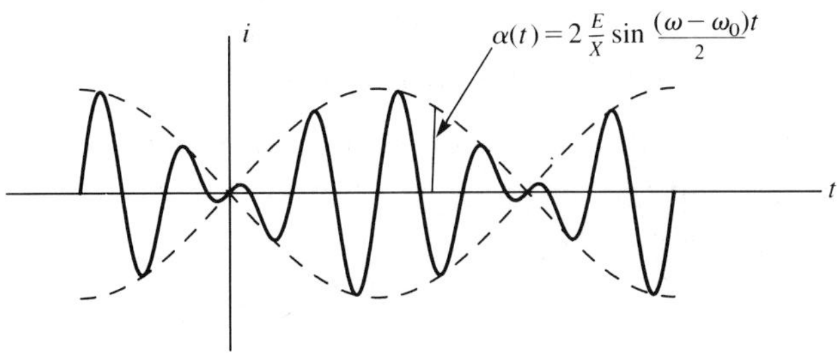

Figure 3.5 *Beats*

The point of interest is when ω is close to ω_0. This type of oscillation can be interpreted to have angular frequency close to ω(and to ω_0) with amplitude $\alpha(t)$, which fluctuates slowly with time with angular frequency $(\omega - \omega_0)/2$. Oscillations of this type are called *beats*. (See Figure 3.5.) Note that the amplitude of the beat is twice the amplitude of the steady-state output when there is damping.

If $R = 0$ and ω is actually equal to ω_0, we have

$$i'' + \omega_0^2 i = -\frac{E\omega_0}{L}\sin\omega_0 t$$

with solution

$$i = c_3\cos(\omega_0 t - c_4) + \frac{E}{2L}t\cos\omega_0 t$$

Regardless of initial conditions, this solution is unbounded as t increases.

In the practical case there is always *some* damping so that the transient current is negligible after a period of time. The amplitude of the steady-state current is dependent upon L, C, R, E, and ω, and is given by the formula

$$\alpha = \frac{E}{Z}$$

We are interested in conditions for which this amplitude is a maximum. Conditions of this type are called *resonance*. Obviously the output is increased whenever the amplitude, E, of the input voltage is increased. For a fixed value of E, and fixed $R \neq 0$, the output is maximized by minimizing the impedance $Z = \sqrt{(\omega L - 1/\omega C)^2 + R^2}$, by making $\omega^2 LC = 1$, that is by making the angular frequency of the applied voltage equal to the undamped natural angular frequency, $\sqrt{1/LC}$, of the circuit. If ω, L, and C are fixed, the amplitude of the steady-state current increases as R decreases.

Since in the special case when $R = 0$, $\omega = \sqrt{1/LC}$ produces an unbounded solution, the entire analysis of current resonance is summarized by calling $\sqrt{1/LC}$ the *resonant angular frequency*.

Example 3.25 Using initial conditions $i(0) = 0$, $i'(0) = 0$, find the current in an RLC circuit in each case.

(a) $L = 1$, $R = 0$, and $C = 1$ with an input voltage, $E = \sin t$

(b) $L = 1, R = 0, C = 1/1.1025$, and $E = (\sin 0.95t)/0.95$
(c) $L = 1, R = 0.2, C = 1/1.01$, and $E = \sin t - 0.05 \cos t$

Solution The three differential equations that must be solved with initial conditions $i(0) = 0, i'(0) = 0$ are
(a) $i'' + i = \cos t$,
(b) $i'' + 1.1025i = \cos 0.95t$, and
(c) $i'' + 0.2i' + 1.01i = \cos t + 0.05 \sin t$.
We recognize case (a) as undamped resonance,

$$i = c_1 \cos t + c_2 \sin t + \tfrac{1}{2}t \sin t$$

Evaluating the constants from the initial conditions, we obtain $c_1 = c_2 = 0$, so that $i = (t \sin t)/2$, which becomes unbounded with time. The complete solution to case (b) is easily found to be

$$i = c_1 \cos 1.05t + c_2 \sin 1.05t + 5 \cos 0.95t$$

The initial conditions give $c_1 = -5$ and $c_2 = 0$, from which

$$i = 5(\cos 0.95t - \cos 1.05t) = 10 \sin(0.05t) \sin t$$

This is the "beats" solution.

The complete solution to case (c) is

$$i = e^{-0.1t}(c_1 \cos t + c_2 \sin t) + 5 \sin t$$

The initial conditions give $c_1 = 0, c_2 = -5$, hence

$$i = (5 - 5e^{-0.1t}) \sin t$$

Note again that the amplitude of the beats solution is exactly twice that of the steady-state solution of the last, damped, case. □

3.8 Application to Mechanical Systems

An elementary mechanical system may be pictured with a coil spring suspended vertically from a fixed support and attached to a mass, m. The mass in turn is connected to a dashpot, a device simulating viscous friction. We assume that when the mass is displaced vertically from its equilibrium position—that is, the position assumed when the mass is at rest—the spring exerts a return force proportional to the displacement y. (Such a spring is said to satisfy Hooke's law.) To simplify computations, the damping force due to friction is assumed to be proportional to the velocity dy/dt. Applying Newton's second law of motion, we have

$$m\frac{d^2 y}{dt^2} = -c\frac{dy}{dt} - ky + F(t)$$

where c and k are the damping and spring constants, respectively, and $F(t)$ describes the instantaneous resultant of all external forces acting on the mass. (See Figure 3.6.)

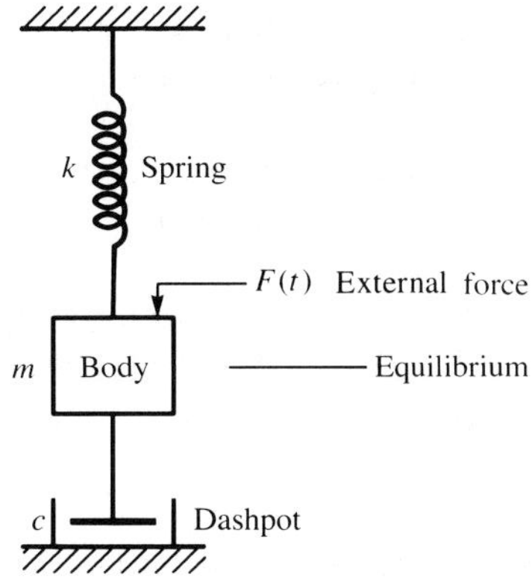

Figure 3.6 *Elementary mechanical system*

This is a differential equation with constant coefficients, which has precisely the same form as the equation for the electric charge

$$L\frac{d^2q}{dt^2} + R\frac{dq}{dt} + \frac{q}{C} = v$$

discussed in the previous section. Indeed, note the correspondence between the electrical and mechanical quantities.

Mass, $m \leftrightarrow$ inductance, L

Fluid friction constant, $c \leftrightarrow$ resistance, R

Spring constant, $k \leftrightarrow$ reciprocal of capacitance, $1/C$

External driving force, $F(t) \leftrightarrow$ driving voltage, v

Displacement, $y \leftrightarrow$ charge, q

Velocity, $y' \leftrightarrow$ current, i

With these correspondences, we can immediately write the displacement as a function of time in response to an external excitation of $-F \sin \omega t$. For $\omega \neq 0$ and $c \neq 0$:

$$y = c_1 e^{-(a+b)t} + c_2 e^{-(a-b)t} + \frac{F}{\omega Z}\cos\left(\omega t - \operatorname{Tan}^{-1}\frac{X}{c}\right) \qquad \textit{Overdamped}$$

$$y = e^{-at}(c_1 + c_2 t) + \frac{F}{\omega Z}\cos\left(\omega t - \operatorname{Tan}^{-1}\frac{X}{c}\right) \qquad \textit{Critically damped}$$

$$y = e^{-at} c_3 \cos(\omega_0 t - c_4) + \frac{F}{\omega Z}\cos\left(\omega t - \operatorname{Tan}^{-1}\frac{X}{c}\right) \qquad \textit{Underdamped}$$

where

$$a = \frac{c}{2m}, \qquad \text{Damping factor}$$

$$b = \sqrt{\left(\frac{c}{2m}\right)^2 - \frac{k}{m}}$$

$$\omega_0 = \sqrt{\frac{k}{m} - \left(\frac{c}{2m}\right)^2}, \qquad \text{Natural angular frequency}$$

$$\omega Z = \sqrt{\omega^2 c^2 + (m\omega^2 - k)^2}$$

$$X = m\omega - \frac{k}{\omega}$$

Analogous to the electrical situation, if the external excitation had been $F \cos \omega t$ and $c \neq 0$, the steady-state response would have been

$$\frac{F}{\omega Z} \sin\left(\omega t - \mathrm{Tan}^{-1} \frac{X}{c}\right)$$

in all cases.

Mechanical resonance is analyzed in a manner that mimics the analysis of the case of the *RLC* circuit. We carry out the details both because of the practical importance of the concept and because our emphasis on displacement (rather than velocity) gives a flavor to the discussion different from that in the electrical case.

The amplitude of the steady-state motion is

$$\alpha = \frac{F}{\omega Z}$$

The method of maximizing this output amplitude is the same as for the electrical case if F, m, c, or k is varied. If ω is varied and F, m, c, and k are fixed, the value of α is maximized by minimizing ωZ, and this quantity is a minimum if $\omega^2 Z^2$ is a minimum.

We therefore take the derivative of $\omega^2 Z^2$ with respect to ω,

$$\frac{d}{d\omega}(\omega^2 Z^2) = \frac{d}{d\omega}\left[(\omega^2 c^2) + m^2\left(\omega^2 - \frac{k}{m}\right)^2\right] = 2\omega c^2 + 4m^2\omega\left(\omega^2 - \frac{k}{m}\right)$$

Setting this quantity equal to zero, we obtain

$$c^2 = 2m^2\left(\frac{k}{m} - \omega^2\right)$$

Hence, the *resonant angular frequency*, ω_r, is

$$\omega_r^2 = \frac{k}{m} - \frac{c^2}{2m^2}$$

If $c^2 > 2mk$, the equation for ω_r does not give a real value and the output amplitude monotonically decreases with increasing ω. If $c^2 < 2mk$, a positive real value of ω_r is obtained. Figure 3.7 shows the variation of the amplitude of the steady-state output with respect to ω for each of these cases. Note that the values of the resonant and natural angular frequencies, ω_r and ω_0, respectively, differ by a small number if the damping is small. Hence, it is sometimes said that resonance is roughly approximated by letting the value of ω be ω_0; this is one

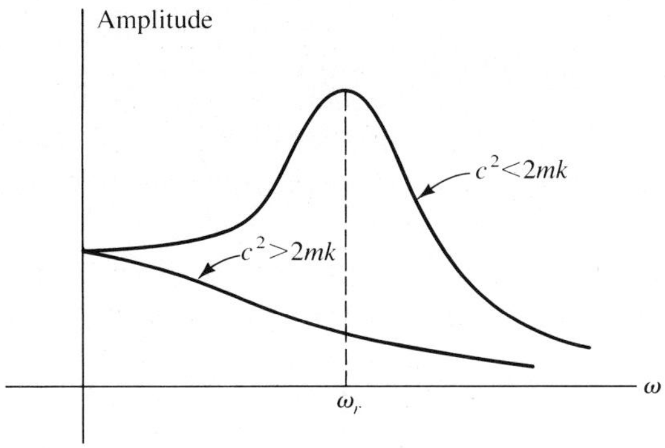

Figure 3.7 *Output amplitude as a function of excitation angular frequency*

of the reasons why the resonant and natural angular frequencies are sometimes confused.

It can readily be shown that the maximum amplitude of the steady-state *velocity* occurs when ω equals the undamped natural angular frequency, $\sqrt{k/m}$. This could be expected owing to the analogy between the current, i, in basic electrical circuits and the velocity, y', in elementary mechanical systems.

As in the electrical analysis, $c = 0$ leads to the interesting, although physically impossible, limiting case of unbounded motion. (See Exercise 7 in this section.)

Resonance is often desirable in an electric circuit where a small incoming signal or driving voltage is to be amplified. In a mechanical system, resonance is often to be avoided. For example, a forcing term $\cos \omega t$ with ω too near ω_r may cause a washing machine to "walk" across the floor, the front end of a car to vibrate violently, a bridge to collapse, or the wings of an aircraft to flap until they sustain structural damage. In either type of system, an analysis of resonance properties is essential.

Exercises for Sections 3.7 and 3.8

1. Find the transient and steady-state current in an *RLC* circuit with $R = 10\,\Omega$, $L = 1\text{H}$, and $C = \frac{1}{9}\text{F}$, if the input voltage is $E = 25 \sin t$ volts. Assume $i(0) = 0$ and $i'(0) = 0$.

2. Repeat Exercise 1 with $R = 1000\,\Omega$, $L = 10$ mH, and $C = 10\,\mu\text{F}$.

3. Find the steady-state and transient oscillations of a mechanical system with $m = 9$, $c = 5$, and $k = 16$ with external excitation $10 \cos t$, if the initial displacement and velocity are zero.

4. Repeat Exercise 3 with $m = 1$, $c = 2$, and $k = 2$ with external force $5 \sin t$.

5. Prove that the overdamped or critically damped transient current passes through zero at most once and hence is nonoscillatory.

6. How does the frequency of an underdamped mechanical vibration with no external excitation depend on the initial conditions?

7. Find the output vibration of an undamped mechanical system with $k = 4$, $m = 1$ if the external excitation is $\sin 2t$. Sketch the graph of the output (transverse) motion versus t. Assume $y(0) = y'(0) = 0$.

8. Sketch the graph of the output amplitude as a function of input frequency for Exercise 7.
9. Sketch the graph of the damped oscillation $y = e^{-\alpha t}\cos t$. Where do the maxima and minima occur? Find the ratio of any two consecutive maxima.
10. Find the resonant angular frequency for the displacement in the mechanical system of Exercise 3.
11. Find the resonant angular frequency for the current in an RLC circuit of $R = 10\ \Omega$, $L = 10$ mH, and $C = 10\ \mu$F.

In Exercises 12–14, use the initial conditions $y(0) = 0$, $y'(0) = 0$ to solve each of the given problems.

12. $y'' + y = \sin t$
13. $y'' + 1.1025y = \sin 0.95t$
14. $y'' + 0.2y' + 1.01y = \sin t - 0.05\cos t$

In a psychological experiment on learning, the amount of new knowledge acquired, u, as a function of time, t, is given by the solution of the initial value problem

$$\frac{d^2u}{dt^2} + a\frac{du}{dt} + bu = a\beta \qquad (a, b \text{ positive constants})$$

$$u(0) = \alpha, \qquad \frac{du}{dt}(0) = \beta$$

Exercises 15–18 refer to this learning model.

15. Analyze the case in which the constants a and b are such that underdamping, analogous to the electrical-mechanical applications, occurs.
16. Can resonance take place in this application? Explain your answer.
17. Solve the initial value problem that results when $\alpha = N$ and $\beta = 0$, and investigate the behavior of the solution as $t \to \infty$.
18. Solve the initial value problem corresponding to $\alpha = 0$ and $\beta = N$, and compare the solution to that obtained in Exercise 17.

The differential equation

$$\frac{d^2g}{dt^2} + 2\alpha\frac{dg}{dt} + \beta^2 g = f(t)$$

occurs in a biomathematical model for the detection of diabetes in humans. Here $g(t)$ is the instantaneous deviation of blood glucose concentration from some predetermined optimum level, α and β are positive constants associated with the kinetics of the body's blood glucose regulatory system, and $f(t)$ is determined by the rate at which the blood glucose concentration is being externally driven at time t.

Exercises 19–22 refer to this biomathematical model.

19. Show that in the absence of external input, that is, with $f(t) \equiv 0$, all solutions of the differential equation tend to zero as $t \to \infty$.
20. Discuss in detail the case of critical damping when $f(t)$ has the form

$$f(t) = F\cos\omega t$$

21. Determine the steady-state response to the input function

$$f(t) = \begin{cases} F, & 0 \leq t \leq 1 \\ 0, & t > 1 \end{cases}$$

22. Find the transient response for the case when $\beta = \alpha$, $g(0) = F$, and $\frac{dg}{dt}(0) = 0$.

3.9 Looking Ahead

It is often the case in mathematics that an operational procedure is used to simplify at least the intermediate steps in the solution of a more complicated problem. An elementary example of this is the utilization of logarithms in the evaluation of nonintegral powers of a given number. In this chapter we saw that the solution of homogeneous linear ordinary differential equations, either with constant coefficients or of Cauchy-Euler form, essentially reduced to the algebraic problem of polynomial root-finding. We will encounter other examples of such simplifying techniques in the next chapter on systems of equations as well as later in this text, when we discuss the application of Laplace transforms to non-homogeneous ordinary differential equations (Chapter 5) and the use of separation of variables in partial differential equations (Chapter 8). Another important example occurs in the analysis of linear integral equations with degenerate kernels (see [6] and [11], for example).

The ideas underlying the method of variation of parameters can be unified into a more general framework by introducing the notion of Green's functions (sometimes called the "impulse response" by engineers). Solution representations using Green's functions have wide applicability in the analysis of ordinary differential equations as well as in partial differential equations. Good references in this regard are [7], [13], and [14].

When numerical techniques are to be applied to differential equations, the continuous problem first must be "discretized." Occasionally, however, problems present themselves in which the independent variable is already in discrete form. Difference equations are discussed in [3], [5], and [9]. Combined differential-difference equations are treated in the classic work [4], for example. Discrete modeling as a general approach for problems in applied mathematics is the subject of [10].

Basic References

In addition to the basic references listed in Chapter 2 (p. 80) we suggest the following:

[1] Rainville, E. D., and P. E. Bedient. 1981. *Elementary Differential Equations.* 6th ed. New York: Macmillan.

[2] Reiss, E. L., et al. 1976. *Ordinary Differential Equations with Applications.* New York: Holt, Rinehart and Winston.

[3] Spiegel, M. R. 1971. *Finite Differences and Difference Equations.* Schaum's Outline Series. New York: McGraw-Hill.

More-Advanced References

[4] Bellman, R., and K. L. Cooke. 1963. *Differential-Difference Equations*. New York: Academic Press.

[5] Bender, C., and S. A. Orszag. 1978. *Advanced Mathematical Methods for Scientists and Engineers*. New York: McGraw-Hill.

[6] Cochran, J. A. 1972. *The Analysis of Linear Integral Equations*. New York: McGraw-Hill.

[7] ———. 1982. *Applied Mathematics: Principles, Techniques, and Applications*. Belmont, Calif.: Wadsworth.

[8] Dickinson, A. B. 1972. *Differential Equations: Theory and Use in Time and Motion*. Reading, Mass.: Addison-Wesley.

[9] Goldberg, S. 1958. *Introduction to Difference Equations*. New York: John Wiley.

[10] Greenspan, D. 1973. *Discrete Models*. Reading, Mass.: Addison-Wesley.

[11] Kanwal, R. P. 1971. *Linear Integral Equations, Theory and Technique*. New York: Academic Press.

[12] Kaplan, W. 1962. *Operational Methods for Linear Systems*. Reading, Mass.: Addison-Wesley.

[13] Roach, G. F. 1982. *Green's Functions: Introductory Theory with Applications*. 2d ed. Cambridge: Cambridge University Press.

[14] Stakgold, I. 1979. *Green's Functions and Boundary Value Problems*. New York: John Wiley.

Chapter 4

Systems of Differential Equations

4.1 Introduction

Although the methods and ideas of the last chapter have been used effectively in an amazing array of applications over the years, they should really be considered as a foundation on which to build rather than as a completed mathematical edifice. To illustrate this point, in this chapter the techniques for solving linear differential equations involving one independent and one dependent variable are modified to accommodate more than one dependent variable. This is most easily done by capitalizing on the conceptual and manipulative advantages of vector-matrix notation.

4.2 Homogeneous Systems with Simple Eigenvalues

We are interested in finding n unknowns $x_1, \ldots, x_n$, each of which is a function of the independent variable t and which satisfy a system of n differential equations. (The use of t rather than x as the independent variable is for the sake of clarity and to conform to common usage. This should cause no confusion, particularly since t is interpreted as time in many applications.) Since linear problems are most amenable to exact solution, we consider systems of the form:

$$\begin{aligned}
x_1' &= a_{11}x_1 + a_{12}x_2 + \cdots + a_{1n}x_n + f_1(t) \\
x_2' &= a_{21}x_1 + a_{22}x_2 + \cdots + a_{2n}x_n + f_2(t) \\
&\vdots \\
x_n' &= a_{n1}x_1 + a_{n2}x_2 + \cdots + a_{nn}x_n + f_n(t)
\end{aligned}$$

This system can be represented more efficiently, using the notation of Chapter 1, as

$$\mathbf{X}' = \mathbf{AX} + \mathbf{F}$$

where

$$\mathbf{X}(t) = \begin{pmatrix} x_1(t) \\ x_2(t) \\ \vdots \\ x_n(t) \end{pmatrix}, \qquad \mathbf{F}(t) = \begin{pmatrix} f_1(t) \\ f_2(t) \\ \vdots \\ f_n(t) \end{pmatrix}$$

and $\mathbf{A}$ is the $n \times n$ matrix $\mathbf{A} = (a_{ij})$. In a general situation the entries of $\mathbf{A}$ and $\mathbf{F}$ may be functions of t. We start with the time-invariant homogeneous case, however, where $\mathbf{A}$ is constant and $\mathbf{F}$ is $\mathbf{0}$.

An obvious solution to $\mathbf{X}' = \mathbf{AX}$ is $\mathbf{X} \equiv \mathbf{0}$. To find other solutions, recall that in Chapter 3 the guess $y = e^{\lambda t}$ for a solution of $y' = ay$ led to the complete solution $y = ce^{at}$, where c is an arbitrary constant. Perhaps then a guess $\mathbf{X} = \mathbf{C}e^{\lambda t}$, where $\mathbf{C}$ is an arbitrary constant vector, would work for the vector matrix equation $\mathbf{X}' = \mathbf{AX}$. Substituting our guess into this differential equation yields

$$\mathbf{C}\lambda e^{\lambda t} = \mathbf{AC}e^{\lambda t}$$

or, since the exponential $e^{\lambda t}$ is nonzero,

$$\mathbf{AC} - \mathbf{C}\lambda = \mathbf{0}$$

which is equivalent to

$$(\mathbf{A} - \lambda\mathbf{I})\mathbf{C} = \mathbf{0}$$

where $\mathbf{I}$ is the $n \times n$ identity matrix.

We see, therefore, that to find a nontrivial solution of the differential equation $\mathbf{X}' = \mathbf{AX}$, we must find a nonzero vector $\mathbf{C}$ satisfying $(\mathbf{A} - \lambda\mathbf{I})\mathbf{C} = \mathbf{0}$. That is, we must find the eigenvectors $\mathbf{C}$ corresponding to an eigenvalue λ of $\mathbf{A}$. We saw in Section 1.9 that the characteristic equation $\det(\mathbf{A} - \lambda\mathbf{I}) = 0$ is just an nth-order polynomial equation, and thus once an eigenvalue λ is found (exactly or by approximation methods), the corresponding $\mathbf{C}$ may be determined by solving a homogeneous system of n linear *algebraic* equations in n unknowns. A fairly complicated calculus problem is thus reduced to familiar algebraic algorithms.

Example 4.1 Find the complete solution of $\mathbf{X}' = \mathbf{AX}$ where $\mathbf{A} = \begin{pmatrix} 1 & -1 & 1 \\ 1 & 1 & -1 \\ 2 & -1 & 0 \end{pmatrix}$. Find a particular solution such that $\mathbf{X}(0) = \begin{pmatrix} -4 \\ 7 \\ 12 \end{pmatrix}$.

Solution

$$\det(\mathbf{A} - \lambda\mathbf{I}) = \det\begin{pmatrix} (1-\lambda) & -1 & 1 \\ 1 & (1-\lambda) & -1 \\ 2 & -1 & -\lambda \end{pmatrix} = -\lambda^3 + 2\lambda^2 + \lambda - 2$$

$$= -(\lambda + 1)(\lambda - 1)(\lambda - 2)$$

so the eigenvalues are $\lambda = 2$, $\lambda = 1$, and $\lambda = -1$. The eigenvector $\mathbf{C}$ corresponding to $\lambda = 2$ must satisfy

$$(\mathbf{A} - 2\mathbf{I})\mathbf{C} = \begin{pmatrix} -1 & -1 & 1 \\ 1 & -1 & -1 \\ 2 & -1 & -2 \end{pmatrix}\begin{pmatrix} c_1 \\ c_2 \\ c_3 \end{pmatrix} = \mathbf{0}$$

The matrix $\begin{pmatrix} -1 & -1 & 1 \\ 1 & -1 & -1 \\ 2 & -1 & -2 \end{pmatrix}$ reduces to $\begin{pmatrix} 1 & -1 & 1 \\ 0 & -2 & 0 \\ 0 & 0 & 0 \end{pmatrix}$, so c_3 is arbitrary, $c_2 = 0$, and $c_1 = -c_3$. One such eigenvector is $\begin{pmatrix} 1 \\ 0 \\ -1 \end{pmatrix}$, so $\mathbf{X}_1 = \begin{pmatrix} 1 \\ 0 \\ -1 \end{pmatrix} e^{2t}$ is a solution of the system. For $\lambda = 1$, $(\mathbf{A} - \lambda\mathbf{I}) = \begin{pmatrix} 0 & -1 & 1 \\ 1 & 0 & -1 \\ 2 & -1 & -1 \end{pmatrix}$ reduces to $\begin{pmatrix} 1 & 0 & -1 \\ 0 & -1 & 1 \\ 0 & 0 & 0 \end{pmatrix}$ and yields the eigenvector $\begin{pmatrix} 1 \\ 1 \\ 1 \end{pmatrix}$ and solution $\mathbf{X}_2 = \begin{pmatrix} 1 \\ 1 \\ 1 \end{pmatrix} e^t$. Similarly $\lambda = -1$ gives rise to $\mathbf{X}_3 = \begin{pmatrix} -1 \\ 3 \\ 5 \end{pmatrix} e^{-t}$. Since the three vectors $\mathbf{X}_1$, $\mathbf{X}_2$, and $\mathbf{X}_3$ are linearly independent, we might suspect that $\mathbf{X} = c_1\mathbf{X}_1 + c_2\mathbf{X}_2 + c_3\mathbf{X}_3$ is the complete solution. This is indeed the case, and therefore

$$\mathbf{X} = c_1\begin{pmatrix} 1 \\ 0 \\ -1 \end{pmatrix} e^{2t} + c_2\begin{pmatrix} 1 \\ 1 \\ 1 \end{pmatrix} e^t + c_3\begin{pmatrix} -1 \\ 3 \\ 5 \end{pmatrix} e^{-t}$$

If

$$\mathbf{X}(0) = \begin{pmatrix} -4 \\ 7 \\ 12 \end{pmatrix}$$

then

$$\begin{aligned} c_1 + c_2 - c_3 &= -4 \\ c_2 + 3c_3 &= 7 \\ -c_1 + c_2 + 5c_3 &= 12 \end{aligned}$$

This system must have a unique solution, since its determinant is not zero, being the determinant of a matrix that has as its columns the linearly independent vectors $\mathbf{X}_1(0)$, $\mathbf{X}_2(0)$, $\mathbf{X}_3(0)$. The solution is $c_1 = 1$, $c_2 = -2$, $c_3 = 3$, so the particular solution we seek is

$$\mathbf{X} = \begin{pmatrix} e^{2t} - 2e^t - 3e^{-t} \\ - 2e^t + 9e^{-t} \\ -e^{2t} - 2e^t + 15e^{-t} \end{pmatrix}$$ □

The methods of Example 4.1 are applicable whenever **A** is a constant matrix with *n distinct* eigenvalues. Indeed, the following theorem forms the theoretical basis for our efforts (compare with Theorem 3.2 and its corollary):

Theorem 4.1 Let $\mathbf{X}' = \mathbf{AX}$ be a first-order homogeneous linear differential system with an $n \times n$ coefficient matrix **A** whose entries are continuous functions on some interval I. The complete solution of the system has the form

$$\mathbf{X}(t) = \sum_{i=1}^{n} c_i \mathbf{X}_i(t)$$

where the $\mathbf{X}_i(t)$ are individually solutions of the system and collectively linearly independent on I. When **A** is a constant matrix with n distinct eigenvalues $\lambda_1, \lambda_2, \ldots, \lambda_n$ and corresponding eigenvectors $\mathbf{C}_1, \mathbf{C}_2, \ldots, \mathbf{C}_n$, the individual $\mathbf{X}_i(t)$ are given by

$$\mathbf{X}_i(t) = \mathbf{C}_i e^{\lambda_i t} \qquad i = 1, 2, \ldots, n$$ ■

Sometimes a given system of homogeneous first-order differential equations cannot be immediately represented in the form $\mathbf{X}' = \mathbf{AX}$ but needs first to be transformed into an equivalent system to which Theorem 4.1 is applicable.

Example 4.2 Determine the complete solution of the homogeneous system

$$\begin{aligned} x_1' + x_1 + 3x_2' &= 0 \\ 3x_1 + x_2' + 2x_2 &= 0 \end{aligned}$$

Solution We begin by solving the two equations simultaneously for x_1' and x_2' to obtain the equivalent system

$$\begin{aligned} x_1' &= 8x_1 + 6x_2 \\ x_2' &= -3x_1 - 2x_2 \end{aligned}$$

or $\mathbf{X}' = \mathbf{AX}$ with $\mathbf{A} = \begin{pmatrix} 8 & 6 \\ -3 & -2 \end{pmatrix}$. The eigenvalues and associated eigenvectors of **A** are then determined to be $\lambda_1 = 3 + \sqrt{7}$, $\mathbf{C}_1 = \begin{pmatrix} -6 \\ 5 - \sqrt{7} \end{pmatrix}$, $\lambda_2 = 3 - \sqrt{7}$, and $\mathbf{C}_2 = \begin{pmatrix} -6 \\ 5 + \sqrt{7} \end{pmatrix}$. Hence, by Theorem 4.1,

$$\mathbf{X}(t) = c_1 \begin{pmatrix} -6 \\ 5 - \sqrt{7} \end{pmatrix} e^{(3+\sqrt{7})t} + c_2 \begin{pmatrix} -6 \\ 5 + \sqrt{7} \end{pmatrix} e^{(3-\sqrt{7})t}$$ □

Exercises for Section 4.2

Find the complete solution for each of the first-order homogeneous linear differential systems in Exercises 1–8.

1. $x_1' = x_1 + 4x_2$
$x_2' = x_1 + x_2$

2. $x_1' = 6x_1 - 3x_2$
$x_2' = 2x_1 + x_2$

3. $\mathbf{X}' = \mathbf{AX}$ for $\mathbf{A} = \begin{pmatrix} -2 & -3 \\ -3 & -2 \end{pmatrix}$

4. $\mathbf{X}' = \mathbf{AX}$ for $\mathbf{A} = \begin{pmatrix} 6 & -1 \\ 3 & 2 \end{pmatrix}$

5. $x_1' = x_1 + x_2 - x_3$
$x_2' = 2x_2$
$x_3' = x_2 - x_3$

6. $x_1' = x_1 + x_2 + x_3$
$x_2' = x_1 + 4x_2 - x_3$
$x_3' = -5x_1 - 8x_2 - 3x_3$

7. $\mathbf{X}' = \mathbf{AX}$ for $\mathbf{A} = \begin{pmatrix} 1 & -3 & 2 \\ 0 & -1 & 0 \\ 0 & -1 & -2 \end{pmatrix}$

8. $\mathbf{X}' = \mathbf{AX}$ for $\mathbf{A} = \begin{pmatrix} 2 & 1 & 5 \\ -3 & -2 & -8 \\ 3 & 3 & 9 \end{pmatrix}$

In each of Exercises 9–12, re-express the given system of equations in the form $\mathbf{X}' = \mathbf{AX}$ and then determine the complete solution.

9. $x_1' + x_2' + 2x_2 = 0$
$x_1' - 3x_1 - 2x_2 = 0$

10. $x_1' - 2x_1 + x_2' - 2x_2 = 0$
$x_1' - 3x_1 + 2x_2' - 2x_2 = 0$

11. $x_1' + 6x_1 + x_2' + 3x_2 = 0$
$x_1' - x_2' + x_2 = 0$

12. $x_1' - 5x_1 + x_2' + 2x_3 = 0$
$x_1' - x_1 - x_2 = 0$
$5x_2' - 11x_2 + 2x_3' - 2x_3 = 0$

13. Solve the initial value problem.

$$x_1' = 2x_1 + 3x_2 \qquad x_1(0) = 0$$
$$x_2' = 2x_1 + x_2 \qquad x_2(0) = 5$$

14. Solve the initial value problem.

$$\mathbf{X}' = \mathbf{AX} \quad \text{for} \quad \mathbf{A} = \begin{pmatrix} 3 & 2 & 2 \\ 1 & 4 & 1 \\ -2 & -4 & -1 \end{pmatrix} \quad \text{and} \quad \mathbf{X}(0) = \begin{pmatrix} -4 \\ 2 \\ 1 \end{pmatrix}$$

15. Show that $\mathbf{X}_1(t) = \begin{pmatrix} 2-t \\ t-1 \end{pmatrix} e^{2t}$ and $\mathbf{X}_2(t) = \begin{pmatrix} t-1 \\ -t \end{pmatrix} e^{2t}$ form a linearly independent set of solutions to the differential system

$$\mathbf{X}' = \mathbf{AX} \quad \text{for} \quad \mathbf{A} = \begin{pmatrix} 1 & -1 \\ 1 & 3 \end{pmatrix}$$

on any interval I.

16. Verify that

$$\mathbf{X}(t) = c_1 \begin{pmatrix} e^{2t} \cos 3t \\ 3e^{2t} \sin 3t \end{pmatrix} + c_2 \begin{pmatrix} e^{2t} \sin 3t \\ -3e^{2t} \cos 3t \end{pmatrix}$$

gives the complete solution to the differential system

$$\mathbf{X}' = \mathbf{AX} \quad \text{for} \quad \mathbf{A} = \begin{pmatrix} 2 & -1 \\ 9 & 2 \end{pmatrix}$$

17. The diffusion of a solute between two compartments with volumes V_1 and V_2, respectively, which are separated by a permeable membrane, is governed by the homogeneous differential system

$$V_1 c_1' = k(c_2 - c_1)$$
$$V_2 c_2' = k(c_1 - c_2)$$

where $c_1(t)$ and $c_2(t)$ are the concentrations present in the individual compartments and k is a (positive) constant of proportionality. Determine the complete solution of this system.

18. Assume that two animal populations $P_1(t)$ and $P_2(t)$ interact according to the differential system

$$P_1' = 3P_1 - 5P_2$$
$$P_2' = 2P_1 - 4P_2$$

Discuss the long-term behavior of these populations.

4.3 Complex and Multiple Eigenvalues

The examples and exercises of Section 4.2 involved characteristic equations that had only *real* distinct roots. If the complex number $a + bi$ were a root of the characteristic equation of $\mathbf{X}' = \mathbf{AX}$, there would be a solution of the form

$$c\mathbf{Q}e^{(a+bi)t}$$

where $\mathbf{Q}$ is a *complex* eigenvector and c is an arbitrary *complex* constant. If, as is usually the case, all the entries of $\mathbf{A}$ are real, the presence of complex exponentials as well as complex vectors and constants may be a nuisance. This is especially true if we have a *real* initial condition $\mathbf{X}(0) = \mathbf{X}_0$. In this situation we would expect (correctly) an answer that could be expressed entirely in terms of real numbers and real-valued functions. Fortunately, this can be done using certain basic notions regarding complex numbers (see page 92).

For $\mathbf{A}$ *real*, if $\lambda = a + bi$ is a complex eigenvalue with corresponding complex eigenvector $\mathbf{C} = \mathbf{C}_1 + i\mathbf{C}_2$ ($\mathbf{C}_1$ and $\mathbf{C}_2$ real), the eigenequation

$$(\mathbf{A} - \lambda\mathbf{I})\mathbf{C} = \mathbf{0}$$

implies, taking conjugates, that

$$\overline{(\mathbf{A} - \lambda\mathbf{I})\mathbf{C}} = (\mathbf{A} - \bar{\lambda}\mathbf{I})\bar{\mathbf{C}} = \mathbf{0}$$

as well. Hence, $\bar{\lambda} = a - bi$ must also be an eigenvalue and have associated with it the complex eigenvector $\bar{\mathbf{C}}$. It follows that both $\mathbf{X}_1 = \mathbf{C}e^{\lambda t}$ and $\mathbf{X}_2 = \bar{\mathbf{C}}e^{\bar{\lambda}t}$ are solutions of $\mathbf{X}' = \mathbf{AX}$. Various linear combinations of $\mathbf{X}_1$ and $\mathbf{X}_2$ would also be solutions. In particular:

$$\operatorname{Re}\mathbf{C}e^{\lambda t} = \operatorname{Re}\mathbf{X}_1 = \frac{\mathbf{X}_1 + \mathbf{X}_2}{2} = \mathbf{C}_1 e^{at}\cos bt - \mathbf{C}_2 e^{at}\sin bt$$

and

$$\operatorname{Im}\mathbf{C}e^{\lambda t} = \operatorname{Im}\mathbf{X}_1 = \frac{\mathbf{X}_1 - \mathbf{X}_2}{2i} = \mathbf{C}_1 e^{at}\sin bt + \mathbf{C}_2 e^{at}\cos bt$$

are *real* solutions, which can be shown to be linearly independent. (The validity of these expressions may be checked using Euler's formula.) Once λ and $\mathbf{C}$ are determined, the expressions involving $\sin bt$ and $\cos bt$ may be calculated directly to find the two real solutions. A better procedure, perhaps, is to calculate, in order, λ, $\mathbf{C}$, $\mathbf{C}e^{\lambda t}$, and, finally—using Euler's formula—$\operatorname{Re}\mathbf{C}e^{\lambda t}$ and $\operatorname{Im}\mathbf{C}e^{\lambda t}$.

Example 4.3 Determine the complete solution of $\mathbf{X}' = \mathbf{AX}$ for $\mathbf{A} = \begin{pmatrix} 2 & -1 & -1 \\ 2 & 1 & -1 \\ 0 & -1 & 1 \end{pmatrix}$.

Solution $\det(\mathbf{A} - \lambda\mathbf{I}) = -\lambda^3 + 4\lambda^2 - 6\lambda + 4 = 0$ yields $\lambda = 2,\ 1+i,$ and $1 - i$. Using $\lambda = 2$,

$$(\mathbf{A} - 2\mathbf{I}) = \begin{pmatrix} 0 & -1 & -1 \\ 2 & -1 & -1 \\ 0 & -1 & -1 \end{pmatrix}$$

which gives rise to the eigenvector $\mathbf{C} = \begin{pmatrix} 0 \\ 1 \\ -1 \end{pmatrix}$. Using $\lambda = 1 + i$, $(\mathbf{A} - \lambda\mathbf{I}) =$ $\begin{pmatrix} 1-i & -1 & -1 \\ 2 & -i & -1 \\ 0 & -1 & -i \end{pmatrix}$, which reduces to $\begin{pmatrix} 1-i & -1 & -1 \\ 0 & 1 & i \\ 0 & 0 & 0 \end{pmatrix}$ with eigenvector $\mathbf{C} = \begin{pmatrix} 1 \\ -i \\ 1 \end{pmatrix}$. Thence $\mathbf{C}e^{\lambda t} = \begin{pmatrix} 1 \\ -i \\ 1 \end{pmatrix} e^{(1+i)t} = \begin{pmatrix} 1 \\ -i \\ 1 \end{pmatrix} e^t(\cos t + i \sin t) =$

$$\begin{pmatrix} e^t \cos t + ie^t \sin t \\ e^t \sin t - ie^t \cos t \\ e^t \cos t + ie^t \sin t \end{pmatrix} = \underbrace{\begin{pmatrix} e^t \cos t \\ e^t \sin t \\ e^t \cos t \end{pmatrix}}_{\operatorname{Re}\mathbf{C}e^{\lambda t}} + i \underbrace{\begin{pmatrix} e^t \sin t \\ -e^t \cos t \\ e^t \sin t \end{pmatrix}}_{\operatorname{Im}\mathbf{C}e^{\lambda t}}$$

so the complete solution may be represented as

$$\mathbf{X}(t) = c_1 \begin{pmatrix} 0 \\ e^{2t} \\ -e^{2t} \end{pmatrix} + c_2 \begin{pmatrix} e^t \cos t \\ e^t \sin t \\ e^t \cos t \end{pmatrix} + c_3 \begin{pmatrix} e^t \sin t \\ -e^t \cos t \\ e^t \sin t \end{pmatrix}$$

If we were satisfied with complex expressions, we could have used

$$\mathbf{X}(t) = c_1 \begin{pmatrix} 0 \\ e^{2t} \\ -e^{2t} \end{pmatrix} + c_2 \begin{pmatrix} 1 \\ -i \\ 1 \end{pmatrix} e^{(1+i)t} + c_3 \begin{pmatrix} 1 \\ +i \\ 1 \end{pmatrix} e^{(1-i)t}$$

Notice that it is unnecessary in either case to calculate the eigenvector corresponding to $\lambda = a - bi$. □

One gap remains—what do we do with multiple roots? If λ is an eigenvalue of multiplicity $k > 1$, occasionally it is possible to find k linearly independent eigenvectors $\mathbf{C}_1, \mathbf{C}_2, \ldots, \mathbf{C}_k$ merely by solving $(\mathbf{A} - \lambda\mathbf{I})\mathbf{C} = \mathbf{0}$. The necessary k linearly independent solutions of $\mathbf{X}' = \mathbf{AX}$ then are $\mathbf{C}_1 e^{\lambda t}, \mathbf{C}_2 e^{\lambda t}, \ldots, \mathbf{C}_k e^{\lambda t}$.

Example 4.4 Solve $\mathbf{X}' = \mathbf{AX}$ for $\mathbf{A} = \begin{pmatrix} 0 & 0 & 0 \\ 0 & 0 & 0 \\ 0 & 0 & 1 \end{pmatrix}$.

Solution $\det(\mathbf{A} - \lambda\mathbf{I}) = \lambda^2(1 - \lambda)$. For $\lambda = 0$, $(\mathbf{A} - \lambda\mathbf{I})\mathbf{C} = \mathbf{0}$ yields $c_3 = 0$, so $\mathbf{C} = \begin{pmatrix} c_1 \\ c_2 \\ 0 \end{pmatrix} = c_1 \begin{pmatrix} 1 \\ 0 \\ 0 \end{pmatrix} + c_2 \begin{pmatrix} 0 \\ 1 \\ 0 \end{pmatrix}$ and the root $\lambda = 0$ has produced two independent eigenvectors. For $\lambda = 1$, $(\mathbf{A} - \lambda\mathbf{I})\mathbf{C} = \mathbf{0}$ yields an eigenvector $\begin{pmatrix} 0 \\ 0 \\ 1 \end{pmatrix}$. The complete solution then has the form

$$\mathbf{X}(t) = c_1 \begin{pmatrix} 1 \\ 0 \\ 0 \end{pmatrix} + c_2 \begin{pmatrix} 0 \\ 1 \\ 0 \end{pmatrix} + c_3 \begin{pmatrix} 0 \\ 0 \\ 1 \end{pmatrix} e^t$$ □

A much more common occurrence is illustrated by the following example:

Example 4.5 Solve

$$\mathbf{X}' = \begin{pmatrix} 1 & -3 & 1 \\ 2 & -1 & -2 \\ 2 & -3 & 0 \end{pmatrix} \mathbf{X}$$

Solution $\det(\mathbf{A} - \lambda\mathbf{I}) = -(\lambda + 1)^2(\lambda - 2)$. For $\lambda = -1$, $(\mathbf{A} - \lambda\mathbf{I}) = \begin{pmatrix} 2 & -3 & 1 \\ 2 & 0 & -2 \\ 2 & -3 & 1 \end{pmatrix}$ reduces to $\begin{pmatrix} 2 & -3 & 1 \\ 0 & 3 & -3 \\ 0 & 0 & 0 \end{pmatrix}$ and yields eigenvectors of the form $\begin{pmatrix} c_3 \\ c_3 \\ c_3 \end{pmatrix}$. Thus $\lambda = -1$ produces only *one* linearly independent solution by way of $(\mathbf{A} - \lambda\mathbf{I})\mathbf{C} = \mathbf{0}$. We must then find another solution corresponding to $\lambda = -1$ using some other method. For a root λ of multiplicity k, a method that works in general is to look for solutions of the forms

1. $\mathbf{C}e^{\lambda t}$
2. $(\mathbf{P} + t\mathbf{Q})e^{\lambda t}$
3. $(\mathbf{R} + t\mathbf{S} + t^2\mathbf{T})e^{\lambda t}$

$\vdots$

finding as many solutions as possible at each stage until k linearly independent solutions have been determined. In these expressions, **C**, **P**, **Q**, **R**, **S**, **T**, and so on, are all constant vectors.

Returning to our example, $\lambda = -1$ has already generated the solution $\begin{pmatrix}1\\1\\1\end{pmatrix}e^{-t}$. Trying $\mathbf{X} = (\mathbf{P} + t\mathbf{Q})e^{-t}$ in the equation $\mathbf{X}' = \mathbf{AX}$, we obtain $(-\mathbf{P} + \mathbf{Q} - \mathbf{AP})e^{-t} = t(\mathbf{AQ} + \mathbf{Q})e^{-t}$ or (equating like terms and dividing by e^{-t}),

$$\mathbf{AP} = -\mathbf{P} + \mathbf{Q}$$

$$\mathbf{AQ} = -\mathbf{Q} \quad \text{or} \quad [\mathbf{A} - (-1)\mathbf{I}]\mathbf{Q} = \mathbf{0}$$

The second equation tells us that either $\mathbf{Q} = \mathbf{0}$ or $\mathbf{Q}$ is an eigenvector corresponding to $\lambda = -1$ (which we have already determined). Taking $\mathbf{Q} = \mathbf{0}$ yields nothing new, so we choose instead

$$\mathbf{Q} = \begin{pmatrix}1\\1\\1\end{pmatrix}$$

We now know $\mathbf{A}$ and $\mathbf{Q}$ and can solve the first equation to obtain $\mathbf{P} = \begin{pmatrix}\frac{1}{2} + p_3\\ p_3\\ p_3\end{pmatrix}$.

Letting $p_3 = 0$ gives rise to $\mathbf{X}(t) = \begin{pmatrix}\frac{1}{2}\\0\\0\end{pmatrix}e^{-t} + \begin{pmatrix}t\\t\\t\end{pmatrix}e^{-t}$. A solution corresponding to $\lambda = 2$ can readily be found, leading to the complete solution

$$\mathbf{X}(t) = c_1\begin{pmatrix}1\\1\\1\end{pmatrix}e^{-t} + c_2\begin{pmatrix}t+\frac{1}{2}\\t\\t\end{pmatrix}e^{-t} + c_3\begin{pmatrix}1\\0\\1\end{pmatrix}e^{2t} \qquad \square$$

Exercises for Section 4.3

Find the complete solution for each of the first-order homogeneous linear differential systems in Exercises 1–8.

1. $x_1' = 6x_1 + 8x_2$
$x_2' = -x_1 + 2x_2$

2. $\mathbf{X}' = \mathbf{AX}$ for $\mathbf{A} = \begin{pmatrix}2 & -1\\9 & 2\end{pmatrix}$

3. $\mathbf{X}' = \mathbf{AX}$ for $\mathbf{A} = \begin{pmatrix}3 & -18\\2 & -9\end{pmatrix}$

4. $x_1' = 2x_1 - x_2$
$x_2' = 4x_1 + 6x_2$

5. $\mathbf{X}' = \mathbf{AX}$ for $\mathbf{A} = \begin{pmatrix}1 & 0 & 1\\1 & 1 & 0\\-2 & 0 & -1\end{pmatrix}$

6. $\mathbf{X}' = \mathbf{AX}$ for $\mathbf{A} = \begin{pmatrix}5 & -5 & -5\\-1 & 4 & 2\\3 & -5 & -3\end{pmatrix}$

7. $x_1' = 5x_1 + 2x_2 + 2x_3$
$x_2' = 2x_1 + 2x_2 - 4x_3$
$x_3' = 2x_1 - 4x_2 + 2x_3$

8. $x_1' = -3x_1 - 3x_3$
$x_2' = 3x_1 - 2x_2 + x_3$
$x_3' = x_1 + x_3$

In each of Exercises 9–12, re-express the given system of equations in the form $\mathbf{X}' = \mathbf{AX}$ and then determine the complete solution.

9. $\begin{aligned} 4x_1' + x_1 + 2x_2' + 7x_2 &= 0 \\ x_1' - x_1 + x_2' + x_2 &= 0 \end{aligned}$

10. $\begin{aligned} x_1' + x_1 + 2x_2' + 3x_2 &= 0 \\ x_1' - 2x_1 + 5x_2' &= 0 \end{aligned}$

11. $\begin{aligned} x_1' + 2x_1 + x_2' + 3x_2 &= 0 \\ x_1' + 3x_2' + x_2 &= 0 \end{aligned}$

12. $\begin{aligned} x_1' - x_1 + 3x_2 - x_3 &= 0 \\ x_2' - 2x_2 - x_3' + 2x_3 &= 0 \\ 2x_1 - 3x_2 - x_3' &= 0 \end{aligned}$

13. Solve the initial value problem.

$$\begin{aligned} x_1' &= 3x_1 - x_2 & x_1(0) &= 1 \\ x_2' &= -4x_1 + x_2 + 2x_3 & x_2(0) &= 3 \\ x_3' &= 8x_1 - 3x_2 & x_3(0) &= 1 \end{aligned}$$

14. Solve the initial value problem.

$$\mathbf{X}' = \mathbf{AX} \quad \text{for} \quad \mathbf{A} = \begin{pmatrix} 0 & 4 & 0 \\ -1 & 0 & 0 \\ 1 & 4 & -1 \end{pmatrix} \quad \text{and} \quad \mathbf{X}(0) = \begin{pmatrix} -2 \\ 0 \\ 0 \end{pmatrix}$$

Two identical pendulums of mass m and length l are suspended from pivots that lie on the same horizontal line and are coupled together by a spring with spring constant k, as shown in Figure 4.1. The motion of this system may be modeled by the differential system

$$\theta'' + \frac{g}{l}\theta = \frac{k}{m}(\psi - \theta)$$

$$\psi'' + \frac{g}{l}\psi = \frac{k}{m}(\theta - \psi)$$

when the oscillations are small and there is no damping. Here g is the constant acceleration due to gravity, $\theta(t)$ and $\psi(t)$ are the angular displacements of the pendulums from the vertical, and the spring is assumed to be unextended when $\theta = 0 = \psi$. One approach to the analysis of this coupled system involves recognizing from the first equation that

$$\frac{k}{m}\psi = \theta'' + \left(\frac{g}{l} + \frac{k}{m}\right)\theta$$

differentiating twice to obtain an expression involving ψ'', and re-expressing ψ'' using the second equation. In this manner we arrive at

$$\theta^{(4)} + 2\left(\frac{g}{l} + \frac{k}{m}\right)\theta'' + \left[\left(\frac{g}{l}\right)^2 + 2\left(\frac{g}{l}\right)\left(\frac{k}{m}\right)\right]\theta = 0$$

a fourth-order linear differential equation with constant coefficients to which the techniques of Chapter 3 are applicable. Alternatively, we can re-express the equations in matrix

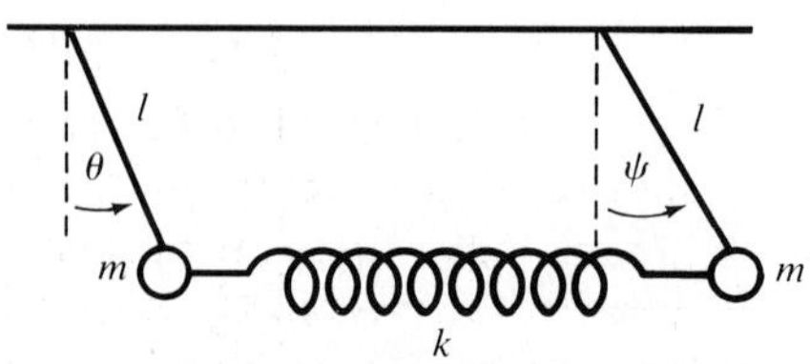

Figure 4.1

form, letting $x_1 = \theta$, $x_2 = \theta'$, $x_3 = \theta''$, and $x_4 = \theta'''$, thereby obtaining the equation

$$\mathbf{X}' = \mathbf{A}\mathbf{X} \quad \text{with} \quad \mathbf{A} = \begin{pmatrix} 0 & 1 & 0 & 0 \\ 0 & 0 & 1 & 0 \\ 0 & 0 & 0 & 1 \\ -\alpha^2\omega^2 & 0 & -\alpha^2 - \omega^2 & 0 \end{pmatrix}$$

(also see Example 4.9). In this formulation $\omega^2 = g/l$, $\alpha^2 = \omega^2 + 2(k/m)$, and, since $\theta = x_1$, we have $\psi = [(\alpha^2 + \omega^2)x_1 + 2x_3]/(\alpha^2 - \omega^2)$. Exercises 15–18 refer to this matrix reformulation of the coupled pendulum mechanical system.

15. Determine the complete solution of the homogeneous differential system.

16. Show that $\theta(t)$ and $\psi(t)$ may be expressed as

$$\theta = c_1 \cos \omega t + c_2 \sin \omega t + c_3 \cos \alpha t + c_4 \sin \alpha t$$
$$\psi = c_1 \cos \omega t + c_2 \sin \omega t - c_3 \cos \alpha t - c_4 \sin \alpha t$$

and hence $(\theta + \psi)$ and $(\theta - \psi)$ are purely sinusoidal with angular frequencies ω and α, respectively.

17. Discuss the "beat" phenomenon, which occurs when the coupling spring is weak, that is, k is small. (Recall the considerations near the end of Section 3.7.)

18. If resistive damping proportional to the pendulum velocities is introduced into the coupled pendulum system, the matrix $\mathbf{A}$ becomes

$$\mathbf{A} = \begin{pmatrix} 0 & 1 & 0 & 0 \\ 0 & 0 & 1 & 0 \\ 0 & 0 & 0 & 1 \\ -\alpha^2\omega^2 & -c(\alpha^2 + \omega^2) & -(\alpha^2 + \omega^2 + c^2) & -2c \end{pmatrix}$$

where c is a positive constant. Without explicitly determining the complete solution in this case, demonstrate that all solutions of the differential system are transient, that is, they approach zero as time increases.

4.4 Nonhomogeneous Equations

Before attacking the nonhomogeneous equation $\mathbf{X}' = \mathbf{A}\mathbf{X} + \mathbf{F}$ with $\mathbf{F} \neq \mathbf{0}$, it is convenient to reorganize our solutions of the homogeneous equation. Here we will make strong use of the definition of matrix differentiation, $\mathbf{B} = \mathbf{A}'$, where $b_{ij} = a'_{ij}$, and the welcome extension of the familiar calculus rule for differentiating products,

$$[\mathbf{A}\mathbf{C}]' = \mathbf{A}'\mathbf{C} + \mathbf{A}\mathbf{C}'$$

Note that in this product formula we must be careful not to commute the order of the matrix product terms.

If $\mathbf{X}_1, \mathbf{X}_2, \ldots, \mathbf{X}_n$ are linearly independent solutions of $\mathbf{X}' = \mathbf{A}\mathbf{X}$, where $\mathbf{A}$ is $n \times n$, then the matrix $\mathbf{\Phi}$, which has $\mathbf{X}_i$ as its ith column, $i = 1, 2, \ldots, n$, is called a *fundamental matrix*. The following are some useful and easily checked properties of such a fundamental matrix:

1. $\det \mathbf{\Phi} = \text{Wronskian }(\mathbf{X}_1, \ldots, \mathbf{X}_n) \neq 0$
2. $\mathbf{\Phi}' = \mathbf{A\Phi}$
3. $c_1\mathbf{X}_1 + c_2\mathbf{X}_2 + \cdots + c_n\mathbf{X}_n = \mathbf{\Phi}\begin{pmatrix} c_1 \\ \vdots \\ c_n \end{pmatrix} = \mathbf{\Phi C}$

Thus the complete solution of the homogeneous equation can be expressed in the form $\mathbf{\Phi C}$ where $\mathbf{C}$ is a column vector of arbitrary constants. The method of variation of parameters, generalized, now suggests a solution of the form $\mathbf{\Phi C}(t)$ for the nonhomogeneous equation.

Rather than using $\mathbf{\Phi C}(t)$, we use $\mathbf{\Phi U}(t)$ in order to emphasize that a *constant* $\mathbf{C}$ is being replaced by a *variable* $\mathbf{U}(t)$. Substituting $\mathbf{X} = \mathbf{\Phi U}$ into $\mathbf{X}' = \mathbf{AX} + \mathbf{F}$, we find $(\mathbf{\Phi U})' = \mathbf{A\Phi U} + \mathbf{F}$, $\mathbf{\Phi' U} + \mathbf{\Phi U}' = \mathbf{A\Phi U} + \mathbf{F}$. Since $\mathbf{\Phi}' = \mathbf{A\Phi}$, it follows that $\mathbf{\Phi U}' = \mathbf{F}$ and, hence, multiplying on both sides by the inverse matrix $\mathbf{\Phi}^{-1}$,

$$\mathbf{U}' = \mathbf{\Phi}^{-1}\mathbf{F}$$

(Note that $\mathbf{\Phi}^{-1}$ exists because $\det \mathbf{\Phi} \neq 0$.) Integration and recombination now lead to

$$\mathbf{U}(t) = \mathbf{C} + \int \mathbf{\Phi}^{-1}\mathbf{F}$$

and

$$\mathbf{X} = \mathbf{\Phi U} = \mathbf{\Phi C} + \mathbf{\Phi}\int \mathbf{\Phi}^{-1}\mathbf{F}$$

Here the first term is the complete solution to the homogeneous equation, whereas the second term is a particular solution of the nonhomogeneous equation. If an initial condition $\mathbf{X}(t_0)$ is given, the solution may be more conveniently written as

$$\mathbf{X}(t) = \mathbf{\Phi}(t)\mathbf{\Phi}^{-1}(t_0)\mathbf{X}(t_0) + \mathbf{\Phi}(t)\int_{t_0}^{t} \mathbf{\Phi}^{-1}(s)\mathbf{F}(s)\,ds$$

Recalling an earlier discussion in Chapter 1, $\mathbf{\Phi}^{-1}$ may be calculated in a variety of ways. The formula

$$\mathbf{\Phi}^{-1} = \frac{\text{Adj } \mathbf{\Phi}}{\det \mathbf{\Phi}}$$

is particularly convenient for 2×2 matrices, since it reduces to

$$\begin{pmatrix} a & b \\ c & d \end{pmatrix}^{-1} = \frac{\begin{pmatrix} d & -b \\ -c & a \end{pmatrix}}{(ad - bc)}$$

There is an alternative formulation of these solutions that not only simplifies calculations, particularly if the order n of the system is large, but also admits important generalizations. We introduce the special fundamental matrix

$$\mathbf{Y}(t) = \mathbf{\Phi}(t)\mathbf{\Phi}^{-1}(t_0)$$

normalized so that $\mathbf{Y}(t_0) = \mathbf{I}$. Substituting into the initial value expression above for $\mathbf{X}(t)$:

$$\mathbf{X}(t) = \mathbf{Y}(t)\mathbf{X}(t_0) + \mathbf{Y}(t)\mathbf{\Phi}(t_0)\int_{t_0}^{t} \mathbf{\Phi}^{-1}(t_0)\mathbf{Y}^{-1}(s)\mathbf{F}(s)\,ds$$

Using $\mathbf{Y}^{-1}(t) = \mathbf{Y}(2t_0 - t)$ and $\mathbf{Y}(t)\mathbf{Y}(t_0 - s) = \mathbf{Y}(t - s)$ (as shown in Exercises 16 and 17 at the end of this section),

$$\begin{aligned}\mathbf{X}(t) &= \mathbf{Y}(t)\mathbf{X}(t_0) + \int_{t_0}^{t} \mathbf{Y}(t)\mathbf{Y}^{-1}(s)\mathbf{F}(s)\,ds \\ &= \mathbf{Y}(t)\mathbf{X}(t_0) + \int_{t_0}^{t} \mathbf{Y}(t)\mathbf{Y}(2t_0 - s)\mathbf{F}(s)\,ds \\ &= \mathbf{Y}(t)\mathbf{X}(t_0) + \int_{t_0}^{t} \mathbf{Y}(t + t_0 - s)\mathbf{F}(s)\,ds\end{aligned}$$

Of course, since $\mathbf{Y}(t)$ *is* a fundamental matrix, if we wish a general expression without a specific initial condition, we may use

$$\mathbf{X}(t) = \mathbf{Y}(t)\mathbf{C} + \mathbf{Y}(t)\int \mathbf{Y}^{-1}(s)\mathbf{F}(s)\,ds$$

or

$$\mathbf{X}(t) = \mathbf{Y}(t)\mathbf{C} + \int \mathbf{Y}(t + t_0 - s)\mathbf{F}(s)\,ds$$

Example 4.6 Find the solution of $\mathbf{X}' = \begin{pmatrix} 3 & 2 \\ 1 & 2 \end{pmatrix}\mathbf{X} + \begin{pmatrix} 4e^{5t} \\ 0 \end{pmatrix}$, where $\mathbf{X}(0) = \begin{pmatrix} 1 \\ -1 \end{pmatrix}$.

Solution $\det(\mathbf{A} - \lambda\mathbf{I}) = (\lambda^2 - 5\lambda + 4) = (\lambda - 4)(\lambda - 1)$. For $\lambda = 4$, $(\mathbf{A} - \lambda\mathbf{I}) = \begin{pmatrix} -1 & 2 \\ 1 & -2 \end{pmatrix}$ and $\mathbf{X}_1 = \begin{pmatrix} 2 \\ 1 \end{pmatrix}e^{4t}$. For $\lambda = 1$, $(\mathbf{A} - \lambda\mathbf{I}) = \begin{pmatrix} 2 & 2 \\ 1 & 1 \end{pmatrix}$ and $\mathbf{X}_2 = \begin{pmatrix} 1 \\ -1 \end{pmatrix}e^{t}$. Hence

$$\mathbf{\Phi}(t) = \begin{pmatrix} 2e^{4t} & e^{t} \\ e^{4t} & -e^{t} \end{pmatrix},$$

$$\mathbf{\Phi}^{-1}(t) = \frac{\begin{pmatrix} -e^{t} & -e^{t} \\ -e^{4t} & 2e^{4t} \end{pmatrix}}{(-3e^{5t})} = \begin{pmatrix} \frac{1}{3}e^{-4t} & \frac{1}{3}e^{-4t} \\ \frac{1}{3}e^{-t} & -\frac{2}{3}e^{-t} \end{pmatrix}$$

$$\mathbf{\Phi}^{-1}(t)\mathbf{F}(t) = \begin{pmatrix} \frac{4}{3}e^{t} \\ \frac{4}{3}e^{4t} \end{pmatrix}, \qquad \int_0^t \mathbf{\Phi}^{-1}(s)\mathbf{F}(s)\,ds = \begin{pmatrix} \frac{4}{3}e^{t} - \frac{4}{3} \\ \frac{1}{3}e^{4t} - \frac{1}{3} \end{pmatrix}$$

$$\mathbf{\Phi}(t)\int_0^t \mathbf{\Phi}^{-1}(s)\mathbf{F}(s)\,ds = \begin{pmatrix} 3e^{5t} - \frac{8}{3}e^{4t} - \frac{1}{3}e^{t} \\ e^{5t} - \frac{4}{3}e^{4t} + \frac{1}{3}e^{t} \end{pmatrix}, \qquad \mathbf{\Phi}^{-1}(0)\mathbf{X}(0) = \begin{pmatrix} 0 \\ 1 \end{pmatrix}$$

$$\mathbf{\Phi}(t)\mathbf{\Phi}^{-1}(0)\mathbf{X}(0) = \begin{pmatrix} e^{t} \\ -e^{t} \end{pmatrix}, \text{ and}$$

$$\mathbf{X}(t) = \mathbf{\Phi}(t)\mathbf{\Phi}^{-1}(0)\mathbf{X}(0) + \mathbf{\Phi}(t)\int_0^t \mathbf{\Phi}^{-1}(s)\mathbf{F}(s)\,ds = \begin{pmatrix} 3e^{5t} - \frac{8}{3}e^{4t} + \frac{2}{3}e^{t} \\ e^{5t} - \frac{4}{3}e^{4t} - \frac{2}{3}e^{t} \end{pmatrix}$$

Alternatively, using the special fundamental matrix

$$\mathbf{Y}(t) = \mathbf{\Phi}(t)\mathbf{\Phi}^{-1}(0) = \begin{pmatrix} 2e^{4t} & e^{t} \\ e^{4t} & -e^{t} \end{pmatrix}\begin{pmatrix} 2 & 1 \\ 1 & -1 \end{pmatrix}^{-1}$$

$$= \begin{pmatrix} \frac{2}{3}e^{4t} + \frac{1}{3}e^{t} & \frac{2}{3}e^{4t} - \frac{2}{3}e^{t} \\ \frac{1}{3}e^{4t} - \frac{1}{3}e^{t} & \frac{1}{3}e^{4t} + \frac{2}{3}e^{t} \end{pmatrix}$$

we calculate, for $t_0 = 0$,

$$\mathbf{Y}(t-s)\mathbf{F}(s) = \begin{pmatrix} \frac{8}{3}e^{4t+s} + \frac{4}{3}e^{t+4s} \\ \frac{4}{3}e^{4t+s} - \frac{4}{3}e^{t+4s} \end{pmatrix}$$

$$\int_0^t \mathbf{Y}(t-s)\mathbf{F}(s)\,ds = \begin{pmatrix} 3e^{5t} - \frac{8}{3}e^{4t} - \frac{1}{3}e^{t} \\ e^{5t} - \frac{4}{3}e^{4t} + \frac{1}{3}e^{t} \end{pmatrix}$$

$$\mathbf{Y}(t)\mathbf{X}(0) = \begin{pmatrix} e^{t} \\ -e^{t} \end{pmatrix} \quad \text{and} \quad \mathbf{X}(t) = \begin{pmatrix} 3e^{5t} - \frac{8}{3}e^{4t} + \frac{2}{3}e^{t} \\ e^{5t} - \frac{4}{3}e^{4t} - \frac{2}{3}e^{t} \end{pmatrix}$$

as before. □

This second approach has the advantage of avoiding one matrix multiplication and the nasty chore of inverting the *nonconstant* matrix $\mathbf{\Phi}(t)$ at the cost of translating the matrix $\mathbf{Y}(t)$ and inverting a *constant* matrix.

In situations (which often occur in applied problems) where all the solutions are transient, we may need only to find a particular "steady-state" solution of the nonhomogeneous equation and do not need to determine completely the solution of the homogeneous equation. To have such transient behavior of the solution of the homogeneous equation, it is necessary and sufficient that the real part of each eigenvalue be negative. If this is the case, we may wish to "guess" some solutions of the nonhomogeneous equation by employing a modified version of the method of undetermined coefficients. As you would expect, some complications arise when there are multiple eigenvalues or forcing terms of the same form as solutions of the homogeneous equation. We do not give a complete discussion of such niceties here; however, a few reasonably well-behaved examples should give a good introduction to this approach.

Example 4.7 Find a particular solution to the nonhomogeneous system

$$\mathbf{X}' = \begin{pmatrix} -4 & -3 \\ 2 & 1 \end{pmatrix}\mathbf{X} + \mathbf{F}(t)$$

in each of the given cases.

(a) $\mathbf{F}(t) = \begin{pmatrix} 2e^{3t} \\ -e^{3t} \end{pmatrix}$ (b) $\mathbf{F}(t) = \begin{pmatrix} \sin 2t \\ 3 \end{pmatrix}$ (c) $\mathbf{F}(t) = \begin{pmatrix} e^{-t} \\ 2e^{-t} \end{pmatrix}$

Solution (a) Assume a particular solution of the nonhomogeneous equation of the form $\mathbf{C}e^{3t}$, where $\mathbf{C}$ is a constant vector. Substitute this guess into the equation:

$$\begin{pmatrix} c_1 \\ c_2 \end{pmatrix} 3e^{3t} = \begin{pmatrix} -4c_1 - 3c_2 \\ 2c_1 + c_2 \end{pmatrix} e^{3t} + \begin{pmatrix} 2e^{3t} \\ -e^{3t} \end{pmatrix}$$

Dividing both sides by e^{3t} and rearranging terms leads to

$$\begin{pmatrix} 7c_1 + 3c_2 \\ -2c_1 + 2c_2 \end{pmatrix} = \begin{pmatrix} 2 \\ -1 \end{pmatrix} \quad \text{and} \quad c_2 = -\tfrac{3}{20}, \qquad c_1 = \tfrac{7}{20}$$

The desired particular solution is then

$$\mathbf{X}(t) = \begin{pmatrix} \frac{7}{20} \\ -\frac{3}{20} \end{pmatrix} e^{3t}$$

(b) Think of $\begin{pmatrix} \sin 2t \\ 3 \end{pmatrix}$ in the form $\begin{pmatrix} \sin 2t \\ 0 \end{pmatrix} + \begin{pmatrix} 0 \\ 3 \end{pmatrix}$ and substitute the guess $\mathbf{X}(t) = \mathbf{A} \sin 2t + \mathbf{B} \cos 2t + \mathbf{C}$ into the differential equation:

$$\begin{pmatrix} a_1 \\ a_2 \end{pmatrix} 2\cos 2t - \begin{pmatrix} b_1 \\ b_2 \end{pmatrix} 2\sin 2t = \begin{pmatrix} -4a_1 - 3a_2 \\ 2a_1 + a_2 \end{pmatrix} \sin 2t + \begin{pmatrix} -4b_1 - 3b_2 \\ 2b_1 + b_2 \end{pmatrix} \cos 2t$$
$$+ \begin{pmatrix} -4c_1 - 3c_2 \\ 2c_1 + c_2 \end{pmatrix} + \begin{pmatrix} \sin 2t \\ 3 \end{pmatrix}$$

Equating coefficients of the three basic linearly independent functions, $\cos 2t$, $\sin 2t$, and 1:

$$\left.\begin{aligned} 2a_1 + 4b_1 + 3b_2 &= 0 \\ 2a_2 - 2b_1 - b_2 &= 0 \end{aligned}\right\} \quad \cos 2t \text{ terms}$$

$$\left.\begin{aligned} 4a_1 + 3a_2 - 2b_1 &= 1 \\ -2a_1 - a_2 - 2b_2 &= 0 \end{aligned}\right\} \quad \sin 2t \text{ terms}$$

$$\left.\begin{aligned} 4c_1 + 3c_2 &= 0 \\ -2c_1 - c_2 &= 3 \end{aligned}\right\} \quad \text{constant terms}$$

We then use row reductions to solve the system:

$$\begin{pmatrix} 2 & 0 & 4 & 3 & 0 & 0 & \cdot & 0 \\ 0 & 2 & -2 & -1 & 0 & 0 & \cdot & 0 \\ 4 & 3 & -2 & 0 & 0 & 0 & \cdot & 1 \\ -2 & -1 & 0 & -2 & 0 & 0 & \cdot & 0 \\ 0 & 0 & 0 & 0 & 4 & 3 & \cdot & 0 \\ 0 & 0 & 0 & 0 & -2 & -1 & \cdot & 3 \end{pmatrix}$$

$$\rightarrow \begin{pmatrix} 2 & 0 & 4 & 3 & 0 & 0 & \cdot & 0 \\ 0 & 2 & -2 & -1 & 0 & 0 & \cdot & 0 \\ 0 & 3 & -10 & -6 & 0 & 0 & \cdot & 1 \\ 0 & -1 & 4 & 1 & 0 & 0 & \cdot & 0 \\ 0 & 0 & 0 & 0 & 4 & 3 & \cdot & 0 \\ 0 & 0 & 0 & 0 & -2 & -1 & \cdot & 3 \end{pmatrix}$$

$$\rightarrow \begin{pmatrix} 2 & 0 & 4 & 3 & 0 & 0 & \cdot & 0 \\ 0 & 1 & -4 & -1 & 0 & 0 & \cdot & 0 \\ 0 & 0 & 6 & 1 & 0 & 0 & \cdot & 0 \\ 0 & 0 & 2 & -3 & 0 & 0 & \cdot & 1 \\ 0 & 0 & 0 & 0 & -2 & -1 & \cdot & 3 \\ 0 & 0 & 0 & 0 & 0 & 1 & \cdot & 6 \end{pmatrix}$$

$$\rightarrow \begin{pmatrix} 2 & 0 & 4 & 3 & 0 & 0 & \cdot & 0 \\ 0 & 1 & -4 & -1 & 0 & 0 & \cdot & 0 \\ 0 & 0 & 2 & -3 & 0 & 0 & \cdot & 1 \\ 0 & 0 & 0 & 10 & 0 & 0 & \cdot & -3 \\ 0 & 0 & 0 & 0 & -2 & -1 & \cdot & 3 \\ 0 & 0 & 0 & 0 & 0 & 1 & \cdot & 6 \end{pmatrix}$$

$$c_2 = 6, \quad c_1 = -\tfrac{9}{2}, \quad b_2 = -\tfrac{3}{10}, \quad b_1 = \tfrac{1}{20}, \quad a_2 = -\tfrac{1}{10}, \quad a_1 = \tfrac{7}{20}$$

$$\mathbf{X}(t) = \begin{pmatrix} \frac{7}{20} \\ -\frac{1}{10} \end{pmatrix} \sin 2t + \begin{pmatrix} \frac{1}{20} \\ -\frac{3}{10} \end{pmatrix} \cos 2t + \begin{pmatrix} -\frac{9}{2} \\ 6 \end{pmatrix}$$

(c) Let $\mathbf{X}(t) = \mathbf{C}e^{-t}$. Then

$$-\begin{pmatrix} c_1 \\ c_2 \end{pmatrix} e^{-t} = \begin{pmatrix} -4c_1 - 3c_2 \\ 2c_1 + c_2 \end{pmatrix} e^{-t} + \begin{pmatrix} e^{-t} \\ 2e^{-t} \end{pmatrix}, \quad \text{and} \quad \begin{cases} 3c_1 + 3c_2 = 1 \\ -2c_1 - 2c_2 = 2 \end{cases}$$

This system of equations is inconsistent, and we must go back to the solution of the homogeneous equation to see why our guess was inappropriate. The characteristic equation of the system is

$$\det\begin{pmatrix} -4-\lambda & -3 \\ 2 & 1-\lambda \end{pmatrix} = \lambda^2 + 3\lambda + 2 = (\lambda + 2)(\lambda + 1)$$

so $\lambda = -1$ is an eigenvalue and there will be a solution of the *homogeneous* equation of the form $\mathbf{C}e^{-t}$. We then produce a new guess by multiplying e^{-t} by a general first-order vector polynomial in t (*not* merely by $t\mathbf{C}$ as we did in the scalar case in Section 3.5):

$$\mathbf{X}(t) = (\mathbf{C}t + \mathbf{B})e^{-t}$$

$$\begin{aligned} \mathbf{X}'(t) &= \begin{pmatrix} c_1 \\ c_2 \end{pmatrix} e^{-t} - \begin{pmatrix} c_1 \\ c_2 \end{pmatrix} te^{-t} - \begin{pmatrix} b_1 \\ b_2 \end{pmatrix} e^{-t} \\ &= \begin{pmatrix} -4c_1 - 3c_2 \\ 2c_1 + c_2 \end{pmatrix} te^{-t} + \begin{pmatrix} -4b_1 - 3b_2 \\ 2b_1 + b_2 \end{pmatrix} e^{-t} + \begin{pmatrix} e^{-t} \\ 2e^{-t} \end{pmatrix} \end{aligned}$$

Equating coefficients of e^{-t} and te^{-t}:

$$\begin{aligned} c_1 \qquad + 3b_1 + 3b_2 &= 1 \\ c_2 - 2b_1 - 2b_2 &= 2 \\ 3c_1 + 3c_2 \qquad &= 0 \\ -2c_1 - 2c_2 \qquad &= 0 \end{aligned}$$

This system has (among others) a solution

$$c_1 = -8, \qquad c_2 = 8 \qquad b_1 = 2, \qquad b_2 = 1$$

so a particular solution to the differential equation is

$$\mathbf{X}(t) = \begin{pmatrix} -8 \\ 8 \end{pmatrix} te^{-t} + \begin{pmatrix} 2 \\ 1 \end{pmatrix} e^{-t} \qquad \square$$

There is a lesson worth noting in part (c) of this example. You do not get an incorrect answer by making an inappropriate guess—you simply do not get any answer at all!

Example 4.8 Solve the initial value problem

$$\begin{aligned} x_1' + x_1 + 3x_2' &= 1 \qquad & x_1(0) = 0 \\ 3x_1 + x_2' + 2x_2 &= t \qquad & x_2(0) = 0 \end{aligned}$$

Solution The given system is equivalent to

$$\mathbf{X}' = \mathbf{AX} + \mathbf{F}$$

with $\mathbf{A} = \begin{pmatrix} 8 & 6 \\ -3 & -2 \end{pmatrix}$ and $\mathbf{F} = \begin{pmatrix} 1 - 3t \\ t \end{pmatrix}$. From Example 4.2, the complete solution of the homogeneous system is

$$\mathbf{X}_c(t) = c_1 \begin{pmatrix} -6 \\ 5 - \sqrt{7} \end{pmatrix} e^{(3+\sqrt{7})t} + c_2 \begin{pmatrix} -6 \\ 5 + \sqrt{7} \end{pmatrix} e^{(3-\sqrt{7})t}$$

As a particular solution of the nonhomogeneous equation, we try

$$\mathbf{X}_p(t) = (\mathbf{C}t + \mathbf{B})$$

Substitution and matching of like terms then leads to

$$\mathbf{B} = \begin{pmatrix} -\frac{1}{2} \\ \frac{1}{2} \end{pmatrix}, \qquad \mathbf{C} = \begin{pmatrix} 0 \\ \frac{1}{2} \end{pmatrix}, \quad \text{and} \quad \mathbf{X}_p(t) = \begin{pmatrix} -\frac{1}{2} \\ \frac{1}{2}(t+1) \end{pmatrix}$$

When we add together $\mathbf{X}_c(t)$ and $\mathbf{X}_p(t)$ and apply the initial conditions, we find $c_1 = (\sqrt{7} - 7)/168$ and $c_2 = -(7 + \sqrt{7})/168$. The desired solution is

$$\mathbf{X}(t) = \frac{1}{28}\begin{pmatrix} 7 - \sqrt{7} \\ -7 + 2\sqrt{7} \end{pmatrix} e^{(3+\sqrt{7})t} + \frac{1}{28}\begin{pmatrix} 7 + \sqrt{7} \\ -7 - 2\sqrt{7} \end{pmatrix} e^{(3-\sqrt{7})t} + \frac{1}{2}\begin{pmatrix} -1 \\ t + 1 \end{pmatrix} \qquad \square$$

Exercises for Section 4.4

In Exercises 1–6, use the matrix formulation of variation of parameters to find the complete solution.

1. $$\begin{aligned} x_1' &= 2x_1 + x_2 + e^{5t} \\ x_2' &= -3x_1 + 6x_2 + e^{5t} \end{aligned}$$

2. $x_1' = 2x_1 + x_2 - e^t$
$x_2' = 3x_1 + 4x_2 - 7e^t$

3. $\mathbf{X}' = \mathbf{AX} + \mathbf{F}$ for $\mathbf{A} = \begin{pmatrix} 0 & 4 \\ -1 & 0 \end{pmatrix}$ and $\mathbf{F} = \begin{pmatrix} -3\cos t \\ 0 \end{pmatrix}$

4. $\mathbf{X}' = \mathbf{AX} + \mathbf{F}$ for $\mathbf{A} = \begin{pmatrix} 2 & 2 \\ -3 & -3 \end{pmatrix}$ and $\mathbf{F} = \begin{pmatrix} 1 \\ 2t \end{pmatrix}$

5. $x_1' = x_1 + x_2 + x_3 + 1$
$x_2' = -x_2$
$x_3' = -2x_1 - x_2 - 2x_3 + 2e^{-t}$

6. $\mathbf{X}' = \mathbf{AX} + \mathbf{F}$ for $\mathbf{A} = \begin{pmatrix} 1 & -1 & 1 \\ 0 & 0 & 1 \\ 0 & -1 & 2 \end{pmatrix}$ and $\mathbf{F} = \begin{pmatrix} 2e^t \\ 6te^t \\ 0 \end{pmatrix}$

7–11. Redo Exercises 1–5 using the matrix version of the undetermined coefficients method.

12. Solve the initial value problem.

$$\begin{aligned} x_1' &= x_2 + t & x_1(0) &= 2 \\ x_2' &= -x_1 + t^2 & x_2(0) &= -1 \end{aligned}$$

13. Solve the initial value problem.

$$\begin{aligned} x_1' &= x_2 + e^{3t} & x_1(0) &= 0 \\ x_2' &= -2x_1 + 3x_2 & x_2(0) &= 0 \end{aligned}$$

14. Solve the initial value problem.

$$\begin{aligned} x_1' + 2x_1 + x_2' + x_2 &= 10\cos t & x_1(0) &= 11 \\ x_1' + 4x_1 + 2x_2' + 3x_2 &= 6 & x_2(0) &= -6 \end{aligned}$$

15. Solve the initial value problem.

$$\mathbf{X}' = \mathbf{AX} + \mathbf{F} \quad \text{for} \quad \mathbf{A} = \begin{pmatrix} 0 & 1 & 0 & 0 \\ 0 & 0 & 1 & 0 \\ 0 & 0 & 0 & 1 \\ 1 & 0 & 0 & 0 \end{pmatrix}, \quad \mathbf{F} = \begin{pmatrix} t^2 \\ 0 \\ 0 \\ 0 \end{pmatrix}, \quad \text{and} \quad \mathbf{X}(0) = \begin{pmatrix} 1 \\ 1 \\ 1 \\ -1 \end{pmatrix}.$$

16. Let $\mathbf{X}_i(t)$ $(i = 1, 2, \ldots, n)$ be n linearly independent solutions of the homogeneous equation $\mathbf{X}' = \mathbf{AX}$, where $\mathbf{A}$ is an $n \times n$ constant matrix. Assume that the fundamental matrix $\mathbf{\Phi}(t)$ has $\mathbf{X}_i$ as its ith column $(i = 1, 2, \ldots, n)$, and form the special matrix $\mathbf{Y}(t) = \mathbf{\Phi}(t)\mathbf{\Phi}^{-1}(t_0)$.

(a) Show that $\mathbf{Y}(t)$ is also a fundamental matrix.

(b) Demonstrate that $\mathbf{Z}(t) \equiv \mathbf{AY}(t) - \mathbf{Y}(t)\mathbf{A}$ is the unique matrix solution of the initial value problem

$$\mathbf{Z}' = \mathbf{AZ}, \qquad \mathbf{Z}(t_0) = \mathbf{0}$$

and hence $\mathbf{Z}(t) \equiv \mathbf{0}$, which implies $\mathbf{AY}(t) \equiv \mathbf{Y}(t)\mathbf{A}$.

(c) Use the result of part (b) to verify that

$$[\mathbf{Y}(t)\mathbf{Y}(2t_0 - t)]' \equiv \mathbf{0}$$

and thus

$$\mathbf{Y}(t)\mathbf{Y}(2t_0 - t) \equiv \mathbf{Y}(t_0)\mathbf{Y}(t_0) = \mathbf{I}$$

or

$$\mathbf{Y}^{-1}(t) \equiv \mathbf{Y}(2t_0 - t)$$

If $t_0 = 0$, this equation has the particularly simple form $\mathbf{Y}^{-1}(t) \equiv \mathbf{Y}(-t)$.

17. Let $\mathbf{Y}(t)$ be the special fundamental matrix for the homogeneous equation $\mathbf{X}' = \mathbf{AX}$ defined in Exercise 16.

(a) Show that if s is fixed and t is variable, both $\mathbf{Y}(t)\mathbf{Y}(t_0 - s)$ and $\mathbf{Y}(t - s)$ satisfy the initial value problem

$$\mathbf{Z}' = \mathbf{AZ}, \qquad \mathbf{Z}(t_0) = \mathbf{Y}(t_0 - s)$$

and hence, by uniqueness,

$$\mathbf{Y}(t)\mathbf{Y}(t_0 - s) \equiv \mathbf{Y}(t - s)$$

(b) Employ the result of part (a) to establish, in alternative fashion, that

$$\mathbf{Y}^{-1}(t) \equiv \mathbf{Y}(2t_0 - t)$$

The differential system

$$\begin{aligned} V_1 c_1' &= h(a - c_1) + k(c_2 - c_1) \\ V_2 c_2' &= k(c_1 - c_2) \end{aligned}$$

arises in the biomathematical analysis of the cellular diffusion of a substance dissolved in a surrounding blood supply. In this model it is assumed that the concentration of the substance in the blood remains at the constant value a and that the substance goes through a two-stage diffusion process, first into a membrane enclosing an individual cell and then into the cell itself. The membrane and cell volumes V_1 and V_2 are constant, whereas the concentrations $c_1(t)$ and $c_2(t)$ in the two volumes vary with time. Both h and k are positive constants.

Exercises 18–20 refer to this biomathematical model. For simplicity, assume $V_1 = V_2 = 1$ throughout.

18. Determine a fundamental matrix for this differential system.

19. Find the complete solution of the nonhomogeneous system and investigate its behavior as $t \to \infty$.

20. Discuss the dependence of the solution of the initial value problem on the parameters a and h, which arises when $c_1(0) = 0 = c_2(0)$.

4.5 Applications

Many real-world problems lend themselves to mathematical modeling leading directly to linear systems of differential equations of the form we have been considering. Moreover, many other models can be put into this form merely by employing a simple trick, which transforms an nth-order linear differential equation into n first-order linear equations.

Example 4.9 Express the equation $y^{(4)} + 4y''' - 3y'' + 7y' - 2y = \sin t$ as a fourth-order system.

Solution Let $u = y'$, $v = y''$, and $w = y'''$. Then $u' = v$, $v' = w$, and $w' = y^{(4)} = -4y''' + 3y'' - 7y' + 2y + \sin t = -4w + 3v - 7u + 2y + \sin t$. The original equation is thus equivalent to

$$\begin{pmatrix} y \\ u \\ v \\ w \end{pmatrix}' = \begin{pmatrix} 0 & 1 & 0 & 0 \\ 0 & 0 & 1 & 0 \\ 0 & 0 & 0 & 1 \\ 2 & -7 & 3 & -4 \end{pmatrix} \begin{pmatrix} y \\ u \\ v \\ w \end{pmatrix} + \begin{pmatrix} 0 \\ 0 \\ 0 \\ \sin t \end{pmatrix}$$

which is in the form $\mathbf{X}' = \mathbf{AX} + \mathbf{F}(t)$. (You should note that the characteristic equation of $\mathbf{A}$ is precisely the characteristic equation $\lambda^4 + 4\lambda^3 - 3\lambda^2 + 7\lambda - 2 = 0$ of the original fourth-order equation.) □

Example 4.10 Assuming arbitrary initial currents and charges on capacitors, find the indicated currents in terms of arbitrary constants for the electrical network shown in Figure 4.2.

Solution Kirchhoff's first law (the algebraic sum of currents at any point in a network is zero) applied at point A mandates that

$$i_1 - i_2 - i_3 = 0$$

Kirchhoff's second law (the algebraic sum of the voltage drops around any loop of a network is zero) for the loops $ABCD$ and $ADEF$ yields

$$\frac{1}{1.8 \times 10^{-3}} \int i_3(t)\,dt + 8i_3' - 18i_2' = 0$$

and

$$\frac{1}{8 \times 10^{-4}} \int i_1(t)\,dt + 18i_2' = 0$$

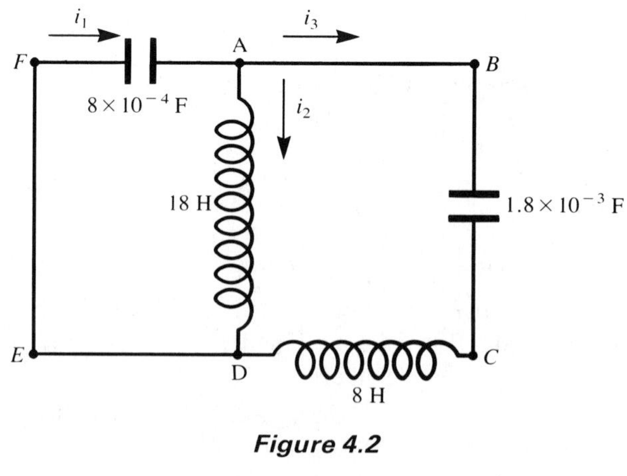

Figure 4.2

Observe that Kirchhoff's second law applied to loop $BCEF$ would merely produce a differential equation that is the sum of the two equations we already have. Differentiating both of these equations to eliminate the antiderivatives and expressing i_1 as $i_2 + i_3$ leads us to the following system of equations:

$$\begin{aligned} 36i_3'' + 2500i_3 - 81i_2'' &= 0 \\ 625i_3 + 9i_2'' + 625i_2 &= 0 \end{aligned}$$

Letting $i_2' = u$ and $i_3' = v$ and adding 9 times the second equation to the first (to eliminate the i_2'' term from the first equation) results in the standard system

$$\begin{pmatrix} i_2 \\ i_3 \\ u \\ v \end{pmatrix}' = \begin{pmatrix} 0 & 0 & 1 & 0 \\ 0 & 0 & 0 & 1 \\ \dfrac{-625}{9} & \dfrac{-625}{9} & 0 & 0 \\ \dfrac{-625}{4} & \dfrac{-(13)(625)}{36} & 0 & 0 \end{pmatrix} \begin{pmatrix} i_2 \\ i_3 \\ u \\ v \end{pmatrix}$$

The characteristic equation of this system is

$$\begin{aligned} (9)(36)\lambda^4 + (153)(625)\lambda^2 + (625)(2500) &= \\ (9\lambda^2 + 2500)(36\lambda^2 + 625) &= 0 \end{aligned}$$

The eigenvalues $\frac{50}{3}i$ and $\frac{25}{6}i$ lead, respectively, to eigenvectors

$$\begin{pmatrix} -3i \\ -9i \\ 50 \\ 150 \end{pmatrix} \quad \text{and} \quad \begin{pmatrix} 24i \\ -18i \\ -100 \\ 75 \end{pmatrix}$$

and corresponding vector solutions

$$\mathbf{X}_1 = \begin{pmatrix} -3i \\ -9i \\ 50 \\ 150 \end{pmatrix} e^{(50i/3)t} \quad \text{and} \quad \mathbf{X}_2 = \begin{pmatrix} 24i \\ -18i \\ -100 \\ 75 \end{pmatrix} e^{(25i/6)t}$$

Since real and imaginary parts of $\mathbf{X}_1$ and $\mathbf{X}_2$ are also solutions, we have a complete solution

$$\mathbf{X} = c_1 \begin{pmatrix} 3\sin\dfrac{50t}{3} \\ 9\sin\dfrac{50t}{3} \\ 50\cos\dfrac{50t}{3} \\ 150\cos\dfrac{50t}{3} \end{pmatrix} + c_2 \begin{pmatrix} -3\cos\dfrac{50t}{3} \\ -9\cos\dfrac{50t}{3} \\ 50\sin\dfrac{50t}{3} \\ 150\sin\dfrac{50t}{3} \end{pmatrix} + c_3 \begin{pmatrix} -24\sin\dfrac{25t}{6} \\ 18\sin\dfrac{25t}{6} \\ -100\cos\dfrac{25t}{6} \\ 75\cos\dfrac{25t}{6} \end{pmatrix}$$

$$+ c_4 \begin{pmatrix} 24\cos\dfrac{25t}{6} \\ -18\cos\dfrac{25t}{6} \\ -100\sin\dfrac{25t}{6} \\ 75\sin\dfrac{25t}{6} \end{pmatrix}$$

and finally

$$\begin{aligned} i_1 &= 12c_1\sin\frac{50t}{3} - 12c_2\cos\frac{50t}{3} - 6c_3\sin\frac{25t}{6} + 6c_4\cos\frac{25t}{6} \\ i_2 &= 3c_1\sin\frac{50t}{3} - 3c_2\cos\frac{50t}{3} - 24c_3\sin\frac{25t}{6} + 24c_4\cos\frac{25t}{6} \\ i_3 &= 9c_1\sin\frac{50t}{3} - 9c_2\cos\frac{50t}{3} + 18c_3\sin\frac{25t}{6} - 18c_4\cos\frac{25t}{6} \end{aligned}$$ □

Sometimes only the natural angular frequencies of such a network are needed, and we can employ various shortcuts to avoid the heavy burden of the straightforward attack illustrated in the solution of Example 4.10. Returning to the two differential equations we first encountered in that example, we outline one such convenient, if mathematically suspect, algebraic approach:

$$\begin{aligned} 36i_3'' + 2500i_3 - 81i_2'' &= 0 \\ 625i_3 + 9i_2'' + 625i_2 &= 0 \end{aligned}$$

is written in operator form

$$\begin{pmatrix} -81D^2 & 36D^2 + 2500 \\ 9D^2 + 625 & 625 \end{pmatrix}\begin{pmatrix} i_2 \\ i_3 \end{pmatrix} = 0$$

Setting the "determinant" of this system equal to zero gives rise to the appropriate characteristic equation:

$$\det\begin{pmatrix} -81D^2 & 36D^2 + 2500 \\ 9D^2 + 625 & 625 \end{pmatrix} = -9(36)D^4 - (153)(625)D^2 - (625)(2500)$$
$$= 0$$

and, of course, the eigenvalues $\lambda = \pm 50i/3, \pm 25i/6$, which directly give us the natural angular frequencies $\frac{50}{3}$ and $\frac{25}{6}$.

Example 4.11 Describe the vibration of the mechanical system illustrated in Figure 4.3.

Solution We assume that the springs obey Hooke's law (see Section 3.8) with the indicated constants. For convenience in setting up the differential equations that model the system, we consider a positive initial displacement x of mass m_1

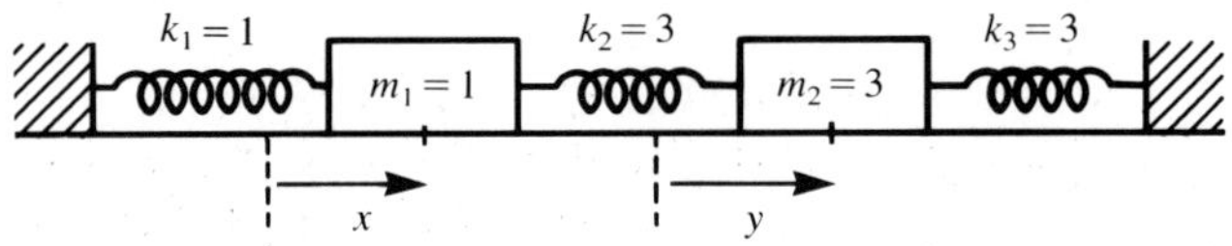

Figure 4.3

and a larger positive displacement y of mass m_2. (The resulting equations are the same no matter what displacements we consider.) Applying Newton's second law to each of the masses, we obtain

$$\begin{aligned} x'' &= -x + 3(y - x) \\ 3y'' &= -3(y - x) - 3y \end{aligned}$$

Letting $x' = u$ and $y' = v$ yields the standard form

$$\begin{pmatrix} x \\ y \\ u \\ v \end{pmatrix}' = \begin{pmatrix} 0 & 0 & 1 & 0 \\ 0 & 0 & 0 & 1 \\ -4 & 3 & 0 & 0 \\ 1 & -2 & 0 & 0 \end{pmatrix} \begin{pmatrix} x \\ y \\ u \\ v \end{pmatrix}$$

which has the characteristic equation $\lambda^4 + 6\lambda^2 + 5 = 0$; eigenvalues $\lambda = \pm\sqrt{5}i$, $\pm i$; and corresponding eigenvectors

$$\begin{pmatrix} 3i \\ -i \\ -3\sqrt{5} \\ \sqrt{5} \end{pmatrix}, \quad \begin{pmatrix} -3i \\ i \\ -3\sqrt{5} \\ \sqrt{5} \end{pmatrix}, \quad \begin{pmatrix} -i \\ -i \\ 1 \\ 1 \end{pmatrix}, \quad \text{and} \quad \begin{pmatrix} i \\ i \\ 1 \\ 1 \end{pmatrix}$$

The complete solution can, therefore, be written in the form

$$\mathbf{X}(t) = c_1 \begin{pmatrix} -3\sin\sqrt{5}t \\ \sin\sqrt{5}t \\ -3\sqrt{5}\cos\sqrt{5}t \\ \sqrt{5}\cos\sqrt{5}t \end{pmatrix} + c_2 \begin{pmatrix} 3\cos\sqrt{5}t \\ -\cos\sqrt{5}t \\ -3\sqrt{5}\sin\sqrt{5}t \\ \sqrt{5}\sin\sqrt{5}t \end{pmatrix}$$

$$+ c_3 \begin{pmatrix} \sin t \\ \sin t \\ \cos t \\ \cos t \end{pmatrix} + c_4 \begin{pmatrix} -\cos t \\ -\cos t \\ \sin t \\ \sin t \end{pmatrix} \qquad \square$$

Since in this last example we focus our attention on x and y, we might prefer the following alternative expressions for the solution:

$$\begin{aligned} x(t) &= -3c_1\sin\sqrt{5}t + 3c_2\cos\sqrt{5}t + c_3\sin t - c_4\cos t \\ y(t) &= c_1\sin\sqrt{5}t - c_2\cos\sqrt{5}t + c_3\sin t - c_4\cos t \end{aligned}$$

The motion is thus a combination, or superposition, of two simple harmonic motions with frequencies $\sqrt{5}/2\pi$ and $1/2\pi$. If the initial conditions are such that only one of these frequencies is present, the system is said to be vibrating in a *normal*, or *principal, mode*. In the normal mode corresponding to the higher frequency, $\sqrt{5}/2\pi$, observe that $c_3 = c_4 = 0$ and $x(t) = -3y(t)$; the masses move in opposite directions and the smaller mass has an amplitude three times as great as the larger mass. In the other principal mode, $x(t) = y(t)$; the masses move in the same direction with the same amplitudes.

The behavior of both electrical and mechanical systems when exterior forcing terms with frequencies near the natural frequencies are present is of crucial practical importance. Although we do not investigate it here, the major themes are exactly those we have already considered in Sections 3.7 and 3.8.

Exercises for Section 4.5

1. Express the equation $3y''' - 7y'' + 2y' + 3y = t^2$ as a third-order system.
2. Can each nth-order system $\mathbf{X}' = \mathbf{AX} + \mathbf{F}$ be expressed as a single nth-order linear differential equation by the method of Example 4.9? Explain your answer.
3. **(a)** Rework Example 4.11 with the new parameter values $m_1 = m_2 = 1$, $k_1 = k_3 = 81$, and $k_2 = 20$.
   (b) Find the solution to this system with the initial values $x(0) = 2$, $y(0) = 0$, $x'(0) = y'(0) = 0$. Discuss this solution from the standpoint of "beats." (See Section 3.7.)
4. Show that no choices of the parameters m_1, m_2, k_1, k_2, and k_3 in Example 4.11 result in the two natural frequencies of motion being equal.
5. **(a)** Write a system of equations describing the motion of the mechanical system in Figure 4.4.
   (b) Put this system of equations in normal form.
6. Referring to Figure 4.2, suppose that at time $t = 0$ the capacitor in loop $ABCD$ has a charge of 10 C, the charge on the other capacitor is zero, and the currents in both loops are zero. Find an expression for $i_1(t)$ and $i_3(t)$.
7. Find the current in each loop of the electrical network in Figure 4.5 if the switch is closed when all currents are zero (E represents a constant voltage).
8. Find the natural angular frequencies of the network in Figure 4.6.

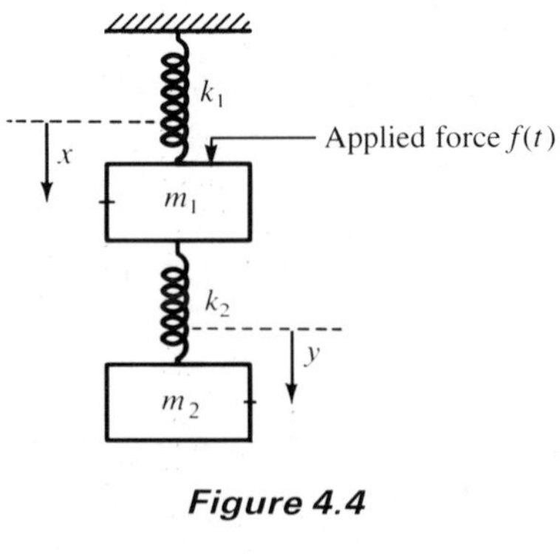

Figure 4.4

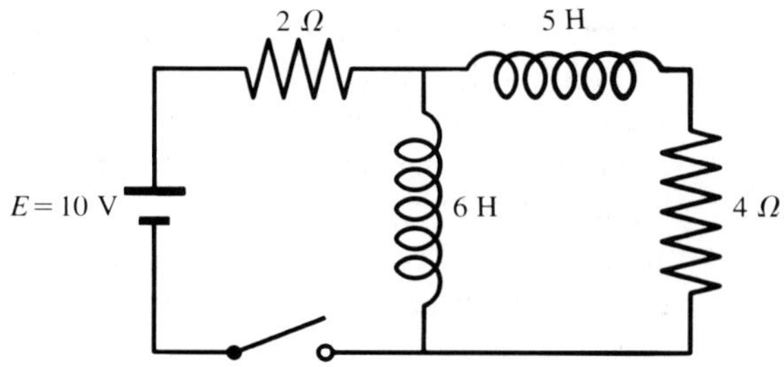

Figure 4.5

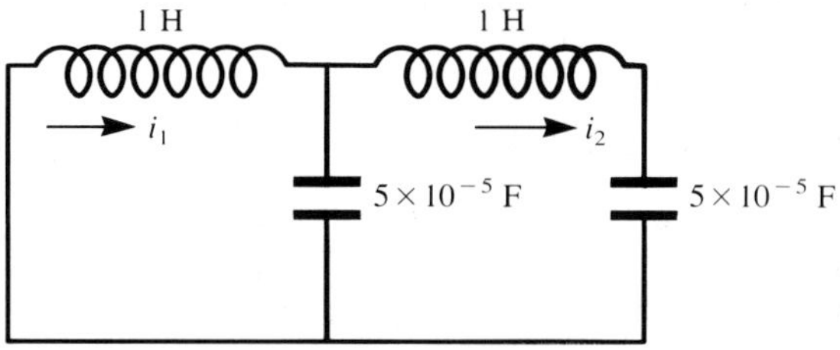

Figure 4.6

The motion of a single body (projectile) in a plane under the sole influence of gravity is a familiar application. If a rectangular coordinate system is employed, the defining equations are

$$m\frac{d^2x}{dt^2} = 0$$

and

$$m\frac{d^2y}{dt^2} = -mg$$

This simple differential system can be investigated using the matrix methods of this chapter. Exercises 9–11 are devoted to this task.

9. Reformulate the equations for planar projectile motion as:
  (a) two separate systems with 2×2 matrices
  (b) one combined differential system with a 4×4 coefficient matrix

10. Determine a fundamental matrix for the combined matrix differential system obtained in Exercise 9(b).

11. Use matrix methods to solve the initial value problem with $x(0) = x_0$, $y(0) = y_0$, $x'(0) = v_0 \cos\alpha$, and $y'(0) = v_0 \sin\alpha$.

In the analysis of homogeneous elastic beam problems, differential equations of the form

$$EIu^{(4)} - Tu'' + ku = f(x)$$

occur. In this expression, E is Young's elastic modulus for the beam material, I is the moment of inertia of the beam cross section (assumed uniform) about the neutral axis, T is the applied axial force (if any), and u is the beam deflection as a function of x. In static problems, $f(x)$ represents the applied transverse load acting to deform the beam and k is a nonnegative constant, which accounts for the presence of a supporting elastic foundation (if there is one). Interestingly, the same equation can also be shown to arise in the case of vibrating beams in which the beam deflection y is a function of both x and t but has the form $y(x,t) = u(x)\sin(\omega t + \theta)$ and only inertial forces are present. In this situation, $f(x) = 0$ and k is now a negative constant proportional to ω^2. For our purposes we note that in both cases the above single equation can be recast in matrix form as

$$\mathbf{X}' = \mathbf{AX} + \mathbf{F}(x) \quad \text{for} \quad \mathbf{A} = \begin{pmatrix} 0 & 1 & 0 & 0 \\ 0 & 0 & 1 & 0 \\ 0 & 0 & 0 & 1 \\ \dfrac{-k}{EI} & 0 & \dfrac{T}{EI} & 0 \end{pmatrix} \quad \text{and} \quad \mathbf{F}(x) = \begin{pmatrix} 0 \\ 0 \\ 0 \\ \dfrac{f(x)}{EI} \end{pmatrix}$$

Exercises 12–16 refer to this matrix formulation of beam problems.

12. Find the complete solution to the *homogeneous* system in the case $T/EI = 2$, $k/EI = 1$.

13. Determine the complete solution when $T = k = 0$ and $f(x) = f =$ constant.

14. Discuss the nature of the solutions to the homogeneous system as functions of the problem parameters (T/EI) and (k/EI). Be sure to consider values of T and k of both signs.

15. Determine a fundamental matrix for the differential system in the case $T = 0$ and $k = -16EI$.

16. Solve the initial value problems that result when $T = 0$, $k = -16EI$, and $\mathbf{F}(x) = \mathbf{0}$, for each given $\mathbf{X}(0)$.

(a) $\mathbf{X}(0) = \begin{pmatrix} 0 \\ 2 \\ 0 \\ -8 \end{pmatrix}$ **(b)** $\mathbf{X}(0) = \begin{pmatrix} 0 \\ 2 \\ 0 \\ 8 \end{pmatrix}$

These solutions can be related to resonant modes that occur in two basic beam vibration problems on the interval $[0, \pi]$.

A general three-component mono-molecular chemical reaction can be described by the matrix differential system

$$\mathbf{X}' = \mathbf{AX} \quad \text{where} \quad \mathbf{A} = \begin{pmatrix} -(k_{21} + k_{31}) & k_{12} & k_{13} \\ k_{21} & -(k_{12} + k_{32}) & k_{23} \\ k_{31} & k_{32} & -(k_{13} + k_{23}) \end{pmatrix}$$

Here the components of $\mathbf{X}(t)$ are the concentrations of the individual chemical species (elements or compounds), and k_{ij} is the (positive) rate constant for the reaction converting the jth species into the ith species.

Exercises 17–20 pertain to this chemical reaction application.

17. **(a)** Show that one eigenvalue of $\mathbf{A}$ is always zero and the other two eigenvalues are either negative or, if complex, have negative real parts.

(b) Infer from the results of part (a) that the solutions of the differential system always remain bounded.

18. Employ the result of Exercise 17 to show that when $k_{ij} = k_{ji}$, the solution $\mathbf{X}(t)$ of any initial value problem has the property that

$$\lim_{t\to\infty} \mathbf{X}(t) = c\begin{pmatrix}1\\1\\1\end{pmatrix}$$

19. Determine a fundamental matrix for the differential system in the case when $k_{ij} = k_{ji}$, $k_{12} = a$, $k_{13} = k_{23} = b$.

20. Solve the initial value problem that occurs when each of the rate constants equals 1 and $\mathbf{X}(0) = \begin{pmatrix}1\\2\\3\end{pmatrix}$.

4.6 Looking Ahead

The study of time-invariant systems of linear ordinary differential equations has occupied our attention in this chapter. Unfortunately, many texts on differential equations give this material relatively modest coverage. Nice accounts of the theory do appear in [1], [4], and [10], however, and both [3] and [5] contain a number of interesting applications. A unified approach to general linear systems theory as well as a substantial introduction to the qualitative considerations important in understanding the behavior of nonlinear and time-variant systems are provided in [11]. (Also see [7] in this regard.) In an excellent text, [12], dynamic systems with both discrete and continuous time are treated in a manner that blends more traditional methods with the modern state-space terminology and techniques commonly encountered in control theory.

The properties of the special fundamental matrix $\mathbf{Y}(t)$ introduced in Section 4.4 suggest a close relationship with an exponential function. Exponentials of matrices (and/or operators) and their role in the solution of constant-coefficient linear systems are discussed in [1], [6], and [10], for example.

The representation of the solution to the nonhomogeneous system in terms of the special fundamental matrix given in Section 4.4 involves a convolution of two matrix functions. Convolutions of ordinary functions occur rather frequently in a variety of contexts, and we encounter them in the next chapter of this text.

A natural generalization of the systems concept would be to partial differential equations. The situation is considerably more complicated, however, when two or more independent variables are present. References such as [8] and [13] touch on this topic.

Basic References

[1] Boyce, W., and R. C. DiPrima. 1977. *Elementary Differential Equations and Boundary Value Problems*. 3d ed. New York: John Wiley.

[2] Braun, M. 1975. *Differential Equations and Their Applications: An Introduction to Applied Mathematics*. New York: Springer-Verlag.

[3] Finizio, N., and G. Ladas. 1982. *Ordinary Differential Equations with Modern Applications*. 2d ed. Belmont, Calif.: Wadsworth.

[4] Ross, S. L. 1980. *Introduction to Ordinary Differential Equations*. 3d ed. New York: John Wiley.

[5] Spiegel, M. R. 1981. *Applied Differential Equations*. 3d ed. Englewood Cliffs, N. J.: Prentice-Hall.

More-Advanced References

[6] Arnold, V. I. 1978. *Ordinary Differential Equations*. Trans. R. A. Silverman. Cambridge, Mass.: MIT Press.

[7] Bender, C., and S. A. Orszag. 1978. *Advanced Mathematical Methods for Scientists and Engineers*. New York: McGraw-Hill.

[8] Chester, C. R. 1971. *Techniques in Partial Differential Equations*. New York: McGraw-Hill.

[9] Coddington, E. A., and N. Levinson. 1955. *Theory of Ordinary Differential Equations*. New York: McGraw-Hill.

[10] Goldberg, J. L., and A. J. Schwartz. 1972. *Systems of Ordinary Differential Equations: An Introduction*. New York: Harper and Row.

[11] Hirsch, M. W., and S. Smale. 1974. *Differential Equations, Dynamical Systems, and Linear Algebra*. New York: Academic Press.

[12] Luenberger, D. G. 1979. *Introduction to Dynamic Systems: Theory, Models, and Applications*. New York: John Wiley.

[13] Zauderer, E. 1983. *Partial Differential Equations of Applied Mathematics*. New York: John Wiley.

Chapter 5

The Laplace Transformation

5.1 Introduction

The Laplace transformation is the backbone of the modern body of knowledge and methods that grew from attempts to develop a calculus of operators. An English electrical engineer, Oliver Heaviside, laid the foundations for this calculus around 1890. In a vastly oversimplified form, operational calculus reduces a problem in differential, partial differential, difference, or integral equations to a form $\mathcal{O}(y) = f$, where $\mathcal{O}$ is an (usually linear) operator and y is an unknown function. Algorithms are then applied to determine the inverse operator $\mathcal{O}^{-1}$ and to find $y = \mathcal{O}^{-1}(f)$.

Many of the problems we solve by using the methods of this chapter can also be solved by the perhaps more familiar techniques discussed in Chapters 2, 3, and 4. The Laplace transform, however, yields new insights, a broad unified approach and, often, a bonus of computational efficiency.

Some examples of problems especially suited to Laplace transform methods are given next. These examples, numbered 1 through 5, include problems with forcing terms that are discontinuous (1), periodic (2), or instantaneous pulses (3), as well as problems involving differences rather than derivatives (5), and those where the qualitative character, rather than a complete or exact expression, of a solution is needed (4).

1. Find the current $i(t)$ in the circuit in Figure 5.1 if the initial current is zero and the applied voltage $E(t)$ is the indicated so-called "unit pulse" function ("unit" refers to the area under the curve being 1).

2. Use the same conditions as in (1); however, now $E(t)$ is the periodic "sawtooth" function indicated in Figure 5.2.

3. Use the same conditions as in (1); however, now the unit pulse is considered to occur *instantaneously* at $t = 2$.

4. In the mechanical system indicated in Figure 5.3, find m and k such that the vibration of M is absorbed, that is, such that the steady state $x(t) \equiv 0$.

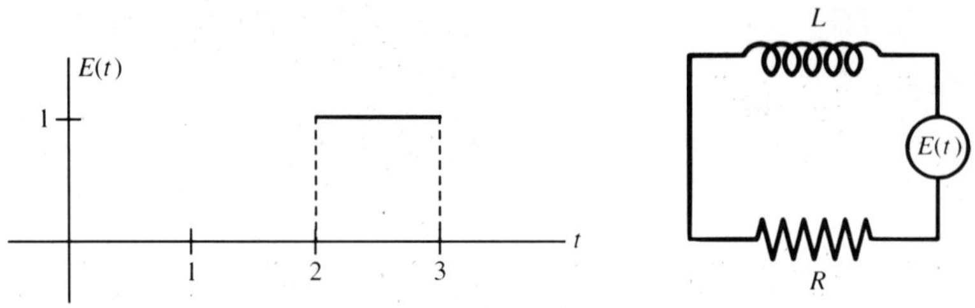

Figure 5.1

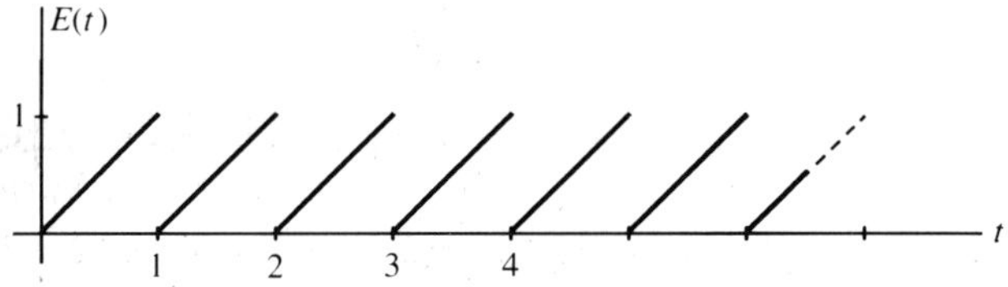

Figure 5.2

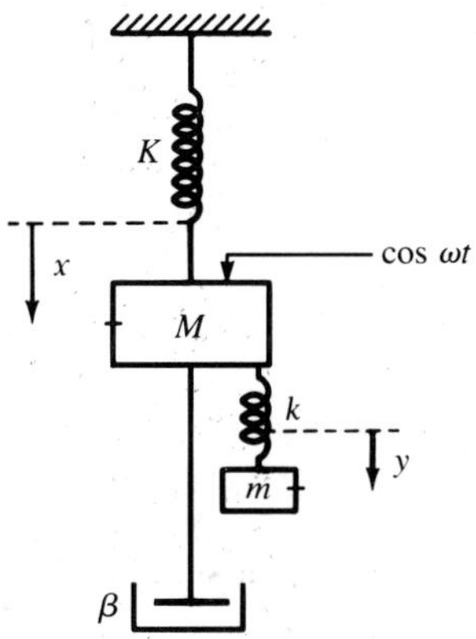

Figure 5.3

5. Devise a replacement scheme for equipment if the number of pieces is to be kept at a high constant value and the "mortality" of the equipment has a negative exponential character.

5.2 Definition of the Laplace Transformation

The general idea of a transform is that of a pairing of two functions, where to be useful, we would expect some sort of uniqueness in the pairing. For example, a function can be thought of as paired with its derivative. We say that the derivative is an "operator," or "transform," which we label $D\{f\}$. This pairing is unique in the sense that two functions with the same derivative differ at most by a constant. Similarly, another transform would be the integral of a function, $I\{f\}$, where

$$I\{f\} = \int_0^x f(t)\,dt$$

This important integral transform can be extended by considering operators of the type

$$T\{f\} = \int_0^\infty K(s, t)f(t)\,dt$$

Here $T\{f\}$ is actually an entire class of transformations, since if K is changed, a new transform is defined. One member of this class is called the Laplace transform.

Definition 5.1 The Laplace transform of a function f, if it exists, is given by the function F defined by

$$F(s) \equiv \int_0^\infty e^{-st}f(t)\,dt$$

The notation $\mathscr{L}\{f\}$ or $\mathscr{L}\{f(t)\}$ is also used.

In general, the variable s may be complex valued, but in this chapter we restrict s to real numbers.

In Chapter 7 you will learn of another kind of integral transform, called the Fourier transform of a function.

Example 5.1 Find the Laplace transform of the function $f(t) = t$.

Solution

$$\mathscr{L}\{t\} = \int_0^\infty te^{-st}\,dt = \lim_{B\to\infty}\int_0^B te^{-st}\,dt$$

Integrating by parts, we obtain

$$\begin{aligned}\mathscr{L}\{t\} &= \lim_{B\to\infty}\left(-\frac{te^{-st}}{s}\bigg|_0^B + \frac{1}{s}\int_0^B e^{-st}\,dt\right)\\ &= \lim_{B\to\infty}\left(-\frac{Be^{-sB}}{s} - \frac{e^{-sB}}{s^2} + \frac{1}{s^2}\right)\\ &= \frac{1}{s^2}, \qquad \text{for } s > 0\end{aligned}$$ □

The result of Example 5.1 is typical in the sense that the Laplace transform of the given function is defined only for values of s greater than some fixed constant a. In this chapter, the key point is not the specific value of a but only that there exists some domain $s > a$ for which the improper integral exists. Later, when an inverse transform is considered, more attention must be paid to these domains.

Note that the values of the function for $t < 0$ have no effect on the transform since the defining integral for $\mathscr{L}\{f\}$ requires a knowledge of $f(t)$ only for $t > 0$.

Example 5.2 Determine the Laplace transform of the function e^{at}.

Solution

$$\begin{aligned}
\mathscr{L}\{e^{at}\} &= \int_0^\infty e^{at}e^{-st}\,dt \\
&= \lim_{B\to\infty}\int_0^B e^{-(s-a)t}\,dt \\
&= \lim_{B\to\infty} -\frac{e^{-(s-a)t}}{s-a}\Bigg|_0^B \\
&= \lim_{B\to\infty}\left(-\frac{e^{-(s-a)B}}{s-a}+\frac{1}{s-a}\right) \\
&= \frac{1}{s-a}, \qquad \text{for } s > a
\end{aligned}$$

□

Example 5.3 Find the Laplace transform of $\sin at$.

Solution Recall from elementary calculus that

$$\int e^{-st}\sin at\,dt = -\frac{e^{-st}(s\sin at + a\cos at)}{s^2+a^2} + C$$

so that

$$\begin{aligned}
\mathscr{L}\{\sin at\} &= \int_0^\infty e^{-st}\sin at\,dt \\
&= \lim_{B\to\infty}\left(-\frac{e^{-sB}(s\sin aB + a\cos aB)}{s^2+a^2}+\frac{a}{s^2+a^2}\right) \\
&= \frac{a}{s^2+a^2}, \qquad s > 0
\end{aligned}$$

□

Example 5.4 Find the Laplace transform of the discontinuous function

$$f(t) = \begin{cases} 1, & 0 < t < 2 \\ -2, & 2 < t < 5 \\ 0 & \text{otherwise} \end{cases}$$

Solution

$$\begin{aligned}
\mathscr{L}\{f\} &= \int_0^2 e^{-st}\,dt - 2\int_2^5 e^{-st}\,dt = -\frac{e^{-st}}{s}\Bigg|_0^2 + 2\frac{e^{-st}}{s}\Bigg|_2^5 \\
&= \frac{1}{s}(1 - 3e^{-2s} + 2e^{-5s})
\end{aligned}$$

□

The short table of Laplace transforms that follows (Table 5.1) lists most of the elementary transforms you should know, plus a few more complicated ones (the last three), which will be needed later. In Section 5.4 we will learn that this little table, along with some general properties, will enable us to find the transforms of surprisingly many functions.

Table 5.1 *Short Table of Laplace Transforms*

Original function $f(t)$	**Laplace transform** $F(s)$	
1	$1/s$	$s > 0$
t^n	$\dfrac{n!}{s^{n+1}}$	$s > 0$, n an integer
t^a	$\dfrac{\Gamma(a+1)^*}{s^{a+1}}$	$s > 0, a > -1$
$\sin bt$	$\dfrac{b}{s^2+b^2}$	$s > 0$
$\cos bt$	$\dfrac{s}{s^2+b^2}$	$s > 0$
$\sinh bt$	$\dfrac{b}{s^2-b^2}$	$s > \lvert b\rvert$
$\cosh bt$	$\dfrac{s}{s^2-b^2}$	$s > \lvert b\rvert$
e^{at}	$\dfrac{1}{s-a}$	$s > a$
$\begin{cases} 1, & 2nc \le t < (2n+1)c \\ 0, & (2n+1)c \le t < (2n+2)c \end{cases}$ $n = 0, 1, 2, \ldots, \quad c > 0$	$\dfrac{1}{s(1+e^{-cs})}$	$s > 0$
$\sqrt{\dfrac{1}{t}}e^{-k^2/4t}$	$\sqrt{\dfrac{\pi}{s}}e^{-k\sqrt{s}}$	$s > 0$
$\sin 2\alpha\sqrt{t}$	$\dfrac{\alpha\sqrt{\pi}}{s^{3/2}}e^{-\alpha^2/s}$	$s > 0$

*See Chapter 9 for a discussion of the gamma function.

5.3 Existence of the Laplace Transform

In order to have an improper integral $\int_0^\infty g(t)\,dt$ exist, we would normally insist on some continuity condition on g together with a condition forcing $g(t)$ to go to zero rapidly as t becomes large. In

$$\int_0^\infty f(t)e^{-st}\,dt$$

the "convergence factor" e^{-st} helps with the second condition. We need two definitions before classifying functions that have Laplace transforms.

Definition 5.2 A function f is said to be *dominated by a function g* if there exist positive constants M and t_0 such that $|f(t)| \le Mg(t)$ for all $t \ge t_0$.

A function is said to be *dominated by an exponential* if there exists a real number $a \ne 0$ such that f is dominated by e^{at}. A function that is dominated by an exponential is said to be of *exponential order*.

Example 5.5 Determine which of the following functions are of exponential order: (a) e^{5t}, (b) t^3, (c) e^{t^2}, (d) $\sin e^{t^2}$.

Solution (a) e^{5t} is dominated by e^{at} for $a > 5$ and hence is of exponential order.

(b) t^3 is dominated by e^{at} for $a > 0$ and hence is of exponential order.

(c) e^{t^2} is not of exponential order, since

$$\lim_{t\to\infty} e^{t^2-at} = \infty$$

regardless of the value of a.

(d) $\sin e^{t^2}$ is of exponential order since it has maximum value 1 and hence is dominated by e^{at} for $a > 0$. □

In passing, note from the example $\sin e^{t^2}$ that a function of exponential order may have a derivative that is not of exponential order.

Recalling the definition of piecewise continuity from Section 1.4, we are now ready for the following theorem.

Theorem 5.1 Let f be piecewise continuous on $[0, t]$ for each $t > 0$, and of exponential order. Then $\mathscr{L}\{f\}$ exists.

Proof Let M_1 be any positive constant. Since f is of exponential order, there are constants a and $t_0 \ge 0$ such that

$$|f(t)| < M_1 e^{at} \le M_1 e^{|a|t}$$

for $t > t_0$. Since f is piecewise continuous, $|f(t)|$ is bounded on $[0, t_0]$; that is,

$$|f(t)| < M_2 \le M_2 e^{|a|t}$$

for $0 \le t \le t_0$. Let M be the larger of M_1 and M_2 so that

$$|f(t)| < Me^{|a|t}$$

for all $t \ge 0$ and

$$\left|\int_0^\infty f(t)e^{-st}\,dt\right| \le \int_0^\infty |f(t)|e^{-st}\,dt < \int_0^\infty Me^{(|a|-s)t}\,dt = \frac{M}{s-|a|}$$

when $s > |a|$. The existence of the Laplace transform for $s > |a|$ is then guaranteed by a standard bounded convergence test from calculus. ■

That the conditions of Theorem 5.1 are not necessary is seen by considering the function $1/\sqrt{t}$, which is not piecewise continuous on $[0, t]$ but whose Laplace transform does exist and is equal to $\sqrt{\pi/s}$.

Exercises for Sections 5.2 and 5.3

1. Verify that the following functions are of exponential order:
   (a) $\sin 3t$ **(b)** e^{100t} **(c)** $\ln(1 + t)$ **(d)** $t^3 e^{7t}$
2. **(a)** Show that if $f(t)$ and $g(t)$ are individually of exponential order, so also is any linear combination of the two functions.
   (b) What can be said about the product fg?
   (c) Can a similar statement be made regarding the quotient f/g? Justify your answer.
3. Demonstrate that if $f(t)$ is of exponential order, then so is $\int_0^t f(u)\,du$.
4. Let G be the transform defined by $G\{f\} = \displaystyle\int_1^5 \frac{f(x)}{sx}\,dx$.
   (a) Find $G\{x^2\}$.
   (b) How do transforms of any two functions differ?
5. Show that $f(t)$ is dominated by an exponential if and only if there is some number a such that $\lim_{t\to\infty} f(t)e^{-at} = 0$.
6. There *are* functions that have Laplace transforms in spite of not being dominated by an exponential. Show that $f(t) = te^{t^2}\sin(e^{t^2})$ is such a function.
7. Verify the formulas for the Laplace transform of $\cos at$, $\cosh at$, and $\sinh at$.

In each of Exercises 8–20, use the definition of the Laplace transform to find $\mathscr{L}\{f\}$ for the given $f(t)$.

8. $f(t) = \begin{cases} 1, & 0 \le t \le 4 \\ 3, & 4 < t \end{cases}$

9. $f(t) = \begin{cases} 0, & 0 \le t < 1 \\ t, & 1 \le t < 2 \\ 1, & 2 \le t \end{cases}$

10. $f(t) = \begin{cases} \sin t, & 0 \le t \le \pi \\ 0, & \pi \le t \end{cases}$

11. $f(t) = \begin{cases} 0, & 0 \le t \le 5 \\ e^{-t}, & 5 < t \end{cases}$

12. $f(t) = t^2 - 2t - 5$

13. $f(t) = (e^t - e^{-t})^2$

14. $f(t) = te^{-t}$

15. $f(t) = \sinh^2 t$

16. $f(t) = \cos^2 at$

17. $f(t) = e^{2t}\cos t$

18. $f(t) = \sin t \cos t$

19. $f(t) = \begin{cases} \dfrac{1}{b - a}, & a \le t \le b \\ 0 & \text{otherwise} \end{cases}$

This includes the waveform displayed in Figure 5.1.

20. $f(t) = \begin{cases} t - a, & a \le t \le a + 1 \\ 0 & \text{otherwise} \end{cases} \quad (a \ge 0)$

This includes any *one* of the sawteeth shown in Figure 5.2.

21. Let $b = a + \varepsilon$ in the result of Exercise 19 and investigate the limiting case $\varepsilon \to 0$. This is one approach to the Laplace transform of the instantaneous unit pulse mentioned in Problem 3 in the list of problems in Section 5.1.

22. Employ the result of Exercise 20 in order formally to express the Laplace transform of the complete sawtooth function of Figure 5.2 as an infinite series. Can you sum this series?

5.4 Some General Properties; The First Shifting Theorem

Before considering some of the interesting and characteristic general properties of the Laplace transform, let us recall the elementary transformation that substitutes for a number its logarithm. This process allows us to replace the operations of multiplying, dividing, and raising to a power by the relatively simpler operations of adding, subtracting, and multiplying, respectively. In a roughly analogous way, the Laplace transform replaces differentiation, integration, and multiplication by an exponential with multiplication by s, division by s, and translation, respectively. These properties, together with linearity, permit us to calculate transforms easily and thus contribute greatly to the manipulative ease and breadth of application, which together make the Laplace transform such a popular and useful device.

As our first basic result, we show that the Laplace transform has the linearity property. In this sense the Laplace transform behaves like the derivative, the integral, and the various other linear operators we have already considered in Chapters 1, 2, and 3.

Theorem 5.2 The Laplace transform is a linear operation; that is,

$$\mathscr{L}\{af(t) + bg(t)\} = a\mathscr{L}\{f(t)\} + b\mathscr{L}\{g(t)\}$$

where a and b are constants and f and g are functions whose Laplace transforms exist.

Proof By definition,

$$\mathscr{L}\{af(t) + bg(t)\} \equiv \int_0^\infty e^{-st}\{af(t) + bg(t)\}\,dt$$

By the linearity property of the integral, this may be written

$$\begin{aligned}\mathscr{L}\{af(t) + bg(t)\} &= a\int_0^\infty e^{-st}f(t)\,dt + b\int_0^\infty e^{-st}g(t)\,dt \\ &= a\mathscr{L}\{f(t)\} + b\mathscr{L}\{g(t)\}\end{aligned}$$ ■

In practical applications, it is often necessary to find Laplace transformations of functions multiplied by an exponential factor. The next theorem shows that if the Laplace transform of the function is known, then multiplying it by an exponential factor does not require a new computation.

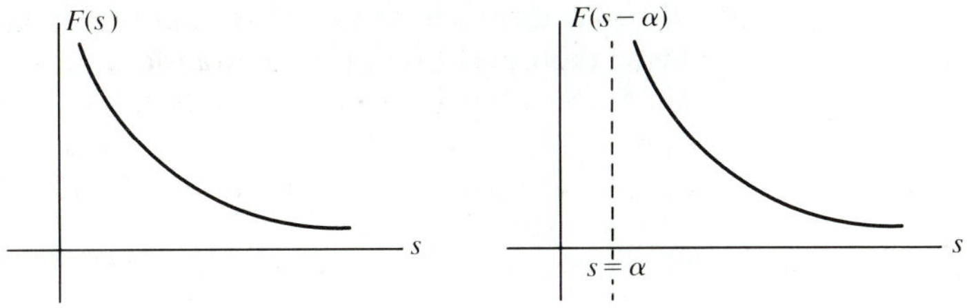

Figure 5.4 *First shifting theorem*

Theorem 5.3 Suppose the Laplace transform of a function $f(t)$ is $F(s)$, $s > \beta$. Then the Laplace transform of $e^{\alpha t}f(t)$ is $F(s - \alpha)$, $s > \alpha + \beta$. (*Note:* The notation $F(s - \alpha)$ denotes the function $F(s)$ "shifted" $|\alpha|$ units to the right if $\alpha > 0$ and $|\alpha|$ units to the left if $\alpha < 0$. Consequently, the theorem is called the *first shifting theorem* or translation property of the Laplace transform. See Figure 5.4.)

Proof

$$\begin{aligned}\mathscr{L}\{e^{\alpha t}f(t)\} &= \int_0^\infty e^{-st}e^{\alpha t}f(t)\,dt \\ &= \int_0^\infty e^{-(s-\alpha)t}f(t)\,dt \\ &= F(s-\alpha), \qquad s > \alpha + \beta\end{aligned}$$ ■

Example 5.6 Find the Laplace transform of te^t.

Solution The Laplace transform of t, from Table 5.1, is $1/s^2$, $s > 0$. Hence, by the first shifting theorem,

$$\mathscr{L}\{te^t\} = \frac{1}{(s-1)^2}, \qquad s > 1$$ □

Example 5.7 Find the Laplace transform of $e^t \sin t$.

Solution The Laplace transform of $\sin t$ is given in Table 5.1 as $1/(s^2 + 1)$, $s > 0$. Hence, by the first shifting theorem,

$$\mathscr{L}\{e^t \sin t\} = \frac{1}{(s-1)^2 + 1}, \qquad s > 1$$ □

Example 5.8 Suppose the Laplace transform of a function $f(t)$ is known to be $e^{-2s}/(s + 2)$, $s > -2$. What is the Laplace transform of $e^{3t}f(t)$?

Solution The Laplace transform of the function $e^{3t}f(t)$ is obtained by letting s be $s - 3$ in the formula for the Laplace transform of $f(t)$. Therefore,

$$\mathscr{L}\{e^{3t}f(t)\} = \frac{e^{-2(s-3)}}{s-1}, \qquad s > 1$$ □

For applications to problems involving differentiation and integration, the next two theorems are of obvious importance.

Theorem 5.4 Let f be continuous and of exponential order, dominated by e^{at}, and let f' be piecewise continuous on every finite interval $0 \le t \le T$ and of exponential order. Then for $s > a$, $\mathscr{L}\{f'(t)\} = s\mathscr{L}\{f(t)\} - f(0^+)$, where by $f(0^+)$ is meant $\lim_{t \to 0^+} f(t)$.

Proof By definition,

$$\mathscr{L}\{f'(t)\} \equiv \lim_{B \to \infty} \int_0^B f'(t)e^{-st}\,dt$$

Subdivide the interval $[0, B]$ at the points t_j where $f'(t)$ has discontinuities. Then

$$\mathscr{L}\{f'(t)\} = \int_0^{t_1} f'(t)e^{-st}\,dt + \int_{t_1}^{t_2} f'(t)e^{-st}\,dt + \cdots + \lim_{B \to \infty} \int_{t_n}^B f'(t)e^{-st}\,dt$$

Each of these can be integrated by parts to give

$$\mathscr{L}\{f'(t)\} = f(t)e^{-st}\Big|_0^{t_1} + s\int_0^{t_1} f(t)e^{-st}\,dt + f(t)e^{-st}\Big|_{t_1}^{t_2}$$

$$+ s\int_{t_1}^{t_2} f(t)e^{-st}\,dt + \cdots + \lim_{B \to \infty} f(t)e^{-st}\Big|_{t_n}^{B} + \lim_{B \to \infty} s\int_{t_n}^{B} f(t)e^{-st}\,dt$$

Since f is continuous, $f(t_k - 0) = f(t_k + 0)$ for $k = 1, 2, 3 \ldots n$. Hence,

$$\mathscr{L}\{f'(t)\} = -f(0^+) + \lim_{B \to \infty} f(B)e^{-sB} + s\int_0^\infty f(t)e^{-st}\,dt$$

Because f is of exponential order, the second term is zero for $s > a$. The third term is recognized as $s\mathscr{L}\{f(t)\}$, and the theorem is proved. ■

With suitable conditions on $f, f', \ldots, f^{(n)}$, Theorem 5.4 may be successively applied to obtain the formula

$$\mathscr{L}\{f^{(n)}(t)\} = s^n\mathscr{L}\{f(t)\} - s^{n-1}f(0^+) - s^{n-2}f'(0^+) - \cdots - f^{(n-1)}(0^+)$$

We will make heavy use of the case $n = 2$:

$$\mathscr{L}\{f''(t)\} = s^2\mathscr{L}\{f(t)\} - sf(0^+) - f'(0^+)$$

Example 5.9 Using the formula for $\mathscr{L}\{f''(t)\}$, find $\mathscr{L}\{\sin 3t\}$.

Solution If we let $f(t) = \sin 3t$, then $f'(t) = 3\cos 3t$ and $f''(t) = -9\sin 3t$. Also $f(0) = 0$ and $f'(0) = 3$. Of course $f(0^+) = f(0)$ and $f'(0^+) = f'(0)$, since f and f' are continuous. Then

$$\mathscr{L}\{-9\sin 3t\} = s^2\mathscr{L}\{\sin 3t\} - 3$$

Solving for $\mathscr{L}\{\sin 3t\}$,

$$\mathscr{L}\{\sin 3t\} = \frac{3}{s^2 + 9} \qquad \square$$

We may now easily approach integration.

Theorem 5.5 Let f be piecewise continuous and of exponential order, dominated by e^{at}. Then for $s > a$

$$\mathscr{L}\left\{\int_0^t f(u)\,du\right\} = \frac{1}{s}\mathscr{L}\{f(t)\}$$

Proof *Case 1: f Continuous.* Let $g(t) = \int_0^t f(u)\,du$. Then $g'(t) = f(t)$ and $g(0^+) = 0$. Since g can be shown to be of exponential order (see Exercise 3, p. 153), we may apply Theorem 5.4 to obtain

$$\mathscr{L}\{g'(t)\} = \mathscr{L}\{f(t)\} = s\mathscr{L}\left\{\int_0^t f(u)\,du\right\}$$

or

$$\mathscr{L}\left\{\int_0^t f(u)\,du\right\} = \frac{1}{s}\mathscr{L}\{f(t)\}$$

Case 2: f Discontinuous. Slight but tedious adjustments in the above argument handle this case, but we will wait until Section 5.9 when the general result becomes a corollary of Theorem 5.9. ■

Example 5.10 Find both $\mathscr{L}\{\int_0^t u^2 e^{3u}\,du\}$ and $\mathscr{L}\{\int_2^t u^2 e^{3u}\,du\}$.

Solution Since $\mathscr{L}\{t^2 e^{3t}\} = \dfrac{2}{(s-3)^3}$, $\mathscr{L}\{\int_0^t u^2 e^{3u}\,du\} = \dfrac{2}{s(s-3)^3}$. Note that $\int_2^t f(u)\,du = \int_0^t f(u)\,du - \int_0^2 f(u)\,du$, so

$$\begin{aligned}\mathscr{L}\left\{\int_2^t u^2 e^{3u}\,du\right\} &= \frac{2}{s(s-3)^3} - \mathscr{L}\left\{\int_0^2 u^2 e^{3u}\,du\right\}\\ &= \frac{2}{s(s-3)^3} - \frac{1}{s}\int_0^2 u^2 e^{3u}\,du\\ &= \frac{2}{s(s-3)^3} - \frac{1}{s}\left[e^{3u}\left(\frac{u^2}{3} - \frac{2}{9}u + \frac{2}{27}\right)\right]\Bigg|_0^2\\ &= \frac{2}{s(s-3)^3} + \frac{2}{27s}(1 - 13e^6)\end{aligned} \qquad \square$$

Example 5.11 If $\mathscr{L}\{f(t)\} = \dfrac{1}{s(s^2+4)}$, find $f(t)$.

Solution Note that $\mathscr{L}\left\{\frac{\sin 2t}{2}\right\} = \frac{1}{s^2+4}$ so, by Theorem 5.5,

$$\mathscr{L}\left\{\int_0^t \frac{\sin 2u}{2}\,du\right\} = \frac{1}{s}\left(\frac{1}{s^2+4}\right)$$

Hence

$$f(t) = \int_0^t \frac{\sin 2u}{2}\,du = -\frac{\cos 2u}{4}\Big|_0^t = \frac{1}{4} - \frac{\cos 2t}{4} \qquad \square$$

The next theorem and corollary provide handy tools for extending our collection of known Laplace transforms and inverse transforms. They also mesh nicely with the extension of Theorem 5.4: to find the transform of $f^{(n)}(t)$ using Theorem 5.4, we multiply $\mathscr{L}\{f\}$ by s^n (and make other adjustments); to find the *inverse* transform of $F^{(n)}(s)$, by the following results, we multiply $f(t)$ by $(-1)^n t^n$.

Theorem 5.6 Let f be piecewise continuous and of exponential order, dominated by e^{at}. If $\mathscr{L}\{f(t)\} = F(s)$, then $\mathscr{L}\{tf(t)\} = -F'(s)$ for $s > a$.

Proof By definition, $F(s) = \int_0^\infty e^{-st} f(t)\,dt$. The conditions on f are sufficient to assure the existence of $\mathscr{L}\{tf(t)\}$ and to allow us to interchange the order of differentiation and integration so that

$$-\frac{dF(s)}{ds} = -\int_0^\infty -te^{-st} f(t)\,dt = \mathscr{L}\{tf(t)\} \qquad \blacksquare$$

Corollary Let f be piecewise continuous and of exponential order, dominated by e^{at}. If $\mathscr{L}\{f(t)\} = F(s)$ then $\mathscr{L}\{t^n f(t)\} = (-1)^n F^{(n)}(s)$ for $s > a$.

Proof This follows inductively from Theorem 5.6. $\blacksquare$

Example 5.12 Find $\mathscr{L}\{t^2 e^{3t}\}$.

Solution Since $\mathscr{L}\{e^{3t}\} = \frac{1}{s-3}$, we can apply Theorem 5.6 to obtain

$$\mathscr{L}\{t^2 e^{3t}\} = \frac{d^2}{ds^2}\left(\frac{1}{s-3}\right) = \frac{2}{(s-3)^3}$$

Notice that the same result is obtained by applying the first shifting theorem to $\mathscr{L}\{t^2\}$. $\square$

Example 5.13 Suppose $\mathscr{L}\{f(t)\} = \ln[(s+2)/(s-3)] = F(s)$. Find $f(t)$.

Solution $F(s)$ does not look like the transform of any of the simple functions we have been considering, but notice that

$$-F'(s) = -\frac{d}{ds}\ln\left(\frac{s+2}{s-3}\right) = -\frac{d}{ds}\Big[\ln(s+2) - \ln(s-3)\Big] = \frac{1}{s-3} - \frac{1}{s+2}$$

which is the transform of $e^{3t} - e^{-2t}$. Thus using Theorem 5.6 with $n = 1$, $tf(t)$ must be $e^{3t} - e^{-2t}$, and $f(t) = \dfrac{e^{3t}}{t} - \dfrac{e^{-2t}}{t}$. □

Table 5.2 summarizes the general properties you should know about the Laplace transform. Note how many other transforms are easily found once the transform of a given f is known.

Table 5.2 *General Properties*

***t* space**	***s* space**
$f(t)$	$F(s) = \int_0^\infty e^{-st}f(t)\,dt$
$g(t)$	$G(s)$
$f(t) + g(t)$	$F(s) + G(s)$
$af(t)$, a a constant	$aF(s)$
$f'(t)$	$sF(s) - f(0^+)$
$f''(t)$	$s^2F(s) - sf(0^+) - f'(0^+)$
$\int_0^t f(u)\,du$	$\frac{1}{s}F(s)$
$e^{at}f(t)$	$F(s-a)$
$t^nf(t)$	$(-1)^nF^{(n)}(s)$

Our primary interest in this chapter is to demonstrate the usefulness of the Laplace transform in various practical applications. As a prelude to this effort, we close this section by giving an example of how to utilize the general properties in Table 5.2 to find the Laplace transform of the solution of a given initial value problem. The solution itself will then result from a determination of the *inverse* Laplace transform, which we discuss in Section 5.6.

Example 5.14 Find the Laplace transform of the solution of the initial value problem

$$y'' - 2y' - 3y = te^{2t}, \qquad y(0) = 4, \qquad y'(0) = -2$$

Solution Transforming both sides of the equation yields

$$s^2\mathcal{L}\{y\} - sy(0) - y'(0) - 2[s\mathcal{L}\{y\} - y(0)] - 3\mathcal{L}\{y\} = -\frac{d}{ds}\left(\frac{1}{s-2}\right)$$

Carrying out the indicated differentiation and substitutions, we easily obtain

$$(s^2 - 2s - 3)\mathscr{L}\{y\} = 4s - 10 + \frac{1}{(s-2)^2}$$

or

$$\mathscr{L}\{y\} = \frac{4s - 10}{s^2 - 2s - 3} + \frac{1}{(s-2)^2(s^2 - 2s - 3)} \qquad \square$$

Notice how our old friend the characteristic polynomial appears in a natural way and how effortlessly the initial conditions are included. We continue this example on p. 166 by finding the solution y of the differential equation.

Exercises for Section 5.4

In each of Exercises 1–25, use the various general properties of the Laplace transform to find $\mathscr{L}\{f\}$ for the given $f(t)$.

1. $f(t) = \cos 2t$

2. $f(t) = (t^2 + 1)^2$

3. $f(t) = \sqrt{t}e^t$

4. $f(t) = e^{-3t} - t + 2t^2 e^t$

5. $f(t) = e^{4t}(t^3 + 2t^2 + t)$

6. $f(t) = e^t \cos 4t$

7. $f(t) = \sin t \sin 2t$

8. $f(t) = \sin^3 3t$

9. $f(t) = t^3 e^{-3t}$ (by two methods)

10. $f(t) = t^2 \sin 2t$

11. $f(t) = \begin{cases} e^{2t}, & 0 \le t \le 4 \\ 3e^{2t}, & 4 < t \end{cases}$

12. $f(t) = \begin{cases} e^{-t}\sin t, & 0 \le t \le \pi \\ 0, & \pi \le t \end{cases}$

13. $f(t) = \begin{cases} 0, & 0 \le t \le 7 \\ e^{-4t}, & 7 < t \end{cases}$

14. $f(t) = te^{-5t} \sin 4t$

15. $f(t) = \int_0^t \cos u \, du$

16. $f(t) = \int_1^t \cos u \, du$

17. $f(t) = t \int_1^t \cos u \, du$

18. $f(t) = \int_0^t u \, du$

19. $f(t) = \int_2^t u \, du$

20. $f(t) = e^{at} \int_2^t u \, du$

21. $f(t) = t \int_0^t e^{-u} \sin 2u \, du$

22. $f(t) = e^{-2t} \int_0^t u \cos u \, du$

23. $f(t) = \int_0^t ue^{-5u} \sin 4u \, du$

24. $f(t) = \int_0^t u^3 e^{-3u} \, du$

25. $f(t) = \int_0^t u^2 \sinh u \, du$

In Exercises 26–29, use the formula for $\mathscr{L}\{f''\}$ to derive the transform of the given function.

26. $f(t) = \cos 2t$

27. $f(t) = \sinh 3t$

28. $f(t) = \cosh t$

29. $f(t) = t^3$, (assuming $\mathscr{L}\{t\} = 1/s^2$)

Determine the Laplace transform of the solution of each of the following initial value problems.

30. $y'' + y = 0, \quad y(0) = 1, \quad y'(0) = 0$

31. $y'' + 3y' - 4y = 0, \quad y(0) = 0, \quad y'(0) = -5$

32. $y'' + 2y' + y = 2te^{-t}, \quad y(0) = -3, \quad y'(0) = 3$

33. $y'' + 4y = \cos 2t, \quad y(0) = -1, \quad y'(0) = -1$

34. $y''' - y' = e^{3t}, \quad y(0) = 0, \quad y'(0) = 5, \quad y''(0) = -2$

35. $y''' + y' = e^{2t}\cos t, \quad y(0) = 2, \quad y'(0) = 0, \quad y''(0) = 1$

36. $y'' + 3y' - 4\int_0^t y\,du = 0, \quad y(0) = 2, \quad y'(0) = 1$

37. $Ri + \dfrac{1}{C}\int_0^t i(x)\,dx = e^{-t}$

38. $i' + 1000i + \dfrac{1}{6.25 \times 10^{-6}}\int_0^t i(x)\,dx = 24, \quad i(0) = 0$

(Note that $i'(0)$ can be found from the integro-differential equation itself.)

39. Show that $\mathscr{L}\{f(at)\} = \dfrac{1}{a}F\left(\dfrac{s}{a}\right)$, where $F(s) \equiv \mathscr{L}\{f\}$.

40. Demonstrate that if f has a jump discontinuity at $t = t_0$, then

$$\mathscr{L}[f'(t)] = -f(0^+) + e^{-st_0}[f(t_0 - 0) - f(t_0 + 0)] + s\mathscr{L}\{f(t)\}$$

In Exercises 41–53, use the various general properties of the Laplace transform to find f for the given $\mathscr{L}\{f\}$. (This type of problem is addressed more systematically in Section 5.6 but now offers good practice on the basic theorems of this section.)

41. $\dfrac{2}{(s+2)^{10}}$

42. $\dfrac{s}{(s+1)^2 + 1}$

43. $\dfrac{1}{s(s+2)}$

44. $\dfrac{3}{s^2 - 4s}$

45. $\dfrac{s-2}{s(s+3)}$

46. $\dfrac{1}{s^2}\left(\dfrac{s-1}{s+1}\right)$

47. $\dfrac{1}{s(s^2+1)}$

48. $\dfrac{1}{s^2(s^2+4)}$

49. $\dfrac{3}{s^2 + 4s + 5}$

50. $\dfrac{s}{(s^2+4)^2}$

51. $\operatorname{Tan}^{-1} s - \dfrac{\pi}{2}$

52. $\ln\left(\dfrac{s+1}{s+2}\right)$

53. $\ln\left(\dfrac{s^2-1}{s^2}\right)$

5.5 Transforming Periodic Functions

Before taking up the question of inverse transforms, it is important to note that in various applications we will, of necessity, encounter periodic functions. (Recall

Problem 2 of Section 5.1.) It turns out that the Laplace transform of such functions can be computed by integrating over just one of the periods.

Theorem 5.7 Let f be piecewise continuous with positive period P, that is, $f(t)$ is such that

$$f(t) \equiv f(t+P) \equiv f(t+2P) \equiv f(t+3P) \equiv \cdots \equiv f(t+nP)$$

Then

$$\mathscr{L}\{f(t)\} = \frac{1}{1-e^{-Ps}} \int_0^P e^{-st} f(t)\,dt, \qquad s > 0$$

Proof

$$\mathscr{L}\{f(t)\} \equiv \int_0^\infty e^{-st} f(t)\,dt$$

We write this integral as the sum of integrals over intervals of length P,

$$\mathscr{L}\{f(t)\} = \int_0^P e^{-st} f(t)\,dt + \int_P^{2P} e^{-st} f(t)\,dt + \int_{2P}^{3P} e^{-st} f(t)\,dt + \cdots + \int_{(n-1)P}^{nP} e^{-st} f(t)\,dt + \cdots$$

In the second integral, make the substitution $u = t - P$; in the third integral, $u = t - 2P$; in the fourth, $u = t - 3P$; in the nth integral, $u = t - (n-1)P$. Then

$$\mathscr{L}\{f(t)\} = \int_0^P e^{-st} f(t)\,dt + e^{-sP} \int_0^P e^{-su} f(u+P)\,du + e^{-2sP} \int_0^P e^{-su} f(u+2P)\,du + \cdots + e^{-(n-1)sP} \int_0^P e^{-su} f(u+(n-1)P)\,du + \cdots$$

Using the fact that $f(u+nP) \equiv f(u)$ and factoring out the integral $\int_0^P e^{-st} f(t)\,dt$,

$$\mathscr{L}\{f(t)\} = \left(\int_0^P e^{-st} f(t)\,dt\right)(1 + e^{-sP} + e^{-2sP} + \cdots + e^{-(n-1)sP} + \cdots)$$

The quantity in the second parentheses is an infinite geometric series with common ratio e^{-sP} and "sum" $1/(1-e^{-sP})$. The series converges because $0 < e^{-sP} < 1$ for $s > 0$. Therefore,

$$\mathscr{L}\{f(t)\} = \frac{1}{1-e^{-sP}} \int_0^P e^{-st} f(t)\,dt, \qquad s > 0$$ ■

Example 5.15 Find the Laplace transform of the function shown in Figure 5.5.

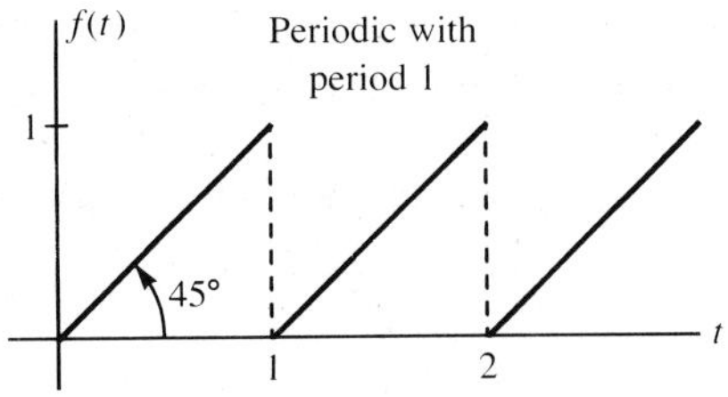

Figure 5.5

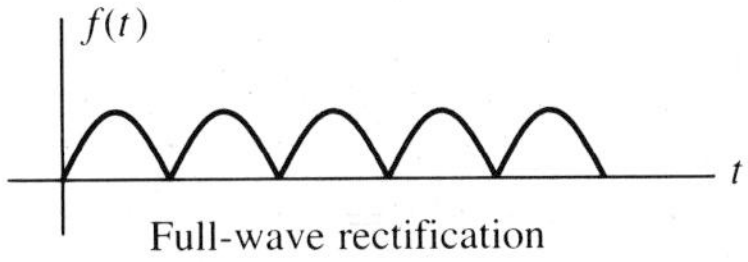

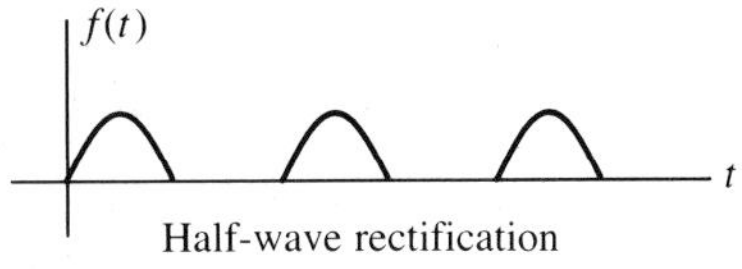

Figure 5.6

Solution Over the interval $0 < t < 1$, the function is defined by the rule $f(t) = t$. Hence,

$$\begin{aligned}\mathscr{L}\{f(t)\} &= \frac{1}{1-e^{-s}}\int_0^1 e^{-st}t\,dt \\ &= \frac{1}{1-e^{-s}}\left(-\frac{t}{s}e^{-st}\bigg|_0^1 + \frac{1}{s}\int_0^1 e^{-st}\,dt\right) \\ &= \frac{1}{1-e^{-s}}\left(-\frac{e^{-s}}{s} - \frac{1}{s^2}(e^{-s}-1)\right) \\ &= \frac{1}{s^2} - \frac{e^{-s}}{s(1-e^{-s})}\end{aligned}$$

□

Example 5.16 Find the transform of the full-wave rectification of $\sin t$ given that the transform of the half-wave rectification of $\sin t$ is

$$\frac{1}{(s^2+1)(1-e^{-\pi s})}$$

(See Figure 5.6, and Exercise 7 of the next exercise set.)

Solution The definition of the half-wave rectified wave is

$$f(t) = \frac{1}{2}(|\sin t| + \sin t)$$

Hence,

$$\mathscr{L}\{|\sin t| + \sin t\} = \frac{2}{(s^2+1)(1-e^{-\pi s})}$$

Since $\mathscr{L}\{\sin t\} = 1/(s^2+1)$ (from Table 5.1), it follows that

$$\begin{aligned}\mathscr{L}\{|\sin t|\} &= \frac{2}{(s^2+1)(1-e^{-\pi s})} - \frac{1}{s^2+1} \\ &= \frac{1}{s^2+1}\,\frac{1+e^{-\pi s}}{1-e^{-\pi s}} \\ &= \frac{1}{s^2+1}\coth\frac{\pi s}{2}\end{aligned}$$

□

Exercises for Section 5.5

1. Employ Theorem 5.7 to verify that

$$\mathscr{L}\{\sin bt\} = \frac{b}{s^2+b^2}, \qquad s>0$$

2. Repeat Exercise 1 for the transform pair

$$\mathscr{L}\{\cos bt\} = \frac{s}{s^2+b^2}, \qquad s>0$$

In Exercises 3–12, find the Laplace transform of the given periodic function. Sketch the graph of the given function in each case.

3. $f(t)$ periodic with period P and

$$f(t) = \begin{cases} 2, & 0<t<P/2 \\ 0, & P/2<t<P \end{cases}$$

4. $f(t)$ periodic with period 4 and

$$f(t) = \begin{cases} c, & 0<t<2 \\ -c, & 2<t<4 \end{cases}$$

5. $f(t)$ periodic with period 3 and

$$f(t) = \begin{cases} 1, & 0<t<1 \\ 0, & 1<t<2 \\ -1, & 2<t<3 \end{cases}$$

6. $f(t)$ periodic with period 2π and

$$f(t) = \begin{cases} 5\sin t, & 0<t<\pi \\ -5\sin t, & \pi<t<2\pi \end{cases}$$

7. $f(t)$ periodic with period 2π and

$$f(t) = \begin{cases} \sin t, & 0<t<\pi \\ 0, & \pi<t<2\pi \end{cases}$$

f is the half-wave rectification of $\sin t$ (recall Example 5.16).

8. $f(t) = |\sin bt|$. This is the full-wave rectification of $\sin bt$ (compare Example 5.16).

9. $f(t) = |\cos bt|$, the full-wave rectification of $\cos bt$.

10. $f(t)$ periodic with period 2 and

$$f(t) = \begin{cases} t, & 0 < t < 1 \\ 2 - t, & 1 < t < 2 \end{cases}$$

11. $f(t)$ periodic with period 3 and

$$f(t) = \begin{cases} t, & 0 < t < 1 \\ 1, & 1 < t < 2 \\ 3 - t, & 2 < t < 3 \end{cases}$$

12. $f(t)$ periodic with period π and

$$f(t) = t^2, \qquad 0 < t < \pi$$

13. Provide an alternative derivation of the Laplace transform for the periodic function in Exercise 5 using the transform determined in Exercise 11.

14. Explain why the Laplace transform of $f(t) = 1$ cannot be obtained from the result of Example 5.15 even though, from the general properties,

$$\mathscr{L}\{f'(t)\} = sF(s) - f(0^+)$$

The function $f(t)$ is called *antiperiodic* if $f(t + a) \equiv -f(t)$ for $t > 0$ and some positive a.

15. Show that any antiperiodic function $f(t)$ is periodic with period $2a$.

16. Show that for an antiperiodic function f, $\mathscr{L}\{f\} = \dfrac{1}{1 + e^{-as}} \displaystyle\int_0^a e^{-st} f(t)\, dt$.

17. Show that the Laplace transform of the half-wave rectification, $\frac{1}{2}[f(t) + |f(t)|]$, of the antiperiodic function f is $\dfrac{\mathscr{L}\{f\}}{1 - e^{-as}}$.

18. Show that the Laplace transform of the full-wave rectification, $|f(t)|$, of the antiperiodic function f is $\mathscr{L}\{f\} \coth(as/2)$.

5.6 The Inverse Laplace Transform

We are now ready to approach the task of "undoing" a Laplace transform, that is, given a function $F(s)$, to find a function $f(t)$ whose Laplace transform is the given F. We begin with a definition and two examples.

Definition 5.3 If there exists a function of t, $f(t)$, such that $\mathscr{L}\{f(t)\} = F(s)$, then $f(t)$ is called the *inverse Laplace transform* of $F(s)$. The notation commonly used is

$$\mathscr{L}^{-1}\{F(s)\} = f(t)$$

Recall that in Example 5.14 we found the Laplace transform of the solution of an initial value problem to be

$$\frac{4s - 10}{s^2 - 2s - 3} + \frac{1}{(s - 2)^2(s^2 - 2s - 3)}$$

In the following example we find the inverse of these two terms and thus find the solution itself.

Example 5.17 (a) If $\mathscr{L}\{f(t)\} = \dfrac{4s - 10}{s^2 - 2s - 3}$, find $f(t)$.

Solution Factoring the denominator and expressing the function in terms of its partial fraction expansion,

$$\frac{4s - 10}{s^2 - 2s - 3} = \frac{4s - 10}{(s - 3)(s + 1)} = \frac{A}{s - 3} + \frac{B}{s + 1}$$

where A and B are constants to be determined. There are several elementary ways to obtain these constants, but perhaps the easiest way to determine the value of A is first to multiply both sides by $s - 3$, simplify, and then let $s = 3$. We find that $A = \frac{1}{2}$. Similarly, $B = \frac{7}{2}$. Thus

$$\mathscr{L}\{y\} = \frac{\frac{1}{2}}{s - 3} + \frac{\frac{7}{2}}{s + 1} \quad \text{and} \quad y = \frac{e^{3t}}{2} + \frac{7e^{-t}}{2}$$

(b) Find $\mathscr{L}^{-1}\{F(s)\}$ where

$$F(s) = \frac{1}{(s - 2)^2(s^2 - 2s - 3)}$$

Solution We again employ partial fractions and expand $F(s)$ as

$$F(s) = \frac{A}{s - 2} + \frac{B}{(s - 2)^2} + \frac{C}{s - 3} + \frac{D}{s + 1}$$

It then follows that $A = -\frac{2}{9}$, $B = -\frac{1}{3}$, $C = \frac{1}{4}$, $D = -\frac{1}{36}$, and

$$\mathscr{L}^{-1}\{F(s)\} = -\frac{2e^{2t}}{9} - \frac{te^{2t}}{3} + \frac{e^{3t}}{4} - \frac{e^{-t}}{36} \qquad \square$$

When we add the results of parts (a) and (b), we finally have the $y(t)$ for Example 5.14. In this example we were fortunate to be able to determine the functions that had the given transforms by using partial fractions and our brief tables of transforms and basic properties. In general this process of finding an inverse Laplace transform takes considerably more skill than finding the transform itself—somewhat akin to the fact that antidifferentiation is usually harder than differentiation. Moreover, not all functions of s have inverse Laplace transforms. Transforms of well-behaved functions of the sort we have been examining have the property of tending to zero as $s \to \infty$. Simple functions lacking this property, such as s, s^2, or $\sin s$, for example, do not have corresponding inverse transforms.

In the following examples we examine a few methods of finding inverse transforms. These techniques are applied to ordinary differential equations and initial value problems in the next section. A general approach to the inversion of Laplace transforms based upon the calculus of residues must be deferred until Chapter 16.

Example 5.18 Find

$$\mathscr{L}^{-1}\left\{\frac{1}{s^2 - 5s + 6}\right\}$$

Solution *Method 1:* As in Example 5.17, use partial fractions to find

$$\mathscr{L}^{-1}\left\{\frac{1}{s^2-5s+6}\right\}=\mathscr{L}^{-1}\left\{\frac{-1}{s-2}+\frac{1}{s-3}\right\}=-e^{2t}+e^{3t}$$

Method 2: Complete the square on the quadratic in the denominator to obtain

$$\frac{1}{s^2-5s+6}=\frac{1}{s^2-5s+(\frac{5}{2})^2-\frac{1}{4}}=\frac{1}{(s-\frac{5}{2})^2-(\frac{1}{2})^2}$$

From Table 5.1, the Laplace transform of $\sinh\dfrac{t}{2}$ is

$$\frac{\frac{1}{2}}{s^2-(\frac{1}{2})^2}$$

Hence,

$$\mathscr{L}^{-1}\left\{\frac{1}{s^2-(\frac{1}{2})^2}\right\}=2\sinh\frac{t}{2}$$

The given function of s is the same as this with s replaced by $(s-\frac{5}{2})$. Thus, from the first shifting theorem,

$$\mathscr{L}^{-1}\left\{\frac{1}{s^2-5s+6}\right\}=2e^{5t/2}\sinh\frac{t}{2}$$

The correspondence of the results of Methods 1 and 2 is easily verified. □

Example 5.19 Find the inverse Laplace transform of $\dfrac{1}{s^2+3s+\frac{13}{4}}$.

Solution The denominator of this function cannot be factored into a product of real linear factors. We complete the square on the quadratic factor,

$$\frac{1}{s^2+3s+\frac{13}{4}}=\frac{1}{s^2+3s+(\frac{3}{2})^2+1}=\frac{1}{(s+\frac{3}{2})^2+1}$$

Using the fact that

$$\mathscr{L}^{-1}\left\{\frac{1}{s^2+1}\right\}=\sin t$$

and the first shifting theorem, we obtain

$$\mathscr{L}^{-1}\left\{\frac{1}{s^2+3s+\frac{13}{4}}\right\}=e^{-(3/2)t}\sin t$$ □

The preceding two examples are representative of the techniques used to find the inverse transform of functions whose denominators are linear or quadratic factors. The first shifting theorem obviously plays an important role.

Example 5.20 Find the inverse transform of the function $\dfrac{s}{s^2+3s+\frac{13}{4}}$.

Solution Since, from the previous example,

$$\mathcal{L}^{-1}\left\{\frac{1}{s^2+3s+\frac{13}{4}}\right\} = e^{-(3/2)t}\sin t$$

and since $sF(s) = f(0^+) + \mathcal{L}\{f'(t)\}$, it follows that

$$\begin{aligned}\mathcal{L}^{-1}\left\{\frac{s}{s^2+3s+\frac{13}{4}}\right\} &= \frac{d}{dt}(e^{-(3/2)t}\sin t)\\ &= e^{-(3/2)t}(\cos t - \tfrac{3}{2}\sin t)\end{aligned}$$ □

Example 5.21 Find the inverse transform of the function $\dfrac{1}{s(s^2-5s+6)}$.

Solution Recall from Theorem 5.5 that division by s corresponds to integration from 0 to t. Since, from Example 5.18,

$$\mathcal{L}^{-1}\left\{\frac{1}{s^2-5s+6}\right\} = -e^{2t}+e^{3t}$$

it follows that

$$\begin{aligned}\mathcal{L}^{-1}\left\{\frac{1}{s(s^2-5s+6)}\right\} &= -\int_0^t e^{2u}\,du + \int_0^t e^{3u}\,du\\ &= -\frac{e^{2t}-1}{2}+\frac{e^{3t}-1}{3}\\ &= \frac{1}{6}-\frac{e^{2t}}{2}+\frac{e^{3t}}{3}\end{aligned}$$ □

We now give a summary of some of the fundamental techniques that are especially useful for finding inverse transforms.

1. If $F(s)$ is of the form $G(s-a)$, find $\mathcal{L}^{-1}\{G(s)\}$ and multiply by e^{at}. (See Theorem 5.3.)

2. If $F(s)$ is of the form $sG(s)$, find $\mathcal{L}^{-1}\{G(s)\}$ and apply Theorem 5.4. If $F(s)$ is of the form $s^2G(s)$, two differentiations will be required. In each case, care must be taken with the constants $f(0^+)$, $f'(0^+)$, and so on, appearing in the formulas.

3. If $F(s) = G(s)/s$, find $\mathcal{L}^{-1}\{G(s)\}$ and perform an integration from 0 to t. (See Theorem 5.5.) If $F(s) = G(s)/s^2$, two such integrations will be required.

4. If $F(s)$ is of the form $G^{(n)}(s)$, where $G(s)$ is the transform of a known function $g(t)$, multiply $g(t)$ by $(-1)^n t^n$. (See the Corollary to Theorem 5.6.)

5. If $F'(s)$ is the transform of a known function $g(t)$, divide $g(t)$ by $-t$. (See Theorem 5.6.)

6. Suppose $F(s)$ is of the form of a rational function of s, for example,

$$F(s) = \frac{N(s)}{D(s)}$$

where the degree of $D(s)$ is greater than the degree of $N(s)$. Factor $D(s)$ into its real quadratic and linear factors (every real polynomial can be so reduced). Then $F(s)$ can be expanded in terms of its partial fractions. Each of the unrepeated linear factors leads to an exponential. Each of the unrepeated quadratic factors leads to an exponential times a trigonometric function. Repeated factors lead to multiplication by powers of t.

In more extensive discussions of the Laplace transformation, you will find a systematic development of a set of formulas by which the inverse transform of a rational function of s may be found quickly. These formulas are obtained by applying the techniques already given to the general cases, and extending the method to the case of repeated quadratic factors.

Note that there are other functions of s that have inverse transforms but are not of the form of a rational function of s. In Section 5.8 we prove the second shifting theorem. This will considerably widen our ability to find inverse transforms.

Example 5.22 Find the inverse transform of the function

$$F(s) = \frac{1}{(s+1)(s-3)^2(s-1)^3}$$

Solution Expanding $F(s)$,

$$\frac{1}{(s+1)(s-3)^2(s-1)^3} = \frac{A}{s+1} + \frac{B}{s-3} + \frac{C}{(s-3)^2} + \frac{D}{s-1} + \frac{E}{(s-1)^2} + \frac{F}{(s-1)^3}$$

It is an easy, but tedious, task to show that

$$A = -\tfrac{1}{128}, \quad B = -\tfrac{7}{128}, \quad C = \tfrac{1}{32}, \quad D = \tfrac{1}{16}, \quad E = \tfrac{1}{16}, \quad F = \tfrac{1}{8}$$

Since $\mathscr{L}^{-1}\{1/s^2\} = t$, and $\mathscr{L}^{-1}\{1/s^3\} = t^2/2$, the use of the first shifting theorem allows us to write

$$\mathscr{L}^{-1}\{F(s)\} = -\frac{e^{-t}}{128} - \frac{7e^{3t}}{128} + \frac{te^{3t}}{32} + \frac{e^t}{16} + \frac{te^t}{16} + \frac{t^2e^t}{16}$$ □

Example 5.23 Find the inverse transform of the function

$$F(s) = \frac{1}{(s-1)(s^2+4)}$$

Solution The denominator is already factored into irreducible real and quadratic factors. (The factor $s^2 + 4$ can be factored, but not into *real* linear factors.) Decomposing the rational function into its partial fraction expansion,

$$\frac{1}{(s-1)(s^2+4)} = \frac{A}{s-1} + \frac{Bs+C}{s^2+4}$$

from which $A = \frac{1}{5}$ and $B = C = -\frac{1}{5}$. Hence

$$\mathscr{L}^{-1}\{F(s)\} = \frac{e^t}{5} - \frac{\cos 2t}{5} - \frac{\sin 2t}{10} \qquad \square$$

Exercises for Section 5.6 (See also Exercises 41–53, p. 161.)

Find the inverse transform of the following functions.

1. $\dfrac{5}{(s-2)^7}$
2. $\dfrac{s}{(s-2)^7}$
3. $\dfrac{s+2}{(s+2)^2+1}$
4. $\dfrac{s}{(s+2)^2+1}$
5. $\dfrac{1}{s}\left(\dfrac{\omega}{s^2+\omega^2}\right)$
6. $\dfrac{1}{s^2}\left(\dfrac{s}{s^2+\omega^2}\right)$
7. $\dfrac{1}{s^2+3s+2}$
8. $\dfrac{1}{s(s^2+3s+2)}$
9. $\dfrac{1}{s^2+4s+4}$
10. $\dfrac{1}{s^2+4s+2}$
11. $\dfrac{s}{s^2+4s+2}$
12. $\dfrac{s^2}{(s+2)^3}$
13. $\dfrac{s}{(s+a)^2+b^2}$
14. $\dfrac{b}{(s+a)^2+b^2}$
15. $\dfrac{s^2}{(s+1)(s+2)(s+3)}$
16. $\dfrac{2s+1}{(s+2)^4}$
17. $\dfrac{1}{(s-1)^2(s+2)}$
18. $\dfrac{1}{(s-3)^2(s^2+3s+6)}$
19. $\dfrac{s^2-3s}{(s-2)(s-1)^2}$
20. $\dfrac{4s-3}{s^3+s^2}$
21. $\dfrac{3s^2+4}{s^4+s^2}$
22. $\ln\dfrac{(s+4)}{(s+2)}$
23. $\dfrac{2s}{(s^2+4)^2}$
24. $\ln\dfrac{s^2}{s^2+6s+10}$
25. $\mathrm{Cot}^{-1}\, s$
26. $\mathrm{Tan}^{-1}\left(\dfrac{3}{s}\right)$
27. $\dfrac{1}{s}\mathrm{Cot}^{-1}\, s$
28. $\dfrac{\int_0^2 e^{-st}t\,dt}{1-e^{-2s}}$
29. $\left(\dfrac{1}{s^2+4}\right)\left(\dfrac{1}{1-e^{-(\pi/2)s}}\right)$ (See Exercise 17, p. 165.)
30. $\left(\dfrac{1}{s^2+4}\right)\coth\dfrac{\pi s}{4}$ (See Exercise 18, p. 165.)

5.7 Solutions to Ordinary Differential Equations

For our purposes, the most immediate and important application of the Laplace transformation is to the solution of ordinary linear differential equations with constant coefficients. The procedure we discuss builds upon the approach suggested in Example 5.14. Consider the equation

$$a_2 y'' + a_1 y' + a_0 y = f(t)$$

where a_2, a_1, and a_0 are constants and $f(t)$ is the driving function or "input." Taking the Laplace transformation of both sides and denoting the Laplace transform of y by Y and that of f by F, we obtain, using Theorem 5.4,

$$a_2[s^2 Y - sy(0) - y'(0)] + a_1[sY - y(0)] + a_0 Y = F(s)$$

Solving for Y,

$$Y(s) = \frac{F(s)}{a_2 s^2 + a_1 s + a_0} + \frac{(a_2 s + a_1)y(0) + a_2 y'(0)}{a_2 s^2 + a_1 s + a_0}$$

which is called the *Laplace transform of the response*. The first term in the expression for Y depends on the driving function and corresponds to a particular solution of the nonhomogeneous equation. The second term depends on the initial conditions and corresponds to the solution of the homogeneous equation. In some applications it is unnecessary to find the actual response; the transform of the response is sufficient.

Note how the initial conditions are considered automatically, as opposed to our previous method where the "arbitrary" constants of the complete solution were found in a separate calculation. We see very clearly and unmistakably the role of the initial conditions in affecting the total solution.

We may summarize the technique of Laplace transforms as applied to ordinary linear differential equations with constant coefficients as follows:

1. Take the Laplace transform of both sides of the given differential equation. The transform of the driving function may prove difficult to find and might require the use of tables.
2. Solve the transformed equation for the transform of the solution. This is a purely algebraic step and should cause no difficulty.
3. Find the inverse transform of $Y(s)$. This may be (and usually is) the most difficult of the steps involved. It is the limitation on the method requiring the most experience to overcome.

Example 5.24 By using the method of Laplace transforms, solve the initial value problem

$$y'' + 2y' + 2y = \sin t, \qquad y(0) = 1, \qquad y'(0) = 1.$$

Solution Taking the transform of both sides, we obtain

$$s^2 Y - s - 1 + 2sY - 2 + 2Y = \frac{1}{s^2 + 1}$$

Solving for Y,

$$Y(s) = \frac{1}{(s^2 + 1)(s^2 + 2s + 2)} + \frac{s + 3}{s^2 + 2s + 2}$$

The partial fraction expansion of the first term is

$$\frac{1}{(s^2+1)(s^2+2s+2)} = \frac{As+B}{s^2+1} + \frac{Cs+D}{s^2+2s+2}$$

from which $A = -\frac{2}{5}$, $B = \frac{1}{5}$, $C = \frac{2}{5}$, and $D = \frac{3}{5}$. Hence

$$\begin{aligned}
\mathscr{L}^{-1}\left\{\frac{1}{(s^2+1)(s^2+2s+2)}\right\} &= \mathscr{L}^{-1}\left\{\frac{1}{5}\cdot\frac{1-2s}{s^2+1}\right\} + \mathscr{L}^{-1}\left\{\frac{1}{5}\cdot\frac{2s+3}{s^2+2s+2}\right\} \\
&= \frac{1}{5}\mathscr{L}^{-1}\left\{\frac{1}{s^2+1}\right\} - \frac{2}{5}\mathscr{L}^{-1}\left\{\frac{s}{s^2+1}\right\} \\
&\quad + \frac{2}{5}\mathscr{L}^{-1}\left\{\frac{s+1}{(s+1)^2+1}\right\} \\
&\quad + \frac{1}{5}\mathscr{L}^{-1}\left\{\frac{1}{(s+1)^2+1}\right\} \\
&= \frac{1}{5}(\sin t - 2\cos t) + \frac{e^{-t}}{5}(2\cos t + \sin t)
\end{aligned}$$

In like manner,

$$\mathscr{L}^{-1}\left\{\frac{s+3}{s^2+2s+2}\right\} = e^{-t}(\cos t + 2\sin t)$$

so that

$$y(t) = \tfrac{1}{5}e^{-t}(7\cos t + 11\sin t) + \tfrac{1}{5}(\sin t - 2\cos t) \qquad \square$$

Example 5.25 The driving voltage to an RC circuit is e^{-t} for $t \geqq 0$. Find the current in the circuit if the initial current is $1/R$. The circuit diagram is shown in Figure 5.7.

Solution The equation governing the circuit is

$$Ri + \frac{1}{C}\int_0^t i(x)\,dx = e^{-t}$$

Taking the Laplace transform of both sides and letting the transform of the current be $I(s)$,

$$RI(s) + \frac{1}{Cs}I(s) = \frac{1}{s+1}$$

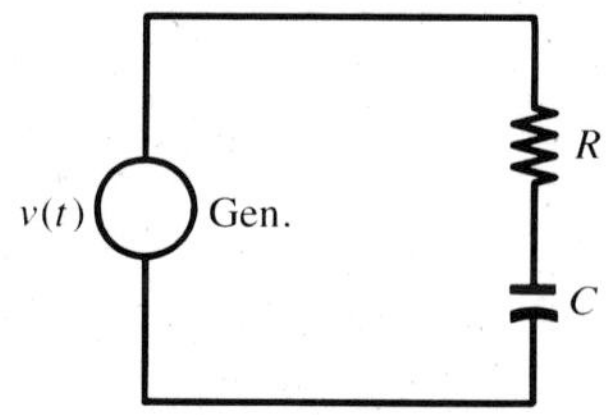

Figure 5.7

Solving for $I(s)$,

$$I(s) = \frac{sC}{(RCs+1)(s+1)} = \frac{1}{R}\frac{s}{\left(s+\dfrac{1}{RC}\right)(s+1)}$$

Since

$$\frac{s}{\left(s+\dfrac{1}{RC}\right)(s+1)} = -\frac{1}{s+\dfrac{1}{RC}}\frac{1}{RC-1} + \frac{1}{s+1}\frac{RC}{RC-1}$$

it follows that

$$i(t) = \frac{-1}{R(RC-1)}e^{-t/RC} + \frac{C}{RC-1}e^{-t} \qquad \square$$

With the use of Laplace transforms, systems of differential equations become systems of algebraic equations.

Example 5.26 Solve the system

$$\begin{aligned} y_1' + y_1 + 3y_2' &= 1 \\ 3y_1 + y_2' + 2y_2 &= t \\ y_1(0) = 0, \qquad y_2(0) &= 0 \end{aligned}$$

Solution Taking the Laplace transform of both sides of both equations, we obtain

$$sY_1 + Y_1 + 3sY_2 = \frac{1}{s}$$

$$3Y_1 + sY_2 + 2Y_2 = \frac{1}{s^2}$$

Using Cramer's rule to solve for Y_1 and Y_2,

$$Y_1(s) = \frac{\det\begin{pmatrix} 1/s & 3s \\ 1/s^2 & s+2 \end{pmatrix}}{\det\begin{pmatrix} s+1 & 3s \\ 3 & s+2 \end{pmatrix}} = \frac{s-1}{s(s^2-6s+2)}$$

$$Y_2(s) = \frac{\det\begin{pmatrix} s+1 & 1/s \\ 3 & 1/s^2 \end{pmatrix}}{\det\begin{pmatrix} s+1 & 3s \\ 3 & s+2 \end{pmatrix}} = \frac{1-2s}{s^2(s^2-6s+2)}$$

$$Y_1(s) = -\frac{1}{2s} + \frac{1}{2}\frac{s-3}{(s-3)^2-7} - \frac{1}{2}\frac{1}{(s-3)^2-7}$$

$$Y_2(s) = \frac{1}{2s} + \frac{1}{2s^2} - \frac{1}{2}\frac{s-3}{(s-3)^2-7} + \frac{1}{(s-3)^2-7}$$

Therefore,

$$y_1(t) = -\tfrac{1}{2} + \tfrac{1}{2}e^{3t}\cosh\sqrt{7}t - \frac{1}{2\sqrt{7}}e^{3t}\sinh\sqrt{7}t$$

$$y_2(t) = \tfrac{1}{2} + \tfrac{1}{2}t - \tfrac{1}{2}e^{3t}\cosh\sqrt{7}t + \frac{1}{\sqrt{7}}e^{3t}\sinh\sqrt{7}t \qquad \square$$

Compare the solution procedure used here with that employed in Example 4.8.

Exercises for Section 5.7

Solve the following initial value problems using the Laplace transform method.

1. $y'' + y' - 2y = 0, \quad y(0) = 0, \quad y'(0) = 2$
2. $y'' + 4y = 0, \quad y(0) = 1, \quad y'(0) = 1$
3. $y'' - 16y = e^{-2t}, \quad y(0) = 1, \quad y'(0) = 0$
4. $4y'' + 4y' + y = 0, \quad y(0) = 2, \quad y'(0) = -1$
5. $y'' + 2y' - 3y = te^t, \quad y(0) = 0, \quad y'(0) = 1$
6. $y'' - 3y' = 2, \quad y(0) = 2, \quad y'(0) = 3$
7. $y' = y, \quad y(0) = 1$
8. $y'' + 2y' + 4y = t, \quad y(0) = 2, \quad y'(0) = 1$
9. $y'' + 3y' + 2y = \cos 2t, \quad y(0) = -1, \quad y'(0) = 1$
10. $y'' + y = \sin 3t, \quad y(0) = 1, \quad y'(0) = 1$
11. $y'' + a^2y = 0, \quad y(0) = 1, \quad y'(0) = 0$
12. $y'' + a^2y = 0, \quad y(0) = 0, \quad y'(0) = 1$
13. Exercise 34 p. 161.
14. Exercise 35 p. 161.
15. Exercise 36 p. 161.
16. Exercise 37 p. 161.
17. Exercise 38 p. 161.
18. An *RLC* circuit has a driving voltage of t^2, with $R = 3\Omega$, $L = 1\text{H}$, and $C = \frac{1}{2}\text{F}$. If $i(0) = 1$ and $i'(0) = -3$, find an expression for the current.
19. An *RLC* circuit has a driving voltage of $\sin t$ volts, with $R = 4\Omega$, $C = \frac{1}{4}\text{F}$, and $L = 1\text{H}$. If $i(0) = 1$ and $i'(0) = -4$, find an expression for the current.
20. A mechanical system with spring constant 5, fluid friction constant 6, and a mass of 1 gm is subjected to an external excitation of $\cos t$. If the initial deflection is 3 cm and the initial velocity is zero, find an expression for the deflection at any time t.
21. The fourth-order equation

$$\theta^{(4)} + (\alpha^2 + \omega^2)\theta'' + \alpha^2\omega^2\theta = 0$$

for the angular displacement $\theta(t)$ occurs in a particular coupled pendulum application (see p. 128 for an explanation of the problem parameters). Determine the solution of the initial value problem that arises when $\theta(0) = \theta_0$, $\theta'(0) = 0$, $\theta''(0) = -(g/l)\theta_0$, and $\theta'''(0) = 0$.

In Exercises 22–26 solve the given system initial value problem using the Laplace transform method.

22.

$$\begin{aligned} &y_1'' - 3y_1' - y_2' + 2y_2 = 14t + 3 \\ &y_1' - 3y_1 + 2y_2 = 1 \\ &y_1(0) = 1, \qquad y_1'(0) = 0, \qquad y_2(0) = 2 \end{aligned}$$

23.

$$\begin{aligned} &y_1' + y_2' = 0 \\ &y_1 - y_2' = 0 \\ &y_1(0) = 0, \qquad y_2(0) = 1 \end{aligned}$$

24.

$$\begin{aligned} &y_1' + y_2 = t^2 + 1 \\ &y_1 - y_2' = -t - 1 \\ &y_1(0) = y_2(0) = 0 \end{aligned}$$

25.

$$\begin{aligned} &y_1'' + y_2'' + y_1 + y_2 = 2 - 6t + t^2 - t^3 \\ &y_1'' - y_2' = 2 + 3t^2 \\ &y_1(0) = 0, \qquad y_2(0) = 3, \qquad y_1'(0) = 3, \qquad y_2'(0) = 0 \end{aligned}$$

26.

$$\begin{aligned} &y_1'' + y_2' - y_2 = 0 \\ &2y_1' - y_1 + y_3' - y_3 = 0 \\ &y_1' + 3y_1 + y_2' - 4y_2 + 3y_3 = 0 \\ &y_1(0) = 0, \qquad y_1'(0) = 1, \qquad y_2(0) = y_3(0) = 0 \end{aligned}$$

27. The coupled system

$$\begin{aligned} c_1' &= -2c_1 + c_2 + c_3 \\ c_2' &= c_1 - 2c_2 + c_3 \\ c_3' &= c_1 + c_2 - 2c_3 \end{aligned}$$

effectively models a particular three-component chemical reaction. (See p. 144.)

(a) Determine the solution of the initial value problem with $c_1(0) = a$, $c_2(0) = b$, $c_3(0) = c$.

(b) Show that if $a + b + c$ is a given constant, the long-term behavior of every solution found in part (a) is effectively the same.

5.8 A "Turn-On" Function; The Second Shifting Theorem

There are many important driving functions that are not combinations of any well-known elementary functions. Figure 5.8 shows several of these very practical functions.

Ordinarily, to describe one of these functions analytically, you would give its "rule" over each subinterval. Sometimes though, and especially when using Laplace transforms to solve differential equations, it is more convenient to give

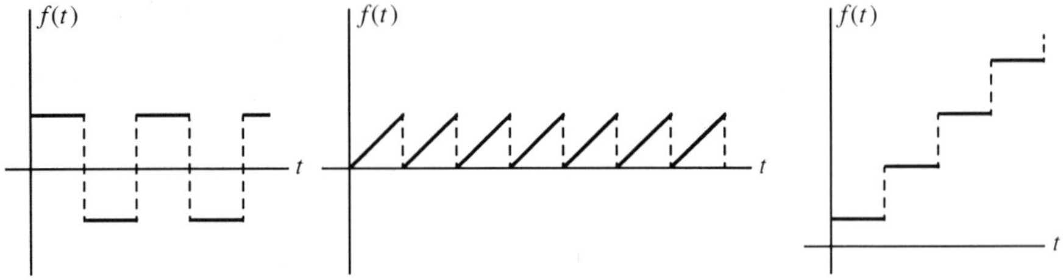

Square wave Sawtooth wave Staircase function

Figure 5.8 *Some important driving functions*

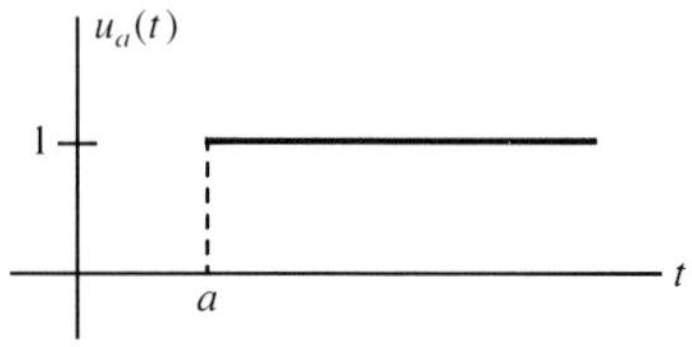

Figure 5.9 *Unit step function*

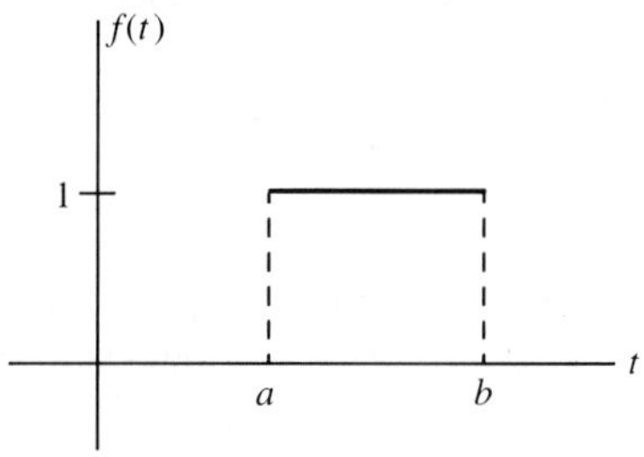

Figure 5.10 *Characteristic function of* $[a, b]$

a single rule, which is valid over the entire interval of interest. To do this, we introduce a function whose only property is that it "turns on" other functions at a specific value of t. This function is called the *unit step function*, $u_a(t)$, and is defined by

$$u_a(t) = \begin{cases} 0, & t < a \\ 1, & t \geqq a \end{cases}$$

(See Figure 5.9.)

A related function, which mathematicians would call a "characteristic function of the interval $[a, b]$" and engineers might call a "pulse" function, is the function whose graph is shown in Figure 5.10. In terms of unit step functions this may be written $u_a(t) - u_b(t)$. It is actually the pulse function rather than the unit step function which allows a compact analytic description of many important driving functions.

Example 5.27 Using unit step functions, describe analytically the function that is 0 except for the line segment connecting $(0, 0)$ to $(2, 6)$. (See Figure 5.11.)

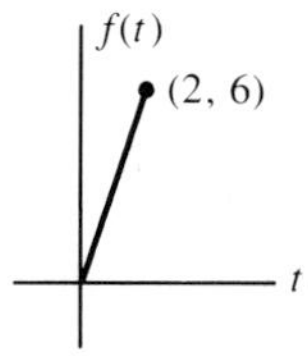

Figure 5.11

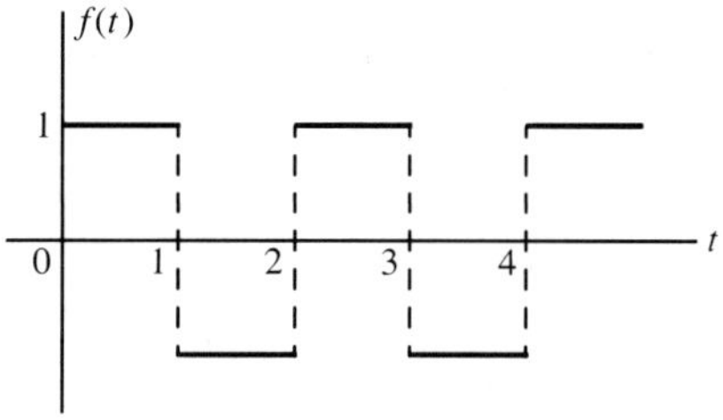

Figure 5.12

Solution Note that on the interval [0, 2] the defining rule for this function is $3t$. Since the function is zero elsewhere, we may describe this by multiplying $3t$ by the pulse function which is "on" from 0 to 2,

$$f(t) = [u_0(t) - u_2(t)]3t \qquad \square$$

Example 5.28 Using combinations of unit step functions, write the rule for the periodic unit square wave shown in Figure 5.12.

Solution The square wave may be thought of as a repeated pulse function alternating between positive and negative unit values. Hence, the function may be written

$$\begin{aligned}[u_0(t) - u_1(t)] - [u_1(t) - u_2(t)] + [u_2(t) - u_3(t)] - \cdots \\ = u_0(t) - 2u_1(t) + 2u_2(t) - 2u_3(t) + \cdots \\ = u_0(t) + 2\sum_{n=1}^{\infty} (-1)^n u_n(t)\end{aligned} \qquad \square$$

The Laplace transform of $u_a(t)$, $a \geq 0$, is of obvious interest and is found by direct appeal to the definition of the transform:

$$\begin{aligned}\mathscr{L}\{u_a(t)\} &= \int_0^\infty u_a(t)e^{-st}\,dt \\ &= \int_a^\infty e^{-st}\,dt = \lim_{B\to\infty} \frac{e^{-st}}{-s}\bigg|_a^B \\ &= \lim_{B\to\infty} \frac{e^{-sB}}{-s} - \frac{e^{-as}}{-s} \\ &= \frac{e^{-as}}{s}, \quad s > 0\end{aligned}$$

Example 5.29 Find the Laplace transform of the unit square wave discussed in the previous example.

Solution It would be possible to find the transform by using the formula for the Laplace transform of a periodic function developed in Theorem 5.7. Alternatively, we can work directly with the function as written in Example 5.28:

$$\begin{aligned}\mathscr{L}\{\text{unit square wave}\} &= \mathscr{L}\{u_0(t)\} - 2\mathscr{L}\{u_1(t)\} \\ &\quad + 2\mathscr{L}\{u_2(t)\} - 2\mathscr{L}\{u_3(t)\} + \cdots \\ &= \frac{1}{s} - 2\frac{e^{-s}}{s} + 2\frac{e^{-2s}}{s} - 2\frac{e^{-3s}}{s} + 2\frac{e^{-4s}}{s} - \cdots\end{aligned}$$

After the first term, this is a geometric series with common ratio equal to $-e^{-s}$. Therefore,

$$\begin{aligned}\mathscr{L}\{\text{unit square wave}\} &= \frac{1}{s} - 2\frac{e^{-s}}{s}\left(\frac{1}{1+e^{-s}}\right), \qquad s > 0 \\ &= \frac{1}{s}\left(\frac{1+e^{-s}-2e^{-s}}{1+e^{-s}}\right) \\ &= \frac{1-e^{-s}}{s(1+e^{-s})} \\ &= \frac{1}{s}\tanh\frac{s}{2}, \qquad s > 0\end{aligned}$$

□

Of more general interest than the Laplace transform of linear combinations of unit step functions is the transform of some well-known functions multiplied by a unit step function. The following theorem provides a formula that is useful in such calculations.

Theorem 5.8 *The Second Shifting Theorem.* Let $F(s) = \mathscr{L}\{f(t)\}$. Then

$$\mathscr{L}\{u_a(t)f(t-a)\} = e^{-as}F(s), \qquad a \geq 0$$

or

$$\mathscr{L}^{-1}\{e^{-as}F(s)\} = u_a(t)f(t-a), \qquad a \geq 0$$

which is the same thing.

Proof By definition

$$e^{-as}F(s) = \int_0^\infty e^{-s(x+a)}f(x)\,dx$$

In the integral, we make the substitution $t = x + a$

$$\begin{aligned}e^{-as}F(s) &= \int_a^\infty e^{-st}f(t-a)\,dt \\ &= \int_0^\infty e^{-st}u_a(t)f(t-a)\,dt \\ &= \mathscr{L}\{u_a(t)f(t-a)\}\end{aligned}$$

■

Theorem 5.8 may be viewed in two ways:

1. When finding the transform of a function times a unit step function, $u_a(t)$, the function must first be expressed in terms of $t - a$. Then the result of the theorem may be applied.

2. The multiplication of the transform by the exponential, e^{-as}, causes a shift of the inverse function to the right and a turn-on at $t = a$. Recall that the first shifting theorem showed how a shift of the transform function occurs when the function of t is multiplied by e^{at}. The first and second shifting theorems are analogous, but the turn-on is unique to the second theorem.

Example 5.30 Find the Laplace transform of $t^2 u_3(t)$.

Solution We express the function t^2 in terms of $t - 3$ (that is, find the Taylor series for t^2 about $t = 3$):

$$t^2 = (t - 3 + 3)^2 = (t - 3)^2 + 6(t - 3) + 9$$

Hence,

$$\begin{aligned}\mathscr{L}\{t^2 u_3(t)\} &= \mathscr{L}\{(t-3)^2 u_3(t)\} + \mathscr{L}\{6(t-3)u_3(t)\} + \mathscr{L}\{9u_3(t)\} \\ &= \frac{2e^{-3s}}{s^3} + 6\frac{e^{-3s}}{s^2} + 9\frac{e^{-3s}}{s}\end{aligned}$$ □

Example 5.31 Find the inverse of the transform function e^{-4s}/s^2.

Solution Since $\mathscr{L}^{-1}\{1/s^2\} = t$, the factor e^{-4s} causes a shift to $t - 4$ and a turn-on at that point:

$$\mathscr{L}^{-1}\left\{\frac{e^{-4s}}{s^2}\right\} = u_4(t)(t - 4)$$ □

We are now in a position to attack three of the five problems outlined in Section 5.1. Problems 1 and 4 are solved in Examples 5.32 and 5.33; the reader is allowed to have some of the fun by solving Problem 2 as Exercise 45.

Example 5.32 Find the response of an RL circuit to a unit pulse function $u_a(t) - u_b(t)$. Assume that the initial current is zero.

Solution The differential equation of the circuit is

$$L\frac{di}{dt} + Ri = u_a(t) - u_b(t)$$

Taking the Laplace transform of both sides and letting $\mathscr{L}\{i(t)\} = I(s)$,

$$LsI(s) + RI(s) = \frac{e^{-as} - e^{-bs}}{s}$$

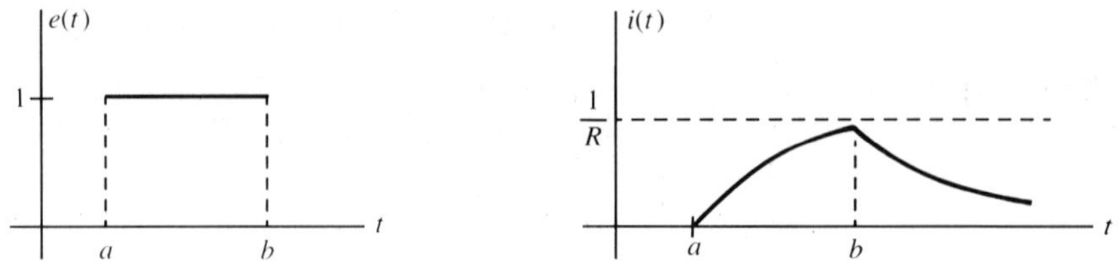

Input Response

Figure 5.13

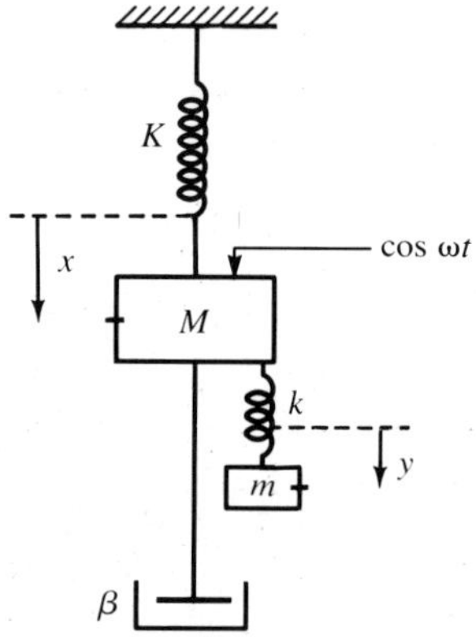

Figure 5.14

from which

$$I(s) = \frac{e^{-as}}{Ls\left(s + \frac{R}{L}\right)} - \frac{e^{-bs}}{Ls\left(s + \frac{R}{L}\right)}$$

$$= \frac{e^{-as}}{R}\left(\frac{1}{s} - \frac{1}{s + \frac{R}{L}}\right) - \frac{e^{-bs}}{R}\left(\frac{1}{s} - \frac{1}{s + \frac{R}{L}}\right)$$

Hence,

$$i(t) = \frac{1}{R}(1 - e^{-R(t-a)/L})u_a(t) - \frac{1}{R}(1 - e^{-R(t-b)/L})u_b(t) \qquad \square$$

(See Figure 5.13.)

Example 5.33 In the mechanical system indicated in Figure 5.14, find m and k such that the vibration of M is absorbed, that is, such that the steady state $x(t) \equiv 0$.

Solution Since we are only interested in the steady state, let us assume that

$$x(0) = x'(0) = y(0) = y'(0) = 0$$

The equations describing the behavior of this system are

$$Mx'' = -Kx + k(y - x) - \beta x' + \cos \omega t$$
$$my'' = -k(y - x)$$

The transforms, X and Y, of x and y then satisfy:

$$[Ms^2 + \beta s + (K + k)]X - kY = \frac{s}{s^2 + \omega^2}$$
$$-kX + (ms^2 + k)Y = 0$$

from which we find that

$$X(s) = \frac{s(ms^2 + k)}{(s^2 + \omega^2)p(s)}$$

where

$$p(s) = (ms^2 + k)(Ms^2 + \beta s + K + k) - k^2$$

Because of the $s^2 + \omega^2$ term in the denominator of the expression for $X(s)$, $x(t)$ will involve terms containing both $\sin \omega t$ and $\cos \omega t$ unless the $ms^2 + k$ factor exactly cancels out $s^2 + \omega^2$, that is, unless $k/m = \omega^2$. This makes good sense—if the natural frequency of the m, k subsystem is equal to the frequency of the applied force, it may resonate and thus damp out the vibration of the mass M. Therefore, we need only show that $p(s)$ does not produce any undamped vibration. To do this we first demonstrate that $p(s)$ has no purely imaginary roots, that is, roots of the form ri where r is real. Indeed, if

$$p(ri) = [(-mr^2 + k)\beta r]i + [(-mr^2 + k)(-Mr^2 + K + k) - k^2] = 0$$

then either $r = 0$ or $(-mr^2 + k) = 0$, since $\beta \neq 0$. If $r = 0$, $p(ri) = 0$ reduces to $k = 0$, which cannot be true. If $(-mr^2 + k) = 0$, we find $k^2 = 0$, another contradiction. We now ask whether $p(s)$ could have a root with positive real part thus producing unbounded motion? This makes no sense physically, but we are *not* dealing with a physical system—we are merely analyzing a mathematical approximation of the system. It is comforting to find out that there are various ways (including using the well-known Routh-Hurwitz stability criterion) to show that the real part of each root is, indeed, negative.

Incidentally, we may also examine Y and find that the mass m does have an undamped vibration so in some sense the m, k system has absorbed the energy produced by the applied force, $\cos \omega t$. □

The principle illustrated in this example has been used to design vibration absorbers for hair clippers, electric razors, and other such familiar devices.

Exercises for Section 5.8

In Exercises 1–11, sketch the graph of the given function for $t \geqq 0$.

1. tu_1

2. $t(u_1 - u_2)$

3. $u_1 - 3u_4 - 4u_5$

4. $(t - 3)u_3$

5. $(t^2 - 4)u_4$

6. $u_\pi \sin t$

7. $u_\pi \sin(t - \pi)$

8. $u_{\pi/2} \cos t$

9. $t(u_0 - u_1) + (u_1 - u_3) + (u_3 - u_4)(4 - t)$

10. $u_1 + u_2 + u_3 + u_4 + \cdots$

11. $u_1 - 2u_2 + 2u_3 - 2u_4 + \cdots$

In Exercises 12–21, represent the given functions in terms of unit step functions and find their Laplace transforms.

12. $f(t) = \begin{cases} 2, & 0 < t < 1 \\ t, & 1 < t \end{cases}$

13. $f(t) = \begin{cases} 3, & 0 < t < 2 \\ t + 1, & 2 < t \end{cases}$

14. $f(t) = \begin{cases} e^{-t}, & 0 < t < 3 \\ 0, & 3 < t \end{cases}$

15. $f(t) = \begin{cases} \sin 3t, & 0 < t < \pi \\ 0, & \pi < t \end{cases}$

16. $f(t) = \begin{cases} t^2, & 0 < t < 3 \\ 9, & 3 < t < 5 \\ 0, & 5 < t \end{cases}$

17.

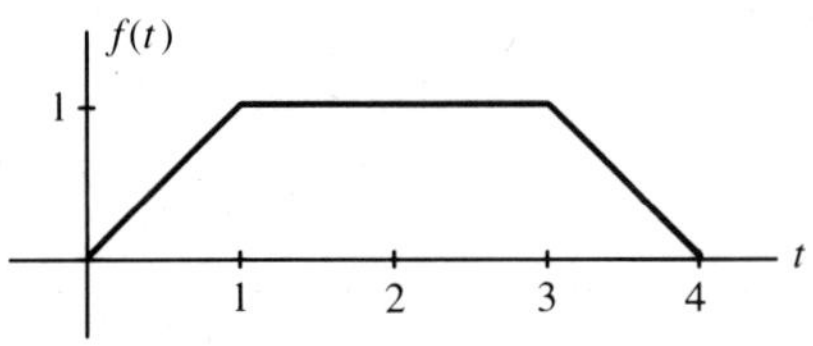

Figure 5.15

18.

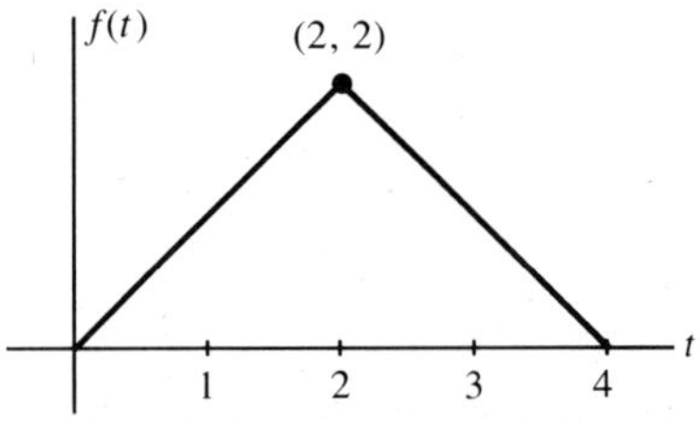

Figure 5.16

19.

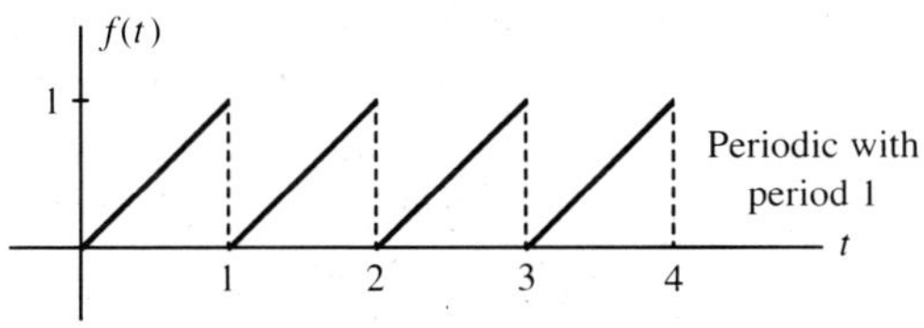

Figure 5.17

20.

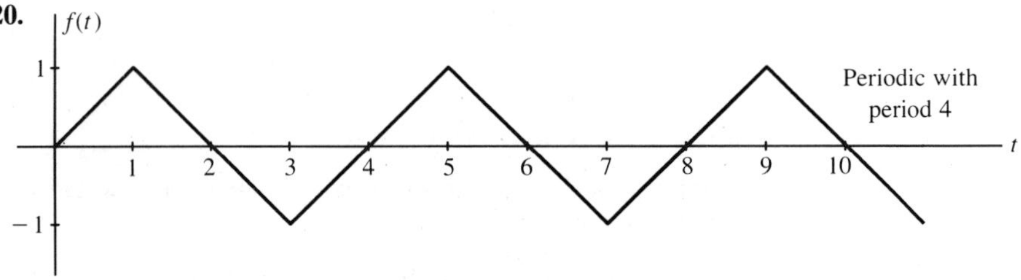

Figure 5.18

21.

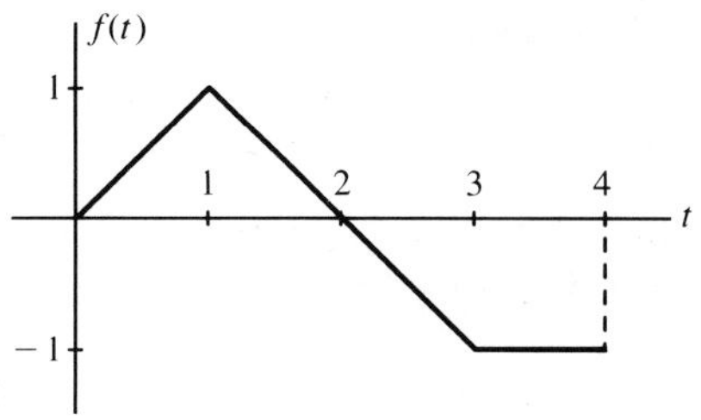

Figure 5.19

In Exercises 22–33, find the inverse Laplace transform of the given function. Sketch the graph of the function of t.

22. $\dfrac{e^{-3s}}{s^2}$ **23.** $\dfrac{e^{-4s} - e^{-s}}{s^3}$ **24.** $\dfrac{e^{-s}}{s^2 + 4}$

25. $\dfrac{e^{-3s}}{s^2 - 9}$ **26.** $\dfrac{e^{-\pi s}}{s}$ **27.** $\dfrac{e^{-s}}{(s-1)(s-2)}$

28. $\dfrac{se^{-2s}}{s^2 + 4}$ **29.** $\dfrac{e^{-s}}{(s+1)^3}$ **30.** $\dfrac{e^{-2s} + e^{-3s}}{s^2 - 5s + 6}$

31. $\dfrac{(s-2)e^{-s}}{s^2 - 4s + 3}$ **32.** $\dfrac{e^{-\pi s}}{s(s^2 + 4)}$ **33.** $\dfrac{e^{-s}}{s^2 - 2s + 2}$

In Exercises 34–42, find the Laplace transform of the given functions.

34. u_3 **35.** tu_2 **36.** $t^2 u_3$

37. $u_\pi \cos t$ **38.** $u_2(t-3)^2$ **39.** $e^t u_3$

40. $u_1 \cos(t-1)$ **41.** $u_{\pi/2} \sin t$ **42.** $u_3 \cos t$

In Exercises 43–45, use the method of Laplace transforms to solve the given problem.

43. Find the current in an RL circuit whose voltage is a constant V_0. Assume no initial current.

44. Find the response of an RC circuit to a single square wave, of height h, between $t = a$ and $t = b$. Assume zero initial charge.

45. Find the current in an RL circuit whose input voltage is the periodic sawtooth function given in Figure 5.17. Assume no initial current. (This is Problem 2 of Section 5.1.)

The small-amplitude forced oscillations of a simple pendulum of length l and mass m are described by the differential equation

$$m\theta'' + \mu\theta' + \frac{mg}{l}\theta = \frac{1}{l}f(t)$$

Here $\theta(t)$ is the angular deviation of the pendulum from the vertical, μ is the coefficient of friction, g is the constant acceleration due to gravity, and $f(t)$ is the applied force.

Exercises 46–48 refer to this oscillating pendulum application and should be solved by the method of Laplace transforms.

46. Show that if $\mu > 0$, all solutions of the reduced equation are transient, that is, they tend to zero for large t.

47. **(a)** Find the motion that results if $\theta(0) = 0 = \theta'(0)$ but

$$f(t) = \begin{cases} mlf_0, & 0 \le t \le 1 \\ 0, & 1 < t \end{cases}$$

(b) Sketch the resulting $\theta(t)$ for some representative values of the problem parameters.

48. An undamped pendulum, initially at rest, is acted upon by the applied force

$$f(t) = mlf_0 t[u_0 - u_a], \qquad a > 0$$

(a) Determine the resulting motion.

(b) Discuss the long-term (steady-state) motion as a function of the parameters a and f_0.

49. Solve the *difference equation* $y(t) = 2y(t-1) + f(t)$, $y(t) = 0$ for $t < 0$ and $f(t) = 0$ for $t < 0$. (*Hint:* After taking transforms, expand $1/(1 - 2e^{-s})$ as $1 + 2e^{-s} + 4e^{-2s} + \cdots$ and take the inverse transform term by term.)

50. Solve the difference equation $y(t) + 2y(t-1) - 3y(t-2) = f(t)$, where $f(t) = 0$ for $t < 0$ and $y(t) = 0$ for $t < 0$. (Use appropriate modifications of the hint for Exercise 49.)

5.9 The Convolution Theorem; Integral Equations

In this last section, we introduce a surprisingly versatile operation, which arises quite naturally in discussing the inverse of the product of two transform functions.

Definition 5.4 Let f and g be piecewise continuous functions. Then the function of t defined by the integral

$$\int_0^t f(t-x)g(x)\,dx$$

is called the *convolution* of f and g and is denoted by $f * g$.

Example 5.34 Find the convolution of the sine and cosine functions.

Solution

$$\begin{aligned}(\sin t) * (\cos t) &= \int_0^t \sin(t-x)\cos x\,dx \\ &= \tfrac{1}{2}\int_0^t [\sin t + \sin(t-2x)]\,dx \\ &= \frac{x \sin t}{2}\bigg|_0^t + \frac{\cos(t-2x)}{4}\bigg|_0^t \\ &= \tfrac{1}{2}t\sin t\end{aligned}$$

□

The fundamental theorem of this section is concerned with the transform of a convolution function.

Theorem 5.9 Let f and g be piecewise continuous and of exponential order. Then $f * g$ has a Laplace transform, which is given by the product of the transform of f and the transform of g.

Proof

$$\begin{aligned}\mathscr{L}\{f * g\} &= \int_0^\infty e^{-st}\left[\int_0^t f(t-x)g(x)\,dx\right]dt \\ &= \int_0^\infty e^{-st}\left[\int_0^\infty u_x(t)f(t-x)g(x)\,dx\right]dt\end{aligned}$$

where the inner integration has been "extended to infinity" by including the factor $u_x(t)$, which is zero for $x > t$. The assumptions that we have made on f and g are sufficient to allow us to interchange the order of integration resulting in

$$\mathscr{L}\{f * g\} = \int_0^\infty g(x)\left[\int_0^\infty u_x(t)e^{-st}f(t-x)\,dt\right]dx$$

and this can now be rewritten using the second shifting theorem (Theorem 5.8) as

$$\mathscr{L}\{f * g\} = \int_0^\infty g(x)\left[e^{-xs}F(s)\right]dx$$

Hence

$$\begin{aligned}\mathscr{L}\{f * g\} &= F(s)\int_0^\infty e^{-xs}g(x)\,dx \\ &= F(s)G(s)\end{aligned}$$ ■

We can now easily supply a proof for the general case of Theorem 5.5—a chore we postponed previously. Let $f_1(t) = 1$ and $f_2(t) = f(t)$. Then by Theorem 5.9,

$$\mathscr{L}\{f_1 * f_2\} = \mathscr{L}\left\{\int_0^t f(u)\,du\right\} = \mathscr{L}\{f_1\}\mathscr{L}\{f_2\} = \frac{1}{s}\mathscr{L}\{f\}$$

Example 5.35 Use the convolution theorem to find

$$\mathscr{L}^{-1}\left\{\frac{1}{s(s^2+1)}\right\}$$

Solution Note that

$$\frac{1}{s(s^2+1)} = \frac{1}{s}\frac{1}{s^2+1} = F(s)G(s)$$

Since

$$\mathscr{L}^{-1}\{F(s)\} = u_0(t)$$

and

$$\mathcal{L}^{-1}\{G(s)\} = \sin t$$

it follows that

$$\begin{aligned}\mathcal{L}^{-1}\left\{\frac{1}{s(s^2+1)}\right\} &= u_0(t) * \sin t \\ &= \int_0^t u_0(t-x)\sin x\,dx \\ &= 1 - \cos t\end{aligned}$$

□

Example 5.36 Find

$$\mathcal{L}^{-1}\frac{s^2}{(s^2+1)^2}$$

Solution Note that the function is the square of $s/(s^2+1)$, whose inverse transform is $\cos t$. Hence

$$\begin{aligned}\mathcal{L}^{-1}\left\{\frac{s^2}{(s^2+1)^2}\right\} &= (\cos t) * (\cos t) \\ &= \int_0^t \cos(t-x)\cos x\,dx \\ &= \frac{\sin t + t\cos t}{2}\end{aligned}$$

□

The convolution theorem may be used to great advantage in solving equations with an unknown function under an integral sign. Equations of this type are called *integral equations*. In order to apply this last result to solve such equations, the integrals occurring must, of course, be of the form of a convolution integral.

Example 5.37 Find the function y, if

$$y(t) = t + \int_0^t \sin(t-x)y(x)\,dx$$

Solution We take the Laplace transform of both sides and denote the transform of y by Y:

$$Y(s) = \frac{1}{s^2} + Y(s)\frac{1}{s^2+1}$$

Solving for Y we obtain

$$\begin{aligned}Y(s) &= \frac{s^2+1}{s^4} \\ &= \frac{1}{s^2} + \frac{1}{s^4}\end{aligned}$$

The inverse transform of $Y(s)$ is

$$y(t) = t + \frac{t^3}{6} \qquad \square$$

Example 5.38 Solve the integral equation

$$y(t) = t + \int_0^t y(x)\,dx + \int_0^t (t - x)y(x)\,dx$$

Solution Taking the transform of both sides we obtain

$$Y(s) = \frac{1}{s^2} + \frac{Y(s)}{s} + \frac{Y(s)}{s^2}$$

Solving for $Y(s)$,

$$\begin{aligned} Y(s) &= \frac{1}{s^2 - s - 1} \\ &= \frac{2}{\sqrt{5}} \frac{\sqrt{5}/2}{(s^2 - s + \frac{1}{4}) - \frac{5}{4}} \end{aligned}$$

Hence, the unknown function is

$$f(t) = \frac{2}{\sqrt{5}} e^{t/2} \sinh \frac{\sqrt{5}}{2} t \qquad \square$$

We return now to Problem 5 of Section 5.1.

Example 5.39 Devise a replacement scheme for equipment if the number of pieces is to be kept a large constant number and the mortality of the equipment has a negative exponential character.

Solution Let $f(t)$ denote the number of pieces of equipment in service at time t, where the number is large enough that we may consider f to be a continuous function rather than taking on only integral values. Without replacement, we would assume $f(t) = f(0)e^{-kt}$. If $g(u)$ is the total replacement up until time u, then $g'(u)\,du$ approximates replacement from u to $u + du$. Of the replacements $g'(u)\,du$, $g'(u)e^{-k(t-u)}\,du$ will be in service at time $t > u$. With replacement, then, we have:

$$f(t) = f(0)e^{-kt} + \int_0^t g'(u)e^{-k(t-u)}\,du$$

Let us assume $g(0) = 0$, that is, the equipment was all new when $t = 0$. We wish $f(t)$ to be a constant, namely, $f(0)$. Thus

$$\frac{f(0)}{s} = \frac{f(0)}{s + k} + \frac{sG(s)}{s + k}, \qquad G(s) = \frac{kf(0)}{s^2}, \quad \text{and} \quad g(t) = kf(0)t \qquad \square$$

Exercises for Section 5.9

1. Find the Laplace transform of $\int_0^t (t-u)\sin 2u\,du$.
2. Find the Laplace transform of $\int_0^t e^{-(t-u)}\cos u\,du$.

In Exercises 3–8 find the inverse transform of the given function using the convolution theorem.

3. $\dfrac{1}{s(s^2+1)}$
4. $\dfrac{1}{s^2(s-1)}$
5. $\dfrac{1}{s^2+6s+8}$
6. $\dfrac{1}{s^2-3s-4}$
7. $\dfrac{1}{s(s^2+4s+8)}$
8. $\dfrac{s}{(s+3)(s^2+4)}$

In Exercises 9–16, solve the given integral equation.

9. $y(t) = t + \int_0^t \sin(t-u)y(u)\,du$
10. $y(t) = \sin t + \int_0^t \sin(t-u)y(u)\,du$
11. $y' + 3y + 2\int_0^t y(u)\,du = u_0(t), \quad y(0) = 0$
12. $y(t) = t^2 + \int_0^t \sin(t-u)y(u)\,du$
13. $y(t) = t + 4\int_0^t (u-t)^2 y(u)\,du$
14. $y(t) = t^2 - 2\int_0^t y(t-u)\sinh 2u\,du$
15. $y(t) = t + \int_0^t y(t-u)e^{-u}\,du$
16. $y(t) = 1 + \int_0^t \sin(t-u)y(u)\,du$
17. Use the convolution theorem to give an expression for a particular solution of $y'' + 2by' + b^2y = f(t)$.
18. Use the convolution theorem to find the complete solution of $y'' + 5y' + 6y = f(t)$.

A simple satellite-tracking system is modeled by the differential equation

$$I\theta''(t) = -k[\theta(t) - \psi(t)]$$

In this expression $\psi(t)$ is an observed input angle, $\theta(t)$ is the system response output angle, I is an appropriate moment of inertia, and k is a positive constant that relates the system's induced compensating torque to the instantaneous pointing error $\theta - \psi$.

Exercises 19–21 are concerned with this servo-control system.

19. If $\psi(t)$ is assumed known, determine the solution of the initial value problem with $\theta(0) = \theta_0, \theta'(0) = \theta_0'$ by Laplace transform methods and show that it can be expressed in convolution form as

$$\theta(t) = \theta_0\cos\omega t + (\theta_0'/\omega)\sin\omega t + \omega\int_0^t \psi(u)\sin\omega(t-u)\,du$$

where $\omega = \sqrt{k/I}$.

20. Find the system response to the sinusoidal input $\psi(t) = a\sin bt$. (Be careful if $b = \omega$.)

21. If $\theta(t)$ is assumed known, the expression appearing in Exercise 19 can be viewed as an integral equation for $\psi(t)$.

(a) Solve for $\psi(t)$ in the case when $\theta(t) = te^{-t}$.

(b) Show that this is the same result as expected from the original differential equation formulation of the problem.

The integral representation

$$P(t) = P_0 s(t) + \int_0^t r(u)s(t-u)\,du$$

arises in the consideration of age-specific population growth. Here $P(t)$ is the quantity of some biological species at time t, with $P_0 = P(0)$, and $s(t)$ and $r(t)$ are the so-called survival and replacement functions for this species.

The remaining exercises refer to this biomathematical example.

22. Discuss the population "growth" in the case $s(t) = e^{-t}$ and $r(t) = k =$ constant. Can such a population ever remain static (unchanging in time)?

23. If $s(t)$ and $P(t)$ are given, the above representation becomes an integral equation for $r(t)$. Determine $r(t)$ in the case when $s(t) = (1 - t)e^{-t}$ and $P(t) = P_0 =$ constant.

24. Carry out an analysis similar to that of Exercise 23 in the case when $s(t) = (1 - 2t + t^2)e^{-t}$ and $P(t) = P_0(1 - te^{-t})$.

5.10 Looking Ahead

In this chapter we have handled problems involving derivatives, differences, differential equations, systems of equations, integrals, and convolutions with one efficient tool—the Laplace transform. There exist other integral transforms, however, which also have wide applicability—Fourier transforms, Mellin transforms, Hankel transforms, Kantorovich-Lebedev transforms, and so on. Fourier transforms and integrals are discussed in Chapters 7 and 16. (See also the classic reference [12].) Good sources for information on integral transforms in general are [5], [11], and [13]. Extensive tables of both Laplace transforms and inverse Laplace transforms appear in [6] and [10]. Laplace transforms are used in the solution of some partial differential equations in Chapter 8. Also see [4] for additional discussions of the application of the Laplace transform to Volterra integral equations with difference kernels.

For understandable reasons, then, the Laplace transform plays a central role in analysis of linear systems. Although in most problems associated with realizable systems, the Laplace transforms encountered have the property of becoming arbitrarily small as $s \to \infty$, it is occasionally of interest to find the inverse transforms of elementary functions such as 1, $\sqrt{s}$, $2s - 5$, and the like, which do not share this feature. Alternatively, we may need to determine the Laplace transform of "functions" such as the instantaneous input of Problem 3 of Section 5.1. Such quests take us beyond the notion of ordinary functions to generalized functions and distributions, an important and exciting new area of mathematics. References [7], [8], [9], and [14] should be consulted by students interested in learning more about these remarkably useful concepts.

Basic References

[1] Boyce, W., and R. C. DiPrima. 1977. *Elementary Differential Equations and Boundary Value Problems.* 3d ed. New York: John Wiley.

[2] Churchill, R. V. 1958. *Operational Mathematics.* 2d. ed. New York: McGraw-Hill.

[3] Kaplan, W. 1962. *Operational Methods for Linear Systems.* Reading, Mass.: Addison-Wesley.

More-Advanced References

[4] Cochran, J. A. 1972. *The Analysis of Linear Integral Equations.* New York: McGraw-Hill.

[5] Davies, B. 1985. *Integral Transforms and Their Applications.* 2d ed. New York: Springer-Verlag.

[6] Erdélyi, A., et al. 1954. *Tables of Integral Transforms. Vol. 1.* New York: McGraw-Hill.

[7] Gel'fand, I. M., and G. E. Shilov. 1964. *Generalized Functions. Vol. I.,* Transl. of Russian edition. New York: Academic Press.

[8] Kaplan, W. 1962. *Operational Methods for Linear Systems.* Reading, Mass: Addison-Wesley.

[9] Lighthill, M. J. 1959. *Introduction to Fourier Analysis and Generalized Functions.* Cambridge: Cambridge University Press.

[10] Oberhettinger, F., and L. Badii. 1973. *Tables of Laplace Transforms.* New York: Springer-Verlag.

[11] Sneddon, I. N. 1972. *The Use of Integral Transforms.* New York: McGraw-Hill.

[12] Titchmarsh, E. C. 1959. *Introduction to the Theory of Fourier Integrals.* 2d ed. Oxford: Clarendon Press.

[13] Tranter, C. J. 1966. *Integral Transforms in Mathematical Physics.* 3d ed. London: Methuen.

[14] Zemanian, A. H. 1965. *Distribution Theory and Transform Analysis.* New York: McGraw-Hill.

Chapter 6

Infinite Series Methods

6.1 Introduction

Our earlier considerations in finding solutions to ordinary differential equations were predicated upon the assumption that the equations under investigation possessed unique solutions which could be expressed as finite linear combinations of elementary functions such as polynomials, exponentials, trigonometric functions, and the like, with which we were already familiar. However, the real world is not always so kind, and in diverse interesting applications we encounter equations that appear (deceptively) simple and yet possess rather complicated and distinctly nonelementary solutions, which, moreover, cannot be obtained by the techniques of Chapters 2, 3, and 4. The equations

$$y'' - xy = 0$$

and

$$xy'' + y' + xy = 0$$

which we meet again later in this chapter, are two cases in point. For such situations new solution approaches are necessary.

In this chapter we consider one such technique, which is based upon assuming the form of the solution of the given differential equation to be an infinite series. Generally, this will be a power series (that is, a Taylor series), but there are important exceptions. The method of infinite series is in many ways the most basic procedure for finding solutions to differential equations, since such a wide set of equations may be solved this way (including linear equations with nonconstant coefficients). You may find the method a bit unsatisfying, however, since ordinarily you will not be able to write the solution in terms of elementary functions. In fact, you may be able to write only the first few terms of the series. If these terms do not suggest a solution involving elementary functions, the first few terms may well serve as a reasonable approximation to the solution. Indeed, in many cases a series generated by a differential equation may be used to *define* and tabulate a new function.

In Sections 6.2 and 6.4 we discuss two separate procedures based on the fundamental assumption that the solution has the form of a Taylor series. The first of these methods emphasizes that the form of the nth coefficient in a Taylor series is

$$\frac{f^{(n)}(a)}{n!}$$

and hence the coefficients may be found by determining $f^{(n)}(a)$ by successive differentiation of the differential equation and application of any initial conditions. Thus, the solution function is essentially known if the values of the derivatives are known at $x = a$. In the second method the coefficients are found directly using the fact that the assumed power series must satisfy the differential equation.

This is a good time to review the basic notions of power series learned in calculus. In what follows we will freely manipulate convergent power series by combining them, differentiating them term by term, expressing them in closed form when possible, and so on. (Complex power series are discussed in Chapter 15.) The key properties of power series that these manipulations depend on, together with some useful specific series, include the following:

1. A power series $\sum_{n=0}^{\infty} a_n(x-a)^n$ either converges only for $x = a$, converges absolutely for $|x - a| < R$ and diverges for $|x - a| > R$ for some positive R, or converges absolutely for each x. We say that the series has a radius of convergence of 0, R, or infinity, respectively, in these three cases.

2. The radius of convergence can be determined by use of the ratio test. If

$$\lim_{n\to\infty}\left|\frac{a_{n+1}(x-a)^{n+1}}{a_n(x-a)^n}\right| = \lim_{n\to\infty}\left|\frac{a_n+1}{a_n}\right||x-a| = \frac{1}{R}|x-a| < 1$$

then the series converges for $|x - a| < R$. If $\lim\left|\frac{a_n+1}{a_n}\right| = 0$, then the series converges for all x.

3. If $f(x) = \sum_{n=0}^{\infty} a_n(x-a)^n$, $g(x) = \sum_{n=0}^{\infty} b_n(x-a)^n$, and $f(x) \equiv g(x)$ for all x such that $|x - a| < R$, then $a_n = b_n$ for each n.

4. Within a positive (possibly infinite) radius of convergence, a power series may be differentiated, integrated, or multiplied by a constant, term by term, that is, if $f(x) = \sum_{n=0}^{\infty} a_n(x-a)^n$ for $|x - a| < R$, then $f'(x)$ exists and $f'(x) = \sum_{n=0}^{\infty} a_n n(x-a)^{n-1}$, $cf(x) = \sum_{n=0}^{\infty} ca_n(x-a)^n$, and $\int_a^x f(u)\,du = \sum_{n=0}^{\infty} a_n \frac{(x-a)^{n+1}}{n+1}$ for $|x - a| < R$.

5. If both $f(x) = \sum_{n=0}^{\infty} a_n(x-a)^n$ and $g(x) = \sum_{n=0}^{\infty} b_n(x-a)^n$ for $|x - a| < R$, then $f(x) + g(x) = \sum_{n=0}^{\infty} (a_n + b_n)(x-a)^n$ and $f(x)g(x) = a_0b_0 + (a_0b_1 + b_0a_1)(x-a) + (b_0a_2 + a_0b_2 + a_1b_1)(x-a)^2 + (a_0b_3 + b_0a_3 + a_1b_2 + b_1a_2)(x-a)^3 + \cdots$ for $|x - a| < R$. The latter formula essentially says you can multiply convergent power series just like "big polynomials." Convergent series may also be divided by the usual polynomial long-division process, but in this case the radius of convergence may be less than that for either of the original series.

6. If $f(x) = \sum_{n=0}^{\infty} a_n(x-a)^n$ for $|x-a| < R$ (R positive or infinite), then $a_n = \dfrac{f^{(n)}(a)}{n!}$.

6.2 Taylor Series

When applying the method of Taylor series, we assume the solution of our given differential equation is of the form

$$y(x) = \sum_{n=0}^{\infty} \frac{y^{(n)}(a)}{n!}(x-a)^n$$

Hence, to write the solution, we need to know the values of the unknown function and its derivatives at $x = a$. Since we will probably use the solution for values of x near to $x = a$, a few terms will usually be sufficient. The first few terms yield an *approximating polynomial* to the infinite series solution. In Section 6.5 we address the important question of when an assumption of this type is valid. For now, if the assumption turns out to be invalid, it will become obvious during the calculation procedure itself. (See Exercise 21 in the next exercise set.)

The values of the derivatives $y^{(n)}(a)$ are found directly from the initial conditions and by successive differentiation of the differential equation. The determination of the various derivatives is time consuming and tedious and, understandably, is the major limitation to this method. On the positive side, the method is general enough to apply to many differential equations, both linear and nonlinear.

Example 6.1 Find the first five nonzero terms of a Taylor series representation of the solution to the initial value problem

$$y'' = y^2 + x, \qquad y(1) = 2, \qquad y'(1) = -1$$

Solution The unknown function and its derivative are given at $x = 1$. Consequently, we want to write a Taylor series in powers of $(x - 1)$. The value of $y''(1)$ is determined from the differential equation to be

$$y''(1) = [y(1)]^2 + 1 = 4 + 1 = 5$$

Differentiating both sides of the differential equation leads to the following expressions for $y'''(x)$ and $y^{(4)}(x)$:

$$y'''(x) = 2y(x)y'(x) + 1$$
$$y^{(4)}(x) = 2[y'(x)]^2 + 2y(x)y''(x)$$

from which

$$y'''(1) = 2(2)(-1) + 1 = -3$$
$$y^{(4)}(1) = 2(-1)^2 + 2(2)(5) = 22$$

The first few terms of the desired Taylor series are, therefore,

$$y(x) = 2 - (x-1) + \tfrac{5}{2}(x-1)^2 - \tfrac{1}{2}(x-1)^3 + \tfrac{11}{12}(x-1)^4 + \cdots \qquad \square$$

It is useful to recognize some of the most often encountered Taylor series, such as:

1. $e^x = 1 + x + \frac{x^2}{2!} + \frac{x^3}{3!} + \cdots = \sum_{n=0}^{\infty} \frac{x^n}{n!}$

2. $\sin x = x - \frac{x^3}{3!} + \frac{x^5}{5!} - \frac{x^7}{7!} + \cdots = \sum_{n=0}^{\infty} \frac{(-1)^n x^{2n+1}}{(2n+1)!}$

3. $\cos x = 1 - \frac{x^2}{2!} + \frac{x^4}{4!} - \frac{x^6}{6!} + \cdots = \sum_{n=0}^{\infty} \frac{(-1)^n x^{2n}}{(2n)!}$

4. $\ln(1 + x) = x - \frac{x^2}{2} + \frac{x^3}{3} - \frac{x^4}{4} + \cdots = \sum_{n=0}^{\infty} (-1)^n \frac{x^{n+1}}{n+1}$

The first three Taylor series are absolutely convergent for all x. (Use the ratio test.) The last series converges absolutely only for $|x| < 1$.

Example 6.2 Find the Taylor series expansion about $x = 0$ of the solution to the differential equation $y'' + y = 0$. Initial conditions on the solution and its derivative are to be left arbitrary.

Solution The solution is assumed to be of the form

$$y(x) = \sum_{n=0}^{\infty} \frac{y^{(n)}(0)}{n!} x^n$$

To determine $y^{(n)}(0)$, we let $y(0)$ and $y'(0)$ be the constants a_0 and a_1, and then the differential equation yields

$$y''(0) = -y(0) = -a_0, \qquad y'''(0) = -y'(0) = -a_1, \qquad y^{(4)}(0) = -y''(0) = a_0,$$
$$y^{(5)}(0) = a_1, \qquad y^{(6)}(0) = -a_0, \qquad \cdots y^{(n)} = -y^{(n-2)}$$

By a simple inductive proof we can show

$$y^{(2n)}(0) = (-1)^n a_0 \quad \text{and} \quad y^{(2n+1)}(0) = (-1)^n a_1$$

and hence, the solution can be written in the form

$$y = a_0 \sum_{n=0}^{\infty} \frac{(-1)^n}{(2n)!} x^{2n} + a_1 \sum_{n=0}^{\infty} \frac{(-1)^n}{(2n+1)!} x^{2n+1}$$

Recognizing the Taylor series for $\cos x$ and $\sin x$, we have

$$y = a_0 \cos x + a_1 \sin x$$ □

Clearly we could have solved the equation of the above example more easily by the standard techniques of previous chapters. This will be true of many examples and exercises that follow. Keep in mind, however, that we are using simple examples to help us learn new procedures, which can then be applied to problems *not* conveniently handled by other methods.

Example 6.3 Find the first three terms of the Taylor expansion about $x = 2$ of an assumed series solution to the differential equation

$$y'' + y' - \ln y = e^x$$

Solution Since $y'' = e^x + \ln y - y'$, we have that $y(2) = a_0$, $y'(2) = a_1$, $y''(2) = e^2 + \ln a_0 - a_1$. Hence, the first three terms are

$$y(x) = a_0 + a_1(x-2) + \frac{e^2 + \ln a_0 - a_1}{2}(x-2)^2 + \cdots \qquad \square$$

This rough approximation may not even converge to a solution, but at least it was easy to obtain!

Example 6.4 Solve the differential equation $y' - y + x^2 + x + 1 = 0$ by assuming a Taylor series solution valid near $x = 0$. Assume $y(0) = 1$.

Solution Let

$$y = \sum_{n=0}^{\infty} \frac{y^{(n)}(0)}{n!} x^n$$

The given differential equation gives

$$y' = y - x^2 - x - 1, \qquad y'' = y' - 2x - 1,$$
$$y''' = y'' - 2, \qquad y^{(4)} = y'''$$

and for $n \geqq 4$,

$$y^{(n)} = y^{(n-1)}$$

Therefore, $y(0) = 1$, $y'(0) = 0$, $y''(0) = -1$, $y'''(0) = -3$, and for $n \geqq 3$, $y^{(n)}(0) = -3$. Consequently, the solution is

$$\begin{aligned} y &= 1 - \frac{x^2}{2} - 3\sum_{n=3}^{\infty} \frac{x^n}{n!} \\ &= 1 - \frac{x^2}{2} - 3\sum_{n=0}^{\infty} \frac{x^n}{n!} + 3\sum_{n=0}^{2} \frac{x^n}{n!} \\ &= 1 - \frac{x^2}{2} - 3\sum_{n=0}^{\infty} \frac{x^n}{n!} + 3 + 3x + \frac{3x^2}{2} \end{aligned}$$

The infinite series can be recognized as the Taylor series for e^x about $x = 0$. Hence

$$y = 4 + 3x + x^2 - 3e^x \qquad \square$$

6.3 Comments on the Form of a Solution

When assuming an infinite series form for the solution, we must choose an appropriate way of stating the answer. There are basically four levels of achievement, each a satisfactory method of stating the solution under certain circumstances.

1. A few of the nonzero terms of the series can be obtained. This is the minimum in the way of an answer, since this requires at most the computation of a few derivatives. (See Examples 6.1 and 6.3.)

2. A recursion (or inductive) formula can be obtained that expresses the nth coefficient or nth derivative in terms of previous derivatives and coefficients. Not only does this yield more information about the coefficients than in level 1, but the recursion formula can often be used to determine the interval of convergence.

3. The recursion formula may be used to find a general formula for the coefficient of x^n or for $f^{(n)}(a)$. An explicit infinite series form of the solution function may then be given. (See Examples 6.2 and 6.4.)

4. Finally, the infinite series form may be expressible in terms of elementary functions (see Examples 6.2 and 6.4) or of known, but perhaps nonelementary functions. The ability to express an answer in terms of elementary functions rather than in terms of an infinite power series, however, is sometimes of little consequence. In these cases, representation of the solution as an infinite series may be precisely what is needed for performing further evaluations and computations.

Example 6.5 Write the power series centered at 0 whose coefficients are related by the inductive formula

$$a_0 = 1, \qquad a_1 = 0, \qquad a_n = -\frac{2}{n}a_{n-2} \qquad \text{for } n \geqq 2$$

Solution Such a power series has the form $\sum_{n=0}^{\infty} a_n x^n$. Because the recursion formula for a_n is in terms of the coefficient of the term two before the given term, we can see that if n is even, a_n can be expressed in terms of a_0, and if n is odd, a_n can be related to a_1. Such a series is often written in rearranged form as

$$\sum_{n=0}^{\infty} a_n x^n = \sum_{k=0}^{\infty} a_{2k} x^{2k} + \sum_{k=0}^{\infty} a_{2k+1} x^{2k+1}$$

The first series on the right is a series of even terms and the second is a series of odd terms.

Since a_1 vanishes, so do all the odd coefficients and the second series vanishes.

The recursion formula for the even coefficients is

$$a_{2k} = -\frac{1}{k}a_{2k-2}$$

Repeatedly applying this formula to $a_{2k-2}, a_{2k-4}, \ldots, a_8, a_6, a_4$, and a_2,

$$\begin{aligned} a_{2k} &= \left(-\frac{1}{k}\right)\left(\frac{-1}{k-1}\right)\left(\frac{-1}{k-2}\right)\cdots\left(-\frac{1}{4}\right)\left(-\frac{1}{3}\right)\left(-\frac{1}{2}\right)\left(-\frac{1}{1}\right)a_0 \\ &= \frac{(-1)^k}{k!}a_0 \end{aligned}$$

Hence, the power series is

$$\sum_{k=0}^{\infty} \frac{(-1)^k}{k!} x^{2k} \qquad \square$$

The preceding example shows how the problem of using the recursion formula to find the general coefficient can sometimes be approached. In many cases, however, the recursion formula is much too complicated to proceed in this manner.

Exercises for Sections 6.1–6.3

Use the method of Taylor series to solve the following initial value problems. Obtain a recursion formula for the derivatives if you can, but if you can not, write at least five nonzero terms of the series. If possible, obtain the general coefficient and try to write the solution in terms of elementary functions.

1. $y' = x + y, \quad y(0) = 1$

2. $y' = y, \quad y(0) = 1$

3. $y'' = y, \quad y(0) = 1, \quad y'(0) = 1$

4. $y' + 2y = x^2, \quad y(0) = 1$

5. $y'' + 4y' + 3y = 0, \quad y(0) = 1, \quad y'(0) = 1$

6. $y' = y^2, \quad y(0) = 2$

7. $y'' = xy, \quad y(1) = 2, \quad y'(1) = -2$

8. $y'' = y^3, \quad y(1) = 1, \quad y'(1) = -1$

9. $y'' = \ln y, \quad y(1) = 1, \quad y'(1) = 1$

10. $y' = e^y, \quad y(0) = 1$

11. $y'' = e^{xy}, \quad y(0) = 1, \quad y'(0) = 0$

In each of the following exercises, find the first three nonzero terms of a Taylor series about $x = a$ by assuming arbitrary initial conditions $y(a) = a_0$, $y'(a) = a_1$.

12. $y'' - xy' + y = 0, \quad a = 0$

13. $y'' + y = 0, \quad a = 0$

14. $y'' - xy' - y = 0, \quad a = 2$

15. $y'' + 3y' + 2(\ln x)y = 0, \quad a = 1$

In each of the following exercises, you are given a recursion formula for the power series solution, $\sum_{n=0}^{\infty} a_n x^n$, to a differential equation. Find the general term and write the solution.

16. $a_n = -\dfrac{1}{n-2} a_{n-1}, \quad n \geqq 4, \quad a_0 = 1, \quad a_1 = 0, \quad a_2 = 1, \quad a_3 = -1$

17. $a_n = \dfrac{n+2}{n} a_{n-2}, \quad n \geqq 2, \quad a_0 = a_1 = 1$

18. $a_n = \dfrac{(n-5)(n-6)}{n^2} a_{n-2}, \quad n \geqq 2, \quad a_0 = a_1 = 1$

19. $a_n = \dfrac{n-5}{n(n-1)} a_{n-2}, \quad n \geqq 2, \quad a_0 = a_1 = 1$

20. $a_n = \dfrac{3}{4n} a_{n-2}, \quad n \geqq 2, \quad a_0 = a_1 = 1$

21. Show that the method of Taylor series cannot be used to solve the differential equation

$$(x^2 - 3x + 2)y'' + 3y' - y = 0 \qquad y(1) = y'(1) = 1$$

(*Hint:* Try it!)

The improper integral

$$y(x) = \int_0^\infty \cos(xt + \tfrac{1}{3}t^3)\,dt$$

arises in a variety of wave diffraction and dispersion problems. The function $y(x)$ can be shown formally to satisfy the differential equation

$$y''(x) = xy$$

which is known as Airy's equation.

22. Show that Airy's equation has two linearly independent solutions that can be expressed in the forms

$$y_1(x) = \sum_{n=0}^{\infty} a_n x^{3n} \quad \text{and} \quad y_2(x) = \sum_{n=0}^{\infty} b_n x^{1+3n}$$

23. Find the first five nonzero terms of the Taylor series representation for the solution of Airy's equation that satisfy the initial conditions $y(1) = 0$ and $y'(1) = 1$.

The Schrödinger equation corresponding to a harmonic oscillator of point mass m and vibrational frequency v is

$$\psi''(s) + \frac{8\pi^2 m}{h^2}(E - 2\pi^2 m v^2 s^2)\psi(s) = 0$$

where E is the total energy of the oscillator and h is Planck's constant. By an appropriate change of both the dependent and independent variable, this equation can be transformed into

$$y''(x) - 2xy'(x) + 2py(x) = 0$$

where $p = (E/hv) - \frac{1}{2}$, which is known as Hermite's differential equation.

24. Show that the substitutions

$$\psi(s) = e^{-x^2/2}y(x)$$

$$s = \sqrt{\frac{h}{mv}}\frac{x}{2\pi}$$

transform Schrödinger's harmonic oscillator equation into the Hermite differential equation.

25. Demonstrate that Hermite's equation has two linearly independent solutions, one consisting solely of even powers of x, the other only of odd powers.

6.4 Method of Undetermined Coefficients

The fundamental idea of assuming a power series solution can be easily adapted to another approach to obtaining the coefficients. This technique is particularly appropriate for linear differential equations. (We call this procedure the method of *undetermined coefficients.* It should not be confused with a related procedure discussed in Chapter 3.)

We assume that the power series can be differentiated term by term as often as necessary and that the series of differentiated terms converges to the derivative of the function. Implicit is the even more fundamental assumption that two power series are equal *if and only if* they have identical coefficients.

The coefficients are found by considering algebraic conditions on the coefficients obtained by substituting the series into the differential equation. The power series expansion of the coefficient functions and the driving function must also be known. Linear differential equations in which the coefficient functions are polynomials yield the prime opportunity for application of this technique. More importantly, such equations occur with surprising regularity in applications.

Example 6.6 Solve the differential equation $y' + xy = x^2 - 2x$ by assuming that the solution may be written in the form of a power series. Use the method of undetermined coefficients.

Solution Since $y = a_0 + a_1x + a_2x^2 + a_3x^3 + a_4x^4 \cdots$, then $y' = a_1 + 2a_2x + 3a_3x^2 + \cdots$, and upon substituting into the differential equation, we obtain

$$a_1 + 2a_2x + 3a_3x^2 + 4a_4x^3 + \cdots + a_0x + a_1x^2 + a_2x^3 + a_3x^4 + \cdots = x^2 - 2x$$

The coefficient a_0 is obtained from a knowledge of $y(0)$ and is left as arbitrary. We now proceed by equating coefficients of like powers of x on both sides of this equation:

$$\begin{array}{lll}
\text{Coefficients of } x^0: & a_1 & = 0 \\
\text{Coefficients of } x^1: & 2a_2 + a_0 & = -2 \\
\text{Coefficients of } x^2: & 3a_3 + a_1 & = 1 \\
\text{Coefficients of } x^3: & 4a_4 + a_2 & = 0 \\
\text{Coefficients of } x^4: & 5a_5 + a_3 & = 0 \\
\vdots & & \vdots \\
\text{Coefficients of } x^{n-1} \text{ for } n \geqq 4: & na_n + a_{n-2} & = 0,
\end{array}$$

From this the necessary coefficient information is

$$a_0 \text{ arbitrary}, \qquad a_1 = 0, \qquad a_2 = -\frac{a_0 + 2}{2},$$

$$a_3 = \tfrac{1}{3}, \qquad a_n = -\frac{1}{n}a_{n-2} \quad \text{for } n \geq 4$$

This recursion formula may also be obtained with the formal use of summation notation. Let $y = \sum_{n=0}^{\infty} a_nx^n$, then $y' = \sum_{n=1}^{\infty} na_nx^{n-1}$. Upon substituting into the differential equation, we obtain

$$\sum_{n=1}^{\infty} na_nx^{n-1} + \sum_{n=0}^{\infty} a_nx^{n+1} = x^2 - 2x$$

In the second series, shift the index back by 2,

$$\sum_{n=1}^{\infty} na_nx^{n-1} + \sum_{n=2}^{\infty} a_{n-2}x^{n-1} = x^2 - 2x$$

Writing the first term of the first series separately and then combining into one series, we obtain

$$1a_1x^0 + \sum_{n=2}^{\infty}(na_n + a_{n-2})x^{n-1} = x^2 - 2x$$

From this point, the procedure is the same as before.

The recursion formula is simple enough that we would expect to be able to obtain the general term. Noting the nature of the recursion formula, we can write the solution in the rearranged form

$$\sum_{k=0}^{\infty} a_{2k}x^{2k} + \sum_{k=0}^{\infty} a_{2k+1}x^{2k+1}$$

The coefficient information for the first series is

$$a_0 \text{ arbitrary,} \qquad a_2 = -\frac{a_0 + 2}{2}, \qquad a_{2k} = -\frac{a_{2k-2}}{2k} \qquad \text{for } k \geqq 2$$

and for the second series is

$$a_1 = 0, \qquad a_3 = \tfrac{1}{3}, \quad \text{and} \quad a_{2k+1} = -\frac{a_{2k-1}}{2k+1} \qquad \text{for } k \geqq 2$$

For the first series

$$\begin{aligned} a_{2k} &= \left(-\frac{1}{2k}\right)\left(-\frac{1}{2k-2}\right)\left(-\frac{1}{2k-4}\right)\cdots\left(-\frac{1}{6}\right)\left(-\frac{1}{4}\right)a_2 \\ &= \frac{(-1)^{k-1}}{2^{k-1}(k!)}\left(-\frac{a_0+2}{2}\right) \qquad \text{for } k \geqq 2 \end{aligned}$$

For the second series

$$\begin{aligned} a_{2k+1} &= \left(-\frac{1}{2k+1}\right)\left(-\frac{1}{2k-1}\right)\left(-\frac{1}{2k-3}\right)\cdots\left(-\frac{1}{7}\right)\left(-\frac{1}{5}\right)a_3 \\ &= \frac{(-1)^{k-1}\{(2k)(2k-2)(2k-4)\ldots(6)(4)(2)\}a_3}{(2k+1)!/3} \\ &= \frac{(-1)^{k-1}2^k k!3}{(2k+1)!}\frac{1}{3} = \frac{(-1)^{k-1}2^k k!}{(2k+1)!} \qquad \text{for } k \geq 2 \end{aligned}$$

Hence, the solution is

$$\begin{aligned} y = a_0 &- \frac{a_0+2}{2}x^2 + \frac{a_0+2}{2}\sum_{k=2}^{\infty}\frac{(-1)^k}{2^{k-1}k!}x^{2k} \\ &+ \frac{1}{3}x^3 + \sum_{k=2}^{\infty}\frac{(-1)^{k-1}2^k k!}{(2k+1)!}x^{2k+1} \end{aligned}$$

□

Example 6.7 Solve the differential equation $y'' + y = e^x$ by assuming a power series solution about the origin.

Solution Letting $y = \sum_{n=0}^{\infty} a_n x^n$, then $y' = \sum_{n=1}^{\infty} a_n n x^{n-1}$ and $y'' = \sum_{n=2}^{\infty} a_n n(n-1)x^{n-2}$. Since the infinite series expansion for e^x is $e^x = \sum_{n=0}^{\infty} \frac{x^n}{n!}$, substituting into the differential equation yields

$$\sum_{n=2}^{\infty} n(n-1)a_n x^{n-2} + \sum_{n=0}^{\infty} a_n x^n = \sum_{n=0}^{\infty} \frac{x^n}{n!}$$

Simplifying slightly,

$$\sum_{n=2}^{\infty} n(n-1)a_n x^{n-2} + \sum_{n=0}^{\infty} \left(a_n - \frac{1}{n!}\right)x^n = 0$$

We now shift the index in the second series to make the exponent the same as in the first series,

$$\sum_{n=2}^{\infty} n(n-1)a_n x^{n-2} + \sum_{n=2}^{\infty} \left(a_{n-2} - \frac{1}{(n-2)!}\right)x^{n-2} = 0$$

and then combine the two series to obtain

$$\sum_{n=2}^{\infty} \left[n(n-1)a_n + a_{n-2} - \frac{1}{(n-2)!}\right]x^{n-2} = 0$$

The right-hand side is identically zero: the coefficient of every power of x on the left-hand side therefore must be zero. As a consequence:

$$\text{for } n \geqq 2$$

$$n(n-1)a_n + a_{n-2} - \frac{1}{(n-2)!} = 0$$

or

$$a_n = \frac{1}{n!} - \frac{1}{n(n-1)}a_{n-2}$$

where the coefficients a_0 and a_1 are arbitrary unless determined by initial conditions.

If we let $a_n = a_n^* + \frac{1}{2(n!)}$ the recursion formula becomes $a_n^* = -\frac{1}{n(n-1)}a_{n-2}^*$. Then the solution may be written in the form

$$y = \sum_{k=0}^{\infty} a_{2k}^* x^{2k} + \sum_{k=0}^{\infty} a_{2k+1}^* x^{2k+1} + \frac{1}{2}\sum_{k=0}^{\infty} \frac{x^k}{k!}$$

where by repeated application of the recurrence relation,

$$a_{2k}^* = \frac{(-1)^k}{(2k)!}a_0^*, \qquad a_{2k+1}^* = \frac{(-1)^k}{(2k+1)!}a_1^*$$

In this form, the solution may be recognized in terms of elementary functions,

$$y = a_0^* \cos x + a_1^* \sin x + \frac{e^x}{2}$$

□

Example 6.8 Solve the differential equation

$$(1 - x^2)y'' - 2xy' + 12y = 0$$

by assuming a solution in power series form valid near the origin. This differential equation is called Legendre's equation of order three. (Legendre's equation of order n has the form

$$(1 - x^2)y'' - 2xy' + n(n + 1)y = 0$$

and is studied extensively in Chapter 9. There is always one polynomial solution of this equation.)

Solution Assuming $y = \sum_{n=0}^{\infty} a_n x^n$, then $y' = \sum_{n=1}^{\infty} a_n n x^{n-1}$, $y'' = \sum_{n=2}^{\infty} a_n n(n-1)x^{n-2}$. Substituting into the given differential equation:

$$(1 - x^2)\sum_{n=2}^{\infty} a_n n(n-1)x^{n-2} - 2x\sum_{n=1}^{\infty} a_n n x^{n-1} + 12\sum_{n=0}^{\infty} a_n x^n = 0$$

Carrying out the indicated multiplications,

$$\sum_{n=2}^{\infty} a_n n(n-1)x^{n-2} - \sum_{n=0}^{\infty} a_n n(n-1)x^n - 2\sum_{n=0}^{\infty} a_n n x^n + 12\sum_{n=0}^{\infty} a_n x^n = 0$$

Combining the last three series,

$$\sum_{n=2}^{\infty} n(n-1)a_n x^{n-2} - \sum_{n=0}^{\infty} a_n\{n^2 - n + 2n - 12\}x^n = 0$$

or

$$\sum_{n=2}^{\infty} n(n-1)a_n x^{n-2} - \sum_{n=0}^{\infty} a_n(n+4)(n-3)x^n = 0$$

Shifting index in the second of these summations,

$$\sum_{n=2}^{\infty} n(n-1)a_n x^{n-2} - \sum_{n=2}^{\infty} a_{n-2}(n+2)(n-5)x^{n-2} = 0$$

This equation implies that a_0 and a_1 are both arbitrary and that for $n \geqq 2$,

$$a_n = \frac{(n+2)(n-5)}{n(n-1)} a_{n-2}$$

It follows that $a_3 = -5a_1/3$, and $a_{2k+1} = 0$ for $k \geqq 2$, and thus one of the solutions to the differential equation is the polynomial of degree three

$$y_1 = a_1(x - \tfrac{5}{3}x^3)$$

The so-called Legendre polynomials are polynomial solutions of Legendre's equation with the multiplicative constant chosen so that their graphs pass through the point (1, 1). In our case we choose $a_1 = -\frac{3}{2}$, so that $y_1 = 5x^3/2 - 3x/2$.

A second solution to the equation is of the form $\sum_{k=0}^{\infty} a_{2k}x^{2k}$, where a_0 is arbitrary and $a_{2k} = \dfrac{(2k+2)(2k-5)}{2k(2k-1)} a_{2k-2}$. Then,

$$a_{2k} = \frac{(2k+2)(2k-5)}{2k(2k-1)} \cdot \frac{(2k)(2k-7)}{(2k-2)(2k-3)} \cdot \frac{(2k-2)(2k-9)}{(2k-4)(2k-5)} \cdots \frac{10\cdot 3}{8\cdot 7} \cdot \frac{8\cdot 1}{6\cdot 5} \cdot \frac{6(-1)}{4\cdot 3} \cdot \frac{4(-3)}{2\cdot 1} a_0$$

Cancelling as many terms as possible in the numerator and denominator,

$$a_{2k} = \frac{(2k+2)(-1)(-3)}{(2k-1)(2k-3)(2)} a_0 = \frac{3(k+1)}{(2k-1)(2k-3)} a_0$$

Hence, a second solution to Legendre's differential equation of order three is

$$y_2 = 3a_0 \sum_{k=0}^{\infty} \frac{k+1}{(2k-1)(2k-3)} x^{2k}$$

It can easily be shown by using the ratio test that this series converges for $|x| < 1$. □

The method of undetermined coefficients may also be used to find power series solutions centered at points other than $x = 0$. In the following example we assume a power series solution about $x = 1$. The example also exhibits how "multi-term" recursion formulas arise. Unfortunately, this kind of recursion formula makes it almost impossible to find a general expression for the nth coefficient.

Example 6.9 Find a solution in power series form, centered at $x = 1$, to the differential equation $xy'' + y' + xy = 0$.

Solution The given differential equation may be written in the form

$$(x-1)y'' + y'' + y' + (x-1)y + y = 0$$

in which all of the coefficients are expressed in terms of powers of $x - 1$. Assume the solution is of the form $y = \sum_{n=0}^{\infty} a_n(x-1)^n$ from which

$$y' = \sum_{n=1}^{\infty} a_n n(x-1)^{n-1} \quad \text{and} \quad y'' = \sum_{n=2}^{\infty} a_n n(n-1)(x-1)^{n-2}$$

Substituting into the differential equation:

$$(x-1)\sum_{n=2}^{\infty} a_n n(n-1)(x-1)^{n-2} + \sum_{n=2}^{\infty} a_n n(n-1)(x-1)^{n-2} + \sum_{n=1}^{\infty} a_n n(x-1)^{n-1} + (x-1)\sum_{n=0}^{\infty} a_n(x-1)^n + \sum_{n=0}^{\infty} a_n(x-1)^n = 0$$

Simplifying,

$$\sum_{n=1}^{\infty} a_n n(n-1)(x-1)^{n-1} + \sum_{n=2}^{\infty} a_n n(n-1)(x-1)^{n-2} + \sum_{n=1}^{\infty} a_n n(x-1)^{n-1} + \sum_{n=0}^{\infty} a_n(x-1)^{n+1} + \sum_{n=0}^{\infty} a_n(x-1)^n = 0$$

Combining terms with like exponents,

$$\sum_{n=1}^{\infty} a_n n^2(x-1)^{n-1} + \sum_{n=2}^{\infty} a_n n(n-1)(x-1)^{n-2} + \sum_{n=0}^{\infty} a_n(x-1)^{n+1} + \sum_{n=0}^{\infty} a_n(x-1)^n = 0$$

Shifting indexes in the various series so that the exponents are all $n-2$,

$$\sum_{n=2}^{\infty} a_{n-1}(n-1)^2(x-1)^{n-2} + \sum_{n=2}^{\infty} a_n n(n-1)(x-1)^{n-2} + \sum_{n=3}^{\infty} a_{n-3}(x-1)^{n-2} + \sum_{n=2}^{\infty} a_{n-2}(x-1)^{n-2} = 0$$

We write the first term separately and combine the other series,

$$(2a_2 + a_1 + a_0)x^0 + \sum_{n=3}^{\infty} \{a_n n(n-1) + (n-1)^2 a_{n-1} + a_{n-2} + a_{n-3}\}(x-1)^{n-2} = 0$$

From this equation, we obtain the following coefficient information: a_0 and a_1 are arbitrary;

$$a_2 = \frac{-a_0 - a_1}{2}$$

and for $n \geqq 3$,

$$a_n = -\frac{(n-1)^2 a_{n-1} + a_{n-2} + a_{n-3}}{n(n-1)}$$

Choosing $a_0 = a_1 = 1$, the next few coefficients are $a_2 = -1$, $a_3 = \frac{1}{3}$, $a_4 = -\frac{1}{4}$. The first few terms of a solution thus are

$$y = 1 + (x-1) - (x-1)^2 + \frac{(x-1)^3}{3} - \frac{(x-1)^4}{4} + \cdots$$ □

The differential equation of the previous example is called *Bessel's equation of order zero*. The technique of solution used in the example is *not* the standard approach to this important equation. A solution valid near the origin is much more popular but requires a slight variation on the procedure of assuming a power series solution. In Chapter 9 we discuss Bessel's equation and Bessel functions extensively.

In certain simple cases, a power series assumption leads directly to a solution without the need for a recursion formula.

Example 6.10 By use of power series find the solution to the differential equation $xy' - y = x^2 e^x$.

Solution Assuming $y = \sum_{n=0}^{\infty} a_n x^n$, then $y' = \sum_{n=1}^{\infty} a_n n x^{n-1}$. The power series for e^x is $\sum_{n=0}^{\infty} (x^n/n!)$. Substituting into the differential equation,

$$\sum_{n=1}^{\infty} a_n n x^n - \sum_{n=0}^{\infty} a_n x^n = x^2 \sum_{n=0}^{\infty} \frac{x^n}{n!}$$

Simplifying,

$$\sum_{n=0}^{\infty} a_n x^n (n-1) = \sum_{n=0}^{\infty} \frac{x^{n+2}}{n!} = \sum_{n=2}^{\infty} \frac{x^n}{(n-2)!}$$

Therefore, $a_0 = 0$ and a_1 is arbitrary. For $n \geqq 2$,

$$a_n = \frac{1}{(n-1)!}$$

and hence the solution is

$$\begin{aligned} y &= a_1 x + \sum_{n=2}^{\infty} \frac{x^n}{(n-1)!} \\ &= a_1 x + x e^x - x \end{aligned}$$ □

Exercises for Section 6.4

By assuming an infinite power series solution about the origin solve the following differential equations, using the method of undetermined coefficients.

1. $y' + y = 0$
2. $y' + y = e^x$
3. $y' + y = e^{-x}$
4. $y' + y = \sin x$
5. $y' - y = \sin x$
6. $y' + xy = 1$
7. $y' + x^2 y = 0$
8. $y'' + y = 0$
9. $y'' + xy = 0$
10. $y'' + x^2 y = 0$
11. $y'' + x^3 y = 0$
12. $(1 - x^2)y'' + 3xy' + 5y = 0$
13. $y'' + y = e^x$
14. $(1 - x^2)y'' + xy' - y = 0$
15. $(1 - x^2)y'' + xy' - y = x$
16. $(x^2 + 1)y'' - 2xy' - 10y = 0$
17. $y'' + xy' - 4y = 0$
18. $y'' + xy' - 4y = x$
19. $(1 + x^2)y'' + 3xy' - 3y = 0$

Use undetermined coefficients to find at least the first three nonzero terms of a power series solution for the following differential equations about the indicated point $x = a$.

20. $y'' - 2(x - 1)y' + 2y = 0, \quad a = 1$
21. $y'' + (x - 2)y = 0, \quad a = 2$
22. $y'' - y' - y = 0, \quad a = 1$
23. $y'' + xy' + 3y = x^2, \quad a = 1$ (*Hint:* $x^2 = (x - 1)^2 + 2(x - 1) + 1$.)
24. $y'' + (x - 1)y = \ln x, \quad a = 1$ (*Hint:* Expand $\ln x$ in powers of $x - 1$.)

If the change of independent variable $x = \cos\theta$ is made in the equation

$$\frac{d^2y}{d\theta^2} + p^2y = 0$$

we obtain

$$(1 - x^2)y'' - xy' + p^2y = 0$$

This latter equation is known as Chebyshev's equation, and its solutions play an important role in the design of microwave filters and in various problems in numerical analysis.

25. Perform the substitution $x = \cos\theta$ and verify that the complete solution of Chebyshev's equation can be expressed as

$$y(x) = \begin{cases} c_1 \cos(p\cos^{-1}x) + c_2 \sin(p\cos^{-1}x), & p \neq 0 \\ c_1 + c_2 \cos^{-1}x, & p = 0 \end{cases}$$

26. Show directly that Chebyshev's equation has two linearly independent solutions, one consisting solely of even powers of x, the other of odd powers only.

27. Solve Chebyshev's equation in the case $p = 3$ by assuming a solution in power series form valid near the origin.

28. Alternatively, employ the result of Exercise 25 to demonstrate that the complete solution of Chebyshev's equation for $p = 3$ has the form

$$y(x) = c_1(4x^3 - 3x) + c_2\sqrt{1 - x^2}(4x^2 - 1)$$

29. Repeat Exercise 28 for the cases $p = 0, 1, 2$.

6.5 Singular Points of a Differential Equation

Although many of the differential equations encountered in applied problems are linear second-order equations with polynomial coefficients, not all such equations have power series solutions valid near any given point. For example, the simple equation

$$x^3y'' + y = 0$$

has no nontrivial solution that may be represented as a power series about the origin. The theory underlying series solutions of ordinary differential equations is designed both to tell us in advance when the method of undetermined coefficients is applicable, as well as to alert us to the possible need for an alternative procedure. It does this by classifying the various values of the independent variable x.

The second-order linear differential equation may be written

$$a_2(x)y'' + a_1(x)y' + a_0(x)y = f(x)$$

or, in standard form,

$$y'' + b_1(x)y' + b_0(x)y = g(x)$$

(Recall Section 3.1.)

Definition 6.1 The point $x = a$ is called an *ordinary* point of the differential equation if both $b_1(x)$ and $b_0(x)$ have Taylor series representations about $x = a$.

Definition 6.2 If a point is not an ordinary point, it is called a *singular* point. If $x = a$ is a singular point but $(x - a)b_1(x)$ and $(x - a)^2 b_0(x)$ have Taylor series representations, the singular point is said to be a *regular singular point*; otherwise it is an *irregular singular point*.

Example 6.11 Locate and classify the singular points of the differential equation

$$(x - 1)^2 x^2 (x - 2)y'' + 3x^3 y' + (x - 2)y = 0$$

Solution Since the coefficients are all polynomials and there is no common factor that can be divided out, the singular points are those values of x for which the coefficient of y'' is 0, hence at $x = 0$, 1, and 2. Dividing by the coefficient of y'', we obtain

$$b_1(x) = \frac{3x}{(x - 1)^2(x - 2)}, \qquad b_0(x) = \frac{1}{(x - 1)^2 x^2}$$

Since

$$(x - 1)b_1(x) = \frac{3x}{(x - 1)(x - 2)}$$

which is undefined at $x = 1$, the singular point at $x = 1$ is irregular. The other two singular points are regular. □

It can be shown that if $x = a$ is an ordinary point of a differential equation, then *every* solution to the equation has a Taylor series representation about $x = a$. If $x = a$ is a singular point, this will generally not be the situation. However, if $x = a$ is a regular singular point, there will be *at least one* solution of the form

$$\sum_{n=0}^{\infty} a_n (x - a)^{n+r}$$

where $a_0 \neq 0$. The *index* r is not usually an integer and, indeed, may be complex. If $x = a$ is an irregular singular point, the situation is usually much more complicated and is not considered here.

The existence of a power series solution about x_0, therefore, is guaranteed *only* when x_0 is an ordinary point. However, since it is often of interest to investigate the behavior of equations and their solutions near regular singular points as well, we consider such problems in the next section.

6.6 Method of Frobenius

Our earlier remarks suggest that it would be advantageous to extend our consideration of series representations to include

$$\sum_{n=0}^{\infty} a_n(x-a)^{n+r}, \qquad a_0 \neq 0$$

In honor of the mathematician who first studied this question, such a series is often called a *Frobenius series.* If we solve a differential equation by assuming a Frobenius series and finding the coefficients by the method of undetermined coefficients, the technique is called the *method of Frobenius.* For a second-order equation, there is *always* at least one solution about a regular singular point which is a Frobenius series. The resulting series will even be a power series if r is a positive integer. (Sometimes you will even be so lucky as to find two solutions this way; but this does not always happen.) To determine the precise form of this series, first substitute the series into the differential equation; second, equate to zero the coefficient of the lowest power of x in the resulting algebraic expression, to obtain the *indicial equation*; third, find the roots r of this equation; fourth, use these roots to find the coefficients a_n of the Frobenius solution.

Example 6.12 Use the method of Frobenius with $a = 0$ to solve the differential equation $2xy'' + 3y' - y = 0$.

Solution Since $x = 0$ is a regular singular point, we assume the solution to be of the form

$$\sum_{n=0}^{\infty} a_n x^{n+r}, \qquad a_0 \neq 0$$

Differentiating twice,

$$y' = \sum_{n=0}^{\infty} a_n(n+r)x^{n+r-1}$$

and

$$y'' = \sum_{n=0}^{\infty} a_n(n+r)(n+r-1)x^{n+r-2}$$

Substituting into the differential equation and carrying out the indicated multiplications,

$$2\sum_{n=0}^{\infty} a_n(n+r)(n+r-1)x^{n+r-1} + 3\sum_{n=0}^{\infty} a_n(n+r)x^{n+r-1} - \sum_{n=0}^{\infty} a_n x^{n+r} = 0$$

Shifting the index in the third series and combining the first two,

$$\sum_{n=0}^{\infty} a_n(n+r)(2n+2r+1)x^{n+r-1} - \sum_{n=1}^{\infty} a_{n-1}x^{n+r-1} = 0$$

Writing the term corresponding to $n = 0$ and combining the remaining terms for $n \geqq 1$ into one series,

$$a_0 r(2r+1)x^{r-1} + \sum_{n=1}^{\infty} \{a_n(n+r)(2n+2r+1) - a_{n-1}\}x^{n+r-1} = 0$$

Equating the coefficient of x^{r-1} to zero yields the indicial equation

$$a_0 r(2r+1) = 0$$

Since $a_0 \neq 0$, we must have either $r = 0$ or $r = -\frac{1}{2}$. We expect, then, two linearly independent solutions to the given differential equation:

$$y_1 = \sum_{n=0}^{\infty} a_n x^n \quad \text{and} \quad y_2 = x^{-1/2} \sum_{n=0}^{\infty} b_n x^n$$

Since $a_n(n+r)(2n+2r+1) - a_{n-1} = 0$ for all $n \geq 1$, we have the following information on the coefficients for the two series:

$$a_0 \neq 0 \text{ (but, otherwise, } a_0 \text{ is arbitrary) and for } n \geq 1,\ a_n = \frac{1}{n(2n+1)} a_{n-1}$$

$$b_0 \neq 0 \text{ (but, otherwise, } b_0 \text{ is arbitrary) and for } n \geq 1,\ b_n = \frac{1}{n(2n-1)} b_{n-1}$$

Finding the general terms of the two series is a matter of a little algebra,

$$a_n = \frac{2^n a_0}{(2n+1)!} \quad \text{and} \quad b_n = \frac{2^n b_0}{(2n)!}$$

The two solutions are

$$y_1 = a_0 \sum_{n=0}^{\infty} \frac{2^n}{(2n+1)!} x^n, \qquad y_2 = b_0 x^{-1/2} \sum_{n=0}^{\infty} \frac{2^n}{(2n)!} x^n$$

One of these solutions, y_1, is a power series, but this is not to be expected in general. □

Example 6.13 Use the method of Frobenius to solve the differential equation $4x^2 y'' + (3x+1)y = 0$ near the origin.

Solution Since the origin is a regular singular point, we assume the solution to be of the form of a Frobenius series

$$y = \sum_{n=0}^{\infty} a_n x^{n+r}$$

Differentiating twice and substituting into the given equation,

$$4x^2 \sum_{n=0}^{\infty} a_n(n+r)(n+r-1)x^{n+r-2} + (3x+1) \sum_{n=0}^{\infty} a_n x^{n+r} = 0$$

Simplifying and expanding,

$$\sum_{n=0}^{\infty} 4a_n(n+r)(n+r-1)x^{n+r} + \sum_{n=0}^{\infty} 3a_n x^{n+r+1} + \sum_{n=0}^{\infty} a_n x^{n+r} = 0$$

Shifting the index in the second series and combining the first and the third series,

$$\sum_{n=0}^{\infty} a_n\{4(n+r)(n+r-1)+1\}x^{n+r} + \sum_{n=1}^{\infty} 3a_{n-1}x^{n+r} = 0$$

Writing the first term of the first series, combining and simplifying once again,

$$a_0\{4r(r-1)+1\}x^r + \sum_{n=1}^{\infty} \{a_n(2n+2r-1)^2 + 3a_{n-1}\}x^{n+r} = 0$$

To assure that $a_0 \neq 0$, we choose r so that

$$4r(r-1)+1=0, \quad \text{or} \quad r=\tfrac{1}{2}$$

Hence, the form of one solution is

$$y = x^{1/2}\sum_{n=0}^{\infty} a_n x^n$$

where the coefficient information in recursive form is

$$a_0 \neq 0 \quad \text{and for } n \geqq 1, \qquad a_n = -\frac{3}{4n^2}a_{n-1} \qquad \square$$

In the above example, the method of Frobenius yielded only one solution since the indicial equation yielded only one value of r. Hence the other solution is not a Frobenius series. It should be mentioned that there are even cases where, though there are two distinct values of r, only one linearly independent Frobenius solution exists (see Exercises 10 and 11). In either of these situations a second independent solution may be obtained by the reduction of order approach of Section 3.3 or by the methods discussed in the next section.

In our Frobenius examples we have considered expansions in terms of powers of x. If we wish a solution valid near $x = a$, we may start with $y(x) = \sum_{n=0}^{\infty} a_n(x-a)^{n+r}$, which adds only a modicum of complication to our calculations. Another method, perhaps more appealing, is to let $t = x - a$ and $y(t) = \sum_{n=0}^{\infty} a_n t^{n+r}$ and then replace t by $x - a$ after all the heavy calculation is over. An opportunity to try one or both of these methods is given in Exercises 20–23.

Exercises for Sections 6.5 and 6.6

Which points are not ordinary points of the following equations? Of the singular points, which are regular and which are irregular?

1. $x^2y'' + xy' + y = e^x$

2. $(x^2-1)^2xy'' + (x+1)y' + xy = 0$

3. $x(x-1)^3y'' + xy' + (x-1)^2y = 0$

4. $xy'' + \dfrac{1}{x}y' + y = 7$

5. $(2-x)y'' + xy' + \dfrac{y}{(x-2)^2} = 0$

In each of the following equations, show that the origin is a regular singular point. Use the method of Frobenius to find at least one solution about the origin.

6. $4xy'' + 2y' + y = 0$

7. $4x^2y'' - 2x(x+2)y' + (x+3)y = 0$

8. $x(x+1)y'' + (5x+1)y' + 3y = 0$

9. $(x^2+x)y'' + (x+1)y' - y = 0$

10. $x^2y'' + xy' + (x^2-1)y = 0$

11. $x^2y'' + xy' + (x^2-4)y = 0$

12. $2x^2y'' + x(1-x)y' - y = 0$

13. $xy'' + y' + xy = 0$

14. $xy'' + y = 0$

15. $4x^2y'' - 2x(x-2)y' - (3x+1)y = 0$

For each of the following equations, find the form of at least one solution about $x = a$. Express the coefficients in terms of a recurrence relation.

16. $x(2x+1)y'' + 3y' - xy = 0, \quad a = 0$

17. $3x^2y'' + (5x - x^2)y' + (2x^2-1)y = 0, \quad a = 0$

18. $2xy'' - (1+x^3)y' + y = 0, \quad a = 0$

19. $x^2(3+x^2)y'' + 5xy' - (1+x)y = 0, \quad a = 0$

20. $x(x-1)y'' - 3y' - 2y = 0, \quad a = 1$

21. $(2x+1)y'' + 4y = 0, \quad a = -\frac{1}{2}$

22. Use the method of Frobenius with $a = 1$ to find at least one solution of Legendre's equation of order 3:

$$(1-x^2)y'' - 2xy' + 12y = 0$$

(Recall Example 6.8.)

23. Use the method of Frobenius with $a = 1$ to find at least one solution of Chebyshev's equation of order 3:

$$(1-x^2)y'' - xy' + 9y = 0$$

(Recall Exercises 27 and 28, p. 206.)

Under appropriate assumptions, one of the quantum-mechanical wave functions ψ associated with the hydrogen atom satisfies the following equation:

$$s^2\psi''(s) + 2s\psi'(s) + \frac{8\pi^2 m}{h^2}\left(Es^2 + Ze^2 s + \frac{h^2}{32\pi^2 m}\right)\psi(s) = 0$$

Here m and $-e$ are the mass and charge of the electron, respectively, Ze is the charge of the hydrogen nucleus, E is the total energy of the system, and h is Planck's constant. Analogous to the situation in Exercise 24, p. 198, an appropriate change of both the dependent and independent variable will transform this equation into Laguerre's differential equation:

$$xy''(x) + (1-x)y'(x) + py(x) = 0$$

where

$$p = \frac{i\pi Ze^2}{h}\sqrt{\frac{2m}{E}} - \frac{1}{2}$$

24. Show that the substitutions

$$\psi(s) = \frac{e^{-(1/2)x}}{\sqrt{x}}\, y(x)$$

$$s = \frac{ihx}{4\pi\sqrt{2mE}}$$

transform the hydrogen atom wave function equation into Laguerre's differential equation.

25. Use the method of Frobenius to find at least one solution of Laguerre's equation in powers of x.

6.7 Frobenius Solutions for Roots Differing by an Integer

In cases where the indicial equation has multiple roots (roots differing by zero) or roots differing by a nonzero integer, special difficulties arise and special care must be taken.

Recalling Exercise 13, p. 105, if the differential equation $a_2x^2y'' + a_1xy' + a_0y = 0$ with indicial equation $a_2r^2 + (a_1 - 1)r + a_0 = 0$ has roots $r_1 = r_2$, then the solutions are of the form $y_1 = x^{r_1}$ and $y_2 = x^{r_1}\ln x$. We might thus expect to have a logarithmic term appear when the roots are equal or differ by an integer. The following theorem verifies this conjecture to a certain extent and also summarizes some of the results of Section 6.6.

Theorem 6.1 Suppose $x = 0$ is a regular singular point of the equation $y'' + b_1(x)y' + b_0(x)y = 0$, where $xb_1(x)$ and $x^2b_0(x)$ have Taylor series representations valid for $|x| < R$, and r_1 and r_2 are solutions of the indicial equation with $r_1 \geq r_2$ (if r_1 and r_2 are complex, the real part of r_1 is greater than or equal to the real part of r_2).* Then there are two linearly independent solutions of the differential equation valid for $0 < |x| < R$ of the following forms:

1. If $r_1 - r_2$ is neither 0 nor an integer,

$$y_1(x) = |x|^{r_1} \sum_{n=0}^{\infty} a_n x^n, \qquad a_0 \neq 0$$

$$y_2(x) = |x|^{r_2} \sum_{n=0}^{\infty} b_n x^n, \qquad b_0 \neq 0$$

2. If $r_1 = r_2$,

$$y_1(x) = |x|^{r_1} \sum_{n=0}^{\infty} a_n x^n, \qquad a_0 \neq 0$$

$$y_2(x) = y_1(x)\ln|x| + |x|^{r_1+1} \sum_{n=0}^{\infty} b_n x^n$$

3. If $r_1 - r_2 =$ a positive integer,

$$y_1(x) = |x|^{r_1} \sum_{n=0}^{\infty} a_n x^n, \qquad a_0 \neq 0$$

$$y_2(x) = cy_1(x)\ln|x| + |x|^{r_2} \sum_{n=0}^{\infty} b_n x^n, \qquad b_0 \neq 0$$

---

* See p. 92.

Notice that in Case 2, the constant b_0, and indeed the entire infinite series, may be zero. In Case 3 the constant c may be zero. The absolute value that appears in the $|x|^{r_1}$ and $|x|^{r_2}$ terms is necessary for complex r_1 and r_2 as well as for real values such as $r = \frac{1}{2}, -\frac{3}{4}$, and so on. ■

There are various shortcuts that skilled practitioners may employ, but with patience a straightforward substitution of the forms above, carefully using the $a_0 \neq 0$ and $b_0 \neq 0$ statements, will produce the desired solutions. In some cases, use of the reduction of order method is less complicated. The following four examples are relatively easy, but typical.

Example 6.14 Find a second independent solution of the equation in Example 6.13.

Solution Since $r_1 = r_2 = \frac{1}{2}$, we look for a solution of the form $y_2 = y_1 \ln x + x^{3/2}\sum_{n=0}^{\infty} b_n x^n$. We restrict ourselves to $x > 0$ and choose $a_0 = 1$ for convenience. Substituting into the equation $L[y] = 4x^2y'' + (3x+1)y = 0$, we obtain

$$L[y_2] = 4x^2\left(y_1''\ln x + \frac{2y_1'}{x} - \frac{y_1}{x^2}\right)$$

$$+ \sum_{n=0}^{\infty} 4\left(n+\frac{3}{2}\right)\left(n+\frac{1}{2}\right)b_n x^{n+(3/2)} + (3x+1)y_1\ln x + \sum_{n=0}^{\infty} 3b_n x^{n+(5/2)}$$

$$+ \sum_{n=0}^{\infty} b_n x^{n+(3/2)} = 0 = (\ln x)L[y_1] + 8xy_1' - 4y_1 + 3b_0x^{3/2} + b_0x^{3/2}$$

$$+ \sum_{n=1}^{\infty}\left[4\left(n+\frac{3}{2}\right)\left(n+\frac{1}{2}\right)b_n + b_n + 3b_{n-1}\right]x^{n+(3/2)} = 0$$

The first few terms of y_1 are $x^{1/2} - \frac{3}{4}x^{3/2} + \frac{9}{64}x^{5/2} - \frac{27}{(36)64}x^{7/2} + \cdots$. Substituting these in the equation above and using the fact that $L(y_1) \equiv 0$, we have $(4b_0 - 6)x^{3/2} + (16b_1 + 3b_0 + \frac{9}{4})x^{5/2} + (36b_2 + 3b_1 - \frac{9}{32})x^{7/2} + \cdots = 0$, so $b_0 = \frac{3}{2}$, $b_1 = -\frac{27}{64}$, $b_2 = \frac{11}{256}, \ldots$ and the first few terms of our solution are

$$y_2 = \ln x\left(x^{1/2} - \frac{3}{4}x^{3/2} + \frac{9}{64}x^{5/2} - \frac{3}{256}x^{7/2} + \cdots\right)$$

$$+ \frac{3}{2}x^{3/2} - \frac{27}{64}x^{5/2} + \frac{11}{256}x^{7/2} + \cdots$$ □

Example 6.15 Solve $x^2y'' + (x^2 - x)y' + y = 0$.

Solution $L(\sum_{n=0}^{\infty} a_n x^{n+r}) = \sum_{n=0}^{\infty} a_n(n+r)(n+r-1)x^{n+r} + \sum_{n=0}^{\infty} a_n(n+r)x^{n+r+1} - \sum_{n=0}^{\infty} a_n(n+r)x^{n+r} + \sum_{n=0}^{\infty} a_n x^{n+r} = a_0[r(r-1) - r + 1] + \sum_{n=1}^{\infty}[a_n(n+r)(n+r-1) + a_{n-1}(n+r-1) - a_n(n+r) + a_n]x^{n+r} = 0$. The indicial equation $r^2 - 2r + 1$ yields the double root $r = 1$. Choosing $a_0 = 1$ we obtain $a_n = -a_{n-1}/n$, $a_1 = -1$, $a_2 = \frac{1}{2}$, $a_3 = -1/3!$, $\ldots$, $a_n = (-1)^n/n!$ and $y_1(x) = x\sum_{n=0}^{\infty}(-1)^n(x^n/n!) = xe^{-x}$.

With such a nice result, the temptation is irresistible to use the formula for the reduction of order method given in Exercise 17, p. 90, that is, $y_2 = y_1 \int \frac{e^{-\int p(x)\,dx}}{y_1^2(x)} dx$. We must first put the differential equation in the standard form $y'' + (1 - (1/x))y' + (1/x^2)y = 0$ and recognize $p(x) = (1 - (1/x))$. Then

$$y_2 = xe^{-x} \int \frac{e^{-(x - \ln x)}}{x^2 e^{-2x}} dx = xe^{-x} \int \frac{e^x}{x} dx$$

$$= xe^{-x} \int \left(\frac{1}{x} + 1 + \frac{x}{2!} + \frac{x^2}{3!} + \cdots + \frac{x^{n-1}}{n!} + \cdots \right) dx$$

$$y_2(x) = xe^{-x} \left[\ln x + x + \frac{x^2}{2 \cdot 2!} + \frac{x^3}{3 \cdot 3!} + \cdots + \frac{x^n}{n \cdot n!} + \cdots \right]$$

Of course, xe^{-x} may be expressed as a series and combined with the other series, but this would not improve things much.* □

Example 6.16 Find the complete solution of $x^2 y'' - (x + 2)y = 0$ near $x = 0$.

Solution Since $x = 0$ is a regular singular point, we assume a solution of the form $y_1 = \sum_{n=0}^{\infty} a_n x^{n+r}$, $a_0 \neq 0$, and the usual manipulations lead to

$$L[y_1] = a_0(r^2 - r - 2)x^r + \sum_{n=1}^{\infty} \{a_n[(n + r)(n + r - 1) - 2] - a_{n-1}\}x^{n+r} = 0$$

The roots of the indicial equation are $r_1 = 2$ and $r_2 = -1$ and the recursion relation is

$$a_n = \frac{a_{n-1}}{(n + r)(n + r - 1) - 2}$$

Choosing $r_1 = 2$ and $a_0 = 1$ we have

$$a_n = \frac{a_{n-1}}{(n + 2)(n + 1) - 2} = \frac{a_{n-1}}{n(n + 3)}$$

so $a_1 = \frac{1}{4}$, $a_2 = \frac{1}{40}$, $a_3 = \frac{1}{720}, \ldots$ and $y_1(x) = x^2 + x^3/4 + x^4/40 + x^5/720 + \cdots$. If we were to try $r = -1$, $y_2(x) = \sum_{n=0}^{\infty} b_n x^{n+r}$, and $b_0 = 1$, we would find $b_n = b_{n-1}/[(n - 1)(n - 2) - 2] = b_{n-1}/n(n - 3)$ with b_3 undefined. If, in desperation, we let $b_0 = b_1 = 0$ and $b_3 = 1$, we would obtain $b_4 = \frac{1}{4}$, $b_5 = \frac{1}{40}$, $b_6 = \frac{1}{720}, \ldots$ and $y_2(x) = x^2 + x^3/4 + x^4/40 + x^5/720 + \cdots = y_1(x)$.

Of course we could have avoided this futile effort by noting that $r_1 - r_2 = 3$, an integer, and by assuming $y_2(x) = cy_1(x) \ln x + x^{-1} \sum_{n=0}^{\infty} b_n x^n$, $b_0 \neq 0$ with $x > 0$ for convenience. If we use the first three terms of $y_1(x) = x^2 + x^3/4 + x^4/40 + \cdots$ found above, then straightforward, if tedious, calculation produces

$$\begin{aligned} L[y_2] = {} & c(2x^2 + \tfrac{3}{2}x^3 + \tfrac{3}{10}x^4 + \cdots - x^3 - \tfrac{1}{4}x^4 + \cdots - 2x^2 \\ & - \tfrac{1}{2}x^3 - \tfrac{1}{20}x^4) \ln x + c(3x^2 + \tfrac{5}{4}x^3 + \tfrac{7}{40}x^4 + \cdots) \\ & + [(-b_0 - 2b_1) + (-b_1 - 2b_2)x - b_2 x^2 - \cdots] = 0 \end{aligned}$$

---

* The integral appearing in this example is actually a tabulated function called the *exponential integral* (see Section 9.4).

As expected, the factor multiplying $\ln x$ is identically zero. Combining the other terms

$$(-b_0 - 2b_1) + (-b_1 - 2b_2)x + (3c - b_2)x^2 + (\tfrac{5}{4}c + 4b_4 - b_3)x^3 + \cdots = 0$$

As usual, take $b_0 = 1$, then $b_1 = -\frac{1}{2}$, $b_2 = \frac{1}{4}$, $c = \frac{1}{12}$, and $b_4 = b_3/4 + \frac{5}{192}$. This last relation is momentarily disquieting, but notice that, in $x^{-1}\sum_{n=0}^{\infty} b_n x^n$, b_3 will be the coefficient of x^2 and the x^2 term is precisely where the series for y_1 starts. Since any multiple of y_1 could be added to y_2 to produce a solution $y_2 + ky_1$, we indeed do have an arbitrary constant b_3 appearing at this stage. If we take $b_3 = 0$, we have $b_4 = \frac{5}{192}$ and no further difficulties arise. Combining all of the above, we have

$$y_2(x) = \frac{1}{12}\left(x^2 + \frac{x^3}{4} + \frac{x^4}{40} + \frac{x^5}{720} + \cdots\right)\ln x + \frac{1}{x} - \frac{1}{2} + \frac{x}{4} + \frac{5x^3}{192} + \cdots$$

where, of course, the complete solution is $y = c_1 y_1 + c_2 y_2$. □

Example 6.17 Find the complete solution of $x(1 - x)y'' - 3y' + 2y = 0$ near $x = 0$.

Solution Proceeding as in the previous example, we let $y_1(x) = \sum_{n=0}^{\infty} a_n x^{n+r}$ with $a_0 = 1$, and then $L[y_1] = a_0 r(r - 4)x^{r-1} + \sum_{n=1}^{\infty} \{(n + r)(n + r - 4)a_n + [2 - (n + r - 1)(n + r - 2)]a_{n-1}\}x^{n+r-1}$. The roots of the indicial equation are $r_1 = 4$ and $r_2 = 0$. For $r_1 = 4$, the recursion relation is $a_n = ((n + 1)/n)a_{n-1}$, so $a_0 = 1, a_1 = 2, a_2 = 3, a_3 = 4, \ldots$, and $y_1(x) = \sum_{n=0}^{\infty} (n + 1)x^{n+4}$. Now consider

$$\begin{aligned} L\left[cy_1(x)\ln x + \sum_{n=0}^{\infty} b_n x^{n+0}\right] &= c\ln x(xy_1'' - x^2 y_1'' - 3y_1' + 2y_1) \\ &\quad + c\left(2y_1' - \frac{4}{x}y_1 - 2xy_1' + y_1\right) \\ &\quad + \sum_{n=0}^{\infty} [b_{n+1}(n + 1)n - b_n n(n - 1) \\ &\quad - 3b_{n+1}(n + 1) + 2b_n]x^n \\ &= c\ln x L[y_1] + c\sum_{n=0}^{\infty} (n + 4)x^{n+3} \\ &\quad + \sum_{n=0}^{\infty} [b_{n+1}(n + 1)(n - 3) \\ &\quad - b_n(n + 1)(n - 2)]x^n \equiv 0 \end{aligned}$$

Since $L[y_1] \equiv 0$, matching of powers of x leads to $(n - 3)b_{n+1} = (n - 2)b_n$ for $n = 0, 1, 2,$ and $c + (n - 3)b_{n+1} = (n - 2)b_n$ for $n \geq 3$. Thus, with $b_0 = 1$, it follows that $b_1 = \frac{2}{3}$, $b_2 = \frac{1}{3}$, $b_3 = 0$, $c = 0$, b_4 is arbitrary, and $b_{n+1} = b_n(n - 2)/(n - 3)$ for $n \geq 4$. For the choice $b_4 = 0$, $y = c_1 y_1 + c_2 y_2 = c_1 \sum_{n=0}^{\infty} (n + 1)x^{n+4} + c_2(1 + \frac{2}{3}x + \frac{1}{3}x^2)$. □

The constant c in the above representation turned out to be zero. This situation occurs whenever the standard Frobenius method actually gives us two linearly independent solutions. Bessel functions of half-integer order provide a notable example of this phenomenon, as we shall see in Chapter 9.

Exercises for Section 6.7

Find the complete solution of each equation near $x = a$. Give at least three nonzero terms of each series.

1. $x^2y'' + 3xy' + (1 - 2x)y = 0, \quad a = 0$

2. $xy'' - y' + 4x^3y = 0, \quad a = 0$

3. $xy'' - (4 + x)y' + 2y = 0, \quad a = 0$

4. $x(1 - x)y'' + 2(1 - x)y' + 2y = 0, \quad a = 0$

5. $x^2y'' + x^2y' - 2y = 0, \quad a = 0$

6. $x(1 - x)y'' + 2y' + 2y = 0, \quad a = 0$

7. $x(1 - x)y'' + 2(1 - x)y' + 2y = 0, \quad a = 1$

8. $xy'' - y' + y = 0, \quad a = 0$

9. $xy'' + (1 - x)y' + 3y = 0, \quad a = 0$
This is Laguerre's equation of order 3. (Recall Exercise 25, p. 212.)

10. $(1 - x^2)y'' - xy' + 9y = 0, \quad a = 1$
This is Chebyshev's equation of order 3. (Recall Exercise 23, p. 211.)

11. $(1 - x^2)y'' - 2xy' + 12y = 0, \quad a = 1$
This is Legendre's equation of order 3. (Recall Exercise 22, p. 211.)

12. $x^2y'' + xy' + x^2y = 0, \quad a = 0$
This is Bessel's equation of order 0. (Recall Exercise 13, p. 211.)

6.8 Looking Ahead

If a given differential equation has polynomial coefficients, it can be conveniently characterized by the number and type of its singular points. This classification scheme includes the point at infinity, which can easily be analyzed by first making the change of independent variable from x to $1/x$ and then considering the origin. Second-order equations with either one or two singular points, both of them regular, are essentially Cauchy-Euler equations (recall the exercises for Section 3.6). The canonical second-order equation with three regular singular points is the hypergeometric equation of Gauss (see [5], [10], for example). Owing to its appearance in various applications involving spherical geometry, we consider at some length the special case of Legendre's equation and its solutions in Chapter 9. The equally important equation, Bessel's equation, which has one regular and one irregular singular point, is also discussed in considerable detail. The classic reference [8] has further information on the classification of linear differential equations.

In the preceding section procedures were proposed for determining the independent second solution of a second-order differential equation in the vicinity of a regular singular point when the Frobenius series representation was inapplicable. (See [6], [7], among other references, for a thorough discussion of these matters.)

Basic References

[1] Boyce, W., and R. C. DiPrima. 1977. *Elementary Differential Equations and Boundary Value Problems*. 3d ed. New York: John Wiley.

[2] Dettman, J. W. 1974. *Introduction to Linear Algebra and Differential Equations*. New York: McGraw-Hill.

[3] Ross, S. L. 1980. *Introduction to Ordinary Differential Equations*. 3d ed. New York: John Wiley.

[4] Spiegel, M. R. 1981. *Applied Differential Equations*. 3d ed. Englewood Cliffs, N. J.: Prentice-Hall.

More-Advanced References

[5] Cochran, J. A. 1982. *Applied Mathematics: Principles, Techniques, and Applications*. Belmont, Calif.: Wadsworth.

[6] Coddington, E. A., and N. Levinson. 1955. *Theory of Ordinary Differential Equations*. New York: McGraw-Hill.

[7] Hildebrand, F. B. 1976. *Advanced Calculus for Applications*. 2d ed. Englewood Cliffs, N. J.: Prentice-Hall, Inc.

[8] Ince, E. L. 1953. *Ordinary Differential Equations*. New York: Dover.

[9] Kaplan, W. 1958. *Ordinary Differential Equations*. Reading, Mass.: Addison-Wesley.

[10] Sneddon, I. N. 1980. *Special Functions of Mathematical Physics and Chemistry*. 3d ed. London: Longman.

Chapter 7

Fourier Expansions

7.1 Introduction

Depending on the particular application on which we are working, we may adopt different standard forms in which to express functions. For instance, an initial value problem focuses our attention on the value of a function and its derivatives at a specific point $x = a$. In this case, power series representations about $x = a$ offer considerable utility. In discussing a musical sound, on the other hand, it is often convenient to break the sound into a basic tone with frequency v and overtones with higher frequencies $2v$, $3v$, $4v$, ... In a similar manner, when analyzing the vibration of a structure such as a column, a drum, or even a bridge, it is helpful to decompose vibrating forces into constituent parts having frequencies related to the natural vibration frequencies of the structure. Likewise, in various modern and classical applications with light rays and beams, it is particularly advantageous to be able to talk about the light spectrum, that is, to analyze the frequencies of components of the light.

These diverse practical applications share a common mathematical feature. They involve expressing a given function in terms of linear combinations (or partial sums) of other functions from some orthogonal set. This, in essence, is the process of Fourier analysis, the subject of this chapter. Representing a function in terms of a standard set of functions is often called an *expansion* in terms of the given set. *Fourier expansions* are always in terms of a set of orthogonal functions and thus are also called *orthogonal expansions*. (You may wish to review the detailed discussion of orthogonal functions in Section 1.5 before proceeding.)

In elementary work, some set of trigonometric functions is usually used to write Fourier expansions. Hence, we will usually assume the terminology "Fourier expansion" to mean a Fourier trigonometric expansion. An example of a different kind of orthogonal set is the set of Legendre polynomials. A Fourier expansion in terms of this set is called a Fourier-Legendre expansion. We have more to say about Fourier-Legendre, Fourier-Bessel, and other generalized Fourier expressions in Chapter 9.

In much of our preceding analysis and, in particular, in our work with Taylor series, we have been forced to restrict our attention to continuous, differentiable, or even infinitely differentiable functions. The results of this chapter will allow us to consider a much broader class of functions and begin to use the extremely fruitful technique known as spectral or Fourier analysis.

7.2 The Space $PC[-L, L]$

From the general discussion of vectors in Chapter 1 we learned that the set of piecewise continuous functions on the interval $[a, b]$ can be considered a normed vector space with an inner product defined by

$$(f, g) = \int_a^b f(x)g(x)\,dx$$

and the corresponding norm

$$\|f\| = \left(\int_a^b [f(x)]^2\,dx\right)^{1/2}$$

In this chapter we consider functions with the *periodicity* property

$$f(x) \equiv f(x + 2L)$$

where the positive quantity $2L$ is called a *period* of the function. Since we can restrict the analysis to any interval of length $2L$, we can discuss the space $PC[-L, L]$ without loss of generality.

As we will be considering orthogonal expansions, it is important to be aware of certain sets of trigonometric functions that are orthogonal on $[-L, L]$. The proofs of the orthogonality of these sets are similar to those in Examples 1.6 and 1.7. The sets are:

1. The set of cosine functions,

$$\left\{\cos\frac{n\pi x}{L}\right\}_{n=0}^{\infty}$$

2. The set of sine functions,

$$\left\{\sin\frac{n\pi x}{L}\right\}_{n=1}^{\infty}$$

3. The complete trigonometric set, which is the union of the two sets described in 1 and 2.

4. Any subset of the complete trigonometric set.

In Section 7.12 we show how other orthogonal sets arise quite naturally in applied mathematics. Most of the results of this chapter apply to expansions in terms of these other orthogonal sets. For simplicity, however, our examples will be in terms of onc of the four sets just listed.

For theoretical purposes, using an orthonormal set instead of an orthogonal one is sometimes easier. This is not a significant restriction to the discussion, since an orthonormal set is easily obtained from one that is merely orthogonal by dividing each function by its norm, assuming that none of these norms is zero.

Example 7.1 Find the orthonormal set corresponding to the complete trigonometric set on $[-L, L]$.

Solution We compute the norm of each of the functions of the orthogonal set,

$$\left\| \cos \frac{0\pi x}{L} \right\| = \|1\| = \left(\int_{-L}^{L} dx \right)^{1/2} = \sqrt{2L}$$

$$\left\| \cos \frac{n\pi x}{L} \right\| = \left(\int_{-L}^{L} \cos^2 \left(\frac{n\pi x}{L} \right) dx \right)^{1/2} = \sqrt{L}, \qquad n \neq 0$$

$$\left\| \sin \frac{n\pi x}{L} \right\| = \left(\int_{-L}^{L} \sin^2 \left(\frac{n\pi x}{L} \right) dx \right)^{1/2} = \sqrt{L}, \qquad n \neq 0$$

Therefore, the orthonormal set is

$$\left\{ \frac{1}{\sqrt{2L}} \right\} \cup \left\{ \frac{\cos(n\pi x/L)}{\sqrt{L}} \right\}_{n=1}^{\infty} \cup \left\{ \frac{\sin(n\pi x/L)}{\sqrt{L}} \right\}_{n=1}^{\infty}$$

□

Sometimes a slightly different inner product is defined for function spaces. This results in different formulas for the norm of a function and for orthogonality.

Definition 7.1 We say that (f, g) is a *generalized inner product* on the space $PC[a, b]$ with respect to the weight function $r(x)$, $(r(x) > 0)$, if

$$(f, g) = \int_a^b f(x)g(x)r(x)\,dx$$

With respect to this inner product, the norm of f takes the form

$$\|f\| = \left(\int_a^b [f(x)]^2 r(x)\,dx \right)^{1/2}$$

A set of functions $\{g_n\}_{n=1}^{\infty}$ is orthogonal on $[a, b]$ with respect to the weight function $r(x)$ if

$$\int_a^b g_m(x)g_n(x)r(x)\,dx = 0 \qquad \text{for } n \neq m$$

If the weight function is 1, the definitions reduce to those previously given.

7.3 Two Kinds of Convergence

Because of the introduction of the integral norms, we are led to consider two types of convergence of a sequence of functions.

Definition 7.2 We say the sequence of functions $\{g_n\}_{n=1}^{\infty}$ *converges in the mean* to the function g if on the interval $[a, b]$

$$\lim_{n\to\infty} \|g - g_n\| = 0$$

We will use the notation $\lim_{n\to\infty} g_n \approx g$ for this type of convergence.* The same sequence is said to *converge pointwise* to g if

$$\lim_{n\to\infty} |g(x) - g_n(x)| \equiv 0$$

for each x in $[a, b]$. As usual, this pointwise convergence will be denoted by $\lim_{n\to\infty} g_n = g$.

Pointwise convergence requires the convergence of many sequences of numbers, one for each value of x. For convergence in the mean, we require the one sequence of numbers

$$d_n = \left(\int_a^b [g(x) - g_n(x)]^2\, dx\right)^{1/2}$$

to converge to zero. Figuratively, $d_n \to 0$ for convergence in the mean and $g_n(x) \to g(x)$ (for each x) for pointwise convergence.

Example 7.2 Consider the sequence of functions $f_n(x) = x^n$ on $[0, 1]$. Analyze for mean convergence. (This sequence of functions converges pointwise to 0 for x in $[0, 1)$ and to 1 for $x = 1$, as is suggested by the sketch in Figure 7.1.)

Solution The sequence converges in the mean to the "zero function" because

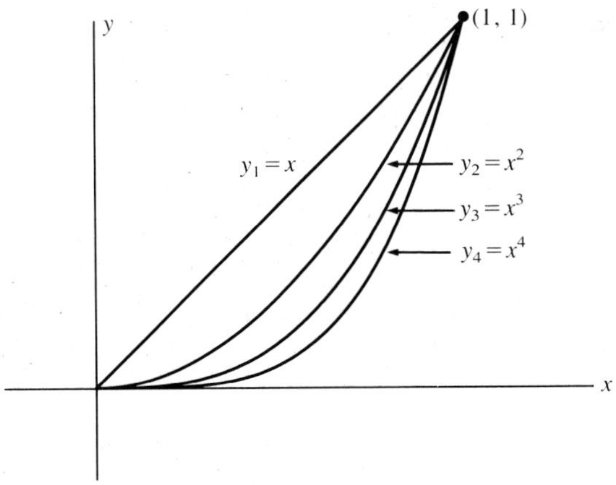

Figure 7.1

*The symbol $\approx$ is used elsewhere in the text with the meaning "is approximately equal to." No serious confusion should arise from these related, but not identical, uses.

$$\|0 - x^n\| = \left(\int_0^1 x^{2n}\,dx\right)^{1/2}$$
$$= \left(\frac{x^{2n+1}}{2n+1}\bigg|_0^1\right)^{1/2}$$
$$= \left(\frac{1}{2n+1}\right)^{1/2}$$

which approaches 0 as n becomes arbitrarily large. □

The next example is introduced only to exhibit the striking difference between the two different types of convergence.

Example 7.3 Let $\{f_n\}_{n=1}^{\infty}$ be a sequence of functions defined on the unit interval as follows:

$f_1(x) = 1$ over the entire interval
$f_2(x) = 1$ on $[0, \frac{1}{2}]$ and 0 otherwise
$f_3(x) = 1$ on $[\frac{1}{2}, 1]$ and 0 otherwise
$f_4(x) = 1$ on $[0, \frac{1}{4}]$ and 0 otherwise
$\vdots$

$$f_{2^n}(x) = 1 \text{ on } \left[0, \frac{1}{2^n}\right] \text{ and 0 otherwise}$$

$$f_{2^n+1}(x) = 1 \text{ on } \left[\frac{1}{2^n}, \frac{2}{2^n}\right] \text{ and 0 otherwise}$$

$$f_{2^n+k}(x) = 1 \text{ on } \left[\frac{k}{2^n}, \frac{k+1}{2^n}\right] \qquad k = 0, 1, 2, \ldots, 2^n - 1$$

$\vdots$

$$f_{2^n+k}(x) = 1 \text{ on } \left[\frac{k}{2^n}, \frac{k+1}{2^n}\right] \qquad k = 0, 1, 2, \ldots, 2^n - 1$$

$\vdots$

Analyze for the two types of convergence. (See Figure 7.2.)

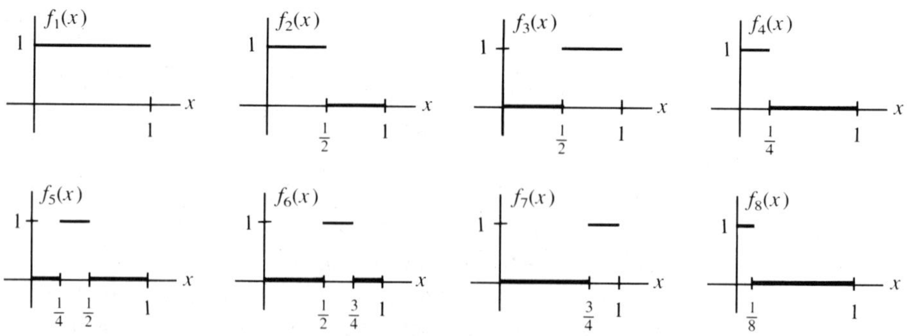

Figure 7.2

Solution The sequence converges pointwise nowhere on the interval since for any value of N and for any fixed x on the interval, there exist $m > N$ and $n > N$ such that $f_m(x) = 0$ and $f_n(x) = 1$.

The sequence converges in the mean to the zero function because

$$\|f_n - 0\|^2 = \int_0^1 [f_n(x)]^2 \, dx$$
$$= \text{length of the interval on which } f_n(x) \text{ is nonzero}$$

As $n \to \infty$, this length approaches 0. □

By definition, a series converges if its sequence of partial sums converges. From Definition 7.2 we then know that the infinite series of functions $\sum_{n=1}^{\infty} g_n$ *converges in the mean* to the function g if

$$\lim_{n\to\infty} \left\| g - \sum_{k=1}^{n} g_k \right\| = 0$$

and *converges pointwise* to the function g if

$$\lim_{n\to\infty} \left| g(x) - \sum_{k=1}^{n} g_k(x) \right| \equiv 0$$

In the first case we write $g \approx \sum_{k=1}^{\infty} g_k$ and in the second case, $g = \sum_{k=1}^{\infty} g_k$.

We are now in a position to define a basis set for the function space $PC[a, b]$.

Definition 7.3 An orthogonal set $\{g_n\}_{n=1}^{\infty}$ on $PC[a, b]$ is a *basis set* for the space $PC[a, b]$ if and only if each function g in the space can be written *uniquely* in the form

$$g \approx \sum_{n=1}^{\infty} c_n g_n$$

The series $\sum_{n=1}^{\infty} c_n g_n$ is called the *Fourier series* of g with respect to the basis set $\{g_n\}_{n=1}^{\infty}$.

It can be shown that if g_n is a member of a basis set, then $\|g_n\| \neq 0$, so we could, if we wished, convert $\{g_n\}_{n=1}^{\infty}$ into an orthonormal set (see Exercise 12, p. 225).

Note that a basis set is defined in terms of mean convergence. In this book we will assume without proof that certain basis sets exist. In particular, the *complete* trigonometric set is a basis set for the function space $PC[-L, L]$; that is, by the very nature of a basis, every piecewise continuous function has a Fourier expansion which converges in the mean to that function. Most detailed treatments of Fourier series *prove* that a set such as the complete trigonometric set is a basis set for $PC[-L, L]$ in the sense of Definition 7.3. The main thing to note is that our assumption is far from being trivial.

It is important also to realize precisely which functions are elements of the trigonometric basis set for a given interval. For example, the function $\sin x$ is a member of the set

$$\{\cos nx\}_{n=0}^{\infty} \cup \{\sin mx\}_{m=1}^{\infty}$$

which is a basis for $PC[-\pi, \pi]$. It is also a member of the set

$$\left\{\cos\frac{nx}{2}\right\}_{n=0}^{\infty} \cup \left\{\sin\frac{mx}{2}\right\}_{m=1}^{\infty}$$

which is a basis for $PC[-2\pi, 2\pi]$. However, it is not a member of the set

$$\{\cos 2nx\}_{n=0}^{\infty} \cup \{\sin 2mx\}_{m=1}^{\infty}$$

which is a basis for $PC[-\pi/2, \pi/2]$.

Exercises for Sections 7.2 and 7.3

1. Show that $\left\{\sin\frac{n\pi x}{L}\right\}_{n=1}^{\infty}$ is an orthogonal set on $[0, L]$. Find the corresponding orthonormal set.
2. Show that $P_0 = 1$, $P_1 = x$, $P_2 = \frac{1}{2}(3x^2 - 1)$ form an orthogonal set on $[-1, 1]$. Find the corresponding orthonormal set. P_0, P_1, and P_2 are the first three members of a set of polynomials called the Legendre polynomials. $P_n(x)$ is the polynomial solution of Legendre's equation of order n.
3. Show that x^2 and x^4 are orthogonal on $[-1, 1]$ with respect to the weight function x.

A function is called *even* if $f(-x) \equiv f(x)$; it is said to be *odd* if $f(-x) \equiv -f(x)$.

4. **(a)** Show that even functions are always orthogonal to odd functions on any symmetric interval $[-L, L]$.
   (b) Generalize this result to the case of an even weight function $r(x)$.
   (*Hint:* Look ahead to the proof of Theorem 7.1 on p. 236.)
5. **(a)** Show that two even functions are always orthogonal on the symmetric interval $[-L, L]$ with respect to the weight function $r(x) = x$.
   (b) Generalize this result to the case of an arbitrary odd weight function.
6. Show that $T_0 = 1$, $T_1 = x$, $T_2 = 2x^2 - 1$ form an orthogonal set on $[-1, 1]$ with respect to the weight function $r(x) = (1 - x^2)^{-1/2}$. T_0, T_1, and T_2 are the first three members of a set of polynomials called the Chebyshev polynomials. $T_n(x)$ is the polynomial solution of Chebyshev's equation of order n.
7. Show that $H_0 = 1$, $H_1 = 2x$, $H_2 = 4x^2 - 2$ form an orthogonal set on $(-\infty, \infty)$ with respect to the weight function $r(x) = e^{-x^2}$. H_0, H_1, and H_2 are the first three members of a set of polynomials called the Hermite polynomials. $H_n(x)$ is the polynomial solution of Hermite's equation of order n.
8. Show that $L_0 = 1$, $L_1 = 1 - x$, $L_2 = 2 - 4x + x^2$ form an orthogonal set on $[0, \infty)$ with respect to the weight function $r(x) = e^{-x}$. L_0, L_1, and L_2 are the first three members of a set of polynomials called Laguerre polynomials. $L_n(x)$ is the polynomial solution of Laguerre's equation of order n.
9. Let $f_n(x) = 0$, if $0 \leqq x \leqq 1/n$ or if $2/n \leqq x \leqq 1$, and let $f_n(x) = \sqrt{n}$ if $1/n < x < 2/n$. Show that the sequence $\{f_n\}_{n=1}^{\infty}$ converges pointwise to the zero function on $[0, 1]$ but does not converge in the mean to the zero function.
10. For the sequence of Example 7.3, sketch f_{63}, f_{100}, f_{200}. Derive a formula for the length of the interval on which $f_n(x) \neq 0$.

11. Is $\cos(n\pi x/4)$ a member of the trigonometric basis set on:
(a) $PC[-1, 1]$ **(b)** $PC[-\pi, \pi]$
(c) $PC[-\frac{1}{4}, \frac{1}{4}]$ **(d)** $PC[-4, 4]$
(e) $PC[-8, 8]$
In general what characteristic must L have if $\cos(n\pi x/4)$ is to be a member of the trigonometric basis for $PC[-L, L]$?

12. Demonstrate that if g_n is a member of a basis set, then $\|g_n\| \neq 0$. (*Hint:* Apply Definition 7.3 to the function $g = kg_n$, where k is an arbitrary constant.)

7.4 Fourier Coefficients

The form of the representation of a function f in terms of an orthogonal set resembles that shown in Chapter 1 for finite dimensional vector spaces. The application to the infinite dimensional vector space $PC[a, b]$ may be considered a purely formal extension of finite dimensional methods.

Assume that the set $\{g_n\}_{n=1}^{\infty}$ is an orthogonal basis for the function space $PC[a, b]$. Then if f is a member of the space, we have convergence in the mean

$$f \approx \sum_{k=1}^{\infty} c_k g_k$$

The restrictions to be imposed on the function f to give meaningful *pointwise* convergence are discussed later. For the formal argument which follows, we assume that the series does converge pointwise to $f(x)$ except possibly at a few isolated points.

Assuming that we may take the inner product termwise (in the case of the integral inner product, this would be assuming term-by-term integration to be permissible), we take the inner product of both sides with respect to g_n to obtain

$$(g_n, f) = \left(g_n, \sum_{k=1}^{\infty} c_k g_k\right) = \sum_{k=1}^{\infty} (g_n, c_k g_k) = \sum_{k=1}^{\infty} c_k (g_n, g_k)$$

Since the set $\{g_n\}_{n=1}^{\infty}$ is orthogonal, $(g_k, g_n) = 0$ for $n \neq k$ and hence only the term $n = k$ remains in the series, $(g_n, f) = c_n(g_n, g_n)$, from which

$$c_n = \frac{(g_n, f)}{\|g_n\|^2}$$

In terms of the integral inner product,

$$c_n = \frac{\int_a^b f(x) g_n(x)\, dx}{\int_a^b [g_n(x)]^2\, dx}$$

The c_n's chosen in this manner are called the *Fourier coefficients of the function f with respect to the orthogonal set* $\{g_n\}_{n=1}^{\infty}$. They are analogous to the components of a vector with respect to a finite basis or more remotely analogous to the Taylor series coefficients for a function.

The name "Fourier coefficient" will also be given to a number c_n defined by

$$c_n = \frac{(g_n, f)}{\|g_n\|^2}$$

as long as all of the $\|g_n\|$ are nonzero, but regardless of whether the orthogonal set $\{g_n\}_{n=1}^{\infty}$ is a basis set or not, and thus regardless of whether $f \approx \sum_{n=1}^{\infty} c_n g_n$ or not. We write

$$f \sim \sum_{n=1}^{\infty} c_n g_n$$

which is read, "the Fourier series of f corresponding to the orthogonal set $\{g_n\}_{n=1}^{\infty}$," *only* to indicate how the c_n's were obtained. No other relationship can be assumed between f and the series. It is possible, for instance, for all the Fourier coefficients of a nonzero function with respect to an orthogonal set to be zero.

Example 7.4 Let $f(x) = x$ be considered as a member of $PC[-\pi, \pi]$. Write the Fourier series of f corresponding to the orthogonal set $\{\cos nx\}_{n=1}^{\infty}$.

Solution The coefficients are

$$c_n = \frac{1}{\pi} \int_{-\pi}^{\pi} x \cos nx \, dx$$

which is easily shown to be zero for all n, either by integrating by parts or using the result that the integral of an odd function over an interval symmetric about the origin is zero. (This is proved in the next section; also see Exercise 4, p. 224.) Hence,

$$x \sim \sum_{n=1}^{\infty} 0 \cdot \cos nx = 0$$

Obviously the Fourier series does not represent the function in any meaningful way. The problem here is that the given orthogonal set is *not* a basis for $PC[-\pi, \pi]$. □

If, in Example 7.4, we had used the complete trigonometric set, which *is* a basis set for $PC[-\pi, \pi]$, we would have obtained (see Example 7.10)

$$x \sim 2 \sum_{n=1}^{\infty} \frac{(-1)^{n+1}}{n} \sin nx$$

The first few partial sums S_n of this series as shown in Figure 7.3 suggest how a series of trigonometric terms can represent such a distinctly nontrigonometric function. This series converges pointwise to x for $-\pi < x < \pi$ and to 0 for $x = -\pi$ and $x = \pi$. The series, of course, converges *in the mean* to $f(x) = x$ on $[-\pi, \pi]$.

Since the complete trigonometric set

$$\left\{\cos \frac{n\pi x}{L}\right\}_{n=0}^{\infty} \cup \left\{\sin \frac{n\pi x}{L}\right\}_{n=1}^{\infty}$$

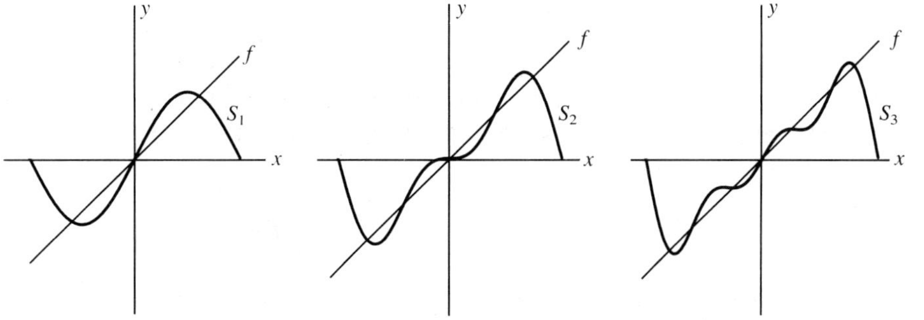

$S_1(x) = 2\sin x$ $\quad$ $S_2(x) = 2\sin x - \sin 2x$ $\quad$ $S_3(x) = 2\sin x - \sin 2x + \frac{2}{3}\sin 3x$

Figure 7.3

is a basis set for the space $PC[-L, L]$, then for *any* $f \in PC[-L, L]$ we know that its Fourier series converges in the mean to f. Such series take on added importance, therefore, so we will find general expressions for these Fourier coefficients:

$$f \sim \sum_{n=0}^{\infty} a_n \cos\frac{n\pi x}{L} + \sum_{n=1}^{\infty} b_n \sin\frac{n\pi x}{L}$$

$$a_0 = \frac{(1, f)}{\|1\|^2} = \frac{1}{2L}\int_{-L}^{L} f(x)\,dx$$

$$a_n = \frac{\left(\cos\frac{n\pi x}{L}, f\right)}{\left\|\cos\frac{n\pi x}{L}\right\|^2} = \frac{1}{L}\int_{-L}^{L} f(x)\cos\frac{n\pi x}{L}\,dx \qquad n \neq 0$$

$$b_n = \frac{\left(\sin\frac{n\pi x}{L}, f\right)}{\left\|\sin\frac{n\pi x}{L}\right\|^2} = \frac{1}{L}\int_{-L}^{L} f(x)\sin\frac{n\pi x}{L}\,dx$$

The formulas for the space $PC[-\pi, \pi]$ are particularly simple as well:

$$f \sim \sum_{n=0}^{\infty} a_n \cos nx + \sum_{n=1}^{\infty} b_n \sin nx$$

$$a_0 = \frac{1}{2\pi}\int_{-\pi}^{\pi} f(x)\,dx$$

$$a_n = \frac{1}{\pi}\int_{-\pi}^{\pi} f(x)\cos nx\,dx \qquad n \neq 0$$

$$b_n = \frac{1}{\pi}\int_{-\pi}^{\pi} f(x)\sin nx\,dx$$

In passing we note that some engineering handbooks write the Fourier series of f in the form

$$f \sim \frac{a_0}{2} + \sum_{n=1}^{\infty} a_n \cos nx + \sum_{n=1}^{\infty} b_n \sin nx$$

This makes the formula for a_0 exactly that given for a_n with $n = 0$.

Example 7.5 Find the Fourier trigonometric series of the function

$$f(x) = \begin{cases} 2, & -\pi \leqq x < 0 \\ 1, & 0 \leqq x \leqq \pi \end{cases}$$

with respect to the complete trigonometric set.

Solution

$$a_0 = \frac{1}{2\pi}\int_{-\pi}^{\pi} f(x)\,dx = \frac{1}{2\pi}\int_{-\pi}^{0} 2\,dx + \frac{1}{2\pi}\int_{0}^{\pi} 1\,dx = \frac{3}{2}$$

$$a_n = \frac{1}{\pi}\int_{-\pi}^{0} 2\cos nx\,dx + \frac{1}{\pi}\int_{0}^{\pi} \cos nx\,dx = 0, \qquad n \geqq 1$$

$$\begin{aligned} b_n &= \frac{1}{\pi}\int_{-\pi}^{0} 2\sin nx\,dx + \frac{1}{\pi}\int_{0}^{\pi} \sin nx\,dx \\ &= \frac{1}{\pi}\left[\frac{-2\cos nx}{n}\bigg|_{-\pi}^{0} - \frac{\cos nx}{n}\bigg|_{0}^{\pi}\right] \\ &= \frac{1}{n\pi}[-2 + 2\cos n\pi - \cos n\pi + 1] \\ &= \frac{\cos n\pi - 1}{n\pi} = \frac{(-1)^n - 1}{n\pi} \end{aligned}$$

Therefore

$$f \sim \frac{3}{2} - \frac{2}{\pi}\left(\sin x + \frac{\sin 3x}{3} + \frac{\sin 5x}{5} + \cdots\right)$$

(See Figure 7.4.) □

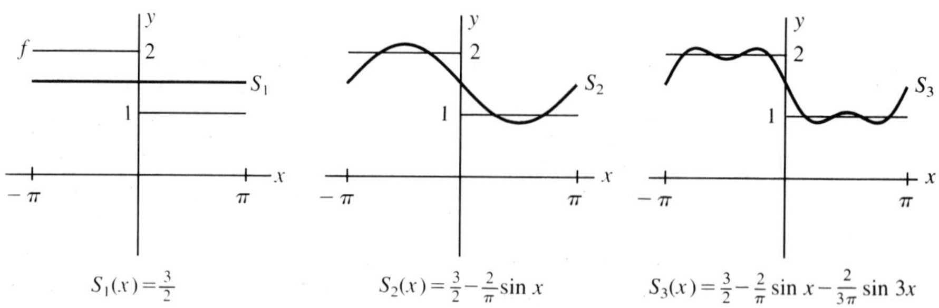

Figure 7.4

Example 7.6 Find the Fourier series of the function

$$f(x)=\begin{cases}0, & -2\leqq x<0\\ x, & 0\leqq x\leqq 2\end{cases}$$

Solution In this case $L=2$, and

$$a_0=\frac{1}{4}\int_0^2 x\,dx=\frac{x^2}{8}\bigg|_0^2=\frac{1}{2}$$

$$a_n=\frac{1}{2}\int_0^2 x\cos\frac{n\pi x}{2}\,dx$$

$$=\frac{1}{2}\left[\frac{4}{n^2\pi^2}\cos\frac{n\pi x}{2}+\frac{2x}{n\pi}\sin\frac{n\pi x}{2}\bigg|_0^2\right]$$

$$=\frac{2(\cos n\pi-1)}{n^2\pi^2}=\begin{cases}0, & n\text{ even}, n\neq 0\\ \dfrac{-4}{n^2\pi^2}, & n\text{ odd}\end{cases}$$

$$b_n=\frac{1}{2}\int_0^2 x\sin\frac{n\pi x}{2}\,dx$$

$$=\frac{1}{2}\left[\frac{4}{n^2\pi^2}\sin\frac{n\pi x}{2}-\frac{2x}{n\pi}\cos\frac{n\pi x}{2}\bigg|_0^2\right]$$

$$=\frac{-2\cos n\pi}{n\pi}=\frac{(-1)^{n+1}2}{n\pi}$$

Therefore,

$$f\sim\frac{1}{2}-\frac{4}{\pi^2}\left(\cos\frac{\pi x}{2}+\frac{1}{9}\cos\frac{3\pi x}{2}+\frac{1}{25}\cos\frac{5\pi x}{2}+\cdots\right)$$
$$+\frac{2}{\pi}\left(\sin\frac{\pi x}{2}-\frac{1}{2}\sin\pi x+\frac{1}{3}\sin\frac{3\pi x}{2}-\cdots\right)$$ □

Example 7.7 Find the Fourier series of the half-wave rectified sine wave shown below and described analytically by

$$f(x)=\begin{cases}0, & -3<x<0\\ \sin\dfrac{\pi x}{3}, & 0\leqq x\leqq 3\end{cases}$$

(See Figure 7.5.)

Solution

$$a_0=\frac{1}{6}\int_0^3\sin\frac{\pi x}{3}\,dx=\frac{1}{\pi}$$

$$a_1=\frac{1}{3}\int_0^3\sin\frac{\pi x}{3}\cos\frac{\pi x}{3}\,dx=0$$

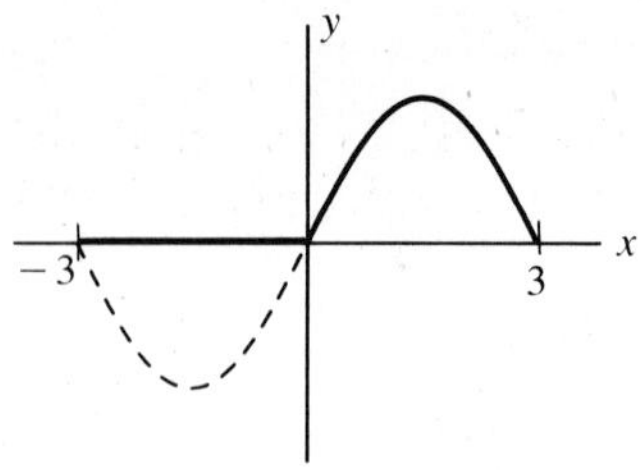

Figure 7.5 *Half-wave rectified sine wave*

For $n \geqq 2$,

$$\begin{aligned}
a_n &= \frac{1}{3}\int_0^3 \sin\frac{\pi x}{3}\cos\frac{n\pi x}{3}\,dx \\
&= \frac{1}{6}\int_0^3 \sin\frac{(n+1)\pi x}{3}\,dx - \frac{1}{6}\int_0^3 \sin\frac{(n-1)\pi x}{3}\,dx \\
&= -\frac{1}{2\pi}\left.\frac{\cos\dfrac{(n+1)\pi x}{3}}{n+1}\right|_0^3 + \frac{1}{2\pi}\left.\frac{\cos\dfrac{(n-1)\pi x}{3}}{n-1}\right|_0^3 \\
&= \frac{1}{2\pi}\left(\frac{-\cos(n+1)\pi + 1}{n+1} + \frac{\cos(n-1)\pi - 1}{n-1}\right) \\
a_n &= 0, \qquad \text{if } n \text{ is odd, } n \neq 1 \\
&= \frac{1}{\pi}\left(\frac{n-1-n-1}{n^2-1}\right) = \frac{-2}{\pi(n^2-1)}, \qquad \text{if } n \text{ is even}
\end{aligned}$$

$$b_1 = \frac{1}{3}\int_0^3 \sin^2\left(\frac{\pi x}{3}\right)dx = \frac{1}{2}$$

For $n \geqq 2$,

$$\begin{aligned}
b_n &= \frac{1}{3}\int_0^3 \sin\frac{\pi x}{3}\sin\frac{n\pi x}{3}\,dx \\
&= \frac{1}{6}\int_0^3 \cos\frac{(n-1)\pi x}{3}\,dx - \frac{1}{6}\int_0^3 \cos\frac{(n+1)\pi x}{3} \\
&= \frac{1}{2}\left.\frac{\sin\dfrac{(n-1)\pi x}{3}}{(n-1)\pi}\right|_0^3 - \frac{1}{2}\left.\frac{\sin\dfrac{(n+1)\pi x}{3}}{(n+1)\pi}\right|_0^3 = 0
\end{aligned}$$

Therefore,

$$f \sim \frac{1}{\pi} + \frac{1}{2}\sin\frac{\pi x}{3} - \frac{2}{\pi}\left(\frac{\cos\dfrac{2\pi x}{3}}{3} + \frac{\cos\dfrac{4\pi x}{3}}{15} + \frac{\cos\dfrac{6\pi x}{3}}{35} + \cdots\right) \qquad \square$$

Exercises for Section 7.4

1. Find the Fourier series of the function $f(x) = x^2$ with respect to the orthogonal set $\{\sin nx\}_{n=1}^{\infty}$ on $[-\pi, \pi]$. Then find the Fourier series with respect to $\{\cos nx\}_{n=0}^{\infty}$ on $[-\pi, \pi]$. What is the Fourier series of x^2 on $[-\pi, \pi]$ with respect to the complete trigonometric set?
2. Show how the Fourier coefficients of $f(x) = x^m$, m a positive integer, with respect to the complete trigonometric set on $[-L, L]$ are related to those with respect to the trigonometric set on $[-\pi, \pi]$.
3. Find the Fourier series of $f(x) = \cos^2 x$ on $[-\pi, \pi]$ with respect to the complete trigonometric set.

In Exercises 4–10, find the Fourier series of the given function with respect to the complete trigonometric set on the indicated interval.

4. $f(x) = \begin{cases} 1, & -\pi < x < 0 \\ 2, & 0 < x < \pi \end{cases}$

5. $|\cos x|$ on $[-\pi, \pi]$

6. $|\sin x|$ on $[-\pi, \pi]$

7. x^2 on $[-1, 1]$

8. $f(x) = \begin{cases} 1, & \text{on } [-\pi, 0) \\ -1, & \text{on } [0, \pi] \end{cases}$

9. $f(x) = \begin{cases} 0, & \text{on } [-2, 0) \\ 1, & \text{on } [0, 2] \end{cases}$

10. $\sin x$ on $[-\pi/2, \pi/2]$

The deflection $u(x)$ of a uniform elastic column of length L subjected to an axial compressive force P, together with a distributed lateral loading $F(x)$, satisfies the ordinary differential equation

$$EIu^{(4)}(x) + Pu''(x) = F(x), \qquad 0 < x < L$$

Here E is the modulus of elasticity of the column material and I the moment of inertia of the column cross section about the reference axis. For special values of the axial load P known as the critical buckling loads, namely,

$$P_n = n^2\pi^2 \frac{EI}{L^2} \qquad (n = 1, 2, \ldots)$$

the special solutions

$$u_n = \sin \frac{n\pi x}{L}$$

satisfy the homogeneous deflection equation. If P is *not* one of these critical values, the method of undetermined coefficients can often be used to find the deflection due to the lateral load F. The key to this approach is to express F, and hence u, as a linear combination of these special solutions of the homogeneous problem.

Exercises 11–14 refer to this practical application.

11. If $P \neq P_n$, find a particular solution for the deflection $u(x)$ when $F(x) = B_n \sin \dfrac{n\pi x}{L}$.
12. Show formally that if $F(x)$ has the Fourier series

$$F(x) \sim \sum_{n=1}^{\infty} B_n \sin \frac{n\pi x}{L}$$

then for $P \neq P_n$ $(n = 1, 2, \ldots)$, the deflection can be expressed as

$$u(x) \sim \frac{L^2}{\pi^2} \sum_{n=1}^{\infty} \frac{B_n}{n^2(P_n - P)} \sin \frac{n\pi x}{L}$$

13. In the case $L = 1$, determine the deflection $u(x)$ when $F(x) = -3$. (Assume $F(x) = 3$ for $-1 < x < 0$, $F(x) = -3$ for $0 < x < 1$.)
14. Let $EI/L^2 = 1$ with $L = \pi$. Find the column deflection in the case $F(x) = x$. (Assume $F(x) = x$ for $-\pi < x < \pi$.)

7.5 Alternate Expressions for Fourier Series

For ease in calculation and application, compact and symmetric notation, or emphasis on particular quantities of interest, different forms may be chosen in which to represent a trigonometric Fourier series. For example, in some situations it is advantageous to combine the cosine and sine terms with the same frequency. In other cases it is helpful to re-express the trigonometric functions in terms of complex exponentials by means of Euler's formula. In this section we discuss these two principal alternative ways for expressing a trigonometric Fourier series. Detailed consideration of Fourier series consisting solely of sine terms or solely of cosine terms, the so-called sine and cosine series, is delayed until Section 7.9.

A form that pairs terms of the same frequency and clearly displays amplitudes and phase shifts corresponding to each frequency (or harmonic) may be obtained from the standard expression

$$f(x) \sim \sum_{n=0}^{\infty} a_n \cos \frac{n\pi x}{L} + \sum_{n=1}^{\infty} b_n \sin \frac{n\pi x}{L}$$

by the substitutions $a_n = A_n \cos \delta_n$ and $b_n = A_n \sin \delta_n$ where $A_n = \sqrt{a_n^2 + b_n^2}$ for $n \geq 1$ and $A_0 = a_0$. It follows that

$$\begin{aligned} f(x) &\sim A_0 + \sum_{n=1}^{\infty} A_n \left(\cos \frac{n\pi x}{L} \cos \delta_n + \sin \frac{n\pi x}{L} \sin \delta_n \right) \\ &\sim A_0 + \sum_{n=1}^{\infty} A_n \cos \left(\frac{n\pi x}{L} - \delta_n \right) \end{aligned}$$

This is called the *shifted cosine* form of the trigonometric Fourier series. Alternatively, making the substitutions $a_n = A_n \sin \gamma_n$ and $b_n = A_n \cos \gamma_n$, we obtain another representation in amplitude and phase form, the *shifted sine* series

$$f(x) \sim A_0 + \sum_{n=1}^{\infty} A_n \sin \left(\frac{n\pi x}{L} + \gamma_n \right)$$

For simplicity the individual phase shifts δ_n and γ_n may be read directly from the triangle in Figure 7.6.

Example 7.8 Re-express the Fourier series of Example 7.6 in shifted cosine form.

Exercises for Section 7.5

In Exercises 1–8, find the trigonometric Fourier series of the given function on the indicated interval and express the result in *shifted cosine* form.

1. $f(x) = \begin{cases} 1, & -\pi < x < 0 \\ 2, & 0 < x < \pi \end{cases}$

2. $|\cos x|$ on $[-\pi, \pi]$

3. $|\sin x|$ on $[-\pi, \pi]$

4. x^2 on $[-1, 1]$

5. $f(x) = \begin{cases} 1, & \text{on } [-\pi, 0) \\ -1, & \text{on } [0, \pi] \end{cases}$

6. $f(x) = \begin{cases} 0, & \text{on } [-2, 0) \\ 1, & \text{on } [0, 2] \end{cases}$

7. $\sin x$ on $[-\pi/2, \pi/2]$

8. $f(x) = \begin{cases} 0, & -3 < x < 0 \\ \sin\dfrac{\pi x}{3}, & 0 \le x \le 3 \end{cases}$

For Exercises 9–16, find the trigonometric Fourier series of the function given in the corresponding Exercises 1–8, but express the result in *complex exponential* form.

9. Exercise 1

10. Exercise 2

11. Exercise 3

12. Exercise 4

13. Exercise 5

14. Exercise 6

15. Exercise 7

16. Exercise 8

17. **(a)** Employ Euler's formula $e^{i\theta} = \cos\theta + i\sin\theta$ and the orthogonality of the complete trigonometric set

$$\left\{\cos\frac{n\pi x}{L}\right\}_{n=0}^{\infty} \cup \left\{\sin\frac{n\pi x}{L}\right\}_{n=1}^{\infty} \qquad \text{on } [-L, L]$$

to show that

$$\int_{-L}^{L} e^{in\pi x/L} e^{-im\pi x/L}\, dx = \begin{cases} 0, & n \ne m \\ 2L, & n = m \end{cases}$$

(b) Use the result of part (a) to demonstrate that if $f(x)$ is represented as a Fourier series in complex exponential form, that is,

$$f(x) \sim \sum_{n=-\infty}^{\infty} c_n e^{in\pi x/L}$$

the Fourier coefficients c_n are given by

$$c_n = \frac{1}{2L}\int_{-L}^{L} f(x) e^{-in\pi x/L}\, dx$$

(*Hint:* Take the inner product on both sides with respect to $e^{-im\pi x/L}$ and proceed as in Section 7.4.)

In Section 3.7, the behavior of the well-known *RLC* circuit was discussed extensively. The governing equation for the current in such a circuit is

$$Li''(t) + Ri'(t) + i(t)/C = v'(t)$$

where $v(t)$ is the driving input voltage. If $R \ne 0$, the response of the circuit to a sinusoidal input voltage is often expressed as a shifted cosine (see p. 110).

18. Find the steady-state current in an RLC circuit in shifted cosine form in each case.
(a) $v(t) = 7 \sin 5t$ **(b)** $v(t) = -2\cos 3t$

19. Find the steady-state current in shifted cosine form when

$$v(t) = \sum_{n=1}^{\infty} B_n \sin n\omega t$$

Write out at least the first three terms of the series. (Assume all $B_n \geq 0$.)

7.6 Subspaces of $PC[-L, L]$

The function space $PC[-L, L]$ has two important subspaces, namely the set of all piecewise continuous even functions $PC_e[-L, L]$ and the set of all piecewise continuous odd functions $PC_o[-L, L]$. As you will see, the work to compute the Fourier coefficients for functions in these two subspaces is considerably reduced. Recall that a function f is even if $f(-x) \equiv f(x)$ and that a function g is odd if $g(-x) \equiv -g(x)$. Typical examples of even functions are the even powers x^0, x^2, x^4, and so on, $|x|$, $\cos x$, and linear combinations of such functions. Representative odd functions are the odd powers x^1, x^3, x^5, and so on, $\tan x$, and $\sin x$.

Theorem 7.1 Let f be an odd function. Then

$$\int_{-L}^{L} f(x)\,dx = 0$$

Proof We split the integral into two parts:

$$\int_{-L}^{L} f(x)\,dx = \int_{-L}^{0} f(x)\,dx + \int_{0}^{L} f(x)\,dx$$

Using the change of variables $u = -x$ in the integral from $-L$ to 0,

$$\begin{aligned}\int_{-L}^{L} f(x)\,dx &= \int_{0}^{L} f(-u)\,du + \int_{0}^{L} f(x)\,dx \\ &= -\int_{0}^{L} f(u)\,du + \int_{0}^{L} f(x)\,dx \\ &= 0\end{aligned}$$

as was to be shown. ■

Theorem 7.2 If f is an even function, then the Fourier coefficients b_n are zero for all n.

Proof

$$b_n = \frac{1}{L}\int_{-L}^{L} f(x) \sin \frac{n\pi x}{L}\,dx$$

Since the integrand is the product of an even function with an odd function, it is odd and hence the integral is zero by the preceding theorem. ■

Theorem 7.3 If f is an odd function, then the Fourier coefficients a_n are zero for all n.

Proof The same reasoning is used as in Theorem 7.2. ■

Example 7.10 Find the Fourier series for the function defined on the interval $[-\pi, \pi]$ by $f(x) = x$.

Solution Since the given function is odd, each a_n equals zero and

$$\begin{aligned} b_n &= \frac{1}{\pi}\int_{-\pi}^{\pi} x \sin nx \, dx \\ &= \frac{2}{\pi}\int_{0}^{\pi} x \sin nx \, dx \\ &= \frac{2}{\pi}\left(\frac{-x}{n}\cos nx + \frac{1}{n^2}\sin nx\right)\Bigg|_0^{\pi} \\ &= (-1)^{n+1}\frac{2}{n} \end{aligned}$$

Therefore,

$$x \sim 2\sum_{n=1}^{\infty} \frac{(-1)^{n+1}}{n} \sin nx$$

(Recall the graphs of partial sums exhibited in Figure 7.3.) □

Example 7.11 Find the Fourier series for the function $f(x) = |x|$ on $[-5, 5]$.

Solution Since the given function is even, all the b_n's $= 0$. Also,

$$\begin{aligned} a_0 &= \frac{1}{10}\int_{-5}^{5} |x| \, dx = \frac{1}{5}\int_{0}^{5} x \, dx = \frac{x^2}{10}\Bigg|_0^5 = \frac{5}{2} \\ a_n &= \frac{2}{5}\int_{0}^{5} x \cos\frac{n\pi x}{5} \, dx \\ &= \frac{2}{n\pi} x \sin\frac{n\pi x}{5}\Bigg|_0^5 + \frac{10}{n^2\pi^2}\cos\frac{n\pi x}{5}\Bigg|_0^5 \\ &= \frac{-20}{n^2\pi^2} \qquad \text{for } n \text{ odd} \\ &= 0 \qquad \text{for } n \text{ even} \end{aligned}$$

Therefore

$$|x| \sim \frac{5}{2} - \frac{20}{\pi^2}\left(\cos\frac{\pi x}{5} + \frac{1}{9}\cos\frac{3\pi x}{5} + \frac{1}{25}\cos \pi x + \cdots\right)$$ □

There are other, not so obvious, examples of subspaces of $PC[-L, L]$ for which even more of the Fourier coefficients can be predetermined equal to zero. The following example is typical.

Example 7.12 Let f be a function in $PC_o[-\pi, \pi]$ that is symmetric with respect to the line $x = \pi/2$. Show that the only possible nonzero Fourier coefficients are b_n for n odd.

Solution Since f is odd, $a_n = 0$ for all n, and

$$b_n = \frac{2}{\pi}\int_0^{\pi} f(x)\sin nx\,dx$$
$$= \frac{2}{\pi}\int_0^{\pi/2} f(x)\sin nx\,dx + \frac{2}{\pi}\int_{\pi/2}^{\pi} f(x)\sin nx\,dx$$

Since f is symmetric with respect to the line $x = \pi/2$,

$$f(x) = f(\pi - x)$$

Hence the integral from $\pi/2$ to π becomes

$$\int_{\pi/2}^{\pi} f(\pi - x)\sin nx\,dx$$

Let $u = \pi - x$ in this integral to obtain

$$b_n = \frac{2}{\pi}\int_0^{\pi/2} f(x)\sin nx\,dx + \frac{2}{\pi}\int_0^{\pi/2} f(u)\sin n(\pi - u)\,du$$
$$= \frac{2}{\pi}[1 + (-1)^{n+1}]\int_0^{\pi/2} f(x)\sin nx\,dx$$

Therefore, if n is even, $b_n = 0$, and if n is odd

$$b_n = \frac{4}{\pi}\int_0^{\pi/2} f(x)\sin nx\,dx$$ □

7.7 Properties of Fourier Coefficients

The Fourier coefficients have some elementary properties, which, while being almost obvious, serve to reduce the amount of computational work. Our first theorem is a case in point.

Theorem 7.4 **1.** Let $f(x) = g(x) + h(x)$. Then the Fourier coefficients of the function f are the sum of the coefficients of g and of h.

2. The Fourier coefficients of $cf(x)$ are c times the Fourier coefficients of f.

Proof The proof follows directly from the formulas for the Fourier coefficients. ■

Theorem 7.5 Let $f(x) - g(x) = c$, where c is constant. Then the Fourier coefficients of f and g differ only in the value of a_0, the difference being precisely c.

Proof Consider the specific case of Fourier expansions on $PC[-\pi, \pi]$, and let $f(x) = g(x) + c$. Let a_n and b_n be the Fourier coefficients of f, and a_n' and b_n' be the Fourier coefficients of g, both with respect to the complete set of trigonometric functions on $[-\pi, \pi]$. Then, for $n \neq 0$,

$$\begin{aligned} a_n &= \frac{1}{\pi}\int_{-\pi}^{\pi} f(x)\cos nx\,dx \\ &= \frac{1}{\pi}\int_{-\pi}^{\pi} g(x)\cos nx\,dx + \frac{c}{\pi}\int_{-\pi}^{\pi}\cos nx\,dx \\ &= a_n' + 0 = a_n' \end{aligned}$$

Similarly,

$$b_n = b_n'$$

But

$$\begin{aligned} a_0 &= \frac{1}{2\pi}\int_{-\pi}^{\pi} f(x)\,dx \\ &= \frac{1}{2\pi}\int_{-\pi}^{\pi} g(x)\,dx + \frac{c}{2\pi}\int_{-\pi}^{\pi} dx \\ &= a_0' + c \end{aligned}$$

■

Theorem 7.6 Let $\{\phi_n\}_{n=1}^{\infty}$ be an orthogonal basis for $PC[a, b]$, and suppose $f(x) = \phi_j(x)$ for some j. Then the Fourier coefficients of f with respect to that basis set are all zero except for c_j which is equal to 1.

Proof See Exercise 8 in the next set of exercises. ■

Example 7.13 Find the Fourier series of $3x + \sin x$ considered as a member of $PC[-\pi, \pi]$.

Solution The Fourier coefficients of the function $f(x) = x$ are found in Example 7.10 to be

$$b_n = (-1)^{n+1}\left(\frac{2}{n}\right), \qquad a_n = 0 \text{ for all } n$$

Hence, the Fourier coefficients of $f(x) = 3x$ are

$$b_n = (-1)^{n+1}\left(\frac{6}{n}\right), \qquad a_n = 0 \text{ for all } n$$

Since $\sin x$ is a member of the trigonometric basis set on $PC[-\pi, \pi]$, the Fourier coefficients of $\sin x$ are

$$b_1 = 1 \qquad \text{and all the rest } 0$$

Thus,

$$3x + \sin x \sim 7 \sin x + \sum_{n=2}^{\infty} \frac{6}{n}(-1)^{n+1} \sin nx \qquad \square$$

Example 7.14 Find the Fourier series for the function $\sin(x/2)$:

(a) on $PC[-2\pi, 2\pi]$ (b) on $PC[-\pi, \pi]$

Solution (a) The trigonometric basis set for $PC_o[-2\pi, 2\pi]$ is

$$\left\{\sin \frac{nx}{2}\right\}_{n=1}^{\infty}$$

Note that $\sin(x/2)$ is a member of this basis set (for $n = 1$), and hence its Fourier series is just $\sin(x/2)$.

(b) The trigonometric basis set for $PC_o[-\pi, \pi]$ is

$$\{\sin nx\}_{n=1}^{\infty}$$

The function is *not* a member of this basis set. We calculate

$$\begin{aligned}
b_n &= \frac{2}{\pi} \int_0^{\pi} \sin \frac{x}{2} \sin nx \, dx \\
&= \frac{1}{\pi} \int_0^{\pi} \cos\left(n - \frac{1}{2}\right) x \, dx - \frac{1}{\pi} \int_0^{\pi} \cos\left(n + \frac{1}{2}\right) x \, dx \\
&= \frac{1}{\pi}\left[\frac{2}{2n-1} \sin\left(\frac{2n-1}{2}\right) x \Big|_0^{\pi} - \frac{2}{2n+1} \sin\left(\frac{2n+1}{2}\right) x \Big|_0^{\pi}\right] \\
&= \frac{2}{\pi}\left[\frac{(-1)^{n+1}}{2n-1} - \frac{(-1)^n}{2n+1}\right] \\
&= (-1)^{n+1}\left(\frac{2}{\pi}\right)\left[\frac{1}{2n-1} + \frac{1}{2n+1}\right] \\
&= (-1)^{n+1}\left(\frac{8}{\pi}\right)\frac{n}{4n^2-1}
\end{aligned}$$

The Fourier series for the given function is

$$f(x) \sim \frac{8}{\pi} \sum_{n=1}^{\infty} (-1)^{n+1} \frac{n}{4n^2 - 1} \sin nx \qquad \square$$

Exercises for Sections 7.6 and 7.7

1. Determine whether the following functions are even, odd, or neither:

(a) x **(b)** x^2

(c) $x^2 + x$ **(d)** $|\sin x|$

(e) $|\cos x|$ (f) $\cos x$
(g) $\sin x$ (h) $\sin 2x$
(i) $\sin x + \cos x$ (j) e^x
(k) $\cosh x$ (l) $\sinh 2x$
(m) x^{2n} (n) x^{2n+1}

2. Prove:
(a) The product of two odd functions is even.
(b) The product of an odd function and an even function is odd.
(c) The product of two even functions is even.

3. Evaluate

$$\int_{-1}^{1} x \tan^8 x \cos^{2/3} x e^{x^2} \sin^2 x^4 \, dx$$

4. Let f be a function in $PC_e[-\pi, \pi]$ that is symmetric with respect to the line $x = \pi/2$. Show that the only possible nonzero Fourier coefficients are a_n for n even. (*Hint:* Recall Example 7.12.)

5. Find the Fourier series of the function which is -1 on $[-\pi, 0)$ and 1 on $[0, \pi]$. Then find the Fourier series of the function which is $-\frac{1}{2}$ on $[-\pi, 0)$ and is $\frac{1}{2}$ on $[0, \pi]$.

6. Find the Fourier series of x and x^2 on $[-\pi, \pi]$ and then $2x + 3x^2$ on $[-\pi, \pi]$.

7. Find the Fourier series of the function $f(x) = 5\cos 2x + \sin(x/2) - 3$ on $[-\pi, \pi]$.

8. Prove Theorem 7.6.

9. Find the Fourier series of $\cos x$:
(a) on $[-\pi, \pi]$ (b) on $[-2\pi, 2\pi]$
(c) on $[-\pi/2, \pi/2]$

10. Find a Fourier series which contains only sine terms and which is the Fourier series of $x - \pi$ on $[0, \pi]$.

11. Find a Fourier series which contains only cosine terms and which is the Fourier series of $x - \pi$ on $[\pi, 2\pi]$.

In Exercises 12–16, find the trigonometric Fourier series of the function displayed in the designated figure. Assume that the basic interval is $[-2, 2]$, and use all available symmetry to simplify the calculations.

12.

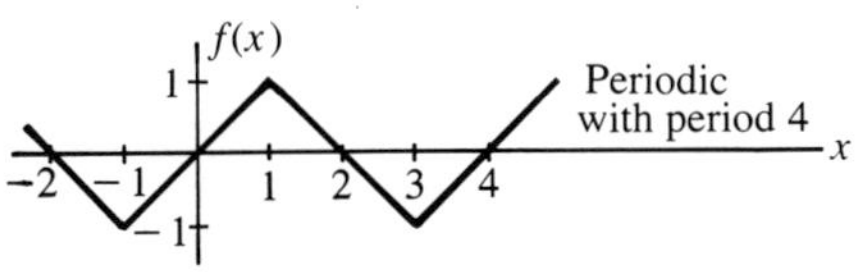

Figure 7.7

13.

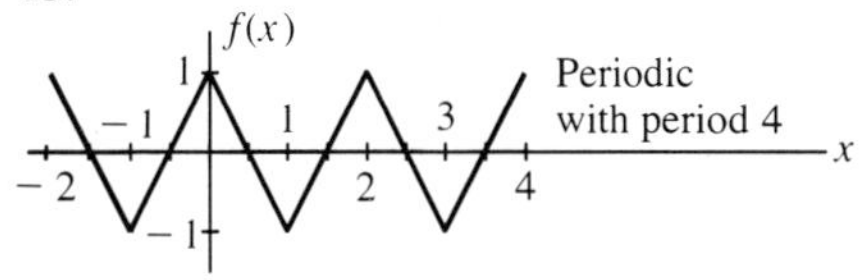

Figure 7.8

14.

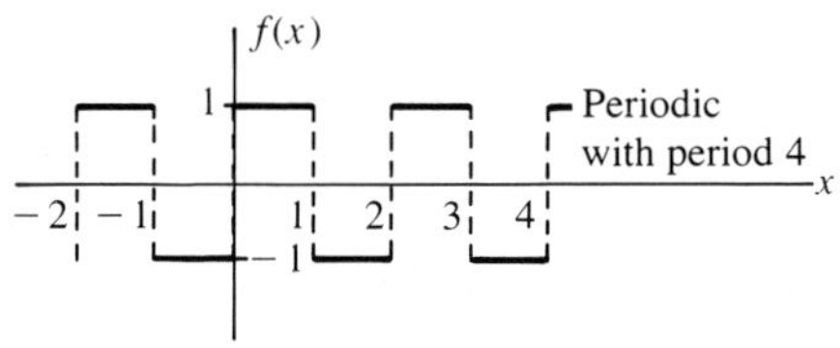

Figure 7.9

15.

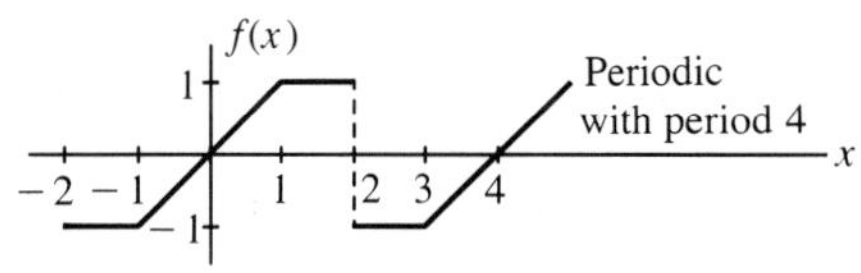

Figure 7.10

16.

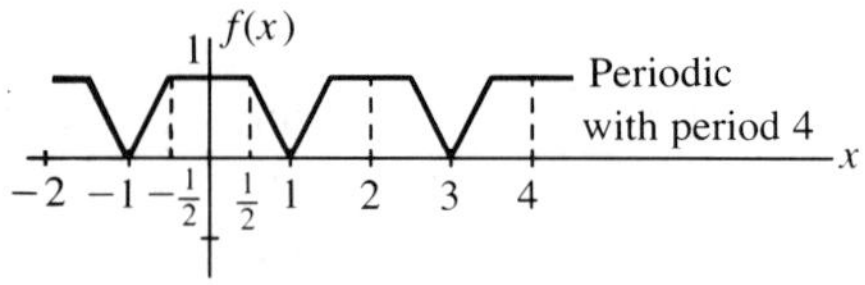

Figure 7.11

7.8 Convergence of a Fourier Series

As we stated earlier, the very definition of a basis assures convergence in the mean of the Fourier series of a function with respect to an orthogonal set which is a basis. The series will converge in the mean to the function itself. Thus for all the examples of the previous sections in which a basis set is used for the expansion, $\approx$ may be substituted for $\sim$.

A very interesting sidetrip that can be taken here passes through a careful analysis of convergence in the mean for generalized Fourier series and leads to some beautiful classical results. We will quote some of them without proof since, even without the supporting analysis, they enhance our understanding and appreciation of Fourier analysis.

1. Using the first N functions of an orthogonal set $\{\phi_n\}_{n=1}^{\infty}$, the Fourier approximation to a function f given by

$$F_N(x) = \sum_{n=1}^{N} c_n \phi_n(x)$$

where $c_n = (f, \phi_n)/\|\phi_n\|^2$ is the usual nth Fourier coefficient, is the best possible approximation in the following sense: The *mean square error*, $E_N = \|F_n - f\|^2$, using F_N as an approximation and the mean square error e_n using any other approximation

$$\sum_{n=1}^{N} b_n \phi_n(x)$$

must satisfy the inequality $e_n \geq E_n$.

2. The Fourier coefficients of a function in $PC[a, b]$ with respect to the orthogonal set $\{\phi_n\}_{n=1}^{\infty}$ satisfy the inequality

$$\sum_{n=1}^{\infty} (c_n \|\phi_n\|)^2 \leq \|f\|^2 \qquad \text{(Bessel's inequality)}$$

with equality (called Parseval's equality) *if and only if* the orthogonal set is a basis set. In terms of the trigonometric basis set on $[-L, L]$, Parseval's equality takes the form

$$\|f\|^2 = L\left[2a_0^2 + \sum_{n=1}^{\infty} (a_n^2 + b_n^2)\right]$$

3. Parseval's equality in the case where the set $\{\phi_n\}_{n=1}^{\infty}$ is an orthonormal basis has a fascinating interpretation. In that case $\|\phi_n\| = 1$ and hence

$$\|f\|^2 = \sum_{n=1}^{\infty} c_n^2$$

Since the c_n's are computed with respect to a basis for $PC[a, b]$, they are often called *components* of f with respect to the basis. Thus, Parseval's equality may be interpreted as the infinite-dimensional analog of the situation in three dimensions

$$|\mathbf{A}|^2 = a_1^2 + a_2^2 + a_3^2$$

where a_1, a_2, and a_3 are the components of $\mathbf{A}$ with respect to some orthonormal basis for the set of three-dimensional geometric vectors. You can probably see why Parseval's equality is sometimes called the "infinite-dimensional Pythagorean theorem."

The situation for pointwise convergence is neither so simple nor so elegant. In fact, one of the unsolved problems of mathematics is the determination of conditions that are both necessary and sufficient for the Fourier series to converge pointwise to the function. For example, there exist continuous functions whose Fourier series do not represent them. However, the class of functions that can be usefully represented by their Fourier series is surprisingly large and general. The *Dirichlet conditions* that *f be bounded with only a finite number of maxima and minima and only a finite number of discontinuities over any one period* constitute a handy set of sufficient conditions. Indeed, we have the following theorem, which we state without proof.

Theorem 7.7 Let f be bounded, piecewise continuous, and have a finite number of maxima and minima on $[-L, L]$. Then the corresponding Fourier series converges pointwise on $[-L, L]$. Its sum is $f(x)$ except at points of discontinuity where it converges to the average of the right- and left-hand limits. ■

The Dirichlet conditions make it clear that f need not be continuous in order to be represented by its Fourier series expansion. This means that the graph of a function may consist of a number of disjointed arcs of different curves, each defined by a different equation and still be represented by a Fourier series. Note that outside the interval $[-L, L]$ the Fourier series converges to the periodic extension of f. Hence if f is already periodic, then its Fourier series represents it everywhere except at its points of discontinuity. Further, the Fourier series of the function will converge to the average value at the discontinuities regardless of the definition (or lack of definition) of the function at these points.

Example 7.15 Sketch the graph of the function to which the Fourier series of f converges if

$$f(x) = \begin{cases} 2, & -\pi < x < 0 \\ 1, & 0 \leqq x \leqq \pi \end{cases}$$

(See Example 7.5.)

Solution The function is extended periodically and at the points of discontinuity, the series averages the right- and left-hand limits. (See Figure 7.12.) □

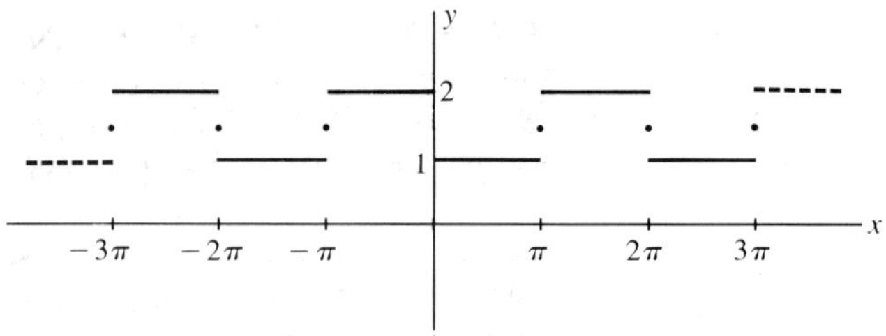

Figure 7.12

Example 7.16 Show that

$$1-\frac{1}{3}+\frac{1}{5}-\frac{1}{7}+\cdots+\frac{(-1)^{n+1}}{2n-1}+\cdots=\frac{\pi}{4}$$

Solution We use the function of Examples 7.15 and 7.5 and the result of Theorem 7.7 to write

$$f(x)=\frac{3}{2}-\frac{2}{\pi}\left(\sin x+\frac{\sin 3x}{3}+\frac{\sin 5x}{5}+\cdots\right)$$

where the equality is valid for all x except at the points of discontinuity.

Using the equality at the point $x = \pi/2$,

$$f\left(\frac{\pi}{2}\right)=1=\frac{3}{2}-\frac{2}{\pi}\left(\sin\frac{\pi}{2}+\frac{\sin(3\pi/2)}{3}+\frac{\sin(5\pi/2)}{5}+\cdots\right)$$

Transposing and simplifying, we obtain

$$\frac{1}{2}=\frac{2}{\pi}\left(1-\frac{1}{3}+\frac{1}{5}-\frac{1}{7}+\cdots\right)$$

Therefore,

$$\frac{\pi}{4}=1-\frac{1}{3}+\frac{1}{5}-\frac{1}{7}+\cdots$$ □

Many other interesting and valuable results may be obtained by evaluating Fourier series at carefully chosen points.

As might be expected, even when the Dirichlet conditions are satisfied, some notable things occur in the case of pointwise convergence. In order to discuss one such phenomenon, we recall the notion of partial sums, S_n, of a Fourier series. The graphs of the partial sums of the Fourier series of a given function are called *approximating curves* to the function. The approximating curves exhibit a peculiarity around points of discontinuity known as the *Gibbs phenomenon.*

Example 7.17 Sketch the first three approximating curves to the function

$$f(x)=\begin{cases}-1, & -\pi<x<0\\ 1, & 0\leqq x\leqq\pi\end{cases}$$

Solution Since the function is odd, $a_n = 0$ for all n. Further,

$$\begin{aligned} b_n &= \frac{2}{\pi}\int_0^\pi \sin nx\,dx \\ &= -\frac{2}{n\pi}\cos nx\Big|_0^\pi \\ &= \frac{4}{n\pi}, \qquad n \text{ odd} \\ &= 0, \qquad n \text{ even} \end{aligned}$$

Therefore,

$$\begin{aligned} S_1 &= \frac{4}{\pi}\sin x \\ S_2 &= \frac{4}{\pi}\left(\sin x + \frac{\sin 3x}{3}\right) \\ S_3 &= \frac{4}{\pi}\left(\sin x + \frac{\sin 3x}{3} + \frac{\sin 5x}{5}\right) \end{aligned}$$

(See Figure 7.13.) □

Note that on either side of the discontinuity the approximating curve tends to "overshoot," a tendency which does not disappear as the peaks are shifted closer and closer to the discontinuity. Although the magnitude of the overshoot decreases, it never disappears completely. As a matter of fact, it can be shown that in the neighborhood of any discontinuity, the overshoot approaches a constant c times the height of the discontinuity jump, where

$$c = \frac{1}{2} - \frac{1}{\pi}\int_0^\pi \frac{\sin u}{u}\,du \doteq 0.09$$

We now return to the mainstream of our considerations and discuss two classical and very useful results concerning Fourier series convergence.

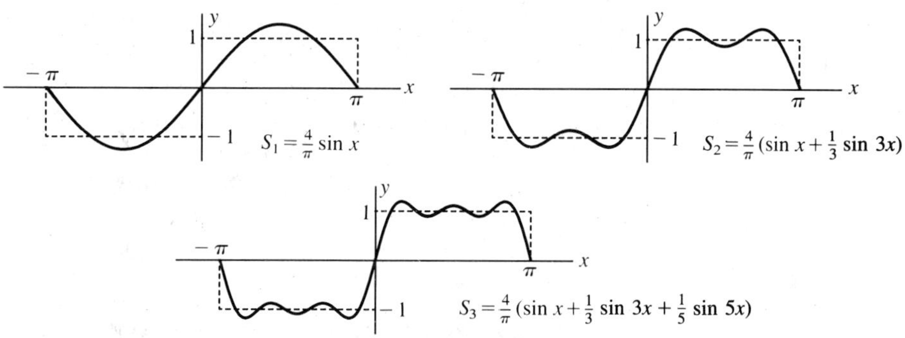

Figure 7.13

Since integration is a smoothing process, the convergence of a Fourier series is enhanced when we take the indefinite integral.

Theorem 7.8 Let f be integrable on $[-L, L]$. Then, for every x on $[-L, L]$

$$\int_{-L}^{x} f(t)\,dt = \int_{-L}^{x} [\text{Fourier series of } f(t)]\,dt$$

where the series is integrated term by term. Note that this result holds even if the Fourier series for f does *not* converge pointwise to f. ■

Example 7.18 Use the previous theorem to compute the Fourier series of the function $f(x) = |x| - \pi$, $-\pi < x < \pi$.

Solution Consider the function

$$g(x) = \begin{cases} -1, & -\pi < x < 0 \\ 1, & 0 < x < \pi \end{cases}$$

First note that

$$\int_{-\pi}^{x} g(t)\,dt = f(x)$$

Since g is odd, we need calculate only the b_n's,

$$\begin{aligned} b_n &= \frac{2}{\pi}\int_0^{\pi} \sin nx\,dx \\ &= -\frac{2}{\pi}\frac{\cos nx}{n}\Big|_0^{\pi} \\ &= 0 \qquad \text{if } n \text{ is even} \\ &= \frac{4}{n\pi} \qquad \text{if } n \text{ is odd} \end{aligned}$$

Hence

$$g(x) = \frac{4}{\pi}\sum_{n=0}^{\infty} \frac{\sin(2n+1)x}{2n+1}$$

where by Theorem 7.7, equality will hold throughout the interval if we define $g(0) = 0$.

From Theorem 7.8,

$$f(x) = -\frac{4}{\pi}\sum_{n=0}^{\infty} \frac{\cos(2n+1)x}{(2n+1)^2} - \frac{\pi}{2}$$

Here we have made use of the fact that

$$\sum_{n=0}^{\infty} \frac{1}{(2n+1)^2} = \frac{\pi^2}{8}$$

which follows from Exercise 10 on p. 248 or Example 7.23. □

Roughly speaking, Theorem 7.8 says that the indefinite integral of a function is equal to the term-by-term indefinite integral of its Fourier series. Even roughly speaking, the same kind of theorem is *not* true for the derivative of a function. It is one of those times in mathematics when things happen a bit differently than perhaps we might suppose, or would wish.

To see this, consider the Fourier series for $f(x) = x$. Its Fourier series is computed in Example 7.10, so that on the open interval $(-\pi, \pi)$,

$$x = 2 \sum_{n=1}^{\infty} (-1)^{n+1} \frac{\sin nx}{n}$$

The derivative of the left-hand side with respect to x is 1 and of the right-hand side is

$$2 \sum_{n=1}^{\infty} (-1)^{n+1} \cos nx$$

This latter series is certainly not a representation of the function $f(x) = 1$, and hence the Fourier series of the derivative of the given function is not equal to the term-by-term derivative of the Fourier series. (Note that the Fourier series of the function $f(x) = 1$ *does* exist.)

The function $f(x) = x$ on the interval $[-\pi, \pi]$ is not a "weird" function in any sense, although its periodic extension does have discontinuities at $x = \pm(2n + 1)\pi$. You should be extremely careful in differentiating both sides of an equality involving Fourier series. The equality is generally not maintained.

Conditions under which a Fourier series *may* be differentiated term by term to yield the derivative of the original function are given in the following theorem.

Theorem 7.9 Let f and its first n derivatives satisfy the Dirichlet conditions on $[-L, L]$ and assume that the periodic extensions of $f, f', \ldots, f^{(n-1)}$ are all continuous. Then the Fourier series for f can be differentiated term by term n times and the resulting series will converge pointwise to $f^{(n)}(x)$ at each point of continuity. ■

In applying this theorem you *must* consider the periodic extension of the function, not just the function as defined on $(-L, L)$. This will explain why an apparently continuous function such as $f(x) = x$ must be considered as discontinuous at $-L$ and L, and hence the differentiated series is not equal to the differentiated function.

Exercises for Section 7.8

1. Sketch the graph of the function to which the Fourier series of f converges on $[-3\pi, 3\pi]$ if the fundamental interval is $[-\pi, \pi]$ and

$$f(x) = \begin{cases} 1, & \text{on } (-\pi, 0) \\ 2, & \text{on } (0, \pi) \end{cases}$$

2. Repeat Exercise 1 for $f(x) = |\cos x|$ on $[-\pi, \pi]$.

3. Repeat Exercise 1 for $f(x) = |\sin x|$ on $[-\pi, \pi]$.
4. Repeat Exercise 1 for $f(x) = |x|$ on $[-\pi, \pi]$.
5. Repeat Exercise 1 for $f(x) = x$ on $[-\pi, \pi]$.
6. Sketch the graph of the function to which the Fourier series of f converges on $[-3, 3]$ if $f(x) = x^2$ on the fundamental interval $[-1, 1]$.
7. Sketch the graph of the function to which the Fourier series of f converges on $[-6, 6]$ if the fundamental interval is $[-2, 2]$ and

$$f(x) = \begin{cases} 0, & \text{on } [-2, 0) \\ 1, & \text{on } [0, 2] \end{cases}$$

8. Using the Fourier series of $f(x) = x^2$ on $PC[-\pi, \pi]$, show

$$\sum_{n=1}^{\infty} \frac{1}{n^2} = \frac{\pi^2}{6}; \qquad \sum_{n=1}^{\infty} \frac{(-1)^{n+1}}{n^2} = \frac{\pi^2}{12}$$

9. Using the Fourier series of $f(x) = x$ on $PC[-\pi, \pi]$, verify that

$$\sum_{n=0}^{\infty} \frac{(-1)^n}{2n+1} = \frac{\pi}{4}$$

10. Use Parseval's equality to demonstrate that

$$\sum_{n=0}^{\infty} \frac{1}{(2n+1)^2} = \frac{\pi^2}{8}$$

(*Hint:* Consider the function

$$f(x) = \begin{cases} -1, & -\pi < x < 0 \\ 1, & 0 < x < \pi \end{cases}$$

11. Show

$$\sum_{n=1}^{\infty} \frac{1}{n^4} = \frac{\pi^4}{90}$$

(*Hint:* Consider $f(x) = x^2$, $-\pi < x < \pi$.)

12. Carefully sketch the first three approximating curves to the following functions:
(a) $f(x) = |\sin x|$, $-\pi < x \leqq \pi$ **(b)** $f(x) = x^2$, $-1 < x < 1$
(c) $f(x) = x$, $-\pi < x < \pi$ **(d)** $f(x) = \begin{cases} 0, & -2 \leqq x \leqq 0 \\ x, & 0 < x < 2 \end{cases}$
13. Starting with the Fourier series for $f(x) = x$ on $[-\pi, \pi]$, use Theorem 7.8 in order to determine the Fourier series for

$$f(x) = 2x^2 - 1 \qquad \text{on } [-\pi, \pi]$$

14. In the same manner as in Exercise 13, find the Fourier series for $x^3 + 3x$ on $[-\pi, \pi]$.
15. In the same manner as in Exercise 13, find the Fourier series for

$$f(x) = \begin{cases} 2x + 2\pi, & \text{on } (-\pi, 0] \\ x + 2\pi, & \text{on } [0, \pi) \end{cases}$$

(*Hint:* See Example 7.5.)

16. Determine for which values of n Theorem 7.9 is applicable to each of the following functions on the interval $[-1, 1]$:

(a) x^2 **(b)** $\sin 3\pi x$
(c) $\cos^2 \pi x$ **(d)** $x^3 - x - 5$
(e) $\sin^4 \dfrac{\pi x}{2}$ **(f)** $(x^2 - 1)^2$

17. Construct a polynomial function $f(x)$ for which Theorem 7.9 is applicable with $n = 4$ on the interval $[-1, 1]$.

7.9 Sine and Cosine Series

Let f be defined on $[0, L]$. The *even extension* of f to $[-L, L]$ is the function f_e such that $f_e(-x) = f(x)$ for x on $(0, L]$. The *odd extension* of f to $[-L, L]$ is a function f_o such that $f_o(-x) = -f(x)$ for x on $(0, L]$. The three functions f, f_e, and f_o all agree on $[0, L]$. (See Figure 7.14.)

Example 7.19 Find f_e and f_o for the function defined on $[0, 1]$ by $f(x) = x$.

Solution

$$f_e(x) = \begin{cases} x & \text{on } [0,1] \\ -x & \text{on } [-1,0) \end{cases} \qquad f_o(x) = \begin{cases} x & \text{on } [0,1] \\ x & \text{on } [-1,0) \end{cases}$$

(See Figure 7.15.) □

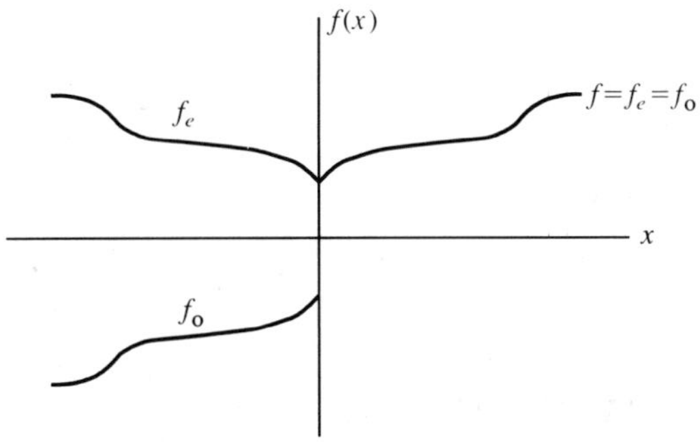

Figure 7.14 *Even and odd extensions*

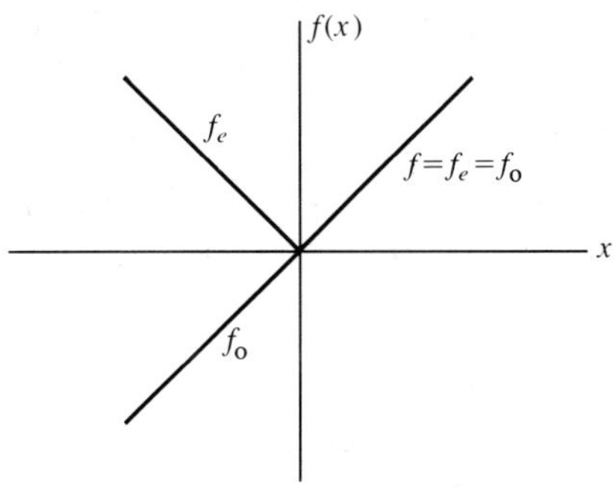

Figure 7.15

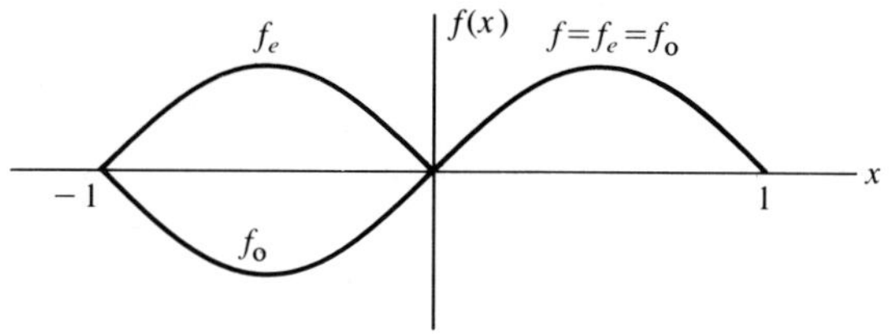

Figure 7.16

Example 7.20 Find f_e and f_o in the case $f(x) = x - x^2$ on $[0, 1]$. Sketch the odd and even periodic extensions.

Solution

$$f_e(x) = \begin{cases} x - x^2 & \text{on } [0,1] \\ -x - x^2 & \text{on } [-1,0) \end{cases} \qquad f_o(x) = \begin{cases} x - x^2 & \text{on } [0,1] \\ x + x^2 & \text{on } [-1,0) \end{cases}$$

(See Figure 7.16.) □

The even extension of f, f_e, is a member of $PC_e[-L, L]$ and thus may be expressed in terms of the set $\{\cos(n\pi x/L)\}_{n=0}^{\infty}$; that is, the set of cosine functions is a basis for $PC_e[-L, L]$. Therefore, the following formulas are valid:

$$f_e \sim \sum_{n=0}^{\infty} a_n \cos\frac{n\pi x}{L}$$

$$a_0 = \frac{1}{L}\int_0^L f_e(x)\,dx = \frac{1}{L}\int_0^L f(x)\,dx$$

$$a_n = \frac{2}{L}\int_0^L f(x)\cos\frac{n\pi x}{L}\,dx, \qquad n \neq 0$$

With appropriate conditions on f, this series will represent f on $[0, L]$ and the even periodic extension of f on $[-L, 0]$. This is true even if the original f is left undefined outside of the interval $[0, L]$. The Fourier series so obtained is called the *Fourier cosine series of f*.

Example 7.21 Let f be defined on $[0, \pi]$ by the rule

$$f(x) = \begin{cases} x, & \text{for } x \text{ on } \left[0, \dfrac{\pi}{2}\right] \\ 0, & \text{for } x \text{ on } \left(\dfrac{\pi}{2}, \pi\right] \end{cases}$$

Write the Fourier cosine series of f.

Solution

$$a_0 = \frac{1}{\pi}\int_0^{\pi/2} x\,dx = \frac{\pi}{8}$$

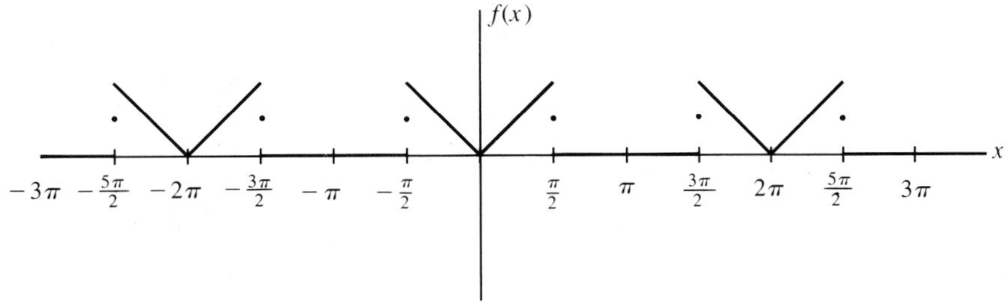

Figure 7.17

$$a_n = \frac{2}{\pi}\int_0^{\pi/2} x\cos nx\,dx = \frac{1}{\pi n^2}\left(2\cos\frac{n\pi}{2} + n\pi\sin\frac{n\pi}{2} - 2\right)$$

$$f(x) \approx \frac{\pi}{8} + \frac{1}{\pi}\left[(\pi - 2)\cos x - \cos 2x - \left(\frac{\pi}{3} + \frac{2}{9}\right)\cos 3x + \cdots\right]$$

Figure 7.17 shows the graph of the function to which this Fourier cosine series converges on $[-3\pi, 3\pi]$. □

In the same way, the function f_o is a member of the subspace $PC_o[-L, L]$ and hence may be written in terms of the basis $\{\sin(n\pi x/L)\}_{n=1}^{\infty}$:

$$f_o \sim \sum_{n=1}^{\infty} b_n \sin\frac{n\pi x}{L}$$

where

$$b_n = \frac{2}{L}\int_0^L f_o(x)\sin\frac{n\pi x}{L}\,dx = \frac{2}{L}\int_0^L f(x)\sin\frac{n\pi x}{L}\,dx$$

The Fourier series so obtained is called the *Fourier sine series of f*. If f satisfies the Dirichlet conditions, the series will actually represent f on $[0, L]$ and will converge to the odd periodic extension of f outside of $[0, L]$ (except at the points of discontinuity).

Example 7.22 Write the Fourier sine series of the function of the previous example.

Solution

$$b_n = \frac{2}{\pi}\int_0^{\pi/2} x\sin nx\,dx$$
$$= \frac{1}{\pi n^2}\left(2\sin\frac{n\pi}{2} - n\pi\cos\frac{n\pi}{2}\right)$$

Therefore,

$$f(x) \approx \frac{1}{\pi}\left(2\sin x + \frac{\pi}{2}\sin 2x - \frac{2}{9}\sin 3x - \frac{\pi}{4}\sin 4x + \cdots\right)$$

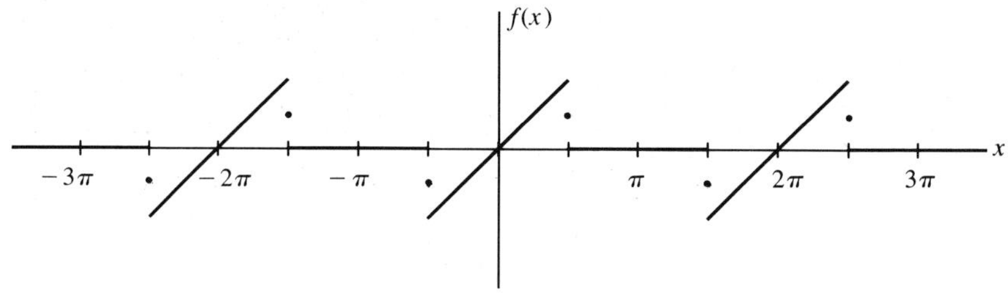

Figure 7.18

Figure 7.18 shows the graph of the function to which the series converges on $[-3\pi, 3\pi]$. □

7.10 Which Fourier Expansion to Use?

Suppose a function is defined on $[-L, L]$ but you desire a Fourier series representation only on $[0, L]$. Or suppose the function is defined only on $[0, L]$ but you may extend it to $[-L, 0]$ any way that you wish so long as you ultimately represent the portion on $[0, L]$ by some Fourier expansion.

1. If f is defined on $[-L, 0]$ you might write the Fourier series for f on the entire interval even though you would want to use only the portion on $[0, L]$.
2. If f is not defined on $[-L, 0]$ then you could use the odd extension of f and the resulting Fourier series is the Fourier sine series of the original function.
3. If f is not defined on $[-L, 0]$ you could use the even extension of f and the resulting Fourier series is the Fourier cosine series of the original function.
4. If f is not defined on $[-L, 0]$ you might define it to be zero there, since perhaps you would conclude that the computation of the Fourier coefficients would be easier.

Remember that in all cases the series will "represent" the *periodic* extension of your choice of functions on $[-L, L]$. Sketches of graphs for these choices are shown in Figure 7.19.

Aside from the motivation of writing the series with the least number of calculations, you might want to choose the expansion that results in the smoothest overall graph. For then it would seem reasonable that the series would converge "faster," that is, fewer terms of the series would be required to give a reasonable approximation to the limit function. Faster convergence may be judged by the order of the coefficients, where we say that a_n is of the *order* of $1/n^p$ if $\lim_{n\to\infty} \dfrac{a_n}{1/n^p} = \text{constant} \neq 0$.

Theorem 7.10 Let f and its first k derivatives satisfy the Dirichlet conditions on $[-L, L]$ and assume that the periodic extensions of $f, f', \ldots, f^{(k-1)}$ are all continuous. Then the Fourier coefficients of f are at least of the order of $1/n^{k+1}$. ■

More concisely, though less accurately, the theorem asserts that the smoother the function, the faster its Fourier expansion converges. Specifically, a function

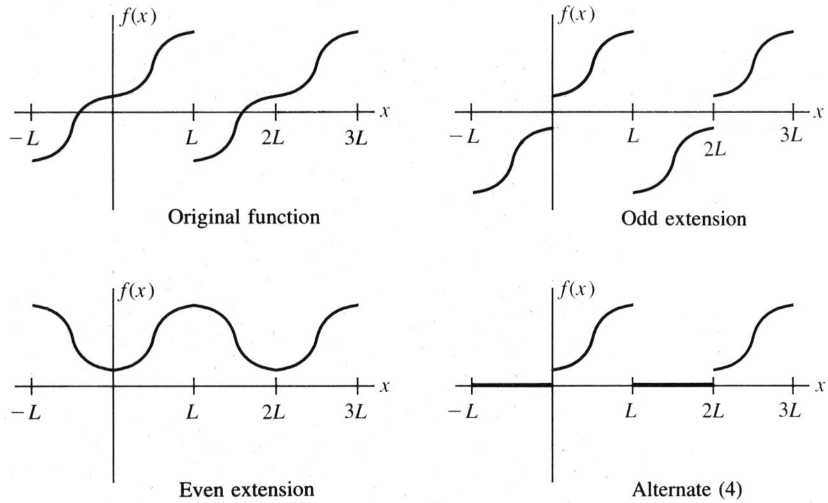

Figure 7.19 *Possible extensions of f*

which is discontinuous will have coefficients of the order of $1/n$; a function which is continuous but has a discontinuous derivative will have coefficients of the order of $1/n^2$, and so forth. So, to decide which Fourier series to use in a practical sense, it is best to sketch the periodic extension, the odd periodic extension, and the even periodic extension of the function. The one with the smoothest graph will yield a series which converges most quickly.

As we will see in Chapter 8 there may also be other criteria besides smoothness that will influence our choice of an expansion for f.

Example 7.23 If you are to represent the function $f(x) = x$ by a Fourier expansion on $[0, \pi]$, would a Fourier sine or cosine series be better?

Solution The graphs of the even and odd periodic extensions are shown in Figure 7.20. Since the graph of the function to which the cosine series converges is continuous and that to which the sine series converges is discontinuous, we should use the Fourier cosine series; both because of the more rapid convergence and also because the cosine series converges to $f(x)$ at $x = \pi$. The Fourier sine series is the same (in this case) as the Fourier series of the function $f(x) = x$ on $[-\pi, \pi]$. This was computed in Example 7.10 and found to be

$$2 \sum_{n=1}^{\infty} \frac{(-1)^{n+1}}{n} \sin nx$$

The Fourier cosine series is easily computed and found to be

$$\frac{\pi}{2} - \frac{4}{\pi} \sum_{n=0}^{\infty} \frac{\cos(2n+1)x}{(2n+1)^2}$$

The nature of the coefficients reinforces Theorem 7.10. □

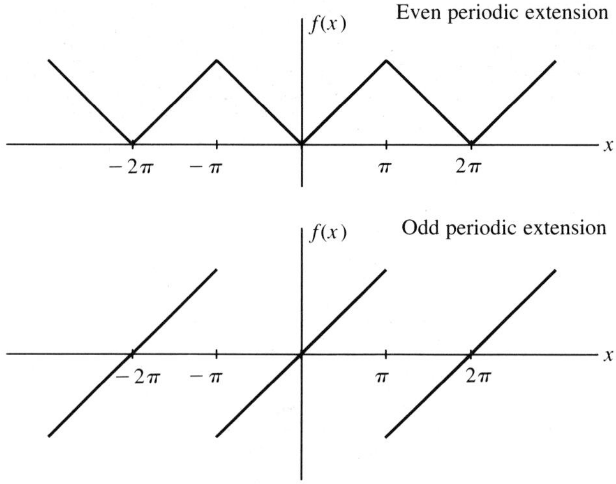

Figure 7.20

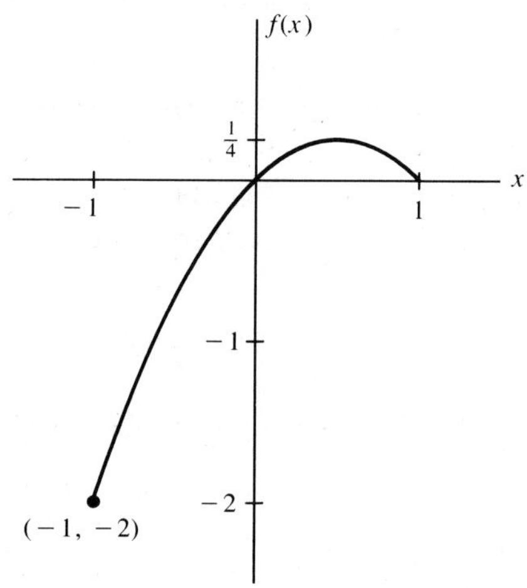

Figure 7.21

Example 7.24 Let $f(x) = x - x^2$ on $[-1, 1]$. (See Figure 7.21.) Sketch the graphs of the functions: (a) to which the Fourier series of the function on $[-1, 1]$ converges on $[-3, 3]$; (b) to which the Fourier cosine series of the function on $[0, 1]$ converges on $[-3, 3]$; (c) to which the Fourier sine series of the function on $[0, 1]$ converges on $[-3, 3]$. Which representation would you rather use?

Solution Figure 7.22 shows the three graphs. The first is the periodic extension of the given graph on $[-1, 1]$, the second is the even periodic extension of the given graph on $[0, 1]$, and the last is the odd periodic extension. (The latter two graphs replace the original definition of f on $[-1, 0]$ before extending periodi-

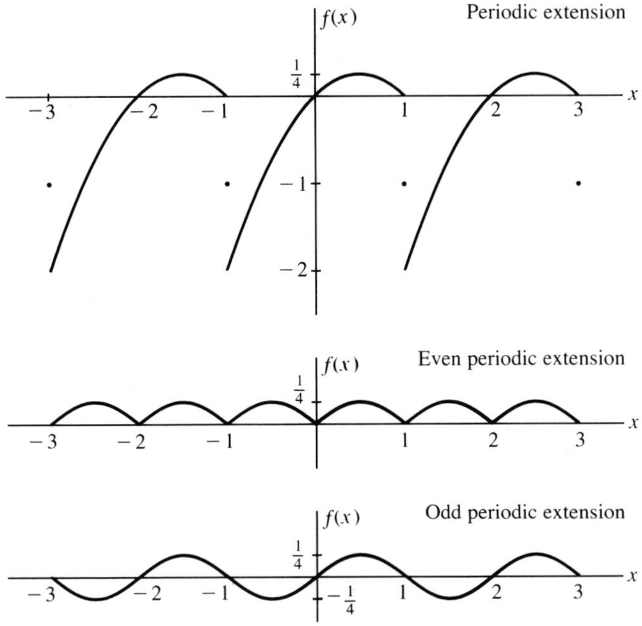

Figure 7.22 Three extensions for the function of Example 7.24

cally and perhaps should be called modifications rather than extensions.) Obviously, the Fourier sine series leads to the smoothest graph. □

Exercises for Sections 7.9 and 7.10

1. Find the functions f_o and f_e on $[-2, 2]$ and sketch their graphs if $f(x)$ is defined on $[0, 2]$ by:

   (a) $f(x) = \begin{cases} x - 1, & 0 < x < 1 \\ 1 - x, & 1 < x < 2 \end{cases}$ (b) $f(x) = x^2 + x$

   (c) $f(x) = \sin \dfrac{\pi x}{2}$ (d) $f(x) = \cos \dfrac{\pi x}{2}$

   (e) $f(x) = e^x$

2. Find the Fourier sine and cosine series of the function $f(x) = \sin x$, $0 < x < \pi$. Sketch the graphs of the functions to which the two series converge on $[-3\pi, 3\pi]$.

3. Find the Fourier sine and cosine series of the function $f(x) = \cos x$, $0 < x < \pi$. Sketch the graphs of the functions to which the two series converge on $[-3\pi, 3\pi]$.

In Exercises 4–10 find both the Fourier sine and cosine series of the given function on the indicated interval.

4. x^2 on $[0, 1]$

5. $\sin \dfrac{x}{2}$ on $[0, \pi]$

6. $\cos^2 \dfrac{x}{4}$ on $[0, \pi]$

7. $x^3 - 4x$ on $[0, 2]$

8. The function displayed in Figure 7.7, on $[0, 2]$.
9. The function displayed in Figure 7.9, on $[0, 1]$.
10. The function displayed in Figure 7.11, on $[0, 1]$.
11. Apply Theorem 7.10 to determine the order of the Fourier coefficients of each of the following functions on the indicated interval:

    (a) x^2 on $[-1, 1]$ **(b)** $f(x) = \begin{cases} 2, & -\pi \le x < 0 \\ 1, & 0 \le x \le \pi \end{cases}$

    (c) $|\cos x|$ on $[-\pi, \pi]$
12. Repeat Exercise 11 for the following functions:

    (a) $\sin^4 \dfrac{\pi x}{2}$ on $[-1, 1]$ **(b)** $x^3 - 4x$ on $[-2, 2]$

    (c) $(x^2 - 9)^2$ on $[-3, 3]$
13. Construct a polynomial function $f(x)$ whose Fourier coefficients for the interval $[-2, 2]$ are at least of the order $1/n^5$.

7.11 An Application

Considerable effort has been expended to explore the structure and computation of Fourier series. We must not forget, however, that such series were developed to solve otherwise intractable physical problems. Fourier's controversial idea of expressing fairly arbitrary functions in terms of a series of sines and cosines, introduced in 1807, came from his experiments in and analysis of heat conduction. He expressed the feeling of many of the great mathematicians of his era when he said "the deep study of nature is the most fruitful source of mathematical discovery." The following application is a natural extension of our consideration in Chapter 3 of the effects of a sinusoidal force applied to a mass-spring system.

Example 7.25 Find the steady-state displacement of the mechanical system shown in Figure 7.23(a) with the periodic applied force shown in Figure 7.23(b).

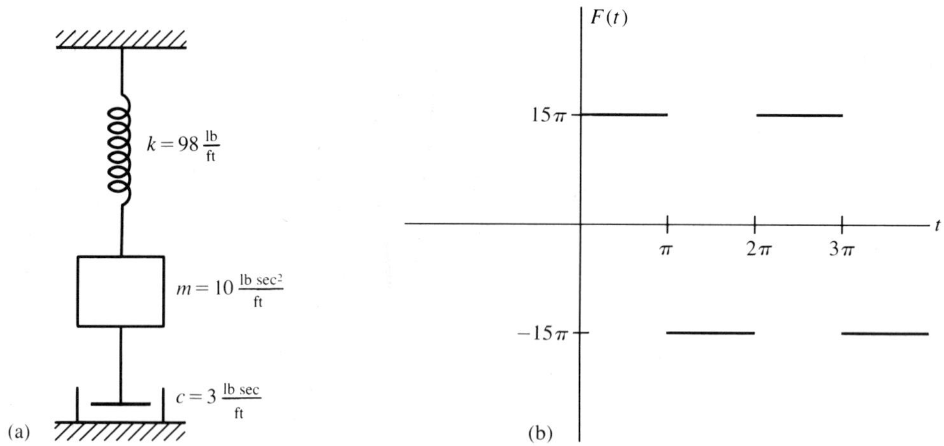

Figure 7.23

Solution The differential equation modeling this mechanical system is $my'' + cy' + ky = F(t)$. Expressing $F(t)$ in a Fourier sine series as in Example 7.18, the equation becomes

$$10y'' + 3y' + 98y = 60\left(\sin t + \frac{\sin 3t}{3} + \frac{\sin 5t}{5} + \cdots + \frac{\sin(2n+1)t}{2n+1} + \cdots\right)$$

We could calculate the response of our system to each of the forcing sine terms using any of several methods at our disposal, such as undetermined coefficients. However, we have already gone through this labor in Section 3.8. Indeed, the steady-state solution to $my'' + cy' + ky = A \sin \omega t$ is

$$y_p = \frac{-A\cos(\omega t - \delta)}{\sqrt{\omega^2 c^2 + (m\omega^2 - k)^2}}, \qquad \delta = \tan^{-1}\frac{m\omega^2 - k}{c\omega}$$

Using this formula to calculate the response to the first few forcing sine terms and adding the results (the method of *superposition*), we obtain the steady-state displacement:

$$\begin{aligned} y_{ss} \approx & -0.68\cos(t + 1.54) - 1.66\cos(3t + 0.73) - 0.08\cos(5t - 1.47) \\ & - 0.02\cos(7t - 1.52) + \cdots \end{aligned}$$

The graph of this displacement is compared with the (scaled) graph of the periodic function forcing in Figure 7.24. □

In earlier examples and exercises we have become accustomed to the response to a forcing term of such a mechanical system being similar in form to the forcing term itself (with perhaps some modification in "amplitude" and some "phase shift"). The response of this system, then, may be somewhat surprising in

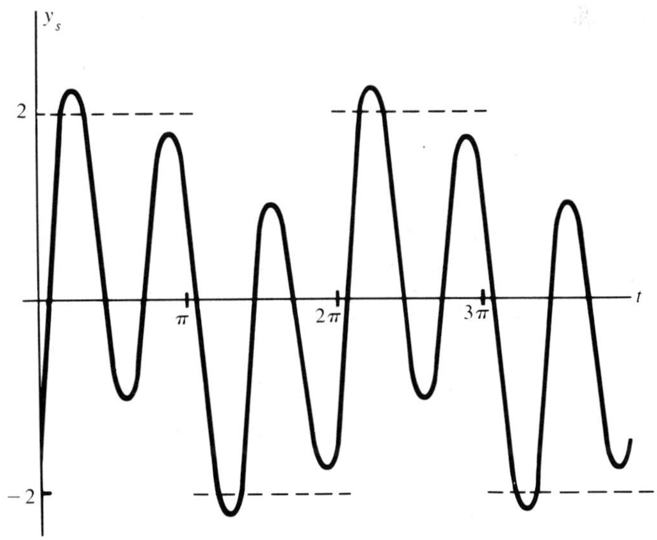

Figure 7.24

that it *appears* to have a frequency about three times that of the forcing term. What is happening here is that in the Fourier expansion of $F(t)$ one of the harmonics, namely, the third one, has a frequency very close to the natural frequency of the mechanical system and, since there is very little damping, the response to this harmonic dominates the other terms making up the total response.

You should note that the response is periodic with period 2π since we have neglected the transient; however, it is not actually periodic with the higher frequency, so our reference to its apparent "frequency" has been a casual one. It is also worth observing that even though the forcing term (and thus the acceleration) is discontinuous, the displacement is continuous. Although we will not demonstrate it, it is true that the velocity is also continuous.

Exercises for Section 7.11

The application discussed in this section is representative of simple oscillation problems with external inputs which are not pure sinusoids but are nevertheless periodic functions. The solution is obtained as a Fourier series in which each term individually gives the response of the system to one component of the Fourier expansion of the applied input. In this way, by means of Fourier series, we can find useful expressions for the solutions of practical problems which are virtually unsolvable by other methods.

Exercises 1–3 refer to the damped mass-spring mechanical system governed by the differential equation

$$my''(t) + cy'(t) + ky(t) = F(t)$$

Exercises 4–6 pertain to the differential equation

$$Li''(t) + Ri'(t) + i(t)/C = v'(t)$$

which models the current in a simple series RLC circuit driven by an input voltage $v(t)$.

Exercises 7–9 are related to the biomathematical model

$$g''(t) + 2\alpha g'(t) + \beta^2 g(t) = f(t)$$

associated with blood glucose concentration levels (see p. 116).

1. Let $m = 5$ lb sec^2/ft, $k = 20$ lb/ft, and $c = 1$ lb sec/ft. Determine the steady-state response of the damped mass-spring oscillator to the periodic external driving force that satisfies

$$F(t) = \begin{cases} 6t^2, & 0 \le t \le \pi \\ 6(2\pi - t)^2, & \pi < t \le 2\pi \end{cases}$$

Express your result as a trigonometric Fourier series in shifted cosine form and explicitly determine at least the first three terms.

2. Repeat Exercise 1 in the case when the applied force is periodic, with period 2π, and satisfies

$$F(t) = \begin{cases} -t, & 0 \le t \le \pi \\ t - 2\pi, & \pi < t \le 2\pi \end{cases}$$

3. Calculate both the resonant frequency ω_r and the natural angular frequency ω_0 for each case.
   (a) the application of Section 7.11
   (b) the configuration of Exercise 1
   Comment on the "appearance" of any resonance-like phenomena in these problems.
4. For the *RLC* circuit with $R = 4\ \Omega$, $L = 3$ H, and $C = \frac{1}{4}$ F, determine the steady-state current when $v(t)$ is the periodic function with period 2π that satisfies

$$v(t) = 3\pi t^2, \qquad -\pi \le t \le \pi$$

   Express your result as a trigonometric Fourier series in shifted cosine form and explicitly determine at least the first three terms.
5. Carry out the same analysis as in Exercise 4 for the case of the periodic voltage function

$$v(t) = |t|, \qquad -\pi < t < \pi$$

   Here $v'(t)$ becomes the familiar periodic square wave function.
6. Sketch the graph of the output current obtained in Exercise 5 and compare with the graph of the input voltage. Discuss the role of resonance in these results, if any.
7. Let $\alpha = 0.025$/min and $\beta = 0.020$/min in the blood glucose biomathematical model. Find the steady state concentration level $g(t)$, in shifted cosine form, that results when $f(t)$ is a pure sinusoid.
8. Repeat Exercise 7 in the case when

$$f(t) \sim \sum_{n=1}^{\infty} B_n \sin nt$$

9. Let β be fixed in the blood glucose model equation and assume that the input $f(t)$ is a pure sinusoid. Determine the phase shift of the output concentration and discuss the effect on this phase shift of varying α.

7.12 The Sturm-Liouville Problem

It may surprise you to learn that orthogonal sets of functions other than the trigonometric set arise in a very natural way in applied mathematics. In this section we will show how orthogonal sets evolve from the boundary value problem on the interval $[a, b]$ of the form

$$\frac{d}{dx}(p(x)y') + (q(x) + \lambda r(x))y = 0$$

$$c_1 y(a) + c_2 y'(a) = 0; \qquad c_1 \text{ and } c_2 \text{ not both zero}$$

$$c_3 y(b) + c_4 y'(b) = 0; \qquad c_3 \text{ and } c_4 \text{ not both zero}$$

This differential equation with the boundary conditions at a and b as given is known as the Sturm-Liouville differential equation with unmixed homogeneous boundary conditions. Notice that if $c_2 = c_4 = 0$, the boundary conditions become $y(a) = y(b) = 0$; that is, the solution function crosses the x-axis at $x = a$ and $x = b$. Notice also that for appropriate choices of coefficients, the differential equation takes on familiar forms, such as $y'' + \lambda y = 0$, Bessel's equation, the Legendre equation, and Euler's equation.

Example 7.26 Show that the boundary value problem $y'' + \lambda y = 0$ with the boundary values $y(0) = y(\pi) = 0$ is a Sturm-Liouville problem.

Solution If, in the general problem, we let $a = 0$, $b = \pi$, $r(x) = p(x) = 1$, and $q(x) = 0$, the given problem is seen to "fit" the general one. □

The form of the equation in the Sturm-Liouville problem is not as specialized as it may at first appear. Indeed, if

$$a_2(x)y'' + a_1(x)y' + [a_0(x) + \lambda w(x)]y = 0$$

multiplication by the "integrating factor"

$$\frac{1}{a_2(x)} \exp \int \frac{a_1(x)}{a_2(x)} dx$$

converts the original equation to the desired Sturm-Liouville "self-adjoint" form

$$\frac{d}{dx}[p(x)y'] + [q(x) + \lambda r(x)]y = 0$$

with $p = \exp \int (a_1/a_2)\, dx$, $q = (a_0/a_2) \exp \int (a_1/a_2)\, dx$, and $r = (w/a_2) \exp \int (a_1/a_2)\, dx$.

Example 7.27 Convert the general Cauchy-Euler equation

$$a_2 x^2 y'' + a_1 x y' + a_0 y = 0$$

where the a_i are constants, into self-adjoint form.

Solution The appropriate integrating factor is

$$\frac{1}{a_2 x^2} \exp \int \frac{a_1 x}{a_2 x^2} dx = \frac{1}{a_2} x^{(a_1/a_2)-2}$$

Hence

$$\frac{d}{dx}(x^{a_1/a_2} y') + \frac{a_0}{a_2} x^{(a_1/a_2)-2} y = 0$$ □

The constant λ in the Sturm-Liouville problem may take on any real value. Thus, one Sturm-Liouville problem is actually many boundary value problems, one for each value of λ. For some values of λ the resulting boundary value problem has no solution other than $y \equiv 0$. Those values of λ for which there is a nontrivial solution to the corresponding boundary value problem are called *eigenvalues*. The associated solutions are called *eigenfunctions*.

Example 7.28 Find the eigenvalues and eigenfunctions of the Sturm-Liouville problem $y'' + \lambda y = 0$; $y(0) = y(\pi) = 0$.

Solution If $\lambda = 0$, then the boundary value problem has the form $y'' = 0$, with the boundary values as given. Solving the differential equation we obtain $y = ax + b$, where a and b are constants to be determined from the boundary conditions. We find that $a = b = 0$, so the solution corresponding to $\lambda = 0$ is $y \equiv 0$ on the interval $[0, \pi]$.

If $\lambda < 0$, say $\lambda = -\alpha^2$, the differential equation becomes $y'' - \alpha^2 y = 0$. The solution is of the form $y = ae^{\alpha x} + be^{-\alpha x}$ where, once more, a and b are determined from the boundary conditions. Again, $a = b = 0$ for all values of α.

If $\lambda > 0$, say $\lambda = \beta^2$, the differential equation becomes $y'' + \beta^2 y = 0$, whose solution is $y = a\cos\beta x + b\sin\beta x$. Applying the boundary conditions,

$$0 = a\cos\beta 0 + b\sin\beta 0$$
$$0 = a\cos\beta\pi + b\sin\beta\pi$$

From the first of these equations $a = 0$, for all β. The second equation is then $b\sin\beta\pi = 0$. If β is nonintegral, $b = 0$; but if β is an integer, b may be any real number, and hence nontrivial solutions exist. The eigenvalues for these eigenfunctions are thus $\lambda = 1, 4, 9, 16, 25, \ldots$. The eigenfunctions themselves are $\sin x$, $\sin 2x$, $\sin 3x$, $\sin 4x$, $\sin 5x$, ... □

The situation occurring in this example is not an isolated event.

Theorem 7.11 Let p', q, and r of the Sturm-Liouville problem be continuous, with p and r positive, on $[a, b]$. Then there exists a countable infinity of real, discrete eigenvalues λ_m, $m = 0, 1, 2, \ldots$ with $\lambda_m \to \infty$ as $m \to \infty$. ■

As we are already well aware, the set of eigenfunctions of Example 7.28 forms an orthogonal set on $[0, \pi]$. The following theorem gives the result in the general case.

Theorem 7.12 Let p', q, and r of the Sturm-Liouville problem be continuous on $[a, b]$. Let y_m and y_n be continuous eigenfunctions corresponding to distinct eigenvalues λ_m and λ_n. Then y_m and y_n are orthogonal with weight function r on $[a, b]$.

Proof By hypothesis y_m and y_n are solutions to the Sturm-Liouville boundary value problem with $\lambda = \lambda_m$ and $\lambda = \lambda_n$, respectively. Hence

$$\frac{d}{dx}(py_m') + (q + \lambda_m r)y_m = 0$$
$$\frac{d}{dx}(py_n') + (q + \lambda_n r)y_n = 0$$

Multiplying the first of these by y_n and the second by $-y_m$ and adding,

$$(\lambda_m - \lambda_n)ry_m y_n = \frac{d}{dx}(py_m y_n' - py_m' y_n)$$

Integrating over the interval $[a, b]$ we obtain

$$(\lambda_m - \lambda_n) \int_a^b r y_m y_n \, dx = p(y_m y_n' - y_m' y_n) \Big|_a^b$$
$$= 0$$

(See the exercises.)

Therefore, since $\lambda_m \neq \lambda_n$,

$$\int_a^b r y_m y_n \, dx = 0$$

which is the desired orthogonality condition. ■

Corollary

1. If $p(a) = 0$, then no boundary condition is needed at $x = a$ in the Sturm-Liouville problem for the orthogonality condition to hold.
2. If $p(b) = 0$, no boundary condition is needed at $x = b$.
3. If $p(a) = p(b)$, then the "regular" boundary conditions can be replaced by $y(a) = y(b)$ and $y'(a) = y'(b)$. (These are known as periodic boundary conditions.)

Proof See Exercise 9 of the next set of exercises. ■

Example 7.29 The differential equation

$$\frac{d}{dx}[(1 - x^2)y'] + \lambda y = 0$$

is called Legendre's differential equation. If $\lambda = n(n + 1)$, where n is a nonnegative integer, the differential equation has a polynomial solution that we denote by P_n, called the *Legendre polynomial* of order n. The first several Legendre polynomials are $P_0 = 1$, $P_1 = x$, $P_2 = \frac{1}{2}(3x^2 - 1)$, and $P_3 = \frac{1}{2}(5x^3 - 3x)$.

Show that the set of *all* Legendre polynomials forms an orthogonal set on $[-1, 1]$.

Solution This is a Sturm-Liouville differential equation with $p(x) = 1 - x^2$, $q(x) = 0$, and $r(x) = 1$. Since $p(1) = p(-1) = 0$, no boundary conditions are necessary for the orthogonality property of the Legendre polynomials to hold, as a direct consequence of Theorem 7.12 and its corollary. □

Exercises for Section 7.12

1. Solve the Sturm-Liouville problem $y'' + \lambda y = 0$, $y'(0) = 0$, $y'(L) = 0$.
2. Solve the Sturm-Liouville problem $y'' + \lambda y = 0$, $y(0) = 0$, $y'(L) = 0$.
3. Find a Sturm-Liouville problem whose eigenfunctions are the modified trigonometric set

$$\left\{\cos\frac{n\pi x}{L}\right\}_{n=0}^{\infty} \cup \left\{\sin\frac{n\pi x}{L}\right\}_{n=1}^{\infty}$$

4. Is $\lambda = 0$ an eigenvalue for the Sturm-Liouville problem $y'' + \lambda y = 0$, $y(0) + y'(0) = 0$, $y(1) = 0$?

5. For what real values of α is $\lambda = 0$ an eigenvalue for the Sturm-Liouville problem $y'' + \lambda y = 0$, $y(0) = 0$, $\alpha y(1) + y'(1) = 0$?
6. Solve the Sturm-Liouville problem $(xy')' + (\lambda/x)y = 0$, $y(1) = 0$, $y(e) = 0$. (Recall the discussion in the exercises for Section 3.6.)
7. Solve the Sturm-Liouville problem $(xy') + (\lambda/x)y = 0$, $y'(1) = 0$, $y'(e) = 0$.
8. Demonstrate that $p(y_m y_n' - y_m' y_n)|_a^b = 0$ in the proof of Theorem 7.12.
9. Establish the corollary to Theorem 7.12.

Chebyshev's differential equation is

$$(1 - x^2)y'' - xy' + p^2 y = 0 \qquad -1 < x < 1$$

If p is a nonnegative integer n, this equation has a polynomial solution, which, when normalized to the value 1 at $x = 1$, is usually denoted by $T_n(x)$ and called the Chebyshev polynomial of order n. The first several Chebyshev polynomials are $T_0 = 1$, $T_1 = x$, $T_2 = 2x^2 - 1$, and $T_3 = 4x^3 - 3x$.

10. Re-express Chebyshev's equation in Sturm-Liouville form and thence show that the set of all Chebyshev polynomials forms an orthogonal set on $[-1, 1]$ with respect to an appropriate weight function.
11. Repeat Exercise 10 for the set of polynomial solutions to the differential equation

$$y'' - 2xy' + 2ny = 0, \qquad -\infty < x < \infty$$

where n is a nonnegative integer. (Properly normalized, these are the Hermite polynomials $H_n(x)$.)

12. Repeat Exercise 10 for the set of polynomial solutions to the differential equation

$$xy'' + (1 - x)y' + ny = 0, \qquad 0 < x < \infty$$

where n is a nonnegative integer. (Properly normalized, these are the Laguerre polynomials $L_n(x)$.)

Bessel's equation of order n is

$$x^2 y'' + xy' + (\lambda x^2 - n^2)y = 0$$

If, for fixed positive n, we append the boundary conditions $y(0) = y(1) = 0$, we obtain a Sturm-Liouville problem on $[0, 1]$, the eigenfunctions and eigenvalues of which are

$$y_m(x) = J_n(\sqrt{\lambda_m}x)$$

and

$$\lambda_m = \alpha_m^2$$

respectively. Here the numbers α_m $(m = 1, 2, 3, \ldots)$ are the zeros of J_n, the Bessel function of the first kind. (Bessel's equation and Bessel functions are discussed extensively in Chapter 9.)

13. Show that the eigenfunctions of this Sturm-Liouville problem form an orthogonal set on $[0, 1]$ with respect to the weight function x.

7.13 Looking Ahead

Fourier expansions and representations of arbitrary functions in terms of more general sets of orthogonal functions turn out to be remarkably useful in applied problem solving. We will encounter them again in each of the next two chapters on partial differential equations and special functions. Applications in numerical analysis and integral equations are explored in [4], [11], and [14]. See also [10].

In Section 7.10 we presented a theorem that essentially stated that the smoother a function is, the faster its Fourier expansion converges. This assertion regarding the rate of decay of Fourier coefficients for large n is one of a number of related results. (See [7], [12].) Analogous connections between the smoothness of kernels in integral equations and the growth properties of their eigenvalues are becoming increasingly important in the numerical solution of such equations.

If the given function is not continuous, its Fourier series need not converge (pointwise) for any x. The classic references on Fourier series, [5] and [16], consider this problem in considerable detail. The situation is even more complicated in the case of generalized Fourier representations with other orthogonal functions (see [3], for example).

Owing partly to their continual recurrence in practical applications, boundary value problems of Sturm-Liouville type have been well studied over the years. As suggested by the theory, examples, and exercises of Section 7.12, under fairly general conditions such problems give rise to a countable infinity of orthogonal eigenfunctions which are then available for use in generalized Fourier representations (see [6], [9]). Other applications involving Sturm-Liouville equations are considered in [8], [15].

Basic References

[1] Churchill, R. V., and J. W. Brown. 1978. *Fourier Series and Boundary Value Problems*. 3d ed. New York: McGraw-Hill.

[2] Spiegel, M. R. 1981. *Applied Differential Equations*. 3d ed. Englewood Cliffs, N. J.: Prentice-Hall.

More-Advanced References

[3] Alexits, G. 1961. *Convergence Problems of Orthogonal Series*. New York: Pergamon Press.

[4] Baker, C. T. H. 1977. *The Numerical Treatment of Integral Equations*. Oxford: Clarendon Press.

[5] Bary, N. K. 1964. *A Treatise on Trigonometric Series. Vols. 1 and 2*. New York: Pergamon Press.

[6] Birkhoff, G., and G. C. Rota. 1978. *Ordinary Differential Equations*. 3d ed. New York: John Wiley.

[7] Boas, R. P. 1967. *Integrability Theorems for Trigonometric Transforms*. New York: Springer-Verlag.

[8] Cochran, J. A. 1982. *Applied Mathematics: Principles, Techniques, and Applications*. Belmont, Calif.: Wadsworth.

[9] Coddington, E. A., and N. Levinson. 1955. *Theory of Ordinary Differential Equations*. New York: Mc Graw-Hill.

[10] Davis, P. J. 1963. *Interpolation and Approximation*. New York: Blaisdell.

[11] Fox, L., and I. B. Parker. 1968. *Chebyshev Polynomials in Numerical Analysis*. London: Oxford University Press.

[12] Hardy, G. H., and W. W. Rogosinski. 1962. *Fourier Series*. 3d ed. Cambridge: Cambridge University Press.

[13] Kaplan, W. 1962. *Operational Methods for Linear Systems*. Reading, Mass.: Addison-Wesley.

[14] Ralston, A., and P. Rabinowitz. 1978. *A First Course in Numerical Analysis*. New York: McGraw-Hill.

[15] Swanson, C. A. 1968. *Comparison and Oscillation Theory of Linear Differential Equations*. New York: Academic Press.

[16] Zygmund, A. 1977. *Trigonometric Series*. Cambridge: Cambridge University Press.

Chapter 8

Partial Differential Equations

8.1 Introduction

In this chapter we consider some elementary techniques of solving equations that involve partial derivatives of some unknown function. A partial differential equation is one which involves several independent variables $x, y, z, \ldots$, a function u of these variables, and the partial derivatives $u_x, u_y, u_{xx}, u_{yy}, u_{xz}, \ldots$ of the function. The order of the highest derivative is called the *order of the differential equation.* Except for a few easy problems, we will limit ourselves to equations in which the unknown function and its partial derivatives appear linearly. Such an equation is called a *linear partial differential equation.*

The most general second-order linear partial differential equation in two independent variables is

$$au_{xx} + bu_{xy} + cu_{yy} + du_x + eu_y + fu + g = 0$$

where a, b, c, d, e, f, and g are known functions of x and y. If g is the zero function, then the equation is called *homogeneous*; otherwise it is said to be *nonhomogeneous.* Linear partial differential equations have a relatively simple general theory, and thus are the kind most often studied. Fortunately, such equations provide reasonably good mathematical models for a wide variety of practical applications.

A function u is called a *solution* to a partial differential equation in some domain D of the space of the independent variables if the equation becomes an identity when u and its partial derivatives are substituted into the equation. This verification that a given function is, or is not, a solution is easy if we know how to differentiate the function.

Various forms of the following three operators receive the bulk of attention in virtually every introductory treatment of partial differential equations:

$$\text{Laplace or potential operator, } \Delta_n \equiv \frac{\partial^2}{\partial x_1^2} + \cdots + \frac{\partial^2}{\partial x_n^2}$$

$$\text{Heat or diffusion operator, } \Delta_{n-1} - k\frac{\partial}{\partial t}$$

$$\text{D'Alembert or wave operator, } \Delta_{n-1} - \frac{1}{c^2}\frac{\partial^2}{\partial t^2}$$

In these equations the space variables appear only in the Laplacian operator Δ, t represents the independent time variable, k is a positive constant inversely proportional to the conductivity/diffusivity of the medium of interest, and c is the wave velocity.

In the constant coefficient case, the general second order equation we have considered earlier may, by way of a linear change of variables, be put into a form involving one of these three operators with $n = 2$. Depending on whether $b^2 - 4ac$ is less than 0, equal to 0, or greater than 0 the operator used will be, respectively, the potential, diffusion, or wave operator. At least partially for this reason, the operators and corresponding equations are called elliptic, parabolic, and hyperbolic, respectively. Even in the nonconstant case, the same operator classification can be used locally, that is, by considering the independent variables only in a neighborhood of a specific point in n-space.

In examining the truly astonishing number of applied problems involving these operators, one may expect, in very crude terms, solutions involving a fixed, equal-potential surface in the elliptic case; smooth, spreading flows in the parabolic case; and disturbance preserving waves in the hyperbolic case.

Example 8.1 Show that $u(x, y) = x^2 - y^2$ and $v(x, y) = e^x \cos y$ are solutions to the two-dimensional Laplace equation

$$u_{xx} + u_{yy} = 0$$

Solution Calculating the partial derivatives of u and v,

$$u_x = 2x, \qquad u_{xx} = 2, \qquad u_y = -2y, \qquad u_{yy} = -2$$

$$v_x = e^x \cos y, \qquad v_{xx} = e^x \cos y$$

$$v_y = -e^x \sin y, \qquad v_{yy} = -e^x \cos y$$

In both cases it is easy to see that the sum of the second partial derivatives is zero and hence both u and v satisfy the differential equation. Any linear combination of u and v is also a solution, but there are many other solutions to this equation as well. □

The set of all solutions to a partial differential equation is called the *solution set*, a set which may be very large and cumbersome. Usually, we are more interested in finding a set of solutions from which all other solutions can be derived.

Even this set may be quite large as is the case with solutions to Laplace's equation (see Example 8.1) whose members are called *harmonic functions* (if their second partial derivatives are continuous). A defining equation of the solution set is called the *complete* or *general solution*. From it any *particular solution* can be obtained.

Some simple equations can be solved in a straightforward manner, as in the following examples.

Example 8.2 Solve the partial differential equation $u_x = y \sin x$.

Solution The complete solution is obtained by antidifferentiating partially with respect to x,

$$u(x, y) = -y \cos x + f(y)$$

where f is an arbitrary function. Any two solutions differ by a function of y. □

Example 8.3 Solve the equation $u_{xx} = x$, considered in three-dimensional space.

Solution The complete solution is found by two antidifferentiations with respect to x,

$$u(x, y, z) = \frac{x^3}{6} + xf(y, z) + g(y, z)$$

where f and g are arbitrary functions of the two variables y and z. □

The approach used in these two examples cannot generally be applied in more complicated cases. In what follows, then, we explore the utility of alternative methods of solution. We also consider the nature of various auxiliary conditions that allow us to determine the unique solutions, if such exist, for the mathematical models associated with physical situations.

8.2 Boundary Value Problems

From the two previous examples one might conjecture that the solution set of an nth-order linear partial differential equation is in some sense a family of functions depending on n arbitrary functions in the same way that such a set for an nth-order linear ordinary differential equation is dependent on n arbitrary constants. This statement is roughly correct, but it will not be a source of concern for us to make it more precise since we rarely can, or have to, find complete solutions.

In practice, we are often asked to find not the complete solution to a partial differential equation but a particular one arising from a given physical situation. In some cases, this particularization is obtained by specifying the value of the solution at points on the boundary; in others, when the time t is one of the variables the values of the function and its derivative(s) at $t = 0$, called *initial conditions*, are prescribed. A differential equation with some specified auxiliary conditions for the solution to satisfy is called a *boundary value problem.*

In the study of ordinary differential equations, the common procedure for the solution of an initial value problem or boundary value problem is to find the complete solution set and then determine values of the arbitrary constants so that the initial values or boundary values are as specified. In the solution of partial differential equations, this is usually not the preferred approach. For even when the complete solution is known, it still may be prohibitively difficult to find the arbitrary functions to match the auxiliary conditions.

Frequently, at least in elementary work, you will find that the most useful procedure in solving boundary value problems is to guess a form for the particular solution immediately. We pick a form to which the auxiliary conditions are easily adapted, and, while the picking of the "best" form for these purposes might be quite difficult in general, it is relatively straightforward in some very important cases.

Example 8.4 The temperature in a two-dimensional region satisfies Laplace's equation $T_{xx} + T_{yy} = 0$ with boundary conditions as shown in Figure 8.1. Formulate the mathematical model and solve the boundary value problem assuming a linear variation.

Solution The mathematical model is the differential equation with the boundary conditions stated in terms of T,

$$T_{xx} + T_{yy} = 0; \qquad T(x, 0) \equiv 0, \qquad T(x, \pi) \equiv 100, \qquad x > 0$$
$$T_x(0, y) \equiv 0, \qquad 0 \leqq y \leqq \pi$$

As a result of the assumed linearity,

$$T(x, y) = ax + by + c$$

Note that any linear function satisfies the equation. Since $T_x(0, y) = 0$, $a = 0$. Applying the boundary conditions at $y = 0$ and $y = \pi$,

$$b = \frac{100}{\pi} \quad \text{and} \quad c = 0$$

Therefore, the solution is

$$T(x, y) = \frac{100}{\pi} y$$

□

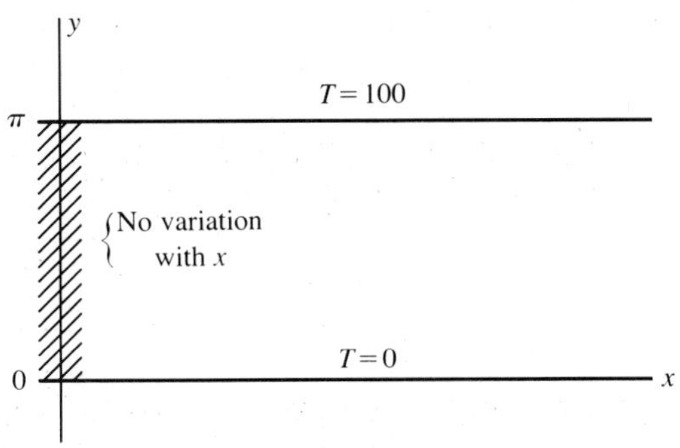

Figure 8.1

It is a theorem, proved in more advanced discussions of the subject of partial differential equations, that the bounded solution to a boundary value problem like that of the previous example is unique. Hence we need concern ourselves solely with guessing or otherwise obtaining one solution to a boundary value problem, since if we do get one, it will be the only one.

Example 8.5 Consider the boundary value problem

$$u_{xx} = u_{tt}; \qquad 0 < x < 2, \quad t > 0$$
$$u(x,0) \equiv 2\sin 3\pi x, \qquad u_t(x,0) \equiv 0 \text{ (initial conditions)}$$
$$u(0,t) \equiv u(2,t) \equiv 0 \text{ (boundary conditions)}$$

Verify that one form of a solution is the product of a trigonometric function in x by a trigonometric function in t, and attempt to match the boundary conditions to this form.

Solution We guess the solution to be of the form

$$u(x,t) = (A\sin\alpha x + B\cos\alpha x)(C\sin\beta t + D\cos\beta t)$$

Partial differentiation of $u(x,t)$ twice with respect to x and twice with respect to t gives

$$u_{xx}(x,t) = -\alpha^2 u(x,t); \qquad u_{tt}(x,t) = -\beta^2 u(x,t)$$

and thus $\alpha = \pm\beta$, since these two partial derivatives must be equal. The trigonometric function in t therefore must have the same angular frequency as the trigonometric function in x. The values of A, B, C, D, and α are determined from auxiliary conditions. The boundary condition $u(0,t) \equiv 0$ gives $B = 0$, since the trivial solution $u \equiv 0$ does not satisfy the initial conditions. For the boundary condition $u(2,t) \equiv 0$, we get

$$u(2,t) = A\sin 2\alpha(C\sin\alpha t + D\cos\alpha t) \equiv 0$$

which means that

$$A\sin 2\alpha = 0$$

which is satisfied for

$$\alpha = \frac{n\pi}{2}, \qquad \text{with } n \text{ an integer}$$

(Note that $A = 0$ would again yield $u \equiv 0$.) Applying the second initial condition gives $C = 0$, so that the solution is of the form

$$u(x,t) = AD\sin\frac{n\pi x}{2}\cos\frac{n\pi t}{2}$$

The other initial condition gives

$$2\sin 3\pi x = AD\sin\frac{n\pi x}{2}$$

from which $AD = 2$ and $n = 6$. Therefore, the solution is

$$u(x, t) = 2 \sin 3\pi x \cos 3\pi t \qquad \square$$

Example 8.6 Heat conduction in a thin rod of length 2 can be modeled with the differential equation

$$u_{xx} - ku_t = 0 \qquad 0 < x < 2, \quad t > 0, \text{ with } k > 0$$

If

$$u(x, 0) \equiv 2 \sin 3\pi x \qquad \text{(initial condition)}$$
$$u(0, t) \equiv u(2, t) \equiv 0 \qquad \text{(boundary conditions)}$$

solve the resulting initial value–boundary value problem by assuming the solution is in the form of a product of a trigonometric function in x and an exponential function in t.

Solution We guess the solution to be of the form

$$u(x, t) = (A \sin \alpha x + B \cos \alpha x)e^{-\beta t}$$

where β is assumed to be nonnegative in order to ensure that the solution to this physically motivated problem remains bounded for increasing time t. Substitution of this expression into the differential equation leads to $\beta = \alpha^2/k$. The values of A, B, and α are then determined from the auxiliary conditions. Carrying out steps precisely as in the previous example, we find $B = 0$, $A = 2$, and $\alpha = 3\pi$. The desired solution in this example is thus

$$u(x, t) = 2(\sin 3\pi x)e^{-9\pi^2 t/k} \qquad \square$$

The guesses that we made for the form of the solutions in Examples 8.5 and 8.6 were suggested by the nature of the differential equations to be solved. An even more striking illustration of the differences that arise in the solutions of seemingly similar problems is given in Example 8.7.

Example 8.7 Verify that

$$u(x, y) = \frac{2u_0}{\pi} \operatorname{Tan}^{-1} \frac{y}{x}$$

$$u(x, t) = \frac{2u_0}{\sqrt{\pi}} \int_{x/\sqrt{4t}}^{\infty} e^{-\tau^2} d\tau$$

and

$$u(x, t) = \begin{cases} 0, & 0 \le t < x \\ u_0, & t > x \end{cases}$$

are solutions of the three initial value–boundary value problems

(a)

$$u_{xx} + u_{yy} = 0, \qquad x > 0, y > 0$$
$$u(0, y) = u_0$$
$$u(x, 0) = 0$$

(b)
$$u_{xx} - u_t = 0, \qquad x > 0, t > 0$$
$$u(0, t) = u_0$$
$$u(x, 0) = 0$$

(c)
$$u_{xx} - u_{tt} = 0, \qquad x > 0, t > 0$$
$$u(0, t) = u_0$$
$$u(x, 0) = 0$$
$$u_t(x, 0) = 0$$

respectively, where u_0 = constant.

Solution Showing that the three given functions satisfy the appropriate differential equations is merely a matter of performing the indicated differentiations. Verification that the auxiliary conditions are also satisfied in each case is accomplished with equal ease. It is interesting to note that the solution to the third problem is discontinuous when $t = x$, whereas the first two solutions are not only continuous but infinitely differentiable in the given region. Differences such as this are characteristic of wave phenomena, on the one hand, as opposed to equilibrium and diffusion phenomena, on the other. □

If nothing else, it should be clear from these various examples that if we are going to be successful in discovering the solutions to a reasonable variety of equations and problems, our "guessing" procedure will need some methodological underpinnings. In Section 8.4 we consider the so-called D'Alembert method, a procedure useful for deriving the solution of problem (c) in Example 8.7. Later in Section 8.9 and again in Chapters 13 and 16 we will consider the application of Laplace transform and other operational techniques to initial value–boundary value problems. Barring extremely clever guessing, it is methods such as these that are needed for the solution of problems (a) and (b) in Example 8.7.

8.3 Superposition

Initially, in the study of homogeneous partial differential equations, it proves beneficial to reason analogously to the theory of ordinary differential equations. In ordinary differential equations, we know that a linear combination of any two solutions to a linear homogeneous equation is also a solution.

Theorem 8.1 Let u_1 and u_2 be solutions to the homogeneous differential equation

$$au_{xx} + bu_{xy} + cu_{yy} + du_x + eu_y + fu = 0$$

where $a, b, c, d, e,$ and f are functions of x and y. Then $Au_1 + Bu_2$ is also a solution for any two constants A and B.

Proof See Exercise 14 in the next set of exercises. ■

We are especially interested in the analog of Theorem 8.1 for nonfinite solution sets.

Example 8.8 Find solutions to $u_x + u_y = 0$ and show that the infinite analog of Theorem 8.1 holds in at least one particular case.

Solution It is easy to show that each of the functions

$$u_n(x, y) = (x - y)^n$$

is a solution to the partial differential equation. Restricting n to the nonnegative integers, we form the infinite series of functions

$$\sum_{n=0}^{\infty} c_n u_n(x, y) = \sum_{n=0}^{\infty} c_n (x - y)^n$$

This series will converge to e^{x-y} if we choose c_n to be $1/n!$. Further, e^{x-y} is also a solution to the given partial differential equation. As a matter of fact, if we choose c_n to be the coefficients of *any* convergent series expansion $f(t) = \sum_{n=0}^{\infty} c_n t^n$ of a function of one real variable, then the infinite series of functions $\sum_{n=0}^{\infty} c_n (x - y)^n$ so obtained converges to $f(x - y)$. The function $f(x - y)$ can easily be verified to be a solution to the given differential equation. □

Use of the principle of superposition will usually be necessary in order to obtain the solutions of the boundary value problems encountered in this chapter. Care must be taken, then, in applying auxiliary conditions since a linear combination of functions will not generally satisfy the same auxiliary conditions as does each function individually. However, there is an important exception to this observation which we will take advantage of frequently; this occurs when the auxiliary conditions require $\alpha u + \beta \partial u/\partial v = 0$ along some curve(s). (Here α and β may vary along the curve(s), and $\partial u/\partial v$ is a derivative of u in some nontangential direction.) If u_1 and u_2 are solutions to a homogeneous boundary value problem whose differential equation is linear, then linear combinations of u_1 and u_2 will satisfy not only the differential equation but also all the *homogeneous auxiliary conditions* of the above form. You will be asked to prove this statement in the exercises.

Exercises for Sections 8.1–8.3

1. Verify that the following functions satisfy Laplace's differential equation $u_{xx} + u_{yy} = 0$:

(a) $u(x, y) = x^2 - y^2$ **(b)** $u(x, y) = \text{Tan}^{-1}\dfrac{y}{x}$

(c) $u(x, y) = ax + by + c$ **(d)** $u(x, y) = \cos x \cosh y$

2. Show that the following functions satisfy the diffusion equation $u_{xx} - u_t = 0$ for $t > 0$:

(a) $\dfrac{1}{\sqrt{t}} e^{-x^2/4t}$ **(b)** $\displaystyle\int_{x/2\sqrt{t}}^{\infty} e^{-\tau^2}\, d\tau$

(c) $ax + b + c(x^2 + 2t)$ **(d)** $e^{-t} \cos x$

3. Find a partial differential equation satisfied by:

(a) $u(x, y) = \sin(x^2 - y^2)$ **(b)** $u(x, y) = \cos(x^2 + y^2) + x^2$

4. Show that $f(w)$ is a solution to $yu_x - xu_y = 0$ if $w = x^2 + y^2$.

5. Verify that $f(w)$ is a solution to $xu_x - yu_y = 0$ if $w = xy$.

6. Verify that $f(w)$ is a solution to $yu_x - u_y = 0$ if $w = 2x + y^2$.
7. Show that $u(x, y) = f(w) + g(v)$ is a solution to $u_{xx} = u_{yy}$ if $w = x + y$ and $v = x - y$.
8. Verify that $u(x, y) = f(v) + xg(v)$ is a solution to $u_{xx} + 2u_{xy} + u_{yy} = 0$ if $v = x - y$.
9. Find the complete solution of:
   (a) $u_y = 0$ **(b)** $u_{xx} + u = 0$
   (c) $u_{xxx} + u_x = 0$ **(d)** $u_{xxy} = 1$
10. Solve the two-dimensional Laplace equation with the boundary conditions $u(0, y) \equiv u(L, y) \equiv 0$, $u_y(x, L) \equiv (\pi/L) \sinh \pi \sin(\pi x/L)$, $u_y(x, 0) \equiv 0$.
11. Find the solution of the two-dimensional Laplace equation with the boundary conditions $u_x(0, y) \equiv (\pi/L) \cosh \pi \sin(\pi y/L)$, $u(L, y) \equiv 0$, $u(x, 0) \equiv u(x, L) \equiv 0$.
12. Determine a solution of the diffusion equation $u_{xx} - ku_t = 0$ that satisfies the auxiliary conditions $u_x(0, t) \equiv u_x(4, t) \equiv 0$, $u(x, 0) = 3 \cos 2\pi x$.
13. By assuming a product of trigonometric functions as a solution, solve $u_{xx} = u_{tt}$ if $u(x, 0) \equiv 3 \sin 10\pi x$, $u_t(x, 0) \equiv 0$, $u(0, t) \equiv u(5, t) \equiv 0$.
14. Prove Theorem 8.1.
15. Prove $u_n(x, y) = (x - y)^n$ satisfies the equation $u_x + u_y = 0$ for every n. Can you find an infinite series of these functions, different from the one in Example 8.8, whose sum is also a solution?
16. Find the solution of $u_{xx} - u = 0$ if $u(0, y) \equiv y$, $u_x(0, y) \equiv \sin y$.
17. Prove the last statement of Section 8.3 for the special case of Laplace's two-dimensional equation and the boundary conditions $u(x, 0) \equiv u(x, L) \equiv 0$.
18. Describe a "physical problem" with boundary values like the ones given in Figure 8.1.

Laplace's equation $\Delta_n u(x_1, x_2, \ldots, x_n) = 0$ appears regularly in the mathematical modeling of equilibrium phenomena. For example, this equation can be used to describe steady-state temperature distributions, electrostatic potentials, and stationary irrotational incompressible fluid flows. When $n = 1$ the equation can be easily and completely solved, a feat that is not possible in higher dimensions. Certain solutions that possess considerable symmetry, therefore, take on special significance when $n = 2, 3, \ldots$.

19. Show that $u(x, y) = \ln(x^2 + y^2)$, $(x \neq 0, y \neq 0)$ is a solution of Laplace's equation in two dimensions. (This solution is proportional to the electrostatic potential of a uniformly charged line.)
20. **(a)** Rewrite Laplace's equation for $n = 2$ in polar coordinates r, θ where

$$x = r \cos \theta$$
$$y = r \sin \theta$$

   (b) Determine the complete solution which is *axisymmetric*, that is, independent of the polar angle θ.
21. Verify that $r^{2-n} = (x_1^2 + x_2^2 + \cdots + x_n^2)^{1-n/2}$ is a solution of Laplace's equation in n dimensions ($n \geq 3$). (In three dimensions this solution is proportional to the electrostatic potential of a point charge.)
22. **(a)** Show that Laplace's equation for $n = 3$ can be rewritten in spherical coordinates ρ, θ, φ, where

$$x = \rho \cos \theta \sin \varphi$$
$$y = \rho \sin \theta \sin \varphi$$
$$z = \rho \cos \varphi$$

   as $\rho^2 u_{\rho\rho} + 2\rho u_\rho + (\text{terms involving derivatives in } \theta, \varphi) = 0$.

(b) Find the complete solution that is *spherically symmetric*, that is, independent of θ and φ.

8.4 Two Methods

For various homogeneous linear partial differential equations, particularly those with constant coefficients, there are two methods that are most commonly used in applications.

Method 1: In the method of *separation of variables* or the *product method*, we seek possible solutions that are products of the functions of the separate variables. For two independent Cartesian variables we let

$$u(x, y) = X(x)Y(y)$$

A good reason to try this substitution is that it often enables us to change a two-dimensional linear partial differential equation into two ordinary differential equations that we can solve. It does seem rather unlikely that there will be a solution of this special form, but, as we will soon see, through superposition we can greatly increase the utility of this method.

Example 8.9 Solve Laplace's equation in two dimensions, $\Delta_2 u = 0$, by assuming a product solution.

Solution We must solve the equation $u_{xx} + u_{yy} = 0$ by substituting $u(x, y) = X(x)Y(y)$:

$$X''(x)Y(y) + X(x)Y''(y) = 0$$

Equivalently,

$$\frac{X''(x)}{X(x)} = -\frac{Y''(y)}{Y(y)}$$

The left member of this equation is independent of y while the right member is independent of x. Thus the two members, known to be equal, are independent of *both* x and y. Therefore they are both equal to a constant, which we denote by λ. Hence we have the two ordinary differential equations

$$\frac{X''}{X} = \lambda \quad \text{and} \quad \frac{Y''}{Y} = -\lambda$$

The solutions will be trigonometric in y and exponential in x if $\lambda > 0$ and just the reverse if $\lambda < 0$. If $\lambda = 0$, both X and Y are linear functions. Hence, if there are solutions to Laplace's equation in the form of a product of a function of x and a function of y, they will be of the forms

$$(Ae^{kx} + Be^{-kx})(C \sin ky + D \cos ky), (A \sin kx + B \cos kx)(Ce^{ky} + De^{-ky})$$

where $k > 0$,

and

$$(Ax + B)(Cy + D) \qquad \square$$

Example 8.10 Solve the boundary value problem $u_{xx} + u_{yy} = 0$, where $u(0, y) \equiv 0$, $u(2, y) \equiv 0$, $u(x, 0) \equiv \sin \pi x - 3 \sin 5\pi x$, $u_y(x, 0) \equiv 0$.

Solution First, we find those product solutions of the form of the previous example which satisfy the homogeneous boundary conditions. Then, by the remarks at the end of Section 8.3, any linear combination of these functions may be used in an attempt to satisfy the remaining, nonhomogeneous, condition. Consider, therefore, the function defined by

$$u(x, y) = (A \sin kx + B \cos kx)(Ce^{ky} + De^{-ky})$$

We choose this from the three forms given earlier because of the appearance of $\sin kx$ terms in the auxiliary condition on $u(x, 0)$. Since $u(0, y) \equiv 0$, we have

$$B(Ce^{ky} + De^{-ky}) \equiv 0$$

so $B = 0$ or $C = D = 0$ (since e^{ky} and e^{-ky} are linearly independent). The second possibility gives us the useless result $u(x, y) \equiv 0$.

With $B = 0$ we now consider $u(2, y) \equiv 0$, that is,

$$A \sin 2k(Ce^{ky} + De^{-ky}) \equiv 0$$

Both $A = 0$ and $C = D = 0$ again lead to $u \equiv 0$ so we must have $k = n\pi/2$ for n an integer. Thus

$$u(x, y) = \sin \frac{n\pi x}{2}(ce^{n\pi y/2} + de^{-n\pi y/2})$$

where $c = AC$ and $d = AD$. Differentiating,

$$u_y(x, y) = \frac{n\pi}{2} \sin \frac{n\pi x}{2}(ce^{n\pi y/2} - de^{-n\pi y/2})$$

$$u_y(x, 0) = \frac{n\pi}{2} \sin \frac{n\pi x}{2}(c - d) \equiv 0$$

which implies $c = d = c_n$. We now have solutions of the form

$$u_n = c_n \sin \frac{n\pi x}{2}(e^{n\pi y/2} + e^{-n\pi y/2})$$

To fit the nonhomogeneous condition we choose $c_2 = \frac{1}{2}$, $c_{10} = -\frac{3}{2}$, and $c_n = 0$ otherwise. Using superposition, the desired solution finally becomes

$$u(x, y) = \sin \pi x\left[\frac{e^{\pi y} + e^{-\pi y}}{2}\right] - 3 \sin 5\pi x\left[\frac{e^{5\pi y} + e^{-5\pi y}}{2}\right] \qquad \square$$

As one might guess by the form of the answer in Example 8.10, calculations are somewhat simplified by choosing the complete solution of $Y'' - k^2 Y = 0$ in the form $C \cosh ky + D \sinh ky$ instead of $Ce^{ky} + De^{-ky}$. (Recall that $\cosh ky = (e^{ky} + e^{-ky})/2$, $\sinh ky = (e^{ky} - e^{-ky})/2$.) The utility of this alternative form for $Y(y)$ should be verified by working back through the example.

The method of separation of variables can be used to find solutions to some equations with nonconstant coefficients. If the substitution $u(x, y) = X(x)Y(y)$ leads to an equation in a form in which the functions X and Y may be separated, then the differential equation is said to be *separable*.

Example 8.11 Solve $x^2u_{xx} + xu_x - u_y = 0$ by assuming a product solution.

Solution Substituting $u(x, y) = X(x)Y(y)$ into the differential equation we obtain

$$x^2X''Y + xX'Y - XY' = 0$$

This leads to the equation in which the variables are separated,

$$\frac{x^2X'' + xX'}{X} = \frac{Y'}{Y}$$

By reasoning similar to that of Example 8.9, the two members of this equation have a common value which is constant and which we denote by λ, thereby obtaining the two ordinary differential equations,

$$x^2X'' + xX' - \lambda X = 0$$
$$Y' - \lambda Y = 0$$

The first of these equations is a "Cauchy-Euler" equation, whose complete solution may be found by a substitution $x = e^v$ (see the exercises for Section 3.6).

If $\lambda > 0$, say $\lambda = k^2$, then $X = c_1x^k + c_2x^{-k}$.
If $\lambda < 0$, say $\lambda = -k^2$, then $X = c_3 \cos(k \ln x) + c_4 \sin(k \ln x)$.
If $\lambda = 0$, then $X = c_4 \ln x + c_5$.

The second of the equations has a complete solution

$$Y = C_1e^{k^2y}, \qquad Y = C_2e^{-k^2y}, \quad \text{or} \quad Y = C_3$$

if $\lambda = k^2, -k^2$, or 0, respectively. Hence linear combinations of terms of the forms

$$e^{k^2y}x^{\pm k}, \qquad e^{-k^2y}\cos(k \ln x), \qquad e^{-k^2y}\sin(k \ln x), \qquad \ln x, \qquad \text{and} \quad 1$$

are solutions of the given partial differential equation. □

There are equations with constant coefficients that are not separable, even though solutions exist that are factorable. You should try separating the variables when assuming a product solution to the innocent-looking equation

$$u_{xx} + u_{xy} + u_{yy} = 0$$

Method 2: A constant-coefficient differential equation of the form

$$Au_{xx} + Bu_{xy} + Cu_{yy} = 0$$

can be put into the form $U_{vw} = 0$ with a linear change of the independent variables $v = x + my$; $w = x + ny$. The function $u(x, y)$ is a composite of the function $U(v, w)$ and the two functions $v(x, y)$ and $w(x, y)$ given by the relationship

$$u(x, y) = U(v(x, y), w(x, y)) = U(x + my, x + ny)$$

The idea is to select real constants m and n so that in the transformed differential equation, the coefficients of U_{vv} and U_{ww} are zero. The method, at least in some particular applications, has come to be called *D'Alembert's method.*

In substituting into the differential equation for the partial derivatives, note that by using the chain rule,

$$\begin{aligned} u_x &= U_v v_x + U_w w_x = U_v + U_w, \qquad \text{since } v_x = w_x = 1 \\ u_{xx} &= U_{vv} v_x + U_{vw} w_x + U_{wv} v_x + U_{ww} w_x \\ &= U_{vv} + 2U_{vw} + U_{ww}, \qquad \text{again since } v_x = w_x = 1 \end{aligned}$$

with similar formulas for u_{xy} and u_{yy}.

Example 8.12 Solve the differential equation $u_{xx} - u_{xy} - 2u_{yy} = 0$.

Solution Using the change of variables, $v = x + my$, $w = x + ny$; and with repeated use of the chain rule,

$$\begin{aligned} u_{xx} &= U_{vv} + 2U_{vw} + U_{ww} \\ u_{xy} &= mU_{vv} + (m + n)U_{vw} + nU_{ww} \\ u_{yy} &= m^2 U_{vv} + 2mnU_{vw} + n^2 U_{ww} \end{aligned}$$

The differential equation becomes

$$U_{vv}(1 - m - 2m^2) + U_{vw}(2 - m - n - 4mn) + U_{ww}(1 - n - 2n^2) = 0$$

We now select m and n to be the distinct real roots of the quadratic equation $1 - t - 2t^2 = 0$; that is, -1 and $\frac{1}{2}$. With this choice of m and n the differential equation can be written $U_{vw} = 0$, whose solution is

$$U(v, w) = g(v) + h(w)$$

where g and h are arbitrary twice differentiable functions. These functions are determined by auxiliary conditions and not by the partial differential equation. Hence, the solution expressed in terms of the original variables x and y has the form

$$u(x, y) = g(x - y) + h(x + \tfrac{1}{2}y)$$ □

Example 8.13 Use D'Alembert's method to solve the *wave equation* $u_{xx} = (1/c^2)u_{tt}$ with the boundary conditions $u(x, 0) \equiv f(x)$ and $u_t(x, 0) \equiv 0$, on $[0, L]$; $u(0, t) \equiv 0 \equiv u(L, t)$ for $t > 0$.

Solution First, we find $u(x, t)$ on $[0, L]$. Using the same technique as in the previous example we can show

$$u(x, t) = g(x + ct) + h(x - ct)$$

for arbitrary functions g and h.

From the conditions at $t = 0$,

$$u(x,0) = g(x) + h(x) \equiv f(x)$$
$$u_t(x,0) = cg'(x) - ch'(x) \equiv 0$$

Therefore, $g'(x) = h'(x)$ and hence $g(x) = h(x) + k$, where k is some constant. Using this in the first of the two equations,

$$h(x) + k + h(x) = f(x)$$

from which

$$h(x) = \frac{f(x) - k}{2}$$

and

$$g(x) = \frac{f(x) + k}{2}$$

Therefore, the solution for $0 \leqq x \leqq L$ is

$$u(x,t) = \frac{f(x + ct) + f(x - ct)}{2}$$

To find $u(x,t)$ for x not on the interval $[0, L]$, note that the condition $u(0,t) \equiv 0$ gives

$$f(ct) + f(-ct) \equiv 0$$

that is,

$$f(ct) = -f(-ct)$$

which says that outside of $[0, L]$, f must be chosen to be the odd extension of the given function as defined on $[0, L]$. Thus the function is determined on $[-L, L]$. The condition $u(L,t) \equiv 0$ gives

$$f(L + ct) \equiv -f(L - ct)$$

Since f is odd, the second member of this equation may be written $f(-L + ct)$ and therefore

$$f(ct + L) = f(ct - L)$$

and this means that f should be chosen to be periodic with period $2L$. Hence, the formula for $u(x,t)$ outside of $[0, L]$ is the same as that on $[0, L]$ with the understanding that the function f is odd and periodic with period $2L$. In Section 8.7 we will see that $f(x + ct)$ may be interpreted as a wave "traveling" to the left and $f(x - ct)$ as a wave "traveling" to the right, each with velocity c. □

Exercises for Section 8.4

In Exercises 1 and 2 find solutions by assuming an exponential form (that is, $u = ce^{ax}e^{by}, \ldots$).

1. **(a)** $u_x + 2u_y - u = 0$ **(b)** $u_{xx} - u_{yy} + u_y = 0$
2. **(a)** $u_x + u_y + u_z + u = 0$ **(b)** $u_{xy} + u_x + u_{yy} = 0$

3. Do the problems of Exercise 1 assuming a product solution.
4. Do the problems of Exercise 2 assuming a product solution.

Solve the equations in Exercises 5 and 6 by D'Alembert's method:

5. **(a)** $u_{xx} - 5u_{xy} + 6u_{yy} = 0$ **(b)** $u_{xy} + 4u_{yy} = 0$
6. **(a)** $u_{xy} - u_{yy} = 0$ **(b)** $u_{xx} = 25u_{yy}$
7. In D'Alembert's method, it may happen that $m = n$. Show that in this case $u = xf(x + my)$ is also a solution to $Au_{xx} + Bu_{xy} + Cu_{yy} = 0$.
8. Solve $u_{xx} + u_{xy} + u_{yy} = 0$ by assuming an exponential solution. Is this a separable differential equation?

Find solutions of the equations given in Exercises 9–15 in any way you can.

9. $x^2u_{xy} + 3y^2u = 0$
10. $u_x + u_y = 0$
11. $u_{xxxx} + yu_{xxy} = 0$
12. $u_{xy} - u = 0$
13. $u_{xx} - (1 + y^2)u_{xy} = 0$
14. $u_{xyz} - xyzu = 0$
15. $u_x + u_y = xu + yu$

An appropriate differential equation for modeling the vibrations, assumed small, of a thin beam of length L and of homogeneous and uniform cross section is

$$EI\frac{\partial^4 u(x,t)}{\partial x^4} + \rho\frac{\partial^2 u(x,t)}{\partial t^2} = -F(x,t), \qquad 0 < x < L, t > 0$$

Here $F(x,t)$ is the applied external transverse force, E is Young's modulus, and ρ, I, and $u(x,t)$ are the beam linear mass density, cross-sectional moment of inertia (about the neutral axis), and deflection (from equilibrium), respectively.

Exercises 16–18 refer to this beam-vibration application.

16. For the case of harmonic excitation in which $F(x,t) = F_0\cos(\omega t - \beta)$, find solutions of the form

$$u(x,t) = \phi(x)\cos(\omega t - \beta)$$

17. If the end of the beam at $x = 0$ is fixed and the end of the beam at $x = L$ is free, the deflection satisfies the boundary conditions

$$u(0,t) \equiv 0 \equiv \frac{\partial u(0,t)}{\partial x}$$

$$\frac{\partial^2 u(L,t)}{\partial x^2} \equiv 0 \equiv \frac{\partial^3 u(L,t)}{\partial x^3}$$

Show that nontrivial solutions of the homogeneous ($F \equiv 0$) beam-vibration problem of the form

$$u(x,t) = \phi(x)\cos(\omega t - \beta)$$

exist *only* if ω is one of the "special" values

$$\omega_n = \left(\frac{\gamma_n}{L}\right)^2\sqrt{\frac{EI}{\rho}}$$

where $\cosh\gamma_n\cos\gamma_n = -1$.

18. Verify that the natural frequencies for the vibration of a simply supported beam are

$$\omega_n = \left(\frac{n\pi}{L}\right)^2 \sqrt{\frac{EI}{\rho}} \qquad n = 1, 2, 3, \ldots$$

(Assume $F \equiv 0$ and $u(x, t) = \phi(x)\cos(\omega t - \beta)$, and let u and $\partial^2 u/\partial x^2$ vanish at both $x = 0$ and $x = L$.)

The *biharmonic* or *bi-Laplacian* equation

$$\Delta_2(\Delta_2 u) = \frac{\partial^4 u}{\partial x^4} + 2\frac{\partial^4 u}{\partial x^2 \partial y^2} + \frac{\partial^4 u}{\partial y^4} = 0$$

is involved in many two-dimensional applications of the theory of elasticity.

19. Find solutions of the biharmonic equation having the form

(a) $u(x, y) = \phi(x)\sin \beta y$ **(b)** $u(x, y) = \psi(y)\cos \alpha x$

20. **(a)** Show that the biharmonic equation is not separable.

(b) How do you reconcile the result of part (a) with the product solutions obtained in Exercise 19?

21. Verify that D'Alembert's method is *not* applicable to the biharmonic equation. (Recall that m and n must be real.)

22. Solve

$$\frac{\partial^4 u}{\partial x^4} - 2\frac{\partial^4 u}{\partial x^2 \partial y^2} + \frac{\partial^4 u}{\partial y^4} = 0$$

by D'Alembert's method.

8.5 More on Separation of Variables and Laplace's Equation

In most cases *finite* linear combinations of product solutions of the forms

$$(A \sinh kx + B \cosh kx)(C \sin ky + D \cos ky)$$
$$(A \sin kx + B \cos kx)(C \sinh ky + D \cosh ky)$$

and

$$(Ax + B)(Cy + D)$$

will be inadequate to solve a given boundary value problem for Laplace's equation in two dimensions. In such situations we must use infinite series representations and take advantage of our knowledge of Fourier series. The following examples illustrate the procedure.

Example 8.14 Determine the solution of the boundary value problem $u_{xx} + u_{yy} = 0$ for $0 < x < L$, $0 < y < K$, where $u_x(0, y) \equiv 0$, $u_x(L, y) \equiv 0$, $u(x, 0) \equiv 0$, and $u(x, K) \equiv x$.

Solution Separation of variables and application of the two homogeneous boundary conditions $u_x(0, y) \equiv 0$ and $u(x, 0) \equiv 0$ lead to product solutions of the form

$$C \cosh kx \sin ky, \qquad C \cos kx \sinh ky, \quad \text{and} \quad Cy$$

The third type of solution satisfies the boundary condition $u_x(L, y) \equiv 0$ for all C, and so does the second type if $k = n\pi/L$, $n = 1, 2, \ldots$. Employing superposition, therefore, we are led to try

$$u(x, y) = C_0 y + \sum_{n=1}^{\infty} C_n \cos \frac{n\pi x}{L} \sinh \frac{n\pi y}{L}$$

as a representation for the solution of the full boundary value problem.

To satisfy the remaining boundary condition, we must have

$$u(x, K) = C_0 K + \sum_{n=1}^{\infty} C_n \cos \frac{n\pi x}{L} \sinh \frac{n\pi K}{L} \equiv x$$

We recognize the series in this expression as essentially in the form of a Fourier cosine series (recall the discussion of Section 7.9). This last condition will be satisfied, therefore, if we choose

$$C_0 K = \frac{1}{L} \int_0^L x \, dx = \frac{L}{2}$$

and, for $n \geq 1$,

$$\begin{aligned} C_n \sinh \frac{n\pi K}{L} &= \frac{2}{L} \int_0^L x \cos \frac{n\pi x}{L} \, dx \\ &= \frac{2L}{n^2\pi^2} [(-1)^n - 1] \\ &= \begin{cases} -\dfrac{4L}{n^2\pi^2}, & n \text{ odd} \\ 0, & n \text{ even} \end{cases} \end{aligned}$$

Thus the formal solution to the given boundary value problem has the representation

$$u(x, y) = \frac{Ly}{2K} - \frac{4L}{\pi^2} \sum_{m=0}^{\infty} \frac{1}{(2m+1)^2} \cos \frac{(2m+1)\pi x}{L} \sinh \frac{(2m+1)\pi y}{L} \Bigg/ \sinh \frac{(2m+1)\pi K}{L}$$

Moreover, this series can be shown to converge pointwise on the rectangle of interest (recall Theorem 7.7). □

In these problems there is nothing special about Cartesian coordinates. Indeed, representations in terms of polar coordinates, for instance, may be more appropriate.

Example 8.15 Find the bounded solution of Laplace's equation $\Delta_2 u = 0$ for the interior of the unit circle $r < 1$, $0 \leq \theta \leq 2\pi$, where on the boundary $u(1, \theta) \equiv f(\theta)$, a specified function of θ.

Solution In polar coordinates Laplace's equation assumes the form

$$r\frac{\partial}{\partial r}\left(r\frac{\partial u}{\partial r}\right) + \frac{\partial^2 u}{\partial \theta^2} = 0$$

If we apply separation of variables and assume a product solution $u(r, \theta) = R(r)\Theta(\theta)$, we arrive at

$$\frac{r(rR')'}{R} = -\frac{\Theta''}{\Theta} = \text{constant}$$

The choice of λ^2 as the constant then leads to the two ordinary differential equations

$$r^2 R'' + rR' - \lambda^2 R = 0$$

and

$$\Theta'' + \lambda^2 \Theta = 0$$

The first of these two equations is of Cauchy-Euler type and can be readily solved (recall Exercises 12–13, pp. 104–105) to yield

$$\begin{aligned} R &= Ar^{\lambda} + Br^{-\lambda}, &\quad \lambda \neq 0 \\ R &= A + B\ln r, &\quad \lambda = 0 \end{aligned}$$

The solution of the equation in Θ is even easier to determine:

$$\begin{aligned} \Theta &= C\sin\lambda\theta + D\cos\lambda\theta, &\quad \lambda \neq 0 \\ \Theta &= C\theta + D, &\quad \lambda = 0 \end{aligned}$$

Now we are looking for a bounded solution $u(r, \theta)$. Moreover, since the points (r, θ) and $(r, \theta + 2\pi)$ are the same, we expect the solution to be periodic in θ with period 2π. These two requirements imply λ is an integer and $B = 0$. Indeed, we can restrict attention to solutions having the form

$$u(r, \theta) = r^n(C_n \sin n\theta + D_n \cos n\theta)$$

where $n = 0, 1, 2, \ldots$.

To satisfy the boundary condition, we first use superposition:

$$u(r, \theta) = D_0 + \sum_{n=1}^{\infty} r^n(C_n \sin n\theta + D_n \cos n\theta)$$

Thence

$$u(1, \theta) = D_0 + \sum_{n=1}^{\infty} (C_n \sin n\theta + D_n \cos n\theta) \equiv f(\theta)$$

a classical Fourier series expression in the variable θ. The familiar formulas of Section 7.4, therefore, immediately lead to

$$D_0 = \frac{1}{2\pi}\int_0^{2\pi} f(\theta)\,d\theta$$

and for $n \geq 1$,

$$C_n = \frac{1}{\pi}\int_0^{2\pi} f(\theta)\sin n\theta\, d\theta$$

$$D_n = \frac{1}{\pi}\int_0^{2\pi} f(\theta)\cos n\theta\, d\theta$$

The desired solution is given by the above representation with these Fourier coefficients. □

Exercises for Section 8.5

The results in the following exercises can all be interpreted in terms of steady-state temperature distributions, electrostatic-magnetostatic potentials, or similar equilibrium phenomena in rectangular and circular two-dimensional regions.

1. Show that the solution of the boundary value problem

$$\begin{aligned} u_{xx} + u_{yy} &= 0, && 0 < x < L, 0 < y < K \\ u(0, y) &= 0, && 0 < y < K \\ u(L, y) &= 0, && 0 < y < K \\ u(x, 0) &= 0, && 0 < x < L \\ u(x, K) &= f(x), && 0 < x < L \end{aligned}$$

can be represented as the infinite series

$$\begin{aligned} u(x, y) &= \sum_{n=1}^{\infty} u_n(x, y) \\ &= \sum_{n=1}^{\infty} C_n \sin\frac{n\pi x}{L}\sinh\frac{n\pi y}{L} \end{aligned}$$

where $C_n \sinh\dfrac{n\pi K}{L} = \dfrac{2}{L}\displaystyle\int_0^L f(x)\sin\frac{n\pi x}{L}\,dx.$

2. Determine the form of the solution if the boundary conditions in Exercise 1 are replaced by

$$\begin{aligned} u(0, y) &= 0 = u(L, y), && 0 < y < K \\ \frac{\partial u(x, 0)}{\partial y} &= 0, && 0 < x < L \\ u(x, K) &= f(x), && 0 < x < L \end{aligned}$$

3. Repeat Exercise 1 under the conditions

$$\begin{aligned} u(0, y) &= g(y), && 0 < y < K \\ u(L, y) &= 0, && 0 < y < K \\ u(x, 0) &= 0 = u(x, K), && 0 < x < L \end{aligned}$$

4. Demonstrate that when the boundary conditions take the form

$$\begin{aligned} u_x(0, y) &= 0 = u_x(L, y), && 0 < y < K \\ u(x, 0) &= 0, && 0 < x < L \\ u_y(x, K) &= f(x), && 0 < x < L \end{aligned}$$

the solution of Laplace's equation for $0 < x < L$, $0 < y < K$ can be represented as the infinite series

$$u(x, y) = \sum_{n=0}^{\infty} u_n(x, y)$$

$$= D_0 y + \sum_{n=1}^{\infty} D_n \cos\frac{n\pi x}{L} \sinh\frac{n\pi}{L} y$$

where $D_0 = \dfrac{1}{L}\displaystyle\int_0^L f(x)\,dx$, and, for $n \geq 1$,

$$D_n \frac{n\pi}{L} \cosh\frac{n\pi K}{L} = \frac{2}{L}\int_0^L f(x)\cos\frac{n\pi x}{L}\,dx$$

5. Determine the form of the solution if the boundary conditions in Exercise 4 are replaced by

$$u(0, y) = g(y), \qquad 0 < y < K$$
$$u_x(L, y) = 0, \qquad 0 < y < K$$
$$u_y(x, 0) = 0 = u_y(x, K), \qquad 0 < x < L$$

6. Solve the boundary value problem

$$u_{xx} + u_{yy} = 0, \qquad 0 < x < 1, 0 < y < 1$$

where $u(0, y) \equiv 0$, $u(1, y) \equiv 0$, $u(x, 0) \equiv 0$, and $u(x, 1) \equiv x$.

7. Determine the solution of the boundary value problem

$$u_{xx} + u_{yy} = 0, \qquad 0 < x < 3, 0 < y < 2$$

where $u(0, y) \equiv y - 1$, $u(3, y) \equiv 0$, $u(x, 0) \equiv 0$, and $u(x, 2) \equiv 0$.

8. Discuss how superposition can be used to solve the boundary value problem

$$u_{xx} + u_{yy} = 0, \qquad 0 < x < L, 0 < y < K$$

where $u(0, y) \equiv f(y)$, $u(L, y) \equiv g(y)$, $u(x, 0) \equiv h(x)$, and $u(x, K) \equiv k(x)$.

9. Determine the solution of the boundary value problem

$$u_{xx} + u_{yy} = 0, \qquad 0 < x < 2, 0 < y < 1$$

where $u_x(0, y) \equiv 0$, $u_x(2, y) \equiv y$, $u(x, 0) \equiv 0$, and $u(x, 1) \equiv 0$.

10. Solve the boundary value problem

$$u_{xx} + u_{yy} = 0, \qquad 0 < x < 1, 0 < y < 2$$

where $u(0, y) \equiv 0$, $u(1, y) \equiv 0$, $u_y(x, 0) \equiv 1 + x$, and $u_y(x, 2) \equiv 0$.

11. Find the bounded solution $u(r, \theta)$ of Laplace's equation in the interior of the unit circle $r < 1$, $0 \leq \theta \leq 2\pi$, where $u(1, \theta) = \cos\theta$.

12. Determine the bounded solution $u(r, \theta)$ of Laplace's equation in the interior of the unit circle $r < 1$, $0 \leq \theta \leq 2\pi$ where

$$u(1, \theta) = \begin{cases} 0, & 0 \leq \theta \leq \pi \\ 2\pi - \theta, & \pi \leq \theta \leq 2\pi \end{cases}$$

13. Solve Laplace's equation $\Delta_2 u = 0$ in the interior of the unit circle where $u(r, \theta)$ is bounded for $r < 1$, $0 \leq \theta \leq 2\pi$ and $u(1, \theta) = 5$ for $0 \leq \theta < \pi/2$, -5 for $\pi/2 \leq \theta < \pi$, 0 for $\pi \leq \theta < 2\pi$.

14. Show that the result of Example 8.15 can be generalized to a circle of arbitrary radius a merely by replacing r by r/a in the solution representation.
15. Determine the form of the solution if the boundary condition in Example 8.15 is replaced by

$$\frac{\partial u(1,\theta)}{\partial r} \equiv g(\theta)$$

Here it turns out that g must satisfy the auxiliary condition

$$\int_0^{2\pi} g(\theta)\, d\theta = 0$$

16. Find a bounded solution $u(r,\theta)$ of Laplace's equation in the interior of the unit circle $r < 1, 0 \leq \theta \leq 2\pi$, where $\partial u/\partial r\,(1,\theta) = \sin 3\theta$.
17. Solve Laplace's equation, $\Delta_2 u = 0$, in the interior of the unit circle, where $u(r,\theta)$ is bounded for $r < 1, 0 \leq \theta \leq 2\pi$, and $\partial u/\partial r = \theta - \pi$ for $0 \leq \theta < 2\pi$ and $r = 1$.

An alternative representation for the solution of Example 8.15 has the form

$$u(r,\theta) = \frac{1}{2\pi}\int_0^{2\pi} \frac{1 - r^2}{1 + r^2 - 2r\cos(\theta - \phi)} f(\phi)\, d\phi$$

This result, known as the *Poisson integral formula*, generalizes the classical *mean-value theorem*

$$u(0,0) = \frac{1}{2\pi}\int_0^{2\pi} f(\phi)\, d\phi$$

for harmonic functions to arbitrary points (r,θ). Moreover, the formula often provides an easier derivation of closed form expressions for given boundary value problems with Laplace's equation in the unit circle.

18. Use the Poisson integral formula to show that the bounded solution of Laplace's equation, $\Delta_2 u = 0$, in the interior of the unit circle that satisfies $u(1,\theta) \equiv C$, where C is constant, is $u(r,\theta) \equiv C$.
19. Verify that

$$u(r,\theta) = \frac{u_0}{2} + \frac{u_0}{\pi}\mathrm{Tan}^{-1}\left(\frac{2r\sin\theta}{1 - r^2}\right)$$

is the bounded solution of Laplace's equation in the interior of the unit circle with the boundary values $u(1,\theta) = u_0$ for $0 \leq \theta < \pi$ and 0 for $\pi \leq \theta < 2\pi$.
20. Derive the solution given in Exercise 19 using the Poisson integral formula.
21. Solve Exercise 11 using the Poisson integral formula.

8.6 The Wave Equation: Two Derivations

You are probably aware that completely different physical phenomena can be effectively modeled by the same mathematical equations. In this section we will show how the wave equation is derived in two widely varied applications.

Consider the problem of a string stretched taut between two points and then

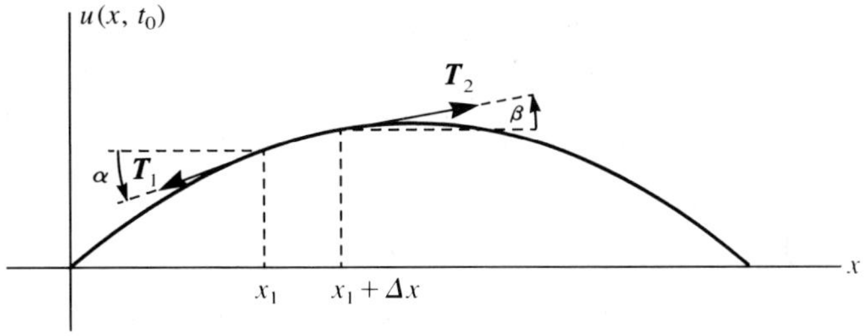

Figure 8.2 *Transverse motion of vibrating string*

given an initial deflection or "plucked." The transverse ("up and down") motion is small, in fact much smaller than the sketch in Figure 8.2 would seem to imply, and the tension in the string is assumed to be large with respect to other forces.

To determine the equation for the transverse vibration we examine a small piece of the string between x_1 and $x_1 + \Delta x$. Since there is no horizontal movement, the horizontal forces are in equilibrium:

$$|\mathbf{T}_1| \cos \alpha = |\mathbf{T}_2| \cos \beta = T = \text{constant}$$

The difference of the vertical forces is equal to the mass of the element times the transverse acceleration,

$$|\mathbf{T}_2| \sin \beta - |\mathbf{T}_1| \sin \alpha = \rho \, \Delta s \, u_{tt}$$

where ρ is the mass per unit length and Δs is the length of the small piece of string under consideration. Since the vibration is small, $\Delta s \approx \Delta x$. Dividing each term of the second equation by the equal quantities of the first equation,

$$\tan \beta - \tan \alpha = \frac{\rho \, \Delta x}{T} u_{tt}$$

Dividing by Δx and using the slope interpretation of the derivative,

$$\frac{u_x(x_1 + \Delta x, t) - u_x(x_1, t)}{\Delta x} = \frac{\rho}{T} u_{tt}$$

Letting Δx approach 0, this becomes

$$u_{xx} = \frac{\rho}{T} u_{tt}$$

which is often written

$$u_{xx} = \frac{1}{c^2} u_{tt}$$

where c turns out to have the dimensions of velocity.

This partial differential equation is called the (one-dimensional) wave equation, and it, along with the auxiliary conditions

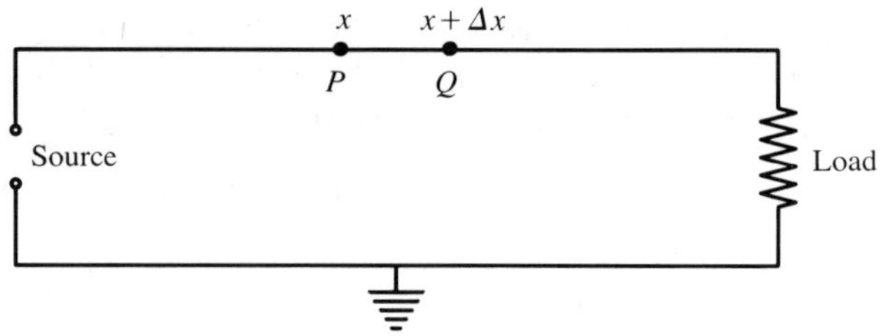

Figure 8.3 *Transmission line*

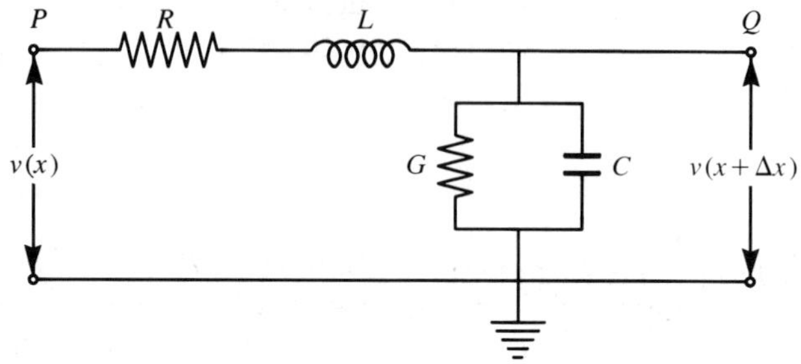

Figure 8.4 *Schematic per unit length*

$$
\begin{aligned}
u(x, 0) &\equiv f(x), && \text{initial deflection} \\
u_t(x, 0) &\equiv g(x), && \text{initial velocity} \\
u(0, t) &\equiv u(L, t) \equiv 0, && \text{both ends fixed}
\end{aligned}
$$

is the mathematical model for the vibrating string problem.

Next consider the flow of electricity in a long cable or transmission line assumed to be imperfectly insulated so that there is both capacitance and leakage of current to ground. Figure 8.3 gives a generalized schematic of the system and Figure 8.4 shows a representative portion of the transmission line for a small length Δx. Assuming R, L, G, and C to be the ordinary electric parameters (resistance, inductance, conductance, and capacitance) per unit length, the values of the lumped components are as shown.

The voltage drop from P to Q is given by

$$v(x + \Delta x) - v(x) = -(R\,\Delta x)i - (L\,\Delta x)i_t$$

and the current loss is given by

$$i(x + \Delta x) - i(x) = -(G\,\Delta x)v - (C\,\Delta x)v_t$$

In both equations, divide by Δx, and let Δx approach 0, to obtain the *transmission line equations*

$$
\begin{aligned}
v_x &= -Ri - Li_t \\
i_x &= -Gv - Cv_t
\end{aligned}
$$

These equations are often written in slightly different form. If you differentiate the first equation with respect to x and the second with respect to t and eliminate the variable i between the two, you get

$$v_{xx} = RGv + RCv_t + LGv_t + LCv_{tt}$$

Reversing the procedure for the current,

$$i_{xx} = RGi + RCi_t + LGi_t + LCi_{tt}$$

Thus, the current and voltage satisfy exactly the same partial differential equation, known as the *wave equation of the transmission line.*

For high frequencies (or for negligible R and G), the terms involving the second derivative with respect to time dominate the right-hand sides and the equations become

$$v_{xx} = LCv_{tt}$$
$$i_{xx} = LCi_{tt}$$

These equations have precisely the same form as the one derived for the vibrating string problem. The actual mathematical model for the two situations will usually be quite different, though, since the vibrating string conditions of zero deflection at the endpoints of the string have no realistic analog on a transmission line.

8.7 Solution by D'Alembert's Method

In Example 8.13 we solved the one-dimensional wave equation with boundary values corresponding to the vibrating string problem. In that example, we saw that the deflection of the string is given by

$$u(x, t) = \frac{f(x + ct) + f(x - ct)}{2}$$

where f is understood to be the odd periodic extension of the initial deflection on $[0, L]$. The graph of $f(x + ct)$ is the same as that of $f(x)$ except that it is translated to the left by ct units. Similarly, the graph of $f(x - ct)$ is the translation of $f(x)$ to the right by ct units.

The functions $f(x + ct)$ and $f(x - ct)$ are said to "travel" since with increasing t, the first wave "moves" to the left and the second one to the right. The physical traveling of the two waves is a very illuminating result and is one of the major reasons for the importance of D'Alembert's method in the solution of the wave equation.

To find $u(x, t)$ on $[0, L]$ at a given instant of time, t_0, we may proceed as follows:

1. Graphically, by plotting the two waves

$$\tfrac{1}{2}f(x + ct_0), \qquad \tfrac{1}{2}f(x - ct_0)$$

and adding the ordinates on $[0, L]$.

2. Analytically, by computing the analytic expressions for $f(x + ct)$ and $f(x - ct)$ on $[0, L]$.

By displaying the deflection (or finding the analytic description of the deflection) at several different times, we can get a good idea of the nature of the vibration.

Example 8.16 Plot solutions to the vibrating string problem corresponding to a triangular initial deflection on $[0, 2]$ given by the formula

$$f(x) = \begin{cases} x, & 0 \leqq x \leqq 1 \\ -x + 2, & 1 \leqq x \leqq 2 \end{cases}$$

Assume $c = 1$ and plot the deflection at $t = 0, \frac{1}{3}, \frac{2}{3}, 1, 1\frac{1}{3}$, and so on. Verify the sketches by computing the deflection analytically.

Solution To sketch the solution curve at a particular value of t, we sum the ordinates of the initial waveform translated to the right by t units with the ordinates of a left-translated waveform (and divide by two). The results are shown in Figure 8.5. Since the initial deflection is defined so explicitly, an analytic solution is possible.

1. At $t = 0$,

$$u(x, 0) = \frac{f(x) + f(x)}{2} = f(x), \qquad \text{on } [0, 2]$$

2. At $t = \frac{1}{3}$,

$$u(x, \tfrac{1}{3}) = \tfrac{1}{2}[f(x + \tfrac{1}{3}) + f(x - \tfrac{1}{3})]$$

where

$$f(x + \tfrac{1}{3}) = \begin{cases} x + \frac{1}{3}, & 0 \leqq x \leqq \frac{2}{3} \\ -x + \frac{5}{3}, & \frac{2}{3} \leqq x \leqq 2 \end{cases}$$

$$f(x - \tfrac{1}{3}) = \begin{cases} x - \frac{1}{3}, & 0 \leqq x \leqq \frac{4}{3} \\ -x + \frac{7}{3}, & \frac{4}{3} \leqq x \leqq 2 \end{cases}$$

Hence,

$$u(x, \tfrac{1}{3}) = \begin{cases} x, & 0 \leqq x \leqq \frac{2}{3} \\ \frac{2}{3}, & \frac{2}{3} \leqq x \leqq \frac{4}{3} \\ -x + 2, & \frac{4}{3} \leqq x \leqq 2 \end{cases}$$

3. At $t = \frac{2}{3}$,

$$f(x + \tfrac{2}{3}) = \begin{cases} x + \frac{2}{3}, & 0 \leqq x \leqq \frac{1}{3} \\ -x + \frac{4}{3}, & \frac{1}{3} \leqq x \leqq 2 \end{cases}$$

$$f(x - \tfrac{2}{3}) = \begin{cases} x - \frac{2}{3}, & 0 \leqq x \leqq \frac{5}{3} \\ -x + \frac{8}{3}, & \frac{5}{3} \leqq x \leqq 2 \end{cases}$$

Hence,

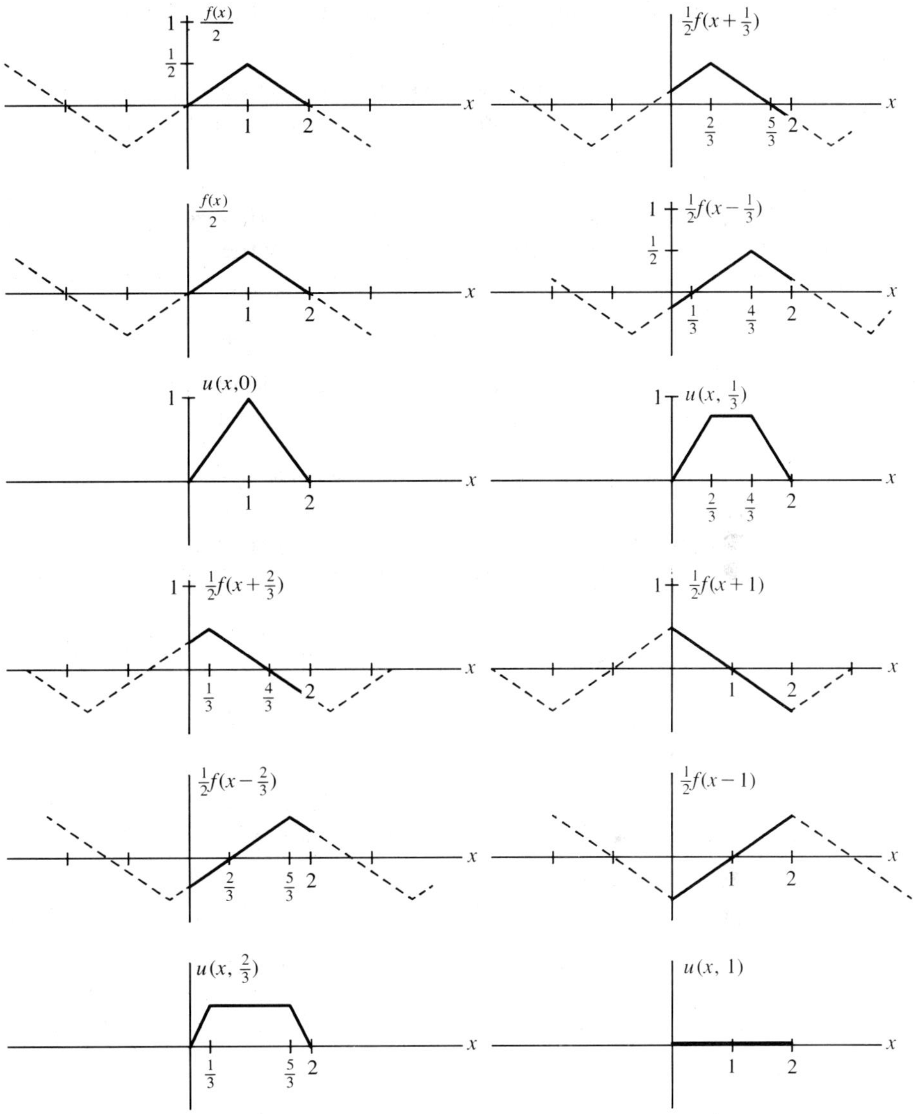

Figure 8.5

$$u(x,\tfrac{2}{3}) = \begin{cases} x, & 0 \leqq x \leqq \frac{1}{3} \\ \frac{1}{3}, & \frac{1}{3} \leqq x \leqq \frac{5}{3} \\ -x+2, & \frac{5}{3} \leqq x \leqq 2 \end{cases}$$

4. At $t = 1$,

$$f(x+1) = -x+1, \qquad 0 \leqq x \leqq 2$$
$$f(x-1) = x-1, \qquad 0 \leqq x \leqq 2$$

Hence

$$u(x,1) = 0, \qquad 0 \leqq x \leqq 2$$

□

Exercises for Sections 8.6 and 8.7

1. The D'Alembert method can be applied to the wave equation on an infinite interval. Verify that

$$u(x,t) = \frac{1}{2}[f(x+ct) + f(x-ct)] + \frac{1}{2c}\int_{x-ct}^{x+ct} g(\tau)\,d\tau$$

solves the initial value problem

$$u_{xx} - \frac{1}{c^2}u_{tt} = 0, \qquad -\infty < x < \infty,\ t > 0$$

$$u(x,0) = f(x)$$

$$u_t(x,0) = g(x)$$

2. Use the representation of Exercise 1 to solve

$$u_{xx} - \frac{1}{c^2}u_{tt} = 0, \qquad -\infty < x < \infty,\ t > 0$$

$$u(x,0) = e^{-x^2}$$

$$u_t(x,0) = 1$$

3. Find the solution of the initial value problem

$$u_{xx} - \frac{1}{c^2}u_{tt} = 0, \qquad -\infty < x < \infty,\ t > 0$$

$$u(x,0) = 0$$

$$u_t(x,0) = \begin{cases} 2, & |x| \le 1 \\ 0, & \text{otherwise} \end{cases}$$

4. The method of D'Alembert is equally applicable if the wave equation is to be analyzed over an interval of semi-infinite extent. Verify that

$$u(x,t) = \begin{cases} \dfrac{1}{2}[f(x+ct) + f(x-ct)] + \dfrac{1}{2c}\displaystyle\int_{x-ct}^{x+ct} g(\tau)\,d\tau, & 0 \le t < \dfrac{x}{c} \\ a\left(t - \dfrac{x}{c}\right) + \dfrac{1}{2}[f(ct+x) - f(ct-x)] + \dfrac{1}{2c}\displaystyle\int_{ct-x}^{ct+x} g(\tau)\,d\tau, & t > \dfrac{x}{c} \end{cases}$$

solves the initial value–boundary value problem

$$u_{xx} - \frac{1}{c^2}u_{tt} = 0, \qquad x > 0,\ t > 0$$

$$u(0,t) = a(t)$$

$$u(x,0) = f(x)$$

$$u_t(x,0) = g(x)$$

5. Use the representation of Exercise 4 to solve

$$u_{xx} - \frac{1}{c^2}u_{tt} = 0, \qquad x > 0,\ t > 0$$

$$u(0,t) = 0$$

$$u(x,0) = 0$$

$$u_t(x,0) = \frac{2x}{1 + x^4}$$

6. Find the solution of the initial value–boundary value problem

$$u_{xx} - \frac{1}{c^2} u_{tt} = 0, \qquad x > 0, t > 0,$$

$$u(0,t) = u_0$$

$$u(x,0) = 0$$

$$u_t(x,0) = 0$$

(This is like problem (c) of Example 8.7.)

7. Use the procedure outlined in Section 8.6 to derive the wave equation for the transmission line from the transmission line equations.

The wave equation of the transmission line

$$v_{xx} = RGv + RCv_t + LGv_t + LCv_{tt}$$

has an important feature. If the substitution $v(x,t) = e^{-Rt/L}V(x,t)$ is made, the equation for the new function V becomes

$$V_{xx} = (LG - RC)V_t + LCV_{tt}$$

Thus, if the transmission line is *balanced*, that is, its electrical parameters satisfy the relation $LG - RC = 0$, the function V satisfies the one-dimensional wave equation, and a general formula for the solution can be obtained.

8. Carry out the substitution $v(x,t) = e^{-Rt/L}V(x,t)$ in the wave equation for the transmission line and verify the form of the differential equation for $V(x,t)$.

9. Give a general relation for the current (voltage) on a balanced ($LG = RC$) transmission line. Express your result in terms of two (damped) traveling waves.

10. Use D'Alembert's method to solve for the current in a balanced ($LG = RC$) transmission line of infinite extent if

$$\left.\begin{aligned} i(x,0) &= f(x) \\ v(x,0) &= g(x) \end{aligned}\right\}, \qquad -\infty < x < \infty$$

(*Hint:* $i_t(x,0) = -\frac{R}{L}f(x) - \frac{1}{L}g'(x)$; also recall the result of Exercise 1.)

In three dimensions the wave equation has the form

$$\Delta_3 u(x,y,z,t) = \frac{1}{c^2}u_{tt}$$

or

$$u_{xx} + u_{yy} + u_{zz} = \frac{1}{c^2}u_{tt}$$

In spherical coordinates the equation becomes

$$u_{\rho\rho} + \frac{2}{\rho}u_\rho + \frac{1}{\rho^2\sin^2\phi}u_{\theta\theta} + \frac{1}{\rho^2}u_{\phi\phi} + \frac{\cos\phi}{\rho^2\sin\phi}u_\phi = \frac{1}{c^2}u_{tt}$$

For spherically symmetric solutions, $u_\theta = 0$, $u_\phi = 0$ and the equation has a particularly simple form:

$$u_{\rho\rho} + \frac{2}{\rho}u_\rho = \frac{1}{c^2}u_{tt}$$

11. Show that in the spherically symmetric case the substitution $u(\rho, t) = v(\rho, t)/\rho$ in the three-dimensional wave equation gives rise to the one-dimensional wave equation for the new function $v(\rho, t)$.
12. Use the result of Exercise 11 to find the complete spherically symmetric solution of the three-dimensional wave equation. Express your answer in terms of traveling waves.
13. Make the appropriate modifications in the representation of Exercise 4 and then use it to solve for the spherically symmetric solution of the three-dimensional wave equation which satisfies the auxiliary conditions $u(1, t) = u_0$, $u(\rho, 0) = 0$, $u_t(\rho, 0) = 1/\rho$ for $\rho > 1$ and $t > 0$.

In Exercises 14–17, plot solutions in a manner similar to Example 8.16 for initial deflection functions as given. Assume zero initial velocity and take $c = 1$.

14. For a string of length 3,

$$f(x) = \begin{cases} x, & 0 < x < 1 \\ 1, & 1 < x < 2 \\ -x + 3, & 2 < x < 3 \end{cases}$$

15. For a string of length 6, $f(x) = x(6 - x)$, $0 < x < 6$.
16. For a string of length 4,

$$f(x) = \begin{cases} 0, & 0 < x < 1 \\ x - 1, & 1 < x < 2 \\ -x + 3, & 2 < x < 3 \\ 0, & 3 < x < 4 \end{cases}$$

17. For a string of length 1, $f(x) = \sin \pi x$, $0 < x < 1$.

18–20. Verify analytically the solutions to Exercises 14–16.

8.8 Product Solutions for the Wave Equation

The one-dimensional wave equation with the homogeneous boundary conditions of the vibrating string problem may also be solved by assuming the solution to be the product of a function of x by a function of t. We substitute $u(x, t) = X(x)T(t)$ into the equation to obtain

$$\frac{X''}{X} = \frac{T''}{c^2 T}$$

which immediately yields the two ordinary differential equations

$$X'' - \lambda X = 0 \quad \text{and} \quad T'' - c^2 \lambda T = 0$$

The homogeneous boundary conditions of u may be easily interpreted in terms of the so-called space function, X,

$$\left.\begin{aligned} u(0,t) &= X(0)T(t) = 0 \\ u(L,t) &= X(L)T(t) = 0 \end{aligned}\right\} \quad \text{for all } t$$

This means $X(0) = X(L) = 0$ if a nontrivial function T is desired. Hence, we are led to the Sturm-Liouville problem

$$X'' - \lambda X = 0$$
$$X(0) = X(L) = 0$$

Up to multiplicative constants, the only nontrivial solutions to this problem are $X_n(x) = \sin(n\pi x/L)$, corresponding to eigenvalues $\lambda_n = -n^2\pi^2/L^2$. (See Section 7.12.) Moreover, when $\lambda_n = -n^2\pi^2/L^2$, the general solution of

$$T'' - c^2\lambda_n T = 0$$

is

$$T_n(t) = C_n \cos\frac{n\pi ct}{L} + D_n \sin\frac{n\pi ct}{L}$$

Forming the product of $X_n(x)$ and $T_n(t)$, we see that each of the functions

$$u_n(x,t) = \sin\frac{n\pi x}{L}\left(C_n \cos\frac{n\pi ct}{L} + D_n \sin\frac{n\pi ct}{L}\right)$$

satisfies the one-dimensional wave equation and the given set of homogeneous boundary conditions. By superposition not only each $u_n(x,t)$ but also any sum of such functions and even any convergent infinite series of such functions satisfies the differential equation and the homogeneous boundary conditions.

For a specific practical problem it remains to match sums of the u_n's to the given nonhomogeneous initial conditions. Such a matching may be possible by inspection.

Example 8.17 Solve the vibrating string problem by the product method for a string of length 5 if the initial deflection is given by $3\sin 2\pi x + \sin 4\pi x - 0.1\sin 20\pi x$. Assume zero initial velocity.

Solution Choose u_{10}, u_{20}, and u_{100} with $C_{10} = 3$, $C_{20} = 1$ and $C_{100} = -0.1$. The sum of these three functions evaluated at $t = 0$ will match the initial deflection function. Since $u_t(x,0) \equiv 0$, it follows that $D_n = 0$, for all n. Hence, the solution is

$$\begin{aligned} u(x,t) = {} & 3(\sin 2\pi x)(\cos 2\pi ct) + (\sin 4\pi x)(\cos 4\pi ct) \\ & -0.1(\sin 20\pi x)(\cos 20\pi ct) \end{aligned}$$

□

This example notwithstanding, in general, no finite sum of the u_n will satisfy arbitrary initial conditions. In such cases we must form an infinite series

$$\begin{aligned} u(x, t) &= \sum_{n=1}^{\infty} u_n(x, t) \\ &= \sum_{n=1}^{\infty} \sin\frac{n\pi x}{L}\left(C_n \cos\frac{n\pi ct}{L} + D_n \sin\frac{n\pi ct}{L}\right) \end{aligned}$$

and choose C_n and D_n so that the initial conditions are satisfied.

To satisfy the condition for initial deflection, $u(x, 0) \equiv f(x)$,

$$u(x, 0) = \sum_{n=1}^{\infty} C_n \sin\frac{n\pi x}{L} \equiv f(x)$$

This condition will be satisfied if the C_n are chosen in such a way that the series converges pointwise to the function f on $[0, L]$. This is the familiar problem of expanding a function by a Fourier sine series on $[0, L]$. Therefore, if we let

$$C_n = \frac{2}{L}\int_0^L f(x) \sin\frac{n\pi x}{L}\, dx$$

the series will converge to f on the interval $[0, L]$ if f satisfies the Dirichlet conditions. (Recall Theorem 7.7.)

The technique for determining the D_n is much the same. Assuming that the series for $u(x, t)$ can be differentiated term by term with respect to t, we equate this series evaluated at $t = 0$ to the initial velocity function,

$$g(x) \equiv u_t(x, 0) = \sum_{n=1}^{\infty} \frac{n\pi c}{L} D_n \sin\frac{n\pi x}{L}$$

and so, $(n\pi c/L)\, D_n$ must be the nth Fourier sine series coefficient of the function g. Hence,

$$D_n = \frac{2}{n\pi c}\int_0^L g(x) \sin\frac{n\pi x}{L}\, dx$$

Example 8.18 Solve Example 8.16 by the product method.

Solution We need only calculate the Fourier sine coefficients of the given triangular function,

$$\begin{aligned} C_n &= \int_0^1 x \sin\frac{n\pi x}{L}\, dx + \int_1^2 (-x + 2) \sin\frac{n\pi x}{L}\, dx \\ &= \frac{8}{n^2\pi^2} \sin\frac{n\pi}{2} \\ &= \begin{cases} \pm\dfrac{8}{n^2\pi^2}, & n \text{ odd} \\ 0, & n \text{ even} \end{cases} \end{aligned}$$

Therefore, since all the $D_n = 0$,

$$u(x, t) = \frac{8}{\pi^2}\sum_{n=0}^{\infty} \frac{(-1)^n}{(2n + 1)^2} \sin\frac{(2n + 1)\pi x}{2} \cos\frac{(2n + 1)\pi t}{2} \qquad \square$$

The frequency of the time factor in the first nonzero u_n is called the *fundamental frequency*. If $u_1 \neq 0$, the fundamental frequency is $c/2L$; all other frequencies are integral multiples of this frequency. The fundamental frequency determines the fundamental *tone* of the vibration; its integral multiples are called *overtones* or harmonics. (Recall the discussion of Section 7.1.)

The solution of the vibrating string problem with zero initial velocity by the product method technique can be identified with that obtained by D'Alembert's method. The solution by the product method gives

$$u(x, t) = \sum_{n=1}^{\infty} C_n \sin\frac{n\pi x}{L} \cos\frac{n\pi ct}{L}$$

where the C_n's are the Fourier sine series coefficients of the initial deflection function on $[0, L]$.

We use the trigonometric identity

$$\sin\frac{n\pi x}{L} \cos\frac{n\pi ct}{L} = \frac{1}{2}\left[\sin\frac{n\pi}{L}(x - ct) + \sin\frac{n\pi}{L}(x + ct)\right]$$

to write

$$u(x, t) = \frac{1}{2}\sum_{n=1}^{\infty} C_n \sin\frac{n\pi}{L}(x - ct) + \frac{1}{2}\sum_{n=1}^{\infty} C_n \sin\frac{n\pi}{L}(x + ct)$$

Both these series are obtainable from the series for $f(x)$ by substituting $x - ct$ and $x + ct$ for x. Hence,

$$u(x, t) = \frac{1}{2}[f(x + ct) + f(x - ct)]$$

as desired.

As suggested in the introductory section of this chapter, wave motion in a higher space dimension is well modeled by the more general wave equation

$$\Delta u = \frac{1}{c^2} u_{tt}$$

where the multidimensional Laplace operator has replaced the $\partial^2/\partial x^2$-term present in the one-dimensional equation. Solving boundary value problems associated with this new equation is naturally more complicated than before. We treat one example here, saving further applications for the exercises.

Example 8.19 The vibration of a square drumhead with fixed edges may be described by the following boundary value problem:

$$\Delta_2 u = \frac{1}{c^2} u_{tt}; \qquad 0 < x < 1, 0 < y < 1, t > 0$$

$$u(0, y, t) \equiv 0, \qquad u(1, y, t) \equiv 0, \qquad u(x, 0, t) \equiv 0, \qquad u(x, 1, t) \equiv 0$$

$$u(x, y, 0) \equiv f(x, y), \qquad u_t(x, y, 0) \equiv g(x, y)$$

Determine a representation for the solution of this problem in the case when the initial velocity $g(x, y) \equiv 0$.

Solution We apply separation of variables with $u(x, y, t) = X(x)Y(y)T(t)$. Substitution into the two-dimensional wave equation then leads to

$$\frac{X''}{X} + \frac{Y''}{Y} = -\frac{T''}{c^2 T}$$

Since the left side of the equation is independent of t, whereas the right side is independent of both x and y, each side must be a constant, which we denote by $-\lambda^2$.

For the "space functions" X and Y, we now have an equation that can be rewritten as

$$\frac{X''}{X} = -\frac{Y''}{Y} - \lambda^2$$

In familiar fashion we can then argue once again that each of the new terms is a constant, which we denote this time by $-\alpha^2$. Combining these several steps, we see that we have obtained the three ordinary differential equations

$$\begin{aligned} X'' + \alpha^2 X &= 0 \\ Y'' + \beta^2 Y &= 0 \\ T'' + c^2\lambda^2 T &= 0 \end{aligned}$$

where $\beta^2 \equiv \lambda^2 - \alpha^2$.

In analogy with our earlier one-dimensional analysis, the homogeneous boundary conditions imply $X(0) = X(1) = 0$ and $Y(0) = Y(1) = 0$. The resulting Sturm-Liouville problems

$$\begin{aligned} X'' + \alpha^2 X &= 0 \qquad & Y'' + \beta^2 Y &= 0 \\ X(0) = X(1) &= 0 \qquad & Y(0) = Y(1) &= 0 \end{aligned}$$

both have the same eigenvalues and comparable eigenfunctions. However, for the sake of complete generality, we must be careful to express them using different subscripts. Therefore, $\alpha^2 = m^2\pi^2$ and $X_m(x) = \sin m\pi x$ with $m = 1, 2, 3, \ldots$, whereas $\beta^2 = n^2\pi^2$ and $Y_n(y) = \sin n\pi y$, with $n = 1, 2, 3, \ldots$. For these values of α and β, the general solution of the equation for T is then the doubly subscripted function

$$T_{mn}(t) = C_{mn} \cos\sqrt{m^2 + n^2}\,\pi ct + D_{mn} \sin\sqrt{m^2 + n^2}\,\pi ct$$

But in this example $u_t(x, y, 0) \equiv 0$, from which it follows that $T'(0) = 0$; hence the coefficients D_{mn} appearing in this expression must be zero for all m and n.

The product solution $u_{mn}(x, y, t) = X_m(x)Y_n(y)T_{mn}(t)$, by construction, satisfies the wave equation along with the various homogeneous auxiliary conditions. In order to satisfy the remaining initial deflection condition, however, we need once more to make use of superposition:

$$\begin{aligned} u(x, y, t) &= \sum_{m=1}^{\infty} \sum_{n=1}^{\infty} u_{mn}(x, y, t) \\ &= \sum_{m=1}^{\infty} \sum_{n=1}^{\infty} C_{mn} \sin m\pi x \sin n\pi y \cos\sqrt{m^2 + n^2}\,\pi ct \end{aligned}$$

This double series will then meet *all* the requirements if the coefficients C_{mn} are such that

$$u(x, y, 0) = \sum_{m=1}^{\infty} \sum_{n=1}^{\infty} C_{mn} \sin m\pi x \sin n\pi y \equiv f(x, y)$$

By formally grouping terms, this latter series can be rewritten as

$$f(x, y) = \sum_{m=1}^{\infty} C_m \sin m\pi x$$

where

$$C_m \equiv \sum_{n=1}^{\infty} C_{mn} \sin n\pi y$$

Each of these new expressions can be recognized as a Fourier sine series. It readily follows then that

$$\begin{aligned} C_{mn} &= \frac{2}{1} \int_0^1 C_m \sin n\pi y\, dy \\ &= 2 \int_0^1 \left[\frac{2}{1} \int_0^1 f(x, y) \sin m\pi x\, dx \right] \sin n\pi y\, dy \\ &= 4 \int_0^1 \int_0^1 f(x, y) \sin m\pi x \sin n\pi y\, dx\, dy \end{aligned}$$

With these coefficients substituted back into $u(x, y, t)$, we have a *double Fourier series representation* for the desired solution. □

Exercises for Section 8.8

In Exercises 1–5, solve the vibrating string problem by the product method if the initial deflection function is as given. Assume zero initial velocity.

1. For a string of length 1, $u(x, 0) = \sin \pi x, \quad 0 < x < 1$.
2. For a string of length 1, $u(x, 0) = x(1 - x), \quad 0 < x < 1$.
3. For a string of length 2, $u(x, 0) = \sin 4\pi x - \sin 100\pi x, \quad 0 < x < 2$.
4. For a string of length 2,

$$u(x, 0) = \begin{cases} x, & 0 < x < 1 \\ -x + 2, & 1 < x < 2 \end{cases}$$

5. For a string of length 3,

$$u(x, 0) = \begin{cases} x, & 0 < x < 1 \\ 1, & 1 < x < 2 \\ -x + 3, & 2 < x < 3 \end{cases}$$

The solution of the vibrating string problem, with given initial velocity but zero initial deflection, by the product method technique can also be identified with that obtained by D'Alembert's method. Here the relevant trigonometric identity is

$$\sin\frac{n\pi x}{L}\sin\frac{n\pi ct}{L} = \frac{1}{2}\left[\cos\frac{n\pi}{L}(x - ct) - \cos\frac{n\pi}{L}(x + ct)\right]$$

6. Verify that if $g(x)$ is considered to be an odd periodic function with period $2L$, the D'Alembert solution of the wave equation $u_{xx} = \frac{1}{c^2}u_{tt}$ with the boundary conditions $u(x, 0) \equiv 0$, $u_t(x, 0) \equiv g(x)$ on $[0, L]$ and $u(0, t) \equiv 0 \equiv u(L, t)$ for $t > 0$ is given by

$$u(x, t) = \frac{1}{2c}\int_{x-ct}^{x+ct} g(\tau)\, d\tau$$

7. Demonstrate the equivalence of the result given in Exercise 6 with the solution obtained for the same problem using the product method. (*Hint:* Note that for the D'Alembert solution, $u_t(x, t) = \frac{1}{2}[g(x + ct) + g(x - ct)]$.)

In Exercises 8–10, solve the vibrating string problem by the product method if the initial velocity is as given. Assume zero initial deflection.

8. For a string of length 1, $u_t(x, 0) = \sin 3\pi x, \quad 0 < x < 1$.
9. For a string of length 2, $u_t(x, 0) = x(2 - x), \quad 0 < x < 2$.
10. For a string of length 4,

$$u_t(x, 0) = \begin{cases} 0, & 0 < x < 1 \\ x - 1, & 1 < x < 2 \\ -x + 3, & 2 < x < 3 \\ 0, & 3 < x < 4 \end{cases}$$

In Exercises 11–14, find the vibration of the square drumhead of Example 8.19 when the initial deflection is as given. Assume no initial velocity.

11. $u(x, y, 0) = f(x, y) = \sin \pi x \sin \pi y$
12. $u(x, y, 0) = f(x, y) = (1 - \cos 2\pi x)(1 - \cos 2\pi y)$
13. $u(x, y, 0) = f(x)f(y)$ where $f(\tau) = \begin{cases} \tau, & 0 < \tau \le \frac{1}{2} \\ 1 - \tau, & \frac{1}{2} \le \tau < 1 \end{cases}$
14. $u(x, y, 0) = f(x, y) = xy(1 - x)(1 - y)$
15. Show that the solution of the boundary value problem

$$\Delta_2 u = \frac{1}{c^2}u_{tt}, \qquad 0 < x < 1, \qquad 0 < y < 1, \qquad t > 0$$

$$u(0, y, t) \equiv 0, \qquad u(1, y, t) \equiv 0, \qquad u(x, 0, t) \equiv 0, \qquad u(x, 1, t) \equiv 0$$

$$u(x, y, 0) \equiv 0, \qquad u_t(x, y, 0) \equiv g(x, y)$$

can be represented as the double Fourier series

$$\begin{aligned} u(x, y, t) &= \sum_{m=1}^{\infty}\sum_{n=1}^{\infty} u_{mn}(x, y, t) \\ &= \sum_{m=1}^{\infty}\sum_{n=1}^{\infty} D_{mn} \sin m\pi x \sin n\pi y \sin\sqrt{m^2 + n^2}\,\pi ct \end{aligned}$$

where

$$\sqrt{m^2 + n^2}\pi c D_{mn} = 4\int_0^1\int_0^1 g(x, y)\sin m\pi x \sin n\pi y \, dx\, dy$$

In Exercises 16–18, find the vibration of the square drumhead of Exercise 15 when the initial velocity is as given. Assume no initial deflection.

16. $u_t(x, y, 0) = g(x, y) = \sin 3\pi x \sin 3\pi y$

17. $u_t(x, y, 0) = g(x, y) = \sin^2 \pi x \sin^2 \pi y$

18. $u_t(x, y, 0) = g(x, y) = xy(1 - x)(1 - y)$

8.9 The Heat Equation

In Section 8.6 we showed how the high-frequency transmission line equations have the same form as the wave equation. If, in the wave equation for the transmission line, we let G and L be negligibly small, which is the same effect as for low frequencies, we obtain

$$v_{xx} = RCv_t$$

and

$$i_{xx} = RCi_t$$

These two equations are known as the *low-frequency transmission line equations*. Both have the form of the heat or diffusion operator. The temperature at any time in a body satisfies a similar equation involving a constant depending on the specific heat, density, and thermal conductivity of the body. We make use of this equation although its derivation is postponed until Chapter 11.

Example 8.20 Find the temperature $u(x, t)$ in a thin bar of length L whose ends are kept frozen and whose sides are well insulated if the bar has some initial heat distribution described by the temperature $u(x, 0) = f(x)$ at each point x.

Solution The mathematical model for this problem is given by the heat equation

$$u_{xx} = ku_t \qquad (k = \text{constant})$$

and the auxiliary conditions

$$u(0, t) = u(L, t) \equiv 0$$
$$u(x, 0) \equiv f(x)$$

Assuming a product solution $u(x, t) = X(x)T(t)$ and substituting into the differential equation, we obtain

$$X''T = kXT'$$

or

$$\frac{X''}{X} = \frac{kT'}{T}$$

As before, this leads to the two ordinary differential equations,

$$X'' + \lambda X = 0$$

$$T' + \frac{\lambda}{k}T = 0$$

The homogeneous boundary conditions are translated into conditions on X such that $X(0) = X(L) = 0$. These, together with the first differential equation, give a now familiar Sturm-Liouville problem which has eigenfunctions $X_n(x) = \sin(n\pi x/L)$ corresponding to the eigenvalues $\lambda_n = n^2\pi^2/L^2$, n an integer. With that choice for λ_n, the second differential equation becomes

$$T' + \left(\frac{n\pi}{\sqrt{k}L}\right)^2 T = 0$$

whose general solution is

$$T_n(t) = B_n e^{-(n\pi/\sqrt{k}L)^2 t}$$

where B_n is an arbitrary constant. Hence, the functions

$$u_n(x,t) = B_n \sin\frac{n\pi x}{L} e^{-(n\pi/\sqrt{k}L)^2 t}$$

satisfy the heat equation and the homogeneous boundary conditions. Generally speaking, to find a function which also satisfies the initial temperature distribution, we must form an infinite series of functions,

$$u(x,t) = \sum_{n=1}^{\infty} u_n(x,t)$$

For $t = 0$, this becomes

$$\begin{aligned} u(x,0) &= \sum_{n=1}^{\infty} u_n(x,0) \equiv f(x) \\ &= \sum_{n=1}^{\infty} B_n \sin\frac{n\pi x}{L} \end{aligned}$$

The B_n should, therefore, be chosen to be the Fourier sine series coefficients of the initial temperature distribution function $f(x)$. □

In a body in which the temperature does not change with time, the so-called steady-state condition is governed by Laplace's equation, $\Delta u = 0$, or in one spatial dimension $u_{xx} = 0$. Sometimes we need this steady-state temperature, since it may form the initial temperature distribution for later times.

The following example is typical of how to handle a boundary value problem in which the boundary conditions are *not* homogeneous. The fundamental approach is to recast the problem in terms of a different dependent variable in which the boundary conditions, or as many of them as possible, are homogeneous.

Example 8.21 Find the temperature in the bar of the previous example if prior to $t = 0$, one end of the bar is kept in boiling water and is then put into water whose temperature is 30°C.

Solution Prior to time $= 0$, under steady-state conditions, the temperature distribution is governed by Laplace's equation, $u_{xx} = 0$, with boundary conditions $u(0, t) \equiv 0$ and $u(L, t) \equiv 100$. The fact that $u_{xx} = 0$ implies the linearity of any solution and hence, prior to $t = 0$, the solution is

$$u(x, t) = \frac{100}{L} x$$

(See Example 8.4.)

For $t \geqq 0$, we must solve the boundary value problem

$$u_{xx} = ku_t; \qquad u(0, t) \equiv 0,\ u(L, t) \equiv 30,\ u(x, 0) \equiv \frac{100}{L} x$$

The boundary conditions of this problem are not homogeneous and hence a sum of solutions satisfying the differential equation will not also satisfy the boundary conditions. Therefore, we introduce another function which satisfies $u_{xx} = 0$ as $t \to \infty$ (called the *steady-state* solution for $t > 0$). It is denoted by $u(x, \infty)$ and is found just like $u(x, t)$ for $t < 0$:

$$u(x, \infty) = \frac{30}{L} x$$

Now, letting $u(x, t) = u(x, \infty) + v(x, t)$, we see that v must satisfy the heat equation with homogeneous boundary conditions,

$$v(0, t) = u(0, t) - u(0, \infty) = 0 - 0 \equiv 0$$
$$v(L, t) = u(L, t) - u(L, \infty) = 30 - 30 \equiv 0$$

The initial condition for the function v is

$$\begin{aligned} v(x, 0) &= u(x, 0) - u(x, \infty) \\ &= \frac{100}{L} x - \frac{30}{L} x \\ &\equiv \frac{70}{L} x \end{aligned}$$

The solution for v is the same as for u in the previous example:

$$v(x, t) = \sum_{n=1}^{\infty} B_n \sin \frac{n\pi x}{L} e^{-(n\pi/\sqrt{k}L)^2 t}$$

where the B_n are the Fourier sine coefficients of the initial temperature distribution function. In this particular case,

$$B_n = \frac{2}{L} \int_0^L \frac{70}{L} x \sin \frac{n\pi x}{L} dx = \frac{140}{n\pi} (-1)^{n+1}$$

Therefore, the temperature function $u(x, t)$ is determined in terms of the steady-state formula $u(x, \infty)$ and the *transient* distribution $v(x, t)$. □

When the domain of interest is not finite, as in the case of an infinite or semi-infinite bar, Fourier series techniques are no longer applicable for heat problems such as in Example 8.21. In these situations classical solution procedures normally involve operational or transform techniques. In Chapter 16 we discuss in detail the utilization of the Fourier integral (transform) in such applications. Here we demonstrate how typical problems may be alternatively approached using the Laplace transform.

Example 8.22 Determine a representation for the temperature $u(x,t)$ in a thin bar of infinite extent whose sides are well insulated if the bar has some initial heat distribution $u(x,0) \equiv f(x)$. Assume that $u(x,t)$ is bounded as $|x| \to \infty$.

Solution The boundary value problem to be solved is

$$\begin{aligned} &u_{xx} = ku_t; \qquad -\infty < x < \infty,\ t > 0 \\ &u(x,t) \text{ bounded as } |x| \to \infty \\ &u(x,0) \equiv f(x) \end{aligned}$$

If we take the Laplace transform with respect to the t variable, considering x for the moment as a parameter, the given partial differential equation and auxiliary conditions give rise to the following ordinary differential equation and boundary conditions for $U(x,s)$, the transformed temperature:

$$\begin{aligned} &U_{xx}(x,s) - ksU(x,s) = -kf(x), \qquad -\infty < x < \infty \\ &U(x,s) \text{ bounded as } |x| \to \infty \end{aligned}$$

(see Exercise 5, p. 307). This latter problem may be readily solved using variation of parameters as discussed in Section 3.6, yielding

$$\begin{aligned} U(x,s) = {} & \frac{1}{2}\sqrt{\frac{k}{s}}\int_0^\infty e^{-\sqrt{ks}(y-x)}f(y)\,dy + \frac{1}{2}\sqrt{\frac{k}{s}}\int_{-\infty}^0 e^{-\sqrt{ks}(x-y)}f(y)\,dy \\ & - \sqrt{\frac{k}{s}}\int_0^x f(y)\sinh\sqrt{ks}\,(x-y)\,dy \end{aligned}$$

(again see Exercise 5). Since $(1/\sqrt{s})e^{-\alpha\sqrt{s}}$ is a known Laplace transform (Table 5.1), the formal inversion process is straightforward and leads to

$$\begin{aligned} u(x,t) = {} & \frac{k}{2}\int_0^\infty \frac{f(y)}{\sqrt{\pi kt}}e^{-k(y-x)^2/4t}\,dy + \frac{k}{2}\int_{-\infty}^0 \frac{f(y)}{\sqrt{\pi kt}}e^{-k(x-y)^2/4t}\,dy \\ & - \frac{k}{2}\int_0^x \frac{f(y)}{\sqrt{\pi kt}}\{e^{-k(y-x)^2/4t} - e^{-k(x-y)^2/4t}\}\,dy \end{aligned}$$

or

$$u(x,t) = \sqrt{\frac{k}{4\pi t}}\int_{-\infty}^\infty f(y)e^{-k(x-y)^2/4t}\,dy$$

□

8.10 A Nontrigonometric Example

Trigonometric functions will not always suffice for the representation of the solution of boundary value problems arising in more complicated applications.

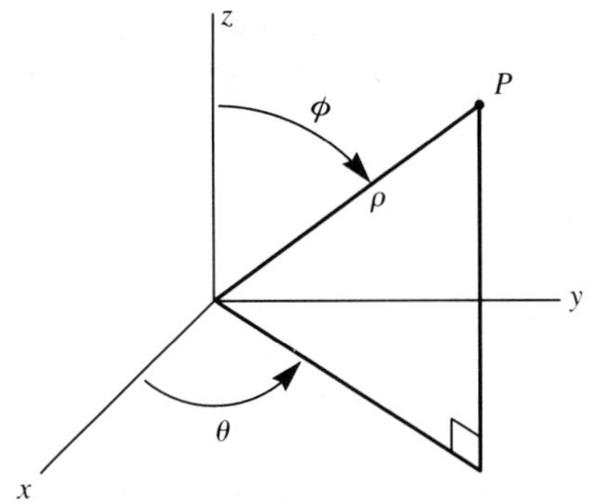

Figure 8.6 *Spherical coordinates*

Example 8.23 Find the steady-state temperature at any point inside a sphere of radius 1 for a given temperature distribution $f(\phi)$ on the surface.

Solution In three dimensions, the steady-state temperature $u(x, y, z)$ satisfies Laplace's equation $\Delta_3 u = 0$. Using the spherical coordinate substitutions

$$x = \rho \sin\phi \cos\theta, \qquad y = \rho \sin\phi \sin\theta, \qquad z = \rho\cos\theta$$

(see Figure 8.6) in

$$\frac{\partial^2 u}{\partial x^2} + \frac{\partial^2 u}{\partial y^2} + \frac{\partial^2 u}{\partial z^2} = 0$$

produces, after a tedious but straightforward computation, the spherical form of Laplace's equation:

$$\rho^2 u_{\rho\rho} + 2\rho u_\rho + u_{\phi\phi} + \frac{\cos\phi}{\sin\phi} u_\phi + \frac{1}{\sin^2\phi} u_{\theta\theta} = 0$$

Owing to the symmetry of the boundary condition, u should be a function solely of ϕ and ρ so the equation reduces to

$$\rho^2 u_{\rho\rho} + 2\rho u_\rho + u_{\phi\phi} + \frac{\cos\phi}{\sin\phi} u_\phi = 0$$

Assuming a solution of the form $R(\rho)\Phi(\phi)$, we find

$$\frac{\rho^2 R''}{R} + \frac{2\rho R'}{R} = -\frac{\Phi''}{\Phi} - \frac{\cos\phi}{\sin\phi}\frac{\Phi'}{\Phi} = \lambda$$

which separates into the two ordinary differential equations

$$\Phi'' + \frac{\cos\phi}{\sin\phi}\Phi' + \lambda\Phi = 0$$

and

$$\rho^2 R'' + 2\rho R' - \lambda R = 0$$

The first equation is Legendre's equation in disguise as can be seen by letting $x = \cos\phi$ yielding, via the chain rule,

$$\Phi' = -\sin\phi \frac{d\Phi}{dx}$$

$$\Phi'' = \sin^2\phi \frac{d^2\Phi}{dx^2} - \cos\phi \frac{d\Phi}{dx}$$

and

$$(1 - x^2)\frac{d^2\Phi}{dx^2} - 2x\frac{d\Phi}{dx} + \lambda\Phi = 0$$

The only solutions of this equation that are finite on $[-1, 1]$ (which corresponds to ϕ on $[0, \pi]$) occur for $\lambda = n(n + 1)$ with n an integer. (See Sections 7.12 and 9.6.) The second equation is a Cauchy-Euler equation (see p. 104) which, for $\lambda = n(n + 1)$, has the complete solution

$$a\rho^n + \frac{b}{\rho^{n+1}}$$

If $n \geq 0$, we are forced to set $b = 0$ in order to have a finite expression at $\rho = 0$.

At this stage we have solutions

$$a_0 P_0(x), a_1 \rho P_1(x), a_2 \rho^2 P_2(x), \ldots, a_n \rho^n P_n(x), \ldots$$

where P_n is the nth Legendre polynomial. Using superposition

$$u(\rho, \phi) = \sum_{n=0}^{\infty} a_n \rho^n P_n(\cos\phi)$$

where we insist that $u(1, \phi) \equiv f(\phi)$, or

$$f(\phi) = \sum_{n=0}^{\infty} a_n P_n(\cos\phi)$$

This is essentially a Fourier expansion of f in terms of the orthogonal functions P_n. Indeed, we know that

$$\int_{-1}^{1} P_n(x) P_m(x)\, dx = 0 \qquad \text{for } m \neq n$$

and it is shown in Chapter 9 that

$$\int_{-1}^{1} P_n^2(x)\, dx = \frac{2}{2n + 1}$$

In terms of the original variable ϕ, these two results become

$$\int_0^{\pi} P_n(\cos\phi) P_m(\cos\phi) \sin\phi\, d\phi = \begin{cases} 0, & n \neq m \\ \dfrac{2}{2n + 1}, & n = m \end{cases}$$

and thus

$$\int_0^{\pi} f(\phi) \sin\phi P_n(\cos\phi)\, d\phi = \int_0^{\pi} \sin\phi P_n(\cos\phi) \sum_{k=0}^{\infty} a_k P_k(\cos\phi) = \frac{2a_n}{2n + 1}$$

It follows that

$$a_n = (n + \tfrac{1}{2}) \int_0^\pi f(\phi) \sin \phi \, P_n(\cos \phi) \, d\phi$$

which completes the formal solution to the problem. □

The properties and uses of Legendre polynomials will be explored in more detail in the next chapter.

Exercises for Sections 8.9 and 8.10

1. A thin bar 2 m long, which is insulated on its longitudinal surface, has a steady-state distribution

$$f(x) = \begin{cases} x, & 0 < x < 1 \\ 2 - x, & 1 < x < 2 \end{cases}$$

At time $t = 0$, the ends of the bar are immersed in ice water at 0°C. Find the temperature distribution $u(x, t)$ for $t > 0$.

2. Repeat Exercise 1, if the distribution prior to $t = 0$ is given by $f(x) = x(2 - x)$, $0 < x < 2$.

3. The ends of a thin bar of unit length are kept at 10° and 90°. What is the steady-state temperature distribution?

4. The ends of the bar in Exercise 3 are suddenly changed (at $t = 0$) to 20° and 70°. Determine the resulting temperature distribution as a function of x and t.

5. **(a)** Show that application of the Laplace transform to the boundary value problem

$$\begin{aligned} &u_{xx} = ku_t \qquad -\infty < x < \infty, \quad t > 0 \\ &u(x, t) \text{ bounded as } |x| \to \infty \\ &u(x, 0) \equiv f(x) \end{aligned}$$

leads, under the assumption that $u(x, t)e^{-st} \to 0$ as $t \to \infty$, to the ordinary differential equation

$$U_{xx}(x, s) - ksU(x, s) = -kf(x), \qquad -\infty < x < \infty$$

and "boundary" conditions $U(x, s)$ bounded as $|x| \to \infty$ for the transformed temperature

$$U(x, s) \equiv \int_0^\infty u(x, t)e^{-st} \, dt$$

(b) Verify that

$$\begin{aligned} U(x, s) = &\frac{1}{2}\sqrt{\frac{k}{s}} \int_0^\infty e^{-\sqrt{ks}(y-x)} f(y) \, dy \\ &+ \frac{1}{2}\sqrt{\frac{k}{s}} \int_{-\infty}^0 e^{-\sqrt{ks}(x-y)} f(y) \, dy \\ &- \sqrt{\frac{k}{s}} \int_0^x f(y) \sinh \sqrt{ks}(x - y) \, dy \end{aligned}$$

is the solution to the resulting "boundary" value problem for U.

In Exercises 6–8, use the integral representation as derived in Example 8.22.

$$u(x,t) = \sqrt{\frac{k}{4\pi t}} \int_{-\infty}^{\infty} f(y) e^{-k(x-y)^2/4t}\, dy$$

for the temperature $u(x,t)$ in a thin bar of infinite extent having an initial heat distribution $u(x,0) \equiv f(x)$.

6. Determine the temperature in the case $f(x) \equiv C$, a constant.

7. If $f(x) \equiv 0$ for $x < 0$ and $f(x) \equiv 1$ for $x > 0$, find the first three terms in a power series representation for the temperature that is valid for small values of $x/\sqrt{t}$.

8. Show that if $f(x)$ is periodic with period $2L$, the above integral representation reduces to a trigonometric Fourier series. (*Hint:* Expand $f(x)$ into a trigonometric Fourier series and integrate termwise.)

9. Use the fact that $P_0(\cos\phi) = 1$ to demonstrate that if $f(\phi) \equiv 1$ in Example 8.23, then $u(\rho, \theta, \phi) \equiv 1$ also.

10. Determine the steady-state temperature in the sphere that results in Example 8.23 if $f(\phi) \equiv \cos\phi$. (Here recall that $P_0(\cos\phi) = 1$, $P_1(\cos\phi) = \cos\phi$.)

In higher dimensions or in nonrectangular two-dimensional regions, application of the method of separation of variables to the heat equation, the wave equation, or Laplace's equation can be fairly complicated. (Example 8.23 is a case in point.) Indeed, in addition to trigonometric functions and exponentials, various nonelementary functions typically begin to occur in the fundamental solutions u_n or u_{mn} provided by the product method. In each of the following exercises, use the product method to determine solutions of the given equation in the specified coordinate system with the desired symmetry.

11. Solutions of Laplace's equation $u_{rr} + \frac{1}{r}u_r + \frac{1}{r^2}u_{\theta\theta} + u_{zz} = 0$ in cylindrical coordinates which are independent of z.

12. Solutions of Laplace's equation

$$u_{\rho\rho} + \frac{2}{\rho}u_\rho + \frac{1}{\rho^2 \sin^2\phi}u_{\theta\theta} + \frac{1}{\rho^2}u_{\phi\phi} + \frac{\cos\phi}{\rho^2 \sin\phi}u_\phi = 0$$

in spherical coordinates that are independent of ϕ.

13. Axisymmetric (that is, θ-independent) solutions of the wave equation

$$u_{rr} + \frac{1}{r}u_r + \frac{1}{r^2}u_{\theta\theta} = \frac{1}{c^2}u_{tt}$$

in polar coordinates. (*Hint:* A simple change of independent variable in the $R(r)$ equation will lead to Bessel's equation of order zero.)

14. Solutions of the heat equation

$$u_{rr} + \frac{1}{r}u_r + \frac{1}{r^2}u_{\theta\theta} = ku_t$$

in polar coordinates that are of the form

$$u(r,\theta,t) = v(r,t)\cos n\theta$$

(*Hint:* You should expect Bessel's equation of order n in this case.)

15. Solutions of Laplace's equation

$$u_{rr} + \frac{1}{r}u_r + \frac{1}{r^2}u_{\theta\theta} + u_{zz} = 0$$

in cylindrical coordinates that are of the form

$$u(r, \theta, z) = v(r, z) \sin n\theta$$

16. Purely radial time-dependent solutions of the heat equation

$$u_{\rho\rho} + \frac{2}{\rho}u_\rho + \frac{1}{\rho^2 \sin^2 \phi}u_{\theta\theta} + \frac{1}{\rho^2}u_{\phi\phi} + \frac{\cos \phi}{\rho^2 \sin \phi}u_\phi = ku_t$$

in spherical coordinates. (*Hint:* Try the substitution $R = v/\rho$.)

8.11 Looking Ahead

As we have seen in a number of examples and exercises, separation of variables applied to particular partial differential equations may lead to nonelementary functions. Perhaps the two most important classes of such functions are Bessel functions and Legendre polynomials, both of which are discussed in the next chapter. When more than one separation parameter is present, the product method gives rise to representations in terms of double (or perhaps triple) Fourier series. If the domain of interest is infinite, series representations may have to give way to integral representations as we encounter again in Chapter 16. The references that follow give extended consideration to these and other related topics. In particular, [4], [5], [10], and [19] contain a wealth of practical applications.

The Poisson integral formula given in the exercises on p. 286 and the integral expression for the solution of Example 8.22 in terms of the fundamental heat kernel are actually solution representations involving multi-dimensional Green's functions. Analogous to the situation in one dimension, similar representations also provide a unified approach to the solution of nonhomogeneous equations. See [8], [11], [15], and [17] for the extension of Green's function techniques to partial differential equations.

The second method discussed in Section 8.4 is the specialization to constant-coefficient equations of a more general procedure applicable to linear hyperbolic and parabolic partial differential equations. The technique involves transformation of the independent variables to so-called "characteristic coordinates." Among others, [9], [12], and [16] discuss this useful approach.

When numerical procedures are needed for the approximate solution of given boundary value problems in partial differential equations, the two most widely used techniques are the finite-element method or some variant of a finite-difference method. Good introductions to this important area of computational science are found in [6], [7], and [13].

Basic References

[1] Berg, P. W., and J. L. McGregor. 1966. *Elementary Partial Differential Equations*. San Francisco: Holden-Day.

[2] Churchill, R. V., and J. W. Brown. 1978. *Fourier Series and Boundary Value Problems.* 3d ed. New York: McGraw-Hill.

[3] Epstein, B. 1975 *Partial Differential Equations, an Introduction.* Huntington, N. Y.: Krieger.

[4] Kovach, L. D. 1984. *Boundary-Value Problems.* Reading, Mass.: Addison-Wesley.

[5] Spiegel, M. R. 1981. *Applied Differential Equations.* 3d ed. Englewood Cliffs, N. J.: Prentice-Hall.

More-Advanced References

[6] Ames, W. F. 1977. *Numerical Methods for Partial Differential Equations.* 2d ed. New York: Academic Press.

[7] Carrier, G. F., and C. E. Pearson. 1976. *Partial Differential Equations: Theory and Technique.* New York: Academic Press.

[8] Cochran, J. A. 1982. *Applied Mathematics: Principles, Techniques, and Applications.* Belmont, California: Wadsworth.

[9] Dennemeyer, R. 1968. *Introduction to Partial Differential Equations and Boundary Value Problems.* New York: McGraw-Hill.

[10] Duff, G. F. D., and D. Naylor. 1966. *Differential Equations of Applied Mathematics.* New York: John Wiley.

[11] Friedman, B. 1956. *Principles and Techniques of Applied Mathematics.* New York: John Wiley.

[12] Garabedian, P. 1964. *Partial Differential Equations.* New York: John Wiley.

[13] Gladwell, I., and R. Wait. 1979. *A Survey of Numerical Methods for Partial Differential Equations.* Oxford: Clarendon Press.

[14] Lin, C. C., and L. A. Segel. 1974. *Mathematics Applied to Deterministic Problems in the Natural Sciences.* New York: Macmillan.

[15] Morse, P. M., and H. Feshback. 1953. *Methods of Theoretical Physics, Parts I and II.* New York: McGraw-Hill.

[16] Sommerfeld, A. 1949. *Partial Differential Equations in Physics.* New York: Academic Press.

[17] Stakgold, I. 1979. *Green's Functions and Boundary Value Problems.* New York: John Wiley.

[18] Weinberger, H. F. 1965. *A First Course in Partial Differential Equations with Complex Variables and Transform Methods.* Waltham, Mass.: Blaisdell.

[19] Williams, W. E. 1973. *Fourier Series and Boundary Value Problems.* London: George Allen & Unwin.

Chapter 9

Special Functions

9.1 Introduction

We often encounter functions that are special in the sense that they are non-elementary in their definition and conception as well as being essential to the efficient resolution of various applied problems; they are the building blocks for useful representations of the desired solutions. A case in point is the application of Section 8.10, where we encountered Legendre polynomials. Since these functions occur regularly in problems involving spherical geometry, we take the time to study them in some detail in this chapter. We also investigate Bessel functions, since they play an especially prominent role in applications with cylindrical geometry. Our discussion begins at a more modest level, however.

9.2 The Gamma Function

There are numerous ways in which special functions arise in problems. Some, like Bessel functions, Legendre polynomials, and Hermite functions, occur as solutions to differential equations of interest. Others, like the Airy functions, the error function, and elliptic functions, appear as definite or indefinite integrals. Still others, such as the Chebyshev polynomials or the beta function, are useful because of some particular regularity or combinatorial property. Perhaps the simplest of functions in this latter category is the *gamma function*, denoted by $\Gamma(x)$, and defined by

$$\Gamma(x) \equiv \int_0^\infty t^{x-1} e^{-t}\, dt, \quad x > 0$$

The gamma function is sometimes called the *generalized factorial function*. To see why, employ integration by parts to show inductively that for $u > 0$,

$$\int_0^\infty e^{-ut}\, dt = \frac{1}{u}$$

$$\int_0^\infty te^{-ut}\,dt = \frac{1!}{u^2}$$

$$\int_0^\infty t^2e^{-ut}\,dt = \frac{2!}{u^3}$$

$$\int_0^\infty t^3e^{-ut}\,dt = \frac{3!}{u^4}$$

$$\int_0^\infty t^4e^{-ut}\,dt = \frac{4!}{u^5}$$

and, in general,

$$\int_0^\infty t^ne^{-ut}\,dt = \frac{n!}{u^{n+1}}$$

In this last expression, we let $u = 1$, to obtain

$$\int_0^\infty t^ne^{-t}\,dt = n!$$

It is not unreasonable then to call the improper integral $\int_0^\infty t^ne^{-t}\,dt$ the *factorial function*. As we have developed the function, the value of n is restricted to positive integers. However, since the improper integral has meaning if n is replaced by any real number $x > -1$, we are tempted to write

$$x! = \int_0^\infty t^xe^{-t}\,dt$$

It is obvious that the "extended" factorial function and the gamma function are one and the same except for a shift of variables; that is,

$$\Gamma(x) = (x-1)!$$

In passing we note that when $x = 0$ in the integral definition of $x!$, we have

$$0! = \int_0^\infty e^{-t}\,dt = 1$$

which supports the definition of 0! as 1 in more elementary discussions of the factorial function.

In beginning courses, $n!$ is defined to be the product of the integers, $n(n-1)(n-2)\ldots(3)(2)(1)$ from which we obtain the *factorial property*, $n! = n(n-1)!$. If x is an integer, $x = n$, we therefore have immediately,

$$\Gamma(n+1) = n! = n(n-1)! = n\Gamma(n)$$

The gamma function has this factorial property for all $x > 0$. Indeed, consider

$$\begin{aligned}\Gamma(x+1) &= \int_0^\infty t^xe^{-t}\,dt \\ &= \lim_{B\to\infty}\int_0^B t^xe^{-t}\,dt\end{aligned}$$

An integration by parts yields

$$\Gamma(x+1) = \lim_{B\to\infty} (-t^x e^{-t})\Big|_{t=0}^{t=B} + x\int_0^\infty t^{x-1}e^{-t}\,dt = x\Gamma(x)$$

which verifies the property.

Tables of values of the gamma function have been compiled. Because of the factorial property, values of the function need be tabulated only over any interval of unit length. The interval usually chosen for this tabulation is $1 \leqq x \leqq 2$. A short list of values is given in Table 9.1.

Table 9.1 *Values of the gamma function for* $1 \leqq x \leqq 2$

x	$\Gamma(x)$
1.0	1.000
1.1	0.951
1.2	0.918
1.3	0.897
1.4	0.887
1.5	0.886
1.6	0.894
1.7	0.909
1.8	0.931
1.9	0.962
2.0	1.000

In Section 9.3 (Example 9.7), we show that $\Gamma(\frac{1}{2}) = \sqrt{\pi}$. Until then, we will assume the validity of that result and use it when necessary.

Example 9.1 Compute $\Gamma(\frac{7}{2})$.

Solution

$$\Gamma(\tfrac{7}{2}) = \tfrac{5}{2}\Gamma(\tfrac{5}{2}) = \tfrac{5}{2}\tfrac{3}{2}\Gamma(\tfrac{3}{2}) = \tfrac{5}{2}\tfrac{3}{2}\tfrac{1}{2}\Gamma(\tfrac{1}{2}) = \tfrac{15}{8}\sqrt{\pi}$$ □

Example 9.2 Evaluate

$$\int_0^\infty e^{-h^2x^2}\,dx$$

using the definition of the gamma function.

Solution Let $y = h^2x^2$. Then $dx = dy/(2h^2x) = dy/(2h\sqrt{y})$, and the given integral becomes

$$\frac{1}{2h}\int_0^\infty y^{-1/2}e^{-y}\,dy = \frac{1}{2h}\Gamma\left(\frac{1}{2}\right) = \frac{\sqrt{\pi}}{2h}$$ □

Example 9.3 In a similar manner evaluate

$$\int_0^1 x^2\left(\ln\frac{1}{x}\right)^3 dx$$

Solution Let $y = -\ln x$. Then $-x^2\,dx = x^3\,dy = e^{-3y}\,dy$, and the given integral becomes

$$-\int_\infty^0 e^{-3y}y^3\,dy = \int_0^\infty e^{-3y}y^3\,dy$$
$$= \tfrac{1}{27}\int_0^\infty e^{-3y}(3y)^3\,dy$$

In this latter integral, let $t = 3y$ to obtain

$$\frac{1}{3^4}\int_0^\infty e^{-t}t^3\,dt = \frac{1}{3^4}\Gamma(4) = \frac{2}{27}$$ □

It is possible to show that for $x > 0$, the defining integral for the gamma function converges to a continuous function all of whose derivatives are also continuous. These derivatives can be obtained by partially differentiating the integrand with respect to x (although such interchanges of the order of integration and partial differentiation are not universally valid).

With the aid of Table 9.1 and with the factorial property, we can graph the gamma function. A knowledge of the sense of concavity of the graph is found from the second derivative. Thus,

$$\Gamma'(x) = \int_0^\infty t^{x-1}(\ln t)e^{-t}\,dt$$
$$\Gamma''(x) = \int_0^\infty t^{x-1}(\ln t)^2 e^{-t}\,dt$$

Since the integrand in the last integral is positive over the interval of integration, $\Gamma''(x) > 0$, and hence the graph of the gamma function is concave up for $x > 0$.

For x near zero, the gamma function is unbounded. To see this, note that

$$\Gamma(x) > \int_0^1 t^{x-1}e^{-t}\,dt > e^{-1}\int_0^1 t^{x-1}\,dt = \frac{1}{ex}$$

which $\to\infty$ as $x \to 0$.

As one would guess from the property $\Gamma(n) = (n-1)!$, where n is a positive integer,

$$\lim_{x\to\infty}\Gamma(x) = \infty$$

The graph of $\Gamma(x)$ for $x > 0$ is shown in Figure 9.1.

We may extend the definition of the gamma function to values of x other than positive real values by postulating the preservation of the factorial property.

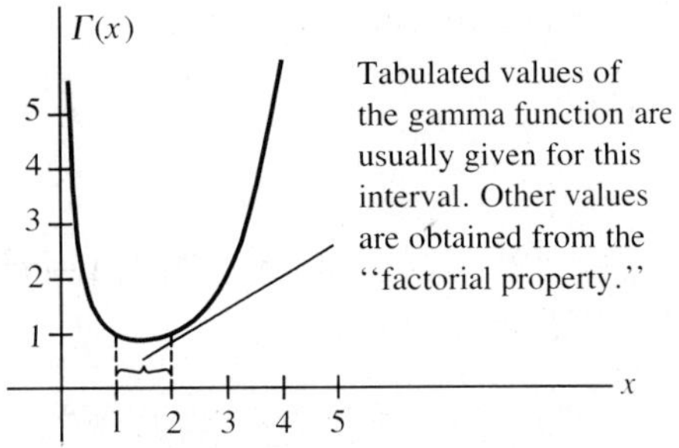

Figure 9.1 *The gamma function*

In this way, all values of the gamma function for negative values of x are written in terms of the values for $0 < x < 1$, and of course, ultimately in terms of the values for $1 < x < 2$. For example,

If $-1 < x < 0$, then

$$\Gamma(x) = \frac{\Gamma(x+1)}{x} \qquad \text{(note that } 0 < x + 1 < 1\text{)}$$

If $-2 < x < -1$, then

$$\Gamma(x) = \frac{\Gamma(x+2)}{x(x+1)} \qquad \text{(note that } 0 < x + 2 < 1\text{)}$$

If $-3 < x < -2$, then

$$\Gamma(x) = \frac{\Gamma(x+3)}{x(x+1)(x+2)} \qquad \text{(note that } 0 < x + 3 < 1\text{)}$$

$$\vdots \qquad\qquad \vdots$$

If $-m < x < -m + 1$, then

$$\Gamma(x) = \frac{\Gamma(x+m)}{x(x+1)\cdots(x+m-1)} \qquad \text{(note that } 0 < x + m < 1\text{)}$$

The preservation of the factorial property does not allow for a definition of the gamma function at the nonpositive integers. Near every negative integer, the function becomes unbounded. The values are alternately positive and negative as determined by the denominator in the defining expression since the numerator is always positive. The graph of the extended function is shown in Figure 9.2.

Example 9.4 Compute $\Gamma(-\frac{3}{2})$.

Solution

$$\Gamma(-\tfrac{3}{2}) = \frac{\Gamma(\frac{1}{2})}{(-\frac{3}{2})(-\frac{1}{2})} = \frac{4\sqrt{\pi}}{3} \qquad \square$$

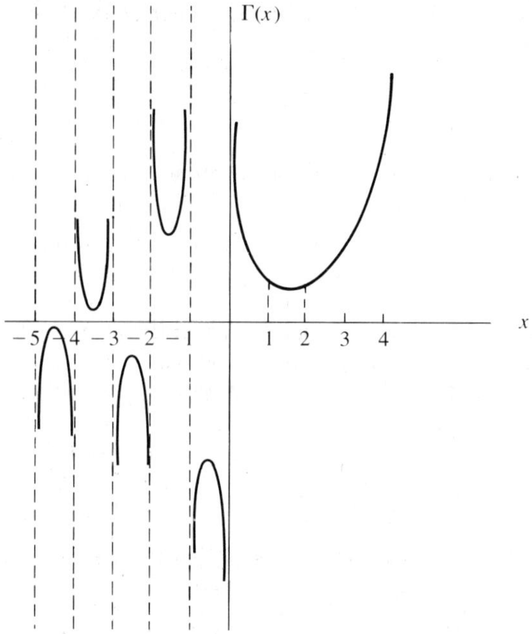

Figure 9.2 *Extended gamma function*

Exercises for Sections 9.1 and 9.2

Evaluate each of the following integrals using the gamma function:

1. $\displaystyle\int_0^1 (\ln x)^{1/3}\,dx$ **2.** $\displaystyle\int_0^\infty x^p e^{-x}\,dx$

3. $\displaystyle\int_0^\infty x^{3/4} e^{-x}\,dx$ **4.** $\displaystyle\int_0^1 x^4\left(\ln\frac{1}{x}\right)^2 dx$

5. $\displaystyle\int_0^\infty \frac{x^a}{a^x}\,dx, \qquad a > 1$ (*Hint:* Recall that $a^x = e^{x\ln a}$)

6. Demonstrate that $\int_0^1 \ln^{x-1}(1/t)\,dt$ provides a definition of $\Gamma(x)$ equivalent to the one introduced in Section 9.2.

7. Show that $2\int_0^\infty e^{-t^2} t^{2x-1}\,dt$ provides a satisfactory definition of $\Gamma(x)$.

8. Verify that $\Gamma(n + \frac{1}{2}) = \dfrac{(2n)!\sqrt{\pi}}{2^{2n} n!}$ for integral n.

9. Demonstrate that the integral on p. 312 defining $x!$ converges for $x > -1$.

10. Let

$$F(x) = \int_0^\infty \frac{te^{-xt}}{t^3 + 1}\,dt$$

Show that this function is well defined for $x \geq 0$. Where is $F'(x)$ meaningful?

11. Let

$$F(x) = \int_0^\infty \frac{\sin xt}{t^2 + 1}\,dt$$

Verify that F is well defined for all real x.

12. Let

$$F(x) = \frac{2}{\pi}\int_0^\infty \frac{dt}{1 + t^2x^2}$$

Show that this improper integral converges to $1/x$ for $x > 0$.

13. Verify that

$$\int_0^\infty \frac{\operatorname{Tan}^{-1} bx - \operatorname{Tan}^{-1} ax}{x}\,dx = \frac{\pi}{2}\ln\frac{b}{a}, \qquad a > 0, \qquad b > 0$$

(*Hint:* Consider the integral $\int_a^b F(x)\,dx$, where F is as given in Exercise 12, and reverse the order of integration.)

14. Make a sketch of the factorial function.

15. The fact that

$$\lim_{x\to\infty} \frac{\Gamma(x + 1)}{\sqrt{2\pi}\, x^{x+1/2} e^{-x}} = 1$$

(usually attributed to Stirling, but actually discovered by D'Moivre) can be used to estimate the gamma function for large values of x. Demonstrate that this limit implies

$$\log_{10}(x + 1) \doteq (x + \tfrac{1}{2})\log_{10} x - 0.43429x + 0.399 \qquad \text{as } x \to \infty$$

16. Use the result of the previous exercise to obtain estimates for
(a) 10! **(b)** 100!

The gamma function is intimately associated with Laplace transform theory; recall that $\mathscr{L}\{t^a\} = \Gamma(a + 1)/s^{a+1}$. Laplace transforms are also often used in an intermediate step in the evaluation of certain improper integrals. The following formal analysis suggests the key idea.

By definition

$$\mathscr{L}\{f(t)\} \equiv F(s) \equiv \int_0^\infty e^{-st} f(t)\,dt$$

If we integrate both sides with respect to s from 0 to ∞ and interchange the order of integration on the right-hand side, we obtain

$$\int_0^\infty F(s)\,ds = \int_0^\infty \frac{f(t)}{t}\,dt$$

Although the steps taken are not always justified, they are in some important cases. For example, we can verify that for $a > 0$

$$\begin{aligned}\int_0^\infty \frac{\sin at}{t}\,dt &= \int_0^\infty \frac{a}{s^2 + a^2}\,ds \\ &= \operatorname{Tan}^{-1}\frac{s}{a}\Big|_{s=0}^{s=\infty} \\ &= \frac{\pi}{2}\end{aligned}$$

Other applications are explored in the remaining exercises.

17. Show that

$$\int_0^\infty \frac{\cos at - \cos bt}{t}\,dt = \ln\frac{b}{a}, \qquad (a, b > 0)$$

18. Demonstrate that

$$\int_0^\infty \frac{e^{-at} - e^{-bt}}{t}\,dt = \ln\frac{b}{a}, \qquad (a, b > 0)$$

19. Establish

$$\int_0^\infty \frac{\sin^2 at}{t^2}\,dt = \frac{\pi a}{2}, \qquad (a > 0)$$

(Here you need to know that $\mathscr{L}\{(\sin^2 t)/t\} = \frac{1}{4}\ln(1 + 4/s^2)$.)

9.3 The Beta Function

In this section we consider a function of *two* variables closely related to the gamma function and usually defined by an improper integral. This function appears frequently in applied mathematics, for example, in the theory of high-energy particle physics. The *beta function*, $B(x, y)$, may actually be defined in several ways, all of which turn out to be equivalent. In each case the function is initially defined only in the first quadrant ($x > 0, y > 0$), with extensions possible by methods similar to those used in Section 9.2.

1. $B(x, y) \equiv \displaystyle\int_0^1 t^{x-1}(1 - t)^{y-1}\,dt$

2. $B(x, y) \equiv 2\displaystyle\int_0^{\pi/2} (\sin\theta)^{2x-1}(\cos\theta)^{2y-1}\,d\theta$

3. $B(x, y) \equiv \displaystyle\int_0^\infty \frac{u^{x-1}}{(1 + u)^{x+y}}\,du$

4. $B(x, y) \equiv \dfrac{\Gamma(x)\Gamma(y)}{\Gamma(x + y)}$

The fact that these four definitions are equivalent is the content of the next theorem.

Theorem 9.1 Definitions 1–4 are equivalent.

Proof If in Definition 1 we let $t = \sin^2\theta$, we obtain Definition 2. If in Definition 1 we let $t = u/(1 + u)$, we obtain Definition 3. The details are left as an exercise. To show the fact that the beta function is expressible in terms of gamma functions, we let $t = s^2$ in the defining integral for $\Gamma(x)$,

$$\Gamma(x) = \int_0^\infty t^{x-1} e^{-t}\, dt$$
$$= 2\int_0^\infty s^{2x-1} e^{-s^2}\, ds$$
$$\Gamma(y) = 2\int_0^\infty t^{2y-1} e^{-t^2}\, dt$$

Therefore,

$$\Gamma(x)\Gamma(y) = 4\int_0^\infty \int_0^\infty s^{2x-1} t^{2y-1} e^{-(s^2+t^2)}\, dt\, ds$$

Using polar coordinates, $s = r\cos\theta$, $t = r\sin\theta$,

$$\Gamma(x)\Gamma(y) = 4\int_0^{\pi/2} \int_0^\infty r^{2(x+y-1)} e^{-r^2} (\cos\theta)^{2x-1} (\sin\theta)^{2y-1}\, r\, dr\, d\theta$$
$$= \left[2\int_0^{\pi/2} (\cos\theta)^{2x-1}(\sin\theta)^{2y-1}\, d\theta\right]\left[2\int_0^\infty r^{2(x+y)-1} e^{-r^2}\, dr\right]$$
$$= B(y,x)\Gamma(x+y)$$

But $B(y,x) = B(x,y)$, as can be seen by letting $u = 1 - t$ in Definition 1. This proves the theorem. ■

Example 9.5 Compute

$$\int_0^\infty \frac{t^3}{(1+t)^5}\, dt$$

Solution This is just the beta function (Definition 3) with $x + y = 5$, $x - 1 = 3$. Therefore, $x = 4$, $y = 1$, and hence the given integral is equal to

$$B(4,1) = \frac{\Gamma(4)\Gamma(1)}{\Gamma(4+1)} = \frac{3!0!}{4!} = \frac{1}{4}$$ □

A more interesting illustration is given in Example 9.6.

Example 9.6 Compute

$$\int_0^2 x^{1/2}(2-x)^{5/2}\, dx$$

Solution The given integral is equal to

$$2^{5/2}\int_0^2 x^{1/2}\left(1 - \frac{x}{2}\right)^{5/2} dx$$

If we let $y = x/2$, this becomes

$$2^4 \int_0^1 y^{1/2}(1-y)^{5/2}\,dy = 2^4 B\left(\frac{3}{2},\frac{7}{2}\right)$$

$$= 2^4 \frac{\Gamma(\frac{3}{2})\Gamma(\frac{7}{2})}{\Gamma(5)}$$

$$= 15\frac{\Gamma(\frac{1}{2})\Gamma(\frac{1}{2})}{\Gamma(5)}$$

$$= \frac{5}{8}\pi \qquad \square$$

Using Definition 3 we can derive a relationship between the gamma function and the sine function. Considering the values of the beta function when $x + y = 1$:

$$B(x, 1-x) = \int_0^\infty \frac{u^{x-1}}{1+u}\,du$$

In Chapter 16 we show that the value of this integral is $\pi/\sin \pi x$, $0 < x < 1$. (See Example 16.6.) Hence,

$$\Gamma(x)\Gamma(1-x) = B(x, 1-x) = \frac{\pi}{\sin \pi x}, \qquad 0 < x < 1$$

This very useful identity for the gamma function can be shown to be valid for any nonintegral x.

Example 9.7 Compute $\Gamma(\frac{1}{2})$.

Solution Using the previous equation along with the definition of the beta function in terms of the gamma function,

$$B\left(\frac{1}{2},\frac{1}{2}\right) = \Gamma\left(\frac{1}{2}\right)\Gamma\left(\frac{1}{2}\right) = \frac{\pi}{\sin \pi/2} = \pi$$

and hence

$$\Gamma\left(\frac{1}{2}\right) = \sqrt{\pi} \qquad \square$$

We employed this result in some examples in Section 9.2.

With the help of the beta function, certain integrals of powers of sines and cosines may be evaluated. For example, let

$$I = \int_0^{\pi/2} \sin^\alpha u \cos^\beta u\,du, \qquad \alpha = 2x - 1, \quad \text{and} \quad \beta = 2y - 1$$

Then,

$$I = \frac{1}{2} B\left(\frac{\alpha+1}{2}, \frac{\beta+1}{2}\right)$$

Example 9.8 Evaluate

$$\int_0^{\pi/2} \sin^6 u \cos^6 u \, du$$

Solution

$$I = \frac{1}{2} B\left(\frac{7}{2}, \frac{7}{2}\right) = \frac{1}{2} \frac{\Gamma(\frac{7}{2})\Gamma(\frac{7}{2})}{\Gamma(7)} = \frac{225\pi}{128 \cdot 6!} = \frac{5\pi}{2048}$$

by virtue of the result of Example 9.1. □

The solution of this example can, of course, also be obtained using more elementary methods; such is not the case, however, when α and β are nonintegral.

9.4 Several Functions Defined by Indefinite Integrals

There is another method of defining functions that is closely allied to the use of improper integrals. Let f be an integrable function over the interval $[c, d]$. Then the indefinite integral

$$\int_c^x f(t) \, dt$$

defines a function of x for every x in $[c, d]$.

Typically this technique is used in elementary calculus to define the natural logarithm, by letting $f(t) = 1/t$. Then

$$\ln x \equiv \int_1^x \frac{1}{t} dt$$

(See Figure 9.3.) The usual properties of this well-known function (and its graph) are directly derivable from this definition. In the exercises you will be asked to

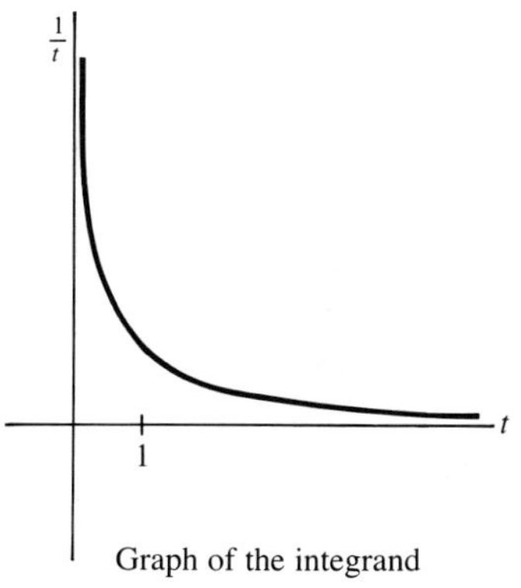

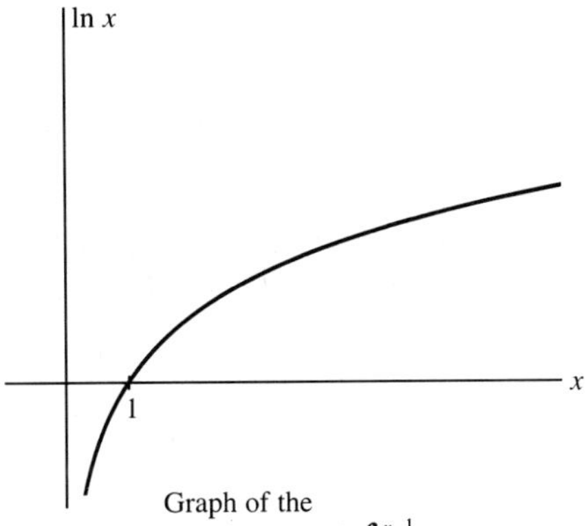

Figure 9.3 *Definition of* ln *x*

carry out some of the familiar details. The values of $\ln x$ are tabulated, of course, and appear in most any set of mathematical tables or can be determined by use of a pocket calculator.

In this section we wish to take at least superficial notice of a few other functions that are defined similarly. These functions are not as familiar as $\ln x$, but, like $\ln x$, their values are tabulated and available in handbooks.

One of the more important functions defined by an indefinite integral, and closely related to the normal distribution in statistics, is called the *error function*,

$$\operatorname{erf} x \equiv \frac{2}{\sqrt{\pi}} \int_0^x e^{-t^2}\, dt$$

Note that

$$\lim_{x\to\infty} \operatorname{erf} x = \frac{2}{\sqrt{\pi}} \int_0^\infty e^{-t^2}\, dt$$

If we substitute $u = t^2$ into this improper integral, then

$$\lim_{x\to\infty} \operatorname{erf} x = \frac{2}{\sqrt{\pi}} \int_0^\infty \frac{e^{-u}u^{-1/2}}{2}\, du = \frac{1}{\sqrt{\pi}}\Gamma\left(\frac{1}{2}\right) = 1$$

which suggests why the "normalizing factor" $2/\sqrt{\pi}$ is used. Table 9.2 shows the approximate values of $\operatorname{erf} x$ for $0 \le x \le 2.7$. For $x > 2.7$ the value of $\operatorname{erf} x$ is, to four significant figures, equal to 1. In Figure 9.4 is a sketch of $\operatorname{erf} x$ and the integrand function e^{-x^2}.

Table 9.2 *Values of erf x*

x	0	0.1	0.2	0.3	0.4	0.5	0.6
erf x	0.000	0.112	0.223	0.329	0.428	0.521	0.604
x	0.7	0.8	0.9	1.0	1.1	1.2	1.3
erf x	0.678	0.742	0.797	0.843	0.880	0.910	0.934
x	1.4	1.5	1.6	1.7	1.8	1.9	2.0
erf x	0.952	0.966	0.976	0.984	0.989	0.993	0.995
x	2.1	2.2	2.3	2.4	2.5	2.6	2.7
erf x	0.996	0.998	0.998	0.999	0.999	0.999	0.999

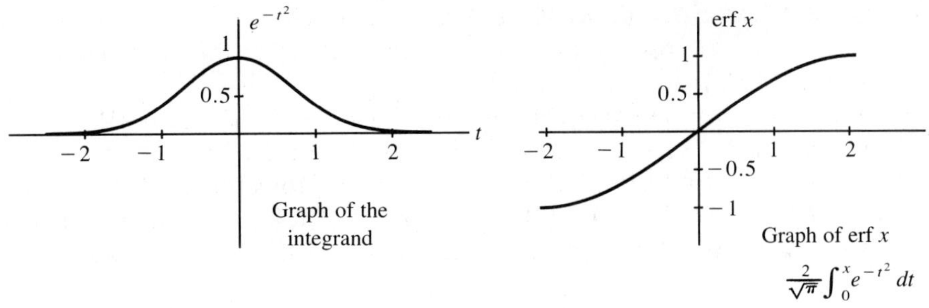

Figure 9.4 *Definition of* erf x

In some applications it is not erf x but a related function that is of interest and whose values are tabulated. In statistics the *cumulative distribution function of the standardized normal or Gaussian distribution* is given by

$$\Phi(x) = \frac{1}{\sqrt{2\pi}}\int_{-\infty}^{x} e^{-t^2/2}\,dt$$

It is easy to show that $\lim_{x\to\infty}\Phi(x) = 1$ and that $\Phi(0) = 1/2$. (See Exercise 19.) Hence, if $x > 0$,

$$\Phi(x) = \frac{1}{\sqrt{2\pi}}\int_{0}^{x} e^{-t^2/2}\,dt + \frac{1}{2}$$

Letting $u^2 = t^2/2$,

$$\begin{aligned}\Phi(x) &= \frac{1}{\sqrt{\pi}}\int_{0}^{x/\sqrt{2}} e^{-u^2}\,du + \frac{1}{2}\\ &= \frac{1}{2}\operatorname{erf}\left(\frac{x}{\sqrt{2}}\right) + \frac{1}{2}\end{aligned}$$

which shows how to find values of the cumulative distribution function when a table for erf x is given. Conversely, it is easy to show

$$\operatorname{erf} x = 2\Phi(x\sqrt{2}) - 1$$

In passing we mention a few other functions defined by indefinite integrals.

$\mathrm{Si}(x) \equiv \int_0^x \frac{\sin t}{t}\,dt$ is called the *sine integral function*

$S(x) \equiv \int_0^x \sin t^2\,dt$ is called a *Fresnel integral*

$E_1(x) \equiv \int_x^\infty \frac{e^{-t}}{t}\,dt$ is called the *exponential integral function*

Exercises for Sections 9.3 and 9.4

1. Using the result of Exercise 8 of the previous exercise set (p. 316) show that:
   (a) $B\left(n, \frac{1}{2}\right) = \frac{2^{2n}(n!)^2}{n(2n)!}$ **(b)** $B\left(n + \frac{1}{2}, \frac{1}{2}\right) = \frac{(2n)!}{2^{2n-1}(n!)^2} \frac{\pi}{2}$

2. Show that

$$\int_0^{\pi/2} \sin^n x\, dx = \frac{1}{2} B\left(\frac{n+1}{2}, \frac{1}{2}\right)$$

   Evaluate in the case $n = 2$.

Express the following integrals in terms of beta functions and then evaluate by using gamma functions.

3. $\displaystyle\int_0^1 \frac{x^4}{(1 - x^2)^{1/2}}\, dx$
4. $\displaystyle\int_0^1 \frac{1}{(1 - x^4)^{1/2}}\, dx$
5. $\displaystyle\int_0^\infty \frac{dx}{1 + x^4}$
6. $\displaystyle\int_0^\infty \frac{x}{(1 + x^3)^2}\, dx$
7. $\displaystyle\int_0^1 \frac{dx}{\sqrt{x(1 - x)}}$
8. $\displaystyle\int_0^2 \frac{x^2}{\sqrt{2 - x}}\, dx$
9. $\displaystyle\int_0^{\pi/2} \sin^5\theta \cos^7\theta\, d\theta$
10. $\displaystyle\int_0^{\pi/2} \frac{d\theta}{\sqrt{\sin\theta}}$
11. $\displaystyle\int_0^{\pi/2} (\sin\theta)^{1/2}(\cos\theta)^{3/2}\, d\theta$

12. Find the first quadrant area bounded by the curve $x^3 + y^3 = 8$ in terms of beta functions.
13. Compute the length of arc of the curve $r^2 = 16\cos 2\theta$ in terms of beta functions.
14. Fill in the details of the proof of Theorem 9.1.
15. Using the indefinite integral definition of $\ln x$, show the following:
    (a) $\ln 1 = 0$ **(b)** $\ln(ab) = \ln a + \ln b$ **(c)** $\ln x^p = p \ln x$
16. Show that $\text{erf}(x)$ is an odd function of x.
17. Determine the points of inflection of the integrand function of $\text{erf}(x)$.
18. Evaluate, using a table of error functions:
    (a) $\displaystyle\int_0^2 e^{-x^2}\, dx$ **(b)** $\text{erf}(0.6)$
    (c) $\displaystyle\frac{2}{\sqrt{\pi}} \int_{1.5}^\infty e^{-x^2}\, dx$ **(d)** $\displaystyle\int_1^{1.5} e^{-x^2}\, dx$
19. Prove that $\lim_{x\to\infty} \Phi(x) = 1$ and that $\Phi(0) = 1/2$.

In statistics, given a random process with a continuous probability density function $f(x)$, it is often of interest to determine the various moments of the process

$$\mu_n = \int_{-\infty}^{\infty} x^n f(x)\, dx, \qquad n = 0, 1, 2, \ldots$$

(μ_0 should be 1 since $f(x)$ is a probability density). The first moment is called the mean and, when $\mu_1 = 0$, the second moment becomes the variance of the process.

20. Show that the mean and variance of the standardized normal, or Gaussian, distribution are 0 and 1, respectively.

21. Demonstrate that for the standardized normal distribution

$$\mu_{2n+1} = 0$$
$$\mu_{2n} = \frac{2^n}{\sqrt{\pi}}\Gamma\left(n + \frac{1}{2}\right)$$

The Hermite polynomials $H_n(x)$ encountered in Chapter 7, pp. 224 and 263, are intimately related to the error function and hence to the standardized normal distribution. The defining relationship is

$$H_n(x) \equiv \frac{1}{2}\sqrt{\pi}(-1)^n e^{x^2}\frac{d^{n+1}}{dx^{n+1}}(\operatorname{erf} x)$$

22. Use the above relationship to generate the Hermite polynomials for $n = 0, 1, 2, 3, 4$ and verify that they satisfy Hermite's equation,

$$y'' - 2xy' + 2ny = 0$$

23. Recast the above defining relationship for the Hermite polynomials in terms of the cumulative distribution function $\Phi(x)$.

9.5 Functions Defined by Differential Equations

This section could have been titled "Functions Defined by Infinite Series." In Chapter 6 we saw that many differential equations may be solved by assuming that the solution function can be written as an infinite power series about some point. The differential equation essentially defines a function since the coefficients in the series representation are determined by forcing the infinite series to be a solution.

While solutions in infinite series form are usually "nonelementary," this does not curtail their usefulness. Tables of values are constructed or computer algorithms devised so that the functions become (in time and with use) as familiar as, for instance, the trigonometric functions. In fact the sine and cosine functions could very easily be considered as being essentially defined by the differential equation $y'' + k^2 y = 0$.

The name given to a solution function is the same as the name of the differential equation from which it arises. Thus the differential equation $y'' + k^2 y = 0$ is the trigonometric differential equation, the differential equation $y'' - k^2 y = 0$ might be called the hyperbolic differential equation, $y'' - xy = 0$ is Airy's equation, $(1 - x^2)y'' - xy' + p^2 y = 0$ is Chebyshev's equation, and so forth.

In the remainder of this chapter we concern ourselves with the solutions of two very special differential equations:

$$(1 - x^2)y'' - 2xy' + n(n + 1)y = 0 \qquad \text{(Legendre's equation)}$$
$$x^2 y'' + xy' + (x^2 - m^2)y = 0 \qquad \text{(Bessel's equation)}$$

These equations, both of which we have seen before, are named after famous mathematicians who were first associated with important practical problems in which the solutions of the equations played key roles.

9.6 Legendre Polynomials

We have previously encountered the Legendre polynomials several times in this text. Most recently we saw in Section 8.10 that they are especially important in applications exhibiting spherical symmetry. Later we shall discuss their relevance to problems in numerical integration. In the meantime, we turn our attention to the gaps in our basic knowledge concerning these "special" and useful functions.

Legendre's equation is a second-order linear differential equation having the general form

$$(1 - x^2)y'' - 2xy' + n(n + 1)y = 0$$

The points $x = 1, -1$ are regular singular points of this differential equation (see Section 6.5) and generally appear as singularities of any solutions as well. If n is a nonnegative integer, however, Legendre's equation has one solution which is well defined for all x including these singular points. Indeed, this solution turns out to be a polynomial and, properly normalized, it is termed the *Legendre polynomial of order n* and designated $P_n(x)$.

One standard method of determining a general representation for $P_n(x)$ is to assume a solution of Legendre's equation in power series form valid near the origin and use the method of undetermined coefficients. Since $x = 0$ is an ordinary point of the equation, substitution of

$$y = \sum_{k=0}^{\infty} a_k x^k$$

into the differential equation eventually leads to a recursion formula for the coefficients:

$$a_{k+2} = -\frac{(n - k)(n + k + 1)}{(k + 2)(k + 1)} a_k, \qquad (k = 0, 1, 2, \ldots)$$

with a_0 and a_1 both arbitrary. (Recall Example 6.8.) Separation into even and odd powers of x then allows us to write the complete solution in the form

$$y(x) = a_0 y_1(x) + a_1 y_2(x)$$

where

$$y_1(x) \equiv 1 - \frac{n(n + 1)}{2!}x^2 + \frac{(n - 2)n(n + 1)(n + 3)}{4!}x^4 - \cdots$$

$$y_2(x) \equiv x - \frac{(n - 1)(n + 2)}{3!}x^3 + \frac{(n - 3)(n - 1)(n + 2)(n + 4)}{5!}x^5 - \cdots$$

If n is even (that is, $n = 0, 2, 4, \ldots$), the infinite series for y_1 terminates and becomes an nth-degree polynomial. In analogous fashion, if n is odd, the expression for

y_2 truncates to an nth-degree polynomial. In either case, by assigning special values to a_0, a_1 (see Exercise 2 in the next set of exercises), we arrive at the compact, though somewhat complicated, expression for $P_n(x)$:

$$P_n(x) \equiv \sum_{j=0}^{N} \frac{(-1)^j (2n-2j)!\, x^{n-2j}}{2^n j!(n-j)!(n-2j)!}$$

where $N = n/2$ or $(n-1)/2$, whichever is an integer. (Note that in this expression we work down from larger powers of x rather than up from smaller.) For lower-order n this formula yields

$$P_0(x) = 1, \qquad P_1(x) = x, \qquad P_2(x) = \tfrac{1}{2}(3x^2 - 1), \quad \text{and} \quad P_3(x) = \tfrac{1}{2}(5x^3 - 3x)$$

all of which we have seen earlier.

9.7 Some Useful Identities for Legendre Polynomials

An alternative derivation of the Legendre polynomials results from the expansion of $(1 - 2xt + t^2)^{-1/2}$ in ascending powers of t. We obtain

$$\frac{1}{\sqrt{1 - 2xt + t^2}} \equiv \sum_{n=0}^{\infty} P_n(x) t^n$$

from which it is easy to verify the form of $P_n(x)$ at least for lower-order n.

The above expression shows that $(1 - 2xt + t^2)^{-1/2}$ is a *generating function* for $P_n(x)$. Such a relation can be tremendously useful in demonstrating the validity of important properties of Legendre polynomials as well as identities between functions of different order and their derivatives.

Example 9.9 Show that, for each n, $P_n(1) = 1$. (This is the rationale for the particular choice of the values for a_0 and a_1.)

Solution We set $x = 1$ in the generating function. The desired result readily follows since the Maclaurin series for $(1 - t)^{-1}$ is

$$\frac{1}{1-t} = \sum_{n=0}^{\infty} t^n$$

□

Identity 9.1

$$(n+1)P_{n+1}(x) \equiv (2n+1)xP_n(x) - nP_{n-1}(x)$$

Proof If we differentiate both sides of the generating function equation with respect to t we obtain

$$\frac{x - t}{(1 - 2xt + t^2)^{3/2}} = \sum_{n=0}^{\infty} P_n(x) n t^{n-1}$$

This may be rewritten in the form

$$(x - t)\sum_{n=0}^{\infty} P_n(x)t^n = (1 - 2xt + t^2)\sum_{n=0}^{\infty} P_n(x)nt^{n-1}$$

Equating coefficients of like powers of t leads to

$$xP_0(x) = P_1(x)$$

and

$$xP_n(x) - P_{n-1}(x) = (n + 1)P_{n+1}(x) - 2xnP_n(x) + (n - 1)P_{n-1}(x)$$

The second equation then simplifies to the desired expression. ■

Identity 9.2

$$(x^2 - 1)P_n'(x) = nxP_n(x) - nP_{n-1}(x)$$

Proof Differentiation of both sides of the generating function equation with respect to x gives rise to

$$\frac{t}{(1 - 2xt + t^2)^{3/2}} = \sum_{n=0}^{\infty} P_n'(x)t^n$$

Combining this expression with the first equation appearing in the proof of Identity 9.1, we find

$$(x - t)\sum_{n=0}^{\infty} P_n'(x)t^n = \sum_{n=0}^{\infty} P_n(x)nt^n$$

This time equating coefficients of t^n leads to

$$xP_n'(x) - P_{n-1}'(x) = nP_n(x)$$

Multiplying by n and adding $nP_n(x)$ to both sides gives

$$\begin{aligned} n(n + 1)P_n(x) &= nP_n(x) + nxP_n'(x) - nP_{n-1}'(x) \\ &= [nxP_n(x) - nP_{n-1}(x)]' \end{aligned}$$

But

$$[(1 - x^2)P_n'(x)]' + n(n + 1)P_n(x) = 0$$

owing to Legendre's equation. The desired identity follows by integration and the fact that $P_n(1) \equiv 1$. (Recall Example 9.9.) ■

From Example 7.29, we know that the Legendre polynomials form an orthogonal set with unit weight function on $[-1, 1]$. If we are to expand an arbitrary function in terms of these polynomials as in the application of Section 8.10, however, we will need to have a value for the norm of these functions.

Example 9.10 Show that

$$\int_{-1}^{1} P_m(x)P_n(x)\,dx = 0, \qquad m \neq n$$
$$= \frac{2}{2n+1}, \qquad m = n$$

Solution When $m \neq n$, the result is a consequence of the orthogonality of the functions. For the case $m = n$, alternately multiply Identity 9.1 by $P_{n+1}(x)$ and $P_{n-1}(x)$ and integrate. Thus,

$$(n+1)\int_{-1}^{1} P_{n+1}^2(x)\,dx = (2n+1)\int_{-1}^{1} xP_n(x)P_{n+1}(x)\,dx$$
$$n\int_{-1}^{1} P_{n-1}^2(x)\,dx = (2n+1)\int_{-1}^{1} xP_n(x)P_{n-1}(x)\,dx$$

If we increase the index by 1 in the second equation, we are then led to the conclusion that

$$(2n+3)\int_{-1}^{1} P_{n+1}^2(x)\,dx = (2n+1)\int_{-1}^{1} P_n^2(x)\,dx$$

This says that the number $(2n+1)\int_{-1}^{1} P_n^2(x)\,dx$ remains the same if the index is increased by 1 and thus that this number must be constant, regardless of n. For $n = 0$ this number is easily found to be 2. □

9.8 Zeros of the Legendre Polynomials and Gauss Integration

The graphs of the first several Legendre polynomials are sketched in Figure 9.5. It appears that $P_n(x)$ has precisely n zeros, symmetrically situated, in $(-1, 1)$, and

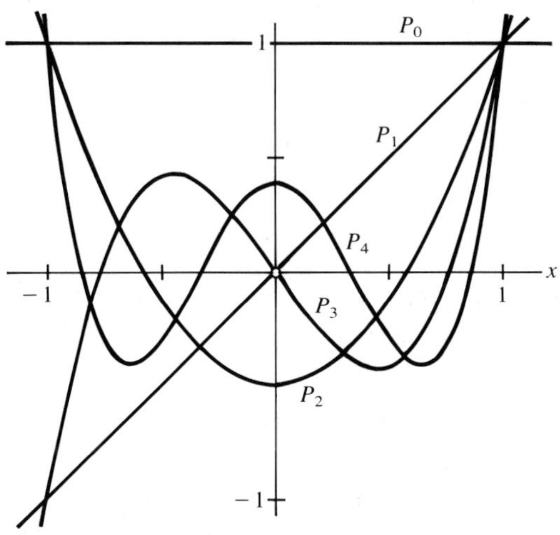

Figure 9.5 *Sketch of $P_n(x)$ for $n = 0, 1, 2, 3, 4$*

this turns out to be the case for arbitrary n. Although it can be shown that the mth zero, $x_{n,m}$, counting from the largest, of $P_n(x)$ satisfies the inequalities.

$$\cos\frac{2m}{2n+1}\pi \le x_{n,m} \le \cos\frac{2m-1}{2n+1}\pi, \qquad m = 1, 2, \ldots, n$$

a general formula for $x_{n,m}$ does not exist. Approximate values of a few zeros are given in Table 9.3.

Table 9.3 *Zeros of $P_n(x)$*

n	Zeros
2	±0.57735
3	±0.77460
	0.00000
4	±0.86114
	±0.33998
5	±0.90618
	±0.53847
	0.00000
6	±0.93247
	±0.66121
	±0.23862

The zeros of the Legendre polynomials are particularly useful in numerical quadrature. If we are interested in numerical evaluation of the integral

$$I = \int_{-1}^{1} f(x)\,dx$$

we know that the approximate formula given by the trapezoidal rule

$$I \approx [f(-1) + f(1)]$$

is exact if $f(x)$ is any linear function, whereas the three-point Simpson's rule

$$I \approx \tfrac{1}{3}[f(-1) + 4f(0) + f(1)]$$

is exact if f is linear or quadratic. It can be shown, however, that for more general functions, improved results can be achieved with an approximation of the form

$$\int_{-1}^{1} f(x)\,dx \approx \sum_{m=1}^{n} A_m f(x_m)$$

if the *sample points* x_m are not equally spaced. Indeed, if for given n, the function is sampled at the zeros of $P_n(x)$, then the *weights* A_m can be selected so that the formula is exact for all polynomial $f(x)$ up to and including those of degree $2n - 1$. With these weights and sample points, this expression is then called the *Gaussian integration formula.*

Example 9.11 Verify that the Gaussian integration formula with unit weights is exact for third-degree polynomials when $n = 2$.

Solution Since the two zeros of $P_2(x)$ are at $\pm 1/\sqrt{3}$, we want to show that

$$\int_{-1}^{1} f(x)\,dx = f\left(\frac{-1}{\sqrt{3}}\right) + f\left(\frac{1}{\sqrt{3}}\right)$$

if $f(x) = ax^3 + bx^2 + cx + d$, where a, b, c, and d are arbitrary constants. Two simple calculations yield

$$\int_{-1}^{1} f(x)\,dx = 2\left(\frac{b}{3} + d\right)$$

and

$$f\left(\frac{-1}{\sqrt{3}}\right) + f\left(\frac{1}{\sqrt{3}}\right) = -\frac{a}{3\sqrt{3}} + \frac{b}{3} - \frac{c}{\sqrt{3}} + d + \frac{a}{3\sqrt{3}} + \frac{b}{3} + \frac{c}{\sqrt{3}} + d$$
$$= 2\left(\frac{b}{3} + d\right)$$

□

Exercises for Sections 9.5–9.8

1. Use the method of Frobenius on Legendre's equation to find a solution in powers of $(x - 1)$. Express the coefficients in terms of a recurrence relation and show that the solution becomes a polynomial when n is a nonnegative integer.
2. Verify that when n is even, the Legendre polynomial $P_n(x)$ has been obtained by choosing
$$a_0 = \frac{(-1)^{n/2} n!}{2^n[(n/2)!]^2}, \qquad a_1 = 0$$
in the expressions of Section 9.6. Similarly when n is odd,
$$a_0 = 0, \qquad a_1 = \frac{(-1)^{(n-1)/2} n!}{2^{n-1}\left[\left(\frac{n-1}{2}\right)!\right]^2}$$
3. Determine the specific form of the Legendre polynomial $P_n(x)$ for $n = 4, 5, 6$.
4. Show that $P_n(-1) \equiv (-1)^n$. (*Hint:* Use the generating function.)
5. Use Identity 9.1 and the fact that $P_0(x) = 1$, $P_1(x) = x$ to demonstrate that $P_n(x)$ is an odd or even function of x according as n is odd or even.
6. Employ the results of earlier exercises to show the following:
   (a) $P'_{2n}(0) = 0$ **(b)** $P_{2n}(0) = (-1)^n \dfrac{(2n)!}{2^{2n}(n!)^2}$
7. Develop alternate proofs that $P_n(1) = 1$ and $P_n(-1) = (-1)^n$ using Identity 9.1.

8. Establish the identity

$$(n+1)P_n(x) \equiv P'_{n+1}(x) - xP'_n(x)$$

9. Use the identity of Exercise 8 to demonstrate that

$$P'_n(1) = \frac{n(n+1)}{2}$$

10. An alternative representation for the Legendre polynomials is given by *Rodrigues' formula*

$$P_n(x) \equiv \frac{1}{2^n n!}\frac{d^n}{dx^n}(x^2-1)^n$$

Verify that this formula gives the familiar expressions for $P_n(x)$ when $n = 0, 1, 2, 3$.

11. Show directly that

$$\int_{-1}^{1} P_n^2(x)\,dx = \frac{2}{2n+1}$$

for $n = 0, 1, 2, 3$ by computing the integral using the polynomial expressions for $P_n(x)$.

12. Show that $P_2(x)$ is the unique polynomial of degree 2 that is orthogonal over $[-1, 1]$ to both $P_0(x) = 1$ and $P_1(x) = x$ and takes on the value $P_2(1) = 1$. (This suggests yet another procedure for determining the Legendre polynomials.)

13. Express the function $f(x) = ax^3 + bx^2 + cx + d$ in terms of Legendre polynomials.

14. The integrals

$$I = \int_{-1}^{1} f(x)P_n(x)\,dx$$

occur in the generalized Fourier representation of functions by series of Legendre polynomials. Use Rodrigues' formula and integration by parts to demonstrate that

$$I = \frac{(-1)^n}{2^n n!}\int_{-1}^{1}(x^2-1)^n\frac{d^n f(x)}{dx^n}\,dx$$

15. Employ the result of Exercise 14 to give an alternative verification of the orthogonality of $P_n(x)$ and $P_m(x)$ over $[-1, 1]$ when $m \neq n$.

16. Use the result of Exercise 14 to show that

$$\int_{-1}^{1} x^n P_n(x)\,dx = \frac{2^{n+1}(n!)^2}{(2n+1)!}$$

17. Assume that the function

$$f(x) = \begin{cases} 0, & -1 < x < 0 \\ x, & 0 < x < 1 \end{cases}$$

is expanded in a series $\sum_{n=0}^{\infty} c_n P_n(x)$ of Legendre polynomials. Find the first three nonzero coefficients.

18. Find the steady-state temperature at a point inside the unit sphere if $u(\rho, \theta, \phi) = \cos 2\phi$ on the surface. (Recall Section 8.10.)

19. Show that $u = 2x^2 - y^2 - z^2$ is a solution of Laplace's equation in three dimensions. Express this function in spherical coordinates using the Legendre polynomials $P_n(\cos\phi)$.

20. Demonstrate that for $n = 1, 2, 3$, the zeros of $P_n(x)$ satisfy the inequalities of Section 9.8.

21. Using the weights $A_1 = \frac{5}{9}$, $A_2 = \frac{8}{9}$, and $A_3 = \frac{5}{9}$, verify that the Gaussian integration formula of Section 9.8 is exact for fifth-degree polynomials when $n = 3$. (Note that the zeros of $P_3(x)$ are at $x = 0, \pm\sqrt{\frac{3}{5}}$).

22. Evaluate $\int_{-1}^{1} e^{-x^2}\,dx$ using the Gaussian integration formula with $n = 3$. (See Exercise 21. The actual value of the integral, exact to five decimal places, is 1.49365.) Compare with the approximate value obtained using the three-point Simpson's rule of Section 9.8.

9.9 Bessel Functions

In Section 8.10 we observed that Legendre's equation and the Legendre polynomials occurred quite naturally when the method of separation of variables was applied to Laplace's equation in spherical coordinates. If the same approach is used on the same equation expressed in *cylindrical coordinates* $x = r\cos\theta$, $y = r\sin\theta$, $z = z$, the equation in the r-variable turns out to be Bessel's equation.

In various exercises and examples we previously encountered Bessel's equation

$$x^2y'' + xy' + (x^2 - m^2)y = 0$$

in situations wherein the parameter m has the integer values $m = 0, 1, 2, \ldots$. In this section we apply the technique of assuming an infinite series solution, valid near the origin, for the case of general real m. Note, however, that since $x = 0$ is a regular singular point (see Section 6.5), it is necessary to assume that the solution is of the form of a Frobenius series rather than a power series about the origin.

Letting $y = x^r \sum_{n=0}^{\infty} c_n x^n$ and substituting into the differential equation, we obtain

$$\sum_{n=0}^{\infty} [(n + r)(n + r - 1)c_n + (n + r)c_n - m^2 c_n]x^{n+r} + \sum_{n=0}^{\infty} c_n x^{n+r+2} = 0$$

or

$$c_0(r^2 - m^2) + c_1[(r + 1)^2 - m^2]x + \sum_{n=2}^{\infty} [((n + r)^2 - m^2)c_n + c_{n-2}]x^{n+r} = 0$$

To determine the coefficients, c_n, we equate each of the coefficients of powers of x in this equation to zero.

Coefficient of x^0: $c_0(r^2 - m^2) = 0$, which implies that $r = \pm m$ with c_0 an arbitrary nonzero constant

Coefficient of x^1: $c_1[(r + 1)^2 - m^2] = 0$, which implies that $c_1 = 0$ (since $r \neq \pm m - 1$)

Coefficient of x^{n+r}: $[(n + r)^2 - m^2]c_n + c_{n-2} = 0$, for $n \geq 2$, which means

$$c_n = \frac{-1}{(n+r)^2 - m^2} c_{n-2}$$

This last result is a recursion relationship valid for $n \geq 2$, and hence all the coefficients are determined. For definiteness it is customary to assign c_0 the value

$$c_0 = \frac{1}{2^m \Gamma(m+1)}$$

if m is not a negative integer, from which the general coefficient formula may be written:

$$c_{2n} = \frac{(-1)^n}{2^{2n+m} n! \Gamma(m+n+1)}$$

$$c_{2n+1} = 0$$

Therefore, one solution to Bessel's equations, called the *Bessel function of the first kind of argument x and order m*, is

$$J_m(x) \equiv \sum_{n=0}^{\infty} \frac{(-1)^n}{2^{2n+m} n! \Gamma(m+n+1)} x^{2n+m}$$

Since Bessel's equation is a second-order differential equation we would naturally expect another solution function, not a constant multiple of the first. Further, it would seem only natural that the "other solution" would be found by letting $r = -m$ and using the same recursion relationship as before. This is valid as long as m is not an integer, but (as is often the "practical" case) if m is an integer, then, as we will now show, the solution J_{-m} is linearly dependent on J_m. Actually, the expression given above for $J_{-m}(x)$ is meaningless when m is a positive integer because in the first m terms the gamma function occurs with a zero or negative integer argument. If we consider $\lim_{u \to m} J_{-u}(x)$ where u is not an integer, however, the terms in question involve $1/\Gamma(v)$ where v is getting close to 0 or a negative integer so $1/\Gamma(v)$ is approaching zero. For positive integral m we thus define

$$J_{-m}(x) = \lim_{u \to m} J_{-u}(x) = \sum_{n=m}^{\infty} \frac{(-1)^n}{2^{2n-m} n! \Gamma(n-m+1)} x^{2n-m}$$

In this power series, use a shift of index by letting $k = n - m$,

$$\begin{aligned} J_{-m}(x) &= \sum_{k=0}^{\infty} \frac{(-1)^{k+m}}{2^{2k+m}(k+m)! \Gamma(k+1)} x^{2k+m} \\ &= (-1)^m J_m(x) \end{aligned}$$

For emphasis, we repeat that this relationship is *not* true if m is *not* an integer.

If m is an integer, a second solution to Bessel's equation that is independent of the first may be found by employing reduction of order or one of the other methods discussed in Section 6.7. This solution will include a term involving the logarithm of x. Consequently the "second" solution to Bessel's equation when m

is an integer is unbounded near the origin and is useful in applications only for $x \neq 0$.

Although $J_{-m}(x)$ is a satisfactory second solution for nonintegral m, what is usually tabulated is

$$Y_m(x) \equiv \frac{(\cos m\pi)J_m(x) - J_{-m}(x)}{\sin m\pi}$$

This function is known either as the Neumann (Weber) function or, more generally, as the *Bessel function of the second kind.* It may seem strange to take such a peculiar combination to be the "second" solution when J_{-m} is so easily available. However, the Neumann functions have the added feature that, while technically undefined for integral m, the limit as m approaches integral values *is* a correct "second" solution for that integer. Indeed, the Wronskian $W[J_m, Y_m]$ is independent of m.

Example 9.12 Show that

$$J_m(x)\,Y_m'(x) - Y_m(x)J_m'(x) = \frac{2}{\pi x}, \qquad x \neq 0$$

Solution $J_m(x)$ and $Y_m(x)$, defined above, are solutions of Bessel's equation. Hence for $x \neq 0$

$$x^2 Y_m'' + x Y_m' + (x^2 - m^2) Y_m = 0$$
$$x^2 J_m'' + x J_m' + (x^2 - m^2) J_m = 0$$

Multiplying the first of these by J_m and the second by $-Y_m$ and adding,

$$x^2 J_m Y_m'' + x J_m Y_m' - x^2 Y_m J_m'' - x Y_m J_m' = 0$$

or

$$[x(J_m Y_m' - Y_m J_m')]' = 0$$

Thus

$$xW[J_m, Y_m] = \text{constant}$$

From the definitions, however, we find

$$xW[J_m, Y_m] = \frac{x}{\sin \pi m} W[J_{-m}, J_m]$$
$$= \frac{x}{\sin \pi m}\left[\frac{2^m x^{-m}}{\Gamma(1-m)} \frac{m x^{m-1}}{2^m \Gamma(m+1)} + \frac{m 2^m x^{-m-1}}{\Gamma(1-m)} \frac{x^m}{2^m \Gamma(m+1)}\right]$$

since only the lead terms in the various series can possibly contribute to this constant. This last expression reduces to the desired result owing to the important gamma function identity $\Gamma(m)\Gamma(1-m) = \pi/\sin \pi m$ (see Section 9.3). □

As we will see later when we consider the oscillatory nature of these functions, some Bessel functions have at least a remote similarity to the sine and cosine

Table 9.4 *Values of $J_0(x)$ and $J_1(x)$*

x	0	0.1	0.2	0.3	0.4	0.5	0.6	0.7	0.8	0.9	1.0
$J_0(x)$	1.000	0.998	0.990	0.978	0.960	0.938	0.912	0.881	0.846	0.808	0.765
$J_1(x)$	0.000	0.050	0.100	0.148	0.196	0.242	0.287	0.329	0.369	0.406	0.440
x	1.5	2.0	2.5	3.0	3.5	4.0	4.5	5.0	5.5	6.0	6.5
$J_0(x)$	0.512	0.224	−0.048	−0.260	−0.381	−0.397	−0.321	−0.178	−0.007	0.151	0.260
$J_1(x)$	0.558	0.577	0.497	0.339	0.137	−0.066	−0.231	−0.328	−0.341	−0.277	−0.154
x	7.0	7.5	8.0	8.5	9.0	9.5	10.0	11.0	12.0	13.0	14.0
$J_0(x)$	0.300	0.266	0.172	0.042	−0.090	−0.194	−0.246	−0.171	0.048	0.207	0.171
$J_1(x)$	−0.005	0.135	0.235	0.273	0.245	0.161	0.043	−0.177	−0.223	−0.070	0.133

functions. We can even notice this now by observing how much the first few terms of the funcitions J_0 and J_1 resemble the cosine and the sine.

$$J_0(x) = 1 - \frac{x^2}{2^2 1!} + \frac{x^4}{2^4 2!} - \frac{x^6}{2^6 3!} + \frac{x^8}{2^8 4!} - \cdots$$

$$J_1(x) = \frac{x}{2} - \frac{x^3}{2^3 1!2!} + \frac{x^5}{2^5 3!2!} - \frac{x^7}{2^7 5!3!} + \frac{x^9}{2^9 7!4!} - \cdots$$

Approximate values of these functions are given in Table 9.4. (See also Figure 9.6.)

Example 9.13 Write $J_{1/2}$ in terms of a function of x times a sine function.

Solution

$$\begin{aligned} J_{1/2}(x) &= \sum_{n=0}^{\infty} \frac{(-1)^n x^{2n+1/2}}{2^{2n+1/2} n!\Gamma(n+\frac{3}{2})} \\ &= \frac{2^{1/2} x^{-1/2}}{\pi^{1/2}} \sum_{n=0}^{\infty} \frac{(-1)^n x^{2n+1}}{(2n+1)!} \qquad \text{(Recall Exercise 8, p. 316)} \\ &= \left(\frac{2}{\pi x}\right)^{1/2} \sin x \end{aligned}$$

□

In similar fashion it can be shown that

$$J_{-1/2}(x) = -Y_{1/2}(x) = \left(\frac{2}{\pi x}\right)^{1/2} \cos x$$

9.10 Modified Bessel Functions

In many applications, Bessel's equation appears in a form slightly different from the one presented in the previous section.

Example 9.14 Find the general solution to the differential equation
$x^2y'' + xy' + (x^2s^2 - m^2)y = 0.$

Solution This equation differs from Bessel's equation only in that sx takes the place of x. Letting $u = sx$ we obtain

$$\left(\frac{u}{s}\right)^2\left(\frac{d^2y}{du^2}s^2\right) + \left(\frac{u}{s}\right)\left(\frac{dy}{du}s\right) + (u^2 - m^2)y = 0$$

which is exactly Bessel's equation. Therefore, the complete solution is

$$y = c_1 J_m(sx) + c_2 Y_m(sx)$$ □

If, in Example 9.14, s is permitted to assume complex values, the solutions to Bessel's modified equation are in terms of Bessel functions with purely imaginary arguments. The procedure we follow is analogous to the case of the real exponential function with complex argument. In that case we express (or define) e^{ix} as a linear combination of trigonometric functions. Here, in order to "simplify" matters, we define the following two functions:

$$I_m(x) \equiv i^{-m} J_m(ix)$$

$$K_m(x) \equiv \frac{\pi}{2} i^{m+1} [J_m(ix) + iY_m(ix)]$$

These are called *modified Bessel functions*. In this manner, solutions are expressed in terms of $I_m(x)$ and $K_m(x)$, each of which have real values for real x.

Exercises for Sections 9.9 and 9.10

1. Show that $J_0(0) = 1$ and $J_m(0) = 0$ for $m > 0$.
2. Establish that $J_{-1/2}(x) = (2/\pi x)^{1/2} \cos x$.
3. Verify that $J_{3/2}(x) = \sqrt{\dfrac{2}{\pi x}}\left[\dfrac{\sin x}{x} - \cos x\right]$.
4. Show that $c_1 I_m + c_2 K_m$ is a solution to $x^2y'' + xy' - (x^2 + m^2)y = 0$.

In Exercises 5–8, solve the given equations in terms of Bessel functions.

5. $x^2y'' + xy' + (x^2 - \frac{1}{4})y = 0$
6. $x^2y'' + xy' - (x^2 + 9)y = 0$
7. $x^2y'' + xy' + \frac{1}{4}(x^2 - 1)y = 0$
8. The coupled system $\begin{aligned} xu' + y &= 0 \\ y' - xu &= 0 \end{aligned}$

   (*Hint:* The change of dependent variable $y = xv$ may be helpful.

In each of Exercises 9–11, write out the first few terms of the specified solution functions.

9. The solution to Exercise 5.
10. That portion of the solution to Exercise 6 which vanishes at $x = 0$.
11. The solution to Exercise 7.

12. Apply L'Hôpital's rule to the definition for the Bessel function of the second kind

$$Y_m(x) \equiv \frac{\cos m\pi J_m(x) - J_{-m}(x)}{\sin m\pi}$$

and show

$$Y_0(x) = \frac{2}{\pi}\left[J_0(x)\ln\frac{x}{2} - \sum_{n=0}^{\infty}\frac{\psi(n+1)(-1)^n(x/2)^{2n}}{(n!)^2}\right]$$

where $\psi(x) \equiv \dfrac{\Gamma'(x)}{\Gamma(x)}$.

13. Repeat Exercise 12 to obtain, for $m \geq 1$,

$$Y_m(x) = \frac{1}{\pi}\left\{2J_m(x)\ln\frac{x}{2} - \sum_{n=0}^{m-1}\frac{(m-n-1)!(x/2)^{2n-m}}{n!}\right.$$
$$\left. - \sum_{n=0}^{\infty}[\psi(n+1) + \psi(n+m+1)]\frac{(-1)^n(x/2)^{2n+m}}{n!(n+m)!}\right\}$$

14. Alternatively, derive the expression for $Y_0(x)$ given in Exercise 12 by applying the Frobenius method, as modified in Section 6.7, to Bessel's differential equation. Here you will need to make use of the fact that $\psi(1) = -\gamma$ (the so-called Euler-Mascheroni constant) and

$$\psi(m) = -\gamma + \sum_{n=1}^{m-1}\frac{1}{n}, \qquad m \geq 2$$

15. Using the tabulated values, make a rough sketch of $J_0(x)$ and $J_1(x)$ for $0 < x < 10$.

16. Show that for small x, $J_0(x) \approx 1 - (x^2/4)$ and use this to determine $J_0(x)$ for $x = 0.1$, 0.2, 0.3, and compare to the tabulated values.

17. Show that $J_n(x)$ is an even function when n is an even integer and an odd function when n is an odd integer.

Bessel functions of particular orders are related to some of the special functions previously encountered. The next few exercises explore some of these relationships.

18. Airy's differential equation is $y'' - xy = 0$. Show that the complete solution of this equation can be expressed as

$$y = \sqrt{x}[c_1 I_{1/3}(t) + c_2 K_{1/3}(t)]$$

where $t = \frac{2}{3}x^{3/2}$.

19. Verify the identity

$$\int_{-1}^{1} e^{xt}P_n(t)\,dt \equiv \sqrt{\frac{2\pi}{x}}I_{n+1/2}(x)$$

by showing that the coefficients of x^n on both sides match. (Recall Exercise 8, p. 316, and Exercise 16, p. 332.)

20. Demonstrate that as $n \to \infty$, $P_n\left(\cos\frac{x}{n}\right)$ satisfies Bessel's equation of order zero, and thus, by matching at $x = 0$, establish that

$$\lim_{n\to\infty} P_n\left(\cos\frac{x}{n}\right) \equiv J_0(x)$$

21. Satisfy yourself as to the validity of the identity

$$\sum_{n=0}^{\infty} \frac{t^n}{n!} P_n(\cos\theta) \equiv e^{t\cos\theta} J_0(t\sin\theta)$$

by testing the values $\theta = 0, \pi/2, \pi$.

If the product method is applied to Laplace's equation in cylindrical coordinates,

$$u_{rr} + \frac{1}{r}u_r + \frac{1}{r^2}u_{\theta\theta} + u_{zz} = 0$$

the resulting ordinary differential equations are

$$\theta'' + \alpha^2\theta = 0$$
$$z'' + \beta^2 z = 0$$
$$r^2R'' + rR' - (\beta^2r^2 + \alpha^2)R = 0$$

This latter equation is recognized as a modified Bessel equation, and thus the modified Bessel function $I_\alpha(\beta r)$ often plays a role in the solution of a given Laplacian boundary value problem.

22. Show that if the boundary conditions are

$$u(r,\theta,0) \equiv 0 \equiv u(r,\theta,1)$$
$$u(1,\theta,z) \equiv f(z)\cos\theta$$

the steady-state temperature within the solid circular cylinder $0 \le r,\ z \le 1$ can be represented as

$$u(r,\theta,z) = 2\cos\theta \sum_{n=1}^{\infty} \left(\int_0^1 f(t)\sin n\pi t\,dt\right)\frac{I_1(n\pi r)}{I_1(n\pi)}\sin n\pi z$$

23. Determine the steady-state temperature within the cylinder of Exercise 22 in the case of insulated ends and a prescribed axisymmetric temperature distribution on the lateral surface, that is,

$$\frac{\partial u(r,\theta,0)}{\partial z} \equiv 0 \equiv \frac{\partial u(r,\theta,1)}{\partial z}$$
$$u(1,\theta,z) \equiv g(z)$$

9.11 Some Fundamental Bessel Function Identities

Because the Bessel functions of the first kind for $m \ge 0$ are bounded near the origin, they are easily the most useful. In the following sections we will restrict our discussion to this type of Bessel function.

Just like Legendre polynomials, Bessel functions of the first kind obey some rather important identities.

Identity 9.3

$$xJ_n'(x) \equiv nJ_n(x) - xJ_{n+1}(x)$$

Proof

$$J_n(x) = \sum_{k=0}^{\infty} \frac{(-1)^k}{k!\Gamma(n+k+1)2^{n+2k}} x^{n+2k}$$

Under the (correct) assumption that term-by-term differentiation is permissible,

$$J_n'(x) = \sum_{k=0}^{\infty} \frac{(n+2k)(-1)^k}{k!\Gamma(n+k+1)2^{n+2k}} x^{n+2k-1}$$

Therefore,

$$xJ_n'(x) = nJ_n(x) + x\sum_{k=1}^{\infty} \frac{(-1)^k}{(k-1)!\Gamma(n+k+1)2^{n+2k-1}} x^{n+2k-1}$$

Shifting the index in the sum by letting $k = s + 1$,

$$xJ_n'(x) = nJ_n(x) - x\sum_{s=0}^{\infty} \frac{(-1)^s}{s!\Gamma(n+s+2)2^{n+2s+1}} x^{n+2s+1}$$

$$= nJ_n(x) - xJ_{n+1}(x) \qquad \blacksquare$$

Identity 9.4

$$xJ_n'(x) \equiv -nJ_n(x) + xJ_{n-1}(x)$$

Proof The details are left as an exercise (see Exercise 1, p. 344) but you should first let $n + 2k = 2(n + k) - n$ in the series expression for $xJ_n'(x)$ and use the fact that $\Gamma(n + k + 1) = (n + k)\Gamma(n + k)$. ■

Identity 9.5 *Differentiation Formula.*

$$J_n'(x) \equiv \frac{J_{n-1}(x) - J_{n+1}(x)}{2}$$

Proof Eliminate J_n between Identities 9.3 and 9.4. ■

Identity 9.6 *Recursion Formula.*

$$J_n(x) \equiv \frac{x}{n}\left(\frac{J_{n+1}(x) + J_{n-1}(x)}{2}\right)$$

or, which is the same thing,

$$J_{n+1}(x) \equiv \frac{2n}{x}J_n(x) - J_{n-1}(x)$$

Proof Eliminate $J_n'(x)$ between Identities 9.3 and 9.4. ■

Example 9.15 Derive a differentiation formula for $J_0(x)$.

Solution From Identities 9.3 and 9.4 it follows immediately that

$$J_0'(x) \equiv J_{-1}(x) = -J_1(x) \qquad \square$$

Example 9.16 Evaluate

$$\int_0^x tJ_0(t)\,dt$$

Solution Note that

$$\frac{d}{dx}(xJ_1(x)) = xJ_1'(x) + J_1(x)$$

Using Identity 9.4 with $n = 1$, $xJ_1'(x) = -J_1(x) + xJ_0(x)$, which means that

$$\frac{d}{dx}(xJ_1(x)) = -J_1(x) + xJ_0(x) + J_1(x) = xJ_0(x)$$

Therefore,

$$\int_0^x tJ_0(t)\,dt = [tJ_1(t)]\Big|_0^x = xJ_1(x)$$

□

Example 9.17 Express J_3 in terms of J_0 and J_1.

Solution Using Identity 9.6 with $n = 2$,

$$\begin{aligned} J_3(x) &= \frac{4}{x}J_2(x) - J_1(x) \\ &= \frac{4}{x}\left[\frac{2}{x}J_1(x) - J_0(x)\right] - J_1(x) \\ &= \left[\frac{8}{x^2} - 1\right]J_1(x) - \frac{4}{x}J_0(x) \end{aligned}$$

□

9.12 Approximations to Bessel Functions

In this section we will show how to approximate the Bessel functions when the argument is either very small (that is, near zero) or very large.

By a rigorous argument, it is possible to show that, for large values of x,

$$J_m(x) \approx \left(\frac{2}{\pi x}\right)^{1/2} \cos\left(x - \frac{2m+1}{4}\pi\right)$$

By a less rigorous, but far easier, argument we can show the same fundamental result; that is, the Bessel functions behave something like a damped cosine function when the value of x is large. To see this, consider Bessel's differential equation

$$x^2y'' + xy' + (x^2 - m^2)y = 0$$

in the form

$$y'' + \frac{1}{x}y' + \left(1 - \frac{m^2}{x^2}\right)y = 0$$

We might anticipate the solution for large x by discarding the term m^2/x^2. The differential equation then becomes

$$y'' + \frac{1}{x}y' + y = 0$$

Letting $u = yx^{1/2}$ (a change in the dependent variable), the differential equation becomes

$$u'' + \left(\frac{1}{4x^2} + 1\right)u = 0$$

Once again, we ignore the term involving x^{-2} to obtain

$$u'' + u = 0$$

whose solution is

$$u = c_1 \cos x + c_2 \sin x$$

and thus the approximate solution to Bessel's equation for large values of x is

$$y = x^{-1/2}(c_1 \cos x + c_2 \sin x) = Ax^{-1/2}\cos(x + \beta)$$

which is what we wanted to show.

For x near zero, we examine the solution itself and discard all terms after the first,

$$J_m(x) \approx \frac{x^m}{2^m\Gamma(m + 1)}$$

This approximation is good *only* for small values of x. Notice that all Bessel functions of positive order are zero for $x = 0$ and that $J_0(0) = 1$.

9.13 Zeros of Bessel Functions

In the previous section, we noted that the Bessel functions of the first kind behave something like damped harmonic functions; indeed for large x

$$J_m(x) \approx \left(\frac{2}{\pi x}\right)^{1/2} \cos\left(x - \frac{2m + 1}{4}\pi\right)$$

Consequently we would suspect that the graphs of these functions should cross the x axis infinitely often and at intervals of length approximately equal to π. This is indeed true but the proof is beyond the scope of our discussion. However, if the order m is small, we can be fairly definitive about the location of the zeros of $J_m(x)$.

Theorem 9.2 If $-\frac{1}{2} \le m \le \frac{1}{2}$, $J_m(x)$ has at least one zero in each of the intervals $[n\pi, (n + 1)\pi]$, $n = 0, 1, 2, \ldots$.

Proof If $u(x) = \sqrt{x}J_m(x)$, then

$$u'' + \left(1 + \frac{\frac{1}{4} - m^2}{x^2}\right)u = 0$$

Also, we know that

$$(\sin x)'' + \sin x = 0$$

We multiply the first of these equations by $\sin x$, the second by $-u(x)$, and add:

$$u'' \sin x - u(\sin x)'' + \left(\frac{\frac{1}{4} - m^2}{x^2}\right) u \sin x = 0$$

or

$$\left(\frac{\frac{1}{4} - m^2}{x^2}\right) u \sin x = -[u' \sin x - u \cos x]'$$

Note that the coefficient of $u \sin x$ is always nonnegative by hypothesis. We now integrate both sides of this equality from $n\pi$ to $(n + 1)\pi$ to obtain

$$\int_{n\pi}^{(n+1)\pi} \left(\frac{\frac{1}{4} - m^2}{x^2}\right) u \sin x \, dx = (-1)^{n+1}[u(n\pi + \pi) + u(n\pi)]$$

If $u(x)$ is of one sign throughout any of the intervals $[n\pi, (n + 1)\pi]$, then the two sides of this last relation are of opposite sign, which is impossible. Thus $u(x)$, and hence $J_m(x)$, must have at least one zero within the interval in question.

■

For certain applications, it will be just as important to know the zeros of the Bessel functions as, say, the zeros of the trigonometric functions. However, unlike the zeros of the trigonometric functions, the zeros of the Bessel functions follow no easily recognized pattern. Hence, it is necessary to have them tabulated. Table 9.5 gives approximate values for the first five zeros of $J_0(x)$ and $J_1(x)$.

Table 9.5

$J_0(x_{0m}) = 0$		$J_1(x_{1m}) = 0$	
m	x_{0m}	m	x_{1m}
1	2.405	1	0
2	5.520	2	3.832
3	8.654	3	7.016
4	11.792	4	10.173
5	14.931	5	13.324

Our original conjecture that the distance between the zeros approaches π as x becomes infinite is true, but we do not give a proof.

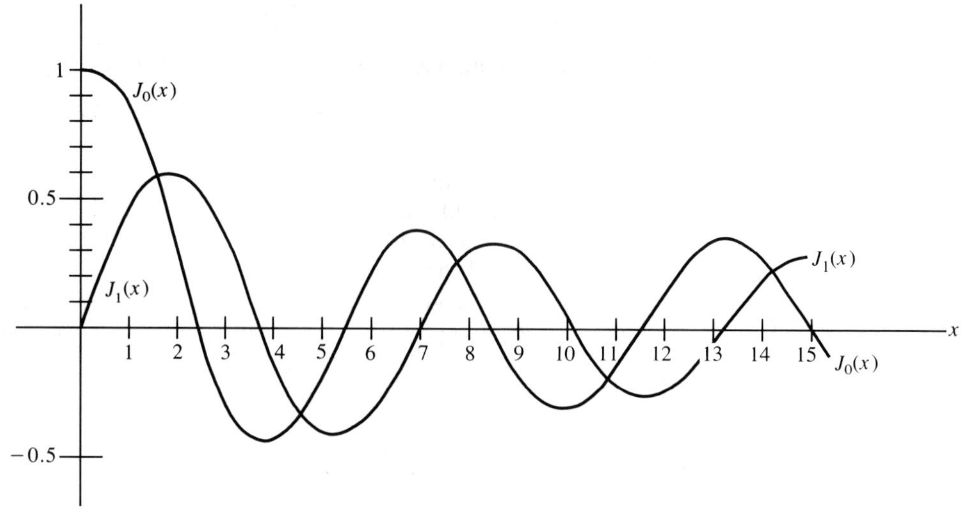

Figure 9.6

With the information on the zeros of the functions and the earlier tabular values (Table 9.4), relatively good graphs of J_0 and J_1 may be sketched. (See Figure 9.6.)

Exercises for Sections 9.11–9.13

1. Complete the proofs of Identities 9.4, 9.5, and 9.6, Section 9.11.
2. Show that when the order n is integral, if b is a zero of $J_n(x)$, then $-b$ is also a zero.
3. Show that $J_2(x) = -\frac{1}{x}J_0'(x) + J_0''(x)$.
4. Express J_4 in terms of J_0 and J_1.
5. Show that

$$J_2'(x) \equiv \frac{1}{2}(J_1(x) - J_3(x)) \equiv \left(1 - \frac{4}{x^2}\right)J_1(x) + \frac{2}{x}J_0(x)$$

6. Show that

$$J_3'(x) \equiv \left(1 - \frac{12}{x^2}\right)J_2(x) + \frac{3}{x}J_1(x)$$

7. Show that

$$\int xJ_n^2(x)\,dx \equiv \frac{1}{2}(x^2 - n^2)J_n^2(x) + \frac{1}{2}[xJ_n'(x)]^2$$

(*Hint:* Use integration by parts along with a knowledge of the differential equation satisfied by $J_n(x)$.)

8. Show that $\int x^n J_{n-1}(x)\,dx = x^n J_n(x) + c$.

9. Show that $\int x^{-n}J_{n+1}(x)\,dx = -x^{-n}J_n(x) + c$.

Evaluate the following ($\int J_0(x)\,dx$ may be left unevaluated).

10. $\int J_2(x)\,dx$ **11.** $\int xJ_1(x)\,dx$ **12.** $\int J_3(x)\,dx$

13. $\int x^2J_0(x)\,dx$ **14.** $\int x^3J_0(x)\,dx$

15. Investigate the accuracy of the "large" and "small" argument approximations of $J_m(x)$ given in Section 9.12. Let $m = 0, 1$ and take $x = 0.1, 0.2, 0.3$ as representative small values and $x = 6, 10, 14$ as representative large values. Compare with the tabular values given in Table 9.4.

16. Compare the values obtained for the lower-order zeros of J_0 and J_1 using the large-argument approximation given in Section 9.12 with the values appearing in Table 9.5. Show that the approximation improves as the order (size) of the zero increases.

The functions that today bear his name were first introduced by Bessel in 1824 in connection with a problem regarding the calculation of anomalies in planetary orbits. The representation he obtained was

$$y(x) = \frac{1}{2\pi}\int_0^{2\pi} \cos(x\sin\theta - n\theta)\,d\theta$$

which can be shown to be an alternative expression for $J_n(x)$ when n is integral.

17. By making use of the identity

$$(x^2\cos^2\theta - n^2)\cos\Theta - x\sin\theta\sin\Theta = \frac{d}{d\theta}[(x\cos\theta + n)\sin\Theta]$$

where

$$\Theta = x\sin\theta - n\theta$$

verify that $y(x)$ as given above satisfies Bessel's equation of order n.

18. Show that $y(0) = 1$ if $n = 0$ and $y(0) = 0$ if $n > 0$, and hence, owing to the result of Exercise 17, deduce that $y(x) = cJ_n(x)$ for nonnegative integral n, where c is a constant perhaps depending on n.

19. Demonstrate that $y(x)$ satisfies Identity 9.5, which implies, in view of the linear independence of the $J_n(x)$ for different n, that the constant c in Exercise 18 does not depend on n; and, since it is unity when $n = 0$, $c = 1$ for all integral n.

9.14 Fourier-Bessel Series

In Chapter 7 you learned that elements of $PC(a, b)$ could be "expanded" in terms of a basis on that space. At that time we considered mainly bases from the set of trigonometric functions but various examples (see especially Section 8.10) did suggest that other functions might be used effectively as the building blocks in certain representations. In this section we see that some sets of Bessel functions can also serve as a basis for function spaces. We do not prove that any given set is a basis, but we show what to expect in the way of orthogonality. Although it is not necessary to do so, we restrict ourselves to integer orders, since this is a case of significant practical importance.

If n is a nonnegative integer, the functions $J_n(sx)$ satisfy Bessel's modified equation

$$x^2 J_n''(sx) + xJ_n'(sx) + (s^2x^2 - n^2)J_n(sx) = 0$$

or

$$[xJ_n'(sx)]' + \left(-\frac{n^2}{x} + s^2 x\right) J_n(sx) = 0$$

This will be a Sturm-Liouville problem with $\lambda = s^2$, $p(x) = x$, $r(x) = x$, and $q(x) = -n^2/x$, if the proper boundary values are chosen. Since $p(0) = 0$, we need only one homogeneous boundary condition for some finite value of x, say R. Thus, we set $J_n(sR) = 0$. This gives

$$s_{n1}R = \alpha_{n1}, s_{n2}R = \alpha_{n2}, \ldots, s_{nm}R = \alpha_{nm}, \ldots$$

where the numbers α_{nm} $(m = 1, 2, 3, \ldots)$ are the zeros of J_n. Hence the values of $s_{n1}, s_{n2}, \ldots, s_{nm} \ldots$ are determined:

$$s_{n1} = \frac{\alpha_{n1}}{R}, s_{n2} = \frac{\alpha_{n2}}{R}, \ldots, s_{nm} = \frac{\alpha_{nm}}{R}, \ldots$$

The squares of s_{nm} are the "eigenvalues" of the Sturm-Liouville problem, and by the result of Section 7.12, the set

$$\{J_n(s_{nm}x)\}_{m=1}^{\infty}$$

is an orthogonal set on $PC(0, R)$ with respect to the weight function x; that is,

$$\int_0^R xJ_n(s_{nm}x)J_n(s_{nq}x)\,dx = 0, \qquad q \neq m$$

Since n is fixed, each of the Bessel functions, along with its zeros, completely determines the following orthogonal sets:

$$\{J_0(s_{0m}x)\}_{m=1}^{\infty}, \{J_1(s_{1m}x)\}_{m=1}^{\infty}, \{J_2(s_{2m}x)\}_{m=1}^{\infty}, \ldots$$

Each of these sets is a basis for the space $PC(0, R)$. Therefore, just like the Fourier expansion of a function in terms of the trigonometric set, we can expect to express rather general functions in terms of these orthogonal sets. You should note, however, precisely which sets of Bessel functions are orthogonal. For example, the set of *all* Bessel functions of integral order is *not* one of the basis sets being considered.

In order to apply the general theory of orthogonal expansions, we must find the norm of each of the functions $J_n(s_{nm}x)$ on $(0, R)$.

$$\|J_n(s_{nm}x)\|^2 = \int_0^R xJ_n^2(s_{nm}x)\,dx$$

which, by a change of variables $y = s_{nm}x$ becomes

$$\|J_n(s_{nm}x)\|^2 = \frac{1}{(s_{nm})^2}\int_0^{s_{nm}R} yJ_n^2(y)\,dy$$

Using Exercise 7 of the previous exercise set to obtain an antiderivative for $yJ_n^2(y)$, this integral is equal to

$$\frac{1}{(s_{nm})^2}\frac{(y^2-n^2)}{2}J_n^2(y)\Big|_0^{s_{nm}R}+\frac{1}{2}\left[\frac{y}{s_{nm}}J_n'(y)\right]^2\Big|_0^{s_{nm}R}$$

Owing to the boundary conditions at 0 and R, the first term vanishes, so that

$$\|J_n(s_{nm}x)\|^2=\tfrac{1}{2}R^2[J_n'(s_{nm}R)]^2$$

By using Identity 9.3 in Section 9.11, with $x = R$, this becomes

$$\|J_n(s_{nm}x)\|^2=\tfrac{1}{2}R^2J_{n+1}^2(s_{nm}R)$$

Thus,

$$\|J_n(s_{nm}x)\|=\frac{1}{\sqrt{2}}R\,|J_{n+1}(s_{nm}R)|$$

For example,

$$\|J_0(s_{0m}x)\|=\frac{R}{\sqrt{2}}|J_1(s_{0m}R)|,\qquad \text{where the } s_{0m}R \text{ are the zeros of } J_0$$

$$\|J_1(s_{1m}x)\|=\frac{R}{\sqrt{2}}|J_2(s_{1m}R)|,\qquad \text{where the } s_{1m}R \text{ are the zeros of } J_1$$

$$\|J_2(s_{2m}x)\|=\frac{R}{\sqrt{2}}|J_3(s_{2m}R)|,\qquad \text{where the } s_{2m}R \text{ are the zeros of } J_2$$

From the formula for the norm, the orthonormal sets corresponding to the orthogonal sets of Bessel functions on $(0, R)$ may now be written

$$\left\{\frac{J_0(s_{0m}x)}{\frac{R}{\sqrt{2}}|J_1(s_{0m}R)|}\right\}_{m=1}^{\infty},\ \left\{\frac{J_1(s_{1m}x)}{\frac{R}{\sqrt{2}}|J_2(s_{1m}R)|}\right\}_{m=1}^{\infty},$$

$$\left\{\frac{J_2(s_{2m}x)}{\frac{R}{\sqrt{2}}|J_3(s_{2m}R)|}\right\}_{m=1}^{\infty},\ldots,\left\{\frac{J_n(s_{nm}x)}{\frac{R}{\sqrt{2}}|J_{n+1}(s_{nm}R)|}\right\}_{m=1}^{\infty},\ldots$$

Example 9.18 Write the first three elements of the set of Bessel functions of order 0 which are orthonormal on $(0, 3)$.

Solution From the table of zeros given on p. 343, $s_{01}R \doteq 2.4$, $s_{02}R \doteq 5.5$, $s_{03}R \doteq 8.7$, and hence $s_{01} \doteq 0.8$, $s_{02} \doteq 1.8$, $s_{03} \doteq 2.9$. Interpolating from the values for $J_1(x)$ given in Table 9.4, $J_1(2.4) \doteq 0.52$, $J_1(5.5) \doteq -0.34$, and $J_1(8.7) \doteq 0.27$. Thus, the first few terms of the orthonormal set are approximately

$$\left\{\frac{J_0(2.4x)}{\frac{3}{\sqrt{2}}(0.52)},\frac{J_0(5.5x)}{\frac{3}{\sqrt{2}}(0.34)},\frac{J_0(8.7x)}{\frac{3}{\sqrt{2}}(0.27)},\ldots\right\}$$

□

An expansion of a function of $PC(0, R)$ in terms of Bessel functions of order n is called a *Fourier-Bessel (order n) expansion.* The results of Chapter 7 apply to this case since we are assuming that each of the orthogonal sets is a basis for $PC(0, R)$. Hence, the Fourier-Bessel series converges in the mean to the function. To determine the coefficients in the expansion, the results of Chapter 7 tell us that if

$$f(x) \approx \sum_{j=1}^{\infty} c_j \phi_j$$

then the Fourier coefficients are given by

$$c_j = \frac{(f, \phi_j)}{\|\phi_j\|^2}$$

In the case of the sets of Bessel functions, the inner product is with respect to the weight function x, so that

$$c_j = \frac{1}{\| J_n(s_{nj}x)\|^2} \int_0^R x f(x) J_n(s_{nj}x)\, dx$$

Using the values of $\| J_n(s_{nj}x)\|$ derived previously, the Fourier-Bessel expansion is written

$$f(x) \approx \sum_{j=1}^{\infty} A_j J_n(s_{nj}x)$$

where

$$A_j = \frac{2}{R^2 J_{n+1}^2(s_{nj}R)} \int_0^R x f(x) J_n(s_{nj}x)\, dx$$

The coefficients A_j are often called the *Fourier-Bessel coefficients.*

Example 9.19 Expand the function defined by $f(x) = 1$ on $PC(0, 2)$ in a series of Bessel functions of order zero.

Solution The Fourier-Bessel coefficients are

$$\begin{aligned} A_j &= \frac{2}{4J_1^2(2s_{0j})} \int_0^2 x J_0(s_{0j}x)\, dx \\ &= \frac{1}{2J_1^2(2s_{0j})s_{0j}^2} \int_0^{2s_{0j}} u J_0(u)\, du \end{aligned}$$

Using the result of Example 9.16, p. 341,

$$\begin{aligned} A_j &= \frac{1}{2J_1^2(2s_{0j})s_{0j}^2} [uJ_1(u)]_0^{2s_{0j}} \\ &= \frac{1}{2J_1^2(2s_{0j})s_{0j}^2} 2s_{0j} J_1(2s_{0j}) \\ &= \frac{1}{s_{0j} J_1(2s_{0j})} \end{aligned}$$

Therefore,

$$1 \sim \sum_{j=1}^{\infty} \frac{J_0(s_{0j}x)}{s_{0j}J_1(2s_{0j})}$$

where $2s_{0j}$ are the zeros of the Bessel function of order zero. □

Example 9.20 Expand the function of $PC(0, 1)$ defined by $f(x) = x$ in a series of Bessel functions of order 1.

Solution For the Bessel functions of order 1, the Fourier-Bessel coefficients with $R = 1$ are

$$\begin{aligned} A_j &= \frac{2}{J_2^2(s_{1j})} \int_0^1 x^2 J_1(s_{1j}x)\,dx \\ &= \frac{2}{J_2^2(s_{1j})(s_{1j})^3} \int_0^{s_{1j}} u^2 J_1(u)\,du \\ &= \frac{2}{J_2^2(s_{1j})(s_{1j})^3} [u^2 J_2(u)]_0^{s_{1j}} \\ &= \frac{2}{s_{1j}J_2(s_{1j})} \end{aligned}$$

(See Exercise 8 of the previous set of exercises.) Thus:

$$x \sim \sum_{j=1}^{\infty} \frac{2}{s_{1j}J_2(s_{1j})} J_1(s_{1j}x)$$

where the s_{1j} are the zeros of the order 1 Bessel function. □

9.15 The Vibrating Membrane

As long ago as 1732, Daniel Bernoulli used what we now recognize as Bessel functions in his investigations of small oscillations of the hanging chain. Since then, there have been a number of other classical problems whose solutions demand the use of Bessel functions. They are also used in more modern analysis, such as in the description of the flux distribution in a nuclear reactor. In this section, we discuss a relatively elementary yet important application.

Consider the vibrations of a taut circular membrane of radius R_1 clamped along its circumference. It is a part of empirical knowledge that the differential equation describing small transverse vibrations is the two-dimensional wave equation

$$u_{xx} + u_{yy} = \frac{1}{c^2} u_{tt}$$

where c is a physical constant of the membrane (recall Example 8.19). We introduce a system of cylindrical coordinates (r, θ) with the origin at the center

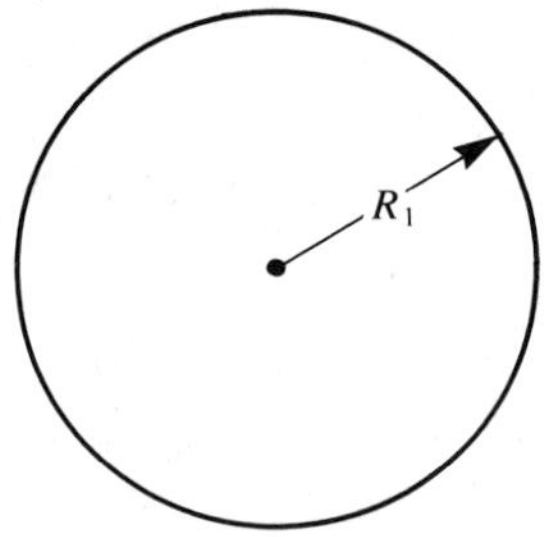

Figure 9.7 *Circular membrane*

of the membrane. (See Figure 9.7.) The u-axis is considered to be perpendicular to the equilibrium plane of the membrane. The wave equation then takes the form

$$u_{rr} + \frac{1}{r}u_r + \frac{1}{r^2}u_{\theta\theta} = \frac{1}{c^2}u_{tt}$$

The circumference of the membrane being clamped means that u must vanish when $r = R_1$ for all $t \geqq 0$. It is customary to simplify the problem by considering only those solutions which are circularly symmetric; that is, those which do not depend on θ. The mathematical model of the problem then becomes

$$u_{rr} + \frac{1}{r}u_r = \frac{1}{c^2}u_{tt}$$

where $u(R_1, t) \equiv 0$, $u(r, 0) \equiv f(r)$, $u_t(r, 0) \equiv g(r)$; the last two conditions are initial conditions on the deflection and velocity. Assuming a product solution we let

$$u(r, t) = R(r)T(t)$$

Substituting into the given differential equation in the usual way, we obtain the two ordinary differential equations

$$T'' + k^2c^2T = 0$$

$$R'' + \frac{1}{r}R' + k^2R = 0$$

and the homogeneous boundary condition $R(R_1) = 0$.

The second of these differential equations is Bessel's modified equation, and hence the solution is

$$R(r) = c_1J_0(kr) + c_2Y_0(kr)$$

where J_0 and Y_0 are the Bessel functions of first and second kind, respectively, of order zero. Since the deflection of the membrane is always finite, whereas Y_0 becomes infinite as r approaches 0, c_2 must be chosen to be zero. At the boundary, $R(R_1) = c_1J_0(kR_1) = 0$, which means that k must be chosen so that kR_1 is a zero of the Bessel function of order zero. Denoting the positive zeros of $J_0(x)$ by α_{01}, $\alpha_{02}, \ldots, \alpha_{0m}, \ldots$, the allowable values of k are

$$k_m = \frac{\alpha_{0m}}{R_1}$$

Corresponding to each of these values of k, the time equation has the solution

$$T_m(t) = a_m \cos k_m ct + b_m \sin k_m ct$$

and, hence, each of the functions

$$u_m(r, t) = (a_m \cos k_m ct + b_m \sin k_m ct) J_0(k_m r)$$

satisfies the two-dimensional wave equation (with the assumed symmetry property with respect to θ), and the boundary condition at the circumference. Further, any sum of these functions is also a solution since the boundary condition is homogeneous. Each u_m is called an eigenfunction of the problem; either k_m or $k_m c$ is called an eigenvalue.

In general, any one of the u_m, or any finite combination of them, would not reduce to $f(r)$ when $t = 0$. We form the infinite series

$$u(r, t) = \sum_{m=1}^{\infty} u_m(r, t)$$

Setting $t = 0$,

$$u(r, 0) = \sum_{m=1}^{\infty} u_m(r, 0) \equiv f(r)$$

$$= \sum_{m=1}^{\infty} a_m J_0(k_m r)$$

Therefore, the a_m must be chosen to be the Fourier-Bessel coefficients of the function f with respect to the orthogonal set $\{J_0(k_m r)\}_{m=1}^{\infty}$, that is,

$$a_m = \frac{2}{R_1^2 J_1^2(k_m R_1)} \int_0^{R_1} r f(r) J_0(k_m r)\, dr$$

The coefficients b_m are obtained similarly from $g(r)$.

This example shows the strikingly similar roles played by the trigonometric functions and the Bessel functions in the solution of the wave equation. The trigonometric differential equation arises in the case of rectangular Cartesian coordinates. Bessel's equation, on the other hand, arises naturally when cylindrical coordinates are used.

Exercises for Sections 9.14 and 9.15

1. Find the Fourier-Bessel expansion of $f(x) = 1$ over the interval $[0, 3]$ in terms of Bessel functions of order 1.
2. Find the Fourier-Bessel expansion of $f(x) = x^2$ over the interval $[0, 1]$ in terms of Bessel functions of order 0.
3. Find the Fourier-Bessel expansion of $f(x) = 1 - x^2$ over the interval $[0, 1]$ in terms of Bessel functions of order 0.

4. Represent $f(x) = x$ on $[0, 2]$ by a Fourier-Bessel series involving Bessel functions of order 3.

5. Make a sketch of $J_0(\alpha_{01}x)$ and $J_0(\alpha_{02}x)$ where α_{01} and α_{02} are the first two zeros of the Bessel function of order 0.

6. Make a sketch of $J_1(\alpha_{11}x)$ and $J_1(\alpha_{12}x)$ where α_{11} and α_{12} are the first two nonzero zeros of the Bessel function of order one.

7. Find the Fourier-Bessel representation for the solution $u(r, t)$ of the vibrating membrane problem of Section 9.15 in the case when $u(R, t) \equiv 0$, $u(r, 0) \equiv 1 - r/R$, $u_t(r, 0) \equiv 0$.

8. Carry out the explicit solution for the vibrating membrane of unit radius when $u(1, t) \equiv 0$, $u(r, 0) \equiv 0$, $u_t(r, 0) \equiv r^2$.

9. Separate variables in Laplace's equation in three dimensions under the assumption of cylindrical symmetry:

$$u_{rr} + \frac{1}{r}u_r + u_{zz} = 0$$

Show that a Fourier-Bessel expansion for $u(r, z)$, which is valid in the half-space $0 \le r \le R$, $z \ge 0$, can be derived in the case that

$$u(0, z) \text{ finite.} \qquad u(r, z) \to 0 \quad \text{as} \quad z \to \infty$$
$$u(R, z) \equiv 0, \qquad u(r, 0) \equiv f(r)$$

(Note the difference between this application and the one considered in Exercises 22 and 23, p. 339.)

10. Determine the series representation for the solution $u(r, z)$ of Laplace's equation as outlined in Exercise 9 when $u(r, 0) \equiv (r/R)^2 - 1$.

11. Make appropriate alterations in the analysis of Exercise 9 in order to obtain a representation for the solution of the following steady-state temperature problem:

$$u_{rr} + \frac{1}{r}u_r + u_{zz} = 0, \qquad 0 < r < R, \quad 0 < z < 1$$

$$u(R, z) \equiv 0$$
$$u_z(r, 0) \equiv 0$$
$$u(r, 1) \equiv g(r)$$

12. Determine the series representation for the solution $u(r, z)$ if, in Exercise 11, $u(r, 1) \equiv g(r) \equiv 100°$.

The analysis of Section 9.15 can be easily modified to determine solutions of the heat equation

$$u_{rr} + \frac{1}{r}u_r + \frac{1}{r^2}u_{\theta\theta} = ku_t$$

in a circular disk of radius R when the temperature vanishes at $r = R$. If we seek solutions of the form

$$u(r, \theta, t) = R(r)T(t)\cos n\theta$$

the pertinent ordinary differential equations are

$$T' + \frac{\beta^2}{k} T = 0$$

$$r^2 R'' + rR' + (\beta^2 r^2 - n^2)R = 0$$

The appropriate infinite series representation is thus

$$u(r, \theta, t) = \cos n\theta \sum_{m=1}^{\infty} a_m J_n(\beta_m r) e^{-(\beta_m^2/k)t}$$

where $\beta_m = \dfrac{\alpha_{nm}}{R}$ in terms of the zeros α_{nm} of $J_n(x)$.

13. Find the series representation for the solution of the two-dimensional heat equation when $u(1, \theta, t) \equiv 0$ and $u(r, \theta, 0) \equiv r^2$.

14. Determine the time-dependent temperature distribution in the circular disk of unit radius when

$$u(1, \theta, t) \equiv 0 \quad \text{and} \quad u(r, \theta, 0) \equiv 2\sin^2 \theta$$

15. Separate variables in the three-dimensional wave equation in cylindrical coordinates,

$$u_{rr} + \frac{1}{r} u_r + \frac{1}{r^2} u_{\theta\theta} + u_{zz} = \frac{1}{c^2} u_{tt}$$

and exhibit product solutions applicable in axisymmetric (θ-independent) situations.

16. Find solutions $u(\rho, t)$ of the three-dimensional wave equation in spherical coordinates

$$u_{\rho\rho} + \frac{2}{\rho} u_\rho = \frac{1}{c^2} u_{tt}$$

that are of the form $u(\rho, t) = (v(\rho)/\sqrt{\rho}) \cos \omega ct$.

9.16 Looking Ahead

In this chapter our attention was focused primarily on Legendre polynomials and Bessel functions of the first kind and their utility in applied problem solving. The classic advanced references on these functions are [10] and [19], respectively.

For annular planar regions or coaxial cylinders, the Bessel function of the second kind, or Neumann function, becomes of increased importance. So also does the Legendre function of the second kind for problems posed in spherical wedges. When the geometry is elliptical or parabolic, completely new special functions arise. In wave diffraction applications, Fresnel integrals appear; Hermite and Laguerre functions are important in various quantum-mechanical considerations; Chebyshev functions play a role in problems as seemingly diverse as microwave filter design and minimum-norm function approximation; Airy functions are among the fundamental building blocks in modern asymptotics. Excellent surveys of these and other special functions are contained in [4] and [9]. (See also [11], [15] and [17].) The renowned monograph [18] concentrates on the particular case of orthogonal polynomials.

In many applications, numerical approximations to certain special functions are needed. Moreover, it often occurs in such situations that truncation of a known power series or Frobenius expansion will yield results of insufficient accuracy. Although additional computational routines based upon more sophisticated approaches are continuously becoming available through IMSL, the FUNPACK effort at Argonne Laboratories, and the work of the NAG contributors in Great Britain, for example, there still remains a substantial amount of art rather than science in this area. Some valuable sourcebooks on both the theoretical and algorithmic levels are [12], [13], and [14].

On occasion an expression which is valid for large values of an independent variable, or parameter, is particularly useful for numerical approximation purposes. An oft-used example is Stirling's formula for the gamma function:

$$\Gamma(x+1) \approx \sqrt{2\pi x}\, e^{x \ln x - x}, \qquad x \to \infty$$

which we exploited in the exercises for Section 9.2. Another example, involving Bessel functions of large argument, was discussed in Section 9.12. The theory that makes possible such asymptotic expressions is detailed in references such as [6], [7], and [8]. Of particular relevance to special functions is [16].

Basic References

[1] Churchill, R. V., and J. W. Brown. 1978. *Fourier Series and Boundary Value Problems*. 3d ed. New York: McGraw-Hill.

[2] Rainville, E. D. 1960. *Special Functions*. New York: Macmillan.

[3] Spiegel, M. R. 1981. *Applied Differential Equations*. 3d ed. Englewood Cliffs, N. J.: Prentice-Hall.

More-Advanced References

[4] Abramowitz, M., and I. A. Stegun. 1965. *Handbook of Mathematical Functions*. New York, Dover; NBS Applied Math Series 55. 1964. Washington, D.C.: U. S. Department of Commerce.

[5] Bender, C. M., and S. A. Orszag. 1978. *Advanced Mathematical Methods for Scientists and Engineers*. New York: McGraw-Hill.

[6] Bleistein, N., and R. A. Handelsman. 1975. *Asymptotic Expansion of Integrals*. New York: Holt, Rinehart, and Winston.

[7] Dingle, R. B. 1973. *Asymptotic Expansions: Their Derivation and Interpretation*. London: Academic Press.

[8] Erdélyi, A. 1961. *Asymptotic Expansions*. New York: Dover.

[9] Erdélyi, A., et al. 1953. *Higher Transcendental Functions*. 3 vols. New York: McGraw-Hill.

[10] Hobson, E. W. 1955. *The Theory of Spherical and Ellipsoidal Harmonics*. New York: Chelsea.

[11] Jahnke, E., and F. Emde. 1945. *Tables of Functions with Formulae and Curves*. 4th ed. New York: Dover.

[12] Luke, Y. L. 1969. *The Special Functions and Their Approximations.* 2 vols. New York: Academic Press.

[13] Luke, Y. L. 1975. *Mathematical Functions and Their Approximations.* New York: Academic Press.

[14] Luke, Y. L. 1979. *Algorithms for the Computation of Mathematical Functions.* New York: Academic Press.

[15] Magnus, W., and F. Oberhettinger. 1949. *Formulas and Theorems for the Functions of Mathematical Physics.* New York: Chelsea.

[16] Olver, F. W. J. 1974. *Asymptotics and Special Functions.* New York: Academic Press.

[17] Sneddon, I. N. 1980. *Special Functions of Mathematics, Physics, and Chemistry.* 3d ed. New York: Longman.

[18] Szego, G. 1982. *Orthogonal Polynomials.* Colloquium Publications, vol. 23, 4th ed. Providence, R. I.: American Mathematical Society.

[19] Watson, G. N. 1958. *A Treatise on the Theory of Bessel Functions.* 2d ed. Cambridge: Cambridge University Press.

Chapter 10

Vector Calculus: Differentiation

10.1 Introduction

Physical quantities such as force, velocity, and electromagnetic field intensity are usually expressed as vectors. In Chapter 1 we discussed the algebraic manipulation of such vector quantities. When the vectors are not constant, however, but depend on parameters such as time and position, an entirely new dimension of analysis is available—vector calculus. Since the subject is so broad, we divide our considerations between this chapter and the next. Here we concentrate on taking derivatives of vector quantities—operations that help us analyze fundamental problems in mechanics, electromagnetics, and fluid dynamics. In Chapter 11 we explore the equally important notions of line, surface, and volume integrals of vector functions.

10.2 Vector Functions and Differentiation

Recall that a function, also called a mapping or a correspondence, from one set to another is a pairing such that each element X of the first set (the domain) is coupled with exactly one element of the second set (the range). The second element is denoted variously by $f(X)$, $G(X)$, and so on. So far in this text, we have concentrated on real functions where both the first and second sets are sets of real numbers. More generally, any set whatsoever could be used as a domain or a range. If the domain is a set of geometric vectors with initial points at the origin and the range is a set of real numbers, the function is called a *real-valued function*, or a *scalar function, of a vector variable*. Such a function could, for example, describe the temperature at each point in the atmosphere. Another common sort of function has a set of real numbers as domain and a set of vectors in three-space as range. Such a *vector-valued function* of one real variable might describe the velocity $\mathbf{V}(t)$ of a rocket at any time t.

In three dimensions a vector function $\mathbf{F}$ of one real variable may be conveniently specified in the form of an equation

$$\mathbf{F}(t) = f_1(t)\mathbf{i} + f_2(t)\mathbf{j} + f_3(t)\mathbf{k}$$

This one vector equation is equivalent to the three *parametric* equations $x = f_1(t)$, $y = f_2(t)$, $z = f_3(t)$ describing the components of $\mathbf{F}(t)$. We will often interpret $\mathbf{F}$ geometrically as mapping a point t from a real *parameter line* into a geometric vector with initial point at the origin and terminal point $(f_1(t), f_2(t), f_3(t))$. As t varies, the tip of this vector traces out a locus in three-space, which, if f_1, f_2, and f_3 are continuous, we recognize intuitively as a curve.

Example 10.1 If $\mathbf{F}(t) = t\cos t\,\mathbf{i} + t\sin t\,\mathbf{j} + t^2\mathbf{k}$, what component functions are equivalent to this function? Sketch the locus of points formed by $\mathbf{F}(t)$ as t varies.

Solution The components are described by $x = t\cos t$, $y = t\sin t$, $z = t^2$. The locus of points formed by (x, y, z) lies on the paraboloid $z = x^2 + y^2$. Representative points on the curve include $(0, 0, 0)$, $(0, \pi/2, \pi^2/4)$, $(-\pi, 0, \pi^2)$, $(0, -3\pi/2, 9\pi^2/4)$, and $(2\pi, 0, 4\pi^2)$. (See Figure 10.1.) □

A vector function of two real variables is a mapping from a set of ordered pairs of real numbers to a set of vectors and, in three dimensions, is specified by an equation of the type

$$\mathbf{F}(u, v) = f_1(u, v)\mathbf{i} + f_2(u, v)\mathbf{j} + f_3(u, v)\mathbf{k}$$

If we let $\mathbf{P} = (u, v)$ and $\mathbf{Q} = (x, y, z)$, then the one equation $\mathbf{Q} = \mathbf{F}(\mathbf{P})$ is equivalent to the three equations $x = f_1(u, v)$, $y = f_2(u, v)$, and $z = f_3(u, v)$. The three real functions f_1, f_2, and f_3 are completely determined by the vector function and conversely the three functions are sufficient to define a vector function of two real variables. As with vector functions of one variable, the geometric interpretation which most interests us is that of a mapping from the uv-plane (called the *parameter plane*) to a set of geometric vectors whose initial point is the origin.

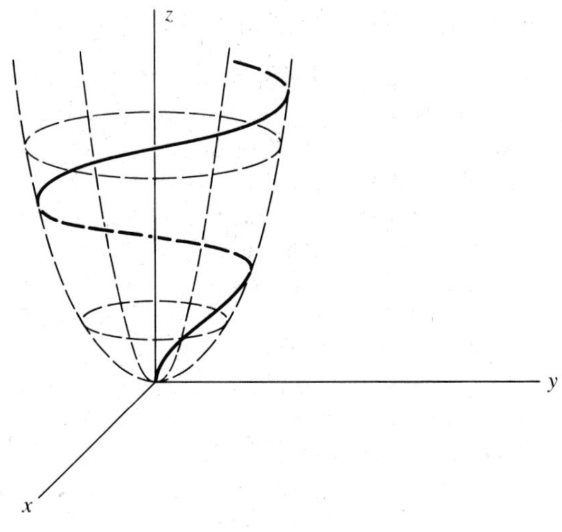

Figure 10.1

Here, as u and v vary, the tip of these vectors $(f_1(u, v), f_2(u, v), f_3(u, v))$ traces out a three-dimensional surface.

Although they are not of interest to us in this text, there are natural generalizations of these notions to vector functions of n real variables.

In what follows, we reluctantly conform to the careless, but almost universal, practice of using the same letter for a function as for the dependent variable determined by the function, that is, $x = x(t)$, $x = x(u, v)$, $y = y(u, v)$, and so on.

The concepts of limit, continuity, and differentiability for vector functions are introduced in much the same manner as they are for real functions.

Definition 10.1 A vector function $\mathbf{F}$ of a real variable t is said to have a *limit* $\mathbf{L}$ as t approaches t_0 if the difference

$$|\mathbf{L} - \mathbf{F}(t)|$$

is arbitrarily small whenever t is sufficiently close to, but not equal to, t_0. We write $\lim_{t \to t_0} \mathbf{F}(t) = \mathbf{L}$. (See Figure 10.2.)

Definition 10.2 A vector function $\mathbf{F}$ of a real variable t is said to be *continuous* at t_0 if

$$\lim_{t \to t_0} \mathbf{F}(t) = \mathbf{F}(t_0)$$

where $\mathbf{F}(t_0)$ is presumed to be defined.

Definition 10.3 A vector function $\mathbf{F}$ of a real variable t is said to have a *derivative* at t_0 (or to be *differentiable* at t_0) if

$$\lim_{t \to t_0} \frac{\mathbf{F}(t) - \mathbf{F}(t_0)}{t - t_0}$$

exists. If this limit does exist it is denoted by $\mathbf{F}'(t_0)$.

As we discuss more fully in Section 10.5, the direction of the vector $\mathbf{F}'(t)$ is tangent to the curve described by $\mathbf{F}(t)$, a fact that has substantial theoretical and practical implications.

These definitions, as well as their geometrical and physical interpretations,

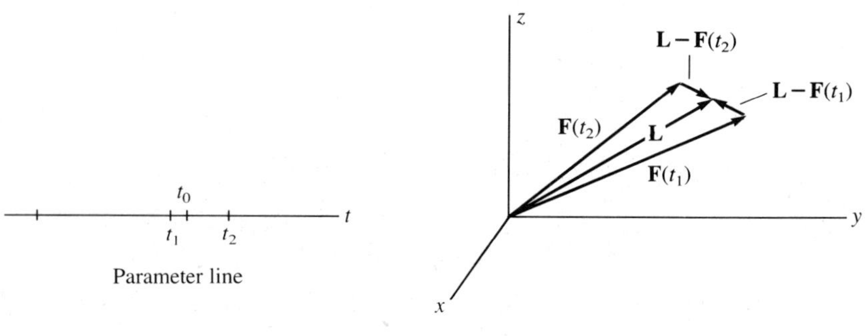

Figure 10.2

are facilitated by use of vector notation. On the other hand, specific calculations often focus on the individual components of the vectors involved, in view of the following theorem.

Theorem 10.1 Let $\mathbf{F}(t) = x(t)\mathbf{i} + y(t)\mathbf{j} + z(t)\mathbf{k}$. Then

(a) $\lim_{t\to t_0} \mathbf{F}(t)$ exists *if and only if* each of the limits $\lim_{t\to t_0} x(t)$, $\lim_{t\to t_0} y(t)$, and $\lim_{t\to t_0} z(t)$ exists. If $\lim_{t\to t_0} \mathbf{F}(t)$ exists, then

$$\lim_{t\to t_0} \mathbf{F}(t) = \left(\lim_{t\to t_0} x(t)\right)\mathbf{i} + \left(\lim_{t\to t_0} y(t)\right)\mathbf{j} + \left(\lim_{t\to t_0} z(t)\right)\mathbf{k}$$

(b) $\mathbf{F}$ is continuous at t_0 *if and only if* each of the component functions is continuous at t_0.

(c) $\mathbf{F}'(t_0)$ exists *if and only if* $x'(t_0)$, $y'(t_0)$, and $z'(t_0)$ exist. If $\mathbf{F}'(t_0)$ exists then

$$\mathbf{F}'(t_0) = x'(t_0)\mathbf{i} + y'(t_0)\mathbf{j} + z'(t_0)\mathbf{k}$$

Proof See the exercises. ■

Definitions and theorems similar to the above hold for vector functions of two or more real variables.

Example 10.2 Let $\mathbf{F}(t) = t^2\mathbf{i} + t\mathbf{j} + t^3\mathbf{k}$. Find $\mathbf{F}'(t)$.

Solution Differentiating each of the components by well-known rules from elementary calculus we obtain

$$\mathbf{F}'(t) = 2t\mathbf{i} + \mathbf{j} + 3t^2\mathbf{k}$$ □

Example 10.3 Find $\mathbf{r}_u \times \mathbf{r}_v$ if $\mathbf{r}(u, v) = (2 + \cos v)\cos u\mathbf{i} + (2 + \cos v)\mathbf{j}$.

Solution

$$\mathbf{r}_u = \frac{\partial \mathbf{r}}{\partial u} = -(2 + \cos v)\sin u\mathbf{i}$$

$$\mathbf{r}_v = \frac{\partial \mathbf{r}}{\partial v} = (-\sin v \cos u)\mathbf{i} - (\sin v)\mathbf{j}$$

$$\begin{aligned} \mathbf{r}_u \times \mathbf{r}_v &= \det\begin{pmatrix} \mathbf{i} & \mathbf{j} & \mathbf{k} \\ -(2 + \cos v)\sin u & 0 & 0 \\ -\sin v \cos u & -\sin v & 0 \end{pmatrix} \\ &= (\sin u \sin v)(2 + \cos v)\mathbf{k} \end{aligned}$$ □

The usual composite function rules hold and, in addition, it is easy to show that if $\mathbf{F}$ and $\mathbf{G}$ are differentiable vector functions of t,

$$(\mathbf{F}\cdot\mathbf{G})' = \mathbf{F}'\cdot\mathbf{G} + \mathbf{F}\cdot\mathbf{G}' \quad \text{and} \quad (\mathbf{F}\times\mathbf{G})' = \mathbf{F}\times\mathbf{G}' + \mathbf{F}'\times\mathbf{G}$$

The order of the terms is, of course, important in the second expression.

Example 10.4 Let $\mathbf{F}(t)$ be a vector function of constant magnitude. Show that $\mathbf{F}(t)$ and $\mathbf{F}'(t)$ are orthogonal for all values of t.

Solution Since $|\mathbf{F}(t)| = c$, it follows that $\mathbf{F}(t)\cdot\mathbf{F}(t) = c^2$. From the general rule for differentiating a dot product, $(\mathbf{F}\cdot\mathbf{F})' = 2\mathbf{F}\cdot\mathbf{F}'$. Hence, differentiating both sides of the equality $\mathbf{F}(t)\cdot\mathbf{F}(t) = c^2$, we obtain $2\mathbf{F}(t)\cdot\mathbf{F}'(t) = 0$. This shows that $\mathbf{F}(t)$ and $\mathbf{F}'(t)$ are orthogonal. □

Exercises for Section 10.2

1. Prove parts (a), (b), and (c) of Theorem 10.1.
2. What vector functions are equivalent to the following sets of parametric equations?
   (a) $x = \sin t,\ y = 2\cos t,\ z = 3$ **(b)** $x = t,\ y = t,\ z = t^2$
   In each case sketch the locus of points formed by (x, y, z) as t varies.
3. What vector function of two real variables is equivalent to the parametric equations $x = \cos u \sin v,\ y = \sin u \sin v,\ z = \cos v$? Is the function periodic? In what way?
4. Find $\mathbf{F}'(t)$ if $\mathbf{F}(t) = (\sin t)\mathbf{i} + (\cos^2 t)\mathbf{j} + t^2\mathbf{k}$.
5. Find an $\mathbf{F}(t)$ such that $\mathbf{F}'(t) = t\mathbf{i} + t^2\mathbf{j} + t^3\mathbf{k}$.
6. Show that if $\mathbf{F}$ and $\mathbf{G}$ are differentiable vector functions of t, then $(\mathbf{F}\cdot\mathbf{G})' = \mathbf{F}\cdot\mathbf{G}' + \mathbf{F}'\cdot\mathbf{G}$.
7. Show that if $\mathbf{F}$ and $\mathbf{G}$ are differentiable vector functions of t, then $(\mathbf{F}\times\mathbf{G})' = \mathbf{F}'\times\mathbf{G} + \mathbf{F}\times\mathbf{G}'$.
8. Find and prove a formula for the derivative of the scalar triple product of three differentiable vector functions of t.
9. Compute $|\mathbf{r}_u \times \mathbf{r}_v|$ when $\mathbf{r}(u, v) = (2\cos u)\mathbf{i} + (2\sin u)\mathbf{j} + v\mathbf{k}$.
10. Compute $|\mathbf{r}_u \times \mathbf{r}_v|$ when $\mathbf{r}(u, v) = x(u, v)\mathbf{i} + y(u, v)\mathbf{j} + z(u, v)\mathbf{k}$.

Netwon's famous second law is actually a vector relationship. If we describe the position of a particle of mass m by the vector function $\mathbf{r}(t)$, then its vector velocity and acceleration are given by $\mathbf{v}(t) \equiv \mathbf{r}'(t)$ and $\mathbf{a}(t) \equiv \mathbf{r}''(t)$, respectively, and Newton's law takes the form

$$\mathbf{F}(t) = m\mathbf{a}(t) = m\mathbf{r}''(t)$$

The following exercises explore some of the implications of this vector expression.

11. In rectilinear motion, $\mathbf{r}(t) = \mathbf{a}f(t) + \mathbf{b}$ where $f(t)$ is a scalar function and $\mathbf{a}$ and $\mathbf{b}$ are constant vectors. Show that the velocity, acceleration, and hence the force vector given by Newton's law are all collinear in this case.
12. The equation $\mathbf{r}(t) = a\cos\omega t\mathbf{i} + a\sin\omega t\mathbf{j}$ with constant ω describes planar motion of a particle along a circle of radius a centered at the origin. Demonstrate that for all values of t:
    (a) $\mathbf{v}(t)$ and $\mathbf{r}(t)$ are orthogonal. **(b)** $\mathbf{a}(t)$ and $\mathbf{v}(t)$ are orthogonal.
    (c) The (centripetal) force given by Newton's law is constant in magnitude and directed toward the origin.
13. Discuss the motion of a particle of mass m which is following a path given by

$$\mathbf{r}(t) = \cosh\omega t(\cos\omega t\mathbf{i} + \sin\omega t\mathbf{j})$$

where ω is constant. In particular, show that the force given by Netwon's law is orthogonal to $\mathbf{r}$ for all t.

14. Describe the motion of a particle of mass m moving under the influence of a constant force. Assume $\mathbf{r}(0) = \mathbf{r}_0$ and $\mathbf{r}'(0) = \mathbf{v}_0$.

15. Solve the vector differential equation $\mathbf{r}'' = -\omega^2\mathbf{r}$, subject to the initial conditions $\mathbf{r}(0) = \mathbf{a}$ and $\mathbf{r}'(0) = \mathbf{b}$. Discuss the motion of a particle modeled by this equation.

10.3 Surfaces

For us, one of the primary applications of vector functions is the representation of surfaces and curves in three dimensions. Now is the time, therefore, for a review of everything from elementary calculus about these concepts. As an aid to refreshing and sharpening your skill in sketching, we have included a number of simple surface and curve sketching problems among the exercises.

Loosely speaking, a *surface* in three-space is a set of points obtained as the image of a mapping from a plane to three-dimensional Euclidean space. We can describe such a set (1) explicitly as $\{(x, y, z) \ 1 \ z = f(x, y)\}$, (2) implicitly as $\{(x, y, z) \ 1 \ f(x, y, z) = c\}$, and (3) vectorially or parametrically as $\{\mathbf{r} \ 1 \ \mathbf{r} = x(u, v)\mathbf{i} + y(u, v)\mathbf{j} + z(u, v)\mathbf{k}\}$. The planar surface given explicitly by the equation $z = f(x, y) = 1 - x - y$ and the sphere defined implicitly by $f(x, y, z) = x^2 + y^2 + z^2 = 4$ are elementary examples of the first two methods of representation. (See the sketches of these surfaces in Figures 10.3 and 10.4.)

The third style of description that uses vector functions of two variables to represent surfaces is quite natural. As suggested earlier, we merely think of the range of the vector function as consisting of position vectors (initial point at the origin). Then the loci of the tips of these vectors describe a surface. Although the vector form of a surface is particulary advantageous for much of our later analysis, it is not always the best for making a sketch. Sometimes we "eliminate the parameters" to obtain a relationship between x, y, and z.

Example 10.5 Identify the surface defined vectorially by

$$\mathbf{r}(u, v) = u\mathbf{i} + v\mathbf{j} + (2 - u + v)\mathbf{k}$$

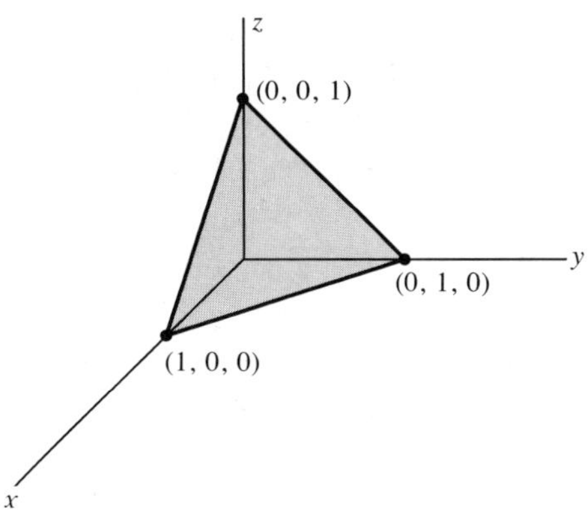

Figure 10.3 $f(x, y) = 1 - x - y$

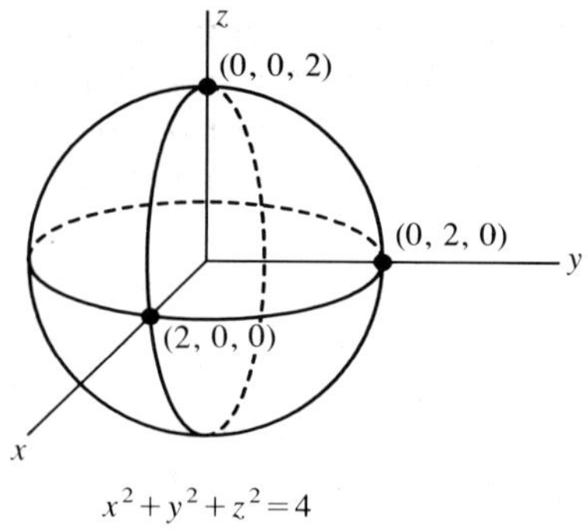

Figure 10.4 $x^2 + y^2 + z^2 = 4$

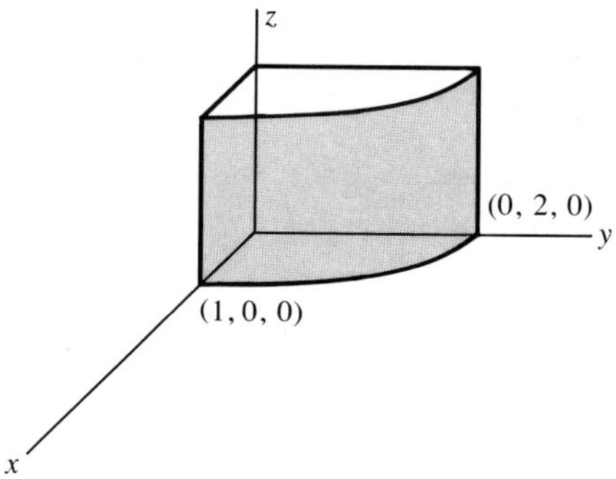

Figure 10.5 $\mathbf{r}(u, v) = (\cos u)\mathbf{i} + (2 \sin u)\mathbf{j} + v\mathbf{k}$

Solution The vector equation is equivalent to the three equations $x = u$, $y = v$, and $z = 2 - u + v$, or $z = 2 - x + y$ which is easily identifiable as a plane. A normal vector to the plane at each point on the plane is $\mathbf{i} - \mathbf{j} + \mathbf{k}$. □

Example 10.6 Identify and sketch the surface defined by the vector function of two variables,

$$\mathbf{r}(u, v) = (\cos u)\mathbf{i} + (2 \sin u)\mathbf{j} + v\mathbf{k}$$

Solution The vector function is equivalent to $x = \cos u$, $y = 2 \sin u$, $z = v$. Elimination of the parameter u from the first two equations is accomplished by dividing the second one by 2 and then squaring both and adding to obtain

$$x^2 + \frac{y^2}{4} = 1$$

which is an elliptic cylinder. Part of this surface is sketched in Figure 10.5. □

One must pay careful attention to domains and ranges when eliminating parameters. For instance, the vector function

$$\mathbf{r}(u, v) = e^u\mathbf{i} + v\mathbf{j} + (e^{2u} + v^2)\mathbf{k}$$

may appear to represent the paraboloid $z = x^2 + y^2$. However, $x = e^u \geq 0$ for all u so $\mathbf{r}$ represents only that part of the paraboloid in the "front" half of our usual graphical representation of three-space.

Without some restrictions on the continuity and differentiability of the functions involved, our "definition" of a surface as the image of a mapping from a plane to three-space would neither allow for the use of calculus techniques nor correspond to our intuitive idea of a surface. For instance, there are no ordered triples (x, y, z) of real numbers satisfying $f(x, y, z) = 0$, where $f(x, y, z) = 1 + x^2 + y^2 + z^2$, and there exist *continuous* functions $\mathbf{r}(u, v)$ with an image that fills a *solid* portion of three-space. We adopt a definition of *smooth surface* which will eliminate such difficulties and be adequate for our purposes.

Definition 10.4 A function $\mathbf{r}(u, v) = x(u, v)\mathbf{i} + y(u, v)\mathbf{j} + z(u, v)\mathbf{k}$ with domain D is said to describe a *smooth surface* in three dimensions if (1) each of the functions x, y, z and each of their first partial derivatives with respect to u and v are continuous over D and (2) the vector $\mathbf{r}_u \times \mathbf{r}_v$ is nonvanishing over D.

The second condition in this definition is equivalent to the nonvanishing, simultaneously, of the three Jacobians

$$\frac{\partial(x, y)}{\partial(u, v)} = \det\begin{pmatrix} \dfrac{\partial x}{\partial u} & \dfrac{\partial y}{\partial u} \\[2ex] \dfrac{\partial x}{\partial v} & \dfrac{\partial y}{\partial v} \end{pmatrix}, \qquad \frac{\partial(y, z)}{\partial(u, v)} = \det\begin{pmatrix} \dfrac{\partial y}{\partial u} & \dfrac{\partial z}{\partial u} \\[2ex] \dfrac{\partial y}{\partial v} & \dfrac{\partial z}{\partial v} \end{pmatrix}$$

and

$$\frac{\partial(x, z)}{\partial(u, v)} = \det\begin{pmatrix} \dfrac{\partial x}{\partial u} & \dfrac{\partial z}{\partial u} \\[2ex] \dfrac{\partial x}{\partial v} & \dfrac{\partial z}{\partial v} \end{pmatrix}$$

(In this form the extension to smooth surfaces in n-dimensions is readily apparent.) The condition ensures the existence of a tangent plane at each point of the surface. Moreover, as we shall see in the next chapter, the vector $\mathbf{r}_u \times \mathbf{r}_v$ is central to a convenient formulation of surface area.

Expressions for surfaces in the explicit and implicit form can be parametrized in such a manner that this definition can be used. For instance, the "top half" of the surface $x^2 + y^2 + z^2 = 4$ can be expressed in the form $z = f(x, y) = \sqrt{4 - x^2 - y^2}$ and *any* such explicit representation $z = f(x, y)$ can be parametrized by taking $x = u$, $y = v$, and $z = f(u, v)$, yielding the equivalent vector representation

$$\mathbf{r}(u, v) = u\mathbf{i} + v\mathbf{j} + f(u, v)\mathbf{k}$$

10.4 Curves

In three-space, a casual definition of a curve would be "a set of points obtained as the image of a mapping from a line to three-dimensional space." As with

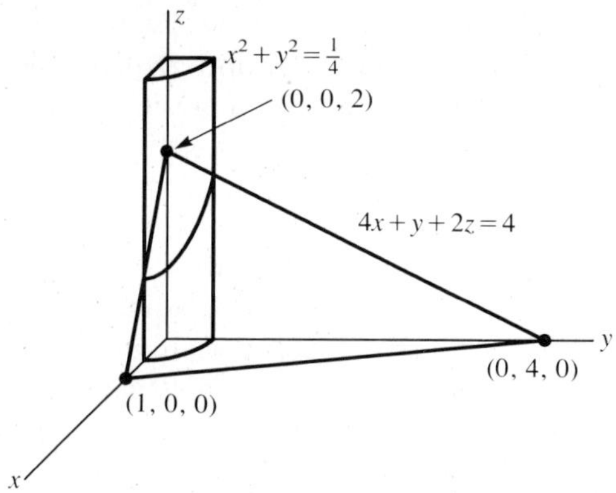

Figure 10.6

surfaces, curves can be described in several ways. For instance, from elementary calculus we recall the definition of a curve as the intersection of two surfaces. Such a method of representation is particularly useful if you are called upon to sketch the curve, for then sometimes it is easier to get an idea of precisely "where the curve is" in space.

Example 10.7 Sketch the curve given by the intersection of $4x + y + 2z = 4$ and $x^2 + y^2 = 1/4$.

Solution $4x + y + 2z = 4$ is a plane and $x^2 + y^2 = 1/4$ is a right circular cylinder whose axis is the z-axis. The first octant sketch of each is shown in Figure 10.6 along with the curve of intersection. □

This example notwithstanding, from an applied analysis point of view it is far more advantageous to use the alternative method of representation for a curve already hinted at in Section 10.2, namely, by means of a vector function of one real variable. If you understand how a vector function of two variables can be used to represent a surface, then it is an easy transition to use vector representations for curves as well. We begin our considerations with some definitions, a few examples, and a discussion of some related terminology. Since it is the area of greatest application, we restrict our attention to three-dimensional vector functions.

Definition 10.5 A function $\mathbf{r}(t) = x(t)\mathbf{i} + y(t)\mathbf{j} + z(t)\mathbf{k}$ defined on some interval $[a, b]$ is said to describe a *smooth curve* if $\mathbf{r}'(t)$ exists, is nonzero, and is continuous on $[a, b]$.

Definition 10.6 A function $\mathbf{r}(t) = x(t)\mathbf{i} + y(t)\mathbf{j} + z(t)\mathbf{k}$ defined on $[a, b]$ is said to describe a *piecewise smooth curve* if $\mathbf{r}(t)$ is continuous on $[a, b]$ and $[a, b]$ can be subdivided into a finite number of closed intervals $[a_i, b_i]$ such that on each (a_i, b_i), $\mathbf{r}'(t)$ exists, is nonzero, and is continuous.

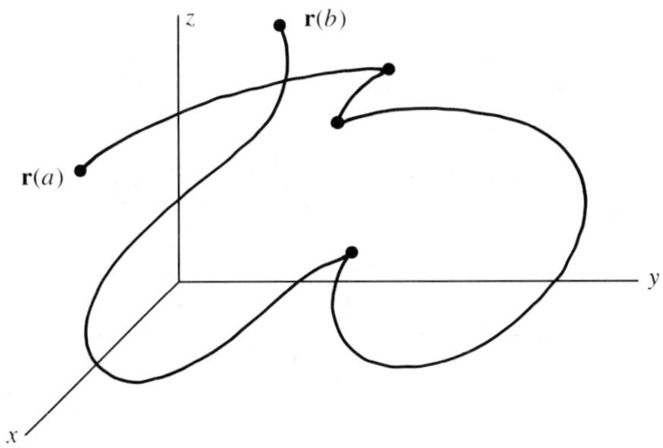

Figure 10.7

A typical piecewise smooth curve is shown in Figure 10.7. The $\mathbf{r}'(t) \neq 0$ restrictions in the preceding two definitions ensure the existence of a continuously turning tangent vector to the given curve, as we shall see in the next section.

Example 10.8 Sketch the plane curve $\mathbf{r}(t) = \sin t\mathbf{i} + \sin t\mathbf{j}$.

Solution The parametric equations are $x(t) = \sin t$ and $y(t) = \sin t$, from which $x = y$. A careless person could conclude that the curve is the entire line $x = y$. However, the parametrization limits x and y between -1 and 1. Thus only a portion of the line $x = y$ is represented by $\mathbf{r}(t)$. □

Example 10.9 Sketch the first octant portion of the curve described vectorially by $\mathbf{r}(t) = \cos t\mathbf{i} + \cos 2t\mathbf{j} + \sin t\mathbf{k}$.

Solution The vector function is equivalent to the three functions of t: $x(t) = \cos t$, $y(t) = \cos 2t$, $z(t) = \sin t$. The parameter t may be eliminated between the first and third of these equations by squaring and adding to obtain $x^2 + z^2 = 1$. Since $\cos 2t = 1 - 2\sin^2 t$, the elimination of the parameter from the second and third equations gives $y = 1 - 2z^2$. Thus, the given curve is the intersection of the two cylinders $x^2 + z^2 = 1$ and $y = 1 - 2z^2$. The curve is sketched in Figure 10.8. □

The vector representation of a curve is usually not unique. Any particular representation will *orient* the curve, that is, give a direction in which the curve is traced out as t increases through the interval $[a, b]$.

Example 10.10 Show that the vector functions $\mathbf{r}_1(t) = (\sin t)\mathbf{i} + (\cos t)\mathbf{j}$ and $\mathbf{r}_2(t) = (\cos t)\mathbf{i} + (\sin t)\mathbf{j}$ both represent the same curve but orient it in opposite directions.

Solution By squaring and adding component functions in each case, we see that both functions represent the circle $\{x^2 + y^2 = 1, z = 0\}$. For the first func-

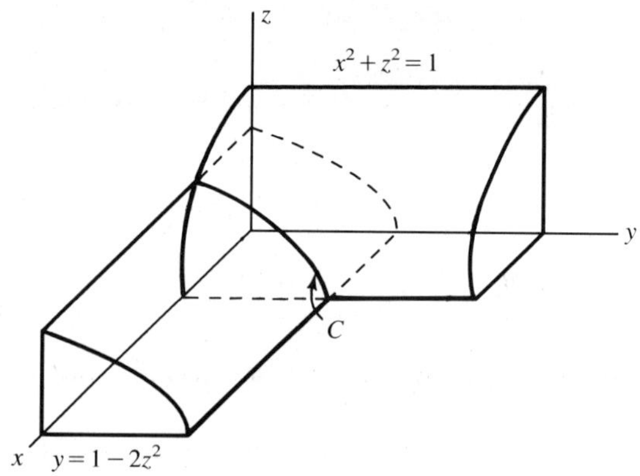

Figure 10.8

tion, as t increases, the tip of the position vector rotates in a clockwise sense, while for the second function increasing t implies counterclockwise rotation of the tip. □

A curve represented by a one-to-one function is said to be *simple.* Note that a simple curve does not intersect itself. If a curve is simple except that its beginning and end points are the same, it is called a *simple closed curve.* A simple portion of a curve defined over some closed subinterval of the domain $[a, b]$ is called an *arc.*

Definition 10.7 A curve is said to be *rectifiable* for $a \leq t \leq b$ and have *arc length* L if the integral

$$L = \int_a^b \left(\frac{d\mathbf{r}}{dt} \cdot \frac{d\mathbf{r}}{dt}\right)^{1/2} dt$$

exists.

It can be shown (much as the comparable result in elementary calculus is demonstrated) that a smooth curve defined on $[a, b]$ is rectifiable. The *arc length function*

$$s(t) = \int_a^t \left(\frac{d\mathbf{r}}{du} \cdot \frac{d\mathbf{r}}{du}\right)^{1/2} du$$

gives the length of the arc of the curve from $\mathbf{r}(a)$ to $\mathbf{r}(t)$. Arc length for a piecewise smooth curve can be found by adding the lengths of the individual smooth arcs of the curve.

Occasionally, we use the arc length s as the parameter in the representation of a curve, as illustrated in the following example.

Example 10.11 Parametrize the curve $\mathbf{r}(t) = 5 \cos t\mathbf{i} + 5 \sin t\mathbf{j}$ in terms of arc length.

Solution This is a circle of radius 5 centered at the origin, and for circles of radius a the relation of angle t to arc length s is known to be $s = at$, or, in this case

$s = 5t$. Therefore,

$$\mathbf{R}(s) = \mathbf{r}(t(s)) = 5\cos\frac{s}{5}\mathbf{i} + 5\sin\frac{s}{5}\mathbf{j} \qquad \square$$

In some situations we may be called upon to develop a vector representation for a curve which is described in alternative fashion. The first example below should be a familiar one from elementary calculus.

Example 10.12 Write a vector equation of a line passing through the points $(-1, 0, 2)$ and $(1, 4, 3)$.

Solution The vector in the direction of the line may be written as $\mathbf{M} = 2\mathbf{i} + 4\mathbf{j} + \mathbf{k}$. Either of the two points may be chosen to give the vector $\mathbf{B}$ from the origin to the point. If $\mathbf{B} = -\mathbf{i} + 2\mathbf{k}$, the vector equation of the line is

$$\begin{aligned}\mathbf{R}(t) = \mathbf{B} + \mathbf{M}t &= -\mathbf{i} + 2\mathbf{k} + (2\mathbf{i} + 4\mathbf{j} + \mathbf{k})t \\ &= (2t - 1)\mathbf{i} + 4t\mathbf{j} + (t + 2)\mathbf{k}\end{aligned}$$

If $\mathbf{B}$ is chosen to be $\mathbf{i} + 4\mathbf{j} + 3\mathbf{k}$, then another equation of the same line (with a different parameter) is

$$\begin{aligned}\mathbf{F}(s) &= \mathbf{i} + 4\mathbf{j} + 3\mathbf{k} + (2\mathbf{i} + 4\mathbf{j} + \mathbf{k})s \\ &= (2s + 1)\mathbf{i} + 4(s + 1)\mathbf{j} + (s + 3)\mathbf{k}\end{aligned}$$

The general construction illustrated by this example is depicted in Figure 10.9.

$\square$

Example 10.13 Write an equation of the circle in the plane $z = 2$, centered at $(0, 0, 2)$, of radius 3, and oriented clockwise as viewed from the origin.

Solution The parametric equations may be written

$$x = 3\sin t, \qquad y = 3\cos t, \qquad z = 2$$

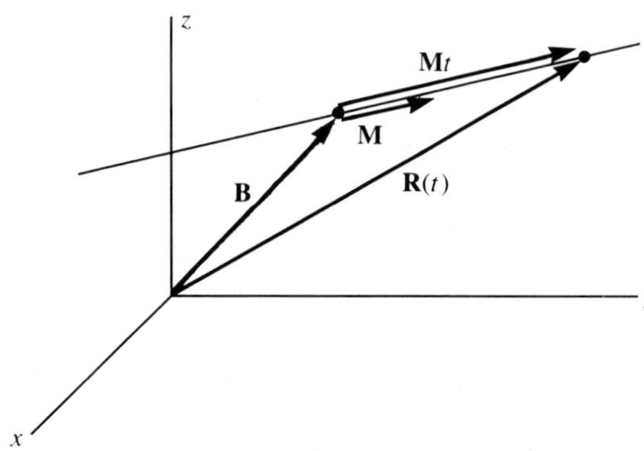

Figure 10.9

and hence a vector function which represents the curve is

$$\mathbf{R}(t) = (3 \sin t)\mathbf{i} + (3 \cos t)\mathbf{j} + 2\mathbf{k}$$ □

Exercises for Sections 10.3 and 10.4

In Exercises 1–4, sketch the specified surfaces:

1. **(a)** $z = x^2 + 3y^2$ **(b)** $y = \sin x$ **(c)** $y = z^2$
2. **(a)** $y^2 + z^2 = x^2$ **(b)** $y = 2$ **(c)** $x + y = 2$
3. **(a)** $x^2 + 4y^2 = 4$ **(b)** $x = y + z$ **(c)** $x^2 - y^2 = 1$
4. **(a)** $z^2 = 4$ **(b)** $z = x^2 + y^2$ **(c)** $x^2 + y^2 = 4 - z$

In Exercises 5–8, sketch the surfaces defined by the given parametric representations:

5. **(a)** $\mathbf{r}(u, v) = u\mathbf{i} + v^2\mathbf{j}$ **(b)** $\mathbf{r}(u, v) = 2u\mathbf{i} + v\mathbf{j} + (1 - 2u - 3v)\mathbf{k}$
6. **(a)** $\mathbf{r}(u, v) = v\mathbf{i} + (u + v - 1)\mathbf{j} + (u - v)\mathbf{k}$
(b) $\mathbf{r}(u, v) = (\cos u \cos v)\mathbf{i} + (\sin u \cos v)\mathbf{j} + (\sin v)\mathbf{k}$
7. **(a)** $\mathbf{r}(u, v) = (u \cos v)\mathbf{i} + (u \sin v)\mathbf{j}$ **(b)** $\mathbf{r}(u, v) = u^2\mathbf{i} + u^2\mathbf{j} + v\mathbf{k}$
8. **(a)** $\mathbf{r}(u, v) = (uv)\mathbf{i} + v\mathbf{j} + 2\mathbf{k}$ **(b)** $\mathbf{r}(u, v) = \cos u \cos v\mathbf{i} + \sin u \sin v\mathbf{j}$

In Exercises 9–11, sketch the curves defined by the given equations:

9. **(a)** $x + y + z = 1$, $y + z = 1$ **(b)** $x^2 + y^2 = 16$, $z = 2$ **(c)** $x^2 + y^2 + z^2 = 4$, $x = y$
10. **(a)** $x^2 + y^2 + z^2 = 4$, $x = 1$ **(b)** $x^2 + y^2 + z^2 = 4$, $x = 3$ **(c)** $x^2 + y^2 = 4$, $z = 3$
11 **(a)** $x^2 + y^2 = 4$, $x = 2$ **(b)** $x^2 = y$, $y = z$ **(c)** $x^2 + y^2 = z$, $x = y$

In Exercises 12–14, sketch the curves defined by the given parametric representations:

12. **(a)** $\mathbf{r}(t) = (\sin t)\mathbf{i} + (\cos t)\mathbf{j} + 2\mathbf{k}$ **(b)** $\mathbf{r}(t) = t^2\mathbf{i} + t\mathbf{j} + t\mathbf{k}$
13. **(a)** $\mathbf{r}(t) = (t)^{1/2}\mathbf{i} + (t)^{1/2}\mathbf{j}$ **(b)** $\mathbf{r}(t) = (2t - 1)\mathbf{i} + t\mathbf{j} + (3t + 1)\mathbf{k}$
14. **(a)** $\mathbf{r}(t) = 2\mathbf{i} + 3t\mathbf{j} + t\mathbf{k}$ **(b)** $\mathbf{r}(t) = t\mathbf{i} + 2t\mathbf{j}$

15. Express the surfaces of Exercises 1–4 in parametric form.

16. Express the curves of Exercises 9–11 in parametric form.

17. Express the curves of Exercises 12–14 as the intersection of two surfaces.

18. Write a vector equation of the line passing through the points $(1, 0, 2)$ and $(5, 1, 1)$.

19. Write a vector equation of the line passing through the points $(1, 0, 0)$ and $(0, 2, 0)$.

20. Write a vector equation of the elliptical helix which lies on a right elliptical cylinder, major semiaxis a, minor semiaxis b, and pitch c units.

21. If two vector functions represent the same line, how do they differ? (See Example 10.12.)

22. Write a vector equation of the parabola $y = x^2$, $z = 0$ oriented from $(2, 4, 0)$ to $(0, 0, 0)$.

23. Write a vector equation of the circle centered at $(0, 2, 2)$, of radius 2, in the plane $y = z$.

24. Write a vector equation of the plane $x - 4y + 2z = 1$.

25. Write a vector equation of the sphere $x^2 + (y - 2)^2 + (z - 3)^2 = 4$.

26. Write a vector equation of the ellipsoid

$$\frac{x^2}{4} + \frac{y^2}{1} + \frac{z^2}{16} = 1$$

10.5 Tangents to Curves

As suggested earlier, it is advantageous to analyze geometrically the derivative of a vector function representing a smooth curve. If $\mathbf{r}(t)$ represents such a curve C, then the tips of the vectors $\mathbf{r}(t_0 + \Delta t)$ and $\mathbf{r}(t_0)$ lie on the curve at points P and Q as shown in Figure 10.10. The difference between the two vectors is thus a secant vector pointing from P to Q. Since division by a positive real number does not affect the direction of a vector, the difference quotient

$$\frac{\mathbf{r}(t_0 + \Delta t) - \mathbf{r}(t_0)}{\Delta t}, \qquad \Delta t > 0$$

is a vector that also has direction from P to Q. As Δt approaches zero, this difference quotient vector approaches a direction tangent to C at P. Hence, if $\mathbf{r}'(t_0) \neq 0$, the direction of $\mathbf{r}'(t_0)$ will be tangential to C at P. The vector

$$\frac{\mathbf{r}'(t_0)}{|\mathbf{r}'(t_0)|}$$

is therefore a unit tangent vector to C, which we will denote by $\mathbf{T}(t_0)$.

The vector $\mathbf{T}(t_0)$ orients C at each point in the direction of increasing t. Notice also that the definition of a smooth curve given in the preceding section ensures that $\mathbf{T}$ exists and is continuous everywhere along C.

Example 10.14 Write a vector equation of the line tangent to the plane curve $\mathbf{r}(t) = \cos t\mathbf{i} + \sin t\mathbf{j}$ when $t = \pi/4$. (See Figure 10.11.)

Solution A vector tangent to C for any value of t is

$$\mathbf{r}'(t) = -\sin t\mathbf{i} + \cos t\mathbf{j}$$

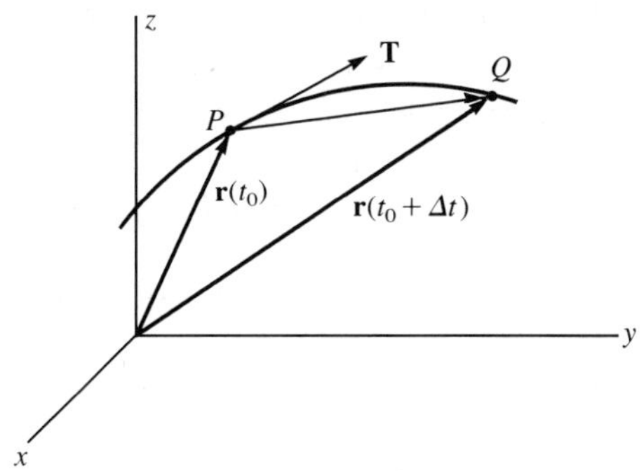

Figure 10.10 Tangent and secant vectors

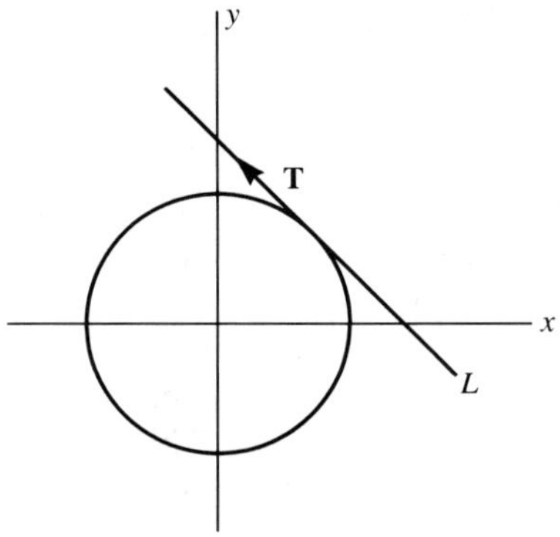

Figure 10.11

and therefore a tangent vector at $\pi/4$ is $-(\sqrt{2}/2)\mathbf{i} + (\sqrt{2}/2)\mathbf{j}$. The point on the curve corresponding to $t = \pi/4$ is $(\sqrt{2}/2, \sqrt{2}/2)$, and the equation of the line is

$$\begin{aligned}\mathbf{q}(w) &= \mathbf{r}\left(\frac{\pi}{4}\right) + \mathbf{r}'\left(\frac{\pi}{4}\right)w \\ &= \left(\frac{\sqrt{2}}{2}\mathbf{i} + \frac{\sqrt{2}}{2}\mathbf{j}\right) + \left(-\frac{\sqrt{2}}{2}\mathbf{i} + \frac{\sqrt{2}}{2}\mathbf{j}\right)w\end{aligned}$$

(We have used the parameter w since t was employed in the description of the curve.) □

Tangent vector calculations are streamlined if C is parametrized by arc length s, that is, C is represented by a vector function $\mathbf{R}(s)$. In this case the difference quotient

$$\frac{\mathbf{R}(s + \Delta s) - \mathbf{R}(s)}{\Delta s}$$

has magnitude given by

$$\frac{\text{chord length}}{\text{arc length}}$$

which approaches 1 as Δs approaches 0. Hence, the unit tangent vector is given merely by

$$\mathbf{T} = \frac{d\mathbf{R}}{ds}$$

This new expression has important practical implications. For example, if C is parametrized by the vector function $\mathbf{r}(t)$, then the arc length is a function of t, and thus using the composite function rule:

$$\frac{d\mathbf{r}}{dt} = \frac{d\mathbf{r}}{ds}\frac{ds}{dt} = \mathbf{T}\frac{ds}{dt}$$

It follows that

$$\left|\frac{d\mathbf{r}}{dt}\right| = \frac{ds}{dt}$$

that is, the magnitude of the first derivative is the first derivative of the arc length function with respect to the parameter t. If t were thought of as time, then ds/dt would be the speed with which the curve is traced out. With this interpretation, $d\mathbf{r}/dt$ at any particular time is a vector in the direction of motion with *magnitude* equal to the speed of that motion. It is natural, then, to call $d\mathbf{r}/dt$ the *velocity* and denote it by $\mathbf{v}(t)$. This idea was mentioned briefly in the exercises on p. 360 and is exploited further in the next section.

In addition to the unit tangent vector, there is another unit vector associated with curve geometry which is of utility in applications. Since $\mathbf{T}(t)$ is of constant magnitude, we have from Example 10.4 that $\mathbf{T}'(t)$ is perpendicular to $\mathbf{T}(t)$ for all t. The unit vector in the direction of $\mathbf{T}'(t)$ we denote by $\mathbf{N}(t)$, that is,

$$\mathbf{N}(t) = \frac{\mathbf{T}'(t)}{|\mathbf{T}'(t)|}$$

This vector is called the *principal normal* to the curve C at each point. Note that since t is an *arbitrary* parameter, $d\mathbf{T}/ds$, for instance, would also be normal to $\mathbf{T}$ and $\mathbf{N}(s) = \mathbf{T}'(s)/|\mathbf{T}'(s)|$.

Example 10.15 Find $\mathbf{T}$ and $\mathbf{N}$ at the point on the curve C corresponding to $t = 2$, for the curve described by

$$\mathbf{r}(t) = t^2\mathbf{i} + 3t\mathbf{j} - t^3\mathbf{k}$$

Solution Since

$$\mathbf{r}(t) = t^2\mathbf{i} + 3t\mathbf{j} - t^3\mathbf{k}$$

we have

$$\mathbf{r}'(t) = 2t\mathbf{i} + 3\mathbf{j} - 3t^2\mathbf{k}$$

so that

$$\mathbf{T}(t) = \frac{\mathbf{r}'(t)}{|\mathbf{r}'(t)|} = \frac{2t\mathbf{i} + 3\mathbf{j} - 3t^2\mathbf{k}}{(4t^2 + 9 + 9t^4)^{1/2}}$$

To compute $\mathbf{N}(t)$ we first find $\mathbf{T}'(t)$:

$$\mathbf{T}'(t) = (2\mathbf{i} - 6t\mathbf{k})(4t^2 + 9 + 9t^4)^{-1/2}$$
$$- \tfrac{1}{2}(4t^2 + 9 + 9t^4)^{-3/2}(8t + 36t^3)(2t\mathbf{i} + 3\mathbf{j} - 3t^2\mathbf{k})$$

Evaluating at $t = 2$,

$$\mathbf{T}(2) = \frac{4\mathbf{i} + 3\mathbf{j} - 12\mathbf{k}}{13}$$

and

$$\mathbf{T}'(2) = \frac{1}{13}\left[2\mathbf{i} - 12\mathbf{k} - \frac{152}{169}(4\mathbf{i} + 3\mathbf{j} - 12\mathbf{k})\right]$$

$$= \frac{1}{13^3}(-270\mathbf{i} - 456\mathbf{j} - 204\mathbf{k})$$

$$= -\frac{6}{13^3}(45\mathbf{i} + 76\mathbf{j} + 34\mathbf{k})$$

from which

$$\mathbf{N}(2) = \frac{\mathbf{T}'(2)}{|\mathbf{T}'(2)|}$$

$$= \frac{1}{\sqrt{8957}}(-45\mathbf{i} - 76\mathbf{j} - 34\mathbf{k}) \qquad \square$$

10.6 Dynamics of a Particle

In the exercises on p. 360, we initiated discussion of particle motion in three-space described by a vector function $\mathbf{r}(t)$ of a single real variable t taken to be time. If $\mathbf{r}(t)$ is continuously differentiable, the particle follows a smooth curve C, and the results of the preceding section allow us to extend significantly our understanding of this practical situation.

The *velocity* and *acceleration* are defined to be the first and second derivatives with respect to time of the vector function:

$$\mathbf{v} \equiv \frac{d\mathbf{r}}{dt} = \frac{dx}{dt}\mathbf{i} + \frac{dy}{dt}\mathbf{j} + \frac{dz}{dt}\mathbf{k}, \qquad \mathbf{a} \equiv \frac{d^2\mathbf{r}}{dt^2} = \frac{d^2x}{dt^2}\mathbf{i} + \frac{d^2y}{dt^2}\mathbf{j} + \frac{d^2z}{dt^2}\mathbf{k}$$

As we have already noted, the velocity vector $\dfrac{d\mathbf{r}}{dt} = \dfrac{ds}{dt}\mathbf{T}$ is tangential to the path of the particle, and the magnitude of $\mathbf{v}$, ds/dt, is the rate of change of arc length, or the *speed*. In terms of components,

$$\frac{ds}{dt} = \left[\left(\frac{dx}{dt}\right)^2 + \left(\frac{dy}{dt}\right)^2 + \left(\frac{dz}{dt}\right)^2\right]^{1/2}$$

To get a better understanding of the vector acceleration, let us start with

$$\mathbf{r}'(t) = \frac{ds}{dt}\mathbf{T}$$

and differentiate using the product rule:

$$\mathbf{a} = \mathbf{r}''(t) = \frac{d^2s}{dt^2}\mathbf{T} + \frac{ds}{dt}\frac{d\mathbf{T}}{dt}$$

But, in terms of the principal normal vector $\mathbf{N}$,

$$\frac{d\mathbf{T}}{dt} = \frac{ds}{dt}\frac{d\mathbf{T}}{ds} = \frac{ds}{dt}\kappa\mathbf{N} \qquad \left(\kappa = \left|\frac{d\mathbf{T}}{ds}\right|\right)$$

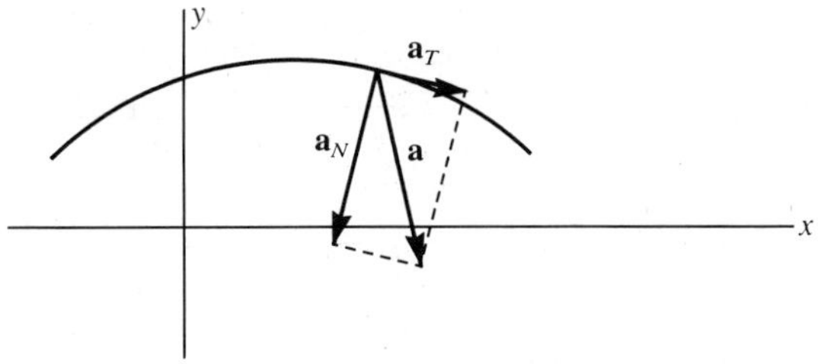

Figure 10.12 Components of acceleration

so the acceleration can be written as a sum of a tangential and a normal component,

$$\mathbf{a} = \frac{d^2s}{dt^2}\mathbf{T} + \left(\frac{ds}{dt}\right)^2 \kappa\mathbf{N} \qquad \left(\kappa = \left|\frac{d\mathbf{T}}{ds}\right|\right)$$
$$= \mathbf{a}_T + \mathbf{a}_N$$

where $\mathbf{a}_T = (\mathbf{a}\cdot\mathbf{T})\mathbf{T}$ and $\mathbf{a}_N = (\mathbf{a}\cdot\mathbf{N})\mathbf{N}$. (See Figure 10.12.) The quantity κ is called the *curvature* and measures, as a function of t, the concavity of the curve described by $\mathbf{r}(t)$. In two dimensions, κ gives rise to the curvature formulas familiar from elementary calculus (see the exercises).

The scalar component of the vector acceleration that is normal to the path of the particle is often called the *centripetal acceleration*, and if this vanishes the value of the acceleration reduces to that obtained for rectilinear motion, d^2s/dt^2. We thus see how it is possible for a particle to move with constant speed and yet have nonzero acceleration.

The quantity $R = 1/\kappa$ is called the *radius of curvature* and, indeed, turns out to be the radius of the circle that, in a natural sense, best approximates the curve at the point in question. In terms of R, the normal component of acceleration can be expressed in a form familiar from elementary physics, namely,

$$\mathbf{a}_N = \frac{v^2\mathbf{N}}{R} \qquad \left(v = |\mathbf{v}| = \frac{ds}{dt}\right)$$

Example 10.16 The path of a particle is described by

$$\mathbf{r}(t) = \cos \pi t\mathbf{i} + \sin \pi t\mathbf{j} + t\mathbf{k}$$

Find the velocity and acceleration vectors at $t = 1$. Also compute the speed, and the unit tangent and unit normal vectors corresponding to that value of t.

Solution Since $\mathbf{r}(t) = \cos \pi t\mathbf{i} + \sin \pi t\mathbf{j} + t\mathbf{k}$,

$$\mathbf{v}(t) = -\pi \sin \pi t\mathbf{i} + \pi \cos \pi t\mathbf{j} + \mathbf{k}$$
$$\mathbf{a}(t) = -\pi^2 \cos \pi t\mathbf{i} - \pi^2 \sin \pi t\mathbf{j}$$

Evaluating at $t = 1$,

$$\mathbf{v}(1) = -\pi\mathbf{j} + \mathbf{k} \quad \text{and} \quad \mathbf{a}(1) = \pi^2\mathbf{i}$$

from which the speed is $\sqrt{1+\pi^2}$. The unit tangent vector

$$\mathbf{T} = \frac{\mathbf{v}(1)}{|\mathbf{v}(1)|} = \frac{-\pi\mathbf{j} + \mathbf{k}}{(\pi^2+1)^{1/2}}$$

There are several ways to obtain $\mathbf{N}$. Since $\mathbf{a}_T = (\mathbf{a}\cdot\mathbf{T})\mathbf{T}$,

$$\mathbf{a}_N = \mathbf{a} - (\mathbf{a}\cdot\mathbf{T})\mathbf{T} = \pi^2\mathbf{i} - 0 = \pi^2\mathbf{i}$$

and hence

$$\mathbf{N} = \frac{\mathbf{a}_N}{|\mathbf{a}_N|} = \mathbf{i}$$

To obtain $\mathbf{N}$, we could also have used the fact that since $\mathbf{T}$ is a constant, $d\mathbf{T}/dt$ points in the direction of $\mathbf{N}$ (see Example 10.4), that is, we could have performed the computation

$$\frac{d}{dt}\left(-\frac{\pi\sin\pi t\mathbf{i} + \pi\cos\pi t\mathbf{j} + \mathbf{k}}{(\pi^2+1)^{1/2}}\right)$$

and evaluated the result at $t = 1$. □

Exercises for Sections 10.5 and 10.6

1. Find a vector equation of the line tangent to the curve $\mathbf{r}(t) = t^2\mathbf{i} + t\mathbf{j} + t^3\mathbf{k}$ at the point $(4, 2, 8)$.
2. Find the unit tangent and unit normal to the curve $\mathbf{r}(t) = t\mathbf{i} + t^2\mathbf{j} + t^3\mathbf{k}$ at $(1, 1, 1)$.
3. Find the unit tangent and unit normal to the curve $\mathbf{r}(t) = (\sin t)\mathbf{i} + (\cos t)\mathbf{j} + t\mathbf{k}$ at $t = 2\pi$.
4. Find a vector equation of the line tangent to the curve of Exercise 3, at $t = 2\pi$.
5. Find the velocity and acceleration at $t = 2$ of a particle that moves according to the law $\mathbf{r}(t) = (\sin^2 \pi t)\mathbf{i} + (\cos \pi t)\mathbf{j} + t^2\mathbf{k}$. What is the speed at that time? What are the tangential and normal vector components of acceleration at that time? Repeat for $t = \frac{1}{4}$.
6. If $\mathbf{r} \times \mathbf{v} \equiv \mathbf{0}$, show that $\mathbf{r}(t)$ has a constant direction.
7. If the curvature $\kappa = 0$ everywhere along a curve $\mathbf{r}(t)$, demonstrate that the curve is a straight line.
8. The velocity vector of a particle is given by $\mathbf{v} = xy\mathbf{i} + y\mathbf{j} + z\mathbf{k}$. Find the path of the particle as a function of time.
9. Solve Exercise 8 except use $\mathbf{v} = t^2\mathbf{i} + t\mathbf{j} - t^3\mathbf{k}$.
10. Solve Exercise 8 except use $\mathbf{v} = -2y\mathbf{i} + 2x\mathbf{j}$.
11. Show that the curvature κ of the plane curve $\mathbf{r}(t) = x(t)\mathbf{i} + y(t)\mathbf{j}$ is given by

$$\kappa = \frac{|x'y'' - x''y'|}{(x'^2 + y'^2)^{3/2}}$$

12. If a planar curve is described in the familiar way by the equation $y = f(x)$, utilize the result of Exercise 11 to obtain a formula for the curvature κ.

For motion of a particle in a plane, we have the option of introducing polar coordinates $x = r\cos\theta$ and $y = r\sin\theta$. In terms of the unit vectors

$$\mathbf{u}_r = \cos\theta\mathbf{i} + \sin\theta\mathbf{j}$$
$$\mathbf{u}_\theta = -\sin\theta\mathbf{i} + \cos\theta\mathbf{j}$$

in the r and θ directions, respectively, the relevant quantities become

$$\mathbf{r} = r\mathbf{u}_r$$
$$\mathbf{v} = \frac{dr}{dt}\mathbf{u}_r + r\frac{d\theta}{dt}\mathbf{u}_\theta$$
$$\mathbf{a} = \left[\frac{d^2r}{dt^2} - r\left(\frac{d\theta}{dt}\right)^2\right]\mathbf{u}_r + \frac{1}{r}\frac{d}{dt}\left(r^2\frac{d\theta}{dt}\right)\mathbf{u}_\theta$$

13. Verify the various expressions for planar motion described in polar coordinates.

14. Establish each of the following polar coordinate expressions.

(a) $\mathbf{a}_T = \dfrac{d}{dt}\left[\left(\dfrac{dr}{dt}\right)^2 + \left(r\dfrac{d\theta}{dt}\right)^2\right]^{1/2}\mathbf{T}$

(b) $\mathbf{a}_N = \dfrac{r\dfrac{d\theta}{dt}\left[\dfrac{d^2r}{dt^2} - r\left(\dfrac{d\theta}{dt}\right)^2\right] - \dfrac{dr}{dt}\left[2\dfrac{dr}{dt}\dfrac{d\theta}{dt} + r\dfrac{d^2\theta}{dt^2}\right]}{\left[\left(\dfrac{dr}{dt}\right)^2 + \left(r\dfrac{d\theta}{dt}\right)^2\right]^{1/2}}\mathbf{N}$

15. If $r = f(\theta)$ is the equation of a planar curve in polar coordinates, demonstrate that its curvature κ is given by

$$\kappa = \frac{|f^2 + 2f'^2 - ff''|}{(f^2 + f'^2)^{3/2}}$$

16. In polar coordinates, the equation $r = 2a\cos\theta$ describes a circle of radius a. Assume that a particle moves along this curve with constant speed. Demonstrate that $\mathbf{a}_T = 0$ and $|\mathbf{a}_N| =$ constant.

17. If a particle moves in a plane with no *radial* acceleration and with a constant angular speed (that is, $d\theta/dt = \omega$), show that

$$r(t) = Ae^{\omega t} + Be^{-\omega t}$$

and hence the path of the particle is given by

$$r(\theta) = c_1\cosh\theta + c_2\sinh\theta$$

18. Demonstrate that if a particle moves in a plane with *only* radial acceleration, the vector $\mathbf{r}$ sweeps out equal areas in equal intervals of time. (We sometimes call such motion *Keplerian* since, to first approximation, this is the motion of the planets as deduced by Kepler.)

19. Verify that in planar spiral motion where

$$r = ae^{b\theta} = c(t + d)^2, \qquad \text{with } a, b, c, d \text{ constant}$$

both the tangential and normal components of the acceleration of the particle are constant.

10.7 Scalar and Vector Fields

Most of our attention, so far, has been devoted to functions of a real variable whose function values lie in a vector space. We now consider a *scalar field*, which is a function whose domain is a subset of three-dimensional geometric vector space and whose range is a set of real members. Physically, this means that to each point P in space is assigned a real number. Temperature of some solid and density of some material are two ready examples of scalar fields. If f is a scalar field, then the equation that describes $f(P)$ is usually dependent upon some choice of coordinate system, but the actual value at P must be the same regardless of which coordinate system is used. We say that f must be *invariant* under changes of the coordinate system.

Perhaps the most important kind of scalar field is a *potential* field. Potential is a measure of work per (some kind of) unit. There are two basic kinds of potential: (a) *gravitational potential*, which measures the work per unit mass required to raise that mass from sea level to a certain altitude, and (b) *electric potential*, which measures the work per unit charge required to transport a test charge from one point to another. The potential at each point is a number (scalar) and not a vector.

A *vector field*, represented by $\mathbf{F}(P)$, is a function whose domain is some subset of the points in three-dimensional space and whose range is some subset of the set of geometric vectors in three dimensions. We write $\mathbf{F}(P) = f(x, y, z)\mathbf{i} + g(x, y, z)\mathbf{j} + h(x, y, z)\mathbf{k}$ to describe the field if the basis $\{\mathbf{i}, \mathbf{j}, \mathbf{k}\}$ is being used. In describing a vector field, a geometric vector is given for each point in space. The vector and the point go together, and we usually do not think of the vector as having initial point at the origin. Velocity fields and force fields are two familiar examples of vector fields.

Example 10.17 Describe the velocity field of a rotating rigid body.

Solution Let $\boldsymbol{\omega}$ be a vector pointing in the direction of the axis of rotation with magnitude equal to the angular speed of the body. The vector $\boldsymbol{\omega}$ is called the *angular rotation vector*. If $\mathbf{r}$ is a vector from some point on the axis of rotation to a point P on the body, then since the magnitude of the velocity is $|\boldsymbol{\omega}||\mathbf{r}|\sin\theta$ and the direction of motion is tangential, the velocity at that point may be represented by

$$\mathbf{v}(P) = \boldsymbol{\omega} \times \mathbf{r}$$

(See Figure 10.13.) For example, if the angular rotation vector is $\boldsymbol{\omega} = 3\mathbf{k}$, then

$$\mathbf{v}(P) = -3y\mathbf{i} + 3x\mathbf{j}$$ □

Example 10.18 Describe the gravitational field due to a particle of mass M at a fixed point P_0. (See Figure 10.14.)

Solution By the law of gravitation, the force at a point P is directed from P to P_0 and has magnitude inversely proportional to the square of the distance from

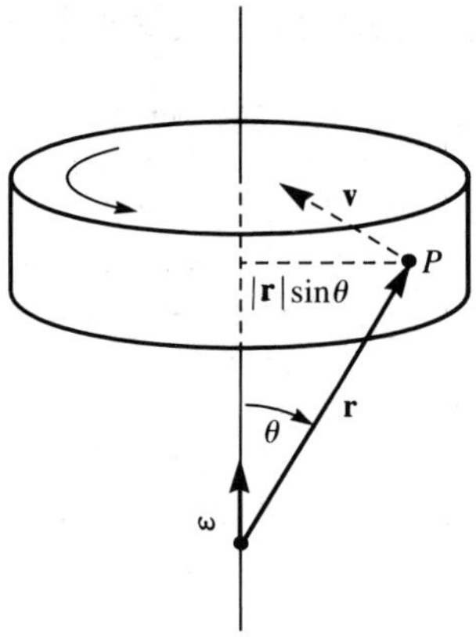

Figure 10.13 Velocity field of rotating body

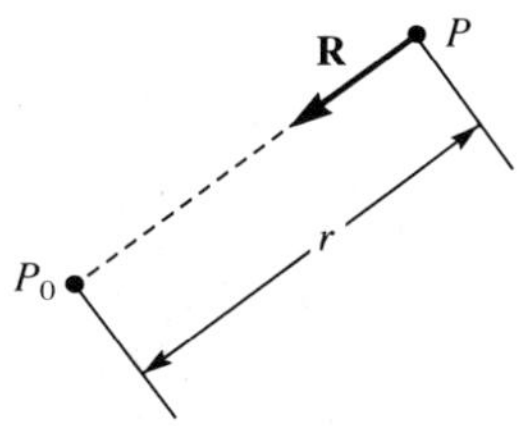

Figure 10.14 Gravitational field

P to P_0. If $\mathbf{R}$ is a unit vector directed from P to P_0, and r is the distance from P to P_0, then the vector $\mathbf{F}$,

$$\mathbf{F} = c\frac{\mathbf{R}}{r^2}, \qquad (c \text{ a constant})$$

describes the force field at every point P. In terms of x, y, z coordinates this would be

$$\mathbf{F}(x, y, z) = c\frac{-(x - x_0)\mathbf{i} - (y - y_0)\mathbf{j} - (z - z_0)\mathbf{k}}{[(x - x_0)^2 + (y - y_0)^2 + (z - z_0)^2]^{3/2}}$$

where c is a constant dependent upon the units chosen. □

This is an example of a *central force field* since the field is directed toward one central point. Inverse square law fields and central force fields are both of prime importance in physics and engineering applications.

Example 10.19 Describe the electric field due to a positive charge of magnitude Q_1 located at the origin.

Solution The electric field $\mathbf{E}$ is defined to be the force per unit charge (newtons per coulomb) at a point. If a charge Q_2 is brought into the neighborhood of the

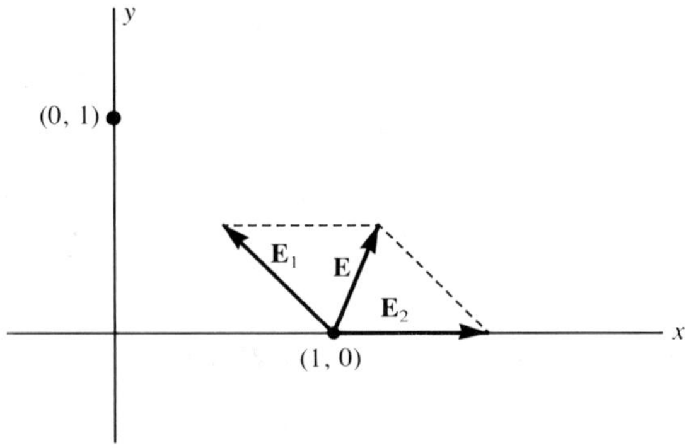

Figure 10.15 *Electric field*

origin it is acted upon by a force directed away from the origin (if Q_2 is positive). From electromagnetic theory we know that this force is given by

$$\mathbf{F} = \frac{Q_1 Q_2}{4\pi\varepsilon_0} \frac{\mathbf{R}}{r^2}, \qquad (\varepsilon_0 = \text{permittivity of vacuum})$$

where if mks units are used the value of ε_0 is $10^{-9}/36\pi$. Since $\mathbf{E}$ is the force per unit charge, it is given by

$$\mathbf{E} = \frac{Q_1}{4\pi\varepsilon_0} \frac{\mathbf{R}}{r^2}$$

□

Example 10.20 Find $\mathbf{E}(1, 0)$ if a charge of -2×10^{-9} coulombs is located one meter above the origin and another of magnitude 10^{-9} is located at the origin. (See Figure 10.15.)

Solution Let $\mathbf{E}_1$ be the electric field at (1, 0) due to the charge at (0, 1) and let $\mathbf{E}_2$ be the field due to the charge at the origin. The magnitude of $\mathbf{E}_1$ is given by

$$|\mathbf{E}_1| = \frac{2 \times 10^{-9}}{4\pi \dfrac{10^{-9}}{36\pi}} \frac{1}{2}$$

$$= 9$$

Since the charge is negative, the field is directed from (1, 0) to (0, 1). Hence the unit vector, $\mathbf{R}$, is given by

$$\mathbf{R} = -\frac{\sqrt{2}}{2}\mathbf{i} + \frac{\sqrt{2}}{2}\mathbf{j}$$

and therefore $\mathbf{E}_1(1, 0)$ is given by

$$\mathbf{E}_1 = \frac{9}{\sqrt{2}}(-\mathbf{i} + \mathbf{j})$$

It is left for you to show that $\mathbf{E}_2 = 9\mathbf{i}$. The total electric field is obtained by adding the fields induced by each charge:

$$\begin{aligned}\mathbf{E} &= \mathbf{E}_1 + \mathbf{E}_2 \\ &\doteq -6.36\mathbf{i} + 6.36\mathbf{j} + 9\mathbf{i} \\ &\doteq 2.64\mathbf{i} + 6.36\mathbf{j}\end{aligned}$$ □

Additional practical applications involving scalar and vector fields appear in the exercise sets throughout the remainder of this chapter.

10.8 Sketching Scalar and Vector Fields

To give an idea of the variation of a scalar field, we sometimes consider *level surfaces.* (In two dimensions, *level curves.*) A level surface of a scalar field f is the set of points with coordinates (x, y, z) such that $f(x, y, z) =$ a constant. Depending upon the application, you will hear level surfaces given more descriptive names. For example, if the scalar function is temperature, the level surfaces are called *isothermal surfaces.* If the scalar function is electric potential, then the constant level surfaces are called *equipotential surfaces.*

Example 10.21 Sketch the level surfaces of the scalar function $f(x, y, z) = z - 4x^2 - y^2$.

Solution (a) The level surfaces of the function are elliptic paraboloids $z - 4x^2 - y^2 = c$. (See Figure 10.16.) □

The sketching of vector fields is considerably more complicated. To give a picture of the variation of a vector field, we must somehow tell how the magnitude *and* the direction of the field are changing. One procedure, although admittedly naive, is to sketch a "certain number" of arrows in the space under consideration. To assist in the sketch, we sometimes calculate the *constant-magnitude curves* for fields in two dimensions and *constant-magnitude surfaces* for fields in three dimensions. These are the curves (or surfaces) on which the vector field has the same magnitude.

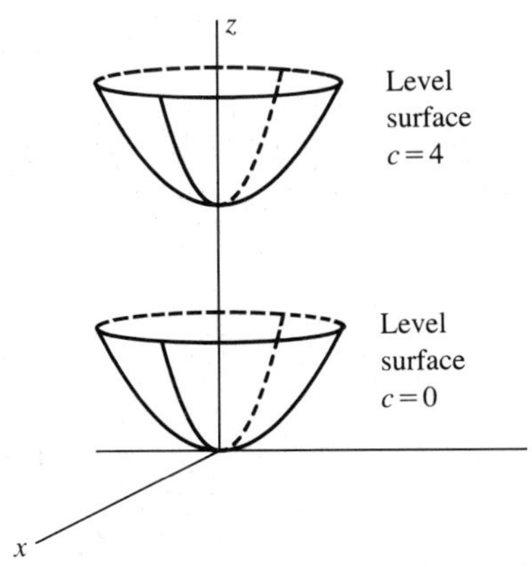

Figure 10.16 Parts of level surfaces for the field $f(x, y, z) = z - 4x^2 - y^2$

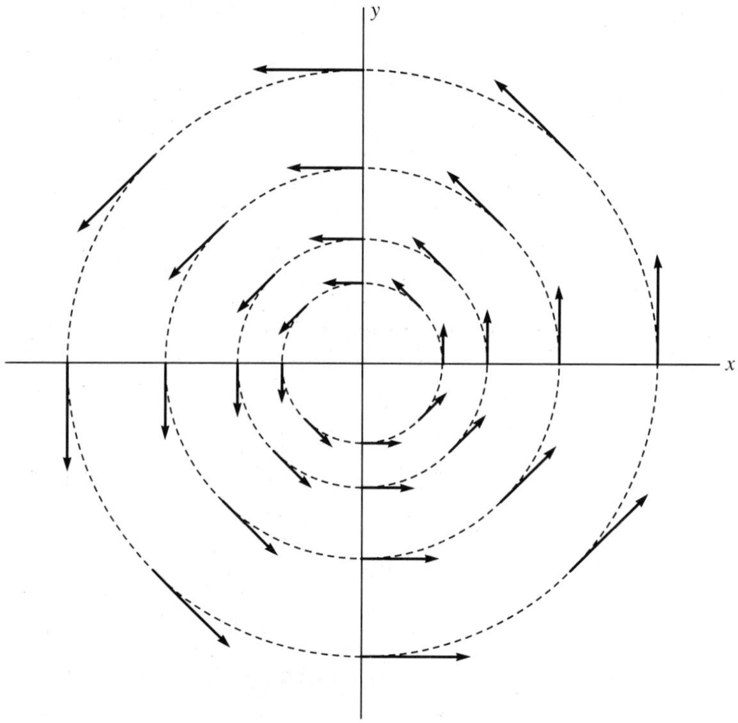

Figure 10.17 *Field of rotating body*

Example 10.22 Sketch the vector field defined by $\mathbf{F}(x, y) = -y\mathbf{i} + x\mathbf{j}$.

Solution First, look back at Example 10.17, and notice that this is a special case of a velocity field of a rotating body. The constant magnitude curves are in the xy-plane with $|\mathbf{F}|^2 = y^2 + x^2 = c^2$. This is a family of circles centered at the origin. In Figure 10.17 you will see some of the constant magnitude curves and how they help in the sketching of the vector field. □

Example 10.23 Sketch the inverse square law central force field

$$\mathbf{v}(P) = \frac{\mathbf{r}}{|\mathbf{r}|^3}$$

where $\mathbf{r}$ is a vector from the origin to a point, P.

Solution Note that this is an inverse *square* law field, even though it might not look like it since the radius in the denominator is taken to the third power. The constant-magnitude curves are given by

$$|\mathbf{v}| = \frac{1}{|\mathbf{r}|^2} = c$$

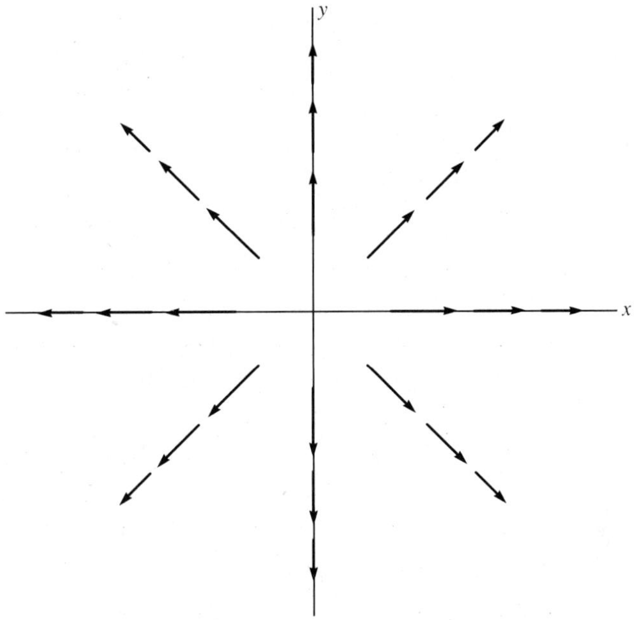

Figure 10.18 Inverse square law field

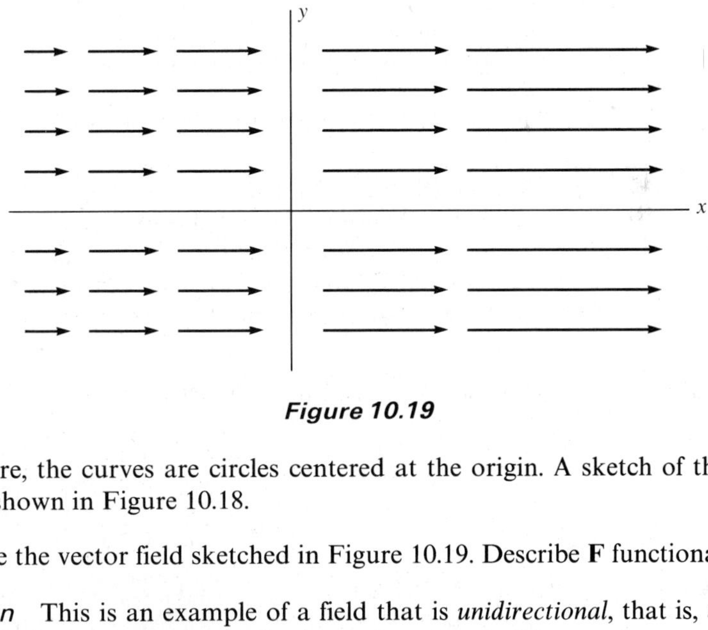

Figure 10.19

Therefore, the curves are circles centered at the origin. A sketch of the vector field is shown in Figure 10.18. □

Example 10.24 Let **F** be the vector field sketched in Figure 10.19. Describe **F** functionally.

Solution This is an example of a field that is *unidirectional*, that is, a field in which all the vectors point in the same direction. In this case the vector field is always in the **i** direction. Since the field does not appear to vary with y, it is a function of x alone and may be described by the expression

$$\mathbf{F} = f(x)\mathbf{i}$$ □

Exercises for Sections 10.7 and 10.8

In Exercises 1–3, sketch constant-magnitude curves for the given vector fields.

1. **(a)** $x^{1/2}\mathbf{i} + y\mathbf{j}$ **(b)** $xy\mathbf{i}$ **(c)** $\sin x\mathbf{i} + \cos x\mathbf{j}$

2. **(a)** $x\mathbf{i}$ **(b)** $y\mathbf{i}$ **(c)** $x\mathbf{j}$

3. **(a)** $y\mathbf{i} + x\mathbf{j}$ **(b)** $-x\mathbf{i} + y\mathbf{j}$ **(c)** $x\mathbf{i} + y\mathbf{j}$

In Exercises 4–6, sketch the specified vector fields in a manner similar to the examples of Section 10.8.

4. The vector fields of Exercise 1.

5. The vector fields of Exercise 2.

6. The vector fields of Exercise 3.

7. Sketch constant-magnitude surfaces for the vector fields defined by each of the following.
(a) $x\mathbf{i} + y\mathbf{j} + z\mathbf{k}$ **(b)** $y\mathbf{i} - z\mathbf{j} + x\mathbf{k}$ **(c)** $x\mathbf{i} + z^{1/2}\mathbf{j} + y\mathbf{k}$

8. Sketch equipotential curves for the following potentials defined on the plane.

(a) $f(x, y) = x^2 + y^2$ **(b)** $f(x, y) = x^2 + 4y^2$
(c) $f(x, y) = x^2 - y^2$ **(d)** $f(x, y) = xy$
(e) $f(x, y) = x^2 - y$

9. Sketch equipotential surfaces for the following potentials.
(a) $w(x, y, z) = x^2 + y^2 - z$ **(b)** $w(x, y, z) = x^2 - z$

10. Write formulas for the following vector fields.
(a) In two dimensions with a positive radial direction and a magnitude equal to the constant C.
(b) In three dimensions with a negative radial direction and unit magnitude.
(c) In two dimensions with a direction that always makes an angle of 60° with the x-axis and a magnitude that is $(x + y)^2$ at the point (x, y).
(d) In two dimensions with a direction that is always orthogonal to the radius vector and a magnitude equal to the distance from the origin.

In the sketching of vector fields, it is often helpful to use families of curves called *field lines*. In two dimensions a curve $y = y(x)$ is a field line of a vector field $\mathbf{F}(x, y) = F_1(x, y)\mathbf{i} + F_2(x, y)\mathbf{j}$ if at each point on the curve, $\mathbf{F}$ is tangent to the curve, that is, the slope of the field lines satisfies

$$\frac{dy}{dx} = \frac{F_2(x, y)}{F_1(x, y)}$$

11. Determine the field lines for the vector fields of Exercise 1.

12. Sketch the vector fields of Exercise 1 using the field lines identified in Exercise 11.

13. Let $\mathbf{r}_1$ and $\mathbf{r}_2$ be vectors from some point on the axis of rotation to two points P and Q on the rotating rigid body of Example 10.17. Show that

$$\mathbf{r}_1 \cdot \frac{d\mathbf{r}_2}{dt} + \mathbf{r}_2 \cdot \frac{d\mathbf{r}_1}{dt} = 0$$

Give a geometrical reason for this result.

14. The acceleration of a point mass moving in an inverse square law central force field, such as in Examples 10.18 and 10.23, is given by

$$\mathbf{a} = \frac{d\mathbf{v}}{dt} = \frac{c}{r^3}\mathbf{r}, \qquad (c = \text{constant})$$

Demonstrate that for this situation,

$$\frac{d}{dt}(\mathbf{r} \times \mathbf{v}) = \mathbf{0}$$

and hence $\mathbf{r} \times \mathbf{v} = \mathbf{C}$, a constant vector.

15. Use the result of Exercise 14 to show that, as the mass moves in the inverse square law central force field, the position vector $\mathbf{r}$ sweeps out equal areas in equal intervals of time. (Compare with Exercise 18 in the preceding set of exercises.)

16. Assume that the acceleration of a point mass moving in the xy-plane, owing to the presence of a particular central force field, is given by

$$\mathbf{a} \equiv \frac{d\mathbf{v}}{dt} \equiv \frac{d^2\mathbf{r}}{dt^2} = -\frac{k^2}{r^6}\mathbf{r}, \qquad (k = \text{constant})$$

Verify that a possible path for this point mass is the circle, given in polar coordinates, $r = \cos\theta$.

17. Show that if the central force field in Exercise 16 is such that

$$\mathbf{a} = -\frac{k^2}{r^3}\mathbf{r}, \qquad (k = \text{constant})$$

elliptical paths $r = 1/(b + \cos\theta)$, with $b > 1$, are possible.

10.9 Variation of a Scalar Field

Let $f(P)$ represent a scalar field and suppose we wish to know the rate of change of this field at a point P. For example we may want to examine the temperature variation in a building near an exit. The measure of the variation will obviously be dependent upon whether we make our computation toward the doorway or away from it, and in fact many other directions could be used to find the rate of change of the temperature at the point. Usually, therefore, the direction of the desired rate of change is specified, in which case we call the value so obtained the *directional derivative* of f at P in the direction given.

The concept of the directional derivative should be a familiar one from elementary calculus. One definition is as follows.

Definition 10.8 The *directional derivative, df/ds,* of a scalar field f at a point P in the direction given by the unit vector $\mathbf{U}$ is the value, if it exists, of

$$\lim_{s\to 0} \frac{f(P + s\mathbf{U}) - f(P)}{s}$$

Note that if $g(s) \equiv f(P + s\mathbf{U})$, then $\dfrac{df}{ds}(P)$ is simply $g'(0)$.*

---

* In these expressions, $P + s\mathbf{U}$ designates the point s units from P in the direction given by the unit vector $\mathbf{U}$.

Example 10.25 Find the directional derivative of the scalar field

$$f(x, y, z) = x^2 + y^2 + xyz$$

at the point $P = (1, 1, 1)$ in the direction given by $\mathbf{U} = \frac{1}{3}\mathbf{i} + \frac{2}{3}\mathbf{j} + \frac{2}{3}\mathbf{k}$.

Solution We form

$$g(s) \equiv f(P + s\mathbf{U}) = \left(1 + \frac{s}{3}\right)^2 + \left(1 + \frac{2s}{3}\right)^2 + \left(1 + \frac{s}{3}\right)\left(1 + \frac{2s}{3}\right)^2$$

Thence

$$g'(s) = \left(\frac{2}{3}\right)\left(1 + \frac{s}{3}\right) + \left(\frac{4}{3}\right)\left(1 + \frac{2s}{3}\right) + \left(\frac{1}{3}\right)\left(1 + \frac{2s}{3}\right)^2 + \left(\frac{4}{3}\right)\left(1 + \frac{s}{3}\right)\left(1 + \frac{2s}{3}\right)$$

and

$$\frac{df}{ds}(P) \equiv g'(0) = \frac{11}{3}$$ □

As you are well aware, there is yet another definition of directional derivative, one based upon the natural generalization of the derivative of a simple real function of one real variable. This is perhaps the most familiar definition and certainly the most advantageous for practical application.

Definition 10.8′ Let $f = f(x, y, z)$ be a continuously differentiable scalar field and define the *gradient vector field*

$$\operatorname{grad} f \equiv \nabla f \equiv \frac{\partial f}{\partial x}\mathbf{i} + \frac{\partial f}{\partial y}\mathbf{j} + \frac{\partial f}{\partial z}\mathbf{k}$$

Then the *directional derivative*, df/ds, of the scalar field f at a point P in the direction of the unit vector $\mathbf{U}$ is given by

$$\frac{df}{ds}(P) \equiv \mathbf{U} \cdot \nabla f(P)$$

In addition to its practicality, this definition has theoretical utility. For example, the extension to n-dimensional scalar fields is clear. Also obvious is the fact that knowledge of the partial derivatives merely in the coordinate directions is sufficient to specify completely *all* the directional derivatives at a given point.

The formula*

$$\frac{df}{ds} = \mathbf{U} \cdot \nabla f = |\nabla f| \cos(\nabla f, \mathbf{U})$$

---

* By $\cos(\mathbf{A}, \mathbf{B})$ is meant the cosine of the angle between the vectors $\mathbf{A}$ and $\mathbf{B}$.

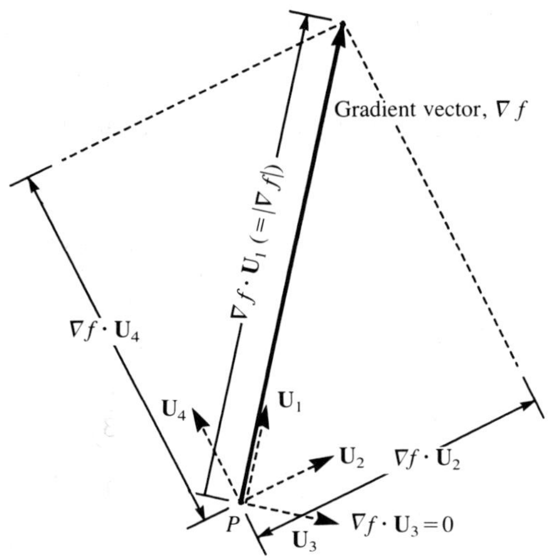

Figure 10.20 Directional derivatives obtained from the gradient vector

associates directions (given by **U**) at a point with values of the directional derivative of f at the point. (See Figure 10.20.) Since df/ds is a maximum when $\cos(\nabla f, \mathbf{U}) = 1$, *the direction of the maximum directional derivative at P is in the direction of the gradient vector at P. The magnitude of the maximum directional derivative is precisely the magnitude of the gradient vector.*

Example 10.26 Let a scalar field be represented by

$$f(x, y, z) = 2xy^2 + z^2xy + x^2$$

Find the value of the directional derivative of f at $(1, 1, 2)$ in the direction of the vector $3\mathbf{i} - 2\mathbf{j} + 5\mathbf{k}$. What is the maximum value of the directional derivative at that point?

Solution In this case

$$\mathbf{U} = \frac{3\mathbf{i} - 2\mathbf{j} + 5\mathbf{k}}{\sqrt{38}}$$

and

$$\nabla f = (2y^2 + z^2y + 2x)\mathbf{i} + (4xy + xz^2)\mathbf{j} + 2xyz\mathbf{k}$$

$$\nabla f(1, 1, 2) = 8\mathbf{i} + 8\mathbf{j} + 4\mathbf{k}$$

Therefore

$$\frac{df}{ds} = \frac{3\mathbf{i} - 2\mathbf{j} + 5\mathbf{k}}{\sqrt{38}} \cdot (8\mathbf{i} + 8\mathbf{j} + 4\mathbf{k})$$

$$= \frac{28}{\sqrt{38}}$$

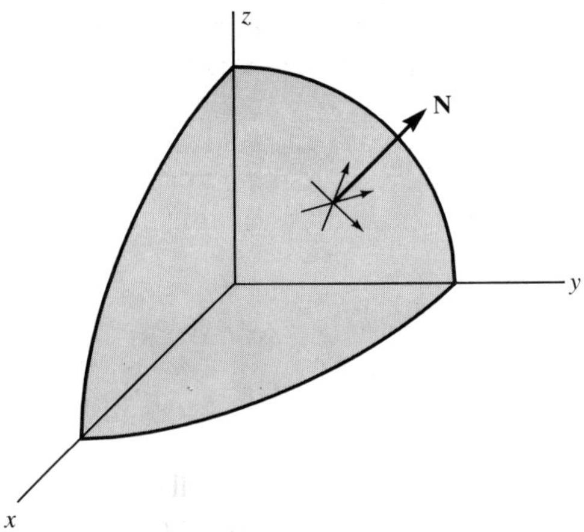

Figure 10.21 *Normal to surface*

Also

$$\left.\frac{df}{ds}\right|_{\max} = |\nabla f| = \sqrt{8^2 + 8^2 + 4^2} = \sqrt{144} = 12 \qquad \square$$

Although the application of the gradient vector to obtain the directional derivative of the scalar field is important, the quantity ∇f is of considerably wider utility. For instance, remember from Section 10.8 that the surfaces $f(x, y, z) = c$ are called level surfaces, which by their very definition are surfaces along which there is no change in the values of the scalar function. Hence, as long as $\mathbf{T}$ is tangential to such a surface,

$$\nabla f \cdot \mathbf{T} = \frac{df}{ds} = 0$$

Therefore the vectors ∇f and $\mathbf{T}$ are perpendicular, and since this must be true for all vectors tangent to the surface at P, ∇f is perpendicular to the level surface. This result leads to one of the more fundamental uses of the gradient vector, that of computing normals to surfaces. (See Figure 10.21.)

Example 10.27 Find a unit vector normal to the surface defined implicity by $z^2 = x^2 + 2y^2$ at $(1, 2, 3)$.

Solution Consider the scalar function $f(x, y, z) = x^2 + 2y^2 - z^2$. The constant level surface $f(x, y, z) = 0$ is the same surface as the one given in the statement of the problem. A unit vector normal to the surface is given by $\nabla f/|\nabla f|$ evaluated at $(1, 2, 3)$. Since

$$\nabla f = 2x\mathbf{i} + 4y\mathbf{j} - 2z\mathbf{k}, \qquad \mathbf{N} = \frac{2\mathbf{i} + 8\mathbf{j} - 6\mathbf{k}}{\sqrt{104}}$$

Another (correct) answer would be the negative of this vector. No confusion should result from this new use of **T** and **N** as unit vectors tangent and normal to a *surface*. □

Vector fields that arise as the gradient of scalar fields are particulary important, since they occur in a variety of applied contexts. Indeed, we give them a special name:

Definition 10.9 Let **F** be a vector field defined in some domain D. If there exists a scalar field, f, such that $\nabla f = \mathbf{F}$ throughout D, then **F** is said to be *conservative* in D. The scalar field f is called a *scalar potential* of **F**.

In Chapter 11 we will consider the appropriateness of the label "conservative" as well as some alternate definitions of the concept, all of which are engendered by consideration of line integrals involving vector fields derived from scalar fields. For the time being, we merely reemphasize that scalar fields are functions that associate with each point of the given domain a (unique) real value.

Example 10.28 A given vector field is represented by $\mathbf{F} = (-\mathbf{r})/|\mathbf{r}|^3$ where $\mathbf{r} = x\mathbf{i} + y\mathbf{j} + z\mathbf{k}$. (This is essentially the inverse square law field of Example 10.23 and Exercises 14, 15, and 17, pp. 382–383.) Show that this field is conservative in any region not including the origin by showing that it has a scalar potential

$$f(x, y, z) = 1/|\mathbf{r}| = (x^2 + y^2 + z^2)^{-1/2}$$

Solution If a potential of the vector field is f, then the three partial derivatives of f must be the components of $(-\mathbf{r})/|\mathbf{r}|^3$. But it is just a matter of elementary differentiation to show

$$f_x = \frac{\partial}{\partial x}(x^2 + y^2 + z^2)^{-1/2} = \frac{-x}{|\mathbf{r}|^3}$$

$$f_y = \frac{-y}{|\mathbf{r}|^3}$$

and

$$f_z = \frac{-z}{|\mathbf{r}|^3}$$

Therefore $\nabla f = \mathbf{F}$ and hence **F** is conservative. □

Example 10.29 A given electrostatic field is represented by

$$\mathbf{E} = (z^2 - 2xy)\mathbf{i} - x^2\mathbf{j} + 2xz\mathbf{k}$$

Show that this field is conservative everywhere by finding an electrostatic potential.

Solution Let v be a scalar potential. Then since $\nabla v = \mathbf{E}$, we have the following three conditions:

$$v_x = z^2 - 2xy; \qquad v_y = -x^2; \qquad v_z = 2xz$$

Partially antidifferentiating the first of these we obtain

$$v(x, y, z) = z^2 x - x^2 y + T(y, z)$$

From this expression for v we may compute v_y,

$$v_y = -x^2 + T_y$$

which by the second of our conditions is equal to $-x^2$. Therefore $T_y = 0$ and hence $T(y, z)$ is a function of z only, say $F(z)$. Thus

$$v(x, y, z) = z^2 x - x^2 y + F(z)$$

We compute v_z from this express to obtain

$$v_z = 2xz + F'(z)$$

which by the third condition is equal to $2xz$. Therefore $F'(z) = 0$ and hence a scalar potential is

$$v(x, y, z) = z^2 x - x^2 y + (\text{a constant}) \qquad \square$$

The process of finding a scalar potential of a conservative vector field is exactly the same as that shown in Section 2.5 for finding a function u (of more than one real variable) if the total differential, du, is known. Hence, the scalar potential is to vector functions what the antiderivative is to real functions.

Exercises for Section 10.9

In Exercises 1–3, find the gradient of the given scalar fields:

1. **(a)** $f(x, y, z) = xyz$ **(b)** $f(x, y, z) = x^2 + y^2 + z^2$
(c) $f(x, y, z) = x \cos y + y \sin z + z \cos x$

2. **(a)** $f(x, y, z) = (x^2 + y^2 + z^2)^{-1/2}$ **(b)** $f(x, y, z) = x^2 y - y^2 z - xyz$
(c) $f(x, y, z) = e^{xyz}$

3. **(a)** $f(x, y, z) = \frac{1}{2}\ln(x^2 + y^2 + z^2)$ **(b)** $f(x, y, z) = \sin^{-1}(z/\sqrt{x^2 + y^2})$
(c) $f(x, y, z) = 2x^2 + 3y^2 + z^2$

4. By using your knowledge of the gradient vector and its relation to constant level curves, find a unit vector normal to the plane curve $y = 3x^2 - 4$, $z = 0$ at $(1, -1, 0)$.

5. Let $f(x, y, z) = xy + yz + xz$. Find each of the following:
(a) df/ds at $(1, 1, 3)$ in the direction toward $(1, 1, 1)$.
(b) The maximum value of df/ds at $(1, 1, 3)$; give the direction in which df/ds is a maximum.
(c) A vector normal to the surface $xy + yz + xz = 7$ at any point of the surface.

In Exercises 6 and 7, assuming $\mathbf{F}$ is conservative, find a scalar potential.

6. **(a)** $\mathbf{F} = \mathbf{i} + \mathbf{j} + \mathbf{k}$ **(b)** $\mathbf{F} = 2x\mathbf{i} + 2y\mathbf{j} + 2z\mathbf{k}$
(c) $\mathbf{F} = \dfrac{x\mathbf{i} + y\mathbf{j}}{(x^2 + y^2)^{1/2}}$

7. (a) $\mathbf{F} = (2xyz + x\sin 2x^2)\mathbf{i} + (x^2z - z\sin yz)\mathbf{j} + (x^2y - y\sin yz)\mathbf{k}$
(b) $\mathbf{F} = (2xy + z^2)\mathbf{i} + (2yz + x^2)\mathbf{j} + (2xz + y^2)\mathbf{k}$

8. Is it possible to take the gradient of a vector field? Explain.

9. By using your knowledge of gradient vectors, find normals to the following surfaces at any point on the surface.
(a) $x^2 + y^2 + z^2 = 2$ (b) $xyz = 1$
(c) $\cos x + \sin y + x\sin z = 1$

10. The magnetic potential is given by $u(x, y, z) = x^2y + xyz$. Find the magnetic induction field vector $\mathbf{B} = \nabla u$.

11. The electrostatic potential of an electric dipole of strength p situated at the origin and oriented in the positive z direction is given by $\phi(x, y, z) = pz/(x^2 + y^2 + z^2)^{3/2}$. Determine the electrostatic field $\mathbf{E} = -\nabla\phi$.

12. Write the equation of the plane tangent to, and the line normal to, the following surfaces at the points indicated.
(a) $x^2 + y^2 = z$ at the point $(1, 1, 2)$ (b) $y = \sin x$ at the point $(\pi/2, 1, 0)$
(c) $xyz = 8$ at the point $(2, 2, 2)$

As we have seen several times earlier, it is sometimes convenient to be able to express a given problem in coordinate systems other than Cartesian. In order to do this with vector quantities, we need to introduce unit vectors in the new coordinate directions. For cylindrical coordinates (r, θ, z) in which $x = r\cos\theta$, $y = r\sin\theta$, $z = z$, we have

$$\mathbf{u}_r = \cos\theta\,\mathbf{i} + \sin\theta\,\mathbf{j}$$
$$\mathbf{u}_\theta = -\sin\theta\,\mathbf{i} + \cos\theta\,\mathbf{j}$$
$$\mathbf{u}_z = \mathbf{k}$$

while for spherical coordinates (ρ, θ, ϕ) in which $x = \rho\sin\phi\cos\theta$, $y = \rho\sin\phi\sin\theta$, $z = \rho\cos\phi$ (see Figure 8.6) we find

$$\mathbf{u}_\rho = \sin\phi\cos\theta\,\mathbf{i} + \sin\phi\sin\theta\,\mathbf{j} + \cos\phi\,\mathbf{k}$$
$$\mathbf{u}_\phi = \cos\phi\cos\theta\,\mathbf{i} + \cos\phi\sin\theta\,\mathbf{j} - \sin\phi\,\mathbf{k}$$
$$\mathbf{u}_\theta = -\sin\theta\mathbf{i} + \cos\theta\mathbf{j}$$

13. Verify that each of the sets $(\mathbf{u}_r, \mathbf{u}_\theta, \mathbf{u}_z)$ for cylindrical coordinates and $(\mathbf{u}_\rho, \mathbf{u}_\theta, \mathbf{u}_\phi)$ for spherical coordinates constitutes a mutually orthogonal collection of unit vectors.

14. Employ the chain rule to show that in cylindrical coordinates the gradient takes the form

$$\nabla f = \frac{\partial f}{\partial r}\mathbf{u}_r + \frac{1}{r}\frac{\partial f}{\partial \theta}\mathbf{u}_\theta + \frac{\partial f}{\partial z}\mathbf{u}_z$$

15. Repeat Exercise 14 in the case of spherical coordinates to demonstrate that

$$\nabla f = \frac{\partial f}{\partial \rho}\mathbf{u}_\rho + \frac{1}{\rho\sin\phi}\frac{\partial f}{\partial \theta}\mathbf{u}_\theta + \frac{1}{\rho}\frac{\partial f}{\partial \phi}\mathbf{u}_\phi$$

16. Use the result of Exercise 14 in order to determine the gradient of the following scalar fields.
(a) $f(r, \theta, z) = (r^2 + z^2)^n$ (b) $f(r, \theta, z) = r^2z\cos\theta\sin\theta$
(c) $f(r, \theta, z) = \cos^{-1}(z/r)$

17. Employ the result of Exercise 15 in order to find the gradient of the following scalar fields.

(a) $f(\rho, \theta, \phi) = e^{\rho \cos \phi}$ **(b)** $f(\rho, \theta, \phi) = \rho \sin \phi \cos \theta$

(c) $f(\rho, \theta, \phi) = \rho^2 \sin^2 \phi \sin \theta \cos \theta + \rho^2 \sin \phi \cos \phi(\sin \theta + \cos \theta)$

18. The electrostatic field associated with a uniform line of charge oriented along the z axis is given in polar coordinates by

$$\mathbf{E}(r, \theta, z) = \frac{\text{const}}{r}\mathbf{u}_r$$

Show that this field is conservative for $r \neq 0$ by finding an electrostatic potential.

19. The field associated with an electric dipole of strength p situated at the origin and oriented in the positive z direction is given in spherical coordinates by

$$\mathbf{E}(\rho, \theta, \phi) = \text{const}\frac{p}{\rho^3}(2 \cos \phi\, \mathbf{u}_\rho + \sin \phi\, \mathbf{u}_\phi)$$

Demonstrate that this field is conservative for $\rho \neq 0$ by finding an electrostatic potential. (Compare with Exercise 11.)

10.10 Variation of a Vector Field: Divergence

As seen in Section 10.9, an understanding of the variation of a *scalar* field depends principally upon knowledge of a single quantity—the derived gradient vector field. The calculus of *vector* fields, however, involves two distinct types of variation. One of these, called the divergence of a vector field, is a scalar quantity. It is discussed in this section and will play a distinguished role in the integration theory of the next chapter. The other variation, called the curl, is a vector variation and is the subject of Section 10.11.

Let $\mathbf{F}$ be a three-dimensional vector field with component functions $\{f, g, h\}$ with respect to the $\{\mathbf{i}, \mathbf{j}, \mathbf{k}\}$ basis. Then we can make the following definition.

Definition 10.10 The *divergence* of the three-dimensional vector field $\mathbf{F}$ is the scalar field whose value at each point is given by the sum $f_x + g_y + h_z$. The divergence of $\mathbf{F}$ is denoted by $\nabla \cdot \mathbf{F}$ (read "del dot $\mathbf{F}$") or div $\mathbf{F}$.

Example 10.30 Find the divergence of the vector field defined by

$$\mathbf{F} = 3xy\mathbf{i} + x^2y\mathbf{j} + y^2z\mathbf{k}$$

Solution Since $f(x, y, z) = 3xy$, $g(x, y, z) = x^2y$, and $h(x, y, z) = y^2z$,

$$\nabla \cdot \mathbf{F} = 3y + x^2 + y^2$$ □

The formula for the divergence might seem naturally associated with the notation $\nabla \cdot \mathbf{F}$ since you can think of the "operator" $\mathbf{i}\frac{\partial}{\partial x} + \mathbf{j}\frac{\partial}{\partial z} + \mathbf{k}\frac{\partial}{\partial z}$ as "dotting" with the vector field to give the scalar quantity indicated. This thinking should not be carried too far, however, since the del operator is certainly not a vector. On the other hand, the notation is exceedingly convenient and clearly suggests the appropriate extension of the divergence concept, like that of the gradient, both to n-dimensions and to other (non-Cartesian) coordinate systems. (See the exercises at the end of this section.)

The word "divergence" is well chosen, as can be seen from the following physical interpretation. Consider the motion of some substance (such as a fluid, a gas, electric charges, alpha particles) spreading through three-space. Then the velocities of the particles at each instant form a vector field, $\mathbf{V} = v_1\mathbf{i} + v_2\mathbf{j} + v_3\mathbf{k}$, where v_1, v_2, and v_3 are the component velocities. At a point P in space we impose a rectangular coordinate system so that without loss of generality P may be considered to have coordinates $(0, 0, 0)$.

We examine the flow of this substance through a small rectangular parallelepiped of dimensions Δx, Δy, Δz with edges parallel to the coordinate axes. (See Figure 10.22.) If ρ is the density of the substance, then the vector quantity $\mathbf{F} = \rho\mathbf{V}$ has component functions ρv_1, ρv_2, and ρv_3.

We compute the change in mass per unit time within the parallelepiped by computing the net outward flow. The normal inward component of $\mathbf{F}$ at the back face is ρv_1 (evaluated at $(0, y, z)$) and the approximate normal outward component at the front face is given by

$$\rho v_1 + \frac{\partial \rho v_1}{\partial x}\Delta x$$

This approximation is good if, for example, the value of Δx is small. Hence, the mass entering the back face per unit time, Δt, is given by

$$\rho v_1(0, y, z)(\Delta y\,\Delta z)\,\Delta t$$

Likewise, the mass leaving the front face is approximately

$$\rho v_1(0, y, z)(\Delta y\,\Delta z)\,\Delta t + \left(\frac{\partial \rho v_1}{\partial x}\Delta x\right)(\Delta y\,\Delta z)\,\Delta t$$

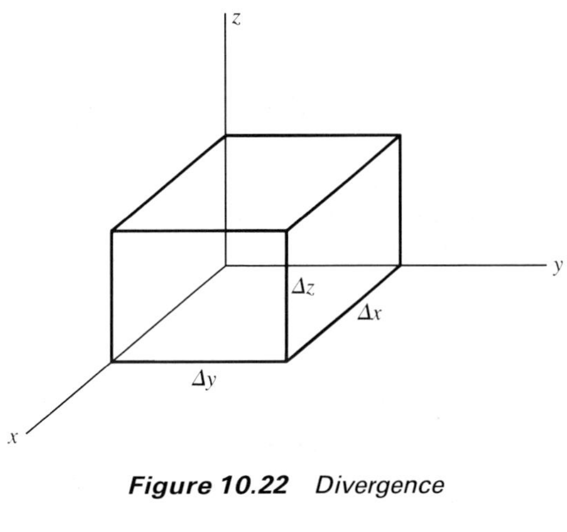

Figure 10.22 Divergence

There are similar expressions for the mass entering and leaving the other four faces. Adding the outward flow of mass we obtain

$$\begin{aligned}\begin{pmatrix}\text{Approximate net change of outward}\\ \text{flow of mass in time interval } \Delta t\end{pmatrix} &= \left(\frac{\partial \rho v_1}{\partial x} + \frac{\partial \rho v_2}{\partial y} + \frac{\partial \rho v_3}{\partial z}\right)\Delta x\,\Delta y\,\Delta z \\ &= \nabla\cdot(\rho\mathbf{V})\,\Delta x\,\Delta y\,\Delta z\end{aligned}$$

which means that the net outward flow of mass per unit volume in time interval Δt is $\nabla \cdot \rho\mathbf{V}$, that is, the divergence of the vector field $\rho\mathbf{V}$ at P.

This net change of mass per unit volume per unit time at a given point may be nonzero, either because of velocities changing through space, or because of a change in density due to a local creation (source) or destruction (sink) of substance. Thus in studying the flow of oxygen through the human body we may have a sink due to conversion into carbon dioxide by metabolism.

Let ϕ represent the algebraic difference between the source density and the sink density (that is, the difference between the rates at which the substance is being created or destroyed per unit time). Then the equation governing the flow of mass is

$$\nabla \cdot \rho\mathbf{V} = \phi - \frac{\partial \rho}{\partial t}$$

In case there are no sources or sinks, we have the *continuity equation for compressible fluid flow* (also called the *conservation of mass equation*),

$$\nabla \cdot \rho\mathbf{V} = -\frac{\partial \rho}{\partial t}$$

If the density is time invariant, that is, the flow is *steady*, then the continuity equation becomes

$$\nabla \cdot \rho\mathbf{V} = 0$$

Finally, if the fluid is *incompressible* (ρ = a constant),

$$\nabla \cdot \mathbf{V} = 0$$

which is called the *condition for incompressibility* (in a field where $\phi = 0$).

As might be expected, the divergence of a vector field has relevance in other practical contexts. In the case of electric fields, the sources and sinks are electric charges and the equation for the condition for incompressibility applies to the electric displacement vector $\mathbf{D}$,

$$\nabla \cdot \mathbf{D} = \phi$$

where ϕ is the charge density. Physically, this means that the electric displacement vector has zero divergence in charge free regions.

For the magnetic field vector, $\mathbf{B}$, you might expect the sources to be magnetic poles. However, there are, in general, no free magnetic poles and hence $\nabla \cdot \mathbf{B} = 0$ always. The two equations which specify the pointwise behavior of the vector fields $\mathbf{B}$ and $\mathbf{D}$ are two from a set of four equations known as Maxwell's equations.

What we have suggested above is how the two divergence equations might be derived from the condition for the conservation of mass. In electromagnetic theory one usually begins with Maxwell's equations themselves as the basis of the development.

Example 10.31 Find the charge density at the point $(1, 1, 2)$ if the electric displacement vector field is

$$\mathbf{D} = x^2 y\mathbf{i} + z^2\mathbf{j} + z^3\mathbf{k}$$

Solution The scalar field $\nabla \cdot \mathbf{D}$ is equal to $(\partial/\partial x)(x^2 y) + (\partial/\partial y)(z^2) + (\partial/\partial z)(z^3)$, which is equal to $2xy + 3z^2$. Hence the value of the divergence of $\mathbf{D}$ at $(1, 1, 2)$ is 14, which by Maxwell's equation is the charge density in coulombs per cubic meter. □

Like conservative vector fields, vector fields whose divergence is zero happen with sufficient regularity in applications that they have also been given a special name.

Definition 10.11 A vector field which has zero divergence throughout a region is said to be *solenoidal* or *incompressible* in that region.

The name solenoidal derives from the fact that a magnetic field, which has zero divergence everywhere, is easily generated by a solenoid. The word incompressible is used since velocity vector fields which have zero divergence satisfy the condition for incompressibility, $\nabla \cdot \mathbf{V} = 0$.

A remarkable situation occurs when a vector field is *both* conservative *and* solenoidal.

Example 10.32 Show that a conservative vector field is solenoidal if the sum of the second partial derivatives of the scalar potential is zero.

Solution If $\mathbf{F}$ is a conservative vector field, then by definition there exists a scalar function f such that $\nabla f = \mathbf{F}$. Hence, in three dimensions,

$$\begin{aligned}\nabla \cdot \mathbf{F} &= \nabla \cdot \nabla f \\ &= \nabla \cdot (f_x\mathbf{i} + f_y\mathbf{j} + f_z\mathbf{k}) \\ &= f_{xx} + f_{yy} + f_{zz}\end{aligned}$$

and therefore the divergence of $\mathbf{F}$ is zero (that is, $\mathbf{F}$ is solenoidal) if the sum of the second partial derivatives of the scalar potential is zero. □

We have already encountered this particular combination of second derivatives in our study of partial differential equations and equilibrium phenomena in Chapter 8. Therefore, instead of writing $\nabla \cdot \nabla f$, as we did in the previous example, we will more often use the notation $\nabla^2 f$ or, as we did earlier, Δf; that is,

$$\nabla^2 f \equiv \Delta f \equiv f_{xx} + f_{yy} + f_{zz}$$

(in three dimensions). Also in keeping with our earlier terminology, we will call this sum of second derivatives either the *Laplacian of f* or the *potential operator applied to f*.

Recall that sufficiently smooth functions f with vanishing Laplacian, that is, functions f that satisfy the important and oft-occurring Laplace differential equation $\Delta f = 0$, are termed *harmonic*. Thus what we have shown in the previous example is that the potential, if it exists, of any solenoidal field is harmonic. For example, the function $f(x, y, z) = (x^2 + y^2 + z^2)^{-1/2}$ is harmonic since it is the potential of the three-dimensional solenoidal field $\mathbf{F} = (-\mathbf{r})/(|\mathbf{r}|^3)$ (see Example 10.28 and the following).

Example 10.33 Show that in three-space the vector field $(-\mathbf{r})/(|\mathbf{r}|^3)$ is solenoidal for $|\mathbf{r}| \neq 0$.

Solution To show that this vector field is solenoidal we must show that the divergence is zero:

$$\begin{aligned}
\nabla \cdot \frac{-\mathbf{r}}{|\mathbf{r}|^3} &= -\frac{\partial}{\partial x}\left[\frac{x}{(x^2 + y^2 + z^2)^{3/2}}\right] - \frac{\partial}{\partial y}\left[\frac{y}{(x^2 + y^2 + z^2)^{3/2}}\right] \\
&\quad - \frac{\partial}{\partial z}\left[\frac{z}{(x^2 + y^2 + z^2)^{3/2}}\right] \\
&= -\left[\frac{y^2 + z^2 - 2x^2}{(x^2 + y^2 + z^2)^{5/2}}\right] - \left[\frac{x^2 + z^2 - 2y^2}{(x^2 + y^2 + z^2)^{5/2}}\right] \\
&\quad - \left[\frac{x^2 + y^2 - 2z^2}{(x^2 + y^2 + z^2)^{5/2}}\right] \\
&= 0
\end{aligned}$$

□

Exercises for Section 10.10

In Exercises 1–3, find the divergence of the given vector fields:

1. **(a)** $xyz(\mathbf{i} + \mathbf{j} + \mathbf{k})$ **(b)** $(3x^2 + 6xz + 3z^2)\mathbf{i}$
   (c) $(z - y)\mathbf{i} + (x - z)\mathbf{j} + (y - x)\mathbf{k}$
2. **(a)** $(x\mathbf{i} + y\mathbf{j} + z\mathbf{k})/\sqrt{x^2 + y^2 + z^2}$
   (b) $(2xy - yz)\mathbf{i} + (x^2 - 2yz - xz)\mathbf{j} - (y^2 + xy)\mathbf{k}$
   (c) $e^{xyz}(yz\mathbf{i} + xz\mathbf{j} + xy\mathbf{k})$
3. **(a)** $4x\mathbf{i} + 6y\mathbf{j} + 2z\mathbf{k}$ **(b)** $\mathbf{k} - (xz\mathbf{i} + yz\mathbf{j})/(x^2 + y^2)$
   (c) $e^x(\sin y\mathbf{i} + \cos y\mathbf{j})$
4. Establish the important identity
$$\nabla \cdot (f\mathbf{F}) \equiv f(\nabla \cdot \mathbf{F}) + \mathbf{F} \cdot \nabla f$$
5. Show directly that the function $f(x, y, z) = (x^2 + y^2 + z^2)^{-1/2}$ is harmonic.
6. Show that the function $e^x \cos y$ is harmonic. What solenoidal field does this potential generate?

7. Find the directional derivative of div $\mathbf{F}$ at (2, 1, 3) in the direction of the outer normal to the sphere $x^2 + y^2 + z^2 = 14$ where $\mathbf{F} = xz^2\mathbf{i} + zy^2\mathbf{j} + x^2z\mathbf{k}$.
8. Show that the steady fluid flow whose velocity vector is given everywhere by $x\mathbf{i}$ is compressible. Sketch this vector field in the xy-plane.
9. Show that the vector field $xz\mathbf{i} + yz\mathbf{j} - z^2\mathbf{k}$ is solenoidal everywhere.
10. An electric displacement vector field is described by $x^2\mathbf{i} + y^2\mathbf{j} + z^2\mathbf{k}$. With this field, are there any regions of space which are charge free?
11. Show that the vector field $y^2z\mathbf{i} + xz^3\mathbf{j} + y^2x^2\mathbf{k}$ is solenoidal everywhere and therefore could describe a magnetic vector field.
12. Review and discuss the concepts of conservation of mass, incompressible fields, solenoidal fields, conservative fields, harmonic functions, and the Laplacian of f.

Expressions for the divergence of an arbitrary vector field $\mathbf{F}$ in terms of cylindrical or spherical coordinate variables are often very useful. There are essentially two ways to derive these relations, which are outlined in the following exercises. In either case, the resulting expressions are:

Cylindrical coordinates (r, θ, z):

$$\mathbf{F} = F_r\mathbf{u}_r + F_\theta\mathbf{u}_\theta + F_z\mathbf{u}_z$$

$$\nabla\cdot\mathbf{F} = \frac{1}{r}\frac{\partial}{\partial r}(rF_r) + \frac{1}{r}\frac{\partial F_\theta}{\partial\theta} + \frac{\partial F_z}{\partial z}$$

Spherical coordinates (ρ, θ, ϕ):

$$\mathbf{G} = G_\rho\mathbf{u}_\rho + G_\theta\mathbf{u}_\theta + G_\phi\mathbf{u}_\phi$$

$$\nabla\cdot\mathbf{G} = \frac{1}{\rho^2}\frac{\partial}{\partial\rho}(\rho^2G_\rho) + \frac{1}{\rho\sin\phi}\frac{\partial G_\theta}{\partial\theta} + \frac{1}{\rho\sin\phi}\frac{\partial}{\partial\phi}(\sin\phi G_\phi)$$

Notice that in these relations the subscripts do *not* denote partial differentiation.

13. Use the defining relations $x = r\cos\theta$, $y = r\sin\theta$, $z = z$ for cylindrical coordinates and the chain rule to show that

$$\frac{\partial}{\partial x} = \cos\theta\frac{\partial}{\partial r} - \frac{\sin\theta}{r}\frac{\partial}{\partial\theta}$$

$$\frac{\partial}{\partial y} = \sin\theta\frac{\partial}{\partial r} + \frac{\cos\theta}{r}\frac{\partial}{\partial\theta}$$

$$\frac{\partial}{\partial z} = \frac{\partial}{\partial z}$$

14. From earlier discussion (p. 389) we know that the cylindrical coordinate unit vectors are related to the classical $\mathbf{i}$, $\mathbf{j}$, $\mathbf{k}$ Cartesian basis by

$$\mathbf{u}_r = \cos\theta\,\mathbf{i} + \sin\theta\,\mathbf{j}$$

$$\mathbf{u}_\theta = -\sin\theta\,\mathbf{i} + \cos\theta\,\mathbf{j}$$

$$\mathbf{u}_z = \mathbf{k}$$

Rewrite $\mathbf{F} = F_r\mathbf{u}_r + \mathbf{F}_\theta\mathbf{u}_\theta + F_z\mathbf{u}_z$ in terms of its Cartesian components and then employ Definition 10.10 for the divergence in conjunction with the result of Exercise 13 to obtain the cylindrical coordinate expression for $\nabla\cdot\mathbf{F}$ given above.

15. Alternatively, verify from first principles that

$$\nabla \cdot \mathbf{u}_r = \frac{1}{r}, \qquad \nabla \cdot \mathbf{u}_\theta = 0, \qquad \nabla \cdot \mathbf{u}_z = 0$$

and then employ the identity of Exercise 4 to establish directly the formula for $\nabla \cdot \mathbf{F}$ in cylindrical coordinates. (Here you will need to make use of the gradient expression given in Exercise 14, p. 389.)

16. Starting from the relations $x = \rho \sin\phi \cos\theta$, $y = \rho \sin\phi \sin\theta$, $z = \rho \cos\phi$, which define spherical coordinates (ρ, θ, ϕ), use the chain rule to demonstrate that

$$\frac{\partial}{\partial x} = \sin\phi\cos\theta\frac{\partial}{\partial\rho} + \frac{\cos\phi\cos\theta}{\rho}\frac{\partial}{\partial\phi} - \frac{\sin\theta}{\rho\sin\phi}\frac{\partial}{\partial\theta}$$

$$\frac{\partial}{\partial y} = \sin\phi\sin\theta\frac{\partial}{\partial\rho} + \frac{\cos\phi\sin\theta}{\rho}\frac{\partial}{\partial\phi} + \frac{\cos\theta}{\rho\sin\phi}\frac{\partial}{\partial\theta}$$

$$\frac{\partial}{\partial z} = \cos\phi\frac{\partial}{\partial\rho} - \frac{\sin\phi}{\rho}\frac{\partial}{\partial\phi}$$

17. Make use of the spherical coordinate formulas

$$\mathbf{u}_\rho = \sin\phi\cos\theta\,\mathbf{i} + \sin\phi\sin\theta\,\mathbf{j} + \cos\phi\,\mathbf{k}$$
$$\mathbf{u}_\phi = \cos\phi\cos\theta\,\mathbf{i} + \cos\phi\sin\theta\,\mathbf{j} - \sin\phi\,\mathbf{k}$$
$$\mathbf{u}_\theta = -\sin\theta\,\mathbf{i} + \cos\theta\,\mathbf{j}$$

and carry out a derivation analogous to that outlined in Exercise 14 to obtain the spherical coordinate expression for $\nabla \cdot \mathbf{G}$ given earlier.

18. Alternatively, verify from first principles that

$$\nabla \cdot \mathbf{u}_\rho = \frac{2}{\rho}, \qquad \nabla \cdot \mathbf{u}_\phi = \frac{\cos\phi}{\rho\sin\phi}, \qquad \nabla \cdot \mathbf{u}_\theta = 0$$

and then carry out a derivation analogous to that outlined in Exercise 15 to establish directly the formula for $\nabla \cdot \mathbf{G}$ in spherical coordinates. (Here you will need the gradient expression given in Exercise 15, p. 389.)

In Exercises 19 and 20, calculate the divergence of each of the given vector fields.

19. **(a)** $\mathbf{F}(r, \theta, z) = \frac{1}{r}\mathbf{u}_r$ **(b)** $\mathbf{F}(r, \theta, z) = r\mathbf{u}_\theta$

(c) $\mathbf{F}(r, \theta, z) = \cos\theta\sin\theta(2rz\mathbf{u}_r + r^2\mathbf{u}_z) + rz(\cos^2\theta - \sin^2\theta)\mathbf{u}_\theta$

20. **(a)** $\mathbf{F}(\rho, \theta, \phi) = \frac{1}{\rho^3}(2\cos\phi\,\mathbf{u}_\rho + \sin\phi\,\mathbf{u}_\phi)$

(b) $\mathbf{F}(\rho, \theta, \phi) = \rho^2\cos\phi(-\cos 2\phi\,\mathbf{u}_\rho + \sin 2\phi\,\mathbf{u}_\phi)$

(c) $\mathbf{F}(\rho, \theta, \phi) = \cos\theta(\sin\phi\,\mathbf{u}_\rho + \cos\phi\,\mathbf{u}_\phi) - \sin\theta\,\mathbf{u}_\theta$

21. The Laplacian of a scalar function f is given by the formula

$$\Delta f \equiv \nabla^2 f \equiv \nabla \cdot \nabla f$$

Combine the above representation for the divergence with the earlier expression for the gradient (Exercise 14, p. 389) to derive the form of the Laplacian in cylindrical coordinates.

22. Use the result of Exercise 15, p. 389 in a repeat of Exercise 21, thus deriving the Laplacian in spherical coordinates (see Exercise 12, p. 308).

10.11 Variation of a Vector Field: Curl

In the previous section we concentrated on one type of important variation of a vector field. In particular we recall that the formula for divergence in Cartesian coordinates can easily be remembered by thinking of ∇ as "dotting" the field. The other possibility is to have del operate in a "crossing" manner.

Definition 10.12 Let **F** be a vector field in three-space with component functions f, g, and h. Then curl **F** $= \nabla \times \mathbf{F}$ is given by

$$\nabla \times \mathbf{F} = (h_y - g_z)\mathbf{i} + (f_z - h_x)\mathbf{j} + (g_x - f_y)\mathbf{k}$$

Symbolically,

$$\nabla \times \mathbf{F} = \det \begin{pmatrix} \mathbf{i} & \mathbf{j} & \mathbf{k} \\ \dfrac{\partial}{\partial x} & \dfrac{\partial}{\partial y} & \dfrac{\partial}{\partial z} \\ f & g & h \end{pmatrix}$$

(This last expression is *not* a determinant but is a device that is often used to enable easy recall of the formula for curl **F**. It is "evaluated" just like a third-order determinant. This device is also helpful in that it reminds us that the concept of the curl of a vector field, like the cross product, is essentially restricted to three dimensions.)

Traditionally, the notion of the curl of a vector field has been quite elusive to the beginner, not so much in evaluating it (for that just involves a few partial derivatives) but in understanding what it is. The following paragraphs and examples are intended to give some feeling for the curl rather than to increase computational skill in working with it.

Let **F** be a vector force field in the xy-plane with component functions with respect to the $\{\mathbf{i}, \mathbf{j}\}$ basis denoted by f and g. We wish to determine the tendency of this force field to rotate the rectangle $ABCD$ shown in Figure 10.23. At point A the horizontal component of the field is given by $f(x_A, y_A)$. At point D the horizontal component is approximately $f(x_A, y_A) + f_y(x_A, y_A)\,\Delta y$. The difference between these two values gives the tendency of the horizontal component of the field to rotate the element, in a clockwise direction if $f_y(x_A, y_A)$ is positive and counterclockwise if negative. Similarly, the change of the vertical component in the horizontal direction, which is g_x, gives a counterclockwise rotation if it is positive and clockwise if negative. Note that it is the variation of the component functions in a direction *perpendicular* to the direction of the field which yields this rotation. Thus, the difference $g_x - f_y$ will be a measure of the vector field's rotational force on the rectangle. In the usual manner, we allow the rectangle to shrink to zero to give a pointwise measure of the rotational force. The direction of the motion follows the convention of the "right-hand rule" and hence we denote the pointwise rotational tendency as $\mathbf{k}(g_x - f_x)$. (Note that this vector is perpendicular to the plane of the rectangle.)

More generally, in three dimensions you can probably convince yourself

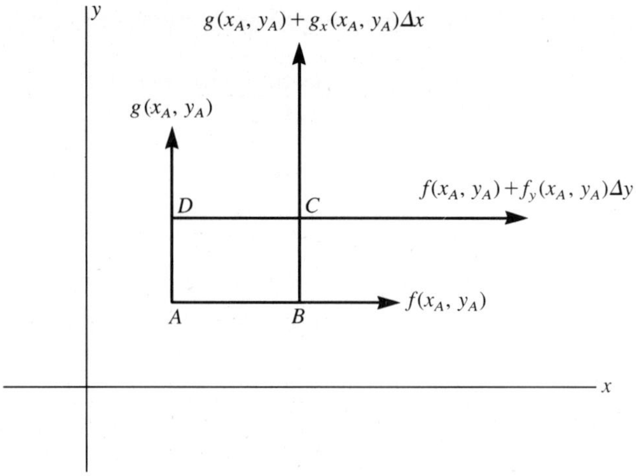

Figure 10.23 *Curl*

that the curl is a measure of the tendency of the vector field to cause a rotation at a point. Its direction is along the axis of rotation and in a direction corresponding to the familiar right-hand rule. This should show you why the words curl and rot are familiar identifiers of $\nabla \times \mathbf{F}$. It should also show the reasonableness of the following definition.

Definition 10.13 A vector field, $\mathbf{F}$, is said to be *irrotational* at those points for which $\nabla \times \mathbf{F} = \mathbf{0}$.

As might be expected, irrotational vector fields arise frequently in practical applications in mechanics, electromagnetics, and elsewhere. Partly this is due to the following result, a result that is fundamental to some of our considerations in the next chapter.

Example 10.34 Show that a conservative vector field is irrotational at each point where it is defined.

Solution If $\mathbf{F}$ represents the vector field, then by the definition of conservative (Definition 10.9), there is a scalar field f such that $\nabla f = \mathbf{F}$. Therefore

$$\nabla \times \mathbf{F} = \nabla \times \nabla f = \begin{pmatrix} \mathbf{i} & \mathbf{j} & \mathbf{k} \\ \dfrac{\partial}{\partial x} & \dfrac{\partial}{\partial y} & \dfrac{\partial}{\partial z} \\ f_x & f_y & f_z \end{pmatrix}$$
$$= (f_{zy} - f_{yz})\mathbf{i} + (f_{xz} - f_{zx})\mathbf{j} + (f_{yx} - f_{xy})\mathbf{k}$$

If we assume that the second mixed partial derivatives of f in each component are equal (a valid assumption if the partials involved exist and are continuous), then each of the components of the curl is zero. Hence, $\nabla \times \mathbf{F} = \mathbf{0}$ and $\mathbf{F}$ is irrotational. □

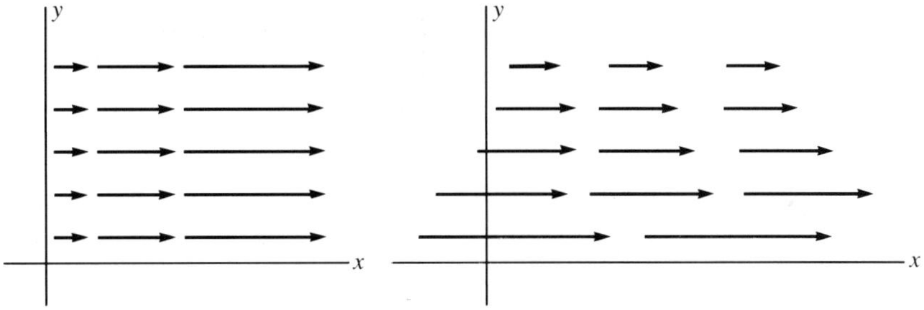

Figure 10.24 *Irrotational and solenoidal fields*

The two fields illustrated in Figure 10.24 are examples of unidirectional fields, one with zero curl and one with zero divergence. Can you tell which is which?

Example 10.35 Show that a vector field **G** that is the curl of another field is solenoidal.

Solution Let $\mathbf{G} = \nabla \times \mathbf{F}$. Then $\nabla \cdot \mathbf{G} = \nabla \cdot \nabla \times \mathbf{F}$. If **F** has component functions f, g, and h, then

$$\begin{aligned}\nabla \cdot \mathbf{G} &= \frac{\partial}{\partial x}(h_y - g_z) + \frac{\partial}{\partial y}(f_z - h_x) + \frac{\partial}{\partial z}(g_x - f_y)\\ &= h_{yx} - g_{zx} + f_{zy} - h_{xy} + g_{xz} - f_{yz}\end{aligned}$$

If we assume the second mixed partial derivatives of the component functions are equal, then $\nabla \cdot \mathbf{G} = \mathbf{0}$ and hence **G** is solenoidal. □

The vector field **F** in the previous example is often called a *vector potential* of the field **G**. In certain applications, if we have a solenoidal field, we postulate the existence of such a potential and find it very useful. For example, in electromagnetics (recall the discussion on p. 392), *Hertz vector potentials* play important roles in various special situations.

In closing this last section of the chapter, it is worth noting that solenoidal vector fields and irrotational vector fields together constitute the fundamental building blocks for the most general of vector fields in three dimensions. Indeed, given an arbitrary smooth vector field **V**, which falls off sufficiently fast for large **r**, it can be shown that **V** has a unique representation (to within constant vectors) as $\mathbf{V}_1 + \mathbf{V}_2$, where $\mathbf{V}_1$ is solenoidal and $\mathbf{V}_2$ is irrotational. The existence of such a representation can be successfully exploited in diverse practical applications.

Exercises for Section 10.11

1. Describe functionally the vector fields of Figure 10.24.

In Exercises 2–4, determine the curl of the given vector fields.

2. **(a)** $x\mathbf{i} + 3xy\mathbf{k}$ **(b)** $z\mathbf{i} + x\mathbf{j} + y\mathbf{k}$
(c) $x\mathbf{i} + y\mathbf{j} + z\mathbf{k}$

3. **(a)** $(x\mathbf{i} + y\mathbf{j} + z\mathbf{k})/\sqrt{x^2 + y^2 + z^2}$
   (b) $(2xy - yz)\mathbf{i} + (x^2 - 2yz - xz)\mathbf{j} - (y^2 + xy)\mathbf{k}$
   (c) $e^{xyz}(yz\mathbf{i} + xz\mathbf{j} + xy\mathbf{k})$
4. **(a)** $xyz(\mathbf{i} + \mathbf{j} + \mathbf{k})$ **(b)** $(z - y)\mathbf{i} + (x - z)\mathbf{j} + (y - x)\mathbf{k}$
   (c) $e^x(\sin y\mathbf{i} + \cos y\mathbf{j})$
5. Are the concepts of irrotational and solenoidal mutually exclusive? Explain.
6. Is the curl of the divergence of a vector field defined? Is the divergence of the curl of a vector field defined? Explain completely.
7. Prove the identity $\nabla\cdot(\mathbf{V} \times \mathbf{W}) = \mathbf{W}\cdot(\nabla \times \mathbf{V}) - \mathbf{V}\cdot(\nabla \times \mathbf{W})$ using components of $\mathbf{V}$ and $\mathbf{W}$ with respect to the $\{\mathbf{i}, \mathbf{j}, \mathbf{k}\}$ basis.
8. Show that the cross product of two conservative fields is solenoidal. (*Hint:* Use the identity of the previous exercise.)
9. Show that a central force field is irrotational everywhere it is defined and differentiable. (*Hint:* a central force field may be described by $f(r)\mathbf{r}$.)
10. Sketch and discuss the following fields with respect to their solenoidal and irrotational properties (these fields were all sketched earlier in the exercises for Sections 10.7–10.8).
    (a) $-y\mathbf{i}$ **(b)** $x\mathbf{j}$ **(c)** $-y\mathbf{i} + x\mathbf{j}$
    (d) $y\mathbf{i} + x\mathbf{j}$ **(e)** $-x\mathbf{i} + y\mathbf{j}$ **(f)** $x\mathbf{i} + y\mathbf{j}$
11. Establish a "law" for unidirectional fields to be solenoidal or irrotational, after examining Figure 10.24. (Also see Exercise 13.)
12. Determine a *necessary* condition for a vector field to have a vector potential.
13. Show that the curl of any unidirectional vector field is perpendicular to that direction.
14. Establish the useful identity

$$\nabla \times (f\mathbf{F}) \equiv f(\nabla \times \mathbf{F}) + \nabla f \times \mathbf{F}$$

As should be expected, the curl of an arbitrary vector field $\mathbf{F}$ can be expressed in cylindrical and spherical coordinates. The necessary intermediate relationships are given in Exercises 13, 14, 16, 17, on p. 395–396. The resulting expressions are:

Cylindrical coordinates (r, θ, z):

$$\mathbf{F} = F_r\mathbf{u}_r + F_\theta\mathbf{u}_\theta + F_z\mathbf{u}_z$$

$$\nabla \times \mathbf{F} = \frac{1}{r}\det\begin{pmatrix} \mathbf{u}_r & r\mathbf{u}_\theta & \mathbf{u}_z \\ \frac{\partial}{\partial r} & \frac{\partial}{\partial \theta} & \frac{\partial}{\partial z} \\ F_r & rF_\theta & F_z \end{pmatrix}$$

Spherical coordinates (ρ, θ, ϕ):

$$\mathbf{F} = F_\rho\mathbf{u}_\rho + F_\phi\mathbf{u}_\phi + F_\theta\mathbf{u}_\theta$$

$$\nabla \times \mathbf{F} = \frac{1}{\rho^2 \sin\phi}\det\begin{pmatrix} \mathbf{u}_\rho & \rho\mathbf{u}_\phi & \rho\sin\phi\,\mathbf{u}_\theta \\ \frac{\partial}{\partial \rho} & \frac{\partial}{\partial \phi} & \frac{\partial}{\partial \theta} \\ F_\rho & \rho F_\phi & \rho\sin\phi\, F_\theta \end{pmatrix}$$

15. Verify each of the following from first principles.

(a) In cylindrical coordinates

$$\nabla \times \mathbf{u}_r = 0, \qquad \nabla \times \mathbf{u}_\theta = \frac{1}{r}\mathbf{u}_z, \qquad \nabla \times \mathbf{u}_z = 0$$

(b) In spherical coordinates

$$\nabla \times \mathbf{u}_\rho = 0, \qquad \nabla \times \mathbf{u}_\phi = \frac{1}{\rho}\mathbf{u}_\theta, \qquad \nabla \times \mathbf{u}_\theta = \frac{\cos\phi}{\rho\sin\phi}\mathbf{u}_\rho - \frac{1}{\rho}\mathbf{u}_\phi$$

16. Combine the results of Exercises 14 and 15 to establish directly the formula for $\nabla \times \mathbf{F}$ in cylindrical coordinates. (The cylindrical coordinate gradient expression is given in Exercise 14, p. 389.)

17. Repeat Exercise 16 to derive the expression for $\nabla \times \mathbf{F}$ in spherical coordinates. (The gradient in spherical coordinates appears in Exercise 15, p. 389.)

In Exercises 18 and 19 determine the curl of each of the given vector fields:

18. **(a)** $\mathbf{F}(r, \theta, z) = \frac{1}{r}\mathbf{u}_r$ **(b)** $\mathbf{F}(r, \theta, z) = r\mathbf{u}_\theta$

(c) $\mathbf{F}(r, \theta, z) = \left(1 - \frac{a^2}{r^2}\right)\cos\theta\,\mathbf{u}_r - \left(1 + \frac{a^2}{r^2}\right)\sin\theta\,\mathbf{u}_\theta$

19. **(a)** $\mathbf{F}(\rho, \theta, \phi) = \frac{1}{\rho^3}(2\cos\phi\,\mathbf{u}_\rho + \sin\phi\,\mathbf{u}_\phi)$

(b) $\mathbf{F}(\rho, \theta, \phi) = \rho^2\cos\phi(-\cos 2\phi\,\mathbf{u}_\rho + \sin 2\phi\,\mathbf{u}_\phi)$

(c) $\mathbf{F}(\rho, \theta, \phi) = \cos\theta(\sin\phi\,\mathbf{u}_\rho + \cos\phi\,\mathbf{u}_\phi) - \sin\theta\,\mathbf{u}_\theta$

10.12 Looking Ahead

The considerations of this chapter can be extended in a variety of ways. For instance, there is much more that can be said about space curves—the binormal vector, torsion, the Frenet-Serret formulas, intrinsic equations, and so on—and surfaces—fundamental forms, Gauss curvature, principal directions, asymptotic lines, and so on. Good references for such differential geometry are [8] and [15]. One of the classics in the field is [18].

Sometimes one is interested in special surfaces such as convex surfaces or surfaces of minimum area spanning given closed space curves (minimal surfaces); other times special curves such as minimum-length curves lying on given fixed surfaces (geodesics) are important. The theory of convex surfaces and sets is fundamental to the study of discrete optimization (see, for example, [14]). Minimal surfaces and geodesics are part of an exciting branch of mathematics called the calculus of variations. Sound introductions to this special field are provided by [5], [9], and [16]. The concepts of arc length and surface area themselves are introduced in Chapter 11.

The fundamental differential operations, the gradient, divergence, and curl, can be expressed in other coordinate systems such as elliptic, parabolic, paraboloidal, spheroidal, and the like, which arise naturally in certain applications. Indeed, there exist convenient formulas for these operations in a general three-dimensional curvilinear coordinate system (see [11], [12], or [13], for example).

In some practical problems, principally in fluid dynamics, elasticity, and relativistic mechanics, the more elementary notions of the vector calculus discussed in this chapter must give way to the more complicated constructs of tensor analysis. In this general setting, in addition to tensors, there are two types of vectors, a new kind of differentiation, and an abundance of nontrivial symbolism. Students interested in this challenging material may profitably consult [6], [10], and [17].

Basic References

In addition to the student's own favorite comprehensive calculus text, we recommend:

[1] Hague, B., and D. Martin. 1970. *An Introduction to Vector Analysis for Physicists and Engineers*. 6th ed. London: Methuen.

[2] Marsden, J. E., and A. J. Tromba. 1981. *Vector Calculus*. 2d ed. San Francisco: W. H. Freeman.

[3] Schey, H. M. 1973. *Div, Grad, Curl, and All That*. New York: W. W. Norton.

[4] Williamson, R. E., et al. 1968. *Calculus of Vector Functions*. 2d ed. Englewood Cliffs, N. J.: Prentice-Hall, Inc.

More-Advanced References

[5] Bliss, G. A. 1925. *Calculus of Variations*. Carus Monograph No. 1. Washington, D.C.: Mathematical Association of America.

[6] Borisenko, A. I., and I. E. Tarapov. 1979. *Vector and Tensor Analysis and Applications*. New York: Dover.

[7] Chorlton, F. 1976. *Vector and Tensor Methods*. Chichester, England: Ellis Horwood.

[8] do Carmo, M. P. 1976. *Differential Geometry of Curves and Surfaces*. Englewood Cliffs, N. J.: Prentice-Hall, Inc.

[9] Ewing, G. W. 1969. *Calculus of Variations with Applications*. New York: W. W. Norton.

[10] Hawkins, G. A. 1963. *Multilinear Analysis for Students in Engineering and Science*. New York: John Wiley.

[11] Hildebrand, F. B. 1976. *Advanced Calculus for Applications*. 2d ed. Englewood Cliffs, N. J.: Prentice-Hall, Inc.

[12] Kaplan, W. 1984. *Advanced Calculus*. 3d ed. Reading, Mass.: Addison-Wesley.

[13] Lass, H. 1950. *Vector and Tensor Analysis*. New York: McGraw-Hill.

[14] Lay, S. R. 1982. *Convex Sets and Their Applications*. New York: John Wiley.

[15] O'Neill, B. 1966. *Elementary Differential Geometry*. New York: Academic Press.

[16] Smith, D. R. 1974. *Variational Methods in Optimization*. Englewood Cliffs, N. J.: Prentice-Hall, Inc.

[17] Sokolnikoff, I. S. 1964. *Tensor Analysis: Theory and Applications*. 2d ed. New York: John Wiley.

[18] Struik, D. J. 1961. *Differential Geometry*. 2d ed. Reading, Mass.: Addison-Wesley.

Chapter 11

Vector Calculus: Integration

11.1 Introduction

In this chapter we will consider in detail the principal integral theorems of the more classical vector field theory in three dimensions. The relationships between the various types of boundary problems, del operations, and integrals are of far-reaching consequence.

We define line, surface, and volume integration based on the familiar definitions of single and multiple Riemann definite integrals. In this process it is also necessary to recall the techniques of evaluating double and triple integrals as iterated integrals. The definitions of line and surface integrals of scalar fields are then extended to integration over curves and surfaces in the presence of vector fields. Here conservative (or irrotational) and solenoidal vector fields play distinguished roles. With the techniques of this chapter we are able to calculate important physical quantities such as mechanical work, fluid circulation, and electrostatic charge distribution when they arise in practical applications.

11.2 Line Integration of Scalar and Vector Fields

The type of integration learned in elementary calculus was along the x-axis. This concept may be generalized to include movement along rather general curves. Our discussion is limited to oriented piecewise smooth curves in three-space. Besides the curve, C, we also consider a scalar valued function, f, assumed to be continuous on the curve.

To define what we mean by "the line integral of a scalar function f along a curve C," we proceed in a manner analogous to that for a function of one real variable. As we show later, there are several types of line integrals and the description given here is the one which defines $\int_C f(P)\,dx\,(P)$ or, as it is more commonly written, $\int_C f(P)\,dx$.

We assume that C is a piecewise smooth simple curve connecting points R and Q. We subdivide C by a finite set of ordered points

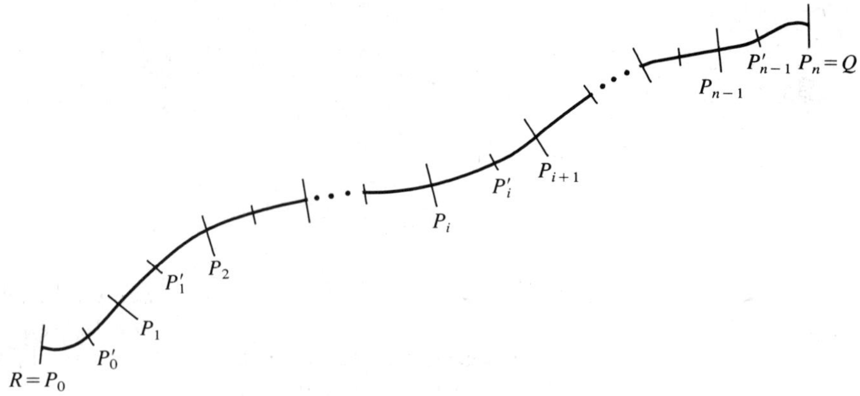

Figure 11.1 *Partitioning a curve*

$$B = \{R = P_0, P_0', P_1, P_1', P_2, \ldots, P_i, P_i', P_{i+1}, \ldots, P_{n-1}, P_{n-1}', P_n = Q\}$$

Figure 11.1 indicates the manner of this subdivision. The finite set of ordered points, which we call a *partition* of C, may be chosen rather arbitrarily. (Indeed, individual points need not be distinct.) Each partition determines a set consisting of the values of x at the points, P_i, P_i':

$$B_x = \{x_0, x_0', x_1, x_1', x_2, \ldots, x_i, x_i', x_{i+1}, \ldots, x_{n-1}, x_{n-1}', x_n\}$$

For a given partition we evaluate each of the quantities $f(P_i')$ and $\Delta x_i = x_{i+1} - x_i$, and then the sum

$$J(B) = \sum_{i=0}^{n-1} f(P_i')\Delta x_i$$

Also associated with each partition is a number called the *x-norm*, which is the largest of the absolute values $|\Delta x_i|$ and is denoted by $\Delta_x(B)$ or $\|\Delta x_i\|$.

To provide some idea of how a partition affects the sum $J(B)$, we consider two examples.

Example 11.1 Find $J(B)$ in the case where the curve C is the parabola in the xy-plane defined by $y = x^2$, the function is $f(x, y) = x + y + 2$, and the partition of the curve is given by $\{(0,0), (\frac{1}{4}, \frac{1}{16}), (\frac{1}{4}, \frac{1}{16}), (\frac{1}{2}, \frac{1}{4}), (1,1)\}$. (See Figure 11.2.)

Solution In this case, the set B_x is given by $\{0, \frac{1}{4}, \frac{1}{4}, \frac{1}{2}, 1\}$ and hence $\Delta x_0 = \frac{1}{4}$, $\Delta x_1 = \frac{3}{4}$, $x_0' = \frac{1}{4}$, and $x_1' = \frac{1}{2}$. Therefore

$$\begin{aligned} J(B) &= f(P_0')\Delta x_0 + f(P_1')\Delta x_1 \\ &= (\tfrac{1}{4} + \tfrac{1}{16} + 2)\tfrac{1}{4} + (\tfrac{1}{2} + \tfrac{1}{4} + 2)\tfrac{3}{4} \\ &= \tfrac{169}{64} \end{aligned}$$

The x-norm of B is $\frac{3}{4}$. □

Example 11.2 Find $J(B)$ if the curve and function are as in the previous example but the partition is defined by the set

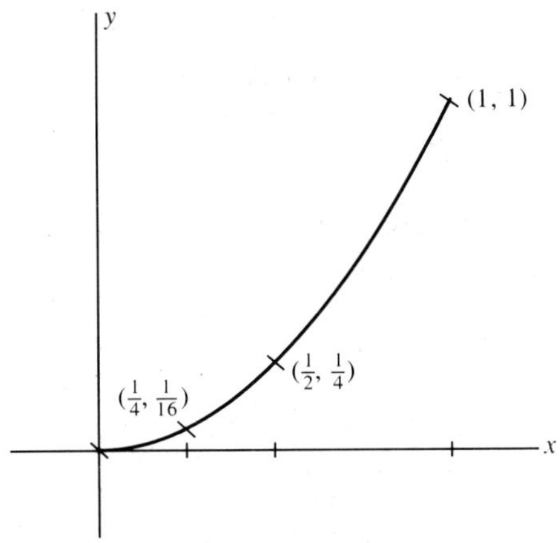

Figure 11.2

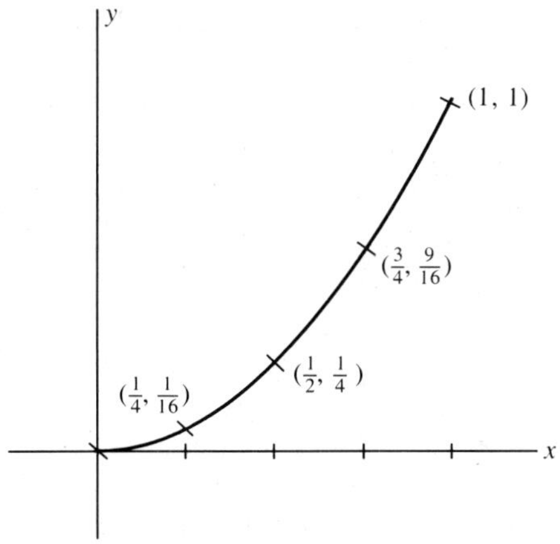

Figure 11.3

$$\{(0, 0), (\tfrac{1}{4}, \tfrac{1}{16}), (\tfrac{1}{4}, \tfrac{1}{16}), (\tfrac{1}{2}, \tfrac{1}{4}), (\tfrac{1}{2}, \tfrac{1}{4}), (\tfrac{3}{4}, \tfrac{9}{16}), (1, 1)\}$$

(See Figure 11.3.)

Solution In this case, the set B_x is given by $\{0, \frac{1}{4}, \frac{1}{4}, \frac{1}{2}, \frac{1}{2}, \frac{3}{4}, 1\}$ and hence $\Delta x_0 = \frac{1}{4}$, $\Delta x_1 = \frac{1}{4}$, $\Delta x_2 = \frac{1}{2}$, $x'_0 = \frac{1}{4}$, $x'_1 = \frac{1}{2}$, and $x'_2 = \frac{3}{4}$. Therefore

$$\begin{aligned} J(B) &= f(P'_0)\Delta x_0 + f(P'_1)\Delta x_1 + f(P'_2)\Delta x_2 \\ &= (\tfrac{1}{4} + \tfrac{1}{16} + 2)\tfrac{1}{4} + (\tfrac{1}{2} + \tfrac{1}{4} + 2)\tfrac{1}{4} + (\tfrac{3}{4} + \tfrac{9}{16} + 2)\tfrac{1}{2} \\ &= \tfrac{187}{64} \end{aligned}$$

The x-norm of the partition is $\frac{1}{2}$. □

The two examples show that the value of $J(B)$ is dependent not only on the number of points chosen for the partition but upon how the selection is made.

Moreover, the importance of the points P_i' and how they affect the sum is apparent. We expect that as different choices of B are made, however, the values of $J(B)$ will be close together if the x-norm of the partition is small. To anticipate this result we make the following definition.

Definition 11.1 The limit (if it exists) of the finite sums $J(B)$, for partitions B of the curve C, as the x-norm approaches 0 is called the *line integral of the scalar function f along C with respect to x* and is denoted by

$$\int_C f(P)\,dx$$

If this development is extended to a simple *closed* curve, we write

$$\oint_C f(P)\,dx$$

Under our fairly reasonable assumptions on the curve C and the function f, the line integral does indeed exist.

Theorem 11.1 If f is continuous on C and C is piecewise smooth, then

$$\lim_{\Delta_x(B)\to 0} J(B)$$

exists and is unique. ■

It is not a difficult matter to define line integrals of the type $\int_C f(P)\,dy$ or $\int_C f(P)\,dz$, since in each case the method of definition is analogous to that for $\int_C f(P)\,dx$. The only difference is that the approximating finite sums are

$$\sum_{i=0}^{n-1} f(P_i')\Delta y_i$$

and

$$\sum_{i=0}^{n-1} f(P_i')\Delta z_i$$

respectively. Also, if the arc length function is known as a function of points on the curve, say $s(P)$, then the differences in arc length $\Delta s_i = s(P_i) - s(P_{i-1})$ may be computed. The limit of the corresponding finite sum

$$\sum_{i=0}^{n-1} f(P_i')\Delta s_i$$

is denoted by $\int_C f(P)\,ds$. Equivalently, $\int_C f(P)\,ds$ can be defined in three-space as the limit of the finite sum

$$\sum_{i=0}^{n-1} f(P_i')\sqrt{\Delta x_i^2 + \Delta y_i^2 + \Delta z_i^2}$$

Example 11.3 Determine an approximation to $\int_C f(P)\,ds$, where the curve and function are as in Example 11.1, but the partition is the one in Example 11.2.

Solution In this case, we use the alternative formulation and calculate

$$\sum_{i=0}^{2} f(P_i')\sqrt{\Delta x_i^2 + \Delta y_i^2} = \left(\frac{37}{16}\right)\frac{\sqrt{17}}{16} + \left(\frac{11}{4}\right)\frac{5}{16} + \left(\frac{53}{16}\right)\frac{\sqrt{13}}{4} \doteq 4.44 \qquad \square$$

In the next section we show how to evaluate line integrals without resorting to the definition. Even so, a knowledge of the definition is essential for approximations to be made with the use of finite sums. Also, the definition of a line integral as a limit of approximating sums makes the following useful properties seem obvious:

1. $\int_C kf(P)\,dx = k\int_C f(P)\,dx$
2. $\int_C [f(P) + g(P)]\,dx = \int_C f(P)\,dx + \int_C g(P)\,dx$
3. $\int_C f(P)\,dx = -\int_{-C} f(P)\,dx$, where by $-C$ is meant the curve which is the same set of points as C but with opposite orientation.
4. If C is comprised of two arcs C_1 and C_2 (disjoint except for end points), then

$$\int_C f(P)\,dx = \int_{C_1} f(P)\,dx + \int_{C_2} f(P)\,dx$$

(See Figure 11.4.)

Similar results are valid for the other line integrals discussed above.

The extension from line integrals of scalar fields to line integrals of vector fields is an easy transition. For example, let f, g, and h be the scalar component functions of a three-dimensional vector field $\mathbf{F}$ and consider how we might write the sum of the three separate line integrals

$$\int_C f(P)\,dx, \quad \int_C g(P)\,dy, \quad \int_C h(P)\,dz$$

as a single line integral involving $\mathbf{F}$. One way begins by letting $d\mathbf{r}$ represent the vector expression $dx\,\mathbf{i} + dy\,\mathbf{j} + dz\,\mathbf{k}$. Then if x, y, and z are considered as functions of arc length, the expression for $d\mathbf{r}$ may be rewritten as

$$\begin{aligned} d\mathbf{r} &= \frac{dx}{ds}\,ds\,\mathbf{i} + \frac{dy}{ds}\,ds\,\mathbf{j} + \frac{dz}{ds}\,ds\,\mathbf{k} \\ &= \mathbf{T}\,ds \end{aligned}$$

Figure 11.4

where $\mathbf{T}$ is the unit vector tangent to C in the direction of increasing arc length. (Recall the discussion of Section 10.5.)

For piecewise smooth curves, $\mathbf{T}$ is well defined except at the endpoints of the smooth component arcs. Hence the sum of the three integrals may be expressed as

$$\int_C f(P)\,dx + \int_C g(P)\,dy + \int_C h(P)\,dz = \int_C f(P)\frac{dx}{ds}\,ds + \int_C g(P)\frac{dy}{ds}\,ds + \int_C h(P)\frac{dz}{ds}\,ds = \int_C \mathbf{F}\cdot\mathbf{T}\,ds = \int_C \mathbf{F}\cdot d\mathbf{r}$$

This integral is the most commonly occurring of all vector line integrals. For obvious reasons, it is called the *line integral of the tangential component of the vector field* $\mathbf{F}$. Other than for a notational device, no new definitions were needed to define this integral since the three line integrals comprising the vector line integral had been defined previously.

Line integrals of vector fields often represent physical quantities. For example, if a particle moves in a force field, $\mathbf{F}$, then the *work*, W, done on the particle is given by

$$W = \int_C \mathbf{F}\cdot d\mathbf{r}$$

This may be interpreted as the integral of the tangential component of the force field acting on the particle. See Figure 11.5.

Example 11.4 Approximate the work done in moving a particle along the curve $y = x^2$ from the point $(0, 0)$ to the point $(1, 1)$ in the presence of the force field $\mathbf{F} = y\mathbf{i} + x\mathbf{j}$. Use the partition of Example 11.2.

Solution In this example, $f(P_i') = y_i'$ and $g(P_i') = x_i'$. Therefore,

$$\sum_{i=0}^{2} f(P_i')\Delta x_i = (\tfrac{1}{16})\tfrac{1}{4} + (\tfrac{1}{4})\tfrac{1}{4} + (\tfrac{9}{16})\tfrac{1}{2} = \tfrac{23}{64}$$

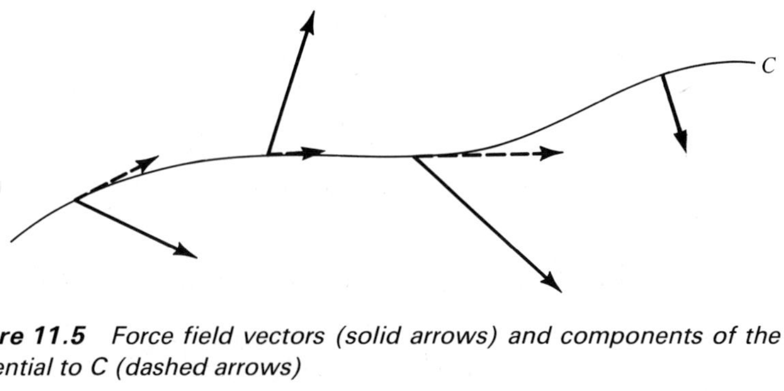

Figure 11.5 *Force field vectors (solid arrows) and components of the field tangential to C (dashed arrows)*

and

$$\sum_{i=0}^{2} g(P_i')\Delta y_i = (\tfrac{1}{4})\tfrac{1}{16} + (\tfrac{1}{2})\tfrac{3}{16} + (\tfrac{3}{4})\tfrac{3}{4} = \tfrac{43}{64}$$

It follows that

$$W \approx \tfrac{66}{64} \doteq 1.03 \qquad \square$$

Exercises for Section 11.2

1. Let C be any smooth curve connecting P_1 to P_2. Show by resorting to the definition of the line integral that

$$\int_C dx(P) = x(P_2) - x(P_1)$$

2. Generalize the result of Exercise 1; that is, let u be any continuous real-valued function defined on C. Prove that

$$\int_C du(P) = u(P_2) - u(P_1)$$

3. Use the definition of the line integral to approximate the work done in moving a particle along the parabola $x = y^2$ from the point $(1, 1)$ to $(9, 3)$ in the presence of the force field $\mathbf{F} = x^2\mathbf{i} + y^2\mathbf{j}$. Use the partition

$$B = \{(1, 1), (\tfrac{9}{4}, \tfrac{3}{2}), (4, 2), (\tfrac{25}{4}, \tfrac{5}{2}), (9, 3)\}$$

4. Approximate $\int_C x^2 y\,dx$ along a straight line segment from $(1, 1)$ to $(5, 5)$. Use the partition

$$B = \{(1, 1), (\tfrac{3}{2}, \tfrac{3}{2}), (2, 2), (\tfrac{5}{2}, \tfrac{5}{2}), (3, 3), (\tfrac{7}{2}, \tfrac{7}{2}), (4, 4), (\tfrac{9}{2}, \tfrac{9}{2}), (5, 5)\}$$

5. Approximate $\int_C x^2 y\,dy$ with the same curve and partition as in the previous exercise.
6. Approximate $\int_C x^2 y\,ds$ with the same curve and partition as in Exercise 4.
7. Approximate $\int_C y^2\,ds$ where C is the unit circle. Use the partition obtained by the eight points equally spaced on C and beginning at $(0, 1)$ and employ both formulations.
8. Repeat Exercise 7 for $\int_C y^2\,dy$.

In the following exercises use the definition of the line integral to approximate the work done in moving a particle along the given curve in the presence of the force field $\mathbf{F} = y\mathbf{i} - x\mathbf{j}$.

9. The parabolic curve and partition of Exercise 3.
10. The straight line segment and partition of Exercise 4.
11. The straight line segment from $(1, -1)$ to $(5, -5)$. Use the partition

$$B = \{(1, -1), (\tfrac{3}{2}, -\tfrac{3}{2}), (2, -2), (\tfrac{5}{2}, -\tfrac{5}{2}), (3, -3), (\tfrac{7}{2}, -\tfrac{7}{2}), (4, -4), (\tfrac{9}{2}, -\tfrac{9}{2}), (5, -5)\}$$

12. The circular curve and partition of Exercise 7. Employ both formulations.

11.3 Evaluation of Line Integrals

The limit defining a line integral of a scalar function f along even a smooth curve C is obviously quite complicated and would be impossible to compute directly

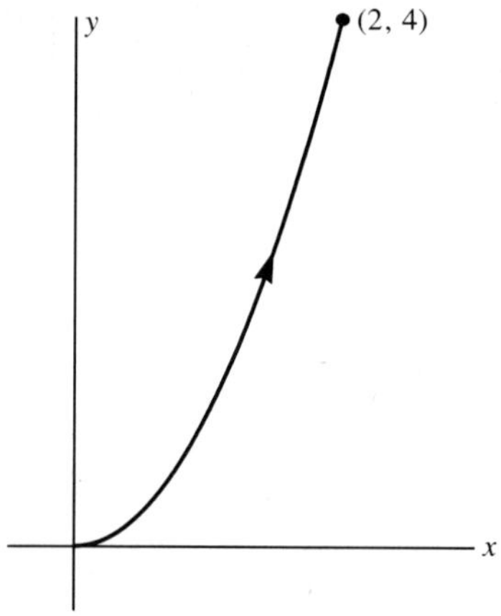

Figure 11.6

except by numerical methods. Often, in practical situations, the curve may be parametrized and thus represented by a well-behaved vector function $\mathbf{r}(t)$ (recall Definition 10.5). Under these circumstances the evaluation of the line integral can be done by appeal to the following theorem.

Theorem 11.2 Let C be a smooth curve described by $\mathbf{r}(t)$ over the interval $a \le t \le b$. If f is continuous on C, then

$$\int_C f(P)\,dx = \int_a^b f[x(t), y(t), z(t)]\frac{dx}{dt}\,dt$$

The value obtained is independent of the method of parametrization. ■

Analogous theorems hold for $\int_C f(P)\,dy$, $\int_C f(P)\,dz$, and $\int_C f(P)\,ds$.

Example 11.5 Evaluate the line integral of the tangential component of the vector field $\mathbf{F} = x^2y\mathbf{i} + (x + y)\mathbf{j}$ along the parabola $y = x^2$ from $(0, 0)$ to $(2, 4)$. (See Figure 11.6.)

Solution The curve C may be represented vectorially as $\mathbf{r}(t) = t\mathbf{i} + t^2\mathbf{j}$. Then

$$\begin{aligned}\int_C \mathbf{F}\cdot\mathbf{T}\,ds &= \int_C \mathbf{F}\cdot d\mathbf{r}\\ &= \int_C x^2y\,dx + \int_C (x + y)\,dy\\ &= \int_0^2 (t^2)(t^2)(1)\,dt + \int_0^2 (t + t^2)(2t)\,dt\\ &= \tfrac{32}{5} + (\tfrac{16}{3} + 8)\\ &= \tfrac{296}{15}\end{aligned}$$

□

Example 11.6 Let C be represented by $\mathbf{r}(t) = (1 - t)\mathbf{i} + (1 - t)\mathbf{j}$, $0 \leqq t \leqq 1$, and let $\mathbf{F}$ be as in the previous example. Evaluate

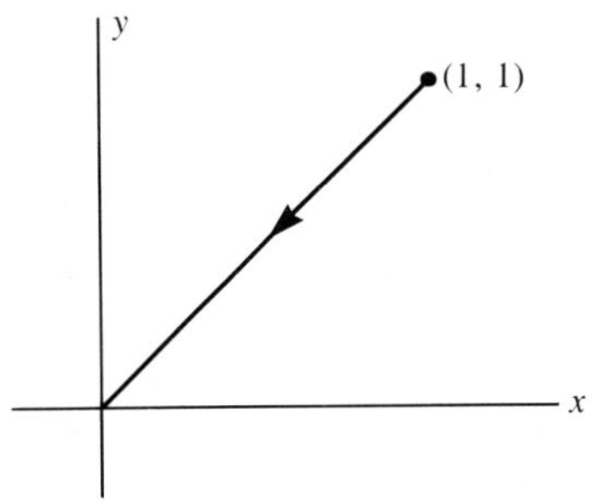

Figure 11.7

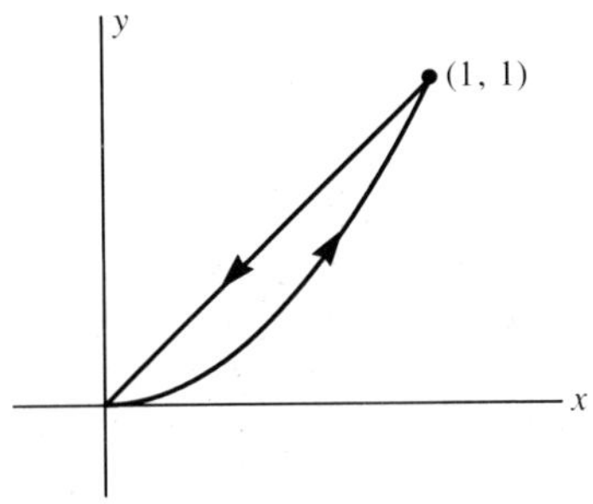

Figure 11.8

$$\int_C \mathbf{F} \cdot d\mathbf{r}$$

(See Figure 11.7.)

Solution

$$\begin{aligned}\int_C \mathbf{F} \cdot d\mathbf{r} &= \int_C x^2 y\,dx + \int_C (x+y)\,dy \\ &= \int_0^1 (1-t)^2(1-t)(-1)\,dt + \int_0^1 (1-t+1-t)(-1)\,dt \\ &= -\tfrac{5}{4}\end{aligned}$$

□

Example 11.7 Evaluate $\int_C \mathbf{F} \cdot d\mathbf{r}$ over the closed curve consisting of $y = x^2$ and $y = x$, taken counterclockwise. Use the same field as in the two previous examples. (See Figure 11.8.)

Solution The given curve is the union of the curve given by $\mathbf{r} = t\mathbf{i} + t^2\mathbf{j}$, $0 \leqq t \leqq 1$ and the curve of the previous example. The value of $\int_C \mathbf{F} \cdot d\mathbf{r}$ over the first of these curves can be shown to be $\frac{41}{30}$. By the additivity property, the value of the integral is the sum of the values of the two separate integrals,

$$\begin{aligned}\oint_C \mathbf{F} \cdot d\mathbf{r} &= \tfrac{41}{30} - \tfrac{5}{4} \\ &= \tfrac{7}{60}\end{aligned}$$

□

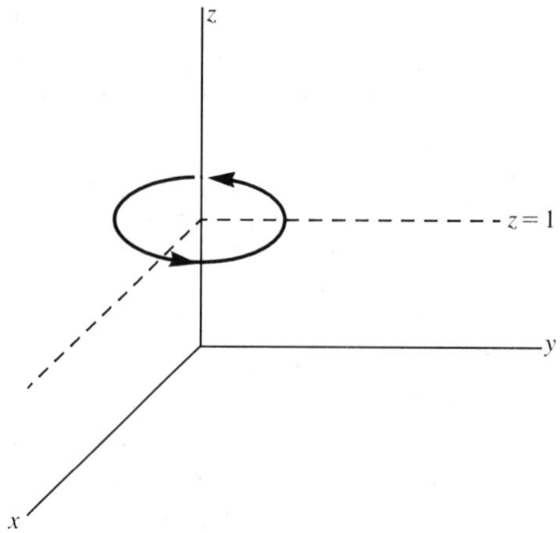

Figure 11.9

It might appear initially that line integrals around closed curves should be zero. Example 11.7 should be evidence enough to keep one from believing that. On the other hand, the next example shows that a line integral around a closed curve could indeed be zero. In the following section, some necessary qualifications for this to be true are examined.

Example 11.8 Let C be the circle of radius 1 in the plane $z = 1$ centered on the z-axis. Let $\mathbf{F} = (y + z)\mathbf{i} + (x + z)\mathbf{j} + (x + y)\mathbf{k}$. Compute $\oint_C \mathbf{F} \cdot d\mathbf{r}$, where the direction around the closed curve is counterclockwise as viewed from above. (See Figure 11.9.)

Solution One convenient parametrization of C is $x = \cos t$, $y = \sin t$, and $z = 1$. Using this, we get

$$\oint_C \mathbf{F} \cdot d\mathbf{r} = \oint_C (y + z)\,dx + \oint_C (x + z)\,dy + \oint_C (x + y)\,dz$$
$$= \int_0^{2\pi} (\sin t + 1)(-\sin t)\,dt + \int_0^{2\pi} (1 + \cos t)(\cos t)\,dt$$

(A third integral doesn't appear since $dz/dt = 0$.) From this, it is easy to show that

$$\oint_C \mathbf{F} \cdot d\mathbf{r} = 0$$ □

Example 11.9 A particle moves in a force field $\mathbf{F} = x\mathbf{i} - z\mathbf{j} + 2y\mathbf{k}$ along the closed curve shown in Figure 11.10. Find the work done in moving the particle completely around the curve.

Solution The curve consists of three arcs, which may be parametrized as follows:

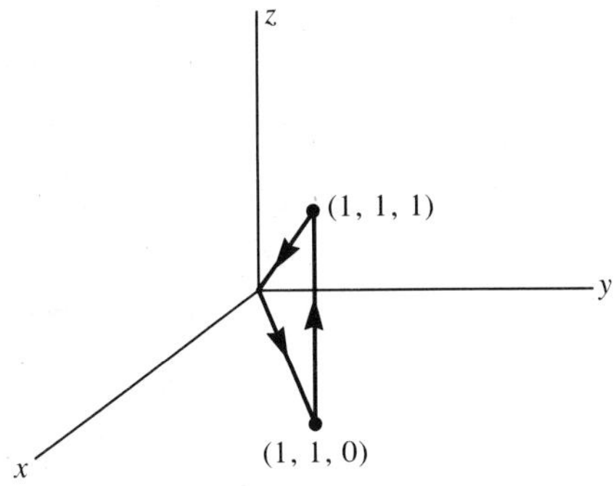

Figure 11.10

$$\begin{aligned}\mathbf{r}(t) &= t\mathbf{i} + t\mathbf{j} & 0 \leqq t \leqq 1 \\ &= \mathbf{i} + \mathbf{j} + (t-1)\mathbf{k} & 1 \leqq t \leqq 2 \\ &= (3-t)\mathbf{i} + (3-t)\mathbf{j} + (3-t)\mathbf{k} & 2 \leqq t \leqq 3\end{aligned}$$

(This is certainly not the only way to represent C and indeed may not be the simplest.) From the definition of work,

$$\begin{aligned}W &= \oint_C \mathbf{F} \cdot d\mathbf{r} \\ &= \oint_C x\,dx - \oint_C z\,dy + \oint_C 2y\,dz \\ &= \int_0^1 t\,dt + \int_1^2 2\,dt - \int_2^3 (3-t)\,dt + \int_2^3 (3-t)\,dt - \int_2^3 2(3-t)\,dt \\ &= \tfrac{1}{2} + 2 - \tfrac{1}{2} + \tfrac{1}{2} - 1 \\ &= \tfrac{3}{2}\end{aligned}$$

□

11.4 Path Independence

As you may have supposed, we have more than passing interest in determining conditions under which the integral of the tangential component of a vector field around a closed curve is zero. Actually our discussion concerns those fields for which such an integration around *any* closed curve is zero. In such fields, the value of a line integral over a curve which is not closed is uniquely determined by the end points. As usual, our discussion is limited to piecewise smooth curves.

Definition 11.2 The line integral $\int_C \mathbf{F} \cdot d\mathbf{r}$ is said to be *path independent* or *independent of the curve* connecting two points P and Q if the value of the integral is the same regardless of the curve connecting the two points. If $\int_C \mathbf{F} \cdot d\mathbf{r}$ is path independent, the usual notation is sometimes replaced by $\int_P^Q \mathbf{F} \cdot d\mathbf{r}$. (See Figure 11.11.)

The idea of path independence is that a curve connecting two points may be "continuously deformed" without affecting the value of the integral. As we

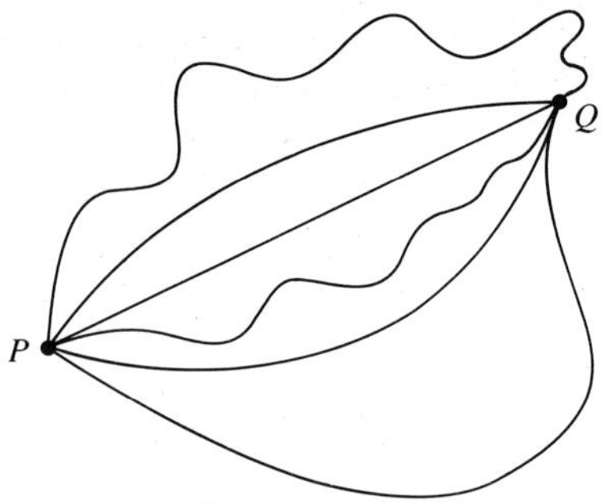

Figure 11.11 *Continuous deformation of curves*

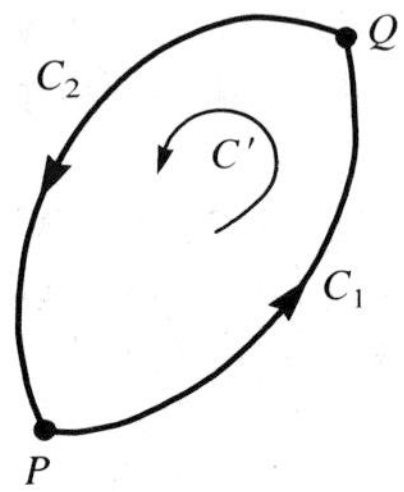

Figure 11.12

will see the only stipulations are that the end points must remain fixed and the curve must stay entirely in a domain in which **F** is continuous. The concepts of path independence and zero-valued line integral around a closed curve can be tied together by using the following theorem.

Theorem 11.3 A line integral $\int_C \mathbf{F} \cdot d\mathbf{r}$ is independent of the curve connecting the end points of C in a domain* D *if and only if* $\oint_{C'} \mathbf{F} \cdot d\mathbf{r} = 0$ for all closed curves C' in D.

Proof Suppose $\int_C \mathbf{F} \cdot d\mathbf{r}$ is path independent. Select two points on a closed curve C' and use them to cut C' into two arcs C_1 and C_2, where C_1 is from P to Q and C_2 is taken from Q to P. (See Figure 11.12.) Then

$$\oint_{C'} \mathbf{F} \cdot d\mathbf{r} = \int_{C_1} \mathbf{F} \cdot d\mathbf{r} + \int_{C_2} \mathbf{F} \cdot d\mathbf{r}$$

Because of the path independence,

$$\int_{C_1} \mathbf{F} \cdot d\mathbf{r} = \int_{-C_2} \mathbf{F} \cdot d\mathbf{r}$$

---

* For our purposes a domain is an open connected set. Loosely speaking, a connected set is one that is all in one piece, or one in which any two points may be connected by a curve that remains in the set. An open set is one for which each of its points may be enclosed in a sphere (3-space) or a circular disk (2-space) lying entirely in the set.

where the curve $-C_2$ is obtained from C_2 by a reversal of sense and thus is oriented from P to Q. Therefore

$$\int_{C_1} \mathbf{F} \cdot d\mathbf{r} = -\int_{C_2} \mathbf{F} \cdot d\mathbf{r}$$

and hence

$$\oint_{C'} \mathbf{F} \cdot d\mathbf{r} = 0$$

The converse of the theorem is proved similarly and is left for the exercises. ■

When a conservative vector field was defined in the previous chapter, you may have wondered about the aptness of the terminology. The previous theorem and the next one (which we state without proof) provide a motive for describing vector fields with scalar potentials as "conservative," since zero work or energy is needed to move a particle around a closed curve in such a field. At the same time, the theorem also gives a well-known method for evaluating a line integral of the tangential component of a conservative vector field. The method looks a bit like the fundamental theorem of calculus, and with good reason. For conservative fields it is the generalization of that basic result to arbitrary curves.

Theorem 11.4 Let $\mathbf{F}$ be a vector field with continuous scalar component functions in a domain D. The line integral $\int_C \mathbf{F} \cdot d\mathbf{r}$ is path independent in D *if and only if* $\mathbf{F}$ is conservative. Further, if U is a scalar potential of $\mathbf{F}$ (that is, $\nabla U = \mathbf{F}$), and P and Q are the endpoints of C, then

$$\int_C \mathbf{F} \cdot d\mathbf{r} = U(Q) - U(P)$$ ■

This important theorem not only tells us that a line integral of the tangential component of a conservative vector field is path independent but also gives us a technique of evaluation based upon finding a scalar potential and computing the "potential difference."

Example 11.10 Show that $\mathbf{F} = yz\mathbf{i} + xz\mathbf{j} + xy\mathbf{k}$ is conservative and then compute $\int_C \mathbf{F} \cdot d\mathbf{r}$ along any curve connecting $(1, -1, 3)$ to $(5, 2, -1)$.

Solution It is easy to show that a scalar potential is the function $U(x, y, z) = xyz$ (see Section 10.9) and hence the value of the line integral is $U(5, 2, -1) - U(1, -1, 3) = -10 - (-3) = -7$. □

The close relationship between path independence and "conservation" can also be extended to include irrotationality.

Theorem 11.5 If a line integral $\int_C \mathbf{F} \cdot d\mathbf{r}$ in three-space is path independent in a domain D, then $\nabla \times \mathbf{F} = \mathbf{0}$ throughout D.

Proof This theorem follows immediately from Theorem 11.4 and the knowledge that the curl of a field with a scalar potential is zero. (Recall Example 10.34.) ■

We later show (in Section 11.8) that the converse of this theorem, although not generally true, *is* valid in certain important situations.

Exercises for Sections 11.3 and 11.4

1. Complete the proof of Theorem 11.3.
2. Evaluate $\int_C \mathbf{F} \cdot d\mathbf{r}$ if $\mathbf{F} = (2x + y)\mathbf{i} + x\mathbf{j}$ and C is as given.
   (a) The straight line segment from $(0, 0)$ to $(3, 2)$.
   (b) The y-axis from $(0, 0)$ to $(0, 2)$ and then the straight line segment from $(0, 2)$ to $(3, 2)$.
3. Evaluate $\int_C \mathbf{F} \cdot d\mathbf{r}$ if $\mathbf{F} = y\mathbf{i} + (x + 2y)\mathbf{j}$ and C is given by $\mathbf{r}(t) = 3t^2\mathbf{i} + 2t\mathbf{j}$ from $t = 0$ to $t = 1$.
4. Evaluate $\int_C \mathbf{F} \cdot d\mathbf{r}$ if $\mathbf{F} = x^2y\mathbf{i} + y^2z\mathbf{j} + z^2x\mathbf{k}$ and C is as given.
   (a) The straight line segment from $(0, 0, 0)$ to $(1, 3, 3)$.
   (b) The polygonal figure connecting $(0, 0, 0)$ to $(1, 0, 0)$ to $(1, 3, 0)$ to $(1, 3, 3)$.
5. Evaluate $\int_C \mathbf{F} \cdot d\mathbf{r}$ if $\mathbf{F} = xy^2\mathbf{i} + yz^2\mathbf{j} + zx^2\mathbf{k}$ and C is given by $\mathbf{r}(t) = t^2\mathbf{i} + t\mathbf{j} + 2t^3\mathbf{k}$ from $t = 0$ to $t = 1$.
6. Find the work done in moving a particle in a force field $\mathbf{F} = x^2y\mathbf{i} - yz\mathbf{j} + z\mathbf{k}$ along the smooth curve $\mathbf{r}(t) = t\mathbf{i} + t^2\mathbf{j} + t^3\mathbf{k}$ from the origin to the point $(2, 4, 8)$. Repeat for the straight line segment connecting the two points.
7. Evaluate $\oint_C \mathbf{F} \cdot d\mathbf{r}$ around the complete unit circle in the plane $z = 0$, taken counterclockwise, if $\mathbf{F} = x^2\mathbf{i} + y\mathbf{j}$.
8. Let C be the unit semicircle $\mathbf{r}(t) = \cos t\mathbf{i} + \sin t\mathbf{j}$, $0 \leq t \leq \pi$. Evaluate $\int_C \mathbf{F} \cdot d\mathbf{r}$ if $\mathbf{F} = x^2y\mathbf{i} + (x + y)\mathbf{j}$. Repeat for the straight line connecting the two endpoints of C.

In Exercises 9–12, evaluate $\oint_C \mathbf{F} \cdot \mathbf{T}\, ds$ for the given vector function $\mathbf{F}$ and curve C.

9. $\mathbf{F} = -3y\mathbf{i} + 3x\mathbf{j} + z\mathbf{k}$ and C is the unit circle centered at $(0, 0, 1)$, in the plane $z = 1$, oriented counterclockwise as viewed from the origin.
10. $\mathbf{F} = y\mathbf{i} + xz^3\mathbf{j} - zy^3\mathbf{k}$ and C is the unit circle centered at $(0, 0, -1)$, in the plane $z = -1$, oriented counterclockwise as viewed from the origin.
11. $\mathbf{F} = \dfrac{-y\mathbf{i} + x\mathbf{j}}{x^2 + y^2}$ and C is the unit circle centered at $(0, 0, 0)$, in the plane $z = 0$, oriented counterclockwise as viewed from above.
12. $\mathbf{F} = (\sin z)\mathbf{i} - (\cos x)\mathbf{j} + (\sin y)\mathbf{k}$ and C is the boundary of the rectangle $0 \leq x \leq 1$, $0 \leq y \leq 2$, $z = 3$ oriented clockwise as viewed from the origin.

In Exercises 13–17, evaluate $\int_C \mathbf{F} \cdot d\mathbf{r}$ for the given vector function $\mathbf{F}$ and curve C.

13. $\mathbf{F} = (y - 7)\mathbf{i} + x\mathbf{j}$ and C is the curve $\mathbf{r}(t) = \sin t^2\mathbf{i} + t\mathbf{j}$ connecting $(0, 0)$ to $(1, \sqrt{\pi/2})$.
14. $\mathbf{F} = (2y^2 - 3x)\mathbf{i} - 4xy\mathbf{j}$ and C is the straight line segment connecting $(0, 0)$ to $(2, 2)$.
15. $\mathbf{F} = yz\mathbf{i} + xz\mathbf{j} + xy\mathbf{k}$ and C is the closed curve $x = y^2$, $y = x^2$ taken counterclockwise in the plane $z = 2$.
16. $\mathbf{F} = (2xy + x^2)\mathbf{i} + x^2\mathbf{j}$ and C is the parabola connecting $(-1, 1)$ to $(1, 1)$ that passes through the origin.

17. $\mathbf{F} = yz\cos(xz)\mathbf{i} + \sin(xz)\mathbf{j} + xy\cos(xz)\mathbf{k}$ and C is any curve connecting $(1, 1, 1)$ to $(2, 3, 4)$.

18. Let $\mathbf{F} = (y^2 + 2xz)\mathbf{i} + (2xy - z)\mathbf{j} + (x^2 - y)\mathbf{k}$. Evaluate $\int_C \mathbf{F}\cdot d\mathbf{r}$ where C is as given.
(a) The ellipse $4x^2 + y^2 = 1$ beginning at $(0, 1)$, oriented clockwise, in the plane $z = 0$.
(b) The curve $\mathbf{r}(t) = \sin^3 \pi t/2\mathbf{i} + t\cos\pi t\mathbf{j} + t^2\cos^2\pi t\mathbf{k}$ from $(0, 0, 0)$ to $(1, -1, 1)$.
(c) The straight line segment from $(0, 0, 0)$ to $(2, 4, 7)$.

19. Let $\mathbf{F} = 3z^2\mathbf{i} + 2y\mathbf{j} + 6xz\mathbf{k}$ and repeat the previous exercise.

In Exercises 20–21 a number of line integrals are given. Some of the line integrals are path independent in suitably chosen domains. Find those in this category and indicate the appropriate domain(s) in each case.

20. **(a)** $\displaystyle\int_C [(2xy + y^2)\mathbf{i} + (x^2 + 1)\mathbf{j}]\cdot d\mathbf{r}$ **(b)** $\displaystyle\int_C [z\mathbf{i} + z^2\mathbf{j} + (x + 2zy)\mathbf{k}]\cdot d\mathbf{r}$

(c) $\displaystyle\int_C \left[\frac{x}{x^2 + y^2}\mathbf{i} + \frac{y}{x^2 + y^2}\mathbf{j}\right]\cdot d\mathbf{r}$

21. **(a)** $\displaystyle\int_C \mathbf{F}\cdot d\mathbf{r}$, where $\mathbf{F} = x\mathbf{i} - y\mathbf{j}$

(b) $\displaystyle\int_C \mathbf{F}\cdot d\mathbf{r}$, where $\mathbf{F} = \frac{y}{y + z}\mathbf{i} + \frac{xz}{(y + z)^2}\mathbf{j} - \frac{xy}{(y + z)^2}\mathbf{k}$

(c) $\displaystyle\int_C \mathbf{F}\cdot d\mathbf{r}$, where $\mathbf{F} = y^2\mathbf{i} + x^2\mathbf{j} + (x^2 + y^2)\mathbf{k}$

In Exercises 22–25, each of the given line integrals is path independent in a suitably chosen domain. Evaluate each integral in *two* different ways.

22. $\displaystyle\int_C \mathbf{F}\cdot d\mathbf{r}$, where $\mathbf{F} = \frac{-y}{x^2 + y^2}\mathbf{i} + \frac{x}{x^2 + y^2}\mathbf{j}$ and C is the right half of the unit circle beginning at $(0, -1)$ and ending at $(0, 1)$.

23. $\displaystyle\int_C \left[\frac{\ln y}{1 - x^2}\mathbf{i} + \frac{\mathrm{Sin}^{-1} x}{y}\mathbf{j}\right]\cdot d\mathbf{r}$, where C is the straight line segment from $(-\frac{1}{2}, 1)$ to $(\frac{1}{2}, 2)$.

24. $\int_C \mathbf{F}\cdot d\mathbf{r}$, where $\mathbf{F}$ is the three-dimensional central force field $-k^2\mathbf{r}/r^3$ and C is an elliptical curve joining $(-1, -1, 0)$ to $(1, 1, 0)$ and lying in the half-space $z \geq 0$.

25. $\int_C f(x^2 + y^2)(x\mathbf{i} + y\mathbf{j})\cdot d\mathbf{r}$, where f is a continuous function of its argument and C is a polygonal curve joining (x_0, y_0) to (x_1, y_1).

In fluid flow problems the line integral

$$\oint_C \mathbf{v}\cdot d\mathbf{r}$$

of the tangential component of the vector velocity around a closed curve C within the fluid is termed the circulation around C. This quantity, calculated for various curves C, is useful in describing the general character of the fluid motion.

For two-dimensional fluid flows determined by the following vector velocities, find the circulation around the circle $x^2 + y^2 = R^2$, oriented counterclockwise.

26. **(a)** $\mathbf{v} = U_0\mathbf{i}$ **(b)** $\mathbf{v} = 2axy\mathbf{i} + a(x^2 - y^2)\mathbf{j}$

27. **(a)** $\mathbf{v} = f(r)\mathbf{u}_r$ **(b)** $\mathbf{v} = r\mathbf{u}_\theta$

28. **(a)** $\mathbf{v} = \dfrac{\theta}{r}\mathbf{u}_r + \dfrac{1}{r}\ln r\mathbf{u}_\theta$ **(b)** $\mathbf{v} = \left(1 - \dfrac{a^2}{r^2}\right)\cos\theta\,\mathbf{u}_r - \left(1 + \dfrac{a^2}{r^2}\right)\sin\theta\,\mathbf{u}_\theta$

29. Could you have predicted the answers determined in Exercises 26–28 on the basis of Theorem 11.4? Does the result for Exercise 28 (a) contradict the theorem? What about the answer for Exercise 28 (b)? Explain.

11.5 Surface Area

In Section 10.3 we discussed surfaces and showed how to represent them in terms of vector functions of two variables. Now we turn to the notion of surface area. The eventual understanding of the more general concept of a surface integral is almost completely dependent upon agreements reached with respect to an elemental piece of surface. There are several approaches to the problem of how to define surface area, some of them based upon practical considerations. First we consider a method applicable to a very general kind of surface and then examine two other approaches that are used in specific cases. We again restrict our attention to surfaces in three dimensions.

Let S be a smooth surface defined by $\mathbf{r}(u, v)$, where the (u, v) are restricted to a domain S'. Then S is the image of S'. The image of a rectangular element with area $\Delta S' = \Delta u \Delta v$ is a curvilinear surface element having area ΔS on S. (See Figure 11.13.) The sides of the curvilinear surface element are parts of coordinate curves, $\mathbf{r}(u_0, v)$, $\mathbf{r}(u_0 + \Delta u, v)$, $\mathbf{r}(u, v_0)$, and $\mathbf{r}(u, v_0 + \Delta v)$. Thus the lengths of the sides are approximately $|\mathbf{r}_u \Delta u|$ and $|\mathbf{r}_v \Delta v|$ and therefore, also approximately,

$$\begin{aligned}\Delta S &= |(\mathbf{r}_u \Delta u) \times (\mathbf{r}_v \Delta v)| \\ &= |\mathbf{r}_u \times \mathbf{r}_v| \Delta u\, \Delta v\end{aligned}$$

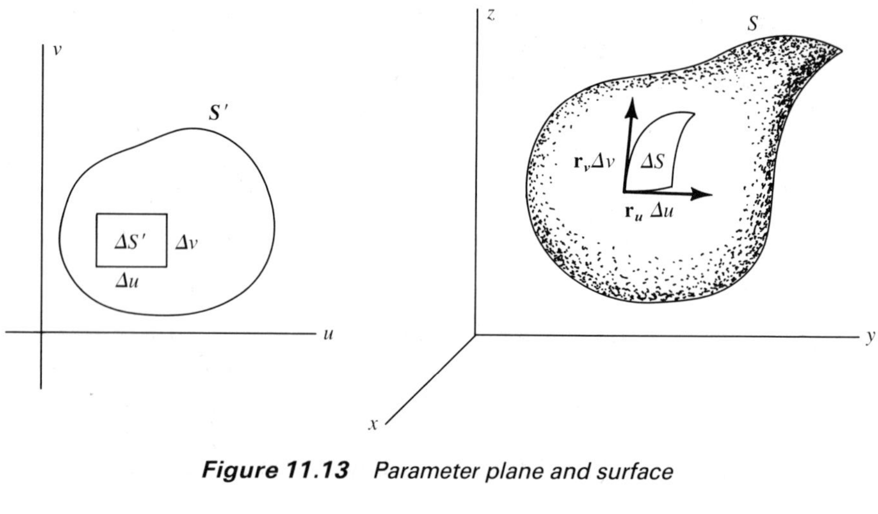

Figure 11.13 Parameter plane and surface

(Recall the nature of the vector cross product, Section 1.6.) It is quite reasonable, then, that we *define* the total surface area to be

$$S \equiv \iint_{S'} |\mathbf{r}_u \times \mathbf{r}_v| \, du \, dv$$

For reasons more apparent in a general setting, the notation for surface area is $\iint_S dS$. This notation will be more meaningful after the discussion of the next section.

Example 11.11 Find the surface area of the portion of the parabolic cylinder $y = x^2$ in the first octant bounded by the planes $z = 2$ and $y = \frac{1}{4}$.

Solution The surface in question may be parametrized by letting $x = u$ and $z = v$ (the xz-plane is the parameter plane). Thus $\mathbf{r}(u, v) = u\mathbf{i} + u^2\mathbf{j} + v\mathbf{k}$ and, by the formula of Example 1.9,

$$\begin{aligned} |\mathbf{r}_u \times \mathbf{r}_v|^2 &= (\mathbf{r}_u \cdot \mathbf{r}_u)(\mathbf{r}_v \cdot \mathbf{r}_v) - (\mathbf{r}_u \cdot \mathbf{r}_v)^2 \\ &= 1 + 4u^2 \end{aligned}$$

Therefore,

$$\begin{aligned} S &= \iint_{S'} (1 + 4u^2)^{1/2} \, du \, dv, \qquad S' = \left\{ (u, v) \middle| \begin{matrix} 0 < u < \frac{1}{2} \\ 0 < v < 2 \end{matrix} \right\} \\ &= \int_0^2 \int_0^{1/2} (1 + 4u^2)^{1/2} \, du \, dv \\ &= \frac{\sqrt{2}}{2} + \tfrac{1}{2} \ln(1 + \sqrt{2}) \end{aligned}$$

□

Often the formula for surface area will be written in a form other than the one we have just given and thus it is important to see how the formula looks in some special cases. If the surface is represented in the form $z = h(x, y)$, where h is always of one sign (say positive), then we may think of the surface S as covering some region S' in the xy-plane; that is, S' is the projection of S on the xy-plane. Let γ be the angle between the upward-directed normal and the z-axis, and consider an element of surface ΔS. (See Figure 11.14.) Then ΔS and $\Delta S'$ are approximately related by

$$\Delta S' = \Delta S \cos \gamma$$

or

$$\Delta S = \Delta S' \sec \gamma$$

In this case we *define* the total surface area to be

$$S \equiv \iint_{S'} \sec \gamma \, dx \, dy$$

where S' is the projection of S on the xy-plane.

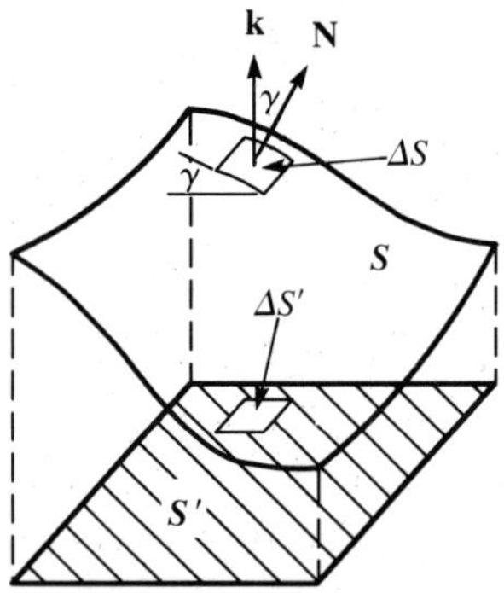

Figure 11.14 *Projection of a surface on the xy-plane*

The fact that the two definitions are equivalent is shown by noting that the explicit equation $z = h(x, y)$ gives the vector representation $\mathbf{r}(u, v) = u\mathbf{i} + v\mathbf{j} + h(u, v)\mathbf{k}$. Hence

$$|\mathbf{r}_u \times \mathbf{r}_v| = \sqrt{h_u^2 + h_v^2 + 1}$$

from which

$$S = \iint_{S'} \sqrt{h_u^2 + h_v^2 + 1}\, du\, dv$$
$$= \iint_{S'} \sqrt{h_x^2 + h_y^2 + 1}\, dx\, dy$$

For the surface defined by $z - h(x, y) = 0$, it is easy to show that the upward-directed unit normal to the surface is

$$\mathbf{N} = \frac{-h_x\mathbf{i} - h_y\mathbf{j} + \mathbf{k}}{\sqrt{h_x^2 + h_y^2 + 1}}$$

It follows that

$$\cos\gamma = \frac{1}{\sqrt{h_x^2 + h_y^2 + 1}}$$

and thus the integrand of the integral over S' is just $\sec\gamma$.

Example 11.12 Find the area of the surface $z^2 = x^2 + y^2$ for $0 < z < 1$. (See Figure 11.15.)

Solution To find a normal to the surface, we consider a constant level surface of the function $\phi(x, y, z) = x^2 + y^2 - z^2$. Then

$$\mathbf{N} = \frac{\nabla\phi}{|\nabla\phi|}$$
$$= \frac{2x\mathbf{i} + 2y\mathbf{j} - 2z\mathbf{k}}{(4x^2 + 4y^2 + 4z^2)^{1/2}}$$
$$= \frac{x\mathbf{i} + y\mathbf{j} - z\mathbf{k}}{z\sqrt{2}}$$

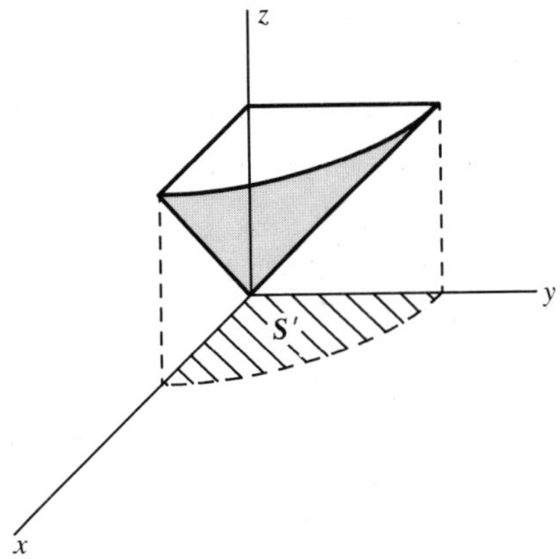

Figure 11.15

Therefore

$$\sec\gamma = \frac{1}{|\mathbf{N}\cdot\mathbf{k}|} = \sqrt{2}$$

The expression for the surface area thus becomes

$$\begin{aligned} S &= \iint_{S'} \sqrt{2}\,dx\,dy \\ &= \sqrt{2}\int_{-1}^{1}\int_{-(1-x^2)^{1/2}}^{(1-x^2)^{1/2}} dy\,dx \\ &= \pi\sqrt{2} \end{aligned}$$

□

If S is represented in the form $y = g(x, z)$ or $x = f(y, z)$, the definition for surface area leads to the integrals $\iint_{S'} \sec\beta\,dx\,dz$ and $\iint_{S'} \sec\alpha\,dy\,dz$ where α, β, and the region of integration are analogous to the case described earlier. Details are relegated to the exercises. In all the cases, the problem of finding the value of the surface area is reduced to a problem of double integration in the parameter plane or (which is the same thing) in the plane onto which the surface projects.

There is another method often used for the computation of surface area, which is quite straightforward but which requires some prior knowledge of the construction of an element of surface. This technique is especially valuable in connection with surfaces such as right circular cylinders or spheres. For example, if θ and z are the parameters for a cylinder of fixed radius a, then one could reason directly that the element of surface area is $a\,d\theta\,dz$. Similarly, one often reasons informally that the element of surface area on a sphere of fixed radius is $a^2\sin\phi\,d\theta\,d\phi$. (See Figure 11.16.) The informal reasoning is done by appealing to the graphs of a right circular cylinder and a sphere but the validity of the reasoning is verified by considering the vector representations

$$\mathbf{r}(\theta, z) = (a\cos\theta)\mathbf{i} + (a\sin\theta)\mathbf{j} + z\mathbf{k}$$

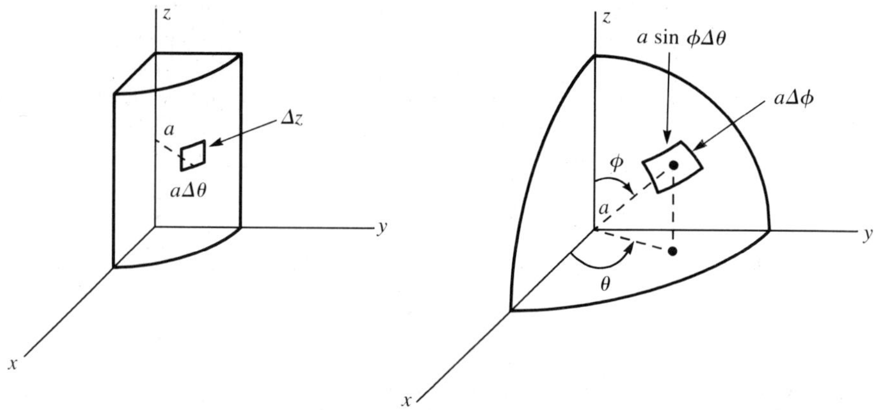

Figure 11.16 *Elements of surface area on a cylinder and a sphere*

for a cylinder and

$$\mathbf{r}(\theta, \phi) = (a \sin \phi \cos \theta)\mathbf{i} + (a \sin \phi \sin \theta)\mathbf{j} + (a \cos \phi)\mathbf{k}$$

for a sphere. The details are again left for the exercises.

Before moving on to the next section, it should be noted that there are many simple surfaces for which one or more of the above formulas are inadequate to determine the entire surface area. As a case in point consider the surface of a sphere with x and y as parameters, the surface in Example 11.12 for both negative and positive z, or the surface of a cube. In each of these situations, however, the surface of interest can be thought of as composed of two or more parts, each one of which may be effectively handled individually as above. These and other examples prompt the following definition.

Definition 11.3 A *piecewise-smooth surface* is *either* a smooth surface together with its piecewise-smooth boundary but perhaps having a finite number of points at which a tangent plane is not uniquely defined, *or* a finite union of such surfaces joined along their common border curves.

The surface area of a piecewise-smooth surface is obviously the sum of the areas of each of the finitely many component surfaces.

11.6 Surface Integrals of Scalar and Vector Fields

Integrals of functions on surfaces (surface integrals) are defined just like the other integration processes except that the element of a partition is an elemental surface area. As before, we begin with scalars and work our way up to vector fields. Let f be a scalar function defined and continuous on a surface S in three-space. After subdividing S into n subregions each of area ΔS_i, we pick a point within each subregion, P_i', and evaluate $f(P_i')$. If the limit of the finite sum

$$\sum_{i=1}^{n} f(P_i')\Delta S_i$$

exists as the largest of the ΔS_i approaches zero, the limit is called the *surface integral of the scalar function f over S* and is denoted by

$$\iint_S f(P)\,dS$$

(Sometimes only one integral sign is used; the reason for the use of two is largely historical.)

In keeping with the tone of this book, we do not concern ourselves with examining the conditions to be imposed on the function and the surface, necessary for the surface integral to exist. Suffice it to say that the following theorem, which is the analog of Theorem 11.2 for line integrals, is known to be true.

Theorem 11.6 Let S be a compact* smooth surface in three-space described by the vector function $\mathbf{r}(u, v)$ with domain S'_{uv}. If f is continuous on S, then $\iint_S f(P)\,dS$ exists and equals

$$\iint_{S'_{uv}} f[x(u, v), y(u, v), z(u, v)]\,|\mathbf{r}_u \times \mathbf{r}_v|\,du\,dv$$ ■

The extension of this result to piecewise-smooth surfaces should be obvious.

Assuming familiarity with the material of the previous section, the only added complication is that now integrand functions other than a constant must be dealt with. If $f(x, y, z)$ is a scalar-valued function and the surface S is represented by

$$\mathbf{r}(u, v) = x(u, v)\mathbf{i} + y(u, v)\mathbf{j} + z(u, v)\mathbf{k}$$

then the first step in the calculations is usually to find and simplify the function $g(u, v)$ defined by

$$g(u, v) = f[x(u, v), y(u, v), z(u, v)]$$

Thus we replace the function f defined over the whole space by the function g defined only for points on the parameter plane.

As in the previous section, the factor $|\mathbf{r}_u \times \mathbf{r}_v|$ may be replaced by $\sec\gamma$ if the xy-plane serves as the parameter plane; or by some other appropriate factor dependent upon the method used to represent the surface. Sometimes, several different approaches are possible.

Example 11.13 Evaluate the surface integral of the function

$$f(x, y, z) = \frac{2y^2 + z}{(4x^2 + 4y^2 + 1)^{1/2}}$$

---

* A set is *compact* if it is *bounded* (that is, contained in some sphere of finite radius) and contains all of its *boundary points* (a boundary point of a set M is a point P such that every sphere of finite radius about P as center contains both points in M and points not in M).

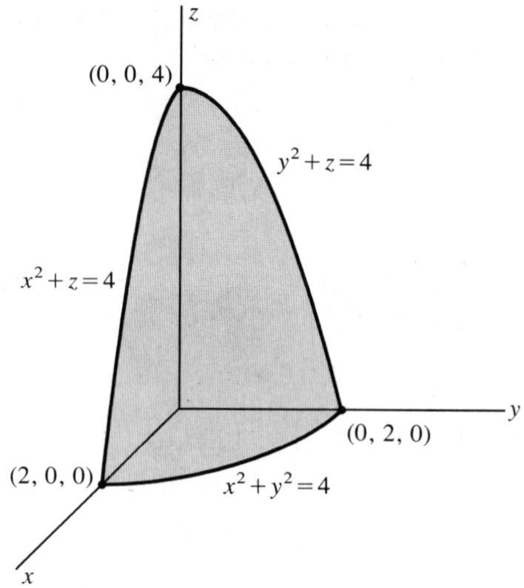

Figure 11.17

over that portion of the surface of the paraboloid

$$x^2 + y^2 + z = 4$$

for which z is positive. (Figure 11.17 shows one-fourth of this surface.)

Solution

$$\iint_S f(x, y, z)\, dS = \iint_{S'_{xy}} \frac{2y^2 + z(x, y)}{(4x^2 + 4y^2 + 1)^{1/2}} \sec\gamma\, dx\, dy$$

A normal to S is found by considering a constant level surface of the function $\phi(x, y, z) = x^2 + y^2 + z$:

$$\mathbf{N} = \frac{\nabla\phi}{|\nabla\phi|} = \frac{2x\mathbf{i} + 2y\mathbf{j} + \mathbf{k}}{(4x^2 + 4y^2 + 1)^{1/2}}$$

$$\sec\gamma = \frac{1}{|\mathbf{N}\cdot\mathbf{k}|} = (4x^2 + 4y^2 + 1)^{1/2}$$

Using this value in the double integral over the region S'_{xy} in the xy-plane, we obtain

$$\iint_S f(x, y, z)\, dS = \int_{-2}^{2} \int_{-(4-x^2)^{1/2}}^{(4-x^2)^{1/2}} (4 - x^2 + y^2)\, dy\, dx$$

In polar coordinates this becomes.

$$\int_0^{2\pi} \int_0^2 (4 + r^2 - 2r^2\cos^2\theta) r\, dr\, d\theta = \int_0^{2\pi} \left(2r^2 + \frac{r^4}{4} - \frac{r^4}{2}\cos^2\theta\right)\Big|_0^2 d\theta$$

$$= \int_0^{2\pi} (12 - 8\cos^2\theta)\, d\theta$$

$$= 24\pi - 8\pi$$

$$= 16\pi$$

□

For surface integrals we have a property analogous to that of independence of path for line integrals:

Definition 11.4 The surface integral

$$\iint_S f(P)\,dS$$

is said to be *independent of the surface* with boundary curve C if the numerical value is the same for each piecewise-smooth surface with the same (total) boundary curve.

Example 11.14 Show that for smooth surfaces S represented in the form $z = h(x, y)$, the surface integral

$$\iint_S (h_x^2 + h_y^2 + 1)^{-1/2}\,dS$$

is independent of the surface with boundary curve C.

Solution Earlier we showed that the surface infinitesimal dS can be replaced by $\sec\gamma\,dx\,dy = \sqrt{h_x^2 + h_y^2 + 1}\,dx\,dy$ for smooth surfaces defined by $z = h(x, y)$. Hence

$$\iint_S (h_x^2 + h_y^2 + 1)^{-1/2}\,dS = \iint_{S'} dx\,dy$$
$$= \text{area of the projection } S' \text{ in the } xy\text{-plane}$$

For fixed boundary curve C, the projections of the surfaces S are all the same, and thus the values of the integrals are all equal to one another. □

It has been suggested that the concept of independence of surface can best be described by a butterfly net in which the surface (the net) may be deformed while holding the boundary curve (the rim) constant. See Figure 11.18. In this fact lies the advantage of knowing when a given surface integral is independent of the surface. The perhaps complicated surface over which the integration is to be performed may be replaced by a simpler surface.

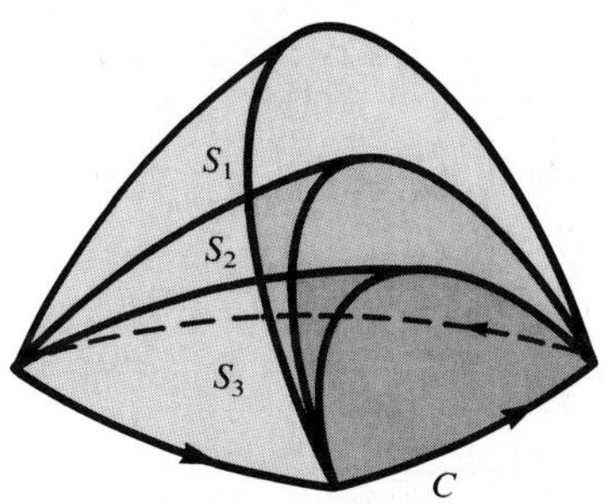

Figure 11.18 *Independence of surface*

Example 11.15 Use the surface independence of

$$\iint_S (ax + by)\,dS \qquad (a, b \text{ constant})$$

where S is the upper unit hemisphere $z = \sqrt{1 - x^2 - y^2}$, to evaluate the given surface integral.

Solution The simplest surface for integration purposes is the unit disk $x^2 + y^2 < 1$, $z = 0$. Thence

$$\int_S (ax + by)\,dS = \int_{-1}^{1} \int_{-(1-x^2)^{1/2}}^{(1-x^2)^{1/2}} (ax + by)\,dy\,dx$$
$$= 0$$

□

In Section 11.8 we determine a condition for a rather general class of surface integrals to be independent of surface. For now we are content with a theorem that resembles the earlier Theorem 11.3 concerning path independence of line integrals.

Theorem 11.7 A surface integral

$$\iint_S f(P)\,dS$$

is *independent of the surface*, S, with boundary curve C in a domain D *if and only if*

$$\iint_{S'} f(P)\,dS = 0$$

for all piecewise-smooth closed* surfaces, S', in D.

Proof This theorem should be proved as an exercise. To do so the closed surface S' may be split into two or more component surfaces (recall Definition 11.3) and the proof of Theorem 11.3 mimicked. ■

With the help of Theorem 11.7 the surface independence of the integral in Example 11.14 can be established in alternative fashion with equal ease. We merely need to note that the projections of the "top" and "bottom" of a closed surface of the given form appear with opposite signs, and hence their "signed" areas sum to zero. Other applications are discussed below and considered in the exercises for this section.

We are now ready to generalize our consideration of surface integrals to include vector fields. Recall that the discussion of *line* integrals in a vector field was merely an agreement on the notation used to represent the integral of the

---

*A surface is said to be *closed* if it consists of the boundary points of some bounded domain.

tangential component of the field. In that case, the vector line integral could be written as the sum of three line integrals of the component functions. Now we make a similar agreement for the surface integral of the component of the field *normal* to the surface. As can be guessed, it is precisely this kind of vector surface integral that is of foremost importance in applications. It is denoted by

$$\iint_S \mathbf{F}\cdot\mathbf{N}\,dS$$

and can be evaluated by finding the dot product of **F** and **N**, and integrating the resulting scalar function by the methods of the previous section.

Example 11.16 Let $\mathbf{F} = y\mathbf{j} + z\mathbf{k}$ and assume S is that portion of the surface of the paraboloid $x^2 + y^2 = 4 - z$ that lies above the xy-plane. If **N** is the upward-directed normal to the surface, evaluate $\iint_S \mathbf{F}\cdot\mathbf{N}\,dS$. (See Figure 11.17.)

Solution By Example 11.13,

$$\mathbf{N} = \frac{2x\mathbf{i} + 2y\mathbf{j} + \mathbf{k}}{(4x^2 + 4y^2 + 1)^{1/2}}$$

and hence

$$\mathbf{F}\cdot\mathbf{N} = \frac{2y^2 + z}{(4x^2 + 4y^2 + 1)^{1/2}}$$

from which

$$\iint_S \mathbf{F}\cdot\mathbf{N}\,dS = 16\pi$$

See Example 11.13 for the technique of evaluating this surface integral. □

Frequently in practical problems a given vector surface integral may be related to a particular physical quantity. In such situations it may then be possible to use Theorem 11.7 to infer surface independence of these integrals purely on physical grounds. This fact explains some of the results to be obtained in Exercises 23–29.

Exercises for Sections 11.5 and 11.6

1. Show that the element of surface area for a sphere or a cylinder can be obtained formally from the definition and is the same as that which we derived heuristically.
2. Justify the formulas for surface area $\iint_{S'_{xz}} \sec\beta\,dx\,dz$ and $\iint_{S'_{yz}} \sec\alpha\,dy\,dz$ without the benefit of the definition of surface area.
3. Using the general formula for surface area, find the area of the surface $x^2 + y^2 = 4$ from $z = 0$ to $z = 3$. Check by using the known result for the area of the curved surface of a right circular cylinder.

4. Find the surface area of the paraboloid of revolution $z = x^2 + y^2$ from $z = 0$ to $z = 2$ in two different ways.
5. Evaluate $\iint_S (x\mathbf{i} + y\mathbf{j} + z\mathbf{k}) \cdot \mathbf{N}\, dS$ over the plane bounded by the triangle whose vertices are the points $(2, 0, 0)$, $(0, 2, 0)$, and $(0, 0, 2)$ where the normal vector points away from the origin.
6. Evaluate $\iint_S \mathbf{F} \cdot \mathbf{N}\, dS$ over the hemisphere $z = (4 - x^2 - y^2)^{1/2}$, where $\mathbf{F} = xz\mathbf{i} + zx\mathbf{j} + xy\mathbf{k}$ and $\mathbf{N}$ is the upward unit normal.

In Exercises 7–10, evaluate $\iint_S \mathbf{F} \cdot \mathbf{N}\, dS$ for the given vector function $\mathbf{F}$ and surface S.

7. $\mathbf{F} = y\mathbf{i} + 2x\mathbf{j} - z\mathbf{k}$ and S is the surface of the plane $2x + y = 6$ in the first octant cut off by the plane $z = 4$.
8. $\mathbf{F} = (x^2 + y^2)\mathbf{i} - 2x\mathbf{j} + 2yz\mathbf{k}$ and S is the surface of the plane $2x + y + 2z = 6$ in the first octant.
9. $\mathbf{F} = 2y\mathbf{i} - z\mathbf{j} + x^2\mathbf{k}$ and S is the surface of the parabolic cylinder $y^2 = 8x$ in the first octant bounded by the planes $y = 4$ and $z = 6$.
10. $\mathbf{F} = 6z\mathbf{i} + (2x + y)\mathbf{j} - x\mathbf{k}$ and S is the *entire* surface of the region bounded by the cylinder $x^2 + y^2 = 9$, $x = 0$, $y = 0$, $z = 0$, $z = 8$.

In Exercises 11–14, let $\mathbf{F} = x\mathbf{i} + y\mathbf{j} + 2\mathbf{k}$. Evaluate $\iint_S \mathbf{F} \cdot \mathbf{N}\, dS$ over the given surface.

11. The surface of the unit cube $0 \leqq x \leqq 1$, $0 \leqq y \leqq 1$, $0 \leqq z \leqq 1$.
12. The portion of the plane $x + 2y + 3z = 6$ which lies in the first octant.
13. The entire surface of the unit sphere centered at the origin.
14. The portion of the cone $x^2 + y^2 - (1 - z)^2 = 0$ between $z = 0$ and $z = 1$.

The vector fields in Exercises 15–16 are such that the given vector surface integrals are surface independent. Perform the indicated surface integration in each case and observe the equality of the results. Also note the relative difficulty of the various integrations.

15. Let $\mathbf{F}$ be the constant vector $\mathbf{F} = a\mathbf{i} + b\mathbf{j} + c\mathbf{k}$. Evaluate $\iint_S \mathbf{F} \cdot \mathbf{N}\, dS$ over each surface.
    - **(a)** The surface of the paraboloid $x^2 + y^2 = 4 - z$ for which $z > 0$.
    - **(b)** The top and curved surface of the right circular cylinder $x^2 + y^2 = 4$, $0 < z < 8$.
    - **(c)** The disk $x^2 + y^2 < 4$, $z = 0$.
16. Let $F = yz\mathbf{i} - xz\mathbf{j} - \mathbf{k}$. Evaluate $\iint_S \mathbf{F} \cdot \mathbf{N}\, dS$ over each surface.
    - **(a)** The conical surface $z^2 = x^2 + y^2$ for $0 < z < 1$.
    - **(b)** The hemisphere $z = 1 - \sqrt{1 - x^2 - y^2}$.
    - **(c)** The disk $x^2 + y^2 < 1$, $z = 1$.

17. Find the surface area of the torus

$$\mathbf{r}(u, v) = (a + b\cos v)\cos u\,\mathbf{i} + (a + b\cos v)\sin u\,\mathbf{j} + b\sin v\,\mathbf{k},$$
$$0 < b < a$$
$$0 \leqq u \leqq 2\pi,\ 0 \leqq v \leqq 2\pi$$

18. Evaluate $\iint_S z\, dS$ where S is the surface defined by $\mathbf{r}(u, v) = u\mathbf{i} + v\mathbf{j} + u^3\mathbf{k}$ for $0 < u < 1$, $0 < v < 1$.
19. Find the surface area of the surface defined by $\mathbf{r}(u, v) = v\cos u\,\mathbf{i} + v\sin u\,\mathbf{j} + v^2\mathbf{k}$, $0 < u < 1$, $0 < v < 1$.
20. Find the surface area of the surface defined by $\mathbf{r}(u, v) = u\mathbf{i} + v\mathbf{j} + u\mathbf{k}$, $0 < u < 1$, $0 < v < 1$.

21. Evaluate $\int\int_S (6x + z - y^2)\,dS$ where S is the surface defined by $\mathbf{r}(u, v) = u\mathbf{i} + v\mathbf{j} + u\mathbf{k}$, $0 \leqq u \leqq 1$, $0 \leqq v \leqq 1$.
22. Evaluate $\int\int_S (6x + y - x^2)\,dS$ where S is the surface defined by $\mathbf{r}(u, v) = u\mathbf{i} + u^2\mathbf{j} + v\mathbf{k}$, $0 \leqq u \leqq 1$, $0 \leqq v \leqq 1$.

In electrostatics, Gauss' law

$$\int\int_S \mathbf{E}\cdot\mathbf{N}\,dS = 4\pi q$$

relates the total charge present within a closed surface to the surface integral of the normal component of the electric field vector engendered by that charge distribution. In Exercises 23–25, find the total charge within the designated surface by applying Gauss' law to the given fundamental electrostatic fields.

23. The field of a point charge $\mathbf{E} = \dfrac{1}{\rho^2}\mathbf{u}_\varrho$ on the sphere $x^2 + y^2 + z^2 = R^2$.
24. The dipole field $\mathbf{E} = \dfrac{p}{\rho^3}(2\cos\phi\,\mathbf{u}_\rho + \sin\phi\,\mathbf{u}_\phi)$ on the sphere $x^2 + y^2 + z^2 = R^2$.
25. The field of a line charge $\mathbf{E} = \dfrac{1}{r}\mathbf{u}_r$ on the cylinder with lateral surface $x^2 + y^2 = R^2$ and ends $z = 0, L$.

In fluid mechanics, the surface integral

$$\int\int_S \mathbf{v}\cdot\mathbf{N}\,dS$$

of the normal component of the vector velocity associated with an incompressible fluid is proportional to the net outward flow of mass across S per unit time. Evaluate $\int\int_S \mathbf{v}\cdot\mathbf{N}\,dS$ for the designated surface and incompressible fluid flows having the vector velocities given in Exercises 26–29.

26. $\mathbf{v} = U_0\mathbf{i}$ for the surface of the cube $|x| < 1$, $|y| < 1$, $|z| < 1$
27. $\mathbf{v} = 2axy\mathbf{i} + a(x^2 - y^2)\mathbf{j}$ for the surface of the cube $|x| < 1$, $|y| < 1$, $|z| < 1$
28. $\mathbf{v} = \dfrac{\theta}{r}\mathbf{u}_r + \dfrac{1}{r}\ln r\mathbf{u}_\theta$ for the surface of the cylinder $x^2 + y^2 < 1$, $|z| < 1$
29. $\mathbf{v} = \left(1 - \dfrac{a^2}{r^2}\right)\cos\theta\,\mathbf{u}_r - \left(1 + \dfrac{a^2}{r^2}\right)\sin\theta\,\mathbf{u}_\theta$ for the surface of the cylinder $x^2 + y^2 < 1$, $|z| < 1$

11.7 Green's Theorem in the Plane

For the remainder of this chapter we study what are called "transformation theorems" or, more classically, "vector integral theorems." Generally speaking, these theorems relate integrals of functions over boundaries of closed sets to integrals of some differential operator of the function over the interior of the set. A special case of this type of transformation is the relationship of a function evaluated at the endpoints of an interval to the derivative of the function integrated over the interval, namely the fundamental theorem of integral calculus

$$f(b) - f(a) = \int_a^b \frac{df}{dx}\,dx$$

The points a and b are the boundary of the interval (a, b) and in this case the differential operator is just the derivative.

We choose to limit our discussion to three main theorems, although others are a part of the exercises and examples. The most fundamental one is called Green's theorem in the plane. It is particularly useful since it immediately generalizes into Stokes' theorem (Section 11.8) and the divergence theorem (Section 11.9).

Theorem 11.8 *Green's Theorem in the Plane.* Let D be a bounded domain in the xy-plane whose boundary is a closed piecewise smooth curve C, and let $\mathbf{F} = f\mathbf{i} + g\mathbf{j}$ be a vector field continuously differentiable on D. Then

$$\oint_C \mathbf{F}\cdot d\mathbf{r} = \iint_D (\nabla \times \mathbf{F})\cdot\mathbf{k}\,dx\,dy$$

where the integration along C is such that D is on the left as one advances in the direction of integration.*

Proof What we must prove is that

$$\oint_C f(P)\,dx + g(P)\,dy = \iint_D [g_x(x, y) - f_y(x, y)]\,dy\,dx$$

the form in which Green's theorem usually appears.

We first prove Green's theorem for the special case in which D is a "standard" domain.† See Figure 11.19. By the ordinary methods of evaluating a double integral as an iterated integral, we have

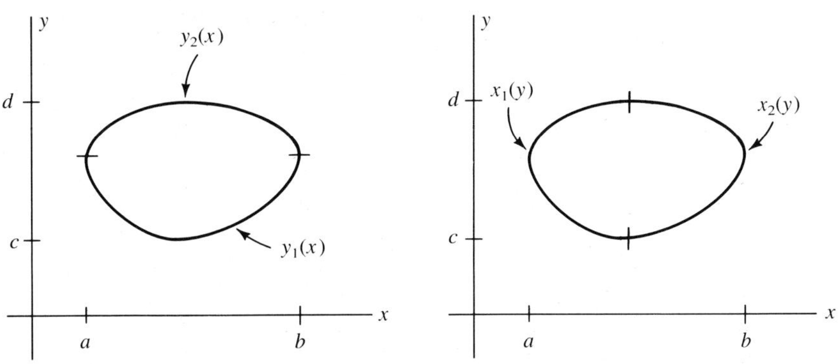

Figure 11.19 *Two views of a standard domain*

*Unless otherwise stated, from now on the direction of integration along a simple closed curve bounding a planar region (surface) will be chosen in this manner.

†A standard domain is a domain D which can be represented in *both* the forms $a < x < b$, $y_1(x) < y < y_2(x)$; $c < y < d$, $x_1(y) < x < x_2(y)$.

$$\iint_D f_y(x, y)\, dy\, dx = \int_a^b \left[\int_{y_1(x)}^{y_2(x)} f_y(x, y)\, dy \right] dx$$

Carrying out the integration on the inside integral we obtain

$$\iint_D f_y(x, y)\, dy\, dx = \int_a^b [f(x, y_2(x)) - f(x, y_1(x))]\, dx$$

which, upon reversing the order of integration in the integral corresponding to the first term, becomes

$$= -\int_b^a f(x, y_2(x))\, dx - \int_a^b f(x, y_1(x))\, dx$$

which is just another way of writing the line integral

$$-\oint_C f(P)\, dx$$

In a similar manner,

$$\begin{aligned} \iint_D g_x(x, y)\, dx\, dy &= \int_c^d \left[\int_{x_1(y)}^{x_2(y)} g_x(x, y)\, dx \right] dy \\ &= \int_c^d [g(x_2(y), y) - g(x_1(y), y)]\, dy \\ &= \oint_C g(P)\, dy \end{aligned}$$

and adding the two results proves the theorem for the case in which D is standard.

Next suppose D is not standard but can be decomposed into two pieces by means of a "cut." Figure 11.20 gives you an idea of how the cut should be introduced. Since the integral over the domain is additive,

$$\begin{aligned} \iint_D [g_x(x, y) - f_y(x, y)]\, dy\, dx &= \iint_{D_1} [g_x(x, y) - f_y(x, y)]\, dy\, dx \\ &\quad + \iint_{D_2} [g_x(x, y) - f_y(x, y)]\, dy\, dx \end{aligned}$$

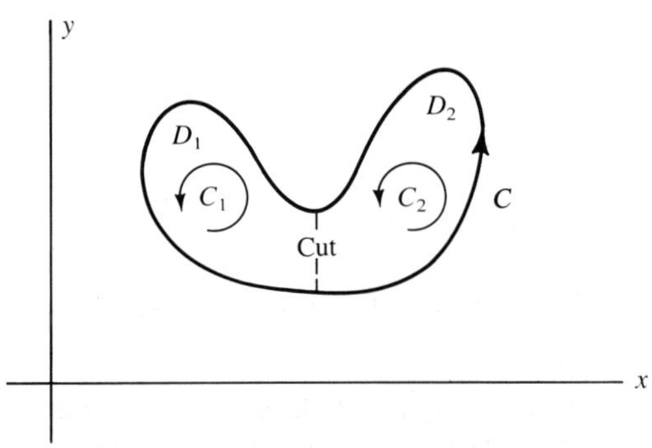

Figure 11.20 *Decomposition of a nonstandard domain*

where D_1 and D_2 are both standard. By applying Green's theorem to each of these separately, we obtain

$$\iint_D [g_x(x, y) - f_y(x, y)]\,dy\,dx = \oint_{C_1} f(P)\,dx + g(P)\,dy + \oint_{C_2} f(P)\,dx + g(P)\,dy$$

The cut is a part of both C_1 *and* C_2 but the integration is in opposite directions and hence

$$\iint_D [g_x(x, y) - f_y(x, y)]\,dy\,dx = \oint_C f(P)\,dx + g(P)\,dy$$

This technique can be employed to prove Green's theorem for domains that can be decomposed by $n - 1$ cuts into n standard domains. ■

Example 11.17 Let $\mathbf{F} = y\mathbf{i} - x\mathbf{j}$ and let C be the circle of radius 1 centered at the origin. Compute

$$\oint_C \mathbf{F} \cdot d\mathbf{r}$$

by using Green's theorem.

Solution C can be represented by $\mathbf{r}(t) = \cos t\mathbf{i} + \sin t\mathbf{j}$. Then D is the domain interior to C and by Green's theorem

$$\oint_C \mathbf{F} \cdot d\mathbf{r} = \iint_D (-1 - 1)\,dx\,dy = -2\pi$$ □

Example 11.18 Evaluate

$$\oint_C \mathbf{F} \cdot d\mathbf{r}$$

where

$$\mathbf{F} = \frac{-y\mathbf{i} + x\mathbf{j}}{x^2 + y^2}$$

and C is the same as in the previous example.

Solution Note that Green's theorem cannot be applied to this closed line integral, since C does not bound a region in which $\mathbf{F}$ is continuously differentiable. ($\mathbf{F}$ is not even defined at (0, 0).) The evaluation can be performed, however, by the methods learned in Section 11.3.

$$\oint_C \mathbf{F} \cdot d\mathbf{r} = \oint_C \frac{-y\,dx + x\,dy}{x^2 + y^2} = \int_0^{2\pi} (\sin^2 t + \cos^2 t)\,dt = 2\pi$$ □

As proved, Green's theorem in the plane applies to what are called *simply connected* domains in the plane. These are domains, D, that have the property

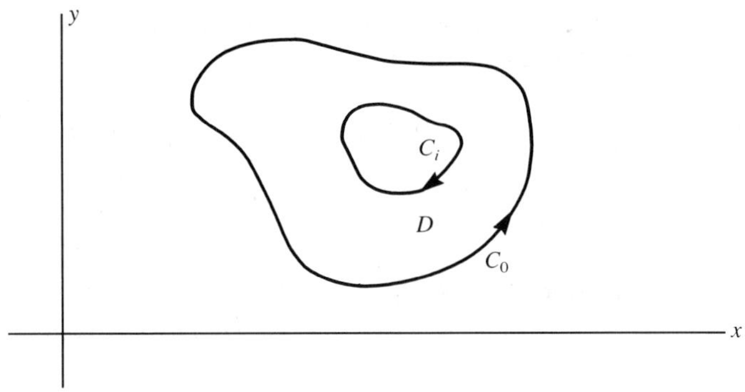

Figure 11.21 *Green's theorem for a doubly connected region*

of being the interior of simple closed curves, C. A property of simply connected domains, often used as the defining property, is that every simple closed curve in D can be shrunk continuously to a point, the point also being in D. Loosely speaking we sometimes say that a simply connected domain is one "without holes in it." The distinction between a general domain and one which is simply connected can be of extreme importance. Later in this chapter we use it to distinguish sharply between an irrotational and a conservative field. A domain that is not simply connected is called *multiply connected.*

To apply Green's theorem to multiply connected domains, proper interpretation must be made of the domain D, the boundary curve C, and the direction of the line integration on C. The case of a doubly connected domain is shown in Figure 11.21. The boundary of D is both the inside and outside curve and the direction of line integration is counterclockwise along the outside curve and clockwise along the inside curve. Hence, the statement of Green's theorem for the domain shown is

$$\iint_D [g_x(x, y) - f_y(x, y)]\, dy\, dx = \oint_{C_0} f(P)\, dx + g(P)\, dy + \oint_{C_i} f(P)\, dx + g(P)\, dy$$

The proof of Green's theorem for a domain such as the one shown in Figure 11.21 is analogous to that of extending Green's theorem to nonstandard regions. The given domain is "cut" into simply connected pieces and the result of Green's theorem is applied to each of the individual pieces. The details are left for Exercise 16 in the next set of exercises.

The application of Green's theorem may be other than as a transformation theorem. For example, the statement of Green's theorem for doubly connected domains can be considered to relate three numerical quantities so that if any two are known, then so is the remaining term. A case in point is in the evaluation of line integrals over "uncommon" closed curves. If we establish a domain D bounded by a "common" curve (such as a circle) and the given one, and if the

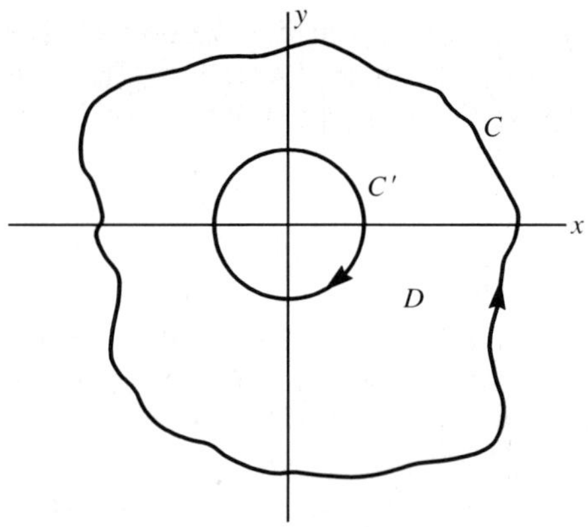

Figure 11.22

double integral over the domain D is known, then the value of the line integral over the "uncommon" curve is known. Naturally f must satisfy the requirements of Green's theorem on D.

Example 11.19 Assume that the area between the curve C in Figure 11.22 and the circle of radius 1 centered at the origin is 5π square units. Compute the value of $\oint_C \mathbf{F} \cdot d\mathbf{r}$ where $\mathbf{F}$ is as in Example 11.17.

Solution By Green's theorem for doubly connected domains,

$$\iint_D (\nabla \times \mathbf{F}) \cdot \mathbf{k}\, dy\, dx = \oint_C \mathbf{F} \cdot d\mathbf{r} + \oint_{C'} \mathbf{F} \cdot d\mathbf{r}$$

where D is the domain between the two curves and C' is the circle of radius 1 centered at the origin. From Example 11.17, the value of the double integral is $-2(\text{Area}) = -2(5\pi) = -10\pi$. Also from that example, the value of the line integral over C' taken in the counterclockwise direction is -2π. Hence, the value of the given line integral is $-10\pi - 2\pi = -12\pi$. □

If the value of the double integral is zero regardless of which domain D is established (such would be the case, for example, if the curl of the vector field is identically zero), the foregoing technique leads to another form of the *principle of deformation of curve* alluded to in Section 11.4. For then, the uncommon curves may be deformed into common curves without affecting the value of the line integral so long as in the deforming process, no points of discontinuity of the function are crossed.

Example 11.20 Evaluate

$$\oint_C \mathbf{F} \cdot d\mathbf{r}$$

where $\mathbf{F}$ is as in Example 11.18 but C is the curve defined by Figure 11.22.

Solution Consider the domain D between the given curve and the circle C' of radius 1 centered at the origin. In this (doubly connected) domain, the function $\mathbf{F}$ satisfies the requirements of Green's theorem and hence,

$$\iint_D (\nabla \times \mathbf{F}) \cdot \mathbf{k}\, dy\, dx = \oint_C \mathbf{F} \cdot d\mathbf{r} + \oint_{C'} \mathbf{F} \cdot d\mathbf{r}$$

where C' is the notation used for the inner circle and the directions of integration are as specified in the figure. Since $\nabla \times \mathbf{F} = \mathbf{0}$ throughout D (show this!), the double integral is zero. But from Example 11.18, the integral along C' is -2π; thus the desired integral has value 2π. □

Whereas Green's theorem can be considered the basic transformation of a double integral over a domain into a line integral around the boundary, other interesting transformations are immediate consequences. The following example shows how to transform the double integral of the Laplacian of a function into a line integral of the "normal derivative" of the function.

Example 11.21 Let w be a continuous scalar function with continuous first and second derivatives in some multiply connected domain D bounded by a curve C. Show

$$\iint_D \nabla^2 w\, dx\, dy = \oint_C \frac{\partial w}{\partial n}\, ds$$

where $\nabla^2 w = w_{xx} + w_{yy}$ and $\partial w/\partial n$ (the "normal derivative") is the directional derivative of w in the direction of the normal pointing to the *exterior* of the domain.

Solution If we let $f = -w_y$ and $g = w_x$, then $g_x - f_y = \nabla^2 w$ and therefore, by Green's theorem,

$$\iint_D \nabla^2 w\, dy\, dx = \oint_C f(P)\, dx + g(P)\, dy$$

But

$$\begin{aligned}
\int_C f(P)\, dx + g(P)\, dy &= \int_C \left(f\frac{dx}{ds} + g\frac{dy}{ds} \right) ds \\
&= \int_C \left(-w_y\frac{dx}{ds} + w_x\frac{dy}{ds} \right) ds \\
&= \int_C \nabla w \cdot \left(\frac{dy}{ds}\mathbf{i} - \frac{dx}{ds}\mathbf{j} \right) ds \\
&= \int_C \nabla w \cdot \mathbf{N}\, ds \\
&= \int_C \frac{\partial w}{\partial n}\, ds
\end{aligned}$$

(The fact that $(dy/ds)\mathbf{i} - (dx/ds)\mathbf{j}$ is the correct expression for $\mathbf{N}$ is left as an exercise.) □

Our statement of Green's theorem related the double integral of the curl of the field to the line integral of the *tangential* component. The next example shows that the double integral of the divergence is related to the line integral of the *normal* component of the field.

Example 11.22 Let **V** represent a continuously differentiable vector field in the plane and let D and C satisfy the requirements of Green's theorem. Show

$$\oint_C \mathbf{V}\cdot\mathbf{N}\,ds = \iint_D \nabla\cdot\mathbf{V}\,dx\,dy$$

where **N** is the unit normal pointing to the exterior of D.

Solution Suppose **V** has component functions v_1 and v_2 and let $\mathbf{F} = -v_2\mathbf{i} + v_1\mathbf{j}$. First, observe that

$$\iint_D (\nabla\times\mathbf{F})\cdot\mathbf{k}\,dx\,dy = \iint_D \left(\frac{\partial v_1}{\partial x} + \frac{\partial v_2}{\partial y}\right)dx\,dy$$

$$= \iint_D \nabla\cdot\mathbf{V}\,dx\,dy$$

Applying Green's theorem to the field **F**,

$$\iint_D (\nabla\times\mathbf{F})\cdot\mathbf{k}\,dx\,dy = \int_C -v_2\,dx + v_1\,dy$$

$$= \oint_C \mathbf{V}\cdot\left(\frac{dy}{ds}\mathbf{i} - \frac{dx}{ds}\mathbf{j}\right)ds$$

$$= \oint_C \mathbf{V}\cdot\mathbf{N}\,ds$$

and the result is proved. □

The relation established in this example is the two-dimensional version of the important divergence theorem, which we consider at length in Section 11.9.

Exercises for Section 11.7

1. Let $\mathbf{r} = x(s)\mathbf{i} + y(s)\mathbf{j}$ represent a smooth, simple, closed curve C. Show that the vector $(dy/ds)\mathbf{i} - (dx/ds)\mathbf{j}$ is a vector normal to C which points to the exterior of the domain of which C is a boundary.
2. Show how to compute the area of a domain in terms of a line integral around its boundary. (*Hint:* Apply Green's theorem to the vector field $-y\mathbf{i}$.)
3. Use the result of Exercise 2 to find the area enclosed by the loop $x = a\sin t\cos t$, $y = a\sin^2 t$, for $a > 0$ and $0 \le t \le \pi$.

In Exercises 4–6, verify Green's theorem for the unit disk $x^2 + y^2 < 1$ and the given vector field:

4. $\mathbf{F} = (x + y)\mathbf{i} + y\mathbf{j}$
5. $\mathbf{F} = x^3y\mathbf{i} + x^2y^2\mathbf{j}$
6. $\mathbf{F} = (e^x\sin y)\mathbf{i} + (e^x\cos y)\mathbf{j}$

In Exercises 7–9, evaluate $\oint_C \mathbf{F}\cdot d\mathbf{r}$ using direct calculations, and then verify the result so obtained using Green's theorem.

7. $\mathbf{F} = y^2\mathbf{i} + x^2\mathbf{j}$; C is the boundary of the square $0 \leqq x \leqq 1, 0 \leqq y \leqq 1$.

8. $\mathbf{F} = (3x^2 + y)\mathbf{i} + 4y^2\mathbf{j}$; C is the boundary of the triangle with vertices at $(0,0)$, $(1,0)$, and $(0,2)$ taken clockwise.

9. $\mathbf{F} = (x^2 + y)\mathbf{i} - xy^2\mathbf{j}$; C is the boundary of the square whose vertices are $(0,0)$, $(1,0)$, $(1,1)$, and $(0,1)$ taken counterclockwise.

10. **(a)** Apply the result of Example 11.21 to the function $\phi(x, y) = x$ on the unit square.
(b) Repeat part (a) for the function $\phi(x, y) = x^2 + y^2$ on the unit disk.

11. Show that if ϕ is harmonic in a bounded domain D of the xy-plane, then

$$\iint_D (\phi_x^2 + \phi_y^2)\,dx\,dy = \int_C \phi\frac{\partial\phi}{\partial n}\,ds$$

where C is the boundary of D. Apply this result to $\phi(x, y) = e^x \cos y$ on the unit square.

12. Evaluate

$$\int_C \frac{\partial\phi}{\partial n}\,ds$$

where $\phi(x, y)$ is equal to the following, and in each case C is the boundary of the unit square:
(a) $x^2 + 2y^2$ **(b)** $x^3y - xy + y^2$ **(c)** $e^{2x} + e^y$

13. Evaluate $\int_C \mathbf{F}\cdot d\mathbf{r}$ for each $\mathbf{F}$ and C.
(a) $\mathbf{F} = -y^3\mathbf{i} + x^3\mathbf{j}$; C is the unit circle centered at the origin.
(b) $\mathbf{F} = \cos x \sin y\mathbf{i} + \sin x \cos y\mathbf{j}$; C is as in part (a).

14. Evaluate

$$\int_C x\frac{\partial x}{\partial n}\,ds$$

where C is the unit square taken counterclockwise.

15. In Example 11.22, how are the fields $\mathbf{F}$ and $\mathbf{V}$ related? Obtain a "div-curl" relationship between the two.

16. Prove Green's theorem for a domain like that of Figure 11.21. Assume that the theorem has been proved for simply connected domains.

17. Assume that the area between the closed curve C in Figure 11.22 and the unit circle is 25 square units. Compute

$$\oint_C \mathbf{F}\cdot d\mathbf{r}$$

if $\mathbf{F} = x^2\mathbf{i} + y^2\mathbf{j}$.

18. Repeat Exercise 17 for the field $\mathbf{F} = y\mathbf{i} - x\mathbf{j}$.

In Exercises 19–20, by means of Green's theorem, rewrite each given surface integral as a line integral and then evaluate both integrals in each case.

19. $\iint_D dx\,dy$, where D is such that $|x| + |y| < 1$

20. $\displaystyle\iint_D \left(\frac{y}{x^2 + y^2} - \frac{x}{x^2 + y^2}\right)dx\,dy$, where D is such that $2 < x^2 + y^2 < 4$

21. Verify Green's theorem for the vector field $\mathbf{F} = (2x^3 - y^3)\mathbf{i} + (x^3 + y^3)\mathbf{j}$ and the specified domain.
(a) The disk $x^2 + y^2 < 9$ **(b)** The annulus $9 < x^2 + y^2 < 16$

The two-dimensional version of the divergence theorem established in Example 11.22 gives us an alternative method for calculating the line integral of the normal component of a given vector $\mathbf{v}$ around a closed planar curve C. For the solenoidal vector velocities associated with incompressible fluid flows (recall the discussion in Section 10.10), the integral should be zero provided the vector field is continuously differentiable within C. Similar remarks apply to electrostatic displacement vectors $\mathbf{D}$ in charge-free regions and to magnetic field vectors $\mathbf{B}$.

In Exercises 22–23, determine which of the given vector fields are *not* solenoidal or *not* continuously differentiable in the plane. For these fields evaluate

$$\oint_C \mathbf{v} \cdot \mathbf{N}\, ds$$

where C is the perimeter of the square $|x| \le 1, |y| \le 1$. Use the divergence theorem in two dimensions wherever it is applicable.

22. **(a)** $\mathbf{v} = U_0\mathbf{i}$ **(b)** $\mathbf{v} = y\mathbf{j}$

23. **(a)** $\mathbf{v} = \dfrac{-y\mathbf{i} + x\mathbf{j}}{x^2 + y^2}$ **(b)** $\mathbf{v} = e^x \cos y\,\mathbf{i} - e^x \sin y\,\mathbf{j}$

In Exercises 24–25, repeat the determinations of Exercises 22–23 for the given vector fields where now C is the perimeter of the unit circle $x^2 + y^2 = 1$.

24. **(a)** $\mathbf{v} = \dfrac{1}{r}\mathbf{u}_\theta$ **(b)** $\mathbf{v} = r^2 \sin 3\theta\, \mathbf{u}_r + r^2 \cos 3\theta\, \mathbf{u}_\theta$

25. **(a)** $\mathbf{v} = (r^2 - 1)\cos\theta\, \mathbf{u}_r - (r^2 + 1)\sin\theta\, \mathbf{u}_\theta$
(b) $\mathbf{v} = e^{r\cos\theta}\,[\cos(r\sin\theta + \theta)\mathbf{u}_r - \sin(r\sin\theta + \theta)\mathbf{u}_\theta]$ (*Hint:* use polar coordinates for the integrations.)

11.8 Stokes' Theorem

In Section 11.4 we presented and used the fact that the value of $\int_C \mathbf{F}\cdot d\mathbf{r}$ is not affected by a deformation of curve if $\mathbf{F}$ has a scalar potential. In this section we determine an analogous result for surface integrals. We show that the value $\iint_S \mathbf{G}\cdot\mathbf{N}\, dS$ will be the same for different surfaces with the same boundary curve if $\mathbf{G}$ has a vector potential; that is, if there exists another vector field $\mathbf{F}$ such that $\nabla \times \mathbf{F} = \mathbf{G}$. The theorem that leads directly to this result is known as Stokes' theorem, one of the more powerful theorems of classical analysis. Its modern-day importance stems from the fact that it can be thought of as the generalization of the fundamental theorem of integral calculus to arbitrary vector fields in three-space.

For the general statement of Stokes' theorem, we need the concept of a simple two-sided or orientable surface.

Definition 11.5 A smooth surface is *simple* if it does not intersect itself. In other words, the function $\mathbf{r}(u, v)$ providing the parametric representation of the surface is one-to-one. (Recall Definition 10.4, p. 363.)

Definition 11.6 A smooth surface S in three dimensions is *orientable* if it is possible to associate with each point (x, y, z) of S a unit vector $\mathbf{N}(x, y, z)$ normal to S in such a way that $\mathbf{N}(x, y, z)$ is a continuous vector field on S.

The notion of orientable surface allows us to exclude from consideration the Möbius strip* and other such surfaces on which it is impossible to specify continuously a normal direction.

Theorem 11.9 *Stokes' Theorem.* Let S be a simple smooth orientable surface in three dimensions bounded by a smooth simple closed curve C. Then, if $\mathbf{F}$ is a vector field that is continuously differentiable in a bounded domain containing S,

$$\oint_C \mathbf{F}\cdot d\mathbf{r} = \iint_S (\nabla \times \mathbf{F})\cdot \mathbf{N}\, dS$$

where $\mathbf{N}$ is a unit vector normal to S and the direction of integration on C is in accordance with the right-hand rule relative to $\mathbf{N}$.† ■

Stokes' theorem tells us that if the integrand of the surface integral is of the form of the normal component of the curl of a vector field, then it may be replaced by the line integral of the tangential component of that field along the path which bounds S. Hence the next theorem is an immediate consequence.

Theorem 11.10 $\iint_S \mathbf{G}\cdot\mathbf{N}\, dS$ is independent of the surface bounded by a closed curve C if $\mathbf{G}$ has a vector potential $\mathbf{F}$ that together with S and $\mathbf{N}$ satisfies the requirements of Stokes' theorem.

Proof See Exercise 1, p. 443. ■

As a result of Theorem 11.7 and the preceding theorem, the following result is obvious.

Theorem 11.11 If $\mathbf{G}$ has a vector potential $\mathbf{F}$ that together with S and $\mathbf{N}$ satisfies the requirements of Stokes' theorem and if S is a closed surface, then

$$\iint_S \mathbf{G}\cdot\mathbf{N}\, dS = 0$$

(Note that $\mathbf{N}$ must be chosen to point always in the same direction relative to S. The usual choice is away from the region bounded by S.) Compare this to Theorem 11.4. ■

Example 11.23 Evaluate the integral of the normal component of curl $\mathbf{F}$ over the hemisphere $z = (4 - x^2 - y^2)^{1/2}$ by Stokes' theorem and then directly by computation of the surface integral. Let $\mathbf{F} = -y\mathbf{i} + x\mathbf{j} + \mathbf{k}$.

---

*This surface can be conveniently realized by taking a strip of paper, giving it a half-twist, and then gluing the ends together.

†The cross product $\mathbf{N} \times \mathbf{T}$, where $\mathbf{T}$ is a tangent vector in the direction of integration on C, points into the surface S rather than away from it.

Solution The surface is bounded by a circle C of radius 2 centered at the origin which may be represented by

$$\mathbf{r}(t) = 2(\cos t\,\mathbf{i} + \sin t\,\mathbf{j})$$

We wish to evaluate $\iint_S (\nabla \times \mathbf{F}) \cdot \mathbf{N}\,dS$. First, by Stokes' theorem,

$$\begin{aligned}\iint_S (\nabla \times \mathbf{F}) \cdot \mathbf{N}\,dS &= \oint_C \mathbf{F} \cdot d\mathbf{r} \\ &= \oint_C -y\,dx + x\,dy + dz \\ &= \int_0^{2\pi} 4(\sin^2 t + \cos^2 t)\,dt \\ &= 8\pi\end{aligned}$$

Evaluating directly,

$$\begin{aligned}\iint_S (\nabla \times \mathbf{F}) \cdot \mathbf{N}\,dS &= \iint_S (2\mathbf{k}) \cdot \mathbf{N}\,dS \\ &= \iint_S 2\cos\gamma|\sec\gamma|\,dx\,dy\end{aligned}$$

Since γ is acute everywhere on the hemisphere,

$$\begin{aligned}\iint_S (\nabla \times \mathbf{F}) \cdot \mathbf{N}\,dS &= 2\,(\text{Area of projection of hemisphere on the } xy\text{-plane}) \\ &= 8\pi\end{aligned}$$

□

Thus Stokes' theorem gives another method of computing surface integrals of normal components of vector fields, *if* the vector field in the integrand has a vector potential. Unfortunately, it is not particularly easy to determine when a vector field has a vector potential (see Exercise 14 for a convenient necessary condition). Nor is it simple to find the vector potential if we know that one exists. It does happen quite often however, as in the previous example, that the form of the vector field in the integrand is already expressed as the curl of another field.

Sometimes it is convenient to use the principle of independence of surface (Theorem 11.10) to integrate over an "easier" surface than the one given. Remember, this theorem cannot be applied unless the integrand is of a very specific form and the vector field has a vector potential.

Example 11.24 Let $\mathbf{F} = 4y\mathbf{i} - x\mathbf{j} + 2z^3\mathbf{k}$, S be the hemisphere of the previous problem, and $\mathbf{N}$ be the outward directed normal to S. Evaluate

$$\iint_S (\nabla \times \mathbf{F}) \cdot \mathbf{N}\,dS$$

in *three* ways.

Solution (a) Integrate the expression as it stands. You can easily show that $\mathbf{N} = (x\mathbf{i} + y\mathbf{j} + z\mathbf{k})/2$ and that $\nabla \times \mathbf{F} = -6\mathbf{k}$, from which $(\nabla \times \mathbf{F})\cdot\mathbf{N} = -6z/2$. Therefore

$$\begin{aligned}\iint_S (\nabla \times \mathbf{F})\cdot\mathbf{N}\,dS &= \iint_S (-3z)\,dS \\ &= \iint_{S'_{xy}} (-3z)|\sec\gamma|\,dx\,dy \\ &= \iint_{S'_{xy}} (-3z)\frac{2}{z}\,dx\,dy \\ &= (-6)\,(\text{Area of projection of hemisphere on the } xy \text{ plane}) \\ &= -24\pi\end{aligned}$$

(b) Use Stokes' theorem to equate the given surface integral to $\oint_C \mathbf{F}\cdot d\mathbf{r}$, where C is the circle $x^2 + y^2 = 4$ in the xy-plane. Using the representation of C as given in the previous example,

$$\begin{aligned}\iint_S (\nabla \times \mathbf{F})\cdot\mathbf{N}\,dS &= \int_C \mathbf{F}\cdot d\mathbf{r} \\ &= \oint_C 4y\,dx - x\,dy + 2z^3\,dz \\ &= \int_0^{2\pi} (-16\sin^2 t - 8\cos^2 t)\,dt \\ &= -24\pi\end{aligned}$$

(c) Use the fact that the vector field whose normal component we are integrating has a vector potential—in this case the vector potential of the integrated field is $\mathbf{F}$. Hence the integral is independent of the surface, so instead of integrating over the given surface, we use the domain in the xy-plane which is interior to the circle $x^2 + y^2 = 4$. Therefore,

$$\begin{aligned}\iint_S (\nabla \times \mathbf{F})\cdot\mathbf{N}\,dS &= \iint_{S'_{xy}} (-6)\,dx\,dy \\ &= -24\pi\end{aligned}$$

□

By Theorem 11.5, if the three-dimensional vector field $\mathbf{F}$ is conservative (or, which is the same thing, if $\oint_C \mathbf{F}\cdot d\mathbf{r} = 0$ for all closed curves C in D), then $\operatorname{curl}\mathbf{F} = \mathbf{0}$ in D. With the assistance of Stokes' theorem we can now show the converse to be true if D is simply connected.*

---

*A domain D in space is *simply connected* if every simple closed curve in D can be continuously shrunk to a point in D *without leaving* D. The interior of a sphere is simply connected; the domain between concentric spheres is simply connected; the interior of a torus is *not* simply connected.

Theorem 11.12 Let D be a simply connected domain in three-space and assume that $\mathbf{F}$ is a vector field that is continuously differentiable and satisfies curl $\mathbf{F} = \mathbf{0}$ in D. Then

$$\oint_C \mathbf{F} \cdot d\mathbf{r} = 0$$

for all simple closed curves C in D.

Proof Let C be a simple closed curve that bounds a surface S in D to which Stokes' theorem is applicable. Hence,

$$\begin{aligned}\oint_C \mathbf{F} \cdot d\mathbf{r} &= \iint_S (\nabla \times \mathbf{F}) \cdot \mathbf{N}\, dS \\ &= \iint_S \mathbf{0} \cdot \mathbf{N}\, dS \\ &= 0\end{aligned}$$

For other closed curves we must first show that the vector field $\mathbf{F}$ is conservative. The details are tedious and we omit them. ■

Theorem 11.12 is not true for other than simply connected domains. See Example 11.18 for a case where curl $\mathbf{F} = \mathbf{0}$ in a doubly connected domain in the plane but the line integral around a closed curve is nonzero.

To summarize, consider these properties of a vector field $\mathbf{F}$:

1. It has a scalar potential throughout D.
2. It is such that the integral $\int_C \mathbf{F} \cdot d\mathbf{r}$ is independent of the curve connecting two fixed points P and Q in D.
3. It is such that the integral $\oint_C \mathbf{F} \cdot d\mathbf{r} = 0$ for all simple closed curves C in D.
4. It has curl $\mathbf{F} = \mathbf{0}$ throughout D.

These properties are all equivalent in a *simply connected domain.* If the domain is not simply connected, only the first three are equivalent. Any of the first three properties may be used to define the concept of a conservative field. In this book we used the first property for that purpose principally because it arose in a natural way *before* the others.

When you study the application of vector analysis in other subject areas, you will find that the main use of Stokes' theorem is not in computing surface integrals or line integrals, but rather in deriving other results. The next example shows how Stokes' theorem is put to good use in the theory of electricity and magnetism.

Example 11.25 Beginning with Ampere's circuital law, derive Maxwell's equation relating the magnetic intensity vector to the current density.

Solution Ampere's circuital law says that $\oint_C \mathbf{H} \cdot d\mathbf{r} = I$, where $\mathbf{H}$ is the magnetic intensity, C is a closed curve, and I is the current crossing *any* surface S bounded by C. (We say that I is the current "linking" C.) If $\mathbf{J}$ is the current density (the current crossing the unit area perpendicular to **J**) then the current across S is given by the surface integral $\int\int_S \mathbf{J}\cdot\mathbf{N}\,dS$. Therefore by Ampere's law

$$\oint_C \mathbf{H}\cdot d\mathbf{r} = \iint_S \mathbf{J}\cdot\mathbf{N}\,dS$$

On the other hand, by Stokes' theorem,

$$\oint_C \mathbf{H}\cdot d\mathbf{r} = \iint_S (\nabla \times \mathbf{H})\cdot\mathbf{N}\,dS$$

Since S is arbitrary, we have Maxwell's equation

$$\nabla \times \mathbf{H} = \mathbf{J}$$

□

Exercises for Section 11.8

1. Prove Theorem 11.10.
2. Prove Theorem 11.11.

In Exercises 3–6, verify Stokes' theorem for the given vector function and surface.

3. $\mathbf{F} = z\mathbf{i} + x\mathbf{j} + y\mathbf{k}$ taken over the hemisphere $x^2 + y^2 + z^2 = a^2$, $z > 0$.
4. $\mathbf{F} = (2y + z)\mathbf{i} + (x - z)\mathbf{j} + (y - x)\mathbf{k}$ taken over the triangle cut from the plane $x + y + z = 1$ by the coordinate planes.
5. $\mathbf{F} = (x\mathbf{i} + y\mathbf{j} + z\mathbf{k}) \times \mathbf{k}$ taken over the hemisphere $x^2 + y^2 + z^2 = a^2$, $z < 0$.
6. $\mathbf{F} = 2x\mathbf{i} + 3y^2\mathbf{j} + xz\mathbf{k}$ taken over that portion of the paraboloid $x^2 + y^2 + z = 2$ with $x^2 + y^2 < 1$.

In Exercises 7–10, evaluate $\int_C \mathbf{F}\cdot d\mathbf{r}$ for the given vector function **F** and curve C:

7. $\mathbf{F} = y\mathbf{i} + z\mathbf{j} + y\mathbf{k}$ and C is the intersection of $x^2 + y^2 + z^2 = 6z$ and $z = x + 3$ oriented clockwise as viewed from the origin.
8. $\mathbf{F} = -y\mathbf{i} + x\mathbf{j} + 2z\mathbf{k}$ and C is the unit circle in the plane $z = 2$, oriented clockwise as viewed from the origin.
9. $\mathbf{F} = y^2\mathbf{i} - x^2\mathbf{j} - (y + z)\mathbf{k}$ and C is the triangle in the xy-plane with vertices at $(0,0,0)$, $(1,0,0)$, and $(1,1,0)$ oriented counterclockwise.
10. $\mathbf{F} = x^2\mathbf{i} + y^2\mathbf{j} + z^2\mathbf{k}$ and C is the intersection of the sphere of radius 2 centered at the origin and the cylinder $z = y^2$. Orient C clockwise as viewed from the origin.

In Exercises 11–12, by means of Stokes' theorem, rewrite the given line integral as a surface integral and then evaluate both integrals in each case:

11. $\oint_C [(yz^2 + 2x)\mathbf{i} + (xz^2 + 1)\mathbf{j} + (2xyz + x^2)\mathbf{k}]\cdot d\mathbf{r}$, where C is the closed curve given parametrically by $x = \cos t$, $y = \sin t$, $z = \sin t$ with $0 \le t \le 2\pi$.
12. $\oint_C [(2y^2 + 3z^2 - x^2)\mathbf{i} + (2z^2 + 3x^2 - y^2)\mathbf{j} + (2x^2 + 3y^2 - z^2)\mathbf{k}]\cdot d\mathbf{r}$, where C is the intersection of the surface $x^2 + y^2 - 2ax + az = 0$ with the plane $z = 0$ oriented counterclockwise.

13. Demonstrate that the following line integrals are 0, where C is any simple closed curve in three-space:

(a) $\oint_C (ye^z\,dx + xe^z\,dy + xye^z\,dz)$ (b) $\oint_C (4x^3y^3z\,dx + 3x^4y^2z\,dy + x^4y^3\,dz)$
(c) $\oint_C (xr^{-3}\,dx + yr^{-3}\,dy + zr^{-3}\,dz)$
where $r = (x^2 + y^2 + z^2)^{1/2}$. (Certain restrictions are needed on the curve C in this last instance. What are they?)

14. As was pointed out in this section, it is sometimes to your advantage to know if a vector field has a vector potential. Can you give a "test" which might be used to give a quick answer, even though the test might fail in some instances? (See Exercise 12 in the exercises for Section 10.11.)
15. Show that the integral of the normal component of the curl of any conservative field over any smooth surface is equal to zero.
16. Use Stokes' theorem to show that $\nabla \times (\nabla f) = \mathbf{0}$.
17. Demonstrate that

$$\oint_C \mathbf{T}\,ds = \mathbf{0}$$

where C is a smooth simple closed curve and $\mathbf{T}$ is a unit vector everywhere tangent to the curve C. (*Hint*: Assume Stokes' theorem is applicable, rewrite $\oint_C \mathbf{T}\,ds$ as $\oint d\mathbf{r}$, and investigate each component separately.)
18. Rework Example 11.22 using Stokes' theorem and thereby show that, in two dimensions, Green's theorem, the divergence theorem, and Stokes' theorem are all equivalent.
19. Verify that $\mathbf{F} = \dfrac{1}{r}\mathbf{u}_\theta$ is irrotational and that $\displaystyle\oint_C \mathbf{F}\cdot d\mathbf{r} = 2\pi$, where C is a circle taken counterclockwise about the origin in the plane $z = 0$. Does this contradict Stokes' theorem? Explain.
20. Prove that if a vector field $\mathbf{F}$ is everywhere orthogonal to a given smooth surface S, then at each point of S, $\nabla \times \mathbf{F}$ either vanishes or is tangent to S.
21. Demonstrate that if a vector field $\mathbf{F}$ vanishes identically everywhere on a given smooth surface S (that is, S is the level surface $\mathbf{F} = 0$), then at each point of S, $\nabla \times \mathbf{F}$ either also vanishes or is tangent to S.
22. Apply Stokes' theorem to the vector field $\mathbf{F} = -y\mathbf{i} + x\mathbf{j}$ in the case of a surface S in the xy-plane bounded by a smooth simple closed curve C and thereby show that the area A within C may be calculated as

$$A = \tfrac{1}{2}\oint_C (x\,dy - y\,dx)$$

(Compare with Exercise 2, p. 436.)
23. Use the result of the previous exercise to determine the area within the ellipse given parametrically by $x = a\cos t$, $y = b\sin t$ $(0 \le t \le 2\pi)$.

In electromagnetic theory the electromotive force $\mathscr{E}$ in a circuit C is given in terms of the electric field E by

$$\mathscr{E} = \oint_C \mathbf{E}\cdot d\mathbf{r}$$

Alternatively, if the circuit is not moving,

$$\mathscr{E} = -\iint_S \frac{\partial \mathbf{B}}{\partial t}\cdot\mathbf{N}\,dS$$

by Faraday's law of induction, where **B** is the magnetic field and S is any capping surface of C.

24. From the above relations, derive Maxwell's equation relating the magnetic and electric fields.

11.9 Divergence Theorem

As you learned in Section 11.7, Green's theorem has two important and fundamental interpretations in the plane:

1. The statement of Theorem 11.8 relates the curl of a vector field to the tangential component of the field around a closed path.
2. Example 11.22, p. 436, shows how the divergence of the field may be related to the normal component of the field around a closed path.

The proper extension of the first of these relationships leads to Stokes' theorem, which is virtually a restatement of Green's theorem, with the difference that the surface and line integrals of Stokes' theorem are not restricted to a plane. In this section we extend the relationship of the normal component and the divergence to three dimensions. The resulting theorem, called the *divergence* theorem, transforms a closed surface integral into a volume integral. It is worth noting that this fundamental result has an obvious and useful generalization to n-space ($n > 3$).

The divergence theorem is usually attributed to Gauss, although it is also called Green's theorem in space. The proof is quite similar to that of Green's theorem.

Theorem 11.13 *Divergence Theorem.* Let D be a bounded domain in three-space bounded by a piecewise smooth surface, S, and let **F** be a vector field with component functions f, g, and h continuously differentiable throughout D. Then

$$\iint_S \mathbf{F}\cdot\mathbf{N}\,dS = \iiint_D (\nabla\cdot\mathbf{F})\,dV$$

where **N** is the unit vector normal to S and pointing away from D.

Proof We outline the proof in the case that D is standard.* In rectangular coordinates the divergence theorem may be stated

$$\iint_S (f\cos\alpha + g\cos\beta + h\cos\gamma)\,dS = \iiint_D (f_x + g_y + h_z)\,dV$$

As in the case of Green's theorem, we prove only one piece of the theorem. The rest of the proof would be repetitious. First we carry out one iteration on the triple integral $\iiint_D h_z\,dV$. Since D is z-standard,

---

* D is said to be "z-standard" if the boundary surface can be decomposed into a lower surface S_1 that has the equation $z = z_1(x, y)$, an upper surface S_2 that has the equation $z = z_2(x, y)$, and a lateral surface S_3 with segments parallel to the z-axis. The surface S_3 many degenerate into a curve (as, for example, with a sphere). Then D is said to be "standard" if it is x-, y-, and z-standard.

$$\iiint_D h_z\,dV = \iint_{D_{xy}} \left(\int_{z_1(x,y)}^{z_2(x,y)} h_z\,dz \right) dx\,dy$$

where D_{xy} is the projection of D onto the xy-plane. Therefore,

$$\iiint_D h_z\,dV = \iint_{D_{xy}} [h(x, y, z_2) - h(x, y, z_1)]\,dx\,dy$$

The surface integral $\iint_S h\cos\gamma\,dS$ can be written as the sum of three surface integrals,

$$\iint_S h\cos\gamma\,dS = \iint_{S_1} h\cos\gamma\,dS + \iint_{S_2} h\cos\gamma\,dS + \iint_{S_3} h\cos\gamma\,dS$$

The direction angle, γ, is acute on S_2, obtuse on S_1, and equal to $\pi/2$ on S_3. It follows that

$$\iint_S h\cos\gamma\,dS = -\iint_{D_{xy}} h(x, y, z_1)\,dx\,dy + \iint_{D_{xy}} h(x, y, z_2)\,dx\,dy$$

which shows that

$$\iiint_D h_z\,dV = \iint_S h\cos\gamma\,dS$$

In like manner

$$\iiint_D g_y\,dV = \iint_S g\cos\beta\,dS$$

and

$$\iiint_D f_x\,dV = \iint_S f\cos\alpha\,dS$$

Adding the last three equations proves the theorem. ■

If the domain D is bounded by two separate surfaces, as in the case of concentric spheres, notice that the normal to the surface must be taken as pointing "away from" D. Further, you should note carefully that the divergence theorem relates a *closed* surface integral to a volume integral. (See the footnote on page 426.)

Example 11.26 Evaluate $\iint_S \mathbf{F}\cdot\mathbf{N}\,dS$ if $\mathbf{F} = xz\mathbf{i} + yz\mathbf{j} + z^2\mathbf{k}$, S is the surface of the sphere $x^2 + y^2 + z^2 = 9$, and $\mathbf{N}$ is the outward directed normal to S.

Solution By the divergence theorem,

$$\begin{aligned}\iint_S \mathbf{F}\cdot\mathbf{N}\,dS &= \iiint_D (\nabla\cdot\mathbf{F})\,dV \\ &= \iiint_D 4z\,dx\,dy\,dz \\ &= 4V\bar{z}\end{aligned}$$

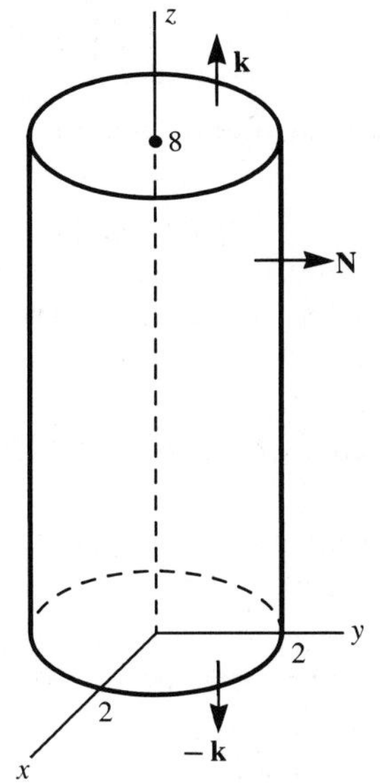

Figure 11.23

where $\bar{z}$ represents the third coordinate of the centroid of the sphere and thus $\bar{z} = 0$. Therefore the given surface integral has the value 0. □

Example 11.27 Evaluate $\iint_S \mathbf{F} \cdot \mathbf{N}\, dS$ if $\mathbf{F} = x\mathbf{i} + y\mathbf{j} + z\mathbf{k}$, S is the complete surface of the finite cylinder $x^2 + y^2 = 4$, $0 \leqq z \leqq 8$, and $\mathbf{N}$ is the outward directed normal. Work the problem in two ways. (See Figure 11.23.)

Solution By the divergence theorem,

$$\iint_S \mathbf{F} \cdot \mathbf{N}\, dS = \iiint_D (\nabla \cdot \mathbf{F})\, dV = 3 \iiint_D dV = 96\pi$$

Directly,

$$\iint_S \mathbf{F} \cdot \mathbf{N}\, dS = \iint_{\substack{\text{Top} \\ (\mathbf{N}=\mathbf{k})}} \mathbf{F} \cdot \mathbf{N}\, dS + \iint_{\substack{\text{Bottom} \\ (\mathbf{N}=-\mathbf{k})}} \mathbf{F} \cdot \mathbf{N}\, dS + \iint_{\substack{\text{curved} \\ \text{surface} \\ (\mathbf{N}=(x\mathbf{i}+y\mathbf{j})/2)}} \mathbf{F} \cdot \mathbf{N}\, dS$$

$$= 32\pi + 0 + 64\pi = 96\pi$$ □

As with Stokes' theorem, a principal use of the divergence theorem is in deriving other very important equations and formulas. The following two examples are illustrations of these derivations.

Example 11.28 Let $T(x, y, z, t)$ represent the temperature of a homogeneous body at time t and at the point (x, y, z). The velocity of heat flow in the body is of the form $\mathbf{V} = -K \operatorname{grad} T$ where K is the thermal conductivity of the body. Using the divergence theorem, derive the *heat equation*

$$T_t = \left(\frac{K}{\sigma\rho}\right)\nabla^2 T$$

where σ is the specific heat of the material of the body and ρ is its density.

Solution Let D represent the domain of the body and S its boundary surface. The amount of heat leaving D per unit time is

$$\iint_S \mathbf{V}\cdot\mathbf{N}\,dS$$

Applying the divergence theorem to this surface integral,

$$\iint_S \mathbf{V}\cdot\mathbf{N}\,dS = -K\iiint_D (\nabla\cdot\nabla T)\,dV = -K\iiint_D \nabla^2 T\,dV$$

The total amount of heat in D is given by $\sigma\rho\iiint_D T\,dV$. The time rate of change of this integral is equal to $\iint_S \mathbf{V}\cdot\mathbf{N}\,dS$. Assuming that the process of differentiation with respect to time and the spatial integration can be interchanged, we obtain

$$-\sigma\rho\iiint_D T_t\,dV = -K\iiint_D \nabla^2 T\,dV$$

Since this equation holds anywhere in the body, and for any part of the body, it follows that

$$T_t = \left(\frac{K}{\sigma\rho}\right)\nabla^2 T$$

□

Example 11.29 Let f and g be scalar functions such that $\mathbf{F} = f\nabla g$ is continuously differentiable in some domain D. Prove Green's first two formulas:

1. $$\iiint_D [f\nabla^2 g + (\nabla f)\cdot(\nabla g)]\,dV = \iint_S f\frac{\partial g}{\partial n}\,dS$$

2. $$\iiint_D (f\nabla^2 g - g\nabla^2 f)\,dV = \iint_S \left(f\frac{\partial g}{\partial n} - g\frac{\partial f}{\partial n}\right)dS$$

Solution We apply the divergence theorem to the vector field $\mathbf{F}$. First, note that $\mathbf{F}\cdot\mathbf{N} = f(\partial g/\partial n)$ from the formula for the directional derivative. Then

$$\nabla\cdot\mathbf{F} = \nabla\cdot(f\nabla g) = f\nabla^2 g + (\nabla f)\cdot(\nabla g)$$

Green's first formula now follows directly from the formula in the divergence theorem.

If we interchange the roles of f and g we obtain a similar formula and subtracting this from Green's first formula, we have

$$\iiint_D (f\nabla^2 g - g\nabla^2 f)\,dV = \iint_S \left(f\frac{\partial g}{\partial n} - g\frac{\partial f}{\partial n}\right)dS$$

which is Green's second formula.

□

Exercises for Section 11.9

Exercises 1–3 are review problems on triple integration.

1. Find the z-coordinate of the centroid of the volume bounded by the planes $x + y + z = 4$, $x = 0$, $y = 0$, and $z = 0$.
2. Find the volume bounded by the cylinder $z = 4/(y^2 + 1)$ and the planes $y = x$, $y = 3$, $x = 0$, and $z = 0$.
3. Find the moment of inertia with respect to the z-axis of the mass in the first octant bounded by the cylinder $x^2 + z^2 = 4$ and the planes $y = x$, $y = 0$, and $z = 0$, if the density is given by $2z$.

In Exercises 4–5, evaluate $\iint_S \mathbf{F}\cdot\mathbf{N}\,dS$ if S is the surface of the sphere $x^2 + y^2 + z^2 = 16$, $\mathbf{N}$ is the unit normal pointing away from the origin, and $\mathbf{F}$ is the given vector function.

4. **(a)** $\mathbf{F} = x\mathbf{i} + y\mathbf{j} + z\mathbf{k}$ **(b)** $\mathbf{F} = xy^2\mathbf{i} + yz^2\mathbf{j} + zx^2\mathbf{k}$
5. **(a)** $\mathbf{F} = xy\mathbf{i} + yz\mathbf{j} + zx\mathbf{k}$ **(b)** $\mathbf{F} = yz\mathbf{i} + zx\mathbf{j} + xy\mathbf{k}$

In Exercises 6–8, verify the divergence theorem for the given vector function and bounded domain.

6. $\mathbf{F} = xy^2\mathbf{i} + x^2y\mathbf{j} + y\mathbf{k}$ over the solid cylinder $x^2 + y^2 < 1$, $|z| < 1$.
7. $\mathbf{F} = (x^2 + y^2)\mathbf{i} + xz\mathbf{j} + yz\mathbf{k}$ over that portion of the solid paraboloid $x^2 + y^2 < z$ with $x > 0$ and $z < 1$.
8. $\mathbf{F} = \nabla(x^4 + y^4 + z^4)$ over the ball $x^2 + y^2 + z^2 < a^2$.

9. Show that if a vector field has a vector potential, the integral of the normal component of the field over a closed surface is zero.

In Exercises 10–12, by means of the divergence theorem, rewrite the given surface integral as a volume integral and then evaluate both integrals in each case.

10. $\iint_S (x\mathbf{i} + y\mathbf{j} - z\mathbf{k})\cdot\mathbf{N}\,dS$, where S is the surface of the unit cube in the first octant.
11. $\iint_S [(x - y)\mathbf{i} + xy^2\mathbf{j} + x^2z\mathbf{k}]\cdot\mathbf{N}\,dS$, where S is the surface of that portion of the solid paraboloid $0 < z < x^2 + y^2$ cut off by the plane $z = 1$.
12. $\iint_S (3xy^2\mathbf{i} + 3x^2y\mathbf{j} + z^3\mathbf{k})\cdot\mathbf{N}\,dS$, where S is the unit sphere.

13. Let $\mathbf{F} = z\mathbf{i} + xy\mathbf{j} + xz\mathbf{k}$. Evaluate $\iint_S \mathbf{F}\cdot\mathbf{N}\,dS$ over the surface of the region bounded by the paraboloid $x^2 + y^2 = 4 - z$ and the planes $x = 0$, $y = 0$, $z = 0$. Work the problem in two distinct ways.

In Exercises 14–16, assume D is a domain in three dimensions with a piecewise-smooth boundary surface S, $\mathbf{N}$ is the unit outward normal to S, V is the volume of the domain, and $(\bar{x}, \bar{y}, \bar{z})$ represents the centroid.

14. Let $\mathbf{F} = x\mathbf{i} + y\mathbf{j} + z\mathbf{k}$ and show $\iint_S \mathbf{F}\cdot\mathbf{N}\,dS = 3V$.
15. Let $\mathbf{F} = y^2\mathbf{i} + 2xy\mathbf{j} - xz\mathbf{k}$ and show $\iint_S \mathbf{F}\cdot\mathbf{N}\,dS = \bar{x}V$.
16. Let $\mathbf{F} = 2xz\mathbf{i} + 3yz\mathbf{j} + 4z^2\mathbf{k}$, and show $\iint_S \mathbf{F}\cdot\mathbf{N}\,dS = 13\bar{z}V$.

17. Demonstrate that if g is harmonic in D, then the integral of the normal derivative of g over the boundary of D is zero.
18. Evaluate $\iint_S \mathbf{F}\cdot\mathbf{N}\,dS$ if $\mathbf{F} = 6z\mathbf{i} + (2x + y)\mathbf{j} - x\mathbf{k}$, where S is the surface of the

region bounded by the cylinder $x^2 + y^2 = 9$ and the planes $x = 0$, $y = 0$, $z = 0$, and $z = 8$.

19. Evaluate $\iint_S \mathbf{F}\cdot\mathbf{N}\,dS$ if $\mathbf{F} = 4xz\mathbf{i} - y^2\mathbf{j} + yx\mathbf{k}$, where S is the surface of the unit cube in the first octant.

20. Use the divergence theorem to show that

$$\iint_S \mathbf{N}\,dS = \mathbf{0}$$

where S is any piecewise-smooth closed surface and $\mathbf{N}$ is the unit vector normal to S pointing in the exterior direction. (*Hint:* Investigate each component of $\mathbf{N}$ separately.)

21. Verify that $\mathbf{F} = \dfrac{1}{\rho^2}\mathbf{u}_\rho$ is solenoidal and that $\displaystyle\iint_S \mathbf{F}\cdot\mathbf{N}\,dS = 4\pi$, where S is a sphere about the origin. Does this contradict the divergence theorem? Explain.

22. **(a)** Rewrite the divergence theorem in a form appropriate for the case of a domain D situated between two closed surfaces S_1 and S_2, one of which lies entirely within the other.

(b) Demonstrate that if $\mathbf{F}$ is solenoidal everywhere between S_1 and S_2, then

$$\iint_{S_1} \mathbf{F}\cdot\mathbf{N}\,dS = \iint_{S_2} \mathbf{F}\cdot\mathbf{N}\,dS$$

23. Apply the divergence theorem to the vector field $\mathbf{F} = x\mathbf{i} + y\mathbf{j} + z\mathbf{k} \equiv \rho\mathbf{u}_\rho$ in the case of a domain D of volume V bounded by a piecewise-smooth surface S and thereby show that

$$V = \tfrac{1}{3}\iint_S \rho\mathbf{u}_\rho\cdot\mathbf{N}\,dS$$

24. Use the result of the previous exercise to determine each volume.

(a) The rectangular parallelepiped $0 \le x \le a$, $0 \le y \le b$, $0 \le z \le c$.

(b) The sphere of radius R.

25. If the total charge present within an arbitrary domain D bounded by a piecewise-smooth surface S is given by

$$q = \iiint_D \rho\,dV$$

demonstrate that Gauss' law,

$$\iint_S \mathbf{E}\cdot\mathbf{N}\,dS = 4\pi q$$

leads to a relationship between the electric field $\mathbf{E}$ and the charge density ρ.

26. In the flow of incompressible fluids with vector velocities $\mathbf{v}$, the surface integral

$$\iint_S \mathbf{v}\cdot\mathbf{N}\,dS$$

is proportional to the net outward flow of mass across S per unit time. Employ the divergence theorem to show that conservation of mass then implies that $\mathbf{v}$ is solenoidal.

11.10 Summary of Techniques of Evaluation

In this section we give a short summary of the techniques used for evaluating line and surface integrals of vector fields in three dimensions. Specifically, we restrict ourselves to line integrals of the tangential component of a vector field and to surface integrals of the normal component of a vector field. In the second instance we assume that all of the vector fields in question are continuously differentiable throughout simply connected domains.

To evaluate $\int_C \mathbf{F}\cdot d\mathbf{r}$, we consider first whether the field is conservative and if the curve is closed.

1. If $\mathbf{F}$ is conservative, it has a scalar potential ϕ and the line integral $\int_C \mathbf{F}\cdot d\mathbf{r}$ is independent of path.
   (a) If C is closed, then $\oint_C \mathbf{F}\cdot d\mathbf{r} = 0$.
   (b) If C is not closed and has endpoints P_1 and P_2, then

$$\int_C \mathbf{F}\cdot d\mathbf{r} = \phi(P_2) - \phi(P_1)$$

2. If $\mathbf{F}$ is not conservative, the line integral is not independent of path and generally must be evaluated by the method of Theorem 11.2.
   (a) If C is closed, it may be possible (and convenient) to find a smooth surface with C as its boundary and then to apply Stokes' theorem equating the given line integral to the surface integral of the normal component of the curl.
   (b) If C is not closed, occasionally a supplementary arc C' may be found that together with C forms a closed path that is the boundary of a smooth surface S. Then, by applying Stokes' theorem,

$$\int_C \mathbf{F}\cdot d\mathbf{r} = \iint_S (\nabla \times \mathbf{F})\cdot\mathbf{N}\,dS - \int_{C'} \mathbf{F}\cdot d\mathbf{r}$$

To evaluate the surface integral of the normal component of a vector function, $\iint_S \mathbf{G}\cdot\mathbf{N}\,dS$, we decide initially (if possible) if the field has a vector potential and if the surface is closed.

1. If $\mathbf{G}$ has a vector potential, there is a field $\mathbf{F}$ such that $\mathbf{G} = \nabla \times \mathbf{F}$ and hence the integral is independent of surface.
   (a) If S is closed, then $\iint_S \mathbf{G}\cdot\mathbf{N}\,dS = 0$.
   (b) If S is not closed, then the surface integral is equal to $\oint_C \mathbf{F}\cdot d\mathbf{r}$, where C is the boundary of S. (We may either evaluate the line integral directly by using Theorem 11.2 or re-apply Stokes' theorem to integrate $\mathbf{G}\cdot\mathbf{N}$ over another surface of which C is the boundary.)
2. If $\mathbf{G}$ does not have a vector potential, then the surface integral is not independent of surface.
   (a) If S is closed, we may apply the divergence theorem to obtain

$$\iint_S \mathbf{G}\cdot\mathbf{N}\,dS = \iiint_D \nabla\cdot\mathbf{G}\,dV$$

where D is the region of which S is the boundary.

(b) If S is not closed,

(i) The surface integral may be evaluated by the method of Theorem 11.6 or one of the ways related to it.

(ii) A supplementary surface, S', may be found, which together with the given one comprises a closed surface. Then by applying the divergence theorem

$$\iint_S \mathbf{G}\cdot\mathbf{N}\,dS = \iiint_D \nabla\cdot\mathbf{G}\,dV - \iint_{S'} \mathbf{G}\cdot\mathbf{N}\,dS$$

where D is the region interior to $S \cup S'$.

Exercises for Section 11.10

In Exercises 1–17, compute $\int_C \mathbf{F}\cdot d\mathbf{r}$ where the vector function $\mathbf{F}$ and the curve C are as given.

1. $\mathbf{F} = xy\mathbf{i} + x^3y^2\mathbf{j}$; C is the line segment from $(1, 1, 1)$ to $(3, 1, 4)$.
2. $\mathbf{F} = x\mathbf{i} + xy\mathbf{j} + x^2yz\mathbf{k}$; C is the circle in the xy-plane of radius 1 centered at $(0, 1, 0)$, taken counterclockwise as viewed from above.
3. $\mathbf{F} = x\mathbf{i} + xy\mathbf{j} + x^2yz\mathbf{k}$; C is the ellipse in the xy-plane, $x^2 + 4(y-1)^2 = 1$, taken counterclockwise as viewed from above.
4. $\mathbf{F} = (2xy + \cos x)\mathbf{i} + (2y + x^2)\mathbf{j}$; C is the curve $x = \cos y$ from $(1, 0)$ to $(1, 2\pi)$.
5. $\mathbf{F} = (y^2 - x^2)\mathbf{i} + 2xy\mathbf{j}$; C is the curve $\mathbf{r}(t) = \sec^2 t\mathbf{i} + \tan^2 t\mathbf{j}$, $0 \leqq t \leqq \pi/4$.
6. $\mathbf{F} = x^2y\mathbf{i} + z\mathbf{j} + y\mathbf{k}$; C is the straight line segment from the origin to $(2, 3, 4)$.
7. $\mathbf{F} = z\mathbf{i}$; C is the curve of intersection of $x^2 + y^2 + z^2 = 4$ and $x = y$.
8. $\mathbf{F} = yz\mathbf{i} + (xz + 1)\mathbf{j} + xy\mathbf{k}$; C is the curve defined by $\mathbf{r}(t) = (1 + \sin^3 t)\mathbf{i} + \tan^2(t/2)\mathbf{j} + (8t/\pi)\mathbf{k}$ from $t = 0$ to $t = \pi/2$.
9. $\mathbf{F} = 2\mathbf{i} + z\mathbf{j} + 3y\mathbf{k}$; C is the intersection of $x^2 + y^2 + z^2 = 1$ and $x = z$ with C oriented counterclockwise as viewed from $(1, 0, 0)$.
10. $\mathbf{F} = \cos x\mathbf{i} + \sin z\mathbf{k}$; C is any curve joining $(0, 1, \pi/2)$ and $(\pi, 0, -\pi/2)$.
11. $\mathbf{F} = xy\mathbf{i} + y^2z\mathbf{j} + xyz^2\mathbf{k}$; C is the line segment from $(0, 0, 0)$ to $(2, 5, 9)$.
12. $\mathbf{F} = 2xy\mathbf{i} + (2yz + x^2)\mathbf{j} + (y^2 + 2z)\mathbf{k}$; C is the curve defined by $\mathbf{r}(t) = t\mathbf{i} + \sin t\mathbf{j} + \sin t\mathbf{k}$ from $(0, 0, 0)$ to $(\pi/2, 1, 1)$.
13. $\mathbf{F} = xy\mathbf{i} + y^2z\mathbf{j} + xyz^2\mathbf{k}$; C is the circle of radius 2 in the xy-plane centered at $(3, 0, 0)$ taken counterclockwise as viewed from above.
14. $\mathbf{F} = y^2\mathbf{i} + x^2\mathbf{j} - (x + z)\mathbf{k}$; C is the boundary of the triangle with vertices $(0, 0, 0)$, $(1, 0, 0)$, and $(1, 1, 0)$ taken in the counterclockwise sense as viewed from above.
15. $\mathbf{F} = yz\mathbf{i} + xz\mathbf{j} + xy\mathbf{k}$; C is the curve $\mathbf{r}(t) = \sin(\pi t/4)\mathbf{i} + \tan(\pi t/4)\mathbf{j} + \sec(\pi t/4)\mathbf{k}$ from $(0, 0, 1)$ to $(\sqrt{2}/2, 1, \sqrt{2})$.
16. $\mathbf{F} = -y\mathbf{i} + 3x\mathbf{j} + z\mathbf{k}$; C is the curve $\mathbf{r}(t) = \cos t\mathbf{i} + \sin t\mathbf{j} + 3\mathbf{k}$.
17. $\mathbf{F} = (2xyz \sin xyz + x^2y^2z^2 \cos xyz)\mathbf{i} + (x^2z \sin xyz + x^3yz^2 \cos xyz)\mathbf{j} + (x^2y \sin xyz + x^3y^2z \cos xyz)\mathbf{k}$; C is the line segment from $(0, 0, 0)$ to $(\frac{1}{2}, \frac{1}{2}, \pi)$.

In Exercises 18–31 compute $\iint_S \mathbf{F}\cdot\mathbf{N}\,dS$ where $\mathbf{F}$ and S are as given and, unless otherwise stated, $\mathbf{N}$ is the unit outward-directed normal.

18. $\mathbf{F} = x^2\mathbf{i} + 2y\mathbf{j} + x^2y\mathbf{k}$; S is the surface of the unit cube lying between $x = 0$, $x = 1$, $y = 0$, $y = 1$, $z = 0$, and $z = 1$.
19. $\mathbf{F} = 3\mathbf{i} + (y^2 - x^2)\mathbf{j} + 2z(2 - y)\mathbf{k}$; S is the complete surface of the sphere $x^2 + y^2 + z^2 = 1$.
20. $\mathbf{F} = x\mathbf{i} + y\mathbf{j} + z\mathbf{k}$; S is the surface of $x^2 + y^2 + z^2 = 4$.
21. $\mathbf{F} = (x^2 + y^2 + z^2)\mathbf{i} + xy\mathbf{j} + xy\mathbf{k}$; S is the surface of the five faces of the unit cube for which $z \neq 0$.
22. $\mathbf{F} = y\mathbf{i} + x\mathbf{j} + y^2\mathbf{k}$; S is the surface of the ellipsoid $x^2 + y^2/4 + z^2/9 = 1$.
23. $\mathbf{F} = (x + z)\mathbf{i} + (y + z)\mathbf{j} + (x + y)\mathbf{k}$; S is the surface of the sphere $x^2 + y^2 + z^2 = 4$.
24. $\mathbf{F} = y\mathbf{j} + z\mathbf{k}$; S is the surface $x^2 + y^2 = 4 - z$, $z > 0$, and $\mathbf{N}$ is the unit upward directed normal to S.
25. $\mathbf{F} = x^2\mathbf{i} + y^2\mathbf{j} + 2z(xy - x - y)\mathbf{k}$; S is the surface of the unit cube.
26. $\mathbf{F} = x\mathbf{i} + y\mathbf{j} + z\mathbf{k}$; S is the surface $x^2 + y^2 - z = 0$ for $0 \leqq z \leqq 4$, and $\mathbf{N}$ is the upward directed unit normal.
27. $\mathbf{F} = \operatorname{curl}[-yx^2\mathbf{i} + (xy^2 - z)\mathbf{j} + x^2\mathbf{k}]$; S and $\mathbf{N}$ are as in Exercise 26.
28. $\mathbf{F} = (x^2/2 + \sin^2 z)\mathbf{i} + (yz + \cos x^2)\mathbf{j} + (x^2 + z)\mathbf{k}$; S is the surface of the six faces of the unit cube.
29. $\mathbf{F} = \operatorname{curl}[z^2\mathbf{i} + x^2\mathbf{j} + y^2\mathbf{k}]$; S is the surface of the sphere $x^2 + y^2 + z^2 = a^2$.
30. $\mathbf{F} = x^2\mathbf{i}$; S is the surface of the plane $x + y + z = 1$ in the first octant, and $\mathbf{N}$ is the unit upward directed normal.
31. $\mathbf{F} = \operatorname{curl}[-y\mathbf{i} + x\mathbf{j} + \mathbf{k}]$; S is the surface of the hemisphere $z = (4 - x^2 - y^2)^{1/2}$, and $\mathbf{N}$ is the unit upward directed normal.

11.11 Looking Ahead

In addition to their importance in their own right and their utility in applied problem solving, several of the notions introduced in this chapter play roles later when we discuss functions of a complex variable. For example, vector line integrals will transform easily into line integrals of complex functions; path independence for integrals of conservative vector fields will appear as the fundamental theorem of integral calculus for analytic functions; and Green's theorem in the plane will provide an easy proof of the renowned Cauchy-Goursat theorem.

As we have suggested earlier, the divergence theorem of Section 11.9 and its consequences (Green's formulas) are not restricted to two- and three-space. Once vectors, planes (hyperplanes), surfaces, and regions have been satisfactorily defined, these results can be established in n-dimensions ($n > 3$) as well. This gives a measure of dimension independence to general conclusions derived using either the divergence theorem or its immediate relatives.

In this regard, it is worth observing that the divergence theorem plays a distinguished role in the theory of second-order partial differential equations. The integral theorems that result are important for representation purposes, uniqueness considerations, transformation into equivalent integral equations, and the like (see [10] and [14], for example). In a specialized setting, the mean-value property, the max-min principle, and Green's representation for harmonic functions (among numerous other results), all follow directly from the divergence theorem. The classic reference for these considerations is [11] (see also [6]).

In two dimensions, the divergence theorem, Stokes' theorem, and Green's theorem all coalesce into a single result. Remarkably, there exists a mathematical theory which unifies these theorems more broadly and shows how an apparently three-dimensional concept such as cross product or curl may be successfully generalized to higher dimension. Differential forms, exterior (wedge) products, and integral theorems based upon these constructs are discussed at length in [8] and [9]. A heuristic introduction to this material is given in [13]; a concise, well-written overview appears in [2].

Basic References

[1] Chambers, L. G. 1969. *A Course in Vector Analysis*. London: Chapman and Hall.

[2] Marsden, J. E., and A. J. Tromba. 1981. *Vector Calculus*. 2d ed. San Francisco: W. H. Freeman.

[3] Mc Lachlan, E. K. 1968. *Calculus of Several Variables*. Belmont, Calif.: Brooks/Cole.

[4] Schey, H. M. 1973. *Div, Grad, Curl, and All That*. New York: W. W. Horton.

[5] Seeley, R. T. 1970. *Calculus of Several Variables: An Introduction*. Glenview, Ill.: Scott, Foresman.

More-Advanced References

[6] Chester, C. R. 1971. *Techniques in Partial Differential Equations*. New York: McGraw-Hill.

[7] Chorlton, F. 1976. *Vector and Tensor Methods*. Chichester, England: Ellis Horwood.

[8] Flanders, H. 1963. *Differential Forms with Applications to the Physical Sciences*. New York: Academic Press.

[9] Fleming, W. 1977. *Functions of Several Variables*. 2d ed. New York: Springer-Verlag.

[10] Gustafson, K. E. 1980. *Partial Differential Equations and Hilbert Space Methods*. New York: John Wiley.

[11] Kellogg, O. D. 1929. *Foundations of Potential Theory*. Berlin: Springer-Verlag; also New York: Dover.

[12] Lass, H. 1950. *Vector and Tensor Analysis*. New York: McGraw-Hill.

[13] Schreiber, M. 1977. *Differential Forms, A Heuristic Introduction*. New York: Springer-Verlag.

[14] Zauderer, E. 1983. *Partial Differential Equations of Applied Mathematics*. New York: John Wiley.

Chapter 12

Complex Functions

12.1 Introduction

There is a fundamental theorem in algebra which states that, counting single roots once, double roots twice, and so on, every polynomial equation of degree n has *precisely* n roots. Although the Babylonians of 3000 years ago may have been satisfied to discard as "imaginary," and hence "impossible," the two roots of

$$x^2 + 1 = 0$$

we regard this as inappropriate, particularly (for instance) if these roots tell us something about the radius of convergence of the power series expansion for the solution of a given differential equation. (See Exercises 16 and 19, p. 205.)

This example is illustrative of the utility of complex numbers. Coupled with their use elsewhere in differential equations, Fourier series, and special functions, complex numbers provide a tremendously convenient tool in applied analysis. But just as real numbers and vectors by themselves are helpful only up to a point—with the full range of applicability needing the notion of real and vector functions—so it is with complex numbers. In this and succeeding chapters we explore the concept of complex functions and sample some of the truly remarkable applications which these new ideas make possible.

12.2 Complex Numbers

From a modern viewpoint, real numbers correspond naturally to points on a Euclidean line, and complex numbers are equally naturally paired with points in the Cartesian plane. In this correspondence the complex numbers may be thought of as an extension of the reals (represented as points along the x-axis). This is not just an attractive abstraction; together with the extension of the reals to n-dimensional vectors as introduced in Chapter 1, it provides us with the numerical, algebraic, and geometrical tools to attack most applied problems.

The fundamental algebraic operations for complex numbers are direct analogs of those for real numbers, and have been employed earlier in Chapters 3, 4, 7, and 9. If we introduce i as standing for $\sqrt{-1}$, then all complex numbers can be represented in the form $z = x + iy$, or more simply (x, y), where x and y are real numbers. Given any two complex numbers $z_1 = x_1 + iy_1$ and $z_2 = x_2 + iy_2$, the following are true.

1. The sum of z_1 and z_2 is given by

$$z_1 + z_2 = (x_1 + x_2) + i(y_1 + y_2)$$

2. The product of z_1 and z_2 is given by

$$z_1 z_2 = (x_1 x_2 - y_1 y_2) + i(x_1 y_2 + x_2 y_1)$$

3. For nonzero z_2, the *reciprocal* of z_2 is given by

$$\frac{1}{z_2} = \frac{x_2}{x_2^2 + y_2^2} - i\frac{y_2}{x_2^2 + y_2^2}$$

4. For nonzero z_2, the quotient of z_1 and z_2 is given by

$$\frac{z_1}{z_2} = z_1\left(\frac{1}{z_2}\right) = \frac{x_1 x_2 + y_1 y_2}{x_2^2 + y_2^2} + i\frac{x_2 y_1 - x_1 y_2}{x_2^2 + y_2^2}$$

The associative, commutative, and distributive properties that are valid for real numbers hold for complex numbers also.

The *real* numbers x and y in the representation $z = x + iy$ are customarily termed the *real and imaginary parts* of z, respectively, that is,

$$x = \operatorname{Re} z \quad \text{and} \quad y = \operatorname{Im} z$$

By convention, if $y = 1$, then "y" may be omitted. Thus $5 + 1i$ is simplified to $5 + i$. Further, if $y = 0$, we delete "iy," and if $x = 0$, we delete "$x +$."

Before leaving this section we note that, since $i = \sqrt{-1}$, computations with complex numbers can be made without continual reference to the definitions of multiplication and division. The algebra of complex numbers becomes identical to that of real binomials if you use the fact that $i^2 = -1$. For example,

$$\begin{aligned}(3 + 4i)(7 + 6i) &= (21 + 28i + 18i + 24i^2)\\ &= 21 + 46i + 24i^2 = -3 + 46i\end{aligned}$$

12.3 The Complex Plane

Complex numbers considered as ordered pairs of real numbers may be represented graphically as points in the Cartesian coordinate plane, in which case it is called the *complex plane*. The x- and y-axes are then called the *real* and *imaginary* axes, respectively. The point with coordinates (x, y) corresponds to the complex number $z = x + iy$. The operation of addition is conveniently interpreted graphically to be the same as the familiar parallelogram law for addition of geometric vectors. (See Figure 12.1.)

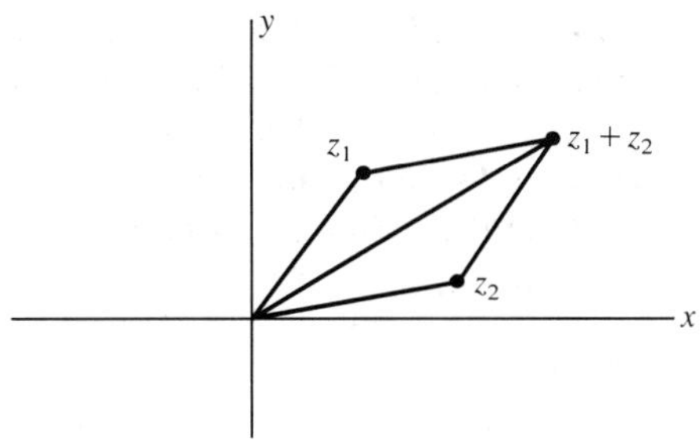

Figure 12.1 *Addition of complex numbers*

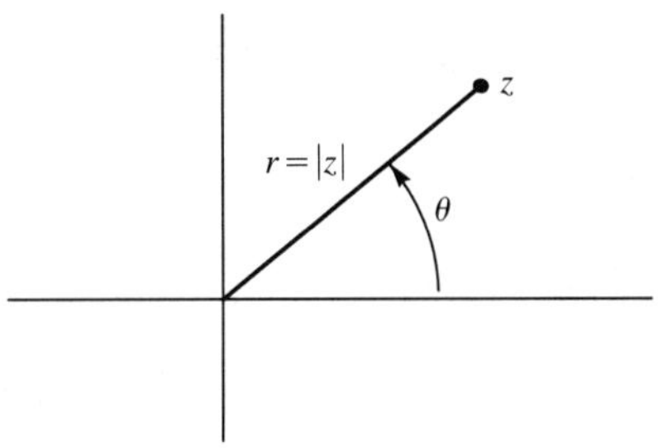

Figure 12.2 *Polar representation*

A complex number $z = x + iy$ may alternatively be represented in *polar form* by use of the transformation equations $x = r\cos\theta$, $y = r\sin\theta$. Then

$$\begin{aligned} z = x + iy &= r\cos\theta + ir\sin\theta \\ &= r(\cos\theta + i\sin\theta) \end{aligned}$$

If we introduce *Euler's formula*

$$e^{i\theta} \equiv \cos\theta + i\sin\theta$$

for the exponential function with purely imaginary exponent (which was used earlier in Chapter 3 and is justified in Section 12.7), this expression can be written more concisely as

$$z = re^{i\theta}$$

In polar representation, the number r is called the magnitude or *modulus* of the complex number z and is denoted by $|z|$; it is always considered to be nonnegative. The real number θ is called the *argument* of z and is denoted by $\arg z$.

If the polar form of a complex number is used, the point in the complex plane corresponding to $z = re^{i\theta}$ is denoted by (r, θ). In this manner the modulus of z is the measure of the distance from the origin to the point and the argument of z is the measure of the counterclockwise angle from the real axis to a line drawn from the origin to the point just as in classical polar coordinates (see Figure 12.2).

Since the complex exponential function enjoys many of the same useful

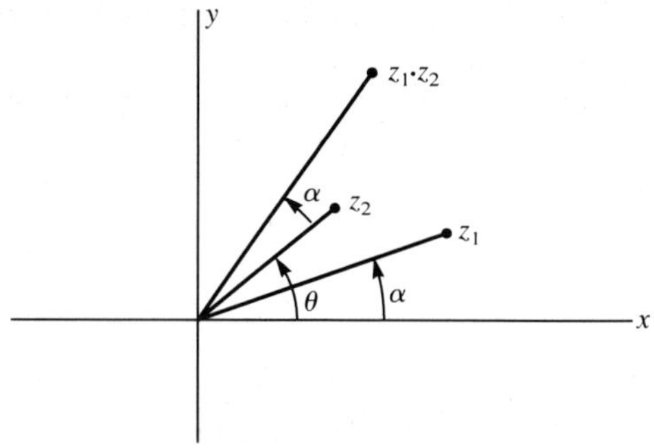

Figure 12.3 *Multiplication of complex numbers*

properties as its real counterpart (and some others also, as we shall see), the polar representation makes it very convenient to multiply or divide complex numbers. For example, if

$$z_1 = r_1 e^{i\theta_1} \quad \text{and} \quad z_2 = r_2 e^{i\theta_2}$$

then

$$\begin{aligned} z_1 z_2 &= (r_1 e^{i\theta_1})(r_2 e^{i\theta_2}) \\ &= r_1 r_2 e^{i(\theta_1 + \theta_2)} \end{aligned}$$

and

$$\begin{aligned} \frac{z_1}{z_2} &= \frac{r_1 e^{i\theta_1}}{r_2 e^{i\theta_2}} \\ &= \left(\frac{r_1}{r_2}\right) e^{i(\theta_1 - \theta_2)} \end{aligned}$$

Algebraically, the modulus of the product (quotient) is the product (quotient) of the individual moduli and the argument of the product (quotient) is the sum (difference) of the arguments of the factors. Geometrically, multiplication or division of z_1 by z_2 results in rotation of the line from the origin to z_1 by an angle equal to an argument of z_2. (See Figure 12.3 for the case of multiplication.)

As the following example suggests, use of the polar form for complex numbers can also substantially simplify the analysis needed in various practical applications.

Example 12.1 Describe the motion in the plane of a particle whose position is expressed in polar form.

Solution We begin by letting

$$z = re^{i\theta}$$

designate the position of the particle and assume that r and θ, and hence z, are functions of time. The complex velocity v of the particle is then given by

$$v \equiv z' = r'e^{i\theta} + r\theta' i e^{i\theta}$$

Note that any term of the form $Ae^{i\theta}$, where A is real, must have the same argument or direction as z while a term of the form $Aie^{i\theta}$ will have the argument $\theta + \pi/2$ (remember the geometric interpretation of complex multiplication). In the expression for z' we then recognize r' as the radial velocity of the particle and $r\theta'$ as its tangential velocity. In similar fashion, the complex acceleration a satisfies

$$\begin{aligned} a \equiv v' = z'' &= (r'' + ir'\theta' + ir\theta'')e^{i\theta} + (r' + ir\theta')e^{i\theta}i\theta' \\ &= (r'' - r(\theta')^2)e^{i\theta} + (2r'\theta' + r\theta'')ie^{i\theta} \end{aligned}$$

Here the radial acceleration term (the term having the same direction as z) is clearly made up of both the time rate of change of the radial velocity as well as the centripetal acceleration $r(\theta')^2$. The tangential component contains not only the expected tangential acceleration term $r\theta''$ but also the *Coriolis acceleration* term, $2r'\theta'$. □

The analysis carried out in Example 12.1 should be contrasted with the discussion given previously in Section 10.6.

Returning now to our definition of θ in the polar representation, we note that a given complex number has infinitely many arguments, each two differing by a multiple of 2π. For example, $\arg(-3) = \pi + 2n\pi$ and $\arg(1 - i) = -\pi/4 + 2n\pi$ with $n = 0, \pm 1, \pm 2, \ldots$. Sometimes, we limit $\arg z$ to a particular 2π span called the *principal values* of $\arg z$ and denoted by $\operatorname{Arg} z$. In this book, the values of $\operatorname{Arg} z$ will be $-\pi < \theta \le \pi$. However, we have to say precisely when $\arg z$ is being limited to its principal values because there are many important occasions in which it is not. For instance, in taking the product of two complex numbers it is often inconvenient to have the argument of the product expressed in terms of a value between $-\pi$ and π.

One final basic definition remains. For each complex number $z = x + iy$, the corresponding number $x - iy$ is called the *conjugate* of z and is denoted by $\bar{z}$. Geometrically, $\bar{z}$ is the reflection of z in the real axis (see Figure 12.4). Note that $z\bar{z} = (x + iy)(x - iy) = x^2 + y^2 = |z|^2$, an important and often-used equality.

In what follows we shall see repeatedly that the interplay of algebraic and geometric notions regarding complex numbers can be extremely fruitful. As a simple illustration we observe, in the next example, that the graphical inter-

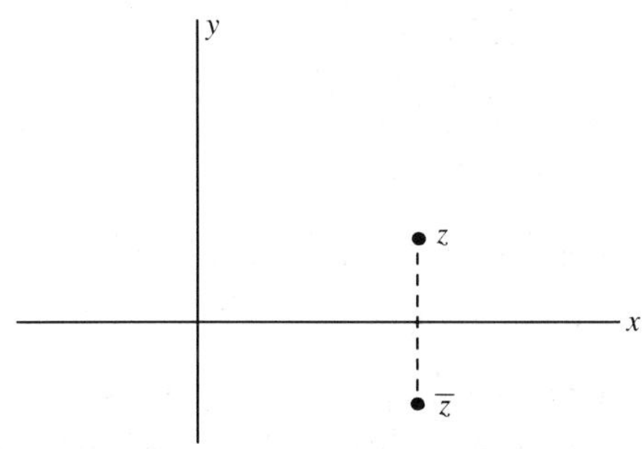

Figure 12.4 *Complex conjugate*

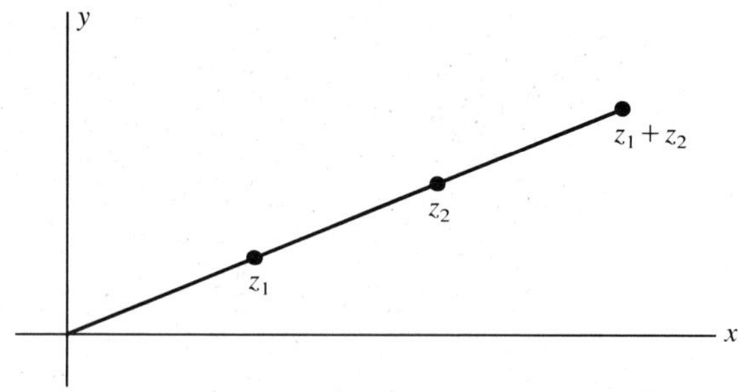

Figure 12.5 *Degenerate case of the triangle inequality*

pretation of complex numbers allows us to accept the truth of an analytic statement on the basis of a geometric argument.

Example 12.2 Show graphically that $|z_1 + z_2| \leqq |z_1| + |z_2|$. When does equality occur?

Solution Since $|z_1|$ and $|z_2|$ are sides of a triangle, Figure 12.1 suggests that the third side is always less than the sum of the other two. Because of the nature of the geometry connected with this inequality, it is called the *triangle inequality*. The degenerate case occurs when $\arg z_1 = \arg z_2$, in which case $|z_1| + |z_2| = |z_1 + z_2|$. (See Figure 12.5.) □

As part of our algebraic-geometric interplay, sets of points in the complex plane are sometimes represented by using magnitude and inequality notation. You should understand that $|z - a|$ is the distance to z from the point a.

Example 12.3 Describe the sets of points $\{z \mid |z| = 1\}$, $\{z \mid |z - 1| = |z - 2|\}$, and $\{z \mid |z - i| < 2\}$.

Solution The equation $|z| = 1$ represents all points that are 1 unit away from the origin, which is a circle of radius 1 centered at the origin. The equation $|z - 1| = |z - 2|$ describes the set of all points that are equidistant from the points $(1, 0)$ and $(2, 0)$. It is the perpendicular bisector of the line segment joining the two points. The inequality $|z - i| < 2$ describes those points which are less than 2 units away from the point $(0, 1)$, that is, the interior of the circle of radius 2 centered at $(0, 1)$. (See Figure 12.6.) □

Along with this set notation comes a certain amount of terminology, much of which repeats or is a simple modification of our earlier work. Given a point a in the complex plane and a real number $\varepsilon > 0$, a *neighborhood* of a is the set of points

$$\{z \mid |z - a| < \varepsilon\}$$

Given a set S of complex numbers, a point a is said to be an *interior point of S* if there is at least one neighborhood (and hence an infinity of neighborhoods) of a, consisting entirely of points of S. A point a is said to be an *exterior point* of S (exterior *to* S) if there is at least one neighborhood of a containing no points of

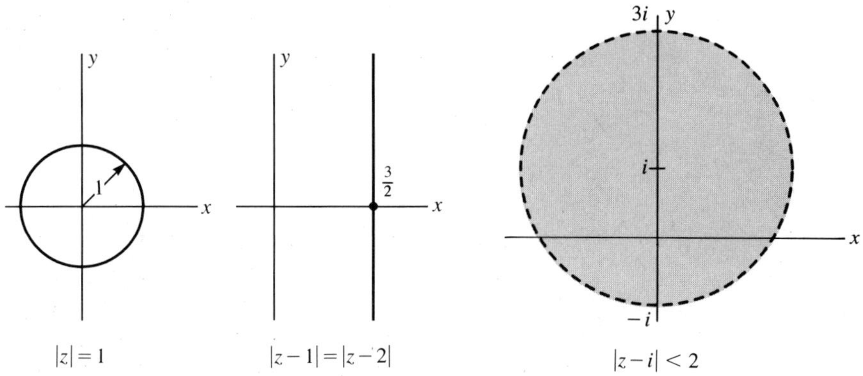

Figure 12.6

S. Points which are neither interior nor exterior to S are termed *boundary points* of S. The set S is called *open* if each point of the set is an interior point. It is *closed* if its complement is open, or equivalently if and only if S contains all of its boundary points.

A set is *bounded* if all its points lie within a circle of sufficiently large radius. An open set is *connected* if any two of its points can be joined by a broken line of finitely many linear segments all of whose points belong to the set. An open connected set is called either a *domain* or a *region*.

Example 12.4 Sketch and describe the sets $\{z \,|\, |z-1| \geqq 1\}$, $\{z \,|\, 0 < |z| < 1\}$, and $\{z \,|\, |\text{Re}\, z| > 1\}$ in terms of these new definitions.

Solution See the sketches in Figure 12.7. The first set is closed, since its complement $\{z \,|\, |z-1| < 1\}$ is open. It is unbounded and connected. The second set is open, is connected, and is bounded. The third set is open, unbounded, but not connected. Only the second set is a domain. □

Exercises for Sections 12.2 and 12.3

1. Find the value of each of the following expressions:
   (a) $(6 + 5i) \times (7 + 2i)$ **(b)** i^3
   (c) $(5e^{i\pi/6})(2e^{-i\pi/2})$ **(d)** $(2e^{i\pi/6})^3$
   (e) $(2e^{i5\pi/6})^3$ **(f)** $(2 + i)^2$
2. Find the value of each of the following expressions:
   (a) $(5 - 2i)^{-1}$ **(b)** $\text{Re}(3 - 2i)^2$
   (c) $|1 + 4i|^2$ **(d)** $\dfrac{i}{(i-1)(i-2)(i-3)}$
3. Write the following complex numbers in the form $z = x + iy$:
   (a) $(1 - \sqrt{3}i)^3$ **(b)** $\dfrac{(1+i)^2}{1-i}$ **(c)** $\dfrac{2 - \sqrt{3}i}{1+i}$
4. Write the following complex numbers in the form $z = x + iy$:
   (a) $\dfrac{3 + 2i}{3i}$ **(b)** $\dfrac{1+i}{2-3i}$ **(c)** $\dfrac{(2-i)^2}{(1+i)^2}$

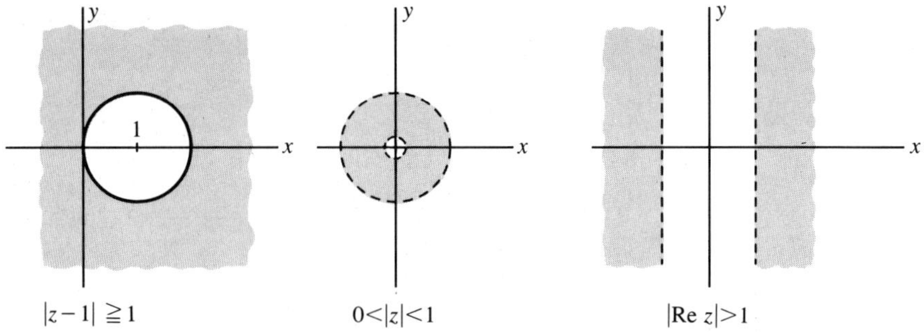

Figure 12.7

5. Write the expression for z in polar form, giving all possible values for $\theta = \arg z$, if:
   (a) $z = 1 - i\sqrt{3}$ **(b)** $z = 2i$ **(c)** $z = 5$
6. Find the moduli and principal arguments of the following complex numbers:
   (a) i^3 **(b)** $\dfrac{1}{1+i}$ **(c)** $\dfrac{3}{(i-\sqrt{3})^2}$
7. Give a graphical description of what it means to multiply a complex number by i.
8. Given the complex number z, interpret geometrically the complex number $(\cos\phi - i\sin\phi)z$.
9. Find the cubes of the following numbers:
   (a) 1 **(b)** $\frac{1}{2}(-1 + i\sqrt{3})$ **(c)** $\frac{1}{2}(-1 - i\sqrt{3})$
10. Verify that $z_n = \dfrac{i^n(1+i)}{\sqrt{2}}$, with $n = 0, 1, 2, 3$, are the 4 distinct roots (solutions) of the equation $z^4 + 1 = 0$.
11. Show that the equation $z^2 - 2z + 2 = 0$ has the solutions $z = 1 \pm i$.
12. Find $\text{Re } z$, $\text{Im } z$, $\bar{z}$, and $|z|$ for the following:
    (a) $z = 1 - 2i$ **(b)** $z = 3 + 4i$ **(c)** $z = -5\dfrac{1-2i}{3+4i}$
13. Show that if $z_1 z_2 = 0$, then either $z_1 = 0$ or $z_2 = 0$.
14. Show that:
    (a) $z + \bar{z} = 2\,\text{Re } z$ **(b)** $z - \bar{z} = 2i\,\text{Im } z$
15. Show that:
    (a) $|\text{Im } z| \leq |z|$ **(b)** $|\text{Re } z| \leq |z|$
16. Demonstrate that:
    (a) $\overline{iz} = -i\bar{z}$ **(b)** $|\overline{z^3}| = |\bar{z}|^3$
17. Sketch and describe the following sets:
    (a) $|z| = 4$ **(b)** $|z| < 4$ **(c)** $|z| \geq 4$
18. Describe the following sets by means of graphs:
    (a) $\{z \mid |z-1| \leqq 1\}$ **(b)** $\{z \mid |z-1| = |z-2|\}$
    (c) $\{z \mid |z-3| \geqq 1\}$ **(d)** $\{z \mid \text{Im } z = \text{Re } z\}$
    (e) $\{z \mid |z-1| = 2|z-2|\}$
19. Describe the following sets by means of graphs:
    (a) $\{z \mid \pi/6 \leq \arg z \leq \pi/2\}$ **(b)** $\{z \mid \arg z = 3\}$

(c) $\left\{z \left| \left|\frac{1}{z}\right| \leq 1 \right.\right\}$ **(d)** $\{z | \operatorname{Re} z = \frac{1}{2}\}$

(e) $\{z | 2 \leq |z - 1| \leq 3\}$

20. Describe the set for which

$$|z - 1| - |z + 1| = c$$

where c is a positive constant. Carefully distinguish between the three cases $c < 2$, $c = 2$, and $c > 2$.

21. If z_1 and z_2 are complex numbers, what is the locus of points given by $\lambda z_1 + (1 - \lambda)z_2$ for $0 \leq \lambda \leq 1$?

22. Under what conditions are the following relations valid:

(a) $|z_1 + z_2| = |z_1| + |z_2|$ **(b)** $|z_1 + z_2| = |z_1| - |z_2|$

23. Show that $|z_1 + z_2|^2 + |z_1 - z_2|^2 = 2(|z_1|^2 + |z_2|^2)$.

24. If $z = x + iy$, write the following in the form $u + iv$ where u and v are both real:

(a) z^2 **(b)** $\frac{1}{z}$ **(c)** $z^2 + z - 1$ **(d)** $\frac{1}{z + i}$

12.4 Functions and Mappings

In the same way that real functions of a real variable provide the raw material for problems in real analysis, complex functions of a complex variable form the essential ingredient for investigations in complex analysis. Moreover, in many applications, it is the nature of the complex function as a geometric transformation or mapping rather than an algebraic expression that is paramount. We shall delay the consideration of some extremely practical illustrations, therefore, until we have a sound understanding of the essence of these new functions.

A *complex function of a complex variable* is a set of ordered pairs of complex numbers (z, w) where to each $z = x + iy$ in a set D of complex numbers (called the domain of the function) there corresponds a unique $w = u + iv$. Such functions are most often specified either by a rule in terms of z or a rule giving coordinates u and v in terms of x and y. In the former case, the notation $w = f(z)$ is usually employed, whence $u = \operatorname{Re} f(z)$ and $v = \operatorname{Im} f(z)$.

Example 12.5 Let $f(z) = z^2$. Give the coordinate method of specifying this function. Repeat for $g(z) = 1/z$.

Solution $f(z) = z^2 = (x + iy)^2 = x^2 - y^2 + 2xyi$. Hence, the function may be written equivalently in terms of $u(x, y) = x^2 - y^2$ and $v = 2xy$. Also

$$g(z) = \frac{1}{z} = \frac{\bar{z}}{z\bar{z}} = \frac{x - iy}{x^2 + y^2}$$

Hence

$$u = \frac{x}{x^2 + y^2}; \qquad v = \frac{-y}{x^2 + y^2}$$

□

Example 12.6 Let $w = u + iv = x^2 + y^2 + 2xyi$. Find $w = f(z)$ in terms of z and $\bar{z}$.

Solution We use the identities $x = (z + \bar{z})/2$ and $y = (z - \bar{z})/2i$ and substitute them into the given rule,

$$\begin{aligned} w &= \left(\frac{z+\bar{z}}{2}\right)^2 + \left(\frac{z-\bar{z}}{2i}\right)^2 + 2i\left(\frac{z+\bar{z}}{2}\right)\left(\frac{z-\bar{z}}{2i}\right) \\ &= z\bar{z} + \frac{z^2 - (\bar{z})^2}{2} \end{aligned}$$

□

The geometric analog of a complex function of a complex variable is called a *mapping*. The domain of a complex mapping is a set of points in the z-plane and the range of values is a set of points in the w-plane. The resolution of innumerable practical problems requires the determination of the images of various point sets under complex mappings.

Example 12.7 Let f be the mapping defined by $f(z) = z^2$. Find the image of the unit square $0 \leqq x \leqq 1, 0 \leqq y \leqq 1$.

Solution Since $w = u + iv = x^2 - y^2 + i2xy$, it follows that $u(x, y) = x^2 - y^2$ and $v(x, y) = 2xy$.

The image of the line segment $y = 0, 0 \leqq x \leqq 1$, is given parametrically by

$$\left.\begin{aligned} u(x,0) &= x^2 \\ v(x,0) &= 0 \end{aligned}\right\} \quad 0 \leqq x \leqq 1$$

which is the set of points on the u-axis between 0 and 1.

The image of the line segment $x = 1, 0 \leqq y \leqq 1$, is

$$\left.\begin{aligned} u(1,y) &= 1 - y^2 \\ v(1,y) &= 2y \end{aligned}\right\} \quad 0 \leqq y \leqq 1$$

which is the portion of the parabola $u = 1 - (v^2/4), 0 \leqq v \leqq 2$.

The image of $y = 1, 0 \leqq x \leqq 1$, is

$$\left.\begin{aligned} u(x,1) &= x^2 - 1 \\ v(x,1) &= 2x \end{aligned}\right\} \quad 0 \leqq x \leqq 1$$

which is the portion of the parabola $u = (v^2/4) - 1, 0 \leqq v \leqq 2$.
The image of the segment $x = 0, 0 \leqq y \leqq 1$, is the set of points on the u-axis between -1 and 0. (See Figure 12.8.)

As you would guess, the inside of the unit square must map either entirely inside or entirely outside the new "triangular" boundary in the uv-plane. We can check one point, say $(x, y) = (\frac{1}{2}, \frac{1}{2})$, to see that "inside" is the proper choice in this case. □

Example 12.8 Find the preimage of $|w| =$ a constant under the mapping $w = xe^{iy}$.

Solution Since $|w| = |x|$ the preimages of the circles centered at the origin in the w-plane are lines parallel to the imaginary axis. (See Figure 12.9.) □

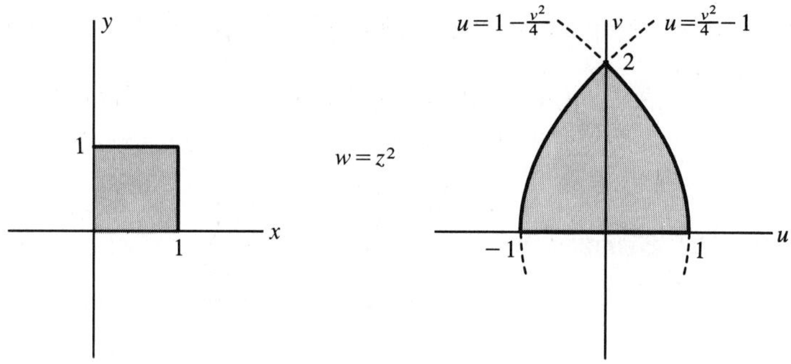

Figure 12.8

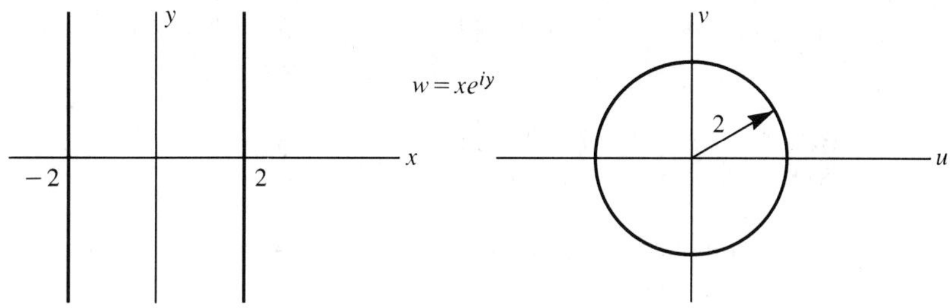

Figure 12.9

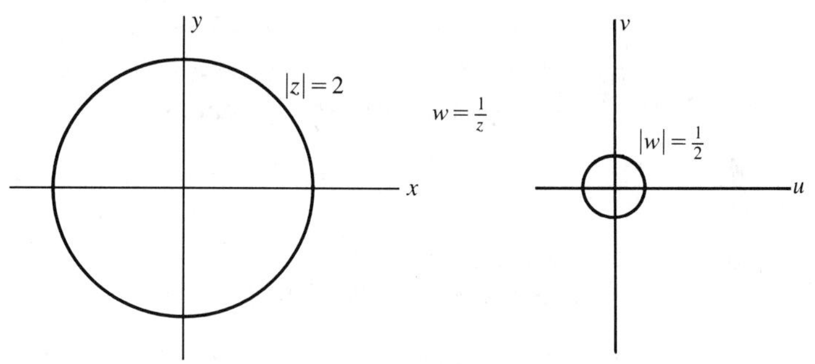

Figure 12.10

Example 12.9 Find the image of the circle $|z| = 2$ under the mapping $w = 1/z$.

Solution Since $|w| = 1/|z|$, the circle $|w| = \frac{1}{2}$ is the image of $|z| = 2$. (See Figure 12.10.) This problem may also be approached by finding the real and imaginary parts of the mapping function. Thus

$$w = \frac{1}{z} = \frac{\bar{z}}{z\bar{z}} = \frac{\bar{z}}{|z|^2} = \frac{x - iy}{x^2 + y^2} = \frac{x - iy}{4}$$

Therefore, $u = x/4$ and $v = -y/4$ from which

$$4 = x^2 + y^2 = 16u^2 + 16v^2$$

Hence, the image of the circle is

$$u^2 + v^2 = \tfrac{1}{4}$$

which is another way of writing $|w| = \tfrac{1}{2}$. □

Exercises for Section 12.4

1. Express the following functions in the form $u(x, y) + iv(x, y)$:
   (a) $z^2 - z + 1$ **(b)** $\dfrac{1}{1 - z}$
   (c) $\dfrac{z - i}{z + i}$ **(d)** $\dfrac{1}{z^2 + i}$
2. Find the images of the following regions under the given mapping:
   (a) $\{z \mid y = x\}$, $w = f(z) = z + 1$
   (b) $\{z \mid y = x^2\}$, $w = f(z) = z^2$
   (c) $\{z \mid |z| < 2\}$, $w = f(z) = z^2$
3. Find the images of the following regions under the given mapping:
   (a) $\left\{z \,\middle|\, |z| < 1, 0 < \arg z < \dfrac{\pi}{2}\right\}$, $f(z) = z^2$
   (b) $\{z \mid \operatorname{Re} z > 1\}$, $f(z) = 1/z$
4. Find the images of the indicated region under the given mapping:
   (a) $\{z \mid y = x\}$, $w = x^2 + iy^2$
   (b) $\{z \mid 0 \leqq x \leqq \pi/2, 0 \leqq y \leqq \pi/2\}$, $w = \sin x \cosh y + i \cos x \sinh y$
5. Find the images of the indicated region under the given mapping:
   (a) $\{z \mid 0 \leqq x \leqq \pi/2, 0 \leqq y \leqq \pi/2\}$, $w = e^x \cos y + ie^x \sin y$
   (b) $\{z \mid |z| < 2\}$, $w = x/(x^2 + y^2) - i[y/(x^2 + y^2)]$
   (c) $\{z \mid 0 < r < 3, 0 < \operatorname{Arg} z < \pi/3\}$, $w = \log_e |z| + i \operatorname{Arg} z$
6. Determine the images of the unit circle $|z| = 1$ under the mappings:
   (a) $w = (1 + i)z$ **(b)** $w = az + b$
   (c) $w = 1/(cz + d)$
7. Determine the images of the following lines under the mapping $w = 1/z$:
   (a) $y - x + 1 = 0$ **(b)** $y - x = 0$
8. Show that the images of the circles $|z| = c$ under the mapping $w = a(z + 1/z)$, with $a > 0$, are confocal ellipses.
9. Verify that the image of the semicircular region $|z| \le 1$, $\operatorname{Im} z < 0$ under the mapping $w = z + 1/z$ is the upper half-plane $v \ge 0$.
10. Show that the size of the angle of intersection of the curves $y = (x/\sqrt{3}) - 1$ and $y = x - \sqrt{3}$ is "preserved" by the mapping $w = z^2$, in the sense that the angle between the tangent lines at the point of intersection in the w-plane is the same as the angle between the lines in the z-plane.
11. Repeat Exercise 10 for the mapping $w = 1/z$.
12. Define what is meant by a complex function of a real variable. Give some examples.
13. Give coordinate expressions $u = u(x, y)$, $v = v(x, y)$ for the complex function $w = z^3 + 3z^2$.
14. Express as a function of z and $\bar{z}$:
    (a) $w = x^2 + y^2$ **(b)** $w = y + ix$

(c) $w = (x^3 - 3xy^2) + i(3x^2y - y^3)$

15. Find $w = u + iv$ if the function in terms of z and $\bar{z}$ is:
(a) $\bar{z}^2 + z$ (b) $z\bar{z} + 5z$ (c) $(z + 1)/(\bar{z}^2 + 1)$

16. Describe the set of points represented by $|az + b| = |cz + d|$ where a, b, c, and d are real numbers. Specifically consider the cases $|a| = |c|$ and $|a| \neq |c|$.

17. Find the preimage of the circle $|w| = 1$ under the mapping $w = (z - 1)/(z + 1)$.

18. Determine the preimage of the circle $|w| = 1$ under the mapping $w = (i - z)/(i + z)$.

19. How should the complex constants a and c be related so that the preimage of the circle $|w| = 1$ under the mapping $w = (az + 1)/(cz + 1)$ is a straight line?

In two-dimensional fluid flows, the complex function

$$\Phi(z) = \phi(x, y) + i\psi(x, y)$$

often termed the complex (velocity) potential, plays an important role. The imaginary part of the function, $\psi(x, y)$, is called the stream function, and the fluid actually flows along the streamlines, that is, the curves where ψ is constant.

20. Determine the streamlines for the two-dimensional fluid flows given by the following complex velocity potentials:
(a) $\Phi(z) = U_0 z$ (b) $\Phi(z) = U_0 z^2$
($U_0 > 0$) Sketch representative streamlines in each case.

21. Find the stream function for the fluid flow with complex velocity potential

$$\Phi(z) = U_0 a\left(\frac{z}{a} + \frac{a}{z}\right)$$

where a and U_0 are positive. Determine and sketch the streamlines $\psi = 0, \pm U_0$. (Assume $|z| \geq a$ throughout.)

22. Repeat Exercise 21 for the complex velocity potential

$$\Phi(z) = \left(ze^{i\alpha} + \frac{e^{-i\alpha}}{z}\right), \qquad \alpha > 0$$

12.5 The Derivative of a Complex Function

Just as in the real-variable case, the applicability of complex function theory is enormously expanded by introduction of the calculus. Indeed, applications in electrostatics, systems analysis, and fluid dynamics, to name just a few important fields, are inextricably intertwined with the notion of the derivative of a complex function. Fortunately, the definitions of limit, continuity, and derivative of a function of a complex variable have forms which correspond closely to those for functions of a real variable. For example, if a complex function $f(z)$ is defined in some neighborhood of a point z_0, with the possible exception of the point z_0 itself, we say that

$$\lim_{z \to z_0} f(z) = w_0$$

if for any $\varepsilon > 0$, there exists a positive number δ such that

$$|f(z) - w_0| < \varepsilon \quad \text{whenever} \quad 0 < |z - z_0| < \delta$$

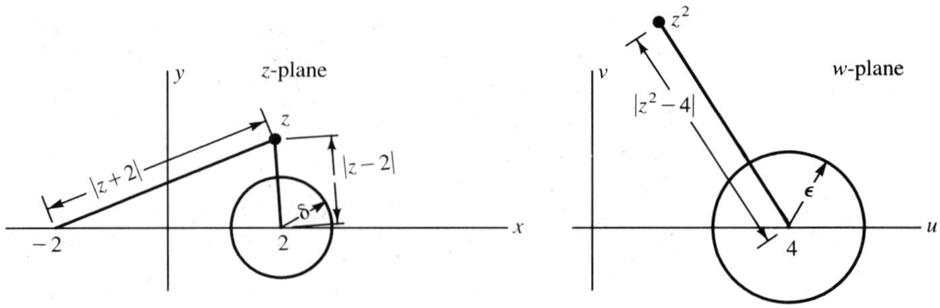

Figure 12.11

One must be wary, however, not to presume undue familiarity with the concept since the *interpretation* of the definition is different in the complex plane. In what follows, we illustrate this with several examples.

Example 12.10 Show that $\lim_{z\to 2} z^2 = 4$.

Solution We must show that for an arbitrary $\varepsilon > 0$, there exists a $\delta > 0$ such that $|z^2 - 4| < \varepsilon$ for each z such that $0 < |z - 2| < \delta$. If, for a given ε, one δ works, then any smaller δ would work, so for convenience we first make the restriction $\delta < 1$. If $0 < |z - 2| < \delta < 1$ then $|z^2 - 4| = |(z - 2)||z + 2| = |z - 2||(z - 2) + 4| < |z - 2|(|z - 2| + 4) < |z - 2|5 < 5\delta$, which will be less than ε if $\delta < \varepsilon/5$. Therefore, choose $\delta < 1$ or $<\varepsilon/5$, whichever is smaller. (See Figure 12.11.) □

To show that $\lim_{z\to z_0} f(z)$ does not exist, it is sufficient to show that the value of the limit is dependent upon the manner in which z_0 is approached.

Example 12.11 Show that $\lim_{z\to 0} (\operatorname{Re} z)/z$ does not exist.

Solution If z approaches 0 along the positive y-axis, then $z = iy$, and $\operatorname{Re} z = 0$. Hence,

$$\lim_{\substack{z\to 0\\ x=0}} \frac{\operatorname{Re} z}{z} = \lim_{y\to 0} \frac{0}{iy} = 0$$

If the approach is along the positive x-axis, then $z = x$, $\operatorname{Re} z = x$, and

$$\lim_{\substack{z\to 0\\ y=0}} \frac{\operatorname{Re} z}{z} = \lim_{x\to 0} \frac{x}{x} = 1$$ □

The following is the formal definition for the derivative in the complex case.

Definition 12.1 If $w = f(z)$ is a complex function of the complex variable $z = x + iy$ defined in some neighborhood of a point z_0, the *derivative* of f at z_0, denoted variously by

$\left.\dfrac{dw}{dz}\right|_{z=z_0}$, $w'(z_0)$, or $f'(z_0)$, is given by

$$f'(z_0) = \lim_{\Delta z \to 0} \frac{f(z_0 + \Delta z) - f(z_0)}{\Delta z}$$

Since the same definition is used for the derivative of a function of a complex variable as for a function of a real variable, the same *general* formulas and theorems are true. (Some specific theorems, such as the mean-value theorem, which depends upon the real numbers being ordered, do not carry over.) However, any real-variables theorem involving differentiation that uses *only* the abstract definition will be true in complex analysis also. In particular:

1. The derivative of a sum is the sum of the derivatives.
2. The derivative of a constant times a function is the constant times the derivative of the function.
3. The product and quotient rules for differentiation are valid.
4. The derivative of a composite function, $w(u(z))$ is given by the chain rule,

$$\frac{dw}{dz} = \frac{dw}{du}\frac{du}{dz}$$

where w and u are both functions which have derivatives.

5. The power rule is the same,

$$\frac{dz^n}{dz} = nz^{n-1}$$

These five general rules are sufficient to enable you to differentiate any rational function of a complex variable. The techniques are the same as in elementary calculus.

The rules given above do not apply to powers of the conjugate of z since, as we will now show, the function $f(z) = \bar{z}$ does not have a derivative anywhere.

Example 12.12 Show that $f(z) = \bar{z}$ does not have a derivative anywhere.

Solution We must show that the limit of the difference quotient does not exist.

$$\lim_{\Delta z \to 0} \frac{f(z + \Delta z) - f(z)}{\Delta z} = \lim_{\Delta z \to 0} \frac{(x + \Delta x) - i(y + \Delta y) - (x - iy)}{\Delta x + i\Delta y}$$

$$= \lim_{\Delta z \to 0} \frac{\Delta x - i\Delta y}{\Delta x + i\Delta y}$$

whose value is dependent upon the path of approach of Δz to 0. For example, the value for the approach through values of $\Delta z = \Delta x$ is 1 while if $\Delta z = i\Delta y$, and $\Delta y \to 0$, the limit is -1. Hence, the limit of the difference quotient does not exist for any value of z. □

When a complex function is defined in terms of two real functions, we will usually want to differentiate it directly, rather than to re-express it in terms of

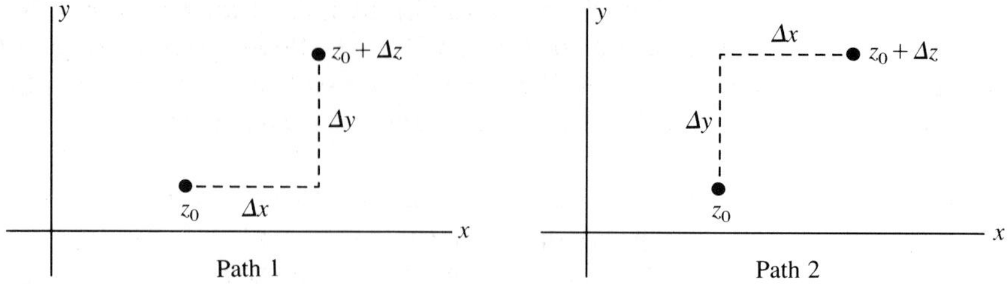

Figure 12.12

the variable z. The following theorem not only gives the formula for this case, but also yields a most interesting and famous set of conditions for the existence of the derivative of a function of a complex variable.

Theorem 12.1 Let $f(z) = u(x, y) + iv(x, y)$ and suppose $f'(z)$ exists at $z_0 = x_0 + iy_0$. Then

$$\begin{aligned} f'(z_0) &= u_x(x_0, y_0) + iv_x(x_0, y_0) \\ &= v_y(x_0, y_0) - iu_y(x_0, y_0) \end{aligned}$$

Proof We approach z_0 in two different ways: one horizontally along the line $\operatorname{Im} z = \operatorname{Im} z_0$ (after we have allowed Δy to go to zero), and the other way vertically along the line $\operatorname{Re} z = \operatorname{Re} z_0$ (after we have allowed Δx to go to zero). (See Figure 12.12.) Since $f'(z_0)$ exists, the limit of the difference quotient is the same in both cases. For the first approach,

$$\begin{aligned} &\lim_{\substack{\Delta z \to 0 \\ \Delta y = 0}} \frac{f(z_0 + \Delta z) - f(z_0)}{\Delta z} \\ &= \lim_{\Delta x \to 0} \frac{u(x_0 + \Delta x, y_0) - u(x_0, y_0) + iv(x_0 + \Delta x, y_0) - iv(x_0, y_0)}{\Delta x} \\ &= \lim_{\Delta x \to 0} \frac{u(x_0 + \Delta x, y_0) - u(x_0, y_0)}{\Delta x} + i \lim_{\Delta x \to 0} \frac{v(x_0 + \Delta x, y_0) - v(x_0, y_0)}{\Delta x} \\ &= u_x(x_0, y_0) + iv_x(x_0, y_0) \end{aligned}$$

For the other approach,

$$\begin{aligned} f'(z_0) &= \lim_{\Delta y \to 0} \frac{u(x_0, y_0 + \Delta y) - u(x_0, y_0) + iv(x_0, y_0 + \Delta y) - iv(x_0, y_0)}{i\Delta y} \\ &= \lim_{\Delta y \to 0} \frac{u(x_0, y_0 + \Delta y) - u(x_0, y_0)}{i\Delta y} + i \lim_{\Delta y \to 0} \frac{v(x_0, y_0 + \Delta y) - v(x_0, y_0)}{i\Delta y} \\ &= \frac{1}{i} u_y + \frac{i}{i} v_y \\ &= v_y(x_0, y_0) - iu_y(x_0, y_0) \end{aligned}$$ ■

Corollary Let $f(z) = u + iv$ and suppose $f'(z_0)$ exists. Then at (x_0, y_0),

$$u_x = v_y; \qquad v_x = -u_y$$ ■

The two partial differential equations of the corollary are called the *Cauchy-Riemann* conditions. We have shown them to be conditions necessary for the existence of $f'(z_0)$, but it is also true that these conditions are sufficient, *if* the partial derivatives of u and v are assumed to be continuous.

Theorem 12.2 Let z_0 be an interior point of a region in which $w = u + iv = f(z)$ is a continuous function of the complex variable $z = x + iy$. If $u(x, y)$ and $v(x, y)$ have continuous first partial derivatives in some neighborhood of the point z_0 and if they satisfy the Cauchy-Riemann equations $u_x = v_y$ and $v_x = -u_y$ at the point itself, then $f'(z_0)$ exists. ■

This result provides a remarkably easy test for the existence of f' at a point, as illustrated by the following three examples.

Example 12.13 Let $f(z) = e^x \cos y + ie^x \sin y$. Where does $f'(z)$ exist?

Solution

$$u_x = e^x \cos y = v_y \quad \text{and} \quad v_x = e^x \sin y = -u_y$$

These equations hold everywhere in the complex plane, and the partial derivatives are continuous; therefore f' exists everywhere. Moreover, $f'(z) = f(z)$. (Why might we have expected this result?) □

Example 12.14 Let

$$g(z) = \frac{x^3}{3} + i\left(y - \frac{y^3}{3}\right)$$

Where does $g'(z)$ exist?

Solution $u_x = x^2$, $v_y = 1 - y^2$, $u_y = 0$, $v_x = 0$. The Cauchy-Riemann conditions are satisfied by those points for which $x^2 = 1 - y^2$, and no others, while the partial derivatives are continuous everywhere. Therefore $g'(z)$ exists for all points on the unit circle. □

Example 12.15 Let $h(z) = x^2 + y^2$. Where does $h'(z)$ exist?

Solution $u_x = 2x$, $u_y = 2y$, $v_x = 0$, $v_y = 0$. The partial derivatives are continuous everywhere but the derivative exists where $2x = 0$ and $2y = 0$, which is at the origin only. □

Although the Cauchy-Riemann conditions are sufficient to guarantee that a function have a derivative at a point if the four partial derivatives are continuous, when these partial derivatives are not continuous, there may be no differentiability.

Example 12.16 Let

$$f(z) = \frac{x^3 - y^3}{x^2 + y^2} + i\frac{x^3 + y^3}{x^2 + y^2}, \qquad z \neq 0$$

$$f(0) = 0$$

Show that the derivative of f does not exist at $z = 0$ even though the Cauchy-Riemann conditions hold at that point.

Solution It is easy to show that $u_x(x, 0) = 1$, $v_y(0, y) = 1$, $v_x(x, 0) = 1$, $u_y(0, y) = -1$ and hence at $z = 0$ the Cauchy-Riemann equations are satisfied. Thus, if the derivative at $z = 0$ exists, the limit of the difference quotient must be $1 + i$ regardless of the path used to approach 0. There are many paths for which this is not so. For example, consider the path $x = 2y$. On this path,

$$\frac{f(0 + \Delta z) - f(0)}{\Delta z} = \frac{\dfrac{(\Delta x)^3 - (\Delta y)^3}{(\Delta x)^2 + (\Delta y)^2} + i\dfrac{(\Delta x)^3 + (\Delta y)^3}{(\Delta x)^2 + (\Delta y)^2}}{\Delta x + i\Delta y}$$

$$= \frac{\dfrac{7(\Delta y)^3}{5(\Delta y)^2} + i\dfrac{9(\Delta y)^3}{5(\Delta y)^2}}{(\Delta y)(2 + i)}$$

$$= \frac{7 + 9i}{5(2 + i)}$$

$$= \frac{23 + 11i}{25} \neq 1 + i$$

The limit of the difference quotient is indeed dependent on the method of approach to zero therefore, and the derivative at $z = 0$ does not exist. We can conclude that some of the partial derivatives must not be continuous at $z = 0$. □

In working with complex functions the existence of f' at isolated points is of little practical use. We are usually interested in those points for which f' exists not only *at* the point but in some neighborhood *about* the point as well. We formalize this in the following definition.

Definition 12.2 A complex function f is said to be *analytic* at the point z_0 if f' exists at z_0 and at every point in some neighborhood of z_0. If f is analytic at each point of a set D, then it is said to be *analytic on D*. If f is analytic for all points in a neighborhood of a but not at a, then the point a is called an isolated *singularity* of the function. There is no requirement that the function be defined at a.

The function $f(z) = 1/(z - 1)$ has a singularity at $z = 1$, since it is obviously nonanalytic at the point, but $f'(z)$ does exist everywhere else.

To determine if a complex function is analytic at a point, it is not sufficient to show the existence of the derivative there. For, in addition to f' existing at the point, it must also exist in some neighborhood about the point. To determine where a function is analytic, we first locate the points where f' exists, after which the points of analyticity are easily determined.

Example 12.17 Determine where the functions of Examples 12.13 through 12.15 are analytic.

Solution Since $f'(z)$ exists everywhere, $f(z)$ is analytic everywhere. The function $g(z)$ is analytic nowhere since $g'(z)$ exists only on the unit circle. At no point where it exists does it have a neighborhood where $g'(z)$ exists. The function $h(z)$ also is analytic nowhere since $h'(z)$ exists only at the origin. □

The seemingly small requirement of demanding the existence of the derivative in a neighborhood rather than just at a point has much more strength than one might realize. Moreover, since it is this feature of analyticity that is important in most practical problems, much of our study of complex variable theory will be directed toward the analysis of analytic functions.

One immediate property of analytic functions concerns their real and imaginary parts.

Theorem 12.3 Both the real and imaginary parts of an analytic function are solutions of Laplace's equation.

Proof See Exercise 19 in the next set of exercises. ■

Recall that a solution of Laplace's equation that has continuous second-order partial derivatives is called a *harmonic function* (see Sections 8.1 and 10.10). If $u + iv$ is analytic and u and v are harmonic, then u and v are called *conjugate* harmonic functions. A conjugate of a given harmonic function may be found easily using the Cauchy-Riemann equations. For example, if we want to determine an analytic function whose real part is $u(x, y) = x^2 - y^2$, we could let the analytic function be $f(z) = x^2 - y^2 + iv(x, y)$. From the Cauchy-Riemann equations we would then calculate $v_x = 2y$ and $v_y = 2x$, which obviously imply that $v(x, y) = 2xy$ (plus a constant). Therefore,

$$f(z) = x^2 - y^2 + 2ixy$$

A more meaningful application of the above ideas is given by the following.

Example 12.18 Find the electrostatic potential $V(x, y)$ in the upper half-plane which takes on the boundary values $V = 0$ and $V = 1$ on the positive and negative x-axes, respectively.

Solution The mathematical model for the problem in the z-plane is

$$V_{xx} + V_{yy} = 0, \qquad -\infty < x < \infty, \qquad y > 0$$

where

$$V(x, 0) = 0, \qquad x > 0$$
$$V(x, 0) = 1, \qquad x < 0$$

The function

$$V(x, y) = \frac{\theta}{\pi}$$

(where $\theta \equiv \operatorname{Arg} z$ is the polar angle of $z = x + iy$ restricted to its principal value) obviously satisfies the boundary conditions. As we shall soon see, θ is the imaginary part of an analytic function, namely the complex logarithm function. By virtue of Theorem 12.3, $V(x, y)$ thus satisfies Laplace's equation and hence is the desired solution. □

In addition to several exercises scattered throughout the rest of this chapter, a number of applications to potential and steady-state temperature problems are given in Section 13.7. For the most part, these depend upon a knowledge of the behavior of compositions of analytic functions. In the remainder of Chapter 12, then, we study the nature of some of the more basic complex functions which we shall be using as building blocks.

Exercises for Section 12.5

1. Show that $\lim_{z \to 1+i} z^2 = 2i$.

2. Show that

$$f(z) = \begin{cases} \dfrac{\operatorname{Im} z^2}{|z|^2}, & z \neq 0 \\ 0, & z = 0 \end{cases}$$

is discontinuous at the origin.

3. Prove the quotient rule for the derivative of a function of a complex variable which can be expressed as the ratio of two other differentiable functions.

4. Find $f'(z)$ in each case:

(a) $f(z) = z^3 + z$ **(b)** $f(z) = \dfrac{z^2 + 2z}{z^4 - 1}$ **(c)** $f(z) = \dfrac{1}{z}$

(d) $f(z) = \dfrac{az + b}{cz + d}$ **(e)** $f(z) = (z^{15} + 3z^{10} + 17)^{18}$

5. Find dw/dz for each w:

(a) $w = \sin x \cosh y + i \cos x \sinh y$

(b) $w = \log_e |z| + i \operatorname{Arg} z$

(c) $w = \cos x \cosh y - i \sin x \sinh y$

6. Where (if anywhere) are the following functions analytic?

(a) $f(z) = |z|^2$ **(b)** $f(z) = x^2 + iy^2$ **(c)** $f(z) = (x - iy)^2$

(d) $f(z) = z\bar{z}$ **(e)** $f(z) = \operatorname{Arg} z$ **(f)** $f(z) = \dfrac{z + 4}{z - i}$

(g) $f(z) = x^3 + 3y + i(y^3 - 3x)$

7. Show that $f(z) = \log_e |z| + i \arg z$ is discontinuous along the negative real axis, if $\arg z$ is limited to its principal values.

8. Show that $f(z) = |z|^{1/2} e^{i(\arg z/2)}$ is discontinuous along the negative real axis if $\arg z$ is limited to its principal values.

9. Find an analytic function whose real part is:

(a) $e^x \cos y$ **(b)** $x^3 - 3xy^2$ **(c)** y

10. Find an analytic function whose real part is:

(a) xy **(b)** $\operatorname{Arg} z$ **(c)** $\log_e |z|$

11. Explain clearly the difference between a function being analytic at a point and having a derivative there.

In Exercises 12–13, show that each of the given functions is harmonic and find a complex analytic function $f(z) = u + iv$:

12. **(a)** $u = x$ **(b)** $v = x$ **(c)** $u = e^x \sin y$
(d) $v = \log_e |z|$

13. **(a)** $u = \dfrac{x}{x^2 + y^2}$ **(b)** $u = \sin x \cosh y$ **(c)** $u = xy + x + y$

14. Show that the conjugate harmonic of a function is unique up to an arbitrary constant.

15. Find the most general harmonic function of the form of a second-degree polynomial in x and y which has no linear or constant terms.

16. Show that if $f(z)$ is analytic with constant modulus, then $f(z)$ is a constant.

17. Prove the following statement. If $f(z)$ is analytic in a domain D, then $f'(z)$ vanishes everywhere in D if and only if $f(z)$ is a constant in that domain.

18. Sketch the families of curves

$$\begin{cases} u = u(x, c) \\ v = v(x, c) \end{cases} \quad \text{and} \quad \begin{cases} u = u(k, y) \\ v = v(k, y) \end{cases}$$

and in each case show that they are orthogonal if:
(a) $f(z) = z$ **(b)** $f(z) = \sin x \cosh y + i \cos x \sinh y$
(c) $f(z) = e^x \cos y + ie^x \sin y$

19. Prove Theorem 12.3. (*Hint:* Use the Cauchy-Riemann equations.)

The velocity fields associated with two-dimensional irrotational fluid flows are conservative. That is, there exists a velocity potential ϕ such that $\mathbf{v} = \nabla\phi$. In what follows we suppress any time dependence and let

$$\mathbf{v} = u(x, y)\mathbf{i} + v(x, y)\mathbf{j}$$

where u and v are assumed to be continuously differentiable.

20. For incompressible fluids $\nabla \cdot \mathbf{v} = 0$. Show that in such cases the velocity potential ϕ must be harmonic.

21. **(a)** Verify that for irrotational incompressible fluid flows there exists a harmonic function $\psi(x, y)$ such that

$$\Phi(x, y) = \phi(x, y) + i\psi(x, y)$$

is analytic.

(b) Demonstrate that for such flows

$$\mathbf{v} \cdot \left(\frac{\partial \psi}{\partial x}\mathbf{i} + \frac{\partial \psi}{\partial y}\mathbf{j}\right) = 0$$

and hence the curves along which ψ = constant are tangent to the given velocity field. (This justifies the terms *stream function* for ψ and *streamlines* for the ψ = constant curves.)

12.6 Integral and Fractional Powers of z

In this and the following sections we will introduce some of the more important complex functions. In most cases, if we can, we will call the complex function by

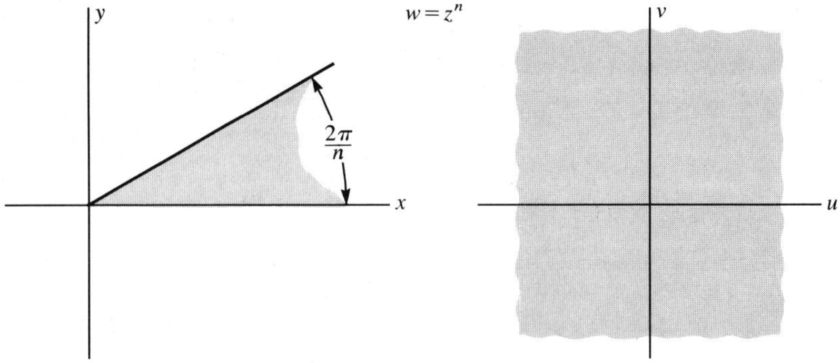

Figure 12.13 Mapping by $w = z^n$

the same name as that to which the function reduces when z is real. However, do not be misled by this practice into believing that you are familiar with the particular function of a complex variable just because you know its properties when z is real. Some properties of these functions are significantly different from their real counterparts.

1. *Integral Powers of z.* This class of functions is denoted by $w = f(z) = z^n$, with n a nonnegative integer. A function of this type is analytic in the entire z-plane. If we write $z = re^{i\theta}$ and $w = Re^{i\phi}$, then the expression for this function becomes

$$Re^{i\phi} = r^n e^{in\theta}$$

or

$$\begin{cases} R = r^n \\ \phi = n\theta \end{cases}$$

Hence the mapping defined by $w = z^n$ maps the sector $0 \leqq \theta < 2\pi/n$ onto the entire w-plane. (See Figure 12.13.)

For $n \neq 1$, the mapping $w = z^n$ is n-to-one except for the correspondence of $w = 0$ with $z = 0$. By this we mean that each point in the w-plane other than $w = 0$ has n distinct preimages in the z-plane.

Example 12.19 Find the preimages of the point

$$w = 32\left(\frac{1+i}{\sqrt{2}}\right)$$

under the mapping $w = z^5$.

Solution We look for values of r and θ that satisfy the equation $r^5 e^{i5\theta} = 32e^{i\pi/4}$. Hence $r^5 = 32$ and $5\theta = \pi/4 + 2k\pi$, from which $r = 2$ and $\theta = \pi/20$(+integral multiples of $2\pi/5$). The *distinct* points which these equations yield are, in polar form, $(2, \pi/20)$, $(2, \pi/20 + 2\pi/5)$, $(2, \pi/20 + 4\pi/5)$, $(2, \pi/20 + 6\pi/5)$, and $(2, \pi/20 + 8\pi/5)$. Figure 12.14 shows the points corresponding to these values.

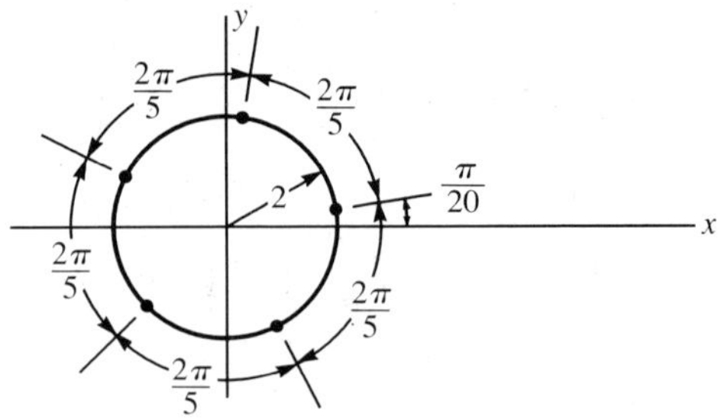

Figure 12.14

They are located on the circle of radius 2 at equally spaced intervals of $2\pi/5$. This, of course, is just an example of D'Moivre's theorem,

$$[r(\cos\theta + i\sin\theta)]^n = r^n[\cos n\theta + i\sin n\theta]$$ □

2. *Fractional Powers of z.* This relation, denoted by $w = z^{1/n}$, is defined by the equation $w^n = z$, with n a nonnegative integer. Classically, it has been called a "multivalued" function, but the more contemporary approach is to agree that *n branch functions* are defined by the equation. Each of the branch functions is defined by limiting $\arg z$ to some interval of length 2π and taking $\arg w = (\arg z)/n$. The rays which emanate from the origin at those intervals separate the domains of the branch functions and are called *branch cuts.* Different choices of branch cuts will result in different branch functions. The branch cuts all meet at the same point (the origin), called a *branch point* of the function. Each of the n branch functions is analytic except at points along the branch cuts.

Example 12.20 Show two ways of defining branch functions from the relation $w = z^{1/3}$.

Solution Let $r = |z|$ and $\theta = \arg z$. By placing branch cuts at $\theta = 0, 2\pi, 4\pi$, and 6π, we obtain the three distinct functions,

$$\begin{aligned} f_1(z) &= r^{1/3}e^{i\theta/3}, && 0 \leqq \theta < 2\pi \\ f_2(z) &= r^{1/3}e^{i\theta/3}, && 2\pi \leqq \theta < 4\pi \\ f_3(z) &= r^{1/3}e^{i\theta/3}, && 4\pi \leqq \theta < 6\pi \end{aligned}$$

By placing branch cuts at $-\pi, \pi, 3\pi$, and 5π, we obtain the three functions

$$\begin{aligned} g_1(z) &= r^{1/3}e^{i\theta/3}, && -\pi \leqq \theta < \pi \\ g_2(z) &= r^{1/3}e^{i\theta/3}, && \pi \leqq \theta < 3\pi \\ g_3(z) &= r^{1/3}e^{i\theta/3}, && 3\pi \leqq \theta < 5\pi \end{aligned}$$

In each of the cases, the range values of the three branch functions taken together completely "cover" the w-plane, although in a somewhat different arrangement, as seen in Figure 12.15. □

Example 12.21 Evaluate $(-i)^{1/3}$ for each of the branch functions of the previous example. Display graphically.

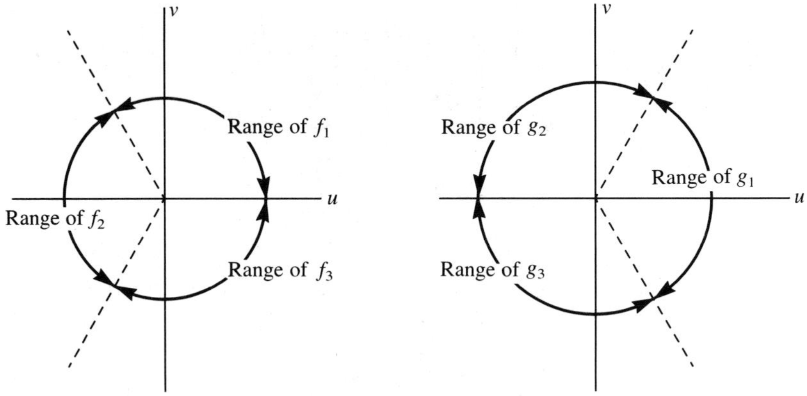

Figure 12.15

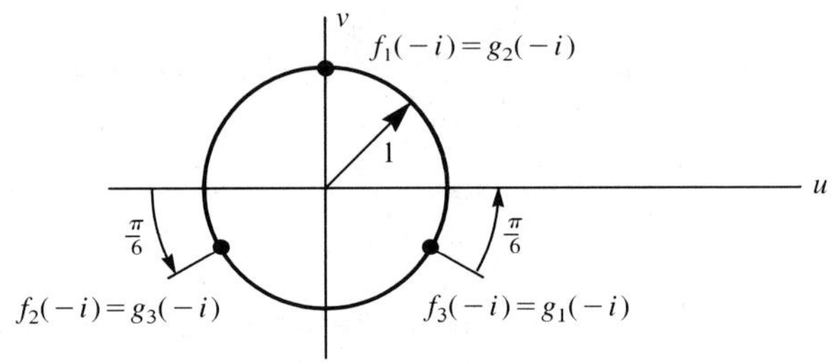

Figure 12.16

Solution

$$f_1(-i) = f_1(e^{i3\pi/2}) = e^{i\pi/2} = i = g_2(e^{i3\pi/2}) = g_2(-i)$$

$$f_2(-i) = f_2(e^{i7\pi/2}) = e^{i7\pi/6} = -\frac{\sqrt{3}}{2} - \frac{i}{2} = g_3(e^{i7\pi/2}) = g_3(-i)$$

$$f_3(-i) = f_3(e^{i11\pi/2}) = e^{i11\pi/6} = \frac{\sqrt{3}}{2} - \frac{i}{2} = g_1(e^{-i\pi/2}) = g_1(-i)$$

(See Figure 12.16.) □

In all cases, the n values of $z^{1/n}$ will form the same set regardless of the method of defining branch functions. From a practical standpoint, if you are asked to evaluate $z^{1/n}$ for a particular value of z_0, you are expected to give all n distinct values. Graphically, the n values are equally spaced at intervals of $2\pi/n$ radians on a circle of radius $R = |z_0|^{1/n}$. This is the geometric form of D'Moivre's theorem.

Each of the n branch functions maps the complete z-plane onto a sector of angular measurement $2\pi/n$ radians in the w-plane. (See Figure 12.17.)

We sometimes think of the z-plane as overlaid by n layers, one layer for the domain of each branch function. Moreover, these layers should be connected in

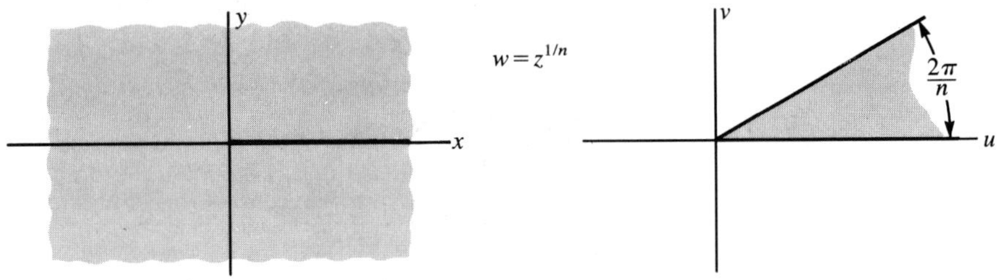

Figure 12.17 Mapping by $w = z^{1/n}$

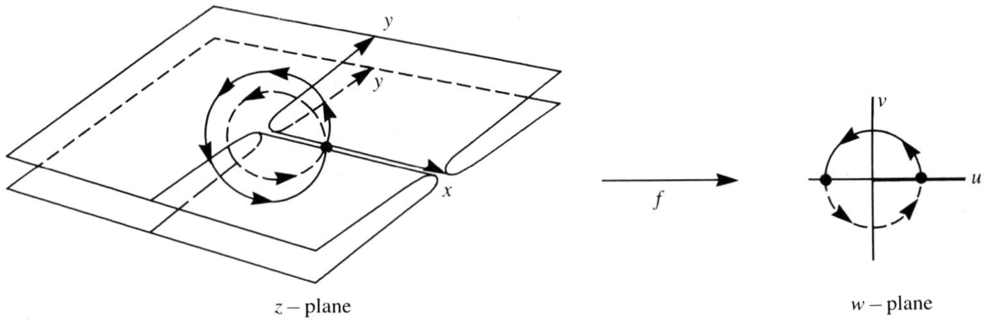

Figure 12.18 Two-sheeted Riemann surface corresponding to $w = z^{1/2}$

a helicoidal arrangement such that as we turn around the origin, at each revolution we move up one layer. After n revolutions the connection should be such (and this is physically impossible) that the top layer continues into the bottom layer. All of the planes together represent the complete domain of the function $w = z^{1/n}$. Such a graphical representation of a domain (or a range) is called an *n-sheeted Riemann surface.* (See Figure 12.18.) With the assistance of Riemann surfaces, the maps of functions which are not one-to-one may be interpreted as being a one-to-one pairing of points. There is an obvious correspondence between points on the w-plane and on the n sheets.

Exercises for Section 12.6

1. Compute and display graphically:
   (a) $(-1)^{1/4}$ **(b)** $(8i)^{1/3}$ **(c)** $(-64)^{1/6}$ **(d)** $(-4i)^{1/2}$
2. Compute and display graphically:
   (a) $\sqrt[3]{1}$ **(b)** $\sqrt[4]{1}$ **(c)** $\sqrt[5]{1}$ **(d)** $\sqrt[6]{1}$
3. Find all fifth roots of $(1 + i)$ and represent them graphically.
4. Solve the following equations and display the roots graphically:
   (a) $z^4 + 256 = 0$ **(b)** $z^2 - 6z + 12 = 0$ **(c)** $z^8 + 15z^4 = 16$
5. Express the following complex numbers in the form $z = x + iy$:
   (a) $\sqrt{i}$ **(b)** $\sqrt{1 - i}$ **(c)** $\dfrac{1}{\sqrt{1 + i}}$ **(d)** $\dfrac{\sqrt{i}}{i}$

Use D'Moivre's theorem

$$[r(\cos\theta + i\sin\theta)]^n = r^n(\cos n\theta + i\sin n\theta)$$

to establish the results in Exercises 6–9.

6. Show that $(\sqrt{3} + i)^8 = -128 - 128\sqrt{3}\,i$.

7. Verify that $(1 - i)^{37} = 2^{18}(i - 1)$.

8. Demonstrate that:
(a) $\cos 2\theta = \cos^2\theta - \sin^2\theta$ **(b)** $\sin 2\theta = 2\sin\theta\cos\theta$

9. Show that:
(a) $\cos 3\theta = \cos^3\theta - 3\cos\theta\sin^2\theta$ **(b)** $\sin 3\theta = 3\cos^2\theta\sin\theta - \sin^3\theta$

10. Define two branch functions for $f(z) = z^{1/2}$. Where is the branch cut? Show that each of the branch functions is discontinuous at its branch cut.

11. Define branch functions for $w = z^{1/5}$ such that the branch cut is along the positive imaginary axis. How many sheets does the Riemann surface of this relation have?

12. Determine and sketch the images of the circular sector $\pi/4 < \arg z < \pi/2$, $|z| < 2$ under the mappings:
(a) $w = z^2$ **(b)** $w = z^3$ **(c)** $w = z^4$

13. Show that the preimage of the lines $v = c$ under the mapping $w = z^{1/2}$ are the parabolas $y^2 = 4c^2(c^2 + x)$.

14. Describe and sketch the image of the rectangular region bounded by the lines $x = 1$, $x = 2$, $y = 1$, $y = 2$ under the mapping $w = z^{1/2}$.

12.7 The Exponential Function

In defining the complex exponential function, we should at least have the one basic property usually associated with its real counterpart, namely, being its own derivative. For the present we denote the complex exponential by $\exp z$ in order to make a clear distinction between the new complex function being defined and the real function e^x (the two notations are often used interchangeably). We seek, then, a function $\exp z$ such that

1. $\exp z$ is analytic for all z,
2. $\dfrac{d}{dz}(\exp z) = \exp z$, and
3. $\exp z$ reduces to the real exponential function e^x when $\operatorname{Im} z = 0$.

The idea of forcing a definition to satisfy some well-known properties is fairly standard practice and should remove the apparent randomness of the definition. It goes almost without saying that other approaches are possible.

Let $\exp z = u + iv$. By using the three properties above, we will obtain expressions for u and v. First, using property 2 and the Cauchy-Riemann conditions,

$$(u + iv) = (u + iv)' = u_x + iv_x = v_y - iu_y$$

Differentiating again,

$$(u + iv) = (u + iv)'' = v_{yx} - iu_{yx} = -u_{yy} - iv_{yy}$$

It follows then that

$$u = u_x \quad \text{and} \quad u = -u_{yy}$$

The first of these equations has the solution

$$u = g(y)e^x$$

Because of the second equation, the function g must satisfy

$$g'' + g = 0$$

an ordinary differential equation whose solution is

$$g(y) = A\cos y + B\sin y$$

Therefore,

$$u(x, y) = e^x(A\cos y + B\sin y)$$

and, since $v = -u_y$,

$$v(x, y) = e^x(A\sin y - B\cos y)$$

When $z = x + i0$, $\exp z = e^x + i0$. Thus,

$$e^x = e^x A \quad \text{and} \quad 0 = e^x(-B)$$

from which $A = 1$ and $B = 0$. Therefore, the complex exponential function is defined to be

$$\exp z \equiv e^x \cos y + ie^x \sin y \equiv e^x(\cos y + i\sin y)$$

Note that when $x = 0$ the above expression gives rise to *Euler's formula*

$$\exp iy \equiv \cos y + i\sin y$$

and thus

$$\exp z = e^x \exp(iy)$$

The exponential function, therefore, has the alternative representation in polar form

$$\exp z = R(\cos\phi + i\sin\phi)$$

where

$$R = |\exp z| = e^x$$
$$\phi = \arg(\exp z) = y$$

From Euler's formula, expressions for the real trigonometric functions in terms of the complex exponential may be derived:

$$\cos\theta = \frac{\exp i\theta + \exp(-i\theta)}{2}$$

$$\sin\theta = \frac{\exp i\theta - \exp(-i\theta)}{2i}$$

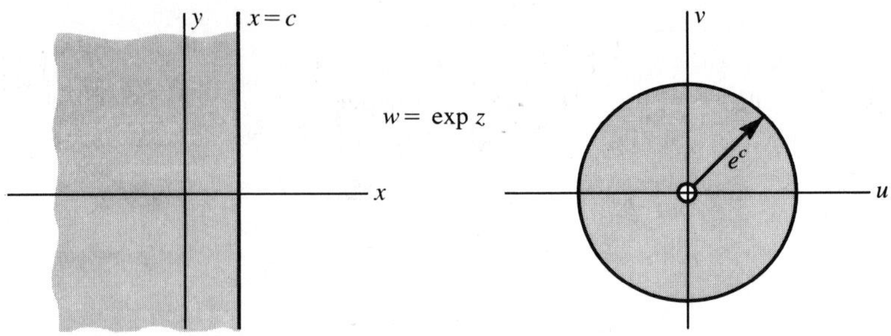

Figure 12.19 *Mapping by* $w = \exp z$

The relationship $\exp(z_1 + z_2) = (\exp z_1)(\exp z_2)$ may also be verified, although it could really not be anticipated considering that this formula was not used as a hypothesis motivating our definition. The proof of this result follows from a straightforward use of the definitions, a few trigonometric identities, and a few real exponential identities.

The complex exponential has some unfamiliar properties as well. One example, which may surprise you, is that of the periodicity of $\exp z$. If the argument of z is increased by $2\pi i$, then

$$\begin{aligned}\exp(z + 2\pi i) &= (\exp z)(\exp 2\pi i)\\ &= \exp z\end{aligned}$$

Therefore, $\exp z$ is periodic with *imaginary* period $2\pi i$.

The periodicity implies that the mapping defined by $w = \exp z$ is many-to-one. The function will assume all its values in *any* horizontal strip of width 2π. In particular, the strip defined by $-\pi < y \leqq \pi$ is called the *fundamental region* of $\exp z$. This choice is consistent with that made for the principal values of $\arg z$. A different choice of principal values of the argument will result in a different fundamental region for $\exp z$.

Since $|w| = e^x$, and $\arg w = y$, the vertical lines $x = c$ are mapped onto circles about the origin and horizontal lines are mapped onto rays through the origin. The region to the left of a vertical line, $x = c$, maps into the interior of the circle of radius e^c. (See Figure 12.19.) The origin in the w-plane has no preimage; that is, $\exp z$ is never 0.

Example 12.22 Find the image of the region $x \geqq 0$, $0 \leqq y \leqq \pi$ under the mapping defined by $w = \exp z$.

Solution If $w = \exp z = Re^{i\phi}$, then $R = e^x$ and $\phi = y$. Therefore, if $x \geqq 0$, $R \geqq 1$ and if $0 \leqq y \leqq \pi$, then $0 \leqq \phi \leqq \pi$. The region in question is shown in Figure 12.20. □

Example 12.23 Find the image of the rectangular region $a \leqq x \leqq b, c \leqq y \leqq d$ under the mapping $w = \exp z$.

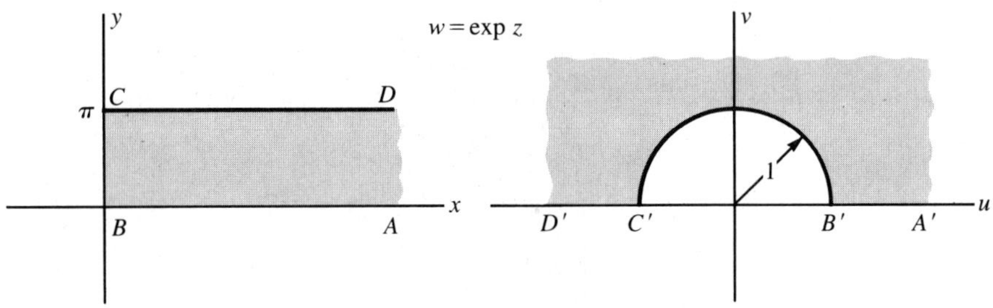

Figure 12.20

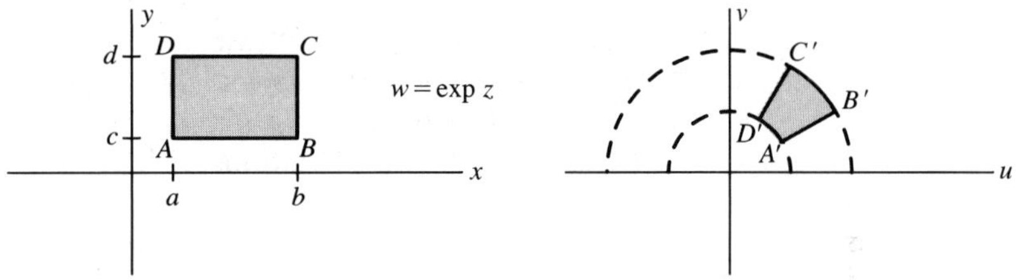

Figure 12.21

Solution The lines $x = a$ and $x = b$ are mapped onto the circles $R = e^a$ and $R = e^b$. The lines $y = c$ and $y = d$ are mapped onto the rays $\phi = c$ and $\phi = d$. The region is shown in Figure 12.21. □

Earlier, we said that $\exp z$ is never 0. It does take on negative values, however.

Example 12.24 Find all values of z such that $\exp z = -1$.

Solution The equation $\exp z = -1$ yields two real equations,

$$\begin{cases} e^x \cos y = -1 \\ e^x \sin y = 0 \end{cases}$$

The second of these is satisfied only if $y = n\pi$, n an integer, for which the first equation becomes $e^x(-1)^n = -1$. This is true only if $x = 0$ and n is an odd integer. Therefore, $\exp z = -1$ for $z = 0 + (2n + 1)\pi i$. □

Exercises for Section 12.7

1. Show that $\exp(z_1 + z_2) = (\exp z_1)(\exp z_2)$.
2. Demonstrate that:
   (a) $e^{3\pi i + z} = -e^z$ **(b)** $(e^z)^n = e^{nz}$ for any integer n

3. Express the following complex numbers in the form $z = x + iy$:
   (a) $\exp\left(-1 + \frac{i\pi}{3}\right)$ **(b)** $\left[\exp\left(2 + \frac{i\pi}{2}\right)\right]\left[\exp\left(-2 + \frac{i2\pi}{3}\right)\right]$
   (c) $\frac{\exp(2 + i\pi/3)}{\exp(-2 - i\pi)}$
4. Find the image of the following under the exponential map $w = e^z$:
   (a) The line $x = 2$ **(b)** The line $y = 2$
   (c) The line $x = y$ **(d)** The region $x \leqq 0$
5. Solve the equations:
   (a) $\exp z = -2$ **(b)** $\exp z = 0$ **(c)** $\exp z = i$ **(d)** $\exp z = 1$
6. Find all complex numbers z such that
   (a) $e^{3z} = 8$ **(b)** $e^{-2iz} = 3$ **(c)** $\exp(z^2) = 1$
7. Find the preimage of the exterior of the unit circle under the mapping $w = e^z$.
8. Show that if $f(z) = \exp z$, then $f(\bar{z}) = \overline{f(z)}$.
9. Show that the exponential function is bounded along the imaginary axis.
10. Use Euler's formula and the nature of the exponential function to prove D' Moivre's formula:

$$[r(\cos\theta + i\sin\theta)]^n = r^n(\cos n\theta + i\sin n\theta)$$

11. Determine the image of the lines $y = cx$ under the exponential map $w = e^{2z}$.
12. Find the image of the semi-infinite strip $x \leqq 0, 0 \leqq y \leqq \pi$ under the exponential map $w = e^z$.
13. Prove that $\frac{d}{dz}(\exp z^n) = nz^{n-1}\exp z^n$.
14. Determine a *nontrivial* electrostatic potential in the region $-\pi/2 < y < \pi/2$ which is such that

$$V\left(x, -\frac{\pi}{2}\right) = 0 = V\left(x, \frac{\pi}{2}\right)$$

   (*Hint:* Consider $\exp z$.)
15. Find the steady-state temperature $T(x, y)$ in the region $x > 0, 0 < y < \pi$ that takes on the boundary values

$$T(x, 0) = 0 = T(x, \pi)$$
$$T(0, y) = 10\sin y$$

12.8 Complex Trigonometric Functions

The representations of the previous section for the real trigonometric functions motivate the following definitions for functions which we denote by $\sin z$ and $\cos z$:

$$\begin{cases} \sin z \equiv \dfrac{\exp(iz) - \exp(-iz)}{2i} \\[2ex] \cos z \equiv \dfrac{\exp(iz) + \exp(-iz)}{2} \end{cases}$$

Since the exponential function is analytic everywhere, so also are these new functions. Moreover, it is an easy application of the definitions to show that

$$\frac{d}{dz}(\sin z) = \cos z \quad \text{and} \quad \frac{d}{dz}(\cos z) = -\sin z$$

a result that should surprise you since, at least superficially, the functions $\sin z$ and $\cos z$ have little relation to the corresponding real functions. The same identities which hold for the real trigonometric functions turn out to be true for their complex counterparts but *must* be proven independently. Once the three identities

$$\sin^2 z + \cos^2 z = 1$$

$$\sin(z_1 \pm z_2) = \sin z_1 \cos z_2 \pm \sin z_2 \cos z_1$$

and

$$\cos(z_1 \pm z_2) = \cos z_1 \cos z_2 \mp \sin z_1 \sin z_2$$

are established, then most of the others follow by techniques common to elementary trigonometry.

Example 12.25 Find $\sin(iy)$ and $\cos(iy)$ where y is real.

Solution

$$\sin(iy) = \frac{\exp i(iy) - \exp(-i(iy))}{2i} = -i\frac{e^{-y} - e^{y}}{2} = i \sinh y$$

$$\cos(iy) = \frac{\exp i(iy) + \exp(-i(iy))}{2} = \frac{e^{-y} + e^{y}}{2} = \cosh y$$

This example shows that the complex trigonometric functions are unbounded. □

Both the sine and cosine functions are periodic with real period 2π.

To interpret the mapping $w = \sin z$, it is convenient to have the real and imaginary parts of the function expressed in terms of real functions. To do this, we use the identity for the sine of the sum of two complex numbers,

$$\sin z = \sin(x + iy) = \sin x \cos iy + \cos x \sin iy$$

Using the result of Example 12.25,

$$\sin z = \sin x \cosh y + i \cos x \sinh y$$

Similarly,

$$\cos z = \cos x \cosh y - i \sin x \sinh y$$

Example 12.26 Determine the image of the semi-infinite strip

$$-\frac{\pi}{2} \leqq x \leqq \frac{\pi}{2}, \qquad y \geqq 0$$

under the mapping $w - \sin z$.

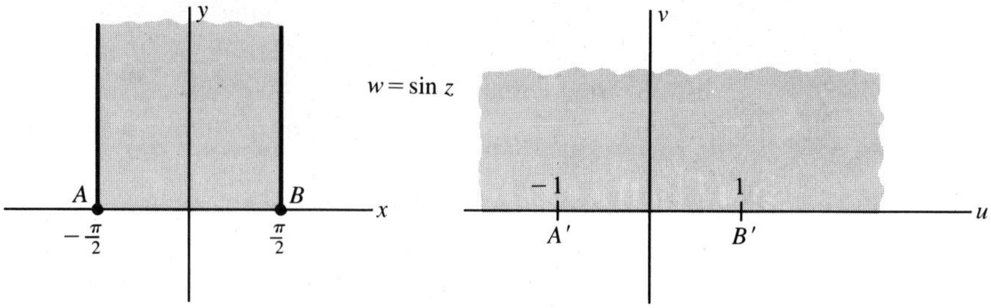

Figure 12.22

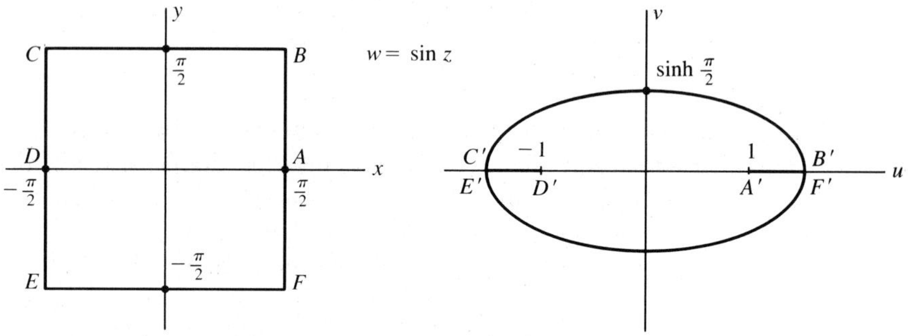

Figure 12.23

Solution When $x = -\pi/2$, $u(-\pi/2, y) = -\cosh y$, $v(-\pi/2, y) = 0$. Therefore the image of the left boundary is the half-line $v = 0$, $u \leqq -1$.

When $x = \pi/2$, $u(\pi/2, y) = \cosh y$, $v(\pi/2, y) = 0$. Hence, the image of the right boundary is the half-line $v = 0$, $u \geqq 1$.

When $y = 0$, $u(x, 0) = \sin x$, $v(x, 0) = 0$, and thus the image of the line segment $-\pi/2 \leqq x \leqq \pi/2$, $y = 0$ is the line segment $-1 \leqq u \leqq 1$, $v = 0$. The given infinite strip is mapped onto the half-plane $v \geqq 0$. (See Figure 12.22.) □

Example 12.27 Determine the image of the square $ABCDEF$ shown in Figure 12.23, under the mapping $w = \sin z$.

Solution The parametric equations of the image of the line $x = \pi/2$ are

$$u = \cosh y$$
$$v = 0$$

Since y varies from $-\pi/2$ to $\pi/2$, the image of $\overline{FAB}$ is the line segment from $(\cosh \pi/2, 0)$ to $(1, 0)$ and then back to $(\cosh \pi/2, 0)$, that is, twice the line segment. (See Figure 12.23.) The parametric equations of the image of the line $y = \pi/2$ are

$$u = \sin x \cosh \frac{\pi}{2}$$

$$v = \cos x \sinh \frac{\pi}{2}$$

Eliminating the parameter x, we obtain

$$\frac{u^2}{\cosh^2 \pi/2} + \frac{v^2}{\sinh^2 \pi/2} = 1$$

This is an ellipse with major semiaxis $\cosh \pi/2$ and minor semiaxis $\sinh \pi/2$. The distance from the origin to either focus is 1. The remainder of the analysis is left as an exercise. □

12.9 Inverse Functions

By properly restricting the domain of the variable so that a function is one-to-one, we can define a function inverse to the one given. We carry out some of the details of this process in the case of the complex logarithm function.

The complex logarithm, $w = \ln z$, is defined as the relation inverse to the complex exponential. Thus $w = \ln z$ means $z = \exp w$. Since the polar representation of z is $re^{i\theta}$, we have

$$z = re^{i\theta} = e^u e^{iv}$$

from which

$$r = e^u \quad \text{and} \quad \theta + 2n\pi = v \quad \text{with } n = 0, \pm 1, \pm 2, \ldots.$$

If we denote the real natural logarithm by $\log_e$, the complex logarithm can be written

$$w = \ln z = \log_e r + i(\theta + 2n\pi) \equiv \log_e |z| + i \arg z$$

where $\arg z$ is *any* argument of z.

The above equation thus defines a many-valued function, or, more correctly, many single-valued branch functions, each determined by how we restrict $\arg z$. For example, we obtain one branch function by imposing $\pi/2 < \arg z \leq 5\pi/2$ and another by requiring $-2\pi < \arg z \leq 0$. If $\arg z$ is limited to its fundamental period, the fundamental branch of $\ln z$ is defined and written as $\operatorname{Ln} z$. Thus,

$$\operatorname{Ln} z \equiv \log_e |z| + i \operatorname{Arg} z$$

Other values of $\ln z$ are related to this fundamental branch by the equation

$$\ln z = \operatorname{Ln} z + 2n\pi i$$

Given any point z_0, the derivative can be determined for any branch of $\ln z$ which contains z_0 as an interior point. We observe that in all cases $\frac{d}{dz}(\ln z) = 1/z$, as can be seen by differentiating $z = \exp w$ implicitly.

The real logarithm of a negative number is undefined, but this is not the case for the complex logarithm which is defined for each $z \neq 0$.

Example 12.28 Compute $\ln(-1)$ and $\mathrm{Ln}(-1)$.

Solution The *principal* polar coordinates of (-1) are $r = 1, \theta = \pi$, and *all* polar coordinates of (-1) are $r = 1, \theta = (2n + 1)\pi$. Therefore,

$$\mathrm{Ln}(-1) = \log_e|-1| + i\pi = 0 + i\pi = i\pi$$
$$\ln(-1) = i\pi + 2n\pi i = (2n + 1)\pi i$$

□

The branch functions of $\ln z$ all map the complete z-plane minus the origin onto a horizontal strip in the w-plane of width 2π. The domains of the various branch functions are separated by the branch cut along the negative real axis.

Since there are infinitely many branches to the complex logarithm, the mapping $w = \ln z$ may be interpreted as a map from an infinite-sheeted Riemann surface onto the w-plane.

12.10 The Power Function

In Section 12.6 we learned that z^n and $z^{1/n}$, with n integral, were defined in essentially the same way as in real analysis. Here we give meaning to the more general power of z, z^c, where c is any complex number. Once again, the method used is similar to the definition of x^b, where x and b are real, namely,

$$z^c \equiv \exp(c \ln z)$$

Perhaps you may recall that this is the technique used to define irrational powers of real numbers. In the case of real numbers, both the real exponential and the real logarithm are single-valued so that x^b is single-valued. However, the complex logarithm is multi-valued and $\exp z$ is periodic so the precise behavior of z^c will depend on the nature of the exponent, c. If c is an integer, we write z^n and if c is rational, the function is written $z^{p/q}$.

We now show that our present definition is consistent with the earlier analysis of Section 12.6.

Case 1: If c is a positive integer, then

$$\begin{aligned} z^n &= \exp(n \ln z) \\ &= \exp(n(\log_e|z| + 2m\pi i + i\,\mathrm{Arg}\,z)) \\ &= \exp(n\log_e|z| + 2mn\pi i + in\,\mathrm{Arg}\,z) \end{aligned}$$

Since n is an integer, so is $2mn$, and because of the periodicity of $\exp z$, the term $2mn\pi i$ drops out so that

$$\begin{aligned} z^n &= \exp(n\log_e|z| + in\,\mathrm{Arg}\,z) \\ &= |z|^n e^{in\,\mathrm{Arg}\,z} \end{aligned}$$

which is the same value previously assigned to z^n.

Case 2: Let $c = 1/n$. Then

$$\begin{aligned} z^{1/n} &= \exp\left(\frac{1}{n}\log_e|z| + \frac{2m\pi i}{n} + i\frac{\operatorname{Arg} z}{n}\right) \\ &= \exp\left(\frac{1}{n}\log_e|z| + i\frac{\operatorname{Arg} z}{n}\right)\exp\left(\frac{2m\pi i}{n}\right) \\ &= [\text{principal value of } z^{1/n}]e^{i2m\pi/n} \end{aligned}$$

which yields distinct values for $m = 0, 1, 2, \ldots, (n-1)$. These are the same n values determined by the definition used in Section 12.6.

Generally, z^c has infinitely many values. For example, if c is a real irrational number, then $\arg z^c = c \operatorname{Arg} z + 2mc\pi$. Since $2mc$ is never an integer, infinitely many values of $\arg z^c$ are obtained. This property sharply distinguishes the complex quantity z^c from the real function x^c, even when z is a real number.

Example 12.29 Calculate $1^{\sqrt{2}}$.

Solution Considered as a problem in real analysis, $1^{\sqrt{2}} = 1$, since "1 to any finite power is 1."

In complex analysis,

$$\begin{aligned} |1^{\sqrt{2}}| &= 1 \\ \arg(1^{\sqrt{2}}) &= \sqrt{2}\cdot 0 + 2\sqrt{2}m\pi \\ &= 2\sqrt{2}m\pi \end{aligned}$$

Therefore

$$\begin{aligned} 1^{\sqrt{2}} &= \exp(2\sqrt{2}m\pi i) \\ &= e^{i2\sqrt{2}m\pi} \end{aligned}$$

where m is an integer. Thus, there are infinitely many values of $1^{\sqrt{2}}$. Graphically, they are all located on a circle of radius 1 centered at the origin. □

In summary, z^c is single-valued if c is an integer, is multiple but finitely many-valued if c is a rational number that is not an integer, and is infinitely many-valued in all other cases.

Example 12.30 Calculate i^i.

Solution

$$\begin{aligned} i^i &= \exp(i \ln i) \\ &= \exp[i(\pi/2\, i + 2m\pi i)] \\ &= e^{-\pi/2 + 2m\pi}, \quad m \text{ an integer} \end{aligned}$$ □

Note that the values of i^i are real, which might be a surprise (although at this stage maybe nothing is!).

To conclude this section, we consider z^c where z is the real irrational number e and c is a complex number. By definition

$$\begin{aligned} e^c &= \exp(c \ln e) \\ &= \exp[c(\log_e e + 0 + 2n\pi i)] \\ &= \exp[c(1 + 2n\pi i)] \end{aligned}$$

which means that e^c is (as is usually the case for the complex power) many-valued. For the case $n = 0$,

$$e^c = \exp c$$

or

$$[\text{principal value of } e^c] = \exp c$$

Hence very often (probably most often), the exponential function is written as e^z instead of $\exp z$. We use the two notations interchangeably.

Exercises for Sections 12.8–12.10

1. Compute:
   (a) $\cos(2 + i)\pi$ **(b)** $\sin(1 + 2i)\pi$ **(c)** $\dfrac{\sin 2\pi i}{\cos 2\pi i}$
2. Show that the vertical strip $-\pi/2 \leq x \leq 0$ is mapped by the sine function onto the left half-plane $u \leq 0$.
3. Show that the equation $\sin z = \pm i \cos z$ has no solutions.
4. Consider the ellipse in the w-plane which is the image of $y = c$, under the mapping $w = \sin z$. Show that the distance from the origin to the focus is independent of c. Show that all horizontal lines map onto a set of confocal ellipses.
5. Determine the image of the region $|x| \leqq c$, $|y| \leqq c$, where $0 < c < \pi/2$, under the mapping $w = \sin z$.
6. Solve the equations:
   (a) $\sin z = 2$ **(b)** $\cos z = \frac{1}{2}$
   (c) $\cos z = i$ **(d)** $\sin z = \cosh 2$
7. Determine the preimage of the real axis under the mapping $w = \sin z$.
8. Show that $\sin z$ and $\cos z$ are periodic.
9. Prove the three identities mentioned at the beginning of Section 12.8.
10. Compute $|\sin z|$ and $|\cos z|$.
11. Show that $\cos \bar{z} = \overline{\cos z}$, and $\sin \bar{z} = \overline{\sin z}$.
12. Calculate and display graphically:
    (a) $\ln 1$ **(b)** $\ln e^i$ **(c)** $\ln i$
13. Solve the equations:
    (a) $\ln z = -1$ **(b)** $\ln z = i$ **(c)** $\ln z = 0$
14. The complex exponential and logarithmic functions are inverse to each other. What differences, if any, are there between the functions $e^{\ln z}$ and $\ln e^z$?
15. Compute:
    (a) $(1 + i)^{-i}$ **(b)** i^{1-i} **(c)** 1^i **(d)** $1^{\sqrt{3}}$

16. Show that $\operatorname{Ln} z$ is discontinuous at its branch cut.
17. Define the function inverse to the sine function. Show that it is many-valued. Write it in terms of the complex logarithm. (*Hint:* Use the definition of $\sin w$.)
18. Demonstrate that $\dfrac{d}{dz}(\sin^{-1} z) = 1/\sqrt{1 - z^2}$.
19. Define $\sinh z$ and $\cosh z$. Find their real and imaginary parts.

In each of the following applications, a specific harmonic function is to be determined. The real or imaginary part of one of the fundamental analytic functions discussed in Sections 12.8–12.10, or its inverse, will suffice in all cases.

20. Find the nontrivial harmonic function $\psi(x, y)$ that vanishes on both the positive x-axis and the ray emanating from the origin at an angle of inclination α, and is positive in between. (ψ serves as the stream function for two-dimensional potential flow within the wedge of angle α.)
21. Determine the nontrivial steady-state temperature $T(x, y)$ that is positive in the semi-infinite strip $0 < x < \pi$, $y > 0$ and goes to zero on the boundary.
22. Find the electrostatic potential $V(x, y)$, harmonic in the upper half-plane, that assumes the boundary values

$$V(x,0) = \begin{cases} \operatorname{Sin}^{-1} x, & |x| < 1 \\ \dfrac{\pi}{2}, & x > 1 \\ -\dfrac{\pi}{2}, & x < -1 \end{cases}$$

on the x-axis.

12.11 Looking Ahead

The material in this chapter provides the foundation for our later consideration of analytic functions both as domain transformations or mapping functions (Chapter 13) and as integrands in complex line integrals (Chapter 14). For the most part our attention has been directed toward single-valued complex functions of a single complex variable. The notions of branch functions and branch cuts that restrict the domain have been utilized in order to preserve this single-valuedness for otherwise multiple-valued algebraic and logarithmic functions. General multiple-valued functions are discussed in greater detail in references such as [8] and [10]. The extended domains of definition of these functions—Riemann surfaces—and their topology are the subject of [6], [13], and the classic [14].

In Sections 12.7 and 12.8 we saw that the exponential function and the complex trigonometric functions are examples of what are termed simply periodic functions, that is, complex periodic functions whose periods are all integer multiples of some fundamental (primitive) period. It turns out that there exist single-valued analytic functions which have two distinct fundamental periods. Doubly periodic functions in general and the important elliptic functions in particular are considered in [5], [10], and [12] (see also [4]). The elliptic functions are inverse functions to elliptic integrals.

Perhaps not surprisingly, Euclidean spaces of even dimension greater than two can be "complexified," like the plane, by combining pairs of real variables x_j, y_j into single complex variables $z_j \equiv x_j + iy_j$. An older reference to the theory of several complex variables is [7]; distinctly modern treatments of the subject are given in [9] and [11]. Readers should understand, however, that the difficulty of the material escalates dramatically as classical complex analytic function theory is left behind.

Basic References

[1] Hawkins, F. M., and J. Q. Hawkins. 1970. *Complex Numbers and Elementary Complex Functions.* New York: Gordon and Breach.

[2] Williams, J. 1972. *Complex Numbers.* London: George Allen and Unwin.

[3] Yaglom, I. M. 1968. *Complex Numbers in Geometry.* New York: Academic Press.

More-Advanced References

[4] Abramowitz, M., and I. A. Stegun. 1965. *Handbook of Mathematical Functions.* New York: Dover; N.B.S. Applied Math Series 55, U. S. Dept. of Commerce, Washington, D. C. (1964).

[5] Ahlfors, L. V. 1979. *Complex Analysis.* 3d ed. New York: McGraw-Hill.

[6] Ahlfors, L. V., and L. Sario. 1960. *Riemann Surfaces.* Princeton, N. J.: University Press.

[7] Bochner, S., and W. T. Martin. 1948. *Several Complex Variables.* Princeton, N. J.: University Press.

[8] Evgrafov, M. A. 1978. *Analytic Functions.* New York: Dover.

[9] Grauert, H., and K. Fritzsche. 1976. *Several Complex Variables.* New York: Springer-Verlag.

[10] Knopp, K. 1947. *Theory of Functions, Part II.* New York: Dover.

[11] Krantz, S. G. 1982. *Function Theory of Several Complex Variables.* New York: John Wiley.

[12] Markushevich. 1977. *Theory of Functions of a Complex Variable.* 2d ed. New York: Chelsea.

[13] Springer, G. 1981. *Introduction to Riemann Surfaces.* 2d ed. New York: Chelsea.

[14] Weyl, H. 1964. *The Concept of a Riemann Surface.* 3d ed. Reading, Mass.: Addison-Wesley.

Chapter 13

Mapping by Analytic Functions

13.1 Introduction

In the previous chapter we posed the problem of finding the electrostatic potential $V(x, y)$ in the upper half-plane, which takes on the value $V = 0$ on the positive x-axis and $V = 1$ on the negative x-axis. Although we did not show how to obtain the solution, we at least verified that the function

$$V(x, y) = \frac{1}{\pi} \operatorname{Arg} z$$

had all the desired properties. In addition, in several exercises we suggested how the real or imaginary part of various other elementary functions could be used to solve certain simple boundary value problems. In this chapter we consider mappings by analytic functions in far greater generality and learn how to find the solutions to more complicated steady-state equilibrium problems. Several other related topics are also introduced which are interesting in their own right.

13.2 Curves in the Complex Plane

Recall that a planar curve may be represented parametrically by two equations $x = x(t)$ and $y = y(t)$. In vector analysis we showed (Chapter 10) how these two equations are equivalent to one vector function of one real variable, $\mathbf{r}(t)$, called the vector representation of the curve. In the same way, the use of a complex function of one real variable,

$$z(t) = x(t) + iy(t)$$

is a convenient method of describing a curve in the complex plane.

The function, $z(t)$, has its domain on some parameter line and its range is a set of complex numbers which give points (on the curve) in the complex plane.

Example 13.1 Use a complex function of one real variable to represent the curve $y = x^2$.

Solution Letting $x = t$, then $y = t^2$, so the function $z(t) = t + it^2$ for $-\infty < t < \infty$ may be used to represent the parabola. □

Example 13.2 Write the equation for the curve whose representation in terms of a complex function is $z(t) = t + (mt + b)i$.

Solution Since $x = t$ and $y = mt + b$, then $y = mx + b$, which is a straight line with slope m and y-intercept b. □

Now consider a mapping from the z-plane to the w-plane. We are often interested in the image not only of a given domain under such a mapping but also of some curve in that domain. The image of a curve, C, represented by $z(t)$, is the curve C^* in the w-plane given by the function $w = w(z(t))$.

Example 13.3 Find the image of the line $z(t) = t + (1 + t)i$ under the mapping $w = z^2$. Sketch both $z(t)$ and $w(z(t))$.

Solution

$$w(z(t)) = (t + (1 + t)i)^2 = (-1 - 2t) + 2t(1 + t)i$$

Therefore $u(t) = -1 - 2t$; $v(t) = 2t(1 + t)$. Eliminating the parameter t, we obtain

$$v = 2\left(\frac{u + 1}{-2}\right)\left(1 + \frac{u + 1}{-2}\right) = \frac{u^2 - 1}{2}$$

(See Figure 13.1.) □

As in vector analysis, the direction of a tangent vector to a smooth curve is defined as the limiting direction of a vector from the point $z(t_0) = z_0$ to another point on the curve $z(t_0 + \Delta t)$. The vector corresponding to the difference $z(t_0 + \Delta t) - z(t_0)$, and the vector corresponding to the difference quotient

$$\frac{z(t_0 + \Delta t) - z(t_0)}{\Delta t}$$

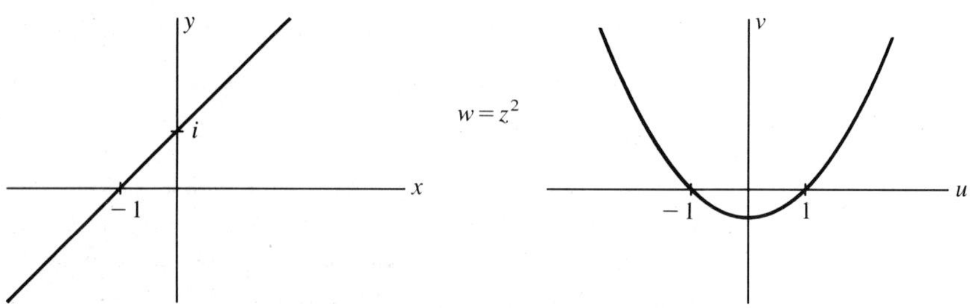

Figure 13.1

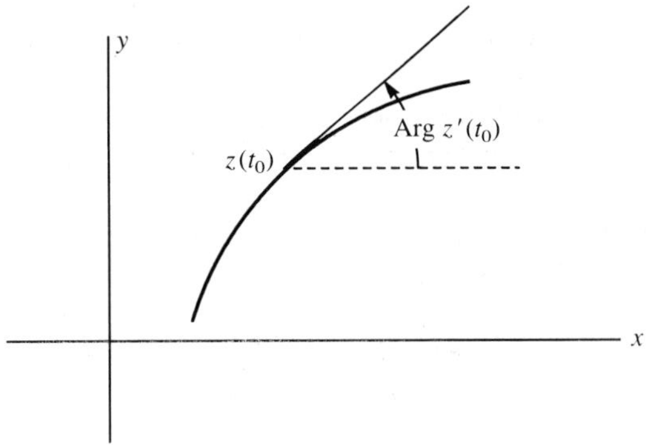

Figure 13.2 *Angle between curve and tangent vector*

have the same direction. Therefore $z'(t_0)$ (that is, dz/dt evaluated at t_0) represents a tangent to the curve at z_0. The angle between this vector and the positive real axis is $\operatorname{Arg} z'(t_0)$. (See Figure 13.2.)

13.3 Conformal Mapping

Generally, the image of a region under a mapping will be significantly distorted. (Recall Figures 12.19–12.23.) A mapping by an analytic function, however, has the important property of not distorting "locally." Specifically, angles between oriented curves are preserved.

Definition 13.1 A mapping defined by $w = f(z)$ is said to be *conformal* at z_0, or *angle preserving*, if the angle between any two oriented smooth curves in the z-plane at z_0 is the same in both magnitude and direction as the angle between the images of these two curves at $w(z_0)$.

Example 13.4 Let C_1 be the parabola defined by $z_1(t) = t + it^2$ and C_2 the straight line given by $z_2(t) = t + it$ with $-\infty < t < \infty$. Find the images of these two curves under the mapping $w = z^2$. Show that the mapping $w = z^2$ preserves the angle between the two curves at the point of intersection corresponding to $t = 1$, but *not* at the point corresponding to $t = 0$.

Solution Let C_1^* and C_2^* be the images of C_1 and C_2 under the mapping. Then

$$C_1^*:\quad w_1(t) = [z_1(t)]^2 = (t^2 - t^4) + i2t^3$$
$$C_2^*:\quad w_2(t) = [z_2(t)]^2 = i2t^2$$

To show that the mapping preserves the angle of intersection at the point P, we must show that the angle between the two tangent lines at P in the z-plane ψ_z is the same as the corresponding angle ψ_w at the image point P^* in the w-plane, or, which is the same thing for our purposes, $\tan\psi_z = \tan\psi_w$ (see Figure 13.3). Now for $t = 1$,

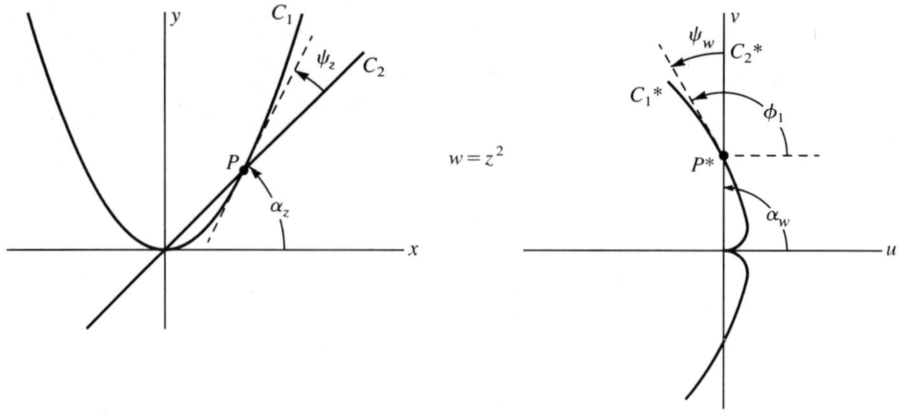

Figure 13.3 *Preservation of angles*

$$\tan \psi_z = \frac{\tan \theta_1 - \tan \theta_2}{1 + \tan \theta_1 \tan \theta_2} = \frac{2 - 1}{1 + (2)(1)} = \frac{1}{3}$$

Similarly, $\tan \psi_w = -\cot \phi_1 = \frac{1}{3}$. In the second case, it is obvious from Figure 13.3 that $\operatorname{Arg}[z_2(0) - z_1(0)] = \alpha_z = \pi/4$ while $\operatorname{Arg}[w_2(0) - w_1(0)] = \alpha_w = \pi/2$, which shows that the angle of intersection is not preserved at the point of intersection corresponding to $t = 0$. □

In Example 13.4 we have shown how a mapping preserves an angle between two specific curves at a point P. To show it to be conformal at P we would have to show this to be true for *every* pair of smooth curves intersecting at P. The following theorem considers the general case.

Theorem 13.1 Let $w = f(z)$ be analytic. Then the mapping defined by $w = f(z)$ is conformal at those points for which $f'(z) \neq 0$.

Proof To prove this theorem we must show that tangents to *all* smooth curves which pass through z_0 are rotated by the same amount. Then the angle between any two curves will be preserved. Let C be any parametrized curve $z(t)$ which passes through z_0. Then $z'(t_0)$ represents a vector tangent to C at z_0. Let C^* be the image of C, $w(z_0)$ the image of z_0, and $w'(z(t_0))$ the image of the tangent vector under the mapping $w = f(z)$. (See Figure 13.4.) To show that the tangent vectors are rotated by a fixed constant, independent of C, we find the relation between $\arg w'$ and $\arg z'$. By the chain rule,

$$w'(z(t_0)) = f'(z_0)z'(t_0)$$

Hence, if $f'(z_0) \neq 0$,

$$\arg w' = \arg f' + \arg z'$$

or

$$\arg w' - \arg z' = \arg f'$$

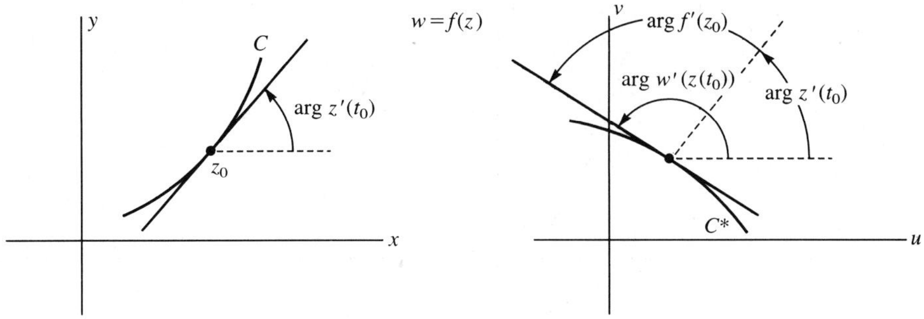

Figure 13.4 *Preservation of angles*

Since arg f' is dependent only on the mapping and the point, the tangents at z_0 are all rotated by a fixed amount. ■

Example 13.5 Where is the mapping given by $w = z^2$ conformal?

Solution Since the mapping is analytic everywhere, and the derivative is nonzero everywhere except at the origin, the mapping is conformal everywhere except at the origin. □

Example 13.6 Where is the mapping defined by $w = (5z + 1)/(2z - 1)$ conformal?

Solution Since

$$\frac{dw}{dz} = \frac{-7}{(2z-1)^2}$$

the function which defines the mapping is analytic everywhere except at the singularity, $z = \frac{1}{2}$. Since dw/dz is never zero, the mapping is conformal everywhere except at $z = \frac{1}{2}$. □

Exercises for Sections 13.1–13.3

In Exercises 1–2 find the images of the given sets of points under the mapping $w = 1/z$:

1. **(a)** The curve $y = x^2 - 1$ **(b)** The circle $|z - 1| = 5$
   (c) The set $\{z | x > 1, y > 0\}$
2. **(a)** The infinite strip $0 < y < \frac{1}{2}$ **(b)** The circle $|z| = 1$
   (c) The circle $|z - 1| = 1$
3. How much does $w = 1/z$ rotate tangents to curves at $z = i$? At $z = 1$? At $z = -i$?
4. What happens to angles between rays emanating from the origin under the mapping $w = z^\alpha$ when:
   (a) $\alpha > 1$ **(b)** $0 < \alpha < 1$

In Exercises 5–6 find the images of the given sets of points under the indicated mappings:

5. **(a)** $x =$ constant and $y =$ constant, under $w = iz$
   (b) $x = y$, under $w = z^2$ **(c)** $x = y$, under $w = iz^2$
   (d) $-\pi/6 < \arg z < \pi/12$, $1 < |z| < 4$, under $w = z^2$

6. **(a)** $x = y$, under $w = x^2 + iy^2$ **(b)** $0 < x, 0 < y < 2$, under $w = iz - 1$
   (c) $x = y$, under $w = \sin z$ **(d)** $x = y$, under $w = e^z$

7. Discuss the mapping functions:
   (a) $w = z^2$ **(b)** $w = z^3$
   In particular, examine the behavior of the images of rays emanating from the origin $z = 0$.
8. Where are the following mappings conformal?
   (a) $w = z + 1/z$ **(b)** $w = z^2 + 1/z^2$ **(c)** $w = z^3 - 1$
   (d) $w = az^2 + bz + c$ **(e)** $w = e^z$ **(f)** $w = \sin z$
   (g) $w = \cos z$ **(h)** $w = 1/z$
9. Determine points in the z-plane at which the mapping by the following functions is not conformal:
   (a) $w = z^2 - z$ **(b)** e^{-z^2} **(c)** $\cos\left(z + \dfrac{1}{z}\right)$
10. How much does the mapping $w = z^2$ rotate tangents to curves at:
    (a) $1 + i$ **(b)** i **(c)** -3 **(d)** $-\frac{1}{2}$
11. Show that the "magnification factor" (equal to $|f'(z)|$) is given by the square root of

$$\det\begin{vmatrix} u_x & u_y \\ v_x & v_y \end{vmatrix}$$

    This is the amount by which infinitesimal lengths are multiplied by the mapping $w = f(z)$.
12. Determine "magnification factors" for each of the following mappings at $z = 1, z = i$, and $z = 1 + i$:
    (a) $w = z^2$ **(b)** $w = z^3$
13. Find the curve(s) along which mapping by the function $w = 3z - z^3$ leaves infinitesimal lengths unchanged.
14. Determine the images of the lines $y = x + 1$ and $y = 3(x - 1)$ under the mapping $w = z^3$. Exhibit that the mapping is conformal at the point of intersection for these two curves.
15. Represent each of the following curves by a complex function of one real variable:
    (a) The line $2x + 3y - 1 = 0$ **(b)** The ellipse $x^2 + 4y^2 = 4$
    (c) The circle $(x - 1)^2 + (y + 1)^2 = 4$ **(d)** The hyperbola $x^2 - y^2 = 1$
16. Find the image of each of the curves of the previous exercise under the mapping $w = z^2$.
17. Represent the tangent lines to the following curves at the indicated points in terms of a complex function of one real variable:
    (a) The parabola $y = x^2$ at the point $(2, 4)$
    (b) The curve $z(t) = t^3 + ie^t$ at $(1, e)$
    (c) The circle $x^2 + y^2 = 4$ at $(0, 2)$

In the consideration of the analytic function $w = f(z)$ as a conformal mapping, the points where $f'(z) = 0$ are exceptional. Such points are generally termed *critical points*. Here the angle between any two smooth curves passing through the point is multiplied by the factor $(n + 1)$ where n is the order of the zero of $f'(z)$ at the critical point.

18. Find the critical point(s) of the analytic function $w = z^4$ and exhibit the nonconformality by sketching the images of various straight lines through the point(s).

19. Determine the critical point(s) of the mapping function $w = 3z - z^3$ and discuss the nature of the nonconformality at these points.

20. Analyze the conformality of the following mappings along the *boundary* of the specified domains:
(a) $w = \exp z$ in the semi-infinite strip $x \geq 0$, $0 \leq y \leq \pi$ (See Figure 12.20.)
(b) $w = \sin z$ in the semi-infinite strip $y \geq 0$, $-\pi/2 \leq x \leq \pi/2$ (See Figure 12.22.)

13.4 Bilinear Transformations

In many applications it is important to use conformal mappings which are defined by functions which are one-to-one, since then an inverse function from the w-plane to the z-plane can be defined. In some cases we are content with the ability to limit the domain of the function sufficiently so that the map is one-to-one in some region of interest. For example, the function $w = z^2$ is not one-to-one in every domain and the inverse relation is not single-valued. The function becomes one-to-one, however, if we restrict consideration to the region $\operatorname{Im} z > 0$, for instance.

A wide and useful class of functions for which no such limitation of domain is necessary is called the set of *bilinear* or *linear fractional* transformations. (The use of the term bilinear in this case is historical even though it is a bit inconsistent with the usual use of the term in other areas of mathematics.)

A bilinear transformation is defined by the equation

$$w = \frac{az + b}{cz + d}, \qquad ad - bc \neq 0$$

where the constants a, b, c, and d may be real or complex. Solving for z, we obtain

$$z = \frac{-dw + b}{cw - a}$$

which shows that the mapping is one-to-one.

It is convenient to associate the "value" $w = \infty$ with $z = -d/c$ and the value $w = a/c$ with $z = \infty$.* The point $z = -d/c$ is an isolated singularity of the bilinear function. Since

$$\frac{dw}{dz} = \frac{ad - bc}{(cz + d)^2}$$

and since $ad - bc \neq 0$, the mapping is conformal everywhere except at this singularity.

There are three special bilinear mappings that "generate" the others.

1. The mapping $w = z + \lambda$ may be regarded as a *translation* in the direction defined by the argument of λ. If $\lambda = 0$ the mapping $w = z$ "fixes" every point.

---

*In complex analysis ∞ is given many of the characteristics of an ordinary complex number. In particular, $z/\infty = 0$ and $z/0 = \infty$ if $z \neq 0$. Geometrically we envisage this single "point" at ∞ as lying at the ends of every line in the plane. This causes lines to take on some of the characteristics of circles; indeed the complex plane may be replaced by a sphere, and lines in the plane by great circles on this sphere.

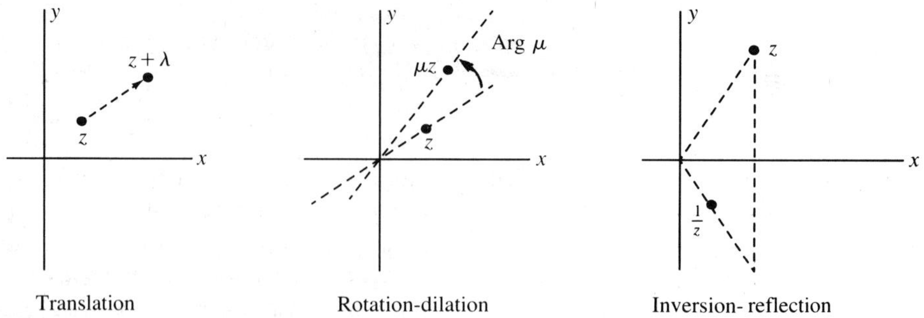

Figure 13.5 Elementary bilinear mappings

(A point is said to be "fixed" if it is mapped onto itself.) The mapping $w = z$ is called the *identity* mapping.

2. The mapping $w = \mu z$ *rotates* points through an angle equal to Arg μ followed by a *dilation* or expansion by the factor $|\mu|$.
3. The mapping $w = 1/z$ is called an *inversion* in the unit circle followed by a *reflection* in the real axis. In polar form the map $w = 1/z$ is $Re^{i\phi} = (1/r)e^{-i\theta}$. Hence, $R = 1/r$ (inversion) and $\phi = -\theta$ (reflection). (See Figure 13.5.)

All bilinear mappings can be obtained by composition of translation, rotation-dilation, and inversion-reflection mappings. To see this, note that if $c \neq 0$, then

$$w = \frac{az + b}{cz + d} = \frac{a}{c} + \frac{bc - ad}{c(cz + d)}$$

Thus, the following composition of the three elementary bilinear mappings gives the most general bilinear map.

$$z \underset{\text{Rotation-dilation}}{\rightarrow} cz \underset{\text{Translation}}{\rightarrow} cz + d \underset{\text{Inversion-reflection}}{\rightarrow} \frac{1}{cz + d} \underset{\text{Rotation-dilation}}{\rightarrow} \frac{\dfrac{bc - ad}{c}}{cz + d} \underset{\text{Translation}}{\rightarrow} \frac{a}{c} + \frac{bc - ad}{c(cz + d)}$$

The case $c = 0$ is left for the exercises.

From a practical point of view, bilinear transformations are important because they preserve lines and circles in the sense of the following theorem.

Theorem 13.2 Under a bilinear transformation each circle or line maps onto a circle or line.

Proof A bilinear map is obtained by composition of translation, rotation-dilation, and inversion-reflection. Obviously, a translation or a rotation-dilation takes a circle onto a circle and a line onto a line. So the theorem will be proved if the map $w = 1/z$ takes a circle onto a circle or a line, and a line onto a circle or a line.

(a) A circle can be represented in the form $|z - a| = R$. The image curve is then

$$\left|\frac{1}{w} - a\right| = R \qquad \text{or} \qquad \left|\frac{1 - aw}{w}\right| = R$$

from which

$$\left|w - \frac{1}{a}\right| = \left|\frac{R}{a}w\right|$$

This is a circle if $R \neq |a|$ and a line if $R = |a|$, as may be easily verified.

(b) A line may be represented in the form $|z - a| = |z - b|$, so that the image curve is

$$\left|\frac{1}{w} - a\right| = \left|\frac{1}{w} - b\right|$$

so that $|aw - 1| = |bw - 1|$, and thus

$$|a|\left|w - \frac{1}{a}\right| = |b|\left|w - \frac{1}{b}\right|$$

This is a line if $|a| = |b|$, and a circle otherwise. ■

Example 13.7 Consider the bilinear transformation defined by

$$w = \frac{z - 1}{z - 2}$$

Find the image of (a) the circle $|z + 1| = 1$; (b) the circle $|z + 1| = 3$; (c) the line $|z - 1| = |z - 2|$.

Solution We first find the inverse transformation by solving for z:

$$z = \frac{2w - 1}{w - 1}$$

(a) The image of $|z + 1| = 1$ is

$$\left|\frac{2w - 1}{w - 1} + 1\right| = 1$$

Simplifying, this becomes $|3w - 2| = |w - 1|$ which is a circle centered at $(\frac{5}{8}, 0)$ with radius $\frac{1}{8}$. (The details of the calculations are left for the reader.)

(b) The image of $|z + 1| = 3$ is

$$\left|\frac{2w - 1}{w - 1} + 1\right| = 3$$

which is $|3w - 2| = |3w - 3|$. This is the perpendicular bisector of the line joining $\frac{2}{3}$ to 1, that is, $u = \frac{5}{6}$ where $w = u + iv$.

(c) The image of $|z-1| = |z-2|$ is

$$\left|\frac{2w-1}{w-1} - 1\right| = \left|\frac{2w-1}{w-1} - 2\right|$$

or $|w| = 1$ which is a circle of radius 1 centered at the origin. □

To determine a specific bilinear transformation, the four constants a, b, c, d must be given. Actually, since

$$\frac{az+b}{cz+d} = \frac{a}{c}\frac{\left(z+\frac{b}{a}\right)}{\left(z+\frac{d}{c}\right)}$$

only three of the constants are independent so three conditions are sufficient to specify completely the bilinear transformation. This informal reasoning leads to the following theorem.

Theorem 13.3 A bilinear transformation is uniquely determined by its effect on three points. ■

If two of the range points specified are 0_w and ∞_w, the bilinear transformation can be determined almost by inspection.

Example 13.8 Find the bilinear transformation that maps 0_z, ∞_z, and a third point, say z_3, onto 0_w, ∞_w, and w_3, respectively.*

Solution The fact that 0 and ∞ are so-called fixed points of the transformation means that the form of the defining equation is $w = cz$. Since the image of z_3 is w_3, $c = w_3/z_3$. Therefore

$$w = \frac{w_3}{z_3}z$$ □

Example 13.9 Find the bilinear transformation that takes 1_z, 2_z, and 3_z onto 0_w, ∞_w, and 1_w, respectively.

Solution The fact that 1_z and 2_z are the preimages of 0_w and ∞_w determines that the transformation has the form

$$w = k\frac{z-1}{z-2}$$

The constant k is determined from the remaining condition. Hence

---

*The subscripts z and w indicate the plane in which the given complex number is being considered.

$$1 = k\frac{3-1}{3-2} = 2k$$

which means $k = \frac{1}{2}$. Therefore

$$w = \frac{1}{2}\left(\frac{z-1}{z-2}\right)$$ □

Example 13.10 Solve the previous example again, except ∞_z rather than 3_z is mapped onto 1_w.

Solution The form of the defining equation is the same. Since ∞_z is mapped onto 1_w,

$$1 = \lim_{z\to\infty} k\frac{z-1}{z-2}$$

and hence $k = 1$. Therefore,

$$w = \frac{z-1}{z-2}$$ □

As these examples suggest, it is an easy problem to map three points in the z-plane onto 0_w, ∞_w, and some third point. In order to find the bilinear transformation which takes any three points z_1, z_2, z_3 in the z-plane onto three general points w_1, w_2, w_3 in the w-plane, we make use of the easily verified identity

$$\frac{w-w_1}{w-w_2}\frac{w_3-w_2}{w_3-w_1} \equiv \frac{z-z_1}{z-z_2}\frac{z_3-z_2}{z_3-z_1}$$

This invariant *cross ratio* is characteristic of bilinear mappings and provides an implicit formula from which the desired transformation can be determined explicitly. We give two examples.

Example 13.11 Find the bilinear transformation that takes the point 0 onto $-i$ and fixes both -1 and 1.

Solution The cross ratio is given in this case by

$$\frac{w+i}{w+1}\frac{1+1}{1+i} = \frac{z-0}{z+1}\frac{1+1}{1-0}$$

Solving for w yields

$$w(1-iz) = -i + z$$

or

$$w = \frac{z-i}{-iz+1}$$ □

Example 13.12 Find the bilinear transformation that takes 0_z, ∞_z, and 2_z onto 1_w, 3_w, and ∞_w, respectively.

Solution Using the invariance of the cross ratio in *formal fashion*, we obtain

$$\frac{w-1}{w-3}\frac{\infty-3}{\infty-1}=\frac{z-0}{z-\infty}\frac{2-\infty}{2-0}$$

Upon simplification, we determine

$$\frac{w-1}{w-3}=\frac{1}{2}z$$

or

$$w=\frac{3z-2}{z-2}$$ □

Exercises for Section 13.4

1. Discuss the transformations defined by:
   (a) $w=(1+i)z$ (b) $w=\dfrac{i}{z}$ (c) $w=\dfrac{1}{z-1}$
2. Find the bilinear transformation that maps:
   (a) $2, i, -2$ onto $1, i, -1$ (b) $-i, 0, i$ onto $-1, i, 1$
3. Find the bilinear transformation that maps:
   (a) $i, 1, \infty$ onto $1, i, \infty$ (b) $1, 2, \infty$ onto $\infty, 5, 2$
   (c) $0, 1, 2$, onto $0, 1, 2$
4. Find the fixed points of the bilinear transformation:
   (a) $w=\dfrac{z+1}{z-1}$ (b) $w=\dfrac{3z-1}{z}$ (c) $w=z$
5. Determine the fixed points of the general bilinear transformation
$$w=\frac{az+b}{cz+d}$$
6. What kind of a bilinear transformation has three fixed points?
7. Does there exist a bilinear transformation with:
   (a) One fixed point (b) No fixed points
8. Determine the general formula for a bilinear transformation specified by three points being mapped onto three points if:
   (a) One of the points z_1 is ∞_z.
   (b) One of the points w_1 is ∞_w.
   (c) A combination of (a) and (b) exists.
9. Find a bilinear transformation whose fixed points are:
   (a) 1 and -1 (b) i and $-i$
10. Find the image of the line $x=5$ under the mapping $(z-1)/(z+1)$.
11. Represent
$$w=\frac{5z-1}{2z+1}$$
as a composition of elementary mappings of translation, inversion-reflection, and rotation-dilation.

12. Show that the bilinear map $w = (az + b)/d$ can be written as a composition of elementary bilinear mappings.

13. Verify that for a general bilinear mapping

$$\frac{w - w_1}{w - w_2} \frac{w_3 - w_2}{w_3 - w_1} \equiv \frac{z - z_1}{z - z_2} \frac{z_3 - z_2}{z_3 - z_1}$$

14. Determine the image of the horizontal strip $0 < y < 1/2c$ with $c > 0$ under the mapping $w = 1/z$.

15. Find the image of the semi-infinite strip $x > 0, 0 < y < 1$ under the transformation $w = i/z$.

16. Determine the image of the first quadrant under the mapping $w = z/(z - 1)$.

17. Find and sketch the image of the vertical strip $-1 < x < 0$ under the mapping $w = (z + 1)/iz$.

18. Determine and sketch the image of the sector $-\pi/4 < \arg z < \pi/4$ under the mapping $w = (z - i)/(z + i)$.

13.5 Special Bilinear Mappings

In this section we particularize the preceding discussion to two specific types of bilinear mappings. These typify the application of bilinear transformations.

A. *Mapping of the upper half-plane onto the unit circle.* The region in the z-plane is described by the inequality $\operatorname{Im} z > 0$, and in the w-plane by $|w| < 1$. There are many ways of accomplishing a map of the upper half-plane onto the unit circle, but clearly, the boundary in the z-plane which is the real axis must correspond to the circle $|w| = 1$. In addition to making this correspondence of boundaries, you must map the upper half-plane onto the interior of the circle. A knowledge of the image of one point is sufficient, since, if the boundaries correspond, the image of the entire upper half-plane will be either the interior or the exterior of the circle $|w| = 1$.

Example 13.13 Show that the bilinear transformation given in Example 13.11 provides a mapping from the upper half-plane onto the interior of the unit circle. What are the images of the lines $y =$ a constant? Write the equation for the image of the line $y = 1$.

Solution Since three points on the real line map onto three points on the unit circle, it is clear from Theorems 13.2 and 13.3 that the entire real axis maps onto the circle $|w| = 1$. It is easy to check that $z = i$ maps onto the origin and hence the upper half-plane is mapped onto the interior of the unit circle. Since ∞_z maps onto $w = i$, the images of all lines in the z-plane pass through $w = i$. Thus, in particular, the lines $y = c$ are mapped onto circles through $w = i$.

To find the image of $y = 1$, we could proceed in one of two ways:

1. If we solve the equation $w = (z - i)/(-iz + 1)$ for z, we obtain $z = (w + i)/(iw + 1)$. Hence

$$1 = y = \operatorname{Im}\left(\frac{w + i}{iw + 1}\right)$$

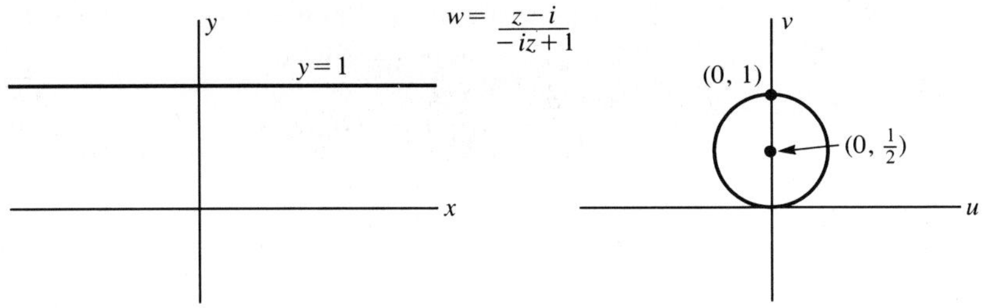

Figure 13.6

In terms of u and v, the equation of the circle is therefore, after a little algebraic manipulation,

$$u^2 + (v - \tfrac{1}{2})^2 = \tfrac{1}{4}$$

(See Figure 13.6.)

2. The equation of the line $y = 1$ can be written $|z| = |z - 2i|$. Using the form of the transformation with z as a function of w, we thus find that the equation of the image of the line is

$$\left|\frac{w+i}{iw+1}\right| = \left|\frac{w+i}{iw+1} - 2i\right|$$

or

$$|w + i| = |3w - i|$$

This is equivalent to the expression obtained by Method 1. □

More generally, to map the upper half-plane onto the unit circle with a bilinear transformation, three points on the real axis in the z-plane should have image points on the circle $|w| = 1$. Moreover, if $z_1 = x_1$, $z_2 = x_2$, $z_3 = x_3$ are such points on the real axis with $x_1 < x_2 < x_3$, then to assure that the image of the upper half-plane is the interior (and not the exterior) of the circle, the image points w_1, w_2, and w_3 must be in cyclic counterclockwise order (see Figure 13.7). This is based, essentially, on the preservation of the sense of an angle under conformal mapping. Alternatively, we may accomplish the same result by assuming that some convenient point z_4 in the upper half-plane is mapped into the interior of the circle.

B. *Mapping of the upper half-plane onto the upper half-plane.* A bilinear transformation that takes the real axis onto the real axis will map the upper half-plane in the z-plane onto either the upper or lower half-plane in the w-plane. Hence, it is sufficient to check the image of a single point after the boundaries are matched to determine whether $\operatorname{Im} z > 0$ is mapped onto $\operatorname{Im} w > 0$ or $\operatorname{Im} w < 0$.

Example 13.14 Determine a bilinear transformation that maps the upper half-plane in such a way that the image of the unit interval is the positive real axis.

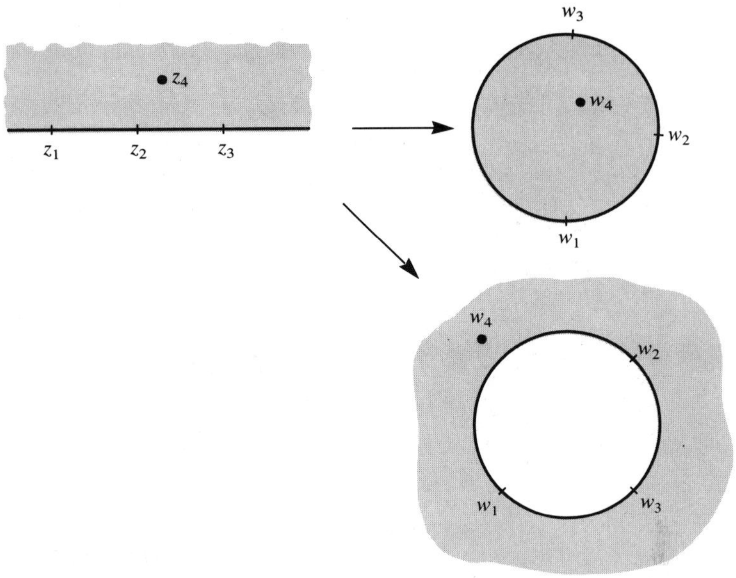

Figure 13.7 Mapping onto the interior and the exterior of the unit circle

Solution Fixing 0 and mapping 1_z onto ∞_w yields $w = kz/(z - 1)$. If we further insist that $\frac{1}{2}$ map onto some positive u, say 1, we have $w = z/(1 - z)$. The image of i_z, namely, $i/(1 - i) = (i - 1)/2$, *is* then in the upper half-plane as desired (and as might have been expected considering the ordering of the images of 0_z, $(\frac{1}{2})_z$, and 1_z). The required transformation is, therefore, $w = z/(1 - z)$. Alternatively, of course, we could have used the invariant cross ratio in deriving the desired mapping. □

Exercises for Section 13.5

1. Discuss the image of the upper half-plane under the mappings:

(a) $w = (1 + i)z$ **(b)** $w = \dfrac{i}{z}$ **(c)** $w = \dfrac{1}{z - 1}$

2. Find the image of the unit disk $|z| \leq 1$ under the same mappings as in Exercise 1.

3. Show that

$$w = \frac{z - z_0}{\bar{z}_0 z - 1}, \qquad |z_0| < 1$$

maps the unit disk onto the unit disk.

4. Find the bilinear mapping that takes $-1, 0, \infty$ onto $-1, \infty, -2$. What is the image of the upper half-plane?

5. Find the bilinear mapping which maps the real axis onto the unit circle and which takes $1, 2, \infty$ onto $i, -1, -i$. What is the image of the upper half-plane?

6. Determine the image of the upper half-plane under the mappings:

(a) $w = \dfrac{z - i}{z + i}$ **(b)** $w = \dfrac{z + i}{z - i}$

7. Show that the general bilinear transformation

$$w = e^{i\alpha}\left(\frac{z-\lambda}{z-\bar{\lambda}}\right)$$

where α is real and $\operatorname{Im}\lambda > 0$, maps the upper half of the z-plane onto the interior of the unit circle $|w| \leq 1$.

8. Find a bilinear transformation that fixes -1 and 1 and that maps the upper half-plane onto the interior of the unit circle.

9. Find an analytic function that maps the region $0 \leqq \arg z \leqq \pi/4$ onto $|w| \leqq 1$.

(*Hint* for Exercises 9 and 10: Use a nonlinear mapping followed by a bilinear mapping.)

10. Find an analytic function that maps $1 \leqq y \leqq x + 2$ onto the interior of the unit circle.

11. Find the image of the unit disk $|z| \leq 1$ under the mappings:

(a) $w = -i\left(\dfrac{z+1}{z-1}\right)$ **(b)** $w = i\left(\dfrac{z-i}{z+i}\right)$

12. Show that the general bilinear transformation

$$w = \frac{\bar{\lambda}z - \lambda e^{i\alpha}}{z - e^{i\alpha}}$$

where α is real and $\operatorname{Im}\lambda > 0$, maps the interior of the unit circle in the z-plane onto the upper half of the w-plane.

13. Use the result of Exercise 12 to generate a two-parameter family of harmonic functions $\phi(x, y)$ (electrostatic potentials, temperature distributions) that are positive in the interior of the unit circle and vanish on the boundary $x^2 + y^2 = 1$.

14. Show that if we follow one bilinear transformation with another, the overall composite mapping is bilinear.

15. Verify that the general bilinear transformation

$$w = \frac{\bar{\lambda}(z-\gamma) - \lambda e^{i\alpha}(z-\bar{\gamma})}{(z-\gamma) - e^{i\alpha}(z-\bar{\gamma})}$$

where α is real and both $\operatorname{Im}\lambda > 0$ and $\operatorname{Im}\gamma > 0$, maps the upper half of the z-plane onto the upper half of the w-plane.

16. Use the result of Exercise 15 to generate a three-parameter family of harmonic functions $\phi(x, y)$ (electrostatic potentials, temperature distributions) that are positive in the upper half-plane and vanish on the x-axis.

13.6 Table of Mappings

In this section we tabulate a number of conformal mappings of the z-plane onto the w-plane which make use of the analytic functions we have discussed in this chapter and the last. More extensive treatments have larger tables for reference. It is quite helpful to have mappings such as these available when making applications as in the next section. Corresponding parts of boundaries are indicated by letters. Each of the mappings should be verified.

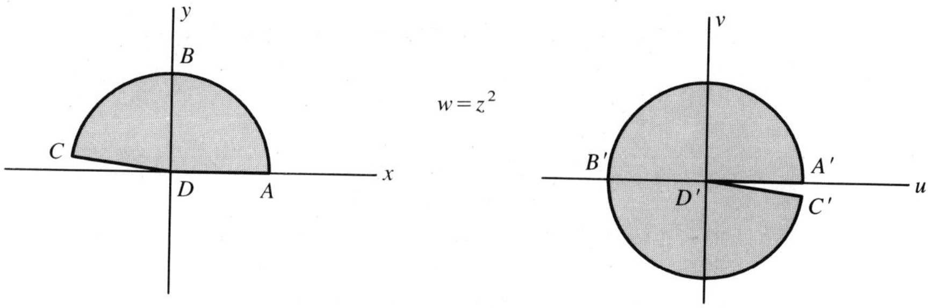

Figure 13.8

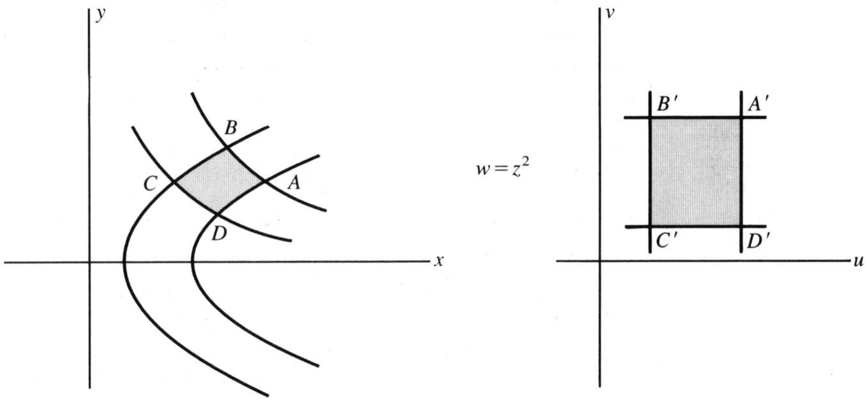

Figure 13.9

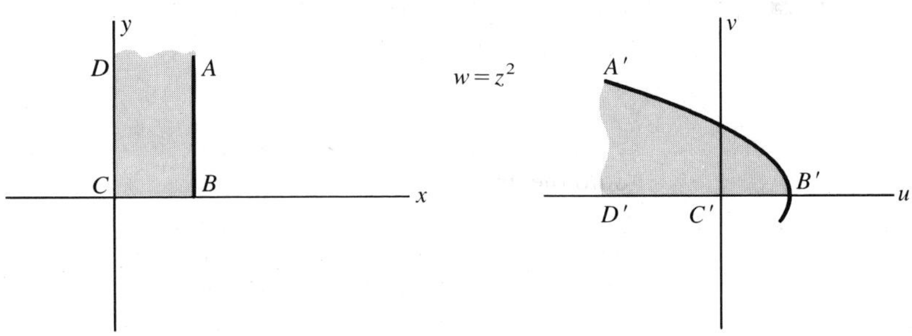

Figure 13.10

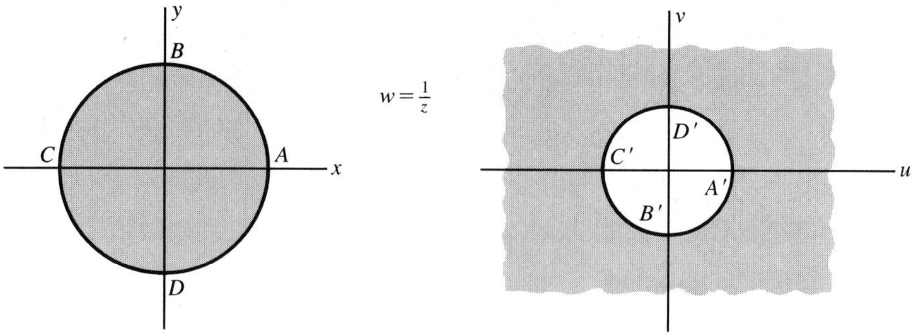

Figure 13.11

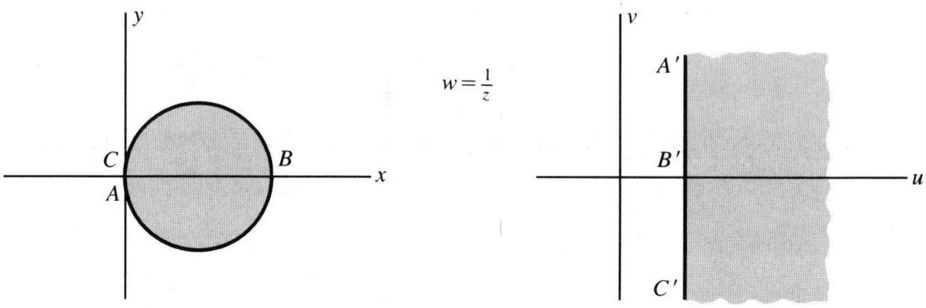

Figure 13.12

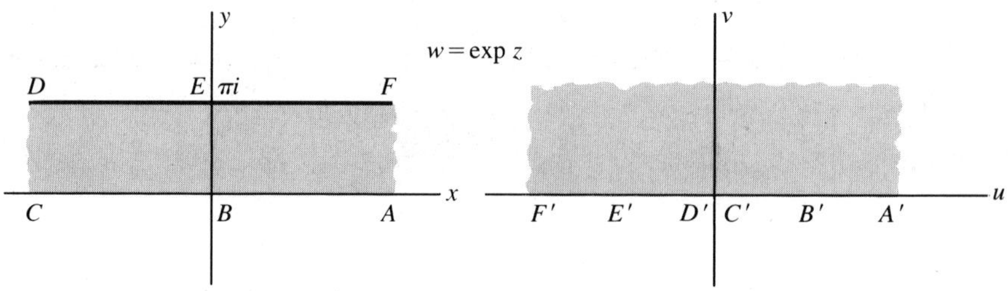

Figure 13.13

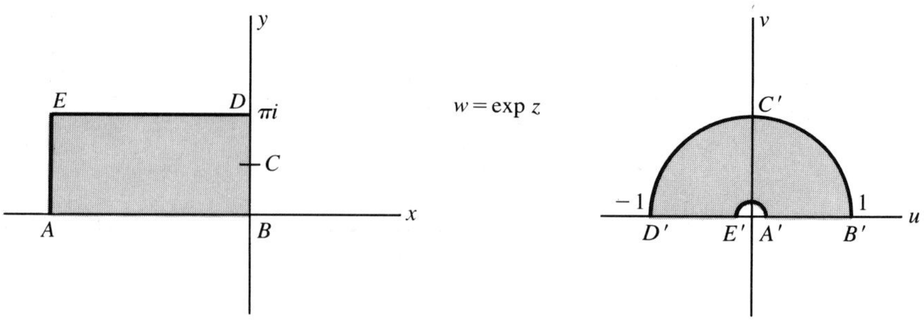

Figure 13.14

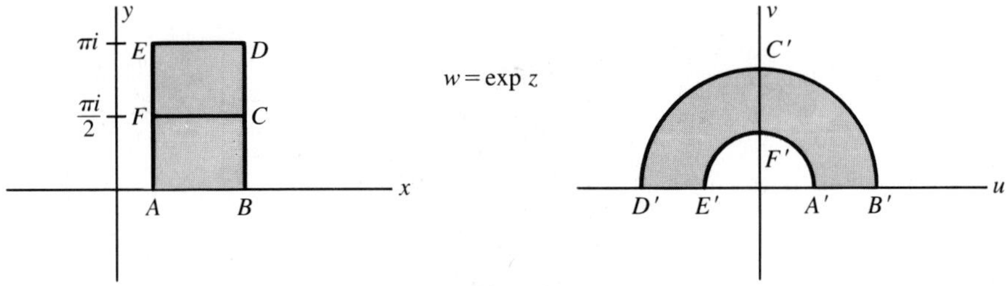

Figure 13.15

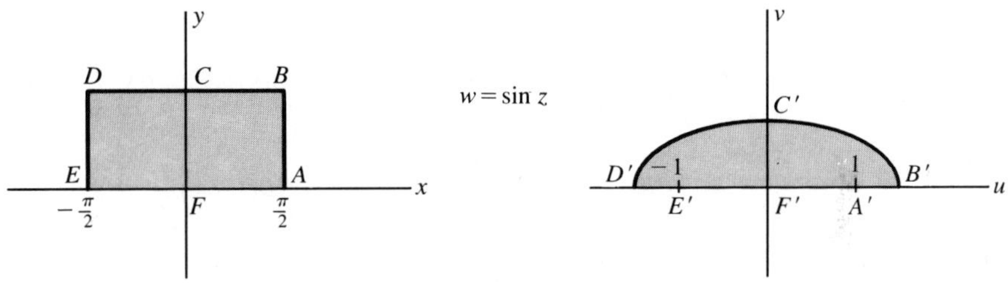

Figure 13.16

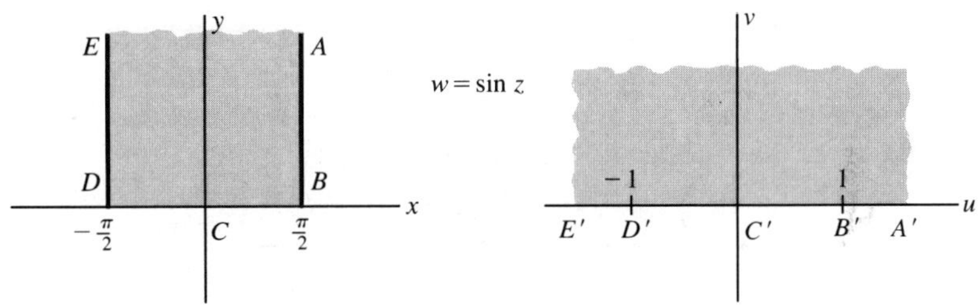

Figure 13.17

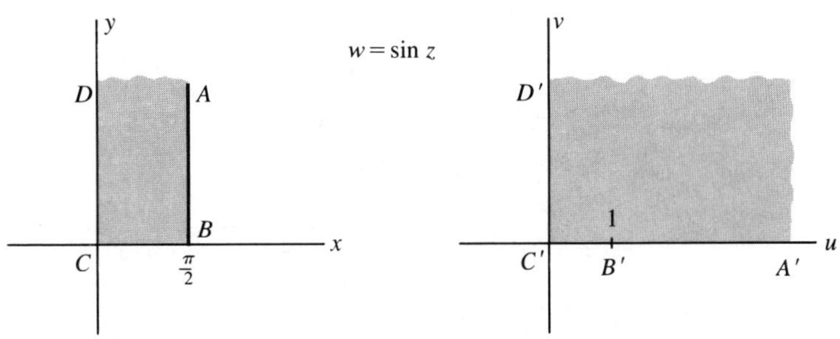

Figure 13.18

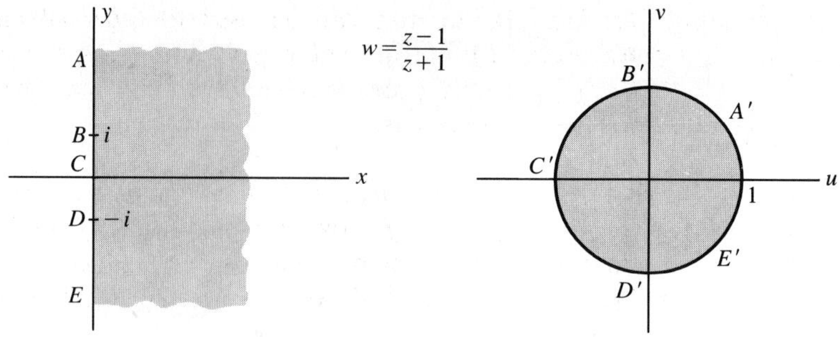

Figure 13.19

Exercises for Section 13.6

In Exercises 1–10 verify the mapping exhibited in the indicated figure.

1. Figure 13.9. In particular, determine the preimages of the lines $u = c$ and $v = d$.

2. Figure 13.10

3. Figure 13.12

4. Figure 13.13

5. Figure 13.14

6. Figure 13.15

7. Figure 13.16

8. Figure 13.17

9. Figure 13.18

10. Figure 13.19

11. Discuss how Figure 13.19 is changed if

$$w = e^{i\alpha}\left(\frac{z-1}{z+1}\right)$$

with α real.

12. Discuss how Figure 13.19 is changed if

$$w = \frac{z-\lambda}{z+\lambda}$$

with $\lambda > 0$.

13. What happens in Exercise 12 if $\lambda < 0$?

14. Determine and sketch the preimage of the unit disk $|w| \le 1$ under the mapping

$$w = \frac{z-e^{i\alpha}}{z+e^{i\alpha}}$$

with α real.

13.7 Application to Boundary Value Problems

Now that we have discussed the theory underlying conformal mappings and analyzed in detail a number of specific transformations, we are ready to apply this knowledge in some practical situations. As we have previously suggested, conformal mappings can be used to great advantage in solving some two-dimensional boundary value problems. We will also need the following theorem, however, whose proof is left to the exercises. The importance of the theorem will become obvious.

Theorem 13.4 Let $f(z) = u(x, y) + iv(x, y)$ be a conformal map from the z-plane to the w-plane with inverse map $f^{-1}(w) = x(u, v) + iy(u, v)$. Let $\phi(x, y)$ be a scalar function defined in some region, R_z, of the z-plane with the composition $\phi(f^{-1}(w))$ denoted by $\Phi(u, v) = \phi(x(u, v), y(u, v))$, being defined in some region, R_w, in the w-plane.

(a) If ϕ is harmonic in the region R_z, then Φ is harmonic in R_w.
(b) If $\phi(x, y) = c$ along some curve C_z, then $\Phi(u, v) = c$ along C_w, where C_w is the image of C_z under f.
(c) If $\partial\phi/\partial n = 0$ along a curve C_z, then $\partial\Phi/\partial n = 0$ along C_w. Recall that the derivatives $\partial\phi/\partial n$ and $\partial\Phi/\partial n$ are the normal derivatives of the scalar functions along the curves. ■

Roughly speaking the theorem says that Laplace's equation and boundary conditions of the type $\phi(x, y) = c$ and $\partial\phi/\partial n = 0$ are invariant under a conformal mapping. As you will see in the examples, therefore, a Dirichlet or Neumann problem with boundary conditions of the kind mentioned may sometimes be solved virtually by inspection if we can find a conformal map to modify the region appropriately. Difficulties can and do arise, however, in expressing the function in terms of the "old" variables; that is, they arise technically in actually finding the inverse of the mapping function.

Example 13.15 Find the steady-state temperature distribution in the plate illustrated in Figure 13.20.

Solution The mathematical model consists of the differential equation

$$T_{xx} + T_{yy} = 0, \qquad x > 0, y > 0, x^2 + y^2 > 1$$

with boundary conditions

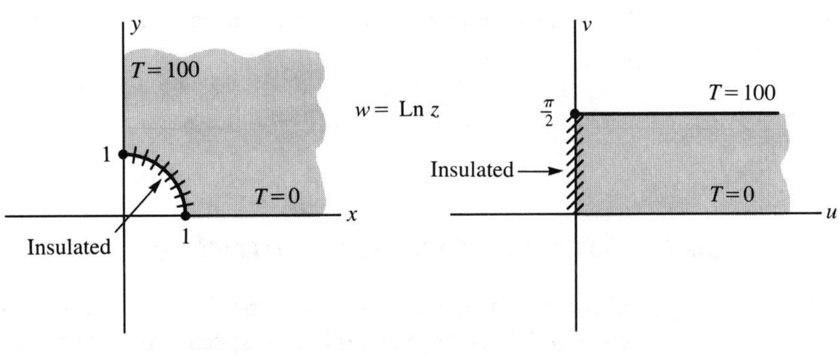

Figure 13.20

$$T(0, y) = 100, \qquad y > 1$$
$$T(x, 0) = 0, \qquad x > 1$$
$$\frac{\partial T}{\partial n} = 0; \qquad \text{on the circular arc } x^2 + y^2 = 1,\ x > 0,\ y > 0$$

By examining the table of mappings in Section 13.6 (especially the inverse mapping associated with Figure 13.15), we see that the function defined by $w = \operatorname{Ln} z$ will map the given region onto the semi-infinite strip shown in the w-plane in Figure 13.20. Owing to Theorem 13.4, then, in the uv-plane the problem has the form

$$T_{uu} + T_{vv} = 0, \qquad u > 0,\ 0 < v < \frac{\pi}{2}$$

where

$$T(u, 0) = 0, \qquad u > 0$$
$$T\left(u, \frac{\pi}{2}\right) = 100, \qquad u > 0$$
$$\frac{\partial T}{\partial u} = 0, \qquad 0 < v < \frac{\pi}{2}$$

The solution can now be determined by inspection. (See Example 8.4, p. 269.)

$$T(u, v) = \frac{200}{\pi} v$$

Since $x = e^u \cos v$ and $y = e^u \sin v$, it follows that $v = \operatorname{Tan}^{-1} y/x$, and hence,

$$T(x, y) = \frac{200}{\pi} \operatorname{Tan}^{-1} \frac{y}{x}$$ □

Sometimes, a sequence of mappings is required. Example 13.16 is typical.

Example 13.16 Find the steady-state temperature in the region bounded by the lines $y = 0$ and $y = \pi$, for $x > 0$, if the boundary conditions are $T = 0$ for $y = 0$ and $y = \pi$, and $T = 50$ for $x = 0$.

Solution We define mappings z_1, z_2, and w to convert the given region sequentially into the one shown in the w-plane as in Figure 13.21. The first of these is

$$z_1 = \sin i\left(z - \frac{i\pi}{2}\right) = \cosh z$$
$$= \cosh x \cos y + i \sinh x \sin y$$

(Note that the subtraction of $i\pi/2$ shifts the region down, the multiplication by i rotates it by 90° and taking the sine opens it up into the upper half-plane as in Figure 13.17.) This is followed by the bilinear transformation

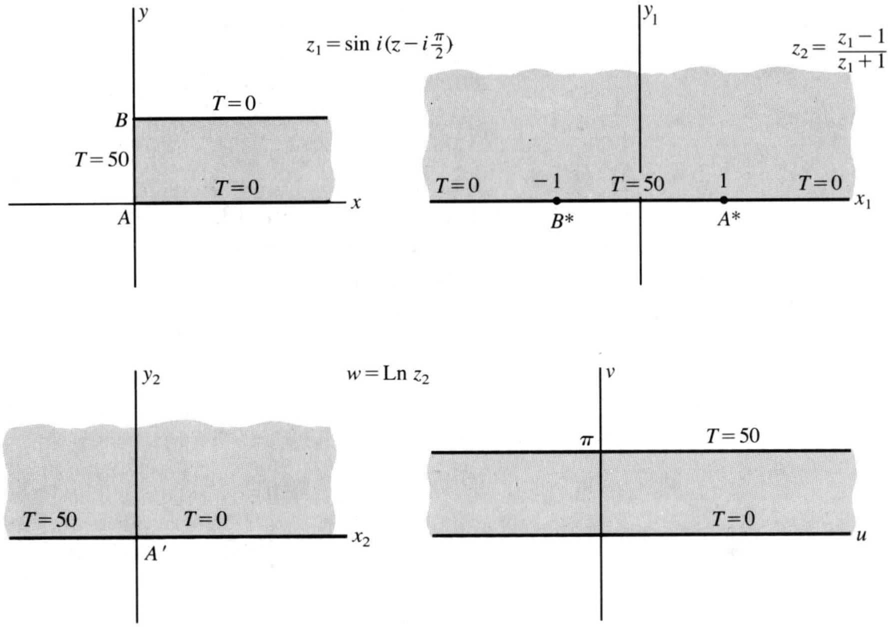

Figure 13.21

$$z_2 = \frac{z_1 - 1}{z_1 + 1}$$

which maps the interval $-1 < \text{Re}\, z_1 < 1$ onto the negative real z_2 axis. Finally,

$$w = \text{Ln}\, z_2$$

takes the transformed region onto the desired infinite strip (recall Figure 13.13). In the w-plane, the solution is readily seen to be

$$T(u, v) = \frac{50v}{\pi}$$

But $v = \text{Im}(\text{Ln}\, z_2) = \text{Arg}\, z_2$, so in the z-plane,

$$T(x, y) = \frac{50}{\pi} \text{Arg}\left[\frac{\cosh x \cos y + i \sinh x \sin y - 1}{\cosh x \cos y + i \sinh x \sin y + 1}\right]$$

Putting the complex number in brackets into standard form leads to

$$\begin{aligned} T(x, y) &= \frac{50}{\pi} \text{Arg}\left[\frac{(\sinh x + i \sin y)^2}{(\cosh x + \cos y)^2}\right] \\ &= \frac{100}{\pi} \text{Arg}(\sinh x + i \sin y) \\ &= \frac{100}{\pi} \text{Tan}^{-1} \frac{\sin y}{\sinh x}, \qquad x > 0, 0 < y < \pi \end{aligned}$$

□

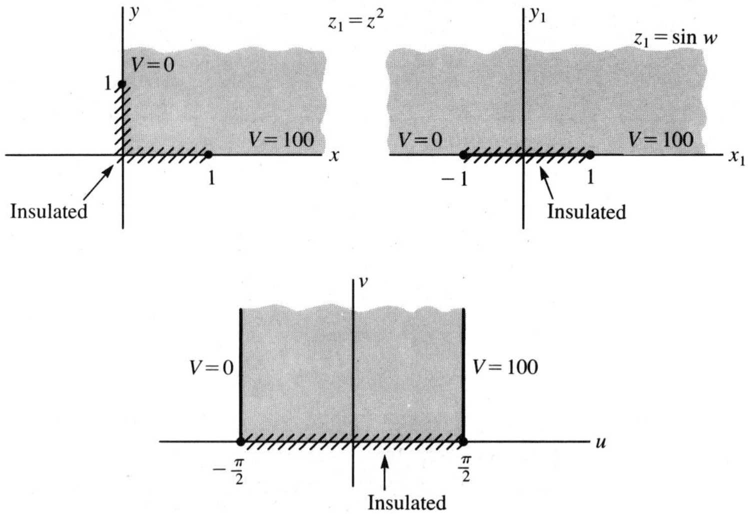

Figure 13.22

Example 13.17 Find the electrostatic potential in the region illustrated in Figure 13.22. What are the equipotential curves?

Solution The mathematical model for the problem in the z-plane is

$$V_{xx} + V_{yy} = 0, \qquad x > 0, y > 0$$

where

$$\begin{aligned} V(x,0) &= 100, & x &> 1 \\ V(0,y) &= 0, & y &> 1 \\ V_x(0,y) &= 0, & 0 &< y < 1 \\ V_y(x,0) &= 0, & 0 &< x < 1 \end{aligned}$$

The mappings $z_1 = z^2$ and $w = \text{Sin}^{-1} z_1$ transform the given region into the one shown for the w-plane. The mathematical model in the z_1-plane is

$$V_{x_1x_1} + V_{y_1y_1} = 0, \qquad y_1 > 0$$

where

$$\begin{aligned} V_{y_1} &= 0, & -1 &< x_1 < 1, y_1 = 0 \\ V(x_1,0) &= 100, & x_1 &> 1 \\ V(x_1,0) &= 0, & x_1 &< -1 \end{aligned}$$

In the w-plane,

$$V_{uu} + V_{vv} = 0, \qquad v > 0, -\frac{\pi}{2} < u < \frac{\pi}{2}$$

where

$$V\left(\frac{\pi}{2}, v\right) = 100, \qquad v > 0$$

$$V\left(-\frac{\pi}{2}, v\right) = 0, \qquad v > 0$$

$$V_v(u, 0) = 0, \qquad -\frac{\pi}{2} < u < \frac{\pi}{2}$$

By inspection,

$$V(u, v) = \frac{100}{\pi}u + 50$$

The equipotential curves in the uv-plane are the lines $u = c$, $-\pi/2 < c < \pi/2$. The preimages of these lines are determined from the equations $x_1 = \sin c \cosh v$ and $y_1 = \cos c \sinh v$. Eliminating v between these two equations gives the equipotential curves in the z_1-plane,

$$\frac{x_1^2}{\sin^2 c} - \frac{y_1^2}{\cos^2 c} = 1$$

This family of hyperbolas is confocal with common foci at $(1, 0)$ and $(-1, 0)$. If we use the definition of the hyperbola ("the difference of the distances of any point on the hyperbola from the two foci is a constant") we obtain

$$\sqrt{(x_1 + 1)^2 + y_1^2} - \sqrt{(x_1 - 1)^2 + y_1^2} = 2 \sin c$$

Substituting the appropriate values for x_1 and y_1 from the relation $z_1 = z^2$,

$$\sqrt{(x^2 - y^2 + 1)^2 + 4x^2y^2} - \sqrt{(x^2 - y^2 - 1)^2 + 4x^2y^2} = 2 \sin c$$

which are the equipotential curves in the z-plane. Since the lines $u = c$ are the equipotential curves in the w-plane, the solution itself can be re-expressed as

$$\begin{aligned} V(x, y) &= 50 + \frac{100}{\pi}u \\ &= 50 + \frac{100}{\pi} \operatorname{Sin}^{-1} \tfrac{1}{2}\left[\sqrt{(x^2 - y^2 + 1)^2 + 4x^2y^2}\right. \\ &\qquad \left. - \sqrt{(x^2 - y^2 - 1)^2 + 4x^2y^2}\right] \quad \square \end{aligned}$$

Exercises for Section 13.7

1. Find the potential distribution in the upper half-plane, if the potential along the x-axis is given by

$$\begin{aligned} V(x, 0) &= 0, & x &< -a \\ &= 100, & -a &< x < a \\ &= 0, & x &> a \end{aligned}$$

2. Find the potential distribution in the first quadrant if the potential along the positive coordinate axes is given by

$$\begin{aligned} V(x,0) &= 0, \quad 0 < x < 2; & V(0,y) &= 0, \quad 0 < y < 1 \\ &= 100, \quad 2 < x & &= 100, \quad 1 < y \end{aligned}$$

(*Hint:* Let $z_1 = z^2$; $z_2 = (z_1 + 1)/(4 - z_1)$; $z_2 = \exp w$.)

3. Find the steady state temperature distribution in the first quadrant if $T = 100°$ along the positive x-axis and $T = 0°$ along the positive y-axis. What are the isotherms? (*Hint:* Let $w = \text{Ln}\, z$.)

4. Find the temperature distribution in the right half-plane for $-a < y < a$, if

$$\begin{aligned} T(0,y) &= 100, \qquad -a < y < a \\ T(x,a) &= T(x,-a) = 0, \qquad x > 0 \end{aligned}$$

(*Hint:* Let $z_1 = i\pi z/2a$; $z_2 = \sin z_1$; $z_3 = (z_2 - 1)/(z_2 + 1)$; $w = \text{Ln}\, z_3$.)

5. Find the temperature distribution in the unit circle if the upper half of the boundary is maintained at 100° and the lower half at 0°. What are the isotherms? (*Hint:* Consider the map $z = (w - i)/(w + i)$.)

6. Find the steady-state temperature in the first quadrant if

$$\begin{aligned} T(x,0) &= 0°, \qquad x > 0 \\ T(0,y) &= 100°, \qquad 5 < y \\ T_x(0,y) &= 0, \qquad 0 < y < 5 \end{aligned}$$

(*Hint:* Let $z_1 = iz/5$; $z_1 = \sin w$.)

7. Show that $w = z + 1/z$ maps the exterior of the circle $|z| = 1$ in the upper half-plane onto the upper half-plane. Use this result to find the steady-state temperature distribution in the part of the upper half-plane exterior to the unit circle, if $T = 100°$ along the linear portion and 0° along the circular portion of the boundary. (*Hint:* Let $z_1 = \frac{1}{2}[z + 1/z]$; $w = (1 + z_1)/(1 - z_1)$.)

8. Prove Theorem 13.4.

As noted previously, the harmonic function $\psi(x, y)$, called the stream function, is central to the analysis of irrotational two-dimensional flows of incompressible fluids—so-called potential flows. Indeed, the fluid streamlines are just the curves $\psi = \text{constant}$.

For uniform flow to the right with speed U_0,

$$\begin{aligned} \psi(x,y) &= \text{Im}(U_0 z) \\ &= U_0 y \end{aligned}$$

and the streamlines are the curves $y = c$. In order to determine the stream functions in more complicated situations, we merely need to transform the given flow (the given flow region) into uniform flow (onto the upper half-plane). For example, for flow in the quadrant $x > 0$, $y > 0$ we use the transformation $w = z^2$, whence

$$\begin{aligned} \psi(x,y) &= \text{Im}(U_0 w) \\ &= \text{Im}(U_0 z^2) \\ &= 2U_0 xy \end{aligned}$$

Notice that in this approach, the boundary of the flow field corresponds naturally to the $\psi = 0$ streamline.

Employ conformal mapping techniques to find the stream function associated with the following two-dimensional potential flows:

9. **(a)** Inside the wedge $0 < \operatorname{Cos}^{-1} \dfrac{x}{\sqrt{x^2 + y^2}} < \pi/\beta, \qquad \beta \geq 1$

(b) Outside the wedge of part (a)

10. Inside the semi-infinite strip $x < 0$, $0 < y < 1$.

11. Outside the circle $x^2 + y^2 = a^2$.
(*Hint:* Dilate/contract the circle to unit radius and use the mapping discussed in Exercise 7.)

12. Around the flat plate of length 4, $y = x \tan \alpha, \quad x^2 + y^2 \leq 4$.
(*Hint:* The transformation $w = \frac{1}{2}(z + \sqrt{z^2 - 4})$ maps the line segment between the points $z = -2$ and $z = 2$ onto the upper half of the unit circle $|w| = 1$. Therefore try $z_1 = \frac{1}{2}(z + \sqrt{z^2 - 4e^{2i\alpha}})$, $w = z_1 + 1/z_1$.)

13. Show that the stream function

$$\psi(x, y) = U_0 \operatorname{Im}\left[ze^{-i\alpha} + \frac{e^{i\alpha}}{z}\right]$$

with $U_0 > 0$ and α real is associated with potential flow outside the circle $x^2 + y^2 = 1$, which as $|z| \to \infty$ approaches uniform flow in a direction making an angle α with the positive x-axis.

14. Discuss why, even though the function $w = e^z$ maps the horizontal strip $0 < y < \pi$ onto the upper half-plane, the harmonic function

$$\psi(x, y) = U_0 \operatorname{Im} e^z = U_0 e^x \sin y$$

is *not* the stream function associated with potential flow in this strip.

13.8 Looking Ahead

The role of analytic functions as mapping functions has been the focus of this chapter. Much of their utility in this capacity is derived from their conformal character—wherever the derivative is nonvanishing, they leave invariant the angles between oriented curves.

It should come as no surprise that invariants of all kinds occur with regularity in practical applications. The conservation of energy and momentum in rigid-body mechanics and the path independence of line integrals of conservative vector fields in fluid dynamics are two familiar examples. The situation is no different in analytic function theory. Indeed, in our discussion in Section 13.4 of the important class of bilinear (linear fractional) transformations, we used the invariant cross ratio to good advantage in determining specific maps. The interplay between conformal invariants and extremal problems forms the central theme of [1].

There is also another type of invariance that we encountered in this chapter. This is the preservation under conformal transformation of Laplace's equation and the associated harmonic character of its smooth solutions in two dimensions. The numerous examples and exercises given were meant to underscore the usefulness of this invariance in applied problem solving. Permanence (invariance) of functional equations involving analytic functions (and hence harmonic functions also) is discussed in a more general context in [9].

The conformal mappings given in Section 13.6 are but a few of the known domain transformations which have proven to be useful in various situations. Additional examples appear in [3]; an entire "dictionary" is contained in [10]. These mappings are typically univalent (one-to-one) on the regions in question. It is a remarkable result, called the Riemann mapping theorem, that any simply connected region that is not the entire complex plane can be mapped by a univalent function onto the interior of the unit circle. This theoretical result notwithstanding, finding the appropriate mapping function for a given domain may be a nontrivial undertaking. See [4], [11], or [14] for consideration of how to determine so-called Schwarz-Christoffel transformations, which are applicable with polygonal domains. Remarks on useful numerical procedures are given in [8]; discussion concerning mappings of multiply-connected domains appears in [2], [6], and [7]. Good references for the many fascinating other aspects of univalent function theory are [5] and [12].

Intermediate and More-Advanced References

[1] Ahlfors, L. V. 1973. *Conformal Invariants: Topics in Geometric Function Theory*. New York: McGraw-Hill.

[2] Bieberbach, L. 1964. *Conformal Mapping*. New York: Chelsea.

[3] Churchill, R. V. et al. 1984. *Complex Variables and Applications*. 4th ed. New York: McGraw-Hill.

[4] Cochran, J. A. 1982. *Applied Mathematics: Principles, Techniques, and Applications*. Belmont, Calif.: Wadsworth.

[5] Duren, P. L. 1983. *Univalent Functions*. New York: Springer-Verlag.

[6] Evgrafov, M. A. 1966. *Analytic Functions*. Philadelphia: Saunders.

[7] Grunsky, H. 1978. *Lectures on Theory of Functions in Multiply Connected Domains*. Göttingen: Vandenhoeck & Ruprecht.

[8] Henrici, P. 1985. *Applied and Computational Complex Analysis, Vol. III*. New York: John Wiley.

[9] Hille, E. 1973. *Analytic Function Theory, Vol. II*. 2d ed. New York: Chelsea.

[10] Kober, H. 1952. *Dictionary of Conformal Representations*. New York: Dover.

[11] Nehari, Z. 1975. *Conformal Mapping*. New York: Dover.

[12] Pommerenke, C. 1975. *Univalent Functions*. Göttingen: Vandenhoeck & Ruprecht.

[13] Schwerdtfeger, H. 1980. *Geometry of Complex Numbers*. New York: Dover.

[14] Silverman, R. A. 1974. *Complex Analysis with Applications*. Englewood Cliffs, N. J.: Prentice-Hall.

Chapter 14

Integration in the Complex Plane

14.1 Introduction

In two-dimensional fluid flow problems, it is convenient to associate with a closed curve C the line integral

$$\Gamma(C) = \oint_C \mathbf{v} \cdot \mathbf{dr}$$

termed the *circulation around* C. (Recall the discussion in the exercises for Section 11.4.) Here $\mathbf{v}$ is the vector velocity that describes the motion of the fluid as a function of space and time and $\mathbf{dr}$ is the infinitesimal arc length vector along the curve C. Frequently, $\mathbf{v}$ can be related to an analytic function in which case the above expression for the circulation becomes nothing more than a simple integration in the complex plane. The nature of such integrals is the subject of this chapter.

While it should not be too unexpected that the concepts of integration and differentiation in the complex plane are related, the strength of the relationship that exists may surprise you. The theory of complex integrals plays a major role in establishing results that concern themselves almost completely with differentiation of complex functions. The relationship between integration and differentiation of an analytic function is both profound and fundamental.

The most important result is the famous Cauchy-Goursat theorem, from which follow the Cauchy integral formula and the formulas for differentiation of an analytic function in terms of a closed line integral. All these results are important in practical applications of which the analysis of fluid circulation is but one example.

14.2 Line Integrals

The line integral of a continuous function, f, of a complex variable is defined analogously to the real line integral of a continuous scalar function (see Section

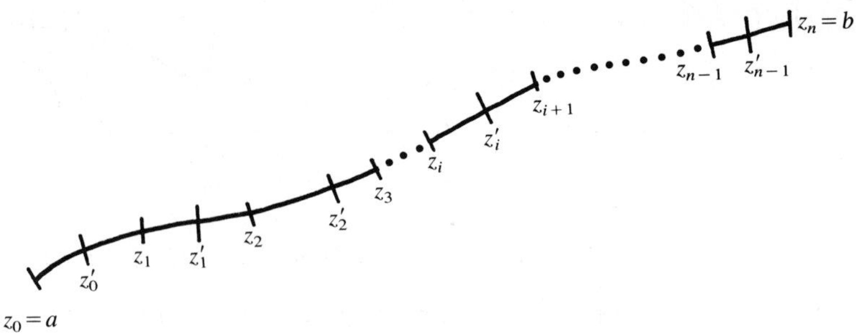

Figure 14.1 *Partitioning of C*

11.2). Let C be a piecewise-smooth simple curve which connects points $z = a$ and $z = b$. Subdivide C with a partition B,

$$B = \{a = z_0, z'_0, z_1, z'_1, z_2, z'_2, \ldots, z_i, z'_i, z_{i+1}, \ldots, z_{n-1}, z'_{n-1}, z_n = b\}$$

(See Figure 14.1.)

For a given partition we evaluate the quantities $f(z'_i)$ at each z'_i, $\Delta z_i = z_{i+1} - z_i$, the finite sum

$$S(B) = \sum_{i=0}^{n-1} f(z'_i)\Delta z_i$$

and the real numbers $|\Delta z_i|$. The largest of the real numbers $|\Delta z_i|$ is called "the norm of the partition B" and is denoted by $\Delta(B)$ or $\|\Delta z_i\|$. The value of $S(B)$ is dependent on the partitioning process, but as the norm of the partition is made smaller and smaller, we would expect the finite sums to approach a limit.

Definition 14.1 The limit of finite sums (if it exists) as the norm of the partition approaches 0,

$$\lim_{\|\Delta z_i\| \to 0} S(B)$$

is called the *line integral of f along C* and is denoted by

$$\int_C f(z)\, dz$$

If C is closed, we write

$$\oint_C f(z)\, dz$$

Example 14.1 Let C be any piecewise-smooth curve connecting the point $z = a$ to the point $z = b$. Evaluate

$$\int_C dz$$

(See Figure 14.2.)

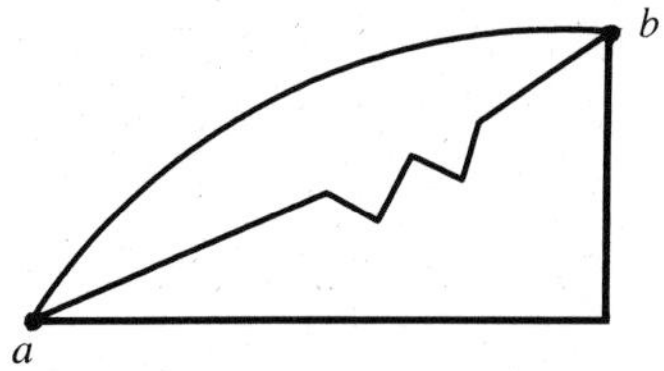

Figure 14.2 *Three paths from z = a to z = b*

Solution For any partition of the curve $\{a = z_0, z'_0, \ldots, z_i, z'_i, z_{i+1}, \ldots, z_n = b\}$,

$$S(B) = (z_1 - a) + (z_2 - z_1) + \cdots + (z_{i+1} - z_i) + \cdots + (b - z_{n-1})$$
$$= b - a$$

and this result is independent of the partition used. Therefore

$$\int_C dz = b - a$$

Note that the value of the integral is dependent only on the end points of the curve and not on the curve itself. □

In complex integration, the integrals

$$\int_C f(z)\,dx, \qquad \int_C f(z)\,dy, \quad \text{and} \quad \int_C f(z)\,ds$$

are also defined but only the latter is of special interest. It is usually denoted by

$$\int_C f(z)\,|dz|$$

Hence, in the previous example, $\int_C |dz|$ is the length of the curve connecting a to b.

Example 14.2 Evaluate

$$\oint_C dz \quad \text{and} \quad \oint_C |dz|$$

where C is a closed circle of radius 1 connecting the point $(1, 0)$ to itself.

Solution From the previous example

$$\oint_C dz = 0$$

since the curve is closed and therefore $a = b$. Since

$$\oint_C |dz|$$

is the length of the path, its value is 2π. □

A complex line integral can be expressed in terms of real line integrals of real functions. In fact, the result of the following theorem is sometimes used as the definition of the line integral of a function of a complex variable.

Theorem 14.1 Let $f(z) = u(x, y) + iv(x, y)$. Then

$$\int_C f(z)\,dz = \int_C (u(x, y)\,dx - v(x, y)\,dy) + i \int_C (u(x, y)\,dy + v(x, y)\,dx)$$

Proof For any partition, B, let $z_k' = x_k' + iy_k'$ and $\Delta z_k = \Delta x_k + i\Delta y_k$. Then

$$\begin{aligned} S(B) &= \sum_{k=0}^{n-1} [u(x_k', y_k') + iv(x_k', y_k')][\Delta x_k + i\Delta y_k] \\ &= \sum_{k=0}^{n-1} u(x_k', y_k')\Delta x_k - v(x_k', y_k')\Delta y_k + i[u(x_k', y_k')\Delta y_k + v(x_k', y_k')\Delta x_k] \end{aligned}$$

which gives the desired result when we take the limit as $\|\Delta z_k\| \to 0$. (Note that both $\|\Delta x_k\| \to 0$ and $\|\Delta y_k\| \to 0$ when $\|\Delta z_k\| \to 0$.) ■

Since the complex line integral may be expressed in terms of real line integrals, the theorems developed for the real case are also true for the complex case. In particular,

1. If f is continuous on C and C is piecewise smooth, then the complex line integral exists (Theorem 11.1).

2. If the smooth curve C connecting $a = x(t_0) + iy(t_0)$, and $b = x(t_1) + iy(t_1)$ is parametrized by $x = x(t)$, $y = y(t)$, and if f is continuous, then

$$\begin{aligned} \int_C f(z)\,dz = &\int_{t_0}^{t_1} \left[u(x(t), y(t))\frac{dx}{dt} - v(x(t), y(t))\frac{dy}{dt} \right] dt \\ &+ i \int_{t_0}^{t_1} \left[u(x(t), y(t))\frac{dy}{dt} + v(x(t), y(t))\frac{dx}{dt} \right] dt \end{aligned}$$

(Theorem 11.2). The substitution in terms of the parameter may sometimes be made directly into the complex function, in which case we have the following formula:

$$\int_C f(z)\,dz = \int_{t_0}^{t_1} f(z(t))\frac{dz}{dt}\,dt$$

where C is represented by $z(t) = x(t) + iy(t)$.

3. The value of the line integral is independent of the parametrization (Theorem 11.2). The value usually *is* dependent on the curve used to connect the two points.

Example 14.3 Evaluate $\int \operatorname{Re} z\,dz$ along:

(a) The straight-line segment from 0 to $1 + i$.
(b) The rectilinear curve from 0 to 1 and then to $1 + i$.

Solution (a) One parametrization of the first curve is $z(t) = (1 + i)t$, $0 \leqq t \leqq 1$. Thus

$$\int_C x\,dz = \int_0^1 t(1 + i)\,dt = \frac{1 + i}{2}$$

(b) The second curve may be written

$$z(t) = \begin{cases} t, & 0 \leqq t \leqq 1 \\ 1 + (t - 1)i, & 1 \leqq t \leqq 2 \end{cases}$$

Therefore,

$$\int_C x\,dz = \int_0^1 t\,dt + i\int_1^2 dt = \tfrac{1}{2} + i$$

□

An especially important line integral is considered in the next example.

Example 14.4 Evaluate

$$\oint_C \frac{1}{z}\,dz$$

where C is the circle $|z| = 1$ traversed once in the counterclockwise direction starting from $z = 1$.

Solution C may be parametrized by $z(t) = \cos t + i \sin t$ from which $dz/dt = -\sin t + i \cos t$. Then

$$\begin{aligned} \int_C \frac{1}{z}\,dz &= \int_0^{2\pi} \frac{-\sin t + i\cos t}{\cos t + i \sin t}\,dt \\ &= i\int_0^{2\pi} dt \\ &= 2\pi i \end{aligned}$$

Alternatively, we could have written the parametrization as $z(t) = e^{it}$. Thence,

$$\begin{aligned} \int_C \frac{1}{z}\,dz &= \int_0^{2\pi} \frac{ie^{it}}{e^{it}}\,dt \\ &= 2\pi i \end{aligned}$$

□

The following properties of the complex line integral are restatements of the same properties for real line integrals as given in Section 11.2:

1. $\displaystyle\int_C kf(z)\,dz = k\int_C f(z)\,dz$
2. $\displaystyle\int_C [f(z) + g(z)]\,dz = \int_C f(z)\,dz + \int_C g(z)\,dz$
3. $\displaystyle\int_C f(z)\,dz = -\int_{-C} f(z)\,dz$

where by $-C$ is meant the curve which is the same set of points as C but with opposite orientation.

4. If C is comprised of two arcs C_1 and C_2 (disjoint except for endpoints), then

$$\int_C f(z)\,dz = \int_{C_1} f(z)\,dz + \int_{C_2} f(z)\,dz$$

Sometimes a numerical estimate of the magnitude of the line integral of f is all that we will need. The following theorem is valuable in making such estimates.

Theorem 14.2 Let L be the length of C and M an upper bound for $|f(z)|$ with z on C. Then

$$\left|\int_C f(z)\,dz\right| \leqq ML$$

Proof For any partition,

$$\begin{aligned} |S(B)| &= \left|\sum_{k=0}^{n-1} f(z_k')\Delta z_k\right| \\ &\leqq \sum_{k=0}^{n-1} |f(z_k')\Delta z_k| \quad \text{(because of the triangle inequality)} \\ &= \sum_{k=0}^{n-1} |f(z_k')||\Delta z_k| \leqq M\sum_{k=0}^{n-1} |\Delta z_k| \leqq ML \end{aligned}$$ ■

By reasoning similar to that used in the proof of Theorem 14.2, we also have the following related theorem.

Theorem 14.3

$$\left|\int_C f(z)\,dz\right| \leqq \int_C |f(z)||dz|$$ ■

Example 14.5 Find an upper bound for

$$\left|\int_C \text{Ln}(z+1)\,dz\right|$$

where C is the straight-line segment from i to $2+i$. (See Figure 14.3.)

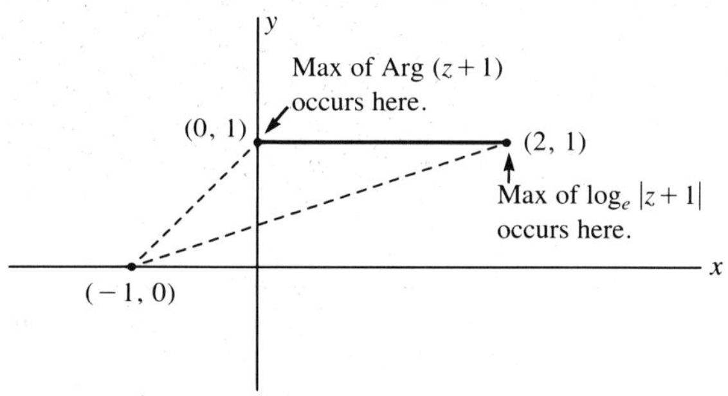

Figure 14.3

Solution

$$\left|\int_C \operatorname{Ln}(z+1)\,dz\right| \leqq [\operatorname{Max}|\operatorname{Ln}(z+1)|]\ [\text{distance from } i \text{ to } 2+i]$$

A bound for $|\operatorname{Ln}(z+1)|$ on the line segment is determined by finding bounds on the real and imaginary parts of $\operatorname{Ln}(z+1)$,

$$|\operatorname{Ln}(z+1)| \leqq |\log_e|z+1|| + |\operatorname{Arg}(z+1)|$$

Since $|z+1|$ represents the distance from $z=-1$, the maximum value of $|z+1|$ is $\sqrt{10}$ (which occurs at $z = 2+i$) and the maximum value of $\operatorname{Arg}(z+1)$ is $\pi/4$ (which occurs at $z = i$). Hence,

$$\begin{aligned}\left|\int_C \operatorname{Ln}(z+1)\,dz\right| &\leqq \left[\log_e\sqrt{10} + \frac{\pi}{4}\right][2] \\ &\doteq (1.15 + 0.79)2 = (1.94)2 \\ &= 3.88\end{aligned}$$

Better estimates are, of course, possible. □

Exercises for Sections 14.1 and 14.2

1. Consider the integral $\int_C z\,dz$ where C is any piecewise-smooth curve connecting a to b. Subdivide C with an arbitrary $n =$ point partition.
   (a) Find S_n if the points of evaluation are the left-hand endpoints of the subintervals in the given partition.
   (b) Find the finite sum if the right-hand endpoints are used, and call it S_n^*.
   (c) Sum $S_n + S_n^*$. Under the assumption that the given integral exists, then

$$\lim_{n\to\infty} (S_n + S_n^*) = 2\int_C z\,dz$$

   Use this technique to evaluate the given integral.
2. Evaluate $\int_C dz$ and $\int_C |dz|$ where C is the perimeter of the square $-1 \leqq x \leqq 1$, $-1 \leqq y \leqq 1$.
3. Same as Exercise 2 except for C use the straight-line segment connecting $1+i$ to $7+6i$.
4. Evaluate $\int_C |z|^2\,dz$ if C is:
   (a) The parabola $y = x^2$ from $(1, 1)$ to $(2, 4)$.
   (b) The straight-line segment connecting the two points.
5. Find a bound on $|\int_C e^z\,dz|$ where C is the part of the circle $|z| = 1$ connecting $(1, 0)$ to $(0, 1)$.
6. Evaluate $\oint_C (z - z_0)^{-1}\,dz$ where C is a circle of any radius centered at z_0 traversed once in the counterclockwise direction. Repeat, if C is the same curve but traversed *twice* in the clockwise direction.
7. Find a better estimate for the integral of Example 14.5 by subdividing the given curve into two equal parts.

In Exercises 8–9, evaluate $\int_C \operatorname{Re} z\, dz$ where C is the specified curve.

8. **(a)** The straight-line segment from 0 to i.
(b) The circular curve from 0 to i which is the right-hand side of the circle $|z - (1/2)i| = 1/2$.
(c) The parabola $y = 2x^2 + 1$ from $(1, 3)$ to $(2, 9)$.

9. **(a)** The circle $|z| = 7$, traversed once in the counterclockwise direction.
(b) The ellipse $4x^2 + y^2 = 4$ traversed once in the clockwise direction.

10. Find the value of $\int_C z^2\, dz$ along the straight-line segment joining 0 to $2 + i$.

11. Evaluate $\int_C \bar{z}\, dz$ if C is:
(a) The parabola $y = x^2$ from $(0, 0)$ to $(1, 1)$.
(b) The straight-line segment connecting the two points.

12. Show that $\oint_C \bar{z}\, dz = 2\pi i$ where C is the unit circle $|z| = 1$ traversed once in the counterclockwise direction.

In Exercises 13–14, use Theorem 14.2 to find upper bounds for the given integrals:

13. **(a)** $\left|\int_C \frac{1}{z^2}\, dz\right|$, where C is the straight-line segment from i to $2 + i$.
(b) $|\int_C e^z\, dz|$, where C is the upper semicircular path connecting 0 to 2π over $|z - \pi| = \pi$.

14. **(a)** $\left|\int_C \frac{\sin z}{z}\, dz\right|$, where C is the straight-line segment from π to 2π.
(b) $|\int_C \operatorname{Ln} z\, dz|$, where C is the straight-line segment from 1 to $1 + i$.
(c) $|\int_C z^{-2}\, dz|$, where C is the straight-line segment from -1 to -2.

15. Evaluate $\int_C \sin \pi z\, dz$ over the straight-line segments from:
(a) 1 to $1 + 3i$ **(b)** i to $3 + i$

16. Find the value of $\int_C \frac{z - 1}{z}\, dz$ if C is:
(a) The upper half of the circle $|z| = 1$ joining $(-1, 0)$ to $(1, 0)$.
(b) The lower half of the circle $|z| = 1$ joining the same two points.

17. Evaluate $\oint_C (1 + z^2)\, dz$, where C is the boundary of the square with vertices at 0, 1, $1 + i$, and i.

18. Evaluate $\oint_C e^{i\pi z}\, dz$, where C is the square boundary of the previous exercise.

19. Evaluate $\oint_C \left(\frac{1}{z - 2} + \frac{1}{(z - 2)^2}\right) dz$, where C is the circle $|z - 2| = 3$ traversed once in the counterclockwise direction.

20. Evaluate $\oint_C (z^2 - z^{-1} + z^{-2})\, dz$, where C is the unit circle traversed once in the clockwise direction.

In Exercises 21–24, use Theorem 14.2 to establish the indicated bounds. Assume a straight-line integration between the designated points.

21. $|\int_C z^2\, dz| < 10$, where C joins i to $2 + i$

22. $\left|\int_C \frac{1}{z^4}\, dz\right| < 2$, where C joins i to $-2 + i$

23. $|\int_C (x^2 + iy^2)\, dz| < 2$, where C joins $-i$ to i

24. $\left|\int_C \frac{\sin z}{z}\,dz\right| < 2\cosh 1$, where C joins i to $2 + i$.

In Section 11.2 we saw that the work done on a particle moving in a force field **F** is given by

$$\text{Work} = \int_C \mathbf{F}\cdot d\mathbf{r}$$

In two dimensions, if we denote the x- and y-components of **F** by $u(x, y)$ and $-v(x, y)$, respectively, we can give an alternative expression in terms of a complex line integral, namely,

$$\text{Work} = \text{Re}\int_C f(z)\,dz$$

where $f(z) = u(x, y) + iv(x, y)$. Use this procedure to calculate the work done on particles moving as follows:

25. Along the straight-line segment from $(0, 0)$ to $(3, 2)$ in the force field $\mathbf{F} = (2x + y)\mathbf{i} + x\mathbf{j}$.

26. Around the complete unit circle traversed counterclockwise with $\mathbf{F} = x^2\mathbf{i} + y\mathbf{j}$.

27. Along the parabola connecting $(-1, 1)$ to $(1, 1)$ that passes through the origin, in the force field $\mathbf{F} = (2xy + x^2)\mathbf{i} + x^2\mathbf{j}$.

28. Around the complete unit circle traversed counterclockwise with

$$\mathbf{F} = \frac{-y\mathbf{i} + x\mathbf{j}}{x^2 + y^2}$$

Employ Theorem 14.2 to find an upper bound for the work done on particles moving as follows:

29. As in Exercise 25.

30. As in Exercise 26.

31. As in Exercise 27.

32. As in Exercise 28.

14.3 The Fundamental Theorem of Complex Integration

The theory of integration of real functions of a real variable is highlighted by the relationship between antidifferentiation and integration. The theory of line integration in the presence of vector fields is highlighted by a similar theorem (Theorem 11.4) stating conditions under which the integral is dependent only on the endpoints of the curve and the fact that the value of the integral is equal to the potential difference of the two endpoints. A similar theorem, called the fundamental theorem of complex line integration, is true for domains in the complex plane.

Before proving the fundamental theorem, we need to review the concept of path independence (analogous to Definition 11.2 for vector fields) and how it relates to integrals around closed curves (analogous to Theorem 11.3). In what follows we introduce the terminology *contour* to denote a piecewise-smooth oriented simple curve.

Definition 14.2 The line integral $\int_C f(z)\,dz$ is *path independent* or *independent of the contour* connecting two points in a domain D if the value of the integral is the same for all contours (with the same orientation) connecting the points. We say that the value of the integral depends only on the endpoints of C.

Theorem 14.4 $\int_C f(z)\,dz$ is path independent in a domain D if and only if $\oint_C f(z)\,dz = 0$ for all closed contours in D.

Proof Prove the theorem as an exercise. (*Hint:* See Theorem 11.3.) ■

The principle of path independence may be interpreted in terms of a continuous deformation of a contour. If the contour is continuously deformed while the endpoints are kept fixed, and if the contour stays entirely within the domain for which the path independence property is valid, then the value of the line integral is unchanged.

Theorem 14.5 *Fundamental Theorem of Complex Line Integration.* Let f be continuous on a domain D. Then $\int_C f(z)\,dz$ is independent of the contour connecting two points in D if and only if there exists a (single-valued) analytic function, F, such that $F'(z) = f(z)$ for all z in D.

Proof Suppose $\int_C f(z)\,dz$ is path independent in D and let z_0^* be in D. Consider any path C, lying in D, which connects the fixed point z_0^* to any other point z_0. Define F by the rule

$$F(z_0) = \int_C f(z)\,dz$$

Because of the path independence, this is a well-defined value dependent only on z_0. We now show $F'(z_0) = f(z_0)$. Let $C_{\Delta z}$ be any contour from z_0 to $z_0 + \Delta z$. (See Figure 14.4.) Forming the difference quotient,

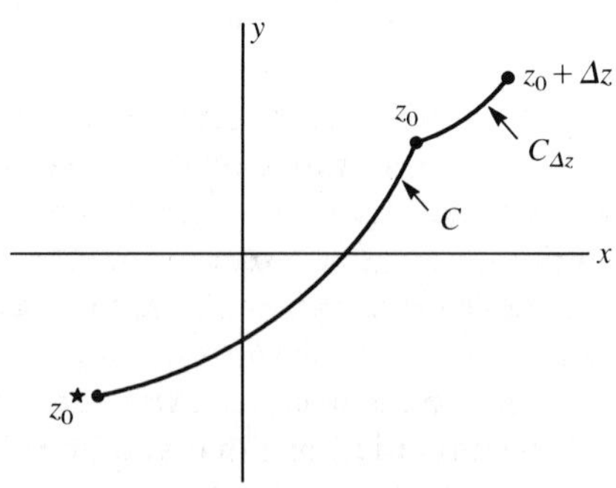

Figure 14.4

$$\frac{F(z_0 + \Delta z) - F(z_0)}{\Delta z} = \frac{1}{\Delta z}\int_{C_{\Delta z}} f(z)\,dz$$

To show that the limit of this difference quotient is $f(z_0)$, we subtract $f(z_0)$ from it and then take the limit,

$$\begin{aligned}\frac{F(z_0 + \Delta z) - F(z_0)}{\Delta z} - f(z_0) &= \frac{1}{\Delta z}\int_{C_{\Delta z}} f(z)\,dz - \frac{f(z_0)}{\Delta z}\int_{C_{\Delta z}} dz \\ &= \frac{1}{\Delta z}\int_{C_{\Delta z}} (f(z) - f(z_0))\,dz\end{aligned}$$

(Recall Example 14.1.) Since f is continuous on $C_{\Delta z}$, the difference $|f(z) - f(z_0)|$ can be made smaller than any arbitrary ε if $|z - z_0|$ is made small enough. Using Theorem 14.2,

$$\left|\frac{F(z_0 + \Delta z) - F(z_0)}{\Delta z} - f(z_0)\right| < \frac{1}{|\Delta z|}\varepsilon L < \frac{1}{|\Delta z|}\varepsilon(|\Delta x| + |\Delta y|) < 2\varepsilon$$

Since ε is arbitrarily small, the limit of the left side as $\Delta z \to 0$ is zero and therefore $F'(z_0) = f(z_0)$.

Conversely, suppose a single-valued antiderivative for f exists throughout D. We denote it by $F(z) = U(x, y) + iV(x, y)$. Therefore $U_x = V_y = \text{Re}\, f$ and $V_x = -U_y = \text{Im}\, f$, and hence

$$\begin{aligned}\int_C f(z)\,dz &= \int_C (U_x\,dx + U_y\,dy) + i\int_C (V_x\,dx + V_y\,dy) \\ &= \int_{t_1}^{t_2} \frac{d}{dt}U(x, y)\,dt + i\int_{t_1}^{t_2} \frac{d}{dt}V(x, y)\,dt\end{aligned}$$

where t_2 and t_1 are the parameter values corresponding to the contour endpoints (x_2, y_2) and (x_1, y_1), respectively. Then

$$\begin{aligned}\int_C f(z)\,dz &= U(x_2, y_2) - U(x_1, y_1) + i(V(x_2, y_2) - V(x_1, y_1)) \\ &= F(z_2) - F(z_1)\end{aligned}$$

which tells us that the value of the integral is dependent only on the difference of the antiderivative values at the endpoints of C. ■

The importance of this theorem and the immediately preceding result should not be underestimated. Taken together, they show that the validity of the fundamental theorem of integral calculus for a complex function f in a domain D is intimately related to the vanishing of $\oint_C f(z)\,dz$ for all closed contours in D. In the next section we make basic connections of these notions to the *analyticity* of f in D.

In the meantime, however, some applications should be helpful in fixing the ideas embodied in Theorem 14.5. Note that in each case we must be sure that the antiderivative F of f is single-valued throughout D. In Example 14.8, this necessitates the use of a single-valued branch of a multiple-valued relation.

Example 14.6 Evaluate $\int_C z^2\,dz$ where C is the contour

$$z(t) = \sin^2 \pi t + i\cos 2\pi t$$

from $t = 0$ to $t = \frac{1}{2}$.

Solution The endpoints of the contour are i and $1 - i$ and an antiderivative of z^2 throughout the plane is $z^3/3$. Therefore

$$\int_C z^2\,dz = \frac{z^3}{3}\bigg|_i^{1-i} = \frac{(1-i)^3}{3} - \frac{i^3}{3} = \frac{-2-i}{3} \qquad \square$$

Example 14.7 Evaluate $\oint_C z^m\,dz$ where C is the unit circle $|z| = 1$ and m is an integer, with $m \neq -1$.

Solution If $m \neq -1$, a single-valued antiderivative throughout the domain excluding the origin is $z^{m+1}/(m+1)$. Therefore $\oint_C z^m\,dz = 0$. $\square$

Example 14.8 Evaluate $\oint_C \frac{1}{z}\,dz$, where C is the unit circle $|z| = 1$ traversed in the counterclockwise direction, by the method of finding antiderivatives.

Solution Care must be taken here since any antiderivative of $1/z$ will involve $\ln z$, which is intrinsically multivalued. Moreover, for any single-valued branch of $\ln z$, there is a ray from the origin that is a branch cut and thus a line of discontinuity. If the integration is over a path enclosing the origin as it is in this case, this line of discontinuity is crossed, and Theorem 14.4 does not apply.

We get around this difficulty by splitting the circle into upper and lower semicircles C_U and C_L, and thus

$$\oint_C \frac{1}{z}\,dz = \int_{C_U} \frac{1}{z}\,dz + \int_{C_L} \frac{1}{z}\,dz$$

Over C_U, we use the antiderivative whose branch cut is, say, the negative imaginary axis,

$$F_1(z) = \log_e|z| + i\arg z; \qquad -\frac{\pi}{2} < \arg z \leqq \frac{3\pi}{2}$$

Over C_L, we place the branch cut on the positive y-axis,

$$F_2(z) = \log_e|z| + i\arg z; \qquad \frac{\pi}{2} < \arg z \leqq \frac{5\pi}{2}$$

Using the result of Theorem 14.4 on each integral,

$$\begin{aligned}\oint_C \frac{1}{z}\,dz &= F_1(-1) - F_1(1) + F_2(1) - F_2(-1)\\ &= \pi i - 0 + 2\pi i - \pi i\\ &= 2\pi i\end{aligned}$$

in agreement with the value obtained previously in Example 14.4. $\square$

Exercises for Section 14.3

1. Evaluate $\int_C \sin z\,dz$, where C is the parabola $y = x^2$ from $z = 0$ to $z = 1 + i$.
2. Evaluate $\int_C e^z\,dz$, where C is the straight-line segment from 0 to πi.
3. Find the value of $\int_C z^2\,dz$, where C is the lower semicircle $|z| = 1$ from -1 to 1.
4. Evaluate $\int_C \pi \cos \pi z\,dz$ over the straight-line segments from:
   (a) 1 to $1 + 3i$ **(b)** i to $2 + i$
5. Evaluate $\int_C z^{-1}\,dz$, where C is the right semicircular contour $|z| = 1$ traversed counterclockwise from $-i$ to i.
6. Find the value of $\oint_C z^{-2}\,dz$, where C is the unit circle $|z| = 1$.
7. Evaluate $\oint_C ze^z\,dz$, where C is the circle $|z - 1| = 2$.
8. Evaluate $\int_C z^{1/2}\,dz$, if C is the upper semicircle $|z| = 1$ from 1 to -1 and for $z^{1/2}$ we choose the branch obtained from the principal values of $\arg z$.
9. Repeat Exercise 8 using the branch of $z^{1/2}$ corresponding to $\pi < \arg z \leqq 3\pi$.
10. Evaluate $\int_C z^2\,dz$, where C is the contour $z(t) = t^3 + \sin^2 \pi t$ from $t = 1$ to $t = 2$.
11. Evaluate $\int_C z^2\,dz$, where C is the straight-line segment connecting 0 to $1 + i$.

In Exercises 12–13, evaluate the given integrals over *straight-line segments* connecting the two points indicated:

12. **(a)** $\int_C \sin 2z\,dz$, from π to 3π **(b)** $\int_C z \sin z\,dz$, from 0 to 1
    (c) $\displaystyle\int_C \frac{1}{z}\,dz$, from 1 to $3 + i$
13. **(a)** $\int_C e^{3zi}\,dz$, from $1 + 2\pi i$ to $1 - 2\pi i$
    (b) $\int_C (z^3 + z^2 + z + 1)\,dz$, from 0 to $3i$
    (c) $\displaystyle\int_C \frac{1}{z^2 + 4}\,dz$, from 0 to i
14. Find the value of $\displaystyle\oint_C \frac{dz}{z^2 + 1}$, where C is the circle $|z - 2| = 3$.
15. Evaluate $\displaystyle\oint_C \frac{dz}{z^2(z^2 + 4)}$, where C is the unit circle.
16. Find the value of $\displaystyle\oint_C \left(\frac{1}{z - 2} + \frac{1}{(z - 2)^2}\right) dz$, where C is the circle $|z - 2| = 3$ traversed once in the counterclockwise direction.
17. Evaluate $\oint_C (z^2 - z^{-1} + z^{-2})\,dz$, where C is the unit circle traversed once in the counterclockwise direction.

In precisely the same fashion as in calculations of the work done in particle motion (see the previous set of exercises), the circulation

$$\Gamma(C) = \oint_C \mathbf{v} \cdot d\mathbf{r}$$

around a closed curve C in a two-dimensional fluid flow as introduced in Section 14.1 can be re-expressed as a complex line integral. If $\mathbf{v} = u(x, y)\mathbf{i} - v(x, y)\mathbf{j}$, the appropriate expression is

$$\Gamma(C) = \operatorname{Re} \oint_C f(z)\,dz$$

where $f(z) = u(x, y) + iv(x, y)$.

In each of the following exercises, a particular velocity field is given in vector notation. Use the given procedure to determine a complex line integral for the circulation around the unit circle $|z| = 1$, and then employ Theorem 14.5 to evaluate that line integral in those cases where the theorem is applicable.

18. **(a)** $\mathbf{v} = U_0\mathbf{i}$ **(b)** $\mathbf{v} = r\mathbf{u}_r$

19. **(a)** $\mathbf{v} = 2axy\mathbf{i} + a(x^2 - y^2)\mathbf{j}$ **(b)** $\mathbf{v} = r\cos 2\theta\,\mathbf{u}_r - r\sin 2\theta\,\mathbf{u}_\theta$

20. $\mathbf{v} = \dfrac{\theta}{r}\mathbf{u}_r + \dfrac{1}{r}\ln r\mathbf{u}_\theta$

21. $\mathbf{v} = \left(1 - \dfrac{a^2}{r^2}\right)\cos\theta\,\mathbf{u}_r - \left(1 + \dfrac{a^2}{r^2}\right)\sin\theta\,\mathbf{u}_\theta$

14.4 The Cauchy-Goursat Theorem

In the previous section we saw that if f has a (single-valued) antiderivative throughout a domain D and if C is any closed path in D, then

$$\oint_C f(z)\,dz = 0$$

In this section we show that if f is analytic in D, then this same result is automatically true for each closed contour C whose interior lies completely within D *without the assumption* that f has an antiderivative. Later, using this theorem, we will be able to *prove* that antidifferentiability of f is a direct consequence of f being analytic in a simply connected domain D.*

The Cauchy-Goursat theorem, sometimes simply called Cauchy's theorem, is one of the cornerstones of mathematical progress. The result was first published by Cauchy in 1825 under the hypothesis not only that f be analytic (that is, f' exists) throughout a simply connected domain, D, but also that f' be continuous. The proof which we give in this section is due to Cauchy. It is a straightforward application of Green's theorem and the Cauchy-Riemann conditions.

A new idea for a proof, which avoided the continuity assumption on f', was introduced by Goursat almost 60 years later. The resulting proof was quite delicate; modified versions are found in most modern texts on functions of a complex variable. The theorem is a dramatic example of the importance of assuming as few hypotheses as necessary. For, by *not* making the assumption of continuity for f', it is possible subsequently to *prove* that f' *is* indeed continuous.

We state this important theorem in its stronger form. For simplicity, however, we establish the result only under the added assumption of the continuity of f'.

Theorem 14.6 *Cauchy-Goursat Theorem.* Let $f(z) = u(x, y) + iv(x, y)$ be analytic throughout a domain D of the complex z-plane. Then

---

*A domain D in the plane is *simply connected* if it has the property that every closed contour lying entirely within the domain encloses *only* points of the domain. Otherwise, it is called *multiply connected*. See Section 11.8 for another characterization.

$$\oint f(z)\,dz = 0$$

for every closed contour C that lies in D and has its interior completely within D.

Proof (Due to Cauchy)

$$\oint_C f(z)\,dz = \oint_C (u(x, y)\,dx - v(x, y)\,dy) + i\oint_C (u(x, y)\,dy + v(x, y)\,dx)$$

If we assume that f' is continuous in D, then Green's theorem in the plane (see Theorem 11.8) may be applied to each of the two real line integrals.

$$\oint_C f(z)\,dz = \iint_R [-v_x(x, y) - u_y(x, y)]\,dx\,dy + i\iint_R [u_x(x, y) - v_y(x, y)]\,dx\,dy$$

where R is the domain interior to C. Because of the Cauchy-Riemann conditions, the integrand of each of the double integrals is zero and hence

$$\oint_C f(z)\,dz = 0$$ ■

Corollary The integral $\int_C f(z)\,dz$ is *path independent* in any simply connected domain throughout which f is analytic.

Proof Obvious.

Corollary If f is analytic in a simply connected domain D, then f has a continuous antiderivative throughout D.

Proof See the next set of exercises. (*Hint:* Consider this theorem and the two theorems of the previous section.)

Example 14.9 Evaluate $\oint_C e^z\,dz$ for any closed contour C in the finite z-plane.

Solution Since e^z is analytic everywhere, $\oint_C e^z\,dz = 0$. □

Example 14.10 Evaluate $\displaystyle\oint_C \frac{dz}{z}$ if C is the circle $|z - 2| = 1$.

Solution The value is zero, since the function is analytic within and on a circle of radius 1 centered at $z = 2$. □

The Cauchy-Goursat theorem and its consequences can also be used in the evaluation of *real* integrals. We give one important illustration here, saving further discussion until Chapter 16.

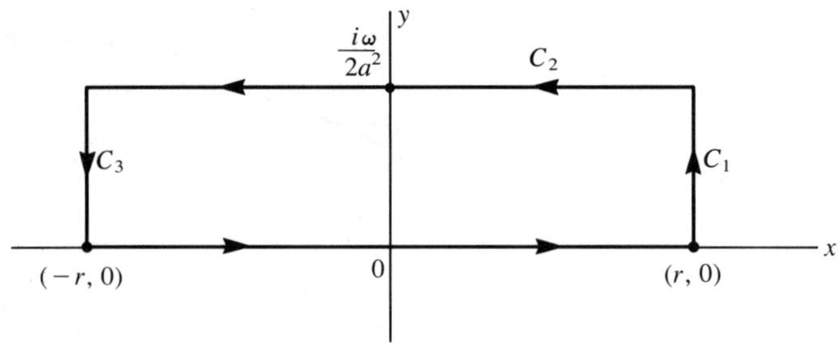

Figure 14.5 *Integration contour C*

Example 14.11 Evaluate $\int_0^\infty e^{-a^2x^2} \cos \omega x \, dx$ by complex integration.

Solution The given improper integral is well defined since

$$\left| \int_0^\infty e^{-a^2x^2} \cos \omega x \, dx \right| \leq \int_0^\infty e^{-a^2x^2} \, dx = \frac{\sqrt{\pi}}{2a}$$

Furthermore,

$$2 \int_0^\infty e^{-a^2x^2} \cos \omega x \, dx = \int_{-\infty}^\infty e^{-a^2x^2} e^{i\omega x} \, dx$$

since $\sin \omega x$ is an odd function. We analyze the contour integral $\oint_C e^{-a^2z^2} e^{i\omega z} \, dz$, therefore, where C is the contour shown in Figure 14.5. By the Cauchy-Goursat theorem, the integration around the entire contour is zero. Hence, using some obvious shorthand notation,

$$\begin{aligned} \int_{-r}^{r} e^{-a^2x^2} e^{i\omega x} \, dx &= -\left(\int_{C_1} + \int_{C_2} + \int_{C_3} \right) e^{-a^2z^2} e^{i\omega z} \, dz \\ &= e^{-\omega^2/4a^2} \int_{-r}^{r} e^{-a^2x^2} \, dx - \left(\int_{C_1} + \int_{C_3} \right) e^{-a^2z^2} e^{i\omega z} \, dz \end{aligned}$$

For the integrals over C_1 and C_3, Theorem 14.2 implies that they are each less than constants times $e^{-a^2r^2}$. As $r \to \infty$, we thus deduce

$$\begin{aligned} \int_0^\infty e^{-a^2x^2} \cos \omega x \, dx &= \frac{1}{2} \int_{-\infty}^\infty e^{-a^2x^2} e^{i\omega x} \, dx = \frac{1}{2} e^{-\omega^2/4a^2} \int_{-\infty}^\infty e^{-a^2x^2} \, dx \\ &= e^{-\omega^2/4a^2} \left(\frac{\sqrt{\pi}}{2a} \right) \end{aligned}$$

□

The integral in the above example turns out to be essentially the Fourier transform of the Gaussian (normal) distribution function (see Sections 16.5 and 16.7). By a simple change of variable the result obtained can also be shown to be equivalent to the final entry in the short table of Laplace transforms given previously on p. 151.

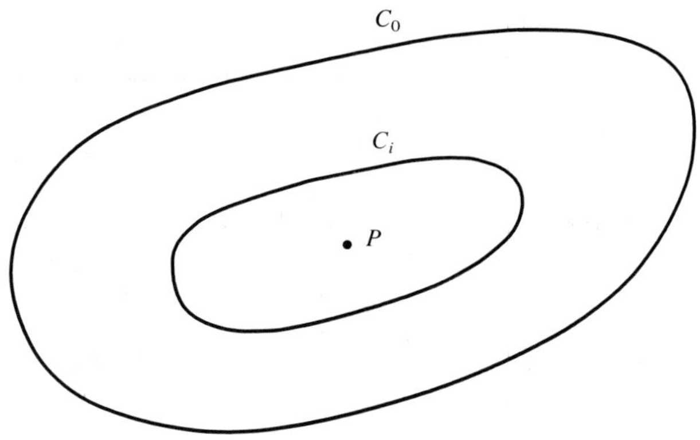

Figure 14.6 *Original curves and domain for Theorem 14.7*

14.5 Principle of Path Deformation

In addition to its utility in the evaluation of real and complex integrals, the Cauchy-Goursat theorem leads to a number of significant fundamental conclusions regarding the nature of analytic functions. In this and the remaining sections of the chapter we consider several of the more important outgrowths. Here we discuss the circumstances under which the shape of a closed contour does not affect the value of the line integral. (We have seen earlier with real line integrals that this property can be advantageous in practical problems.) As one immediate consequence, if the integrand is analytic, rather arbitrary closed contours may be replaced by circles (in fact *any* other closed contour is allowable).

We state and prove a theorem for the case in which f is analytic between two closed contours. The extension to other types of domains should be obvious.

Theorem 14.7 Let C_0 and C_i be two closed contours in a domain D (see Figure 14.6) and suppose f is analytic at all points between and on C_0 and C_i. Then*

$$\oint_{\substack{C_0 \\ (\text{ccw})}} f(z)\,dz = \oint_{\substack{C_i \\ (\text{ccw})}} f(z)\,dz$$

Proof A picture of the situation is shown in Figure 14.6, where we have assumed at least the possibility of f being nonanalytic at P. We make two cuts C_1 and C_2, as shown in Figure 14.7, thereby establishing contours C' and C'', which separately bound simply connected domains throughout which f is analytic. Therefore the Cauchy-Goursat theorem may be applied to each one.

---

* The notation *ccw* and *cw* for counterclockwise and clockwise orientation of the curves, respectively, is introduced at this stage merely for emphasis. As was the case in Chapter 11, the counterclockwise direction for integration along closed contours will be assumed if there is no statement or notation to the contrary.

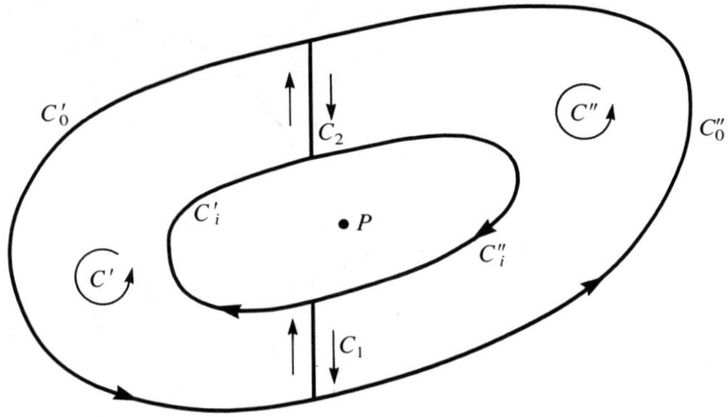

Figure 14.7 *Domain after insertion of cuts*

$$\oint_{\substack{C' \\ \text{(ccw)}}} f(z)\,dz = \oint_{\substack{C_0' \\ \text{(ccw)}}} f(z)\,dz + \int_{\substack{C_1 \\ \text{(up)}}} f(z)\,dz + \int_{\substack{C_i' \\ \text{(cw)}}} f(z)\,dz + \int_{\substack{C_2 \\ \text{(up)}}} f(z)\,dz = 0$$

$$\oint_{\substack{C'' \\ \text{(ccw)}}} f(z)\,dz = \int_{\substack{C_0'' \\ \text{(ccw)}}} f(z)\,dz + \int_{\substack{C_2 \\ \text{(down)}}} f(z)\,dz + \int_{\substack{C_i'' \\ \text{(cw)}}} f(z)\,dz + \int_{\substack{C_1 \\ \text{(down)}}} f(z)\,dz = 0$$

where the sense of the integration over each curve is as indicated. Adding the members of the last two equalities, we obtain

$$\oint_{\substack{C_0 \\ \text{(ccw)}}} f(z)\,dz + \oint_{\substack{C_i \\ \text{(cw)}}} f(z)\,dz = 0$$

which means that

$$\oint_{\substack{C_0 \\ \text{(ccw)}}} f(z)\,dz = \oint_{\substack{C_i \\ \text{(ccw)}}} f(z)\,dz \qquad \blacksquare$$

The content of the previous theorem is a special case of the principle of *path deformation*: If a contour C_0 is obtained from C_i by a continuous deformation, the value of the line integral will not change as long as none of the deformed contours passes through a point where $f(z)$ is not analytic. The principle, in slightly different form, was discussed in Section 14.3.

The extension of this idea to a situation that would require more cuts is shown in Figure 14.8. It is easy to show that

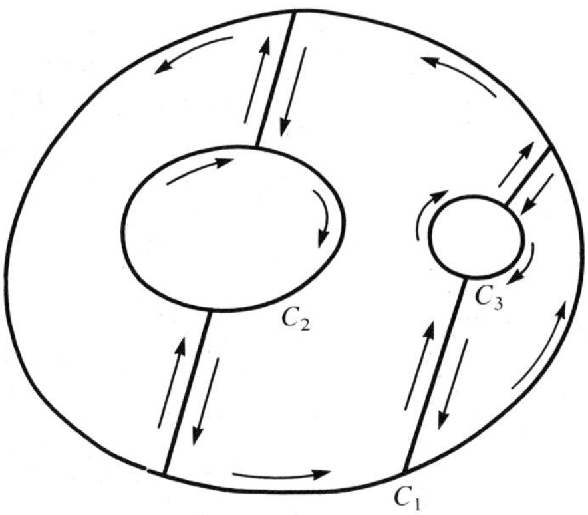

Figure 14.8 *Domain requiring four cuts*

$$\oint_{\substack{C_1 \\ \text{(ccw)}}} f(z)\,dz = \oint_{\substack{C_2 \\ \text{(ccw)}}} f(z)\,dz + \oint_{\substack{C_3 \\ \text{(ccw)}}} f(z)\,dz$$

Example 14.12 Let C be *any* closed contour, traversed in the counterclockwise direction, that encloses the origin. Evaluate $\oint_C \frac{1}{z}\,dz$.

Solution By the principle of path deformation,

$$\oint_C \frac{1}{z}\,dz = \oint_{C_1} \frac{1}{z}\,dz$$

where C_1 is the unit circle $|z| = 1$. By Example 14.4 or 14.8 the value of the integral is $2\pi i$. □

Example 14.13 Evaluate

$$\oint_C \frac{2z - 1}{z^2 - z}\,dz$$

where C is a contour, traversed in the counterclockwise sense, that contains the points 0 and 1 in its interior.

Solution By the usual partial fractions expansion,

$$\oint_C \frac{2z - 1}{z^2 - z}\,dz = \oint_C \frac{dz}{z} + \oint_C \frac{dz}{z - 1}$$

The first integral has value $2\pi i$ from the previous example. In the second integral, we replace the contour C by a circle C_2 of radius 1 about $z = 1$. This circle may be represented by $z(t) = 1 + e^{it}$. With this parametrization it is easy to show that

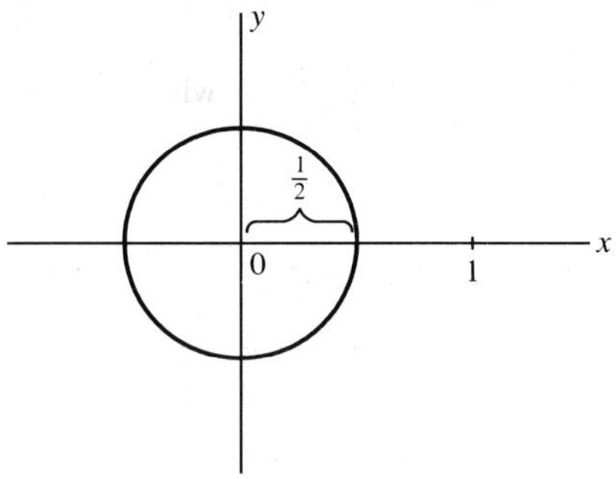

Figure 14.9

the value of the second integral is also $2\pi i$ and hence

$$\oint_C \frac{2z-1}{z^2-z}\,dz = 4\pi i \qquad \square$$

Example 14.14 Evaluate

$$\oint_C \frac{2z-1}{z^2-z}\,dz$$

where C is the circle $|z| = \frac{1}{2}$ traversed in the counterclockwise direction. (See Figure 14.9.)

Solution As in the previous example,

$$\oint_C \frac{2z-1}{z^2-z}\,dz = \oint_C \frac{dz}{z} + \oint_C \frac{dz}{z-1}$$

The first integral has value $2\pi i$ as before, but the second integral is zero since the integrand is analytic within and on C. Therefore

$$\oint_C \frac{2z-1}{z^2-z}\,dz = 2\pi i \qquad \square$$

Exercises for Sections 14.4 and 14.5

1. Show that each of the following integrals is zero, where C is the circle $|z| = 1$:

(a) $\oint_C z^3 e^{-z}\,dz$ **(b)** $\oint_C \frac{z}{z^2+9}\,dz$

(c) $\oint_C \frac{dz}{\cos z}$ **(d)** $\oint_C \ln(z+2)\,dz$

(e) $\oint_C \frac{e^z - e^{-z}}{e^z + e^{-z}}\,dz$

2. Find the value of $\oint_C \frac{z+1}{z^2}\,dz$, where C is as follows:

(a) $|z| = \frac{1}{2}$ **(b)** $|z - 1| = 3$ **(c)** $|z - 1| = \frac{1}{2}$ **(d)** $|z + i| = \frac{1}{2}$

3. Evaluate $\oint_C \frac{z\,dz}{z - 2}$, where C is as follows and where the curves are traversed in the clockwise sense:

(a) $|z| = 1$ **(b)** $|z - 1| = 2$ **(c)** $|z - 1| = \frac{1}{2}$ **(d)** $|z - i| = 3$

4. Find the value of $\oint_C \frac{dz}{1 - z^2}$, where C is as follows and where the curves are traversed in the clockwise sense:

(a) $|z| = \frac{1}{2}$ **(b)** $|z| = 2$ **(c)** $|z - 1| = 1$ **(d)** $|z + 1| = 1$

5. Evaluate $\oint_C \frac{z + 2}{z^2 + z}\,dz$, where C is as follows:

(a) $|z| = \frac{1}{2}$ **(b)** $|z - 1| = 3$ **(c)** $|z - 1| = \frac{1}{2}$ **(d)** $|z - i| = \frac{1}{2}$

6. When does the Cauchy-Goursat theorem apply to $\oint_C z^m\,dz$? Explain.

7. Evaluate $\oint_C \frac{dz}{z}$, where $C = C_1 \cup C_2$, C_1 is the circle $|z| = 1$ traversed counterclockwise and C_2 is the circle $|z| = 2$ traversed clockwise.

8. Evaluate

$$\oint_C \frac{dz}{(z - 1)(z - 2)}$$

where C is as follows:

(a) $|z| = \frac{1}{2}$ **(b)** $|z - 1| = \frac{1}{2}$ **(c)** $|z| = \frac{3}{2}$
(d) $|z - i| = 1$ **(e)** $|z - 1| = 2$ **(f)** $|z| = 4$

9. Find the value of $\oint_C \frac{2z^2 - z + 1}{(z - 1)^2(z + 1)}\,dz$, where $C = C_1 \cup C_2$, C_1 is the circle $|z + 1| = 1$ traversed counterclockwise and C_2 is the circle $|z - 1| = 1$ traversed clockwise.

10. For which of the following complex functions $f = u(x, y) + iv(x, y)$ is

$$\oint_C (u\,dx - v\,dy) = 0$$

for every closed contour C in the finite z-plane?

(a) $f = y^3 - 3x^2y + i(x^3 - 3xy^2)$ **(b)** $f = \text{Tan}^{-1}(y/x) - \frac{i}{2}\ln(x^2 + y^2)$

(c) $f = \cos x \cosh y + i \sin x \sinh y$ **(d)** $f = \frac{y + ix}{x^2 + y^2}$

11. For each of the functions given in Exercise 10, characterize the simply connected domains in which

$$\int_C f\,dz$$

is path independent.

12. Evaluate

$$\oint_C \frac{dz}{z^2(z^2 + 4)}$$

where C is as follows:

(a) $|z| = \frac{3}{2}$ **(b)** $|z - 2i| = 1$ **(c)** $|z + 2| = 3$

13. Which of the following domains are simply connected?
(a) The disk $|z - i| < 2$
(b) The semi-infinite strip $x < 0, 0 < y < 1$
(c) The annulus $1 < |z + 1| < 2$
(d) The exterior of the ellipse $x^2 + 9y^2 = 1$
(e) The set of points in the plane except those for which both $x \geq 0$ and $y = 0$
(f) The sector $-\pi/4 < \arg z < \pi/4$

14. For which of the following complex functions is

$$\oint_{C_1} f(z)\,dz = \oint_{C_2} f(z)\,dz$$

where C_1 and C_2 are the circles $|z - i| = 3$ and $|z| = 1$, respectively:
(a) $f(z) = e^{-z}(2 + z)$ **(b)** $f(z) = \text{Ln}(z + 2)$
(c) $f(z) = z^{-2}$ **(d)** $f(z) = \dfrac{\sin z}{z^2 - 7z + 10}$

15. Carry out the demonstration of the second corollary to Theorem 14.6.

16. State and prove a theorem concerning the *integration-by-parts* formula:

$$\int_C f'(z)g(z)\,dz = f(z)g(z)\Big|_{z_i}^{z_f} - \int_C f(z)g'(z)\,dz$$

where z_i and z_f are the initial and terminal points of the contour C, respectively.

14.6 Cauchy's Integral Formula

This is the second major consequence of Cauchy's theorem. The formula yields a means of representing an analytic function in terms of a line integral.

Theorem 14.8 *Cauchy Integral Formula.* Let f be analytic throughout a domain D of the complex z-plane, and let z_0 be a point of D. If C is a closed contour traversed in the counterclockwise sense which lies in D, has its interior completely within D, and encloses z_0, then

$$\frac{1}{2\pi i}\oint_C \frac{f(z)}{z - z_0}\,dz = f(z_0)$$

(See Figure 14.10.)

Proof

$$\oint_C \frac{f(z)}{z - z_0}\,dz = \oint_C \frac{f(z) - f(z_0)}{z - z_0}\,dz + f(z_0)\oint_C \frac{dz}{z - z_0}$$

The value of

$$\oint_C \frac{dz}{z - z_0} = 2\pi i$$

by the principle of path deformation and the fact that the value is $2\pi i$ if the contour is a circle. The proof will be completed, therefore, if we show that the

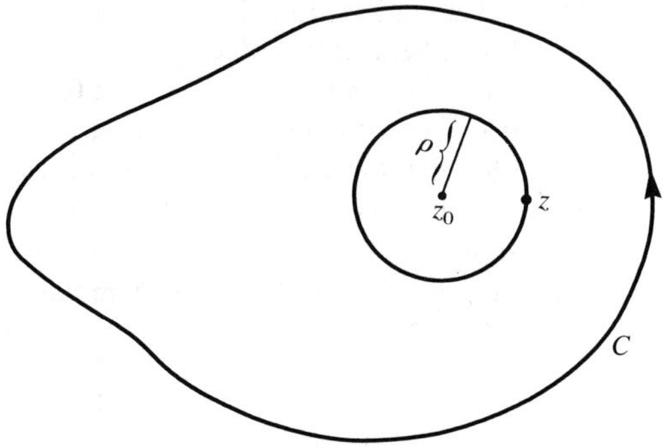

Figure 14.10

value of the first integral is zero. Toward this end, we replace C by a circle of radius ρ centered at z_0. By the continuity of f, there exists a δ such that $|f(z) - f(z_0)| < \varepsilon$ if $|z - z_0| < \delta$. Thus, if we choose ρ to be less than δ, we have

$$\left|\oint_{|z-z_0|=\rho} \frac{f(z) - f(z_0)}{z - z_0}\,dz\right| \leqq \oint_{|z-z_0|=\rho} \frac{|f(z) - f(z_0)|}{|z - z_0|}|dz|$$
$$\leqq \frac{\varepsilon}{\rho}\oint_{|z-z_0|=\rho} |dz| = 2\pi\varepsilon$$

Since ε can be made arbitrarily small, the value of the integral must be zero. ■

Example 14.15 Find the value of

$$\oint \frac{z^2 + 1}{z^2 - 1}\,dz$$

where C is:

(a) $|z - 1| = 1$ (b) $|z - \frac{1}{2}| = 1$ (c) $|z + 1| = 1$ (d) $|z - i| = 1$

Solution (a)

$$\oint_C \frac{z^2 + 1}{z^2 - 1}\,dz = \oint_C \frac{f(z)}{z - 1}\,dz$$

where $f(z) = (z^2 + 1)/(z + 1)$. Since f is analytic within and on the simply connected domain bounded by C and since $z = 1$ is in D,

$$\oint_C \frac{z^2 + 1}{z^2 - 1}\,dz = 2\pi i f(1) = 2\pi i$$

(b) This answer is the same as part (a) by the principle of path deformation.

(c) In this case we can write

$$\oint_C \frac{z^2 + 1}{z^2 - 1}\,dz = \oint_C \frac{g(z)}{z + 1}\,dz$$

where $g(z) = (z^2 + 1)/(z - 1)$. The function g is analytic within the circle $|z + 1| = 1$, and thus

$$\oint_C \frac{z^2+1}{z^2-1}\,dz = 2\pi i\, g(-1) = -2\pi i$$

(d) The value is zero by the Cauchy-Goursat theorem, since the integrand is analytic within and on the circle of radius 1 centered at $z = i$. □

The Cauchy integral formula tells us that *all* values of an analytic function in a simply connected domain are determined once the functional values on the curve which bounds the domain are defined. (Actually, this statement remains valid for multiply connected domains as well—we just need to recall how we used cuts in our analysis of path deformation.) The same statement is true for real harmonic functions.

14.7 Derivatives of an Analytic Function

We are now in a position to explore, at least briefly, the remarkable relationship that exists between integration and differentiation of analytic functions. Indeed, in what follows we use the Cauchy integral formula to demonstrate that an analytic function has derivatives of all orders. The actual formulas for the derivatives are not important computationally; rather, the theorem's importance is in the statement of existence. Note that the only assumption necessary for f is that f' exist throughout some domain D. No continuity assumptions are necessary for f'. This is particularly crucial since we will then be able to show that f' *is* continuous, a situation significantly different from that for functions of a real variable.

We do not include a proof of the theorem here since, in technique, it is similar to that used in the previous two theorems. Formally, the formulas may be obtained by "differentiating inside the integral sign."

Theorem 14.9 *Generalized Cauchy Integral Formula.* If f is analytic throughout a domain D, then it has derivatives of all orders which themselves are analytic in D. The specific formula for the nth-derivative at the point z_0 in D is given by the following line integral around a closed contour C in D which encloses z_0 and has its interior completely within D:

$$f^{(n)}(z_0) = \frac{n!}{2\pi i}\oint_C \frac{f(z)}{(z-z_0)^{n+1}}\,dz$$

(Do not forget the assumed positive orientation for the contour C!) ■

Example 14.16 Evaluate

$$\oint_C \frac{z^4+1}{(2z+1)^3}\,dz$$

where C is the unit circle $|z| = 1$.

Solution

$$\oint_C \frac{z^4+1}{(2z+1)^3}\,dz = \frac{1}{8}\oint_C \frac{z^4+1}{(z+\frac{1}{2})^3}\,dz = \frac{1}{8}\oint_C \frac{f(z)}{(z+\frac{1}{2})^3}\,dz$$

where $f(z) = z^4 + 1$. Since

$$\oint_C \frac{f(z)}{(z+\frac{1}{2})^3}\,dz = \frac{2\pi i}{2!} f''(-\tfrac{1}{2}) = \pi i(12z^2)|_{z=-1/2} = 3\pi i$$

the given integral is equal to $3\pi i/8$. □

With the use of Theorem 14.9 several very important corollaries can be established.

Corollary 1 (*Morera*) Let f be continuous on a domain D. If $\oint_C f(z)\,dz = 0$ around all closed contours interior to D, then f is analytic throughout D.

Proof If $\oint_C f(z)\,dz = 0$ around all closed contours in D, then the line integral $\int_C f(z)\,dz$ is path independent and hence f has an antiderivative F throughout D. But now we know that F is analytic in D and has analytic derivatives of all orders; in particular since F' is equal to f, f must actually be analytic in D. ■

Corollary 2 (*Cauchy's Inequality*) Let C be a circle of radius r with center at z_0 and let $M = \max_{z \text{ on } C} |f(z)|$. Then

$$|f^{(n)}(z_0)| \leqq \frac{n!M}{r^n} \qquad (n = 0, 1, 2, \ldots)$$

Proof

$$|f^{(n)}(z_0)| = \frac{n!}{2\pi}\left|\oint_C \frac{f(z)\,dz}{(z-z_0)^{n+1}}\right| \leqq \frac{n!}{2\pi} M \frac{1}{r^{n+1}} 2\pi r = \frac{n!M}{r^n}$$ ■

Cauchy's inequality, like many things in mathematics, is important because of results to which it leads. For example, we have the following corollary which is often called Liouville's theorem.

Corollary 3 (*Liouville's Theorem*) If f is analytic and bounded in absolute value for all finite z, then $f(z)$ is a constant.

Proof By Cauchy's inequality as applied to the first derivative, $|f'(z)| \leq M/r$ for arbitrary z. If M is bounded, say by M_1, then $|f'(z)| \leq M_1/r$, which goes to 0 as r becomes unbounded. Thus $f'(z) = 0$, which means $f(z)$ equals a constant. ∎

Liouville's theorem may be used to establish the fundamental theorem of algebra. To prove this theorem by algebraic methods is quite difficult but it is an easy consequence of the results of this section.

Theorem 14.10 *Fundamental Theorem of Algebra.* If $P(z)$ is a nonconstant polynomial in z,

$$P(z) = \sum_{k=0}^{n} a_k z^k, \qquad n \geqq 1, a_n \neq 0$$

then $P(z) = 0$ for at least one value of z.

Proof Suppose $P(z) \neq 0$ for all z. Then, the function

$$f(z) = \frac{1}{P(z)}$$

is everywhere analytic. In the exterior of increasingly large circles $|z| = M$, $|f(z)|$ is bounded, since $P(z)$ behaves more and more like $a_n z^n$. (Technically speaking, there exists an M such that $|f(z)| < \dfrac{2}{|a_n||z|^n} \leq 1$, say, for $|z| > M$.) For $|z| \leq M$, $|f(z)|$ is also bounded, since by continuity $P(z)$ is bounded away from zero on any bounded domain. We conclude that $|f(z)|$ is bounded for all finite z and, by Liouville's theorem, $f(z)$ is a constant. But this would mean $P(z)$ is also a constant, which is an obvious contradiction to the hypothesis. Hence, $P(z) = 0$ for at least one value of z. ∎

Exercises for Sections 14.6 and 14.7

1. Evaluate $\oint_C \frac{e^z}{z}\,dz$, where C is the circle $|z| = 2$.
2. Find the value of $\oint_C \frac{dz}{4z^2 + 1}$, where C is the circle $|z| = 1$.
3. Evaluate the following integrals over the contour C formed by the lines $x = \pm 1$, $y = \pm 1$:

   (a) $\oint_C \frac{\sin z}{z}\,dz$ **(b)** $\oint_C \frac{\cos z}{z}\,dz$ **(c)** $\oint_C \frac{e^z}{2z - i}\,dz$

   (d) $\oint_C (\sin z + e^z)\,dz$ **(e)** $\oint_C \frac{\cosh z}{z}\,dz$
4. Evaluate $\oint_C \frac{dz}{z^2(z^2 + 9)}$, where C is:

   (a) $|z - 1| = 2$ **(b)** $|z - 1| = 4$

5. Evaluate

$$\oint_C \frac{z^2 - 1}{z^2 + 1}\,dz$$

where C is as follows:

(a) $|z - 1| = 1$ **(b)** $|z - i| = 1$ **(c)** $|z + i| = 1$
(d) $|z - 2i| = 2$ **(e)** $|z| = 2$

In Exercises 6–7, evaluate the given integrals over $|z| = 1$:

6. **(a)** $\oint_C \frac{e^{2z^3}}{z}\,dz$ **(b)** $\oint_C \frac{e^z - 1}{z}\,dz$ **(c)** $\oint_C \frac{\sec z}{z}\,dz$
(d) $\oint_C \frac{z^2 - 1}{z^2 + 2}\,dz$ **(e)** $\oint_C \frac{z^3}{\left(z - \frac{i}{2}\right)}\,dz$

7. **(a)** $\oint_C \frac{\cos z}{z^2}\,dz$ **(b)** $\oint_C \frac{\cos z}{z^3}\,dz$ **(c)** $\oint_C \frac{z^8 + 1}{(z - \frac{1}{2})^7}\,dz$
(d) $\oint_C \frac{e^z}{z^4}\,dz$ **(e)** $\oint_C \frac{\sin z}{z^4}\,dz$

8. Find the value of $\oint_C \frac{3z^2 + 7z + 1}{a - z}\,dz$, where C is the circle $|z - i| = 2$, when:
(a) $a = 0$ **(b)** $a = i$ **(c)** $a = 2$

9. Evaluate $\oint_C \frac{3z^2 + 7z + 1}{(a - z)^2}\,dz$, where C is the circle $|z - i| = 2$, when:
(a) $a = 0$ **(b)** $a = i$ **(c)** $a = 2$

10. Find the value of $\oint_C \frac{3z^2 + z}{z^2 - 1}\,dz$, where C is as follows:
(a) $|z| = 2$ **(b)** $|z - 1| = 1$

11. Evaluate $\oint_C \frac{dz}{z^2(z^2 + 4)}$, where C is as follows:
(a) $|z + 2i| = 1$ **(b)** $|z| = 1$

12. Exhibit a function of a real variable for which $f'(x_0)$ exists but f' is not continuous at x_0. Repeat for complex functions.

13. Show that

$$\oint_C \frac{dz}{(z - z_0)^n} = 0 \qquad (n = 2, 3, \ldots)$$

for any closed contour C as long as the point z_0 does not lie on C.

14. Starting from the Cauchy integral formula, show that if $f(z)$ is analytic inside and on the circle C_R given by $|z - z_0| = R$, then

$$f(z_0) = \frac{1}{2\pi}\int_0^{2\pi} f(z_0 + Re^{it})\,dt$$

(*Hint:* Parametrize C_R by $z = z_0 + Re^{it}$, $0 \le t \le 2\pi$.)

15. Let $f(z)$ and $g(z)$ be analytic inside and on the closed contour C. Use Cauchy's integral formula to demonstrate that if $f(z) = g(z)$ for all z on C, then $f(z)$ and $g(z)$ are identical for all z in the interior of C.

16. Let $f(z)$ be analytic inside and on the closed contour C. Verify from the generalized Cauchy integral formula that

$$\oint_C \frac{f'(z)}{z - z_0}\,dz = \oint_C \frac{f(z)}{(z - z_0)^2}\,dz$$

for all z_0 not on C.

17. Extend the result of Exercise 16 to show that if $f(z)$ is analytic inside and on the closed contour C,

$$\oint_C \frac{f^{(n)}(z)}{z - z_0}\,dz = n!\oint_C \frac{f(z)}{(z - z_0)^{n+1}}\,dz$$

18. Let $P(z)$ be a polynomial which is nonvanishing on the closed contour C and has a single zero z_0 of multiplicity n within C. Prove that

$$\frac{1}{2\pi i}\oint_C \frac{P'(z)}{P(z)}\,dz = n$$

(*Hint:* Rewrite $P(z)$ as $(z - z_0)^n Q(z)$.)

19. Generalize the result of Exercise 18 to show that if $P(z)$ is a polynomial that is nonvanishing on the closed contour C, then

$$\frac{1}{2\pi i}\oint_C \frac{P'(z)}{P(z)}\,dz = N$$

where N is the total number of zeros of P (counting multiplicity) that lie inside C.

Occasionally the evaluation of a given contour integral can be simplified by making a change of variables in the integrand. For example, if in the integral

$$I = \oint_C \frac{dz}{z^2(z - 1)^3}$$

where C is the circle $|z| = 2$, we set $z = 1/w$, then

$$\begin{aligned} I &= \oint_{C'} \frac{(-1/w^2)\,dw}{(1/w)^2[(1/w) - 1]^3} \\ &= \oint_{C'} \frac{w^3\,dw}{(w - 1)^3} \end{aligned}$$

where C' is the circle $|w| = \frac{1}{2}$, traversed in the clockwise (negative) direction. It is now easy to see that $I = 0$, a result that would be considerably more difficult to derive if the integral were left in its original form.

20. Carry out the substitution $z = 1/w$ and then evaluate

$$I = \oint_C \frac{z + i}{z^3 + 2z^2}\,dz$$

where C is the circle $|z| = 4$.

21. Repeat Exercise 20 for the case of the integral

$$I = \oint_C \frac{5z^2 + 2z + 1}{(z - i)^3}\,dz$$

where C is the circle $|z| = 3$.

22. Evaluate the integral in Exercise 21 in alternative fashion using the generalized Cauchy integral formula.

Exercise 14 has an obvious analog for harmonic functions called the *mean-value theorem.* (Recall the earlier discussion on the Poisson integral formula in the exercises for Section 8.5.) There is also a uniqueness result for harmonic functions corresponding to Exercise 15. The proof in the latter case usually employs the following result, which is yet another consequence of the Cauchy integral formula:

> *Maximum-minimum principle*: A function harmonic in a bounded domain, and continuous up to and including the boundary, takes on its maximum and minimum values on the boundary.

23. Use the maximum-minimum principle to establish a uniqueness result for harmonic functions that is analogous to Exercise 15.
24. Show by example that the maximum-minimum principle need not hold in unbounded domains. (*Hint:* Consider the imaginary part of a mapping to the upper half-plane.)
25. Discuss the implications of the maximum-minimum principle and the result of Exercise 23 for applications in electrostatics and steady-state heat flow.

14.8 Looking Ahead

The principal results of this chapter, the Cauchy-Goursat theorem and the Cauchy integral formula, are useful in two major ways—*directly*, through immediate application in various special situations, and *indirectly* (and perhaps more importantly), through the utility of their numerous consequences. In Section 14.5 we discussed path deformation; this valuable principle is applied over and over again in each of the next two chapters. The Cauchy inequalities and Liouville's theorem were derived in Section 14.7; other straightforward outgrowths of the Cauchy integral formula include the maximum modulus theorem and Schwarz's lemma (see [8], [9], for example). All these results have relevance in the theory of entire functions, about which we will have more to say in Section 15.10. The Cauchy integral formula itself is used as the basis for the development of power series and Laurent series representations of analytic functions in the next chapter; it also undergirds all of the residue calculus and its application (as we shall see in Chapter 16) to the evaluation of real and complex integrals.

The remark at the end of Section 14.6 (p. 544) suggests an interpretation of the Cauchy integral formula as a representation of an analytic function within a domain in terms of its boundary values on the curve(s) that surrounds the domain. For the interior of the circle, a representation called the Poisson integral formula can readily be derived for harmonic functions from this recognition (see [6], for example; also recall the alternative expression given earlier in Section 8.5, p. 286). Owing to the Riemann mapping theorem (see Section 13.8), analogous representations can be determined, and the Dirichlet problem solved for Laplace's equation with arbitrary continuous boundary values in general simply connected regions. (This represents a significant extension of what was possible using only Theorem 13.4.) Additional discussion of this useful technique appears in [2], [4], [10], and [11].

The concept of a complex line integral and the associated Cauchy theory presented in this chapter can be generalized in a number of respects. More advanced texts in analytic function theory such as [1], [3], [4], or [7] typically include extensions involving such notions as chains, cycles, winding numbers, homology, homotopy, and the like.

Intermediate and More-Advanced References

[1] Ahlfors, L. V. 1979. *Complex Analysis*. 3d ed. New York: McGraw-Hill.

[2] Carrier, G. F., et al. 1966. *Functions of a Complex Variable*. New York: McGraw-Hill.

[3] Curtiss, J. H. 1978. *Introduction to Functions of a Complex Variable*. New York: Marcel Dekker.

[4] DePree, J. D., and C. C. Oehring. 1969. *Elements of Complex Analysis*. Reading, Mass.: Addison-Wesley.

[5] Evgrafov, M. A. 1978. *Analytic Functions*. New York: Dover.

[6] Nehari, Z. 1968. *Introduction to Complex Analysis*. rev. ed. Boston: Allyn and Bacon.

[7] Nevanlinna, R., and V. Paatero. 1982. *Introduction to Complex Analysis*. 2d ed. New York: Chelsea.

[8] Saff, E. B., and A. D. Snider. 1976. *Fundamentals of Complex Analysis for Mathematics, Science, and Engineering*. Englewood Cliffs, N. J.: Prentice-Hall.

[9] Silverman, H. 1975. *Complex Variables*. Boston: Houghton Mifflin.

[10] Silverman, R. A. 1974. *Complex Analysis with Applications*. Englewood Cliffs, N. J.: Prentice-Hall.

[11] Weinberger, H. F. 1965. *A First Course in Partial Differential Equations with Complex Variables and Transform Methods*. Waltham, Mass.: Blaisdell.

Chapter 15

Complex Series

15.1 Introduction

In Chapters 13 and 14 we have shown that an analytic function of a complex variable has immediate utility in conformal mapping and consequently in solving the Dirichlet problem in two dimensions. In this chapter, which may seem a bit more theoretically oriented, we discuss how convergent power series and analytic functions are identified. One of our major goals is to relate complex series techniques to the theory of integration of analytic functions developed previously. Once we have done this, we will be able to use "residue calculus" methods in many applications, a number of which are considered in Chapter 16.

15.2 Elementary Definitions

Many of the results for complex functions and complex series are precisely the same as for real functions and real series. In a number of cases, even the proofs have the same form. We can afford, then, to be a bit brief and incomplete in this initial material. (Now is the time for refreshing any needed background in sequences and series.) We begin by making several definitions.

Definition 15.1 A *complex sequence* is a complex function whose domain is the set of positive integers. The range values are called "terms" of the sequence. We denote them by $z_1, z_2, z_3, \ldots, z_n, \ldots$.

As in the real case, the sequences of most interest are those whose terms have a limiting value in the sense of the following definition.

Definition 15.2 A complex sequence $\{z_n\}$ is said to *converge to the complex number c*, if for each $\varepsilon > 0$, there exists a positive integer N such that if $n > N$, then $|z_n - c| < \varepsilon$. We write

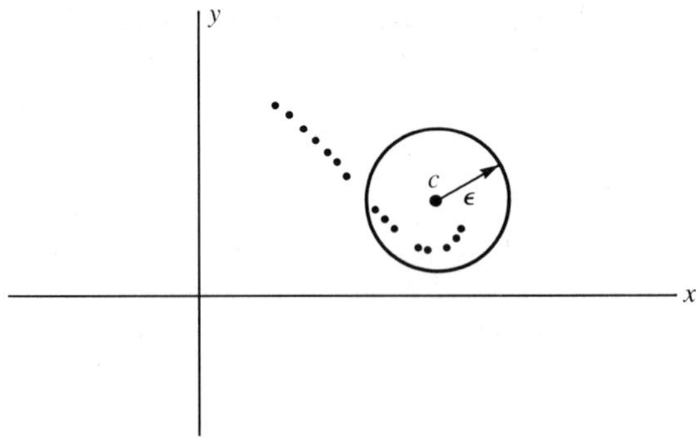

Figure 15.1 *Convergence of a complex sequence*

$$\lim_{n\to\infty} z_n = c$$

A sequence that does not converge is said to *diverge.* (See Figure 15.1.)

Example 15.1 Determine if the sequence whose general term is $z_n = (2n + 1)/ni$ converges.

Solution Since

$$\lim_{n\to\infty} \frac{2n + 1}{ni} = -2i$$

the sequence converges to $-2i$. □

Terms of a sequence are sometimes written $z_n = x_n + iy_n$. The limit of each of the real sequences $\{x_n\}_{n=1}^{\infty}$ and $\{y_n\}_{n=1}^{\infty}$ may often be found easily. The knowledge of these two limits is sufficient to determine the limit of the complex sequence $\{z_n\}_{n=1}^{\infty}$.

Theorem 15.1 A sequence of terms $z_n = x_n + iy_n$ converges to $c = a + ib$ if and only if $\lim_{n\to\infty} x_n = a$ and $\lim_{n\to\infty} y_n = b$.

Proof Prove the theorem as an exercise. ■

The definition of a complex infinite series is essentially the same as in the real case.

Definition 15.3 An *infinite series* is a sequence of *partial sums*

$$S_n = \sum_{k=1}^{n} z_k$$

where the terms z_k are complex numbers.* (The term "partial sum" refers to the sum of the first n terms of the formal expression $z_1 + z_2 + z_3 + \cdots$.)

*Sometimes the series will start with the term $k = 0$. There should be no difficulty whenever this occurs.

The limit $\lim_{n\to\infty} S_n$, if it exists, is called the *sum* of the series and is denoted by

$$\sum_{k=1}^{\infty} z_k$$

In this case, the infinite series is said to *converge*. If $\lim_{n\to\infty} S_n$ does not exist, the infinite series is said to *diverge*. It is sloppy, but standard, practice to use $\sum_{k=1}^{\infty} z_k$ to refer to the series itself as well as to the sum of the series.

Theorem 15.1 has its analog for series in the following theorem.

Theorem 15.2 Let $z_k = x_k + iy_k$. Then $\sum_{k=1}^{\infty} z_k$ converges to $a + ib$ if and only if $\sum_{k=1}^{\infty} x_k = a$ and $\sum_{k=1}^{\infty} y_k = b$. ■

Theorems 15.1 and 15.2 allow the application of virtually all the theory of real sequences and real series to the complex case.

To determine if a series converges, we analyze it directly or else apply one of the standard "tests."

Example 15.2 Consider

$$\sum_{k=0}^{\infty} z^k = 1 + z + z^2 + \cdots + z^k + \cdots$$

Show that this series converges for $|z| < 1$ and diverges otherwise.

Solution $S_n = 1 + z + z^2 + \cdots + z^n$ and $zS_n = z + z^2 + z^3 + \cdots + z^{n+1}$. Therefore

$$S_n - zS_n = 1 - z^{n+1}$$

and

$$S_n = \frac{1 - z^{n+1}}{1 - z}$$

Thus

$$\begin{aligned} \lim_{n\to\infty} S_n &= \frac{1}{1 - z}, \qquad \text{if} \quad |z| < 1 \\ &= \infty, \qquad \text{if} \quad |z| > 1 \end{aligned}$$

If $|z| = 1$ but $z \neq 1$, then $z = e^{i\theta}$ and

$$S_n = \frac{1 - e^{i(n+1)\theta}}{1 - e^{i\theta}}$$

which has no limit. If $z = 1$, $S_n = n$ and $\lim_{n\to\infty} S_n = \infty$. □

The series in the previous example is called the *geometric series*. It is very important, since many series that are not strictly geometric can be analyzed using the known results for geometric series.

Example 15.3 Find the value of $\sum_{k=0}^{\infty}(-z)^k$.

Solution By letting the $(-z)$ of this example take the place of z in the previous one, the series is seen to converge to $1/(1+z)$ for $|z| < 1$. □

Definition 15.4 A series $\sum_{k=1}^{\infty} z_k$ is said to *converge absolutely* if the series of positive terms $\sum_{k=1}^{\infty} |z_k|$ converges.

The importance of absolute convergence stems from the following two theorems whose proofs we omit.

Theorem 15.3 If the series $\sum_{k=1}^{\infty} z_k$ is absolutely convergent, it is convergent. ■

Theorem 15.4 *Comparison Test.* If $\sum_{k=1}^{\infty} b_k$ is a convergent series of nonnegative real terms and if $|z_k| < b_k$ for all k greater than some N, then $\sum_{k=1}^{\infty} z_k$ is (absolutely) convergent. ■

The following two tests for convergence of a series $\sum_{k=1}^{\infty} z_k$ are the same as for real series.

Ratio Test. If

$$\lim_{k\to\infty}\left|\frac{z_{k+1}}{z_k}\right| = L$$

then the series converges if $L < 1$ and diverges if $L > 1$.

Root Test. If

$$\lim_{k\to\infty}|z_k|^{1/k} = L$$

then the series converges if $L < 1$ and diverges if $L > 1$.

Although other tests are available, these two will be sufficient for most elementary applications. They both have the nice feature of being easy to apply.

Exercises for Sections 15.1 and 15.2

In Exercises 1–10, examine the given sequences for convergence or divergence.

1. $z_n = (-1)^n + \dfrac{1}{2^n}$

2. $z_n = \dfrac{(-1)^n}{2^n}$

3. $z_n = \dfrac{\cos n\pi}{n}$

4. $z_n = \sqrt{\sqrt{n}}$

5. $z_n = i^n$

6. $z_n = \dfrac{i^n}{n}$

7. $z_n = \dfrac{2n^2+n}{3n^2-1}$

8. $z_{2n} = 2, \quad z_{2n+1} = \dfrac{2n+1}{1+n^2}$

9. $z_{2n} = \dfrac{2n+1}{3n}, \quad z_{2n+1} = \dfrac{2n}{3n-5}$

10. $z_{2n} = \dfrac{1}{2n}, \quad z_{2n+1} = \dfrac{1}{3n}$

In Exercises 11–20, examine the given series for convergence or divergence.

11. $\sum_{n=1}^{\infty} \frac{n+1}{n2^n}$

12. $\sum_{n=0}^{\infty} \frac{1}{n!}$

13. $\sum_{n=1}^{\infty} \frac{n}{\sqrt{n}}$

14. $\sum_{n=1}^{\infty} \frac{1}{\sqrt{n}}$

15. $\sum_{n=0}^{\infty} \frac{1}{2^n}$

16. $\sum_{n=0}^{\infty} 2^n$

17. $\sum_{n=1}^{\infty} \frac{2^n}{n!}$

18. $\sum_{n=1}^{\infty} \frac{2n+1}{n^2}$

19. $\sum_{n=1}^{\infty} \frac{2n+1}{n^3}$

20. $\sum_{n=1}^{\infty} n^p$

21. Show that if $\sum_{n=1}^{\infty} a_n$ and $\sum_{n=1}^{\infty} b_n$ both converge, then $\sum_{n=1}^{\infty} (a_n + b_n)$ converges.

22. Demonstrate that if $\sum_{n=1}^{\infty} a_n$ converges, but $\sum_{n=1}^{\infty} b_n$ diverges, then $\sum_{n=1}^{\infty} (a_n + b_n)$ also diverges.

23. **(a)** Show that the partial sums of $\sum_{n=0}^{\infty} \frac{1}{2^n}$ are all less than 2.

(b) Find an upper bound for the partial sums of $\sum_{n=0}^{\infty} \frac{1}{a^n}$ where $a > 1$.

24. Determine the values of α for which the comparison test (Theorem 15.4) implies the convergence of

$$\sum_{n=1}^{\infty} \frac{(-1)^n \sqrt{n}}{(n^2 + 2n + 5)^\alpha}$$

In Exercises 25–32, examine the given series for absolute convergence.

25. $\sum_{n=1}^{\infty} \frac{(-1)^n}{\sqrt{n}}$

26. $\sum_{n=0}^{\infty} (-1)^n \frac{2n^2 - 3}{5n^3 + 1}$

27. $\sum_{n=0}^{\infty} \frac{(-1)^n}{n^2 + 2}$

28. $\sum_{n=1}^{\infty} (-1)^n \frac{n^2 - n^4}{n - 3n^6}$

29. $\sum_{n=0}^{\infty} (-1)^n \frac{3^n + 4^n}{4^n + 5^n}$

30. $\sum_{n=1}^{\infty} n^2(-\tfrac{1}{2})^n$

31. $\sum_{n=1}^{\infty} (-1)^n \frac{e^n}{n^p}$

32. $\sum_{n=1}^{\infty} (-1)^n \frac{1}{n\sqrt{n}}$

15.3 Series of Variable Terms

If the terms of a series are variable, we apply the definitions, theorems, and tests of the previous section and attempt to determine the values of z for which the series converges. The set of values of z for which a series of variable terms converges is called the *domain of convergence* of the series. Usually, we are content to find the domain of absolute convergence.

Example 15.4 Find the domain of absolute convergence of the series

$$\sum_{k=1}^{\infty} e^{kz}$$

Solution Since

$$\left|\frac{e^{(k+1)z}}{e^{kz}}\right| = |e^z|$$

this series converges for those values of z for which $|e^z| < 1$. Since $|e^z| = e^x$, the solution set for this inequality is $\{z \mid \operatorname{Re} z < 0\}$, that is, the left half-plane. □

Example 15.5 Find the domain of absolute convergence of the series

$$\sum_{k=1}^{\infty} \frac{z^k}{k^2(1+z)^k}$$

Solution Applying the ratio test, we find the domain of absolute convergence to be the set of z for which $|z/(1+z)| < 1$, or $|z| < |1+z|$. This inequality is satisfied for $\operatorname{Re} z > -\frac{1}{2}$. □

Example 15.6 Determine the domain of absolute convergence for the series

$$\sum_{k=1}^{\infty} \left(\frac{3kz}{5k+1}\right)^k$$

Solution Applying the root test, the domain of absolute convergence is given by $|3z/5| < 1$, that is, $|z| < \frac{5}{3}$. □

15.4 Power Series

For our purposes, the most important series of variable terms are series of nonnegative powers of $(z - a)$. These are called *power series* with center $z = a$.

The convergence behavior for power series is more easily characterized than is that for a more general series of variable terms. Indeed, the comparison test (Theorem 15.4) shows that if the power series

$$\sum_{n=0}^{\infty} c_n(z-a)^n$$

converges absolutely at some point z_0, then it converges absolutely for all z such that $|z - a| \leq |z_0 - a|$. Examples 15.2 and 15.3 show how power series converge in a circle of finite radius. Two other possibilities exist, however, and these are exhibited in the following two examples.

Example 15.7 Determine the domain of convergence of the power series

$$\sum_{n=0}^{\infty} \frac{z^n}{n!}$$

Solution By the ratio test

$$\left|\frac{z^{n+1}}{(n+1)!}\,\frac{n!}{z^n}\right| = \left|\frac{z}{n+1}\right| \to 0$$

as $n \to \infty$, and thus the series converges for all finite z. □

Example 15.8 Determine the domain of convergence of $\sum_{n=0}^{\infty} n!z^n$.

Solution The ratio of succeeding terms is

$$\frac{(n+1)!z^{n+1}}{n!z^n} = (n+1)z$$

whose absolute value $\to\infty$ as $n \to \infty$, unless $z = 0$. Hence the series converges only for $z = 0$. □

The following fundamental theorem characterizes the general situation.

Theorem 15.5 Every power series

$$\sum_{n=0}^{\infty} c_n(z-a)^n$$

behaves in one of the following three ways:

1. It converges for $z = a$ only.
2. It converges absolutely for all values of z.
3. There exists a positive number R such that the series converges absolutely for $|z - a| < R$ and diverges for $|z - a| > R$. ■

The circle $|z - a| = R$ is called the *circle of convergence* and R is called the *radius of convergence* for a given power series. The radius R is a function of the power series *coefficients* c_n. The ratio test may be conveniently used to find the radius of convergence, but the behavior on the circle $|z - a| = R$ must be determined by some other method. The precise analysis of what does happen on the circle of convergence can be very subtle, and will not be considered in this text. A series can diverge everywhere on the circle (as in the case $\sum_{n=0}^{\infty} z^n$) or converge everywhere (as in the case $\sum_{n=1}^{\infty} z^n/n^2$), but these are only two of many possibilities.

Example 15.9 Find the circle of convergence of the power series

$$\sum_{n=1}^{\infty} \frac{(z+4)^{2n-1}}{2n-1}$$

Solution The ratio of two succeeding terms is

$$\left|\frac{(z+4)^{2n+1}(2n-1)}{(2n+1)(z+4)^{2n-1}}\right|$$

which has a limit of $|z + 4|^2$. Hence, the circle of convergence is $|z + 4| = 1$. (See Figure 15.2.) □

At each point where a power series converges there is a natural correspondence of that point with the sum of the series there. This correspondence

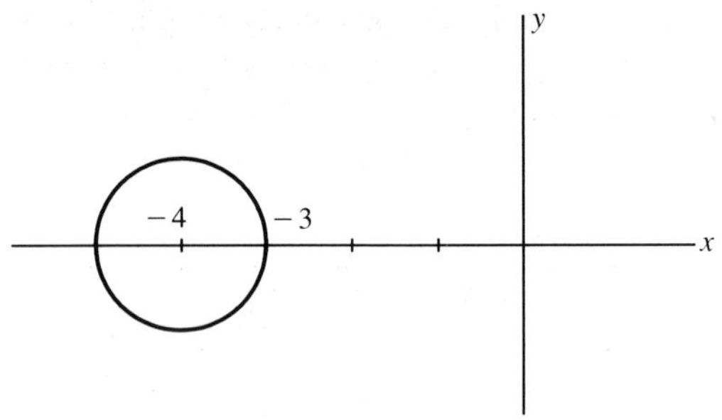

Figure 15.2

defines a complex function of a complex variable. A function defined by a power series in this manner is very well behaved.

Theorem 15.6 Suppose $\sum_{n=0}^{\infty} c_n(z-a)^n$ converges for $|z-a|<R$. Then, in this domain the power series represents an analytic function. The derivative (an antiderivative) is found by term-by-term differentiation (integration) and the derived (integrated) series converges in the same domain. ■

Example 15.10 Let $f(z)=\sum_{n=1}^{\infty} nz^{n-1}$. What is the domain of f? Find the infinite series for f'. Determine the power series representation for antiderivatives of f.

Solution By the ratio test, this series converges for $|z|<1$ and this is the domain of f. Moreover, within this circle,

$$f'(z)=\sum_{n=2}^{\infty} n(n-1)z^{n-2}$$

Similarly, if $F(z)$ is an antiderivative of f, then

$$F(z)=\sum_{n=1}^{\infty} z^n + \text{constant}, \qquad |z|<1$$

□

Example 15.11 Repeat Example 15.10 for the function

$$f(z)=\sum_{n=0}^{\infty} \frac{(-1)^n(z-\pi)^{2n}}{(2n)!}$$

Solution This series converges for all finite z by the ratio test. Hence

$$f'(z)=\sum_{n=1}^{\infty} \frac{(-1)^n(z-\pi)^{2n-1}}{(2n-1)!}$$

and

$$F(z)=\sum_{n=0}^{\infty} \frac{(-1)^n(z-\pi)^{2n+1}}{(2n+1)!} + \text{constant}$$

for each F such that $F'=f$. □

It should be clearly understood that the results given above are true only for a power series, not for more general series of variable terms. Other series do not necessarily converge in circles about a point; indeed, much more complicated patterns of convergence are possible.

Roughly speaking, this section demonstrates that every power series is an analytic function. In the next section, we will show that every analytic function can be written as a power series. The concepts of convergent power series and analytic functions will thus be united. Sometimes, this characterization of analytic functions in terms of power series representations proves more useful than the definition we used initially.

Exercises for Sections 15.3 and 15.4

In Exercises 1–10, determine the domain of absolute convergence.

1. $\sum_{n=1}^{\infty} \frac{\sin nz}{n^2}$

2. $\sum_{n=0}^{\infty} e^{n^2 z}$

3. $\sum_{n=1}^{\infty} \left(\frac{z}{5+3z}\right)^n \frac{1}{n^2}$

4. $\sum_{n=1}^{\infty} \left(\frac{4z}{z+2}\right)^n \frac{1}{n^3}$

5. $\sum_{n=0}^{\infty} 3z^n$ and $\sum_{n=0}^{\infty} (3z)^n$

6. $\sum_{n=1}^{\infty} \left(\frac{2z+5}{z}\right)^n \frac{1}{n^4}$

7. $\sum_{n=0}^{\infty} n^2(z-a)^n$

8. $\sum_{n=0}^{\infty} (z-a)^{n^2}$

9. $\sum_{n=0}^{\infty} \frac{(-1)^n z^n}{2^n}$

10. $\sum_{n=0}^{\infty} \left(\frac{z-4}{3}\right)^n$

11. Can there exist a power series $\sum_{n=0}^{\infty} c_n z^n$ that converges for $z = -3 + i$ and diverges when $z = 2 - 2i$? Explain.

12. Can there exist a power series $\sum_{n=0}^{\infty} c_n(z+1)^n$ that converges for $z = -3 + i$ and diverges when $z = 2 - 2i$? Explain.

13. Can there exist a power series $\sum_{n=0}^{\infty} c_n(z - i/3)^n$ that converges for $z = -3 + i$ and diverges when $z = 2 - 2i$? Explain.

14. Demonstrate that the two power series $\sum_{n=0}^{\infty} c_n z^n$ and $\sum_{n=0}^{\infty} nc_n z^{n-1}$ have the same radius of convergence.

15. **(a)** Show that $w(z) = \sum_{n=0}^{\infty} \frac{z^n}{n!}$ satisfies the differential equation $\frac{dw}{dz} = w$ for all finite z.

(b) Discuss the relationship of w to the analytic function e^z.

16. **(a)** Verify that $w(z) = \sum_{n=0}^{\infty} (-1)^n \frac{z^{2n+1}}{(2n+1)!}$ satisfies the differential equation $\frac{d^2 w}{dz^2} + w = 0$ for all finite z.

(b) Discuss the relationship of w to the analytic functions $\cos z$ and $\sin z$.

In Exercises 17–22, apply Theorem 15.6 to the given infinite series, finding the domain for $f(z)$ and power series representations for both the derivative and antiderivatives of f.

17. $f(z) = \sum_{n=0}^{\infty} (-1)^n z^{2n}$

18. $f(z) = \sum_{n=0}^{\infty} (-1)^n \frac{z^{2n}}{(2n)!}$

19. $f(z) = \sum_{n=1}^{\infty} \frac{n^3 z^{3n}}{27^n}$

20. $f(z) = \sum_{n=0}^{\infty} \frac{(z+2)^n}{(n+1)2^{2n}}$

21. $f(z) = \sum_{n=1}^{\infty} n^n (z-1)^n$

22. $f(z) = \sum_{n=1}^{\infty} \frac{(3+i)^n}{n^2}(z-i)^n$

23. If the method of Frobenius of Section 6.6, with $a = 1$, is applied to Legendre's equation

$$(1 - x^2)y'' - 2xy' + p(p+1)y = 0$$

the power series

$$P_p(x) = \sum_{n=0}^{\infty} \frac{\Gamma(p+1+n)}{\Gamma(p+1-n)(n!)^2}\left(\frac{x-1}{2}\right)^n$$

is obtained. When p is an integer, this series gives rise to the familiar Legendre polynomials discussed earlier in considerable detail (see Section 9.6).

(a) Verify that the infinite series reduces to the first four Legendre polynomials if $p = 0, 1, 2, 3$, respectively.

(b) Replace x by the complex variable z in the above series and determine the domain of absolute convergence in the case of general p.

24. In analogy with Exercise 23, one solution of Chebyshev's equation

$$(1 - x^2)y'' - xy' + p^2 y = 0$$

has the power series representation

$$T_p(x) = 1 + \sum_{n=1}^{\infty} \frac{p\Gamma(p+n)\Gamma(\frac{1}{2})}{\Gamma(p+1-n)\Gamma(n+\frac{1}{2})n!}\left(\frac{x-1}{2}\right)^n$$

(a) Demonstrate that when $p = 0, 1, 2, 3$, respectively, this infinite series gives rise to the first four of the so-called Chebyshev polynomials

$$T_0 = 1, \quad T_1(x) = x, \quad T_2(x) = 2x^2 - 1, \quad \text{and } T_3(x) = 4x^3 - 3x$$

(b) Show that if x is replaced by the complex variable z in the above series, the radius of convergence is $R = 2$.

25. Prove that if m is a nonnegative integer, the power series for the Bessel function of the first kind

$$J_m(z) = \sum_{n=0}^{\infty} \frac{(-1)^n}{\Gamma(m+n+1)n!}\left(\frac{z}{2}\right)^{2n+m}$$

converges absolutely for all z.

15.5 Power Series Representation of a Function

We are now ready to turn our attention to the analog for complex anlaysis of the familiar Taylor expansion of a real function. As might be expected, there are some very important theoretical and applied consequences of the results which we obtain.

Assume f is analytic inside and on the circle C centered at a. To represent

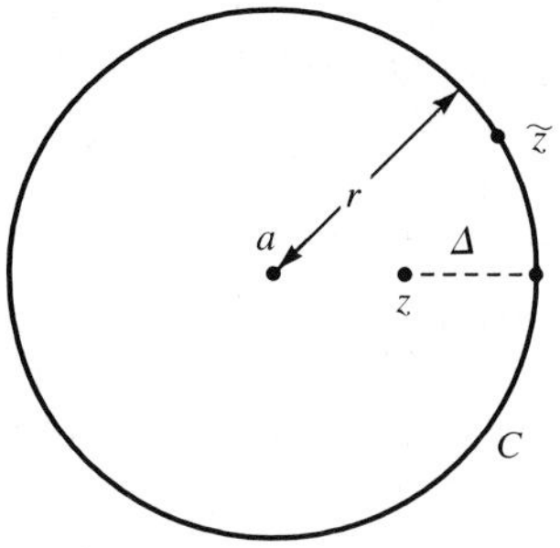

Figure 15.3

$f(z)$ as a series in powers of $(z - a)$, let z be a point of this circular neighborhood. (See Figure 15.3.) By Cauchy's integral formula, the value of the function at z is given by

$$f(z) = \frac{1}{2\pi i} \oint_C \frac{f(\zeta)}{\zeta - z}\, d\zeta$$

The essence of Taylor's theorem is to convert Cauchy's integral representation of $f(z)$ to a power series representation in the same domain.

Theorem 15.7 *Taylor Series.* Let f be analytic in a neighborhood of $z = a$. Then $f(z)$ may be alternatively expressed as

$$f(z) = \sum_{n=0}^{\infty} A_n (z - a)^n$$

where

$$A_n = \frac{1}{2\pi i} \oint_C \frac{f(\zeta)}{(\zeta - a)^{n+1}}\, d\zeta$$

The series is unique and converges to $f(z)$ in the largest circle centered at a such that $f(z)$ is analytic throughout the interior. (We sometimes say the series converges "out to the nearest singularity of f.")

Proof Note that

$$\frac{1}{\zeta - z} = \frac{1}{(\zeta - a) - (z - a)} = \frac{1}{(\zeta - a)\left(1 - \dfrac{z - a}{\zeta - a}\right)}$$

By the familiar expression for a geometric progression with "common ratio" $(z - a)/(\zeta - a)$,

$$\frac{1}{\zeta - z} = \frac{1}{\zeta - a}\left(1 + \frac{z - a}{\zeta - a} + \left(\frac{z - a}{\zeta - a}\right)^2 + \cdots + \left(\frac{z - a}{\zeta - a}\right)^n + \frac{\left(\dfrac{z - a}{\zeta - a}\right)^{n+1}}{1 - \dfrac{z - a}{\zeta - a}}\right)$$

$$= \frac{1}{\zeta - a} + \frac{z-a}{(\zeta-a)^2} + \frac{(z-a)^2}{(\zeta-a)^3} + \cdots + \frac{(z-a)^n}{(\zeta-a)^{n+1}} + \frac{\left(\dfrac{z-a}{\zeta-a}\right)^{n+1}}{\zeta - z}$$

Therefore, since

$$f(z) = \frac{1}{2\pi i}\oint_C \frac{f(\zeta)}{\zeta - z}\,d\zeta$$

then

$$\begin{aligned} f(z) = {} & \frac{1}{2\pi i}\oint_C \frac{f(\zeta)}{\zeta - a}\,d\zeta + \frac{z-a}{2\pi i}\oint_C \frac{f(\zeta)}{(\zeta-a)^2}\,d\zeta \\ & + \frac{(z-a)^2}{2\pi i}\oint_C \frac{f(\zeta)}{(\zeta-a)^3}\,d\zeta + \cdots + \frac{(z-a)^n}{2\pi i}\oint_C \frac{f(\zeta)}{(\zeta-a)^{n+1}}\,d\zeta \\ & + \frac{(z-a)^{n+1}}{2\pi i}\oint_C \frac{f(\zeta)}{(\zeta-a)^{n+1}(\zeta-z)}\,d\zeta \end{aligned}$$

The last term is denoted by R_n and is called the *remainder*. Our proof will be complete if we can show that $|R_n|$ approaches 0 as $n \to \infty$. Since f is analytic on C, $|f(z)|$ is bounded, say by M. For a point z not on the circle, there is a minimum value for $|\zeta - z|$, which we denote by Δ. The ratio $|(z-a)/(\zeta-a)|$ has a constant value, ρ, which is less than 1. Therefore,

$$\begin{aligned} |R_n| &= \frac{1}{2\pi}\left|\oint_C \frac{f(\zeta)}{\zeta - z}\left(\frac{z-a}{\zeta-a}\right)^{n+1} d\zeta\right| \\ &\leq \frac{1}{2\pi}\oint_C \left|\frac{f(\zeta)}{\zeta - z}\right|\left|\frac{z-a}{\zeta-a}\right|^{n+1} |d\zeta| \\ &\leq \frac{1}{2\pi}\frac{M}{\Delta}\rho^{n+1}2\pi r \end{aligned}$$

which tends to 0 as $n \to \infty$. ■

The formula for the coefficients, A_n, in the Taylor series takes on a more familiar look as a result of the following corollary.

Corollary With the hypothesis the same as in the previous theorem,

$$A_n = \frac{f^{(n)}(a)}{n!}$$

Proof Since f is analytic at $z = a$, the formula for the derivative given by Theorem 14.9 may be used to give the desired result. ■

Example 15.12 Find the Taylor series for the function $f(z) = 1/(1 - z)$ about $z = 0$. Where does the series converge? Where does the series represent the function?

Solution Since $f(z) = (1 - z)^{-1}$ and $f^{(n)}(z) = n!(1 - z)^{-n-1}$, $f^{(n)}(0) = n!$ and therefore

$$f(z) = \sum_{n=0}^{\infty} z^n$$

The series converges and represents the function in a circular neighborhood about $z = 0$ extending to the circle $|z| = 1$ which passes through the singularity of f. The series diverges (and consequently could not represent $f(z)$) for $|z| > 1$. (Recall Example 15.2.) □

Example 15.13 Find the Taylor series for $\sin z$ about $z = 0$.

Solution Since $f(z) = \sin z$, $f(0) = 0$, $f'(0) = 1$, $f''(0) = 0$, $f'''(0) = -1$, and so on. Therefore the Taylor series for $\sin z$ is

$$\sum_{n=0}^{\infty} \frac{(-1)^n z^{2n+1}}{(2n+1)!}$$

Since $\sin z$ is analytic everywhere, the series converges and represents $\sin z$ for all finite z. □

Example 15.13 shows that the Taylor series for the complex sine function has the same *form* as the real series for the real sine function. This is also true for other real functions obtained by restricting the complex variable of an analytic function to real values since the form of the coefficients is the same in the real as in the complex case.

Example 15.14 Find the Taylor series for $1/(1 - z)$ about $z = i$.

Solution It is easy to show that $f^{(n)}(z) = n!(1 - z)^{-n-1}$. Therefore $f^{(n)}(i) = n!(1 - i)^{-n-1}$, and thus the Taylor series is

$$\sum_{n=0}^{\infty} (1 - i)^{-n-1}(z - i)^n$$

The series represents the function within a circle passing through the nearest (and only) singularity of f. The domain of representation is $|z - i| < \sqrt{2}$. (See Figure 15.4.) □

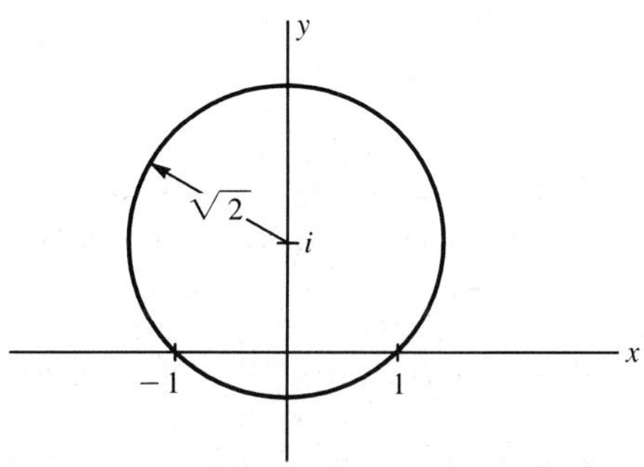

Figure 15.4

Even though the formula for the coefficients is simplified when expressed in terms of the derivative, there are other, and usually more practical, methods for obtaining a Taylor series representation. The uniqueness of the representation allows us to find the series in any convenient way. One method often used is to write the given function in a form in which the results for geometric series may be applied.

Example 15.15 Write the Taylor series for $1/(1 + z^2)$ about $z = 0$.

Solution Substitute $-z^2$ for z in the known expansion of $1/(1 - z)$ to obtain

$$\frac{1}{1 + z^2} = \sum_{n=0}^{\infty} (-z^2)^n = \sum_{n=0}^{\infty} (-1)^n z^{2n}$$

The series represents the function for $|z^2| < 1$, that is, inside a circle of radius 1 centered at the origin. □

Example 15.16 Find the series for $1/(1 + z)$ about $z = 1$.

Solution

$$\frac{1}{1 + z} = \frac{1}{2 + (z - 1)} = \frac{1}{2\left(1 + \dfrac{z - 1}{2}\right)}$$

Substitute $-(z - 1)/2$ for the z of the geometric series to obtain

$$\frac{1}{1 + z} = \frac{1}{2} \sum_{n=0}^{\infty} (-1)^n \left(\frac{z - 1}{2}\right)^n$$

The representation is valid for $|(z - 1)/2| < 1$, that is, $|z - 1| < 2$. □

If the function is rational, you may wish to find the partial fraction decomposition before using other methods.

Example 15.17 Find the Taylor series for $1/[(1 + z)(2 + z)]$ about $z = 0$.

Solution By elementary methods,

$$\frac{1}{(1 + z)(2 + z)} = \frac{1}{1 + z} - \frac{1}{2 + z}$$

The Taylor series for $1/(1 + z)$ is $\sum_{n=0}^{\infty} (-1)^n z^n$ and the representation is valid for $|z| < 1$. Also,

$$\frac{1}{2 + z} = \frac{1}{2\left(1 + \dfrac{z}{2}\right)} = \frac{1}{2} \sum_{n=0}^{\infty} (-1)^n \left(\frac{z}{2}\right)^n$$

valid for $|z| < 2$. Therefore, the Taylor series for the given function is

$$\frac{1}{(1+z)(2+z)} = \sum_{n=0}^{\infty}\left[(-1)^n z^n - \frac{1}{2}(-1)^n\left(\frac{z}{2}\right)^n\right] = \sum_{n=0}^{\infty}(-1)^n z^n\left(1 - \frac{1}{2^{n+1}}\right)$$

which is a valid representation for $|z| < 1$. □

In this last example we have made use of the fundamental fact that convergent power series having the same center can be termwise *added, subtracted,* and *multiplied by a constant.* Two power series can also be *multiplied* together in the following sense.

Cauchy Product. If $f(z) = \sum_{n=0}^{\infty} a_n(z-a)^n$ is analytic for $|z-a| < R_1$ and $g(z) = \sum_{n=0}^{\infty} b_n(z-a)^n$ is analytic for $|z-a| < R_2$, then

$$f(z)g(z) = \sum_{n=0}^{\infty} c_n(z-a)^n$$

where

$$c_n \equiv \sum_{k=0}^{n} a_k b_{n-k}$$

and the series representation is absolutely convergent for $|z-a| < \min(R_1, R_2)$. These results are, of course, extensions of the corresponding properties of real series which we discussed in Chapter 6.

Example 15.18 Find the Taylor series for $1/(1+z)^2$ about $z = 0$.

Solution The Taylor series for $1/(1+z)$ is $\sum_{n=0}^{\infty}(-1)^n z^n$. Therefore

$$\begin{aligned}\frac{1}{(1+z)^2} &= \sum_{n=0}^{\infty}(-1)^n z^n \sum_{n=0}^{\infty}(-1)^n z^n \\ &= \sum_{n=0}^{\infty}\left[\sum_{k=0}^{n}(-1)^k(-1)^{n-k}\right]z^n \\ &= \sum_{n=0}^{\infty}\left[(-1)^n \sum_{k=0}^{n} 1\right]z^n \\ &= \sum_{n=0}^{\infty}(-1)^n(n+1)z^n\end{aligned}$$

The representation is valid for $|z| < 1$.

This result could also be obtained by differentiating both sides of the equality

$$\frac{1}{1+z} = \sum_{n=0}^{\infty}(-1)^n z^n, \qquad |z| < 1$$

□

The next example is an immediate application of Theorem 15.6.

Example 15.19 Write the Taylor series for $\operatorname{Tan}^{-1} z$ about $z = 0$.

Solution Antidifferentiating both sides of the equality

$$\frac{1}{1+z^2} = \sum_{n=0}^{\infty}(-1)^n z^{2n}$$

and noting that $\mathrm{Tan}^{-1} 0 = 0$, we obtain

$$\mathrm{Tan}^{-1} z = \sum_{n=0}^{\infty} \frac{(-1)^n z^{2n+1}}{2n+1}$$

The representation is valid for $|z| < 1$. □

The previous examples show why power series are best discussed in the complex plane. The reason that the series representation of the function $f(x) = 1/(1 + x^2)$ has an interval of convergence restricted to $(-1, 1)$ is inexplainable when only real functions are considered. In the complex plane, we see that the power series for $1/(1 + z^2)$ converges out to the nearest singularity of which there are two, located at i and $-i$. The barrier of the circle $|z| = 1$ is invisible when only real functions are discussed.

Exercises for Section 15.5

In Exercises 1–24, find the Taylor series of the given function about $z = a$ and determine the radius of convergence R.

1. $\cos z, \quad a = 0$

2. $\sin z, \quad a = 0$

3. $e^{-z}, \quad a = 0$

4. $\dfrac{1}{z}, \quad a = 1$

5. $\dfrac{1}{1 - z}, \quad a = i$

6. $f(z) = \begin{cases} \dfrac{\sin z}{z}, & z \neq 0, \\ 1 & z = 0, \end{cases} \quad a = 0$

7. $\dfrac{1}{z^2}, \quad a = 1$

8. $\dfrac{1}{1 - z^3}, \quad a = 0$

9. $\sin z^2, \quad a = 0$

10. $e^{z^2}, \quad a = 0$

11. $z^2, \quad a = 1$

12. $\dfrac{1}{3z + 1}, \quad a = 2$

13. $z - 1, \quad a = -1$

14. $z^2 - 8z + 2, \quad a = 0$

15. $z^2 - 8z + 2, \quad a = 1$

16. $\dfrac{1}{2z + i}, \quad a = 2$

17. $\dfrac{1}{z^2 - 4}, \quad a = 1$

18. $\dfrac{1}{z^2 + 4}, \quad a = 1$

19. $\dfrac{1}{(1 + z)^4}, \quad a = 0$

20. $\dfrac{1}{(z + 3)(z + 2)}, \quad a = -1$

21. $\dfrac{1}{z^4 - 1}, \quad a = 0$

22. $\cos z, \quad a = \dfrac{\pi}{3}$

23. $\displaystyle\int_0^z \frac{d\zeta}{1 + \zeta^4}, \quad a = 0$

24. $\displaystyle\int_0^z \exp(1 + \zeta^2)\, d\zeta, \quad a = 0$

In Exercises 25–30, find the first few terms of the Taylor series of the given function about $a = 0$.

25. $e^{1/(1+z)}$

26. $\cos(1 - z)$

27. $e^{\sin z}$

28. $(1 + z)^{1/2}$

29. $e^z \ln(1 - z)$

30. $\dfrac{z^2 + 1}{(z^2 - 4)(z + 3)}$

31. Using the definition

$$\ln z = \int_1^z \frac{d\zeta}{\zeta}$$

determine the Taylor series for $\operatorname{Ln} z$ about $z = 1$.
(*Hint:* Rewrite $1/\zeta$ as $1/(1 + \zeta - 1)$ and expand.)

In Exercises 32–36, find the Taylor series expansion about the origin $z = 0$ for the given special function:

32. The error function

$$\operatorname{erf} z = \frac{2}{\sqrt{\pi}} \int_0^z e^{-\zeta^2}\, d\zeta$$

33. The exponential integral

$$Ein(z) = \int_0^z \frac{(1 - e^{-\zeta})}{\zeta}\, d\zeta$$

34. The sine integral

$$Si(z) = \int_0^z \frac{\sin \zeta}{\zeta}\, d\zeta$$

35. The Fresnel integral

$$S(z) = \int_0^z \sin \zeta^2\, d\zeta$$

36. The incomplete beta function

$$B_z(a, b) = \int_0^z \zeta^{a-1}(1 - \zeta)^{b-1}\, d\zeta, \qquad a > 0, b > 0$$

(*Hint:* Expand $(1 - \zeta)^{b-1}$ by the binomial theorem.)

15.6 Laurent Series

The Taylor series of a function is a suitable representation of a function around points where f is analytic. There are two major deficiencies, however:

1. No power series representation is possible about a singularity.
2. Even if the power series is written about a point where the function is analytic, the representation is valid only within a circle out to the nearest singularity.

A more general type of series expansion of a function is of the form

$$\sum_{n=-\infty}^{\infty} A_n(z - a)^n$$

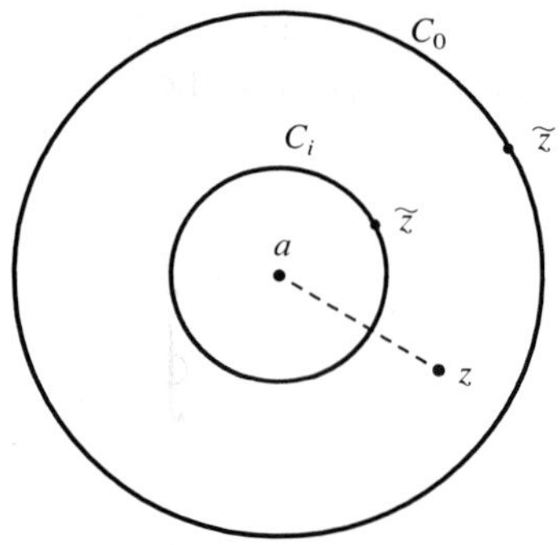

Figure 15.5

and is called a *Laurent series*. In this section we show how to represent a function in terms of such a series about a point where the function is nonanalytic. The representation takes a form such that if f is analytic at a, the Laurent series reduces to the Taylor series, that is, $A_n = 0$ for $n < 0$.

Theorem 15.8 *Laurent Series.* Let f be analytic on two concentric circles centered at a and in the annulus between them. (See Figure 15.5.) Then $f(z)$ can be represented by a Laurent series with coefficients

$$A_n = \frac{1}{2\pi i}\oint_C \frac{f(\zeta)}{(\zeta - a)^{n+1}}\,d\zeta$$

(Recall the assumed positive orientation for integration around closed contours.) Here C is any closed contour that lies in the annulus and encloses the inner circle. The series converges and represents $f(z)$ inside the annulus obtained by continuously increasing C_0 and decreasing C_i until each of the two circles reaches a singularity of f.

Proof The proof parallels that of Taylor's theorem (Theorem 15.7) so most of the details are left to the exercises. Let z be a point in the annulus. By Cauchy's integral formula applied to doubly connected domains (section 14.6 and Figure 14.6).

$$\begin{aligned} f(z) &= \frac{1}{2\pi i}\oint_{C_0} \frac{f(\zeta)}{\zeta - z}\,d\zeta - \frac{1}{2\pi i}\oint_{C_i} \frac{f(\zeta)}{\zeta - z}\,d\zeta \\ &= g(z) - h(z) \end{aligned}$$

an equality that has the effect of defining $g(z)$ and $h(z)$. Since z lies inside of C_0, the first integral is the same as the type considered in the proof of Taylor's theorem. Therefore,

$$g(z) = \frac{1}{2\pi i}\oint_{C_0} \frac{f(\zeta)}{\zeta - z}\,d\zeta = \sum_{n=0}^{\infty} A_n(z - a)^n$$

By the principle of deformation of path, C_0 may be deformed to any closed contour that encloses the inner circle. Hence, the form of the coefficients is verified for $n \geqq 0$. The series which defines $g(z)$ converges within a circle obtained by enlarging C_0 out to the "next" singularity of f.

It remains to show that A_n has the correct form for $n < 0$. Note that

$$
\begin{aligned}
-h(z) &= \frac{-1}{2\pi i}\oint_{C_i} \frac{f(\zeta)}{\zeta - z}\,d\zeta \\
&= \frac{1}{2\pi i}\oint_{C_i} \frac{f(\zeta)}{(z-a)\left(1 - \dfrac{\zeta - a}{z - a}\right)}\,d\zeta \\
&= \frac{1}{2\pi i}(z-a)^{-1}\oint_{C_i} f(\zeta)\,d\zeta + \frac{(z-a)^{-2}}{2\pi i}\oint_{C_i} f(\zeta)(\zeta - a)\,d\zeta \\
&\quad + \frac{(z-a)^{-3}}{2\pi i}\oint_{C_i} f(\zeta)(\zeta - a)^2\,d\zeta + \cdots \\
&\quad + \frac{(z-a)^{-m}}{2\pi i}\oint_{C_i} f(\zeta)(\zeta - a)^{m-1}\,d\zeta \\
&\quad + \frac{1}{2\pi i}\oint_{C_i} \frac{f(\zeta)}{z - \zeta}\left(\frac{\zeta - a}{z - a}\right)^m d\zeta, \qquad \text{where } m > 0
\end{aligned}
$$

By a technique similar to that used in Theorem 15.7 it can be shown that the last term goes to zero as $m \to \infty$. Thus, for $m > 0$, we define

$$A_m \equiv \frac{1}{2\pi i}\oint_{C_i} f(\zeta)(\zeta - a)^{m-1}\,d\zeta$$

so that if $n = -m$,

$$A_n = \frac{1}{2\pi i}\oint_{C_i} \frac{f(\zeta)}{(\zeta - a)^{n+1}}\,d\zeta, \qquad n < 0$$

By the principle of deformation of path, C_i may be deformed to any closed contour C with the prescribed properties. Hence A_n has the correct form for $n < 0$.

The series for $h(z)$ will converge outside of the circle C_i since $|(\zeta - a)/(z - a)| < 1$ only if $|\zeta - a| < |z - a|$. The circle C_i may be shrunk until a singularity of f is reached. The domain common to g and h is the open annulus characterized in the statement of the theorem. ■

If the only singularity of f is at $z = a$, then the Laurent series converges to $f(z)$ for all $z \neq a$. The coefficients of the Laurent expansion cannot be expressed in terms of the derivatives of f at a, however, since f is not analytic there.

Just as for Taylor series, the Laurent representation *in a given annulus* is unique. This allows the computation of the A_n by methods other than appeal to the formula.

Example 15.20 Find all Laurent series for $f(z) = z^2 e^{1/z}$ with center at $z = 0$.

Solution The only singularity for f is at $z = 0$, the center of the series. The Laurent expansion for $e^{1/z}$ about 0, obtained from the Taylor series for e^z, is

$$\sum_{n=0}^{\infty} \frac{1}{n!z^n}$$

valid for $|z| > 0$. Hence the expansion for $z^2 e^{1/z}$ is

$$\sum_{n=0}^{\infty} \frac{1}{n!z^{n-2}} = z^2 + z + \frac{1}{2} + \frac{1}{6z} + \frac{1}{24z^2} + \cdots$$

The representation is valid for $|z| > 0$. □

Example 15.21 Find all Laurent expansions for $f(z) = 1/(z-2)$:
(a) With center at $z = 0$ (b) With center at $z = 2$

Solution (a) With center at $z = 0$, there is a (Taylor) representation valid for $0 \leqq |z| < 2$,

$$\frac{1}{z-2} = \frac{-1}{2\left(1 - \dfrac{z}{2}\right)} = -\frac{1}{2}\sum_{n=0}^{\infty}\left(\frac{z}{2}\right)^n$$

For $|z| > 2$,

$$\frac{1}{z-2} = \frac{1}{z\left(1 - \dfrac{2}{z}\right)} = \frac{1}{z}\sum_{n=0}^{\infty}\left(\frac{2}{z}\right)^n = \sum_{n=0}^{\infty}\frac{2^n}{z^{n+1}}$$

(b) With center at $z = 2$, the Laurent representation for the function given is $1/(z-2)$, that is, the original expression for the function is the desired Laurent series with $z = 2$ as the center. □

Example 15.22 Find all Laurent expansions for $\dfrac{1}{(z-1)(z-2)}$:
(a) With center at $z = 0$ (b) With center at $z = 1$

Solution (a) By a partial fraction expansion,

$$\frac{1}{(z-1)(z-2)} = -\frac{1}{z-1} + \frac{1}{z-2}$$

Since the function has singularities at 1 and 2, there are Laurent expansions about 0 valid for $0 \leqq |z| < 1$, $1 < |z| < 2$, and $|2| < z$. (See Figure 15.6.) For $0 \leqq |z| < 1$,

$$\frac{-1}{z-1} = \sum_{n=0}^{\infty} z^n \quad \text{and} \quad \frac{1}{z-2} = -\frac{1}{2}\sum_{n=0}^{\infty}\left(\frac{z}{2}\right)^n$$

The first of these is valid for $0 \leqq |z| < 1$ and the second for $0 \leqq |z| < 2$. Hence

$$f(z) = \sum_{n=0}^{\infty}\left(1 - \frac{1}{2^{n+1}}\right)z^n, \qquad \text{for } 0 \leqq |z| < 1$$

For $1 < |z|$

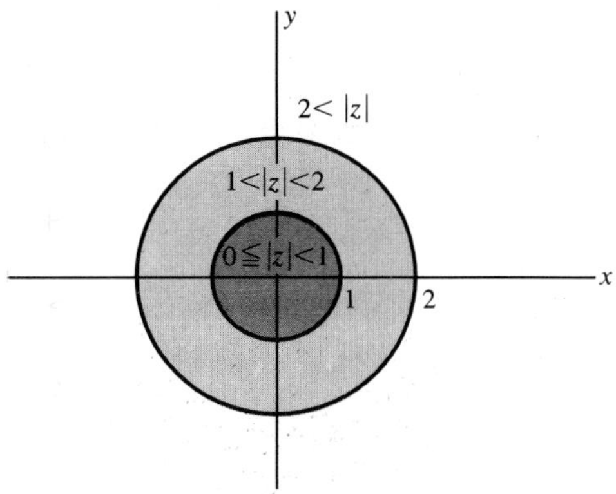

Figure 15.6

$$\frac{-1}{z-1} = -\sum_{n=0}^{\infty} z^{-n-1}$$

and thus

$$f(z) = -\sum_{n=0}^{\infty}\left(\frac{1}{z^{n+1}} + \frac{z^n}{2^{n+1}}\right), \qquad \text{for } 1 < |z| < 2$$

For $|z| > 2$,

$$\frac{1}{z-2} = \sum_{n=0}^{\infty} \frac{2^n}{z^{n+1}}$$

which leads to

$$f(z) = \sum_{n=0}^{\infty} \frac{(2^n - 1)}{z^{n+1}}, \qquad \text{for } 2 < |z|$$

(b) About $z = 1$, there will be two representations, one valid for $0 < |z - 1| < 1$ and the other for $1 < |z - 1|$ (see Figure 15.7),

$$\frac{1}{z-2} = \frac{1}{(z-1)-1} = \frac{-1}{1-(z-1)} = -\sum_{n=0}^{\infty} (z-1)^n$$

Therefore

$$f(z) = -\frac{1}{z-1} - \sum_{n=0}^{\infty} (z-1)^n, \qquad \text{for } 0 < |z-1| < 1$$

For $|z - 1| > 1$, we have

$$\frac{1}{z-2} = \frac{1}{(z-1)\left(1 - \dfrac{1}{z-1}\right)} = \frac{1}{z-1}\sum_{n=0}^{\infty} (z-1)^{-n}$$

Hence

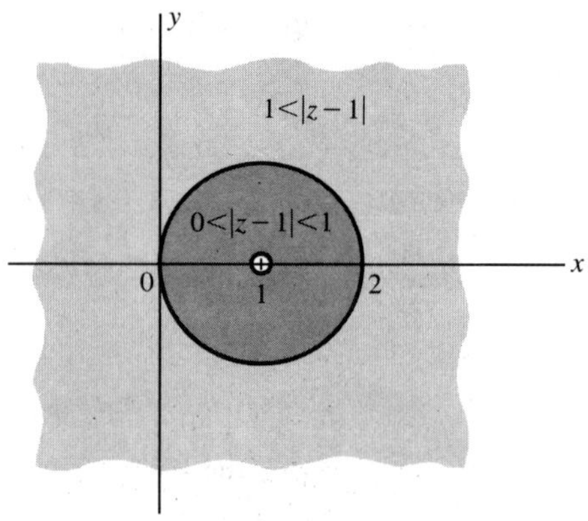

Figure 15.7

$$f(z) = \sum_{n=0}^{\infty} (z-1)^{-n-1} - (z-1)^{-1}$$
$$= \sum_{n=1}^{\infty} \frac{1}{(z-1)^{n+1}}$$
□

Exercises for Section 15.6

Find all Laurent series of the following functions about the specified centers. Give the domain of representation in each case.

1. $z^3 e^{1/z}$, $a = 0$

2. $\dfrac{1}{z^2+1}$, $a = 1$

3. $\dfrac{1}{z^4-1}$, $a = 0$

4. $\dfrac{1}{z^2}$, $a = 1$

5. $\dfrac{1}{z^3}$, $a = 1$

6. $\dfrac{1}{z^4}$, $a = 1$

7. $\dfrac{1}{z^2(z^2+1)}$, $a = 0$

8. $\dfrac{e^z}{(z-1)^2}$, $a = 1$

9. $\dfrac{3z^2+z}{z^3-z}$, $a = -1$

10. $\dfrac{\sin 3z}{z^2}$, $a = 0$

11. $z^3 \cos\left(\dfrac{1}{2z}\right)$, $a = 0$

12. $\dfrac{z}{(z-1)(z-2)}$, $a = 0$

13. $\dfrac{z}{(z-1)(z-2)}$, $a = 1$

14. $\dfrac{z}{(z-1)(z-2)}$, $a = 2$

In Exercises 15–18 find the first *three* nonzero terms of the Laurent series for the given function in the specified domain.

15. $\csc z$ for $0 < |z| < \pi$

16. $\dfrac{1}{e^z-1}$ for $0 < |z| < 2\pi$

17. $\dfrac{e^z}{z^2(1+z^2)}$ for $0 < |z| < 1$ **18.** $\cot z$ for $0 < |z| < \pi$

19. Find all Laurent expansions of $1/(z^2 - 1)$ about:
(a) 0 **(b)** 1 **(c)** -1

20. Find all Laurent expansions of $1/z$ with center at:
(a) 0 **(b)** 1 **(c)** 2

In Exercises 21–23, find all Laurent series for $1/[z(z-i)(z+1)]$ with the indicated center:

21. 0 **22.** i **23.** 1

24. In the proof of Laurent's theorem, show that the last integral goes to zero as $m \to \infty$.

25. If p is an integer, the following is the Laurent series for a special function:

$$Q_p(z) = \sum_{n=0}^{\infty} \frac{\Gamma(\frac{1}{2})\Gamma(p+1+2n)}{\Gamma(p+\frac{3}{2}+n)n!}\left(\frac{1}{2z}\right)^{p+1+2n}$$

(a) Determine the domain of representation for this series.
(b) Show that for real z, $Q_p(x)$ satisfies Legendre's equation

$$(1-x^2)y'' - 2xy' + p(p+1)y = 0$$

(Q_p is known as the Legendre function of the second kind.)

26. A second solution of Chebyshev's equation,

$$(1-x^2)y'' - xy' + p^2y = 0$$

is given by

$$V_p(x) = \frac{1}{p}[(x+\sqrt{x^2-1})^p - (x-\sqrt{x^2-1})^p]$$

If we replace x by z and consider large $|z|$, we find

$$V_p(z) = \frac{1}{p}\left(1-\frac{1}{z^2}\right)^{1/2}\left[(2z)^p \sum_{n=0}^{\infty} \frac{\Gamma(1-p+2n)}{\Gamma(1-p+n)n!}\frac{1}{(2z)^{2n}} - (2z)^{-p}\sum_{n=0}^{\infty}\frac{\Gamma(1+p+2n)}{\Gamma(1+p+n)n!}\frac{1}{(2z)^{2n}}\right]$$

an expression which, when the factor $(1 - 1/z^2)^{1/2}$ is expanded and multiplied in, gives rise to a Laurent series in the case of integral p.
(a) Determine the domain of representation for the resulting Laurent series.
(b) Find the first few terms of the Laurent series for $p = 3$. (Recall Exercise 28, p. 206.)

27. Employ the definition

$$\ln z \equiv \int_1^z \frac{d\zeta}{\zeta}$$

to derive the Laurent series

$$\operatorname{Ln}\left(\frac{z+1}{z}\right) = \sum_{n=1}^{\infty}\frac{(-1)^{n+1}}{nz^n}, \qquad |z| > 1$$

28. Repeat Exercise 27 starting from the Laurent series for $\dfrac{1}{z+1} - \dfrac{1}{z}$.

29. Establish the Laurent series

$$\operatorname{Ln}\left(\frac{z+1}{z-1}\right) = 2\sum_{n=0}^{\infty} \frac{1}{(2n+1)z^{2n+1}}, \qquad |z| > 1$$

30. Show that the principal branch of $\tan^{-1} z$ has the Laurent series

$$\operatorname{Tan}^{-1} z = \frac{\pi}{2} - \sum_{n=0}^{\infty} \frac{(-1)^n}{(2n+1)z^{2n+1}}, \qquad |z| > 1$$

15.7 Poles and Zeros

The ideas of the last section allow us to be more definitive about the structure of analytic functions. Recall that the complex number $z = a$ is said to be a *zero* of the function $f(z)$ if $f(a) = 0$. A zero of the function $f(z)$ is said to be *isolated* if $f(a) = 0$ and there is a neighborhood of a such that $f(z) \neq 0$ in that neighborhood except at $z = a$. It is possible to show that if the function is analytic, then its zeros are isolated unless, of course, $f(z) \equiv 0$.

With the Laurent expansion of a function we can further classify the zeros. An analytic function is said to have a *zero of the Nth order* at $z = a$ if for some integer $N > 0$, $A_n = 0$ for all $n < N$, but $A_N \neq 0$ in the function's Laurent expansion with center at a. Equivalently, f has a zero of the Nth order if $f^{(n)}(a) = 0$ for $n < N$ and $f^{(N)}(a) \neq 0$. When $N = 1$, the zero is said to be *simple.*

Singularities are also classified. If f has an isolated singularity at $z = a$, then in the neighborhood of $z = a$ the function has a Laurent representation

$$f(z) = g(z) + h(z) = \sum_{n=0}^{\infty} A_n(z-a)^n + \sum_{n=-\infty}^{-1} A_n(z-a)^n$$

The series for $h(z)$ is called the *principal part of* $f(z)$ at $z = a$. The principal part is used to classify singularities:

1. If the principal part is zero, then the singularity is said to be *removable.* The singularity is removed by defining $f(a)$ to be A_0.
2. If the principal part has finitely many terms, the singularity is called a *pole.* In this case, the function may be written

$$f(z) = g(z) + A_{-1}(z-a)^{-1} + A_{-2}(z-a)^{-2} + \cdots + A_{-N}(z-a)^{-N}$$

   The value of N is called the *order* of the pole. First-order poles are said to be *simple.*
3. If the principal part has infinitely many nonzero terms, the singularity is called *essential.*

Example 15.23 Classify the poles and zeros of the functions:

(a) $f(z) = e^{1/z}$ (b) $f(z) = \dfrac{-9}{z(z-2)^5} + \dfrac{3}{(z-2)^4}$

(c) $f(z) = 2z + 6z^3$

Solution (a) The Laurent expansion for the function $e^{1/z}$ with center $z = 0$ is given by

$$\sum_{n=0}^{\infty} \frac{z^{-n}}{n!}$$

and so the singularity at $z = 0$ is essential. There are no zeros since the exponential function is never zero.
(b) By inspection, there is a simple pole at $z = 0$ and a pole of order 5 at $z = 2$. The function has simple zeros at $z = -1, 3$.
(c) The given function obviously has no poles in the finite z-plane. There are simple zeros at 0, $i/\sqrt{3}$, and $-i/\sqrt{3}$. □

Exercises for Section 15.7

Classify the poles and zeros of the following functions in the finite complex plane.

1. $1 - z^4$

2. z^4

3. $\sin z$

4. $\dfrac{1 + z^2}{1 - z}$

5. $\tan^2 z$

6. $z^3 \sin^2 (1/z)$

7. $e^{2z} - e^z$

8. $\dfrac{(z - 6)^4}{z}$

9. $\dfrac{z - 1}{z^2 + 1}$

10. $z + 5$

11. $z^2 + 1$

12. $\dfrac{1 + z}{z^2}$

13. $ze^{1/z}$

14. $\cos z$

15. $z + \dfrac{1}{z}$

16. z

17. $\dfrac{\sin z}{z}$

18. $\dfrac{\sin z}{z^2}$

19. $\dfrac{15}{z - 1} - \left(\dfrac{1}{z + 1}\right)^3 + 64\left(\dfrac{1}{z - 2}\right)^2$

20. $\dfrac{\sin z^2}{z}$

21. $\dfrac{z + 1}{z^3 - z^2}$

22. $\dfrac{1 - e^{-z}}{z^3}$

23. $\cot z$

24. $\tanh z$

25. $\dfrac{1 + e^{2z}}{\cos z + \sin z}$

15.8 Residues

If f has an isolated singularity at $z = a$, then in the neighborhood of $z = a$, the function has the Laurent representation

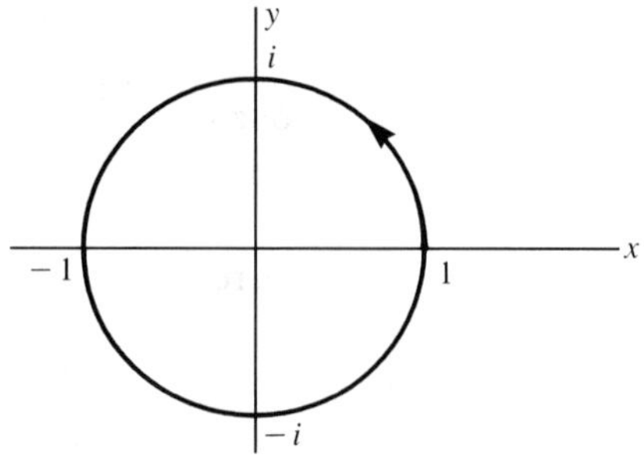

Figure 15.8

$$f(z) = \sum_{n=-\infty}^{\infty} A_n(z-a)^n$$

where the coefficients A_n have been derived previously. In particular,

$$A_{-1} = \frac{1}{2\pi i} \oint_C f(z)\,dz$$

where C is a closed contour enclosing no singularities of f other than the one at $z = a$. From this we have

$$\oint_C f(z)\,dz = 2\pi i A_{-1}$$

which gives another, and extraordinarily useful, technique for evaluating complex line integrals over closed curves. Once the value of A_{-1} is known, the value of the integral is determined.

Definition 15.5 The number A_{-1}, which is the coefficient of $1/(z-a)$ in the Laurent expansion of the function with center at $z = a$, is called the *residue* of the function at $z = a$.

Example 15.24 Evaluate $\oint_C z^{-4} \sin z\,dz$ where C is the unit circle $|z| = 1$. (See Figure 15.8.)

Solution The only pole of the integrand is of order 3 at $z = 0$. (Prove this!) The Laurent series for $z^{-4} \sin z$ about $z = 0$ is

$$z^{-3} - \frac{z^{-1}}{3!} + \frac{z}{5!} - \frac{z^3}{7!} + \cdots$$

Therefore $A_{-1} = -\frac{1}{6}$ and hence

$$\oint_C z^{-4} \sin z\,dz = -\frac{\pi i}{3}$$ □

Example 15.25 Evaluate $\oint_C \cot z\,dz$ where C is the circle $|z| = 1$ traversed clockwise.

Solution This is an example where the computation of the entire Laurent series would be tedious. Since only A_{-1} is required, a "long division" process may be used on the quotient of two series for $\sin z$ and $\cos z$.

$$\cot z = \frac{\cos z}{\sin z} = \frac{1 - \dfrac{z^2}{2} + \dfrac{z^4}{6} + \cdots}{z - \dfrac{z^3}{6} + \dfrac{z^5}{5!} + \cdots} = \frac{1}{z} + \cdots$$

which means that the residue of $\cot z$ at 0 is 1. Therefore

$$\oint_C \cot z\, dz = 2\pi i$$

if C is traversed in a counterclockwise sense, or

$$\oint_{\substack{C \\ (\mathrm{cw})}} \cot z\, dz = -2\pi i \qquad \square$$

Some special techniques are available for computing A_{-1} other than by finding the complete Laurent expansion. In particular, if the pole is simple, the residue may be found easily. When the function has a simple pole at $z = a$, the Laurent expansion has the form

$$f(z) = \frac{A_{-1}}{z-a} + A_0 + A_1(z-a) + A_2(z-a)^2 + \cdots$$

If we multiply both sides of this equality by $z - a$ and take the limit as z approaches a we obtain

$$\lim_{z \to a} (z-a)f(z) = A_{-1} = \operatorname*{Res}_{z=a} f(z)$$

Example 15.26 Let

$$f(z) = \frac{(z-1)(z-2)}{z(z-3)}$$

Find the residue of $f(z)$ at $z = 0$ and at $z = 3$.

Solution Since the poles are both simple, the formula just derived may be applied.

$$\operatorname*{Res}_{z=0} f(z) = \lim_{z \to 0} z f(z) = \frac{(-1)(-2)}{(-3)} = -\frac{2}{3}$$

$$\operatorname*{Res}_{z=3} f(z) = \lim_{z \to 3} (z-3) f(z) = \frac{(2)(1)}{3} = \frac{2}{3} \qquad \square$$

The expression for the residue at simple poles can be simplified even further in the case where $f(z)$ can be expressed as the quotient of two analytic functions. This method is especially appropriate if it is inconvenient or impractical to factor the denominator into linear factors. Suppose f has a simple pole at $z = a$ and that $f(z) = h(z)/g(z)$ where h and g have no common zeros. The fact that f has a simple pole at $z = a$ means $g(a) = 0$ and $g'(a) \neq 0$. Therefore,

$$\operatorname*{Res}_{z=a} f(z) = \lim_{z\to a}(z-a)\frac{h(z)}{g(z)}$$

$$= \lim_{z\to a}\frac{h(z)}{\dfrac{g(z)-g(a)}{z-a}}$$

$$= \frac{h(a)}{g'(a)}$$

Example 15.27 Find

$$\operatorname*{Res}_{z=i}\frac{1}{z^4-1}$$

Solution The pole at $z = i$ is simple. Hence

$$\operatorname*{Res}_{z=i}\frac{1}{z^4-1} = \left.\frac{1}{4z^3}\right|_{z=i} = -\frac{1}{4i} = \frac{i}{4} \qquad \square$$

Example 15.28 Find $\operatorname*{Res}_{z=0}\cot z$ by the foregoing technique.

Solution

$$\operatorname*{Res}_{z=0}\cot z = \operatorname*{Res}_{z=0}\frac{\cos z}{\sin z} = \frac{\cos 0}{\dfrac{d}{dz}(\sin z)|_{z=0}} = 1$$

Actually, this is the residue at *any* pole of $\cot z$. □

Variations of the above techniques may be used for poles of higher order. Rather than deriving formulas applicable to these more infrequent cases, we will be content with an example.

Example 15.29 Find

$$\operatorname*{Res}_{z=1}\frac{z^2}{(z-1)^3(z+2)}$$

Solution Since there is a third-order pole at $z = 1$, the Laurent series for $f(z)$ with center at 1 is of the form

$$\frac{z^2}{(z-1)^3(z+2)} = \frac{A_{-3}}{(z-1)^3} + \frac{A_{-2}}{(z-1)^2} + \frac{A_{-1}}{z-1} + A_0 + A_1(z-1) + \cdots$$

If we multiply both sides by $(z-1)^3$ we have

$$\frac{z^2}{z+2} = A_{-3} + A_{-2}(z-1) + A_{-1}(z-1)^2 + A_0(z-1)^3 + \cdots$$

Differentiating both sides twice and taking the limit as $z \to 1$, we obtain

$$\lim_{z\to 1} \frac{d^2}{dz^2}\left(\frac{z^2}{z+2}\right) = 2A_{-1}$$

and therefore

$$\operatorname*{Res}_{z=1} f(z) = \frac{4}{27}$$ □

Exercises for Section 15.8

In Exercises 1–15 find the residues at each of the isolated singularities of the given functions.

1. $\dfrac{e^z}{(z-1)^2}$
2. $\dfrac{\sin z}{z+2i}$
3. $\tan z$
4. $z^4 e^{1/z}$
5. $\sec z$
6. $e^{1/z}$
7. $\dfrac{1}{z^4+1}$
8. $\dfrac{1}{z^7+i}$
9. $\dfrac{3+4z}{z^3+3z^2+2z}$
10. $\dfrac{1}{z^2}$
11. $\dfrac{\cos z}{z^3}$
12. $\dfrac{1}{e^z-1}$
13. $\dfrac{z}{(z^2-1)^2}$
14. $\dfrac{z^2}{1-z^4}$
15. $\dfrac{e^z}{z^2(z+1)^3}$

Evaluate the following integrals over the unit circle $|z| = 1$.

16. $\displaystyle\oint_C \frac{z+1}{z(z+2)}\,dz$
17. $\displaystyle\oint_C \frac{z+1}{2z+1}\,dz$
18. $\displaystyle\oint_C \frac{dz}{\sin z}\,dz$
19. $\displaystyle\oint_C \frac{dz}{\cos z}$
20. $\displaystyle\oint_C \frac{e^z}{z}\,dz$
21. $\displaystyle\oint_C \frac{z^2+4}{(z-2)^2}\,dz$
22. $\displaystyle\oint_C z^2 e^{1/z}\,dz$
23. $\displaystyle\oint_C \frac{e^z}{(2z-1)^2}\,dz$
24. $\displaystyle\oint_C e^z \csc z\,dz$
25. $\displaystyle\oint_C \tan 2z\,dz$
26. $\displaystyle\oint_C \frac{dz}{z^2 \sin z}$

27. Evaluate

$$\oint_C \frac{z^3 + 1}{(z + 1)(z^2 + 9)} dz$$

where C is the square with vertices at 2, $2i$, -2, and $-2i$.

In the previous chapter we saw that the circulation around a closed contour C in a two-dimensional fluid flow can be expressed in terms of a complex line integral as

$$\Gamma(C) = \operatorname{Re} \oint_C f(z)\, dz$$

(see p. 533). When the fluid is incompressible and the flow is irrotational, the function $f(z)$ is analytic in the flow field.

For each of the following functions associated with a particular potential flow, determine the residues at all isolated singularities and then evaluate the circulation around the unit circle $|z| = 1$.

28. Flow around a circle (cylinder) of radius a, $|a| < 1$:

$$f(z) = U_0\left(1 - \frac{a^2}{z^2}\right) + \frac{Ai}{2\pi z}, \qquad (A \text{ real})$$

29. Flow around a flat plate of unit length centered at the origin and making an angle α with the positive x-axis:

$$f(z) = 2(1 + e^{-2i\alpha}) + \frac{4z}{\sqrt{4z^2 - e^{2i\alpha}}}(1 - e^{-2i\alpha}), \qquad (\alpha \text{ real})$$

30. Uniform flow combined with a source flow and a sink flow:

$$f(z) = U_0 + \frac{Ba}{a^2 - z^2} \qquad (|a| \neq 1,\ B \text{ real})$$

31. Flow around a so-called Joukowski airfoil:

$$f(z) = U_0\left[1 - \frac{1}{k^2(z)}\right]k'(z) + \frac{Ai}{2\pi k(z)}k'(z), \qquad (A \text{ real})$$

where

$$k(z) \equiv \frac{1}{4}[3 + 2z + \sqrt{4z^2 - 1}]$$

(*Hint:* In evaluating the circulation ignore the singularities of $k'(z)$ and use only the residue arising from the zero of $k(z)$.)

15.9 The Residue Theorem

We come now to the final result of this chapter. Virtually all the previous techniques of integration around closed curves of functions with isolated singularities are special cases of the following theorem. It is the major elementary application of residues.

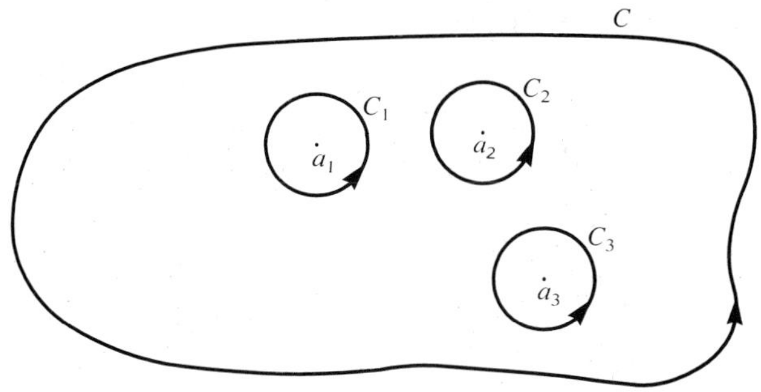

Figure 15.9

Theorem 15.9 *Residue Theorem.* Let f be analytic inside and on a closed contour C *except for finitely many isolated singular points* $a_1, a_2, a_3, \ldots, a_m$ *inside* C. (See Figure 15.9.) Then

$$\oint_C f(z)\,dz = 2\pi i \sum_{j=1}^{m} \operatorname*{Res}_{z=a_j} f(z)$$

Proof The method of proof is to enclose each of the singularities a_j by a circle C_j such that any two such circles do not intersect. Then f is analytic in the domain which is interior to C but exterior to the C_j, and thus the Cauchy-Goursat theorem for multiply connected domains may be applied. (See Figure 14.8.)

$$\oint_C f(z)\,dz + \sum_{j=1}^{m} \oint_{\substack{C_j \\ (\text{cw})}} f(z)\,dz = 0$$

so that

$$\oint_C f(z)\,dz = \sum_{j=1}^{m} \oint_{C_j} f(z)\,dz$$

Each of the integrals around the closed paths C_j is recognizable as $2\pi i \operatorname{Res}_{z=a_j} f(z)$ and the theorem is proved. ■

Example 15.30 Evaluate

$$\oint_C \frac{z^2 - 2}{z(z^2 - 1)}\,dz$$

where C is as follows (see Figure 15.10):

(a) The circle $|z| = 2$
(b) The circle $|z - 1| = 1/2$
(c) The circle $|z| = 1$
(d) The circle $|z - 2| = 1/2$

Solution (a) The integrand has singularities at 0, 1, and -1. The path of integration encloses all the singularities. Therefore

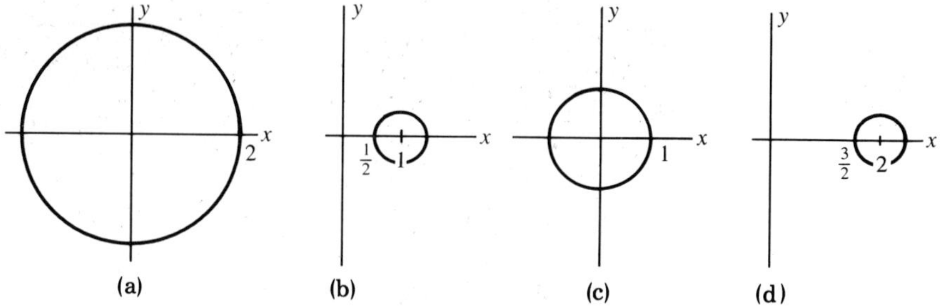

Figure 15.10

$$\oint_{|z|=2} \frac{z^2-2}{z(z^2-1)}\,dz = 2\pi i \sum_{j=1}^{3} \operatorname*{Res}_{z=a_j} f(z)$$

$$= 2\pi i\left[\operatorname*{Res}_{z=0} f(z) + \operatorname*{Res}_{z=1} f(z) + \operatorname*{Res}_{z=-1} f(z)\right]$$

$$= 2\pi i(2 - \tfrac{1}{2} - \tfrac{1}{2}) = 2\pi i$$

(b) The circle $|z| = 1/2$ encloses only the singularity at $z = 1$. Therefore

$$\oint_{|z|=1/2} \frac{z^2-2}{z(z^2-1)}\,dz = 2\pi i \operatorname*{Res}_{z=1} f(z)$$

$$= -\pi i$$

(c) The path $|z| = 1$ passes through two of the singularities of the integrand and therefore the integral is undefined.
(d) The circle $|z - 2| = 1/2$ encloses no singularities and hence, from the Cauchy-Goursat theorem, the value of the integral is zero. □

Exercises for Section 15.9

Evaluate the following integrals around the indicated circles.

1. $\oint_C \frac{dz}{\sin z}, \quad |z| = 6$

2. $\oint_C \frac{dz}{\cos z}, \quad |z| = 10$

3. $\oint_C \frac{e^z\,dz}{z(z+2)}, \quad |z| = 3$

4. $\oint_C \frac{dz}{1+3z^2}, \quad |z| = 1$

5. $\oint_C \frac{z\,dz}{4+z^2}, \quad |z| = 3$

6. $\oint_C \frac{(z+2)^3}{z^4+4z^3+5z^2}\,dz, \quad |z| = 1$

7. $\oint_C \frac{z^2+1}{(z+1)(z-3)}\,dz, \quad |z| = 5$

8. $\oint_C \tan z\,dz, \quad |z| = 1$

9. $\oint_C \tan z\,dz, \quad |z| = 3$

10. $\oint_C \frac{1}{e^z}\,dz, \quad |z| = 1$

11. $\oint_C \frac{dz}{e^z - e^{-z}}, \quad |z| = 4$

12. $\oint_C \frac{\sinh \frac{1}{2}\pi z}{z^2+1}\,dz, \quad |z| = 2$

13. $\oint_C \frac{e^{z^2-2z-1}}{z^4}dz, \quad |z| = 1$

14. $\oint_C \frac{dz}{z^3(z+4)}, \quad |z+2| = 3$

15. $\oint_C \frac{3+2z}{z^4+1}dz, \quad |z| = 2$

16. Evaluate the integral

$$\oint_C \frac{dz}{(z+1)(z+2)(z+3)}$$

where C is as follows:
(a) $|z| = 4$ **(b)** $|z| = \frac{1}{2}$ **(c)** $|z| = \frac{3}{2}$ **(d)** $|z| = \frac{5}{2}$

17. Evaluate the integral

$$\oint_C \frac{(z+2)^2}{2z^3+z}dz$$

where C is as follows:
(a) $|z| = \frac{1}{2}$ **(b)** $|z| = 2$ **(c)** $|z-1| = 3$ **(d)** $|z-1| = 1$

18. Evaluate the integral

$$\oint_C \frac{6z^2+3z+1}{z^2(z^3+1)}dz$$

where C is as follows:
(a) $|z| = \frac{1}{2}$ **(b)** $|z-1| = 3$ **(c)** $|z-2| = 3$
(d) $|z-3| = 1$ **(e)** $|z+1| = \frac{1}{2}$

19. Evaluate the integral

$$\oint_C \frac{(z^2+2)^3}{z^2-2z}dz$$

where C is as follows:
(a) $|z| = 1$ **(b)** $|z| = 3$ **(c)** $|z| = 4$ **(d)** $|z-1| = 3$
(e) $|z-2| = 1$

The evaluation of a number of the integrals encountered above can be simplified by first making a change of variable in the integral as suggested earlier in the exercises on page 548. Carry out the substitution $z = 1/w$ in each of the following five integrals and then evaluate.

20. The integral of Exercise 7

21. The integral of Exercise 14 for $C\colon |z| = 5$

22. The integral of Exercise 15

23. The integral of Exercise 16 for $C\colon |z| = 4$

24. The integral of Exercise 18 for $C\colon |z| = 2$

25. Let $f(z)$ be analytic inside and on a closed contour C. Furthermore, assume that $f(z)$ is nonzero on C but has a single simple zero z_0 inside C. Prove that

$$\frac{1}{2\pi i}\oint_C \frac{f'(z)}{f(z)}dz = 1$$

26. Generalize the result of Exercise 25 to show that if the single zero z_0 has multiplicity (order) n,

$$\frac{1}{2\pi i}\oint_C \frac{f'(z)}{f(z)}\,dz = n$$

(*Hint:* Recall Exercise 18, p. 548.)

27. Let $f(z)$ be analytic inside and on a closed contour C except for a single pole z_0 of order (multiplicity) m. Demonstrate that if $f(z)$ is nonzero inside and on C, then

$$\frac{1}{2\pi i}\oint_C \frac{f'(z)}{f(z)}\,dz = -m$$

(*Hint:* Rewrite $f(z)$ as $(z - z_0)^{-m}g(z)$.)

28. Combine the results of Exercises 26 and 27 to establish that if:
(i) $f(z)$ is analytic inside and on a closed contour C except for poles inside C at the points $b_1, b_2, \ldots, b_n$, and
(ii) $f(z)$ has zeros $a_1, a_2, \ldots, a_m$ inside C but none on C itself,
then

$$\frac{1}{2\pi i}\oint_C \frac{f'(z)}{f(z)}\,dz = N - P$$

$$= \sum_{k=1}^{m} \alpha_k - \sum_{k=1}^{n} \beta_k$$

where α_k and β_k are the orders (multiplicities) of the zeros a_k and poles b_k, respectively.

29. Generalize Exercise 28 to show that under the same conditions on $f(z)$, if $\phi(z)$ is analytic inside and on C

$$\frac{1}{2\pi i}\oint_C \phi(z)\frac{f'(z)}{f(z)}\,dz = \sum_{k=1}^{m} \alpha_k\phi(a_k) - \sum_{k=1}^{n} \beta_k\phi(b_k)$$

15.10 Looking Ahead

The conclusion arrived at in Exercises 26–28 above (and hinted at in Exercises 18 and 19, p. 548) constitutes one of the most useful counting theorems in modern analysis. The result is known as the Argument Principle, since the given contour integral is also equal to the (net) increase in the argument (phase) of the image point $w = f(z)$, measured in multiples of 2π, as z makes one complete circuit around the contour C in the positive direction. More detailed treatments of the argument principle appear in many places (see [12] and [15], for example); application of the result to stability considerations for linear systems is given in [6] (see also [10]).

The equivalence of analytic functions and (convergent) power series established in this chapter raises a natural question: if the only information we have about an analytic function is its representation as a power series, can we compute the values of the function outside its circle of convergence when we know it is analytic in a larger region? The answer to this query is a resounding yes, and the comprehensive collection of concepts that makes possible such calculations is of significant theoretical and practical utility. References such as [13] and [14] provide solid introductions to the basic notions of analytic continuation; [4] and [8] present more advanced discussions.

Sometimes the power series expansion of an analytic function is convergent for all finite z. Such a function is called entire since its power series representation is valid in the entire complex plane. In certain applications it is useful to have an alternative expression for such a function, one that displays the position and orders (multiplicities) of the function's zeros. Product factorizations with this property are covered in [7] and [9] along with still other representations that are particularly advantageous for analytic functions with only pole singularities—so-called meromorphic functions. The general theory of entire functions is considered in substantial detail in [5], [11], and the classic [16].

Basic References

[1] Hyslop, J. M. 1950. *Infinite Series*. 4th ed. rev. Edinburgh: Oliver and Boyd.

[2] Knopp, K. 1956. *Infinite Sequences and Series*. New York: Dover.

[3] Markushevich, A. I. 1967. *Infinite Series*. Boston: D. C. Heath.

More-Advanced References

[4] Ahlfors, L. V. 1979. *Complex Analysis*. 3d ed. New York: McGraw-Hill.

[5] Boas, R. 1954. *Entire Functions*. New York: Academic Press.

[6] Cochran, J. A. 1982. *Applied Mathematics: Principles, Techniques, and Applications*. Belmont, Calif.: Wadsworth.

[7] DePree, J. D., and C. C. Oehring. 1969. *Elements of Complex Analysis*. Reading, Mass.: Addison-Wesley.

[8] Hille, E. 1973. *Analytic Function Theory, Vol. II*. 2d ed. New York: Chelsea.

[9] Knopp, K. 1947. *Theory of Functions, Part II*. New York: Dover.

[10] Kyrala, A. 1972. *Applied Functions of a Complex Variable*. New York: John Wiley.

[11] Markushevich, A. I. 1966. *Entire Functions*. New York: American Elsevier.

[12] Nehari, Z. 1968. *Introduction to Complex Analysis*. rev. ed. Boston: Allyn and Bacon.

[13] Saff, E. B., and A. D. Snider. 1976. *Fundamentals of Complex Analysis for Mathematics, Science, and Engineering*. Englewood Cliffs, N. J.: Prentice-Hall.

[14] Silverman, H. 1975. *Complex Variables*. Boston: Houghton Mifflin.

[15] Silverman, R. A. 1974. *Complex Analysis with Applications*. Englewood Cliffs, N. J.: Prentice-Hall.

[16] Valiron, G. 1949. *Lectures on the General Theory of Integral Functions*. New York: Chelsea.

Chapter 16

Applications of Residue Theory

16.1 Introduction

We have now come to the final chapter of this book, in which we consider the utility of the residue calculus. Surprisingly, the most immediate application of the principal result of the last chapter, the residue theorem, is in the evaluation of *real* integrals. In the next several sections we explore the use of Theorem 15.9 in carrying out the evaluations when the integrand has one of four distinct forms. Subsequently, we show how these same ideas can be applied in problems involving operational methods such as the Fourier and Laplace transforms. Applications here abound and we consider several representative practical situations. In all cases, the trick is to express the integral of interest in terms of a complex integration around a closed contour. The general theory is somewhat limited, however, since as our examples suggest, success is often more a matter of technique than anything else.

16.2 Rational Trigonometric Functions

Real integrals whose integrands are rational functions of $\sin\theta$ and $\cos\theta$ and whose limits of integration are from 0 to 2π may be evaluated by using the residue theorem after an initial substitution $z = e^{i\theta}$. The relevant relations are

$$\cos\theta = \frac{1}{2}(e^{i\theta} + e^{-i\theta}) = \frac{1}{2}\left(z + \frac{1}{z}\right)$$

$$\sin\theta = \frac{1}{2i}(e^{i\theta} - e^{-i\theta}) = \frac{1}{2i}\left(z - \frac{1}{z}\right)$$

$$d\theta = \frac{d(e^{i\theta})}{ie^{i\theta}} = \frac{dz}{iz}$$

After substitution, the integrand is a rational function of z and the path of integration in the complex plane is the circle $|z| = 1$.

Example 16.1 Evaluate

$$\int_0^{2\pi} \frac{d\theta}{5 - 3\cos\theta}$$

Solution We first write $\cos\theta$ in terms of $e^{i\theta}$:

$$\int_0^{2\pi} \frac{d\theta}{5 - 3\cos\theta} = \int_0^{2\pi} \frac{d\theta}{5 - 3\dfrac{e^{i\theta} + e^{-i\theta}}{2}}$$

Letting $z = e^{i\theta}$, the real integral is transformed into an integral in the complex plane over the unit circle:

$$\begin{aligned}\int_0^{2\pi} \frac{d\theta}{5 - 3\cos\theta} &= \oint_{|z|=1} \frac{\dfrac{dz}{iz}}{5 - \frac{3}{2}(z + z^{-1})} \\ &= \frac{1}{i}\oint_{|z|=1} \frac{dz}{5z - \frac{3}{2}z^2 - \frac{3}{2}} \\ &= -\frac{2}{i}\oint_{|z|=1} \frac{dz}{3z^2 - 10z + 3} \\ &= -\frac{2}{i}\oint_{|z|=1} \frac{dz}{(3z - 1)(z - 3)}\end{aligned}$$

The integrand has simple poles at $z = 3$ and $z = \frac{1}{3}$. Hence

$$\begin{aligned}\int_0^{2\pi} \frac{d\theta}{5 - 3\cos\theta} &= 2\pi i\left(-\frac{2}{i}\right)\left(\operatorname*{Res}_{z=1/3} \frac{1}{(3z - 1)(z - 3)}\right) \\ &= -4\pi\left(-\frac{1}{8}\right) = \frac{\pi}{2}\end{aligned}$$

(see Figure 16.1). □

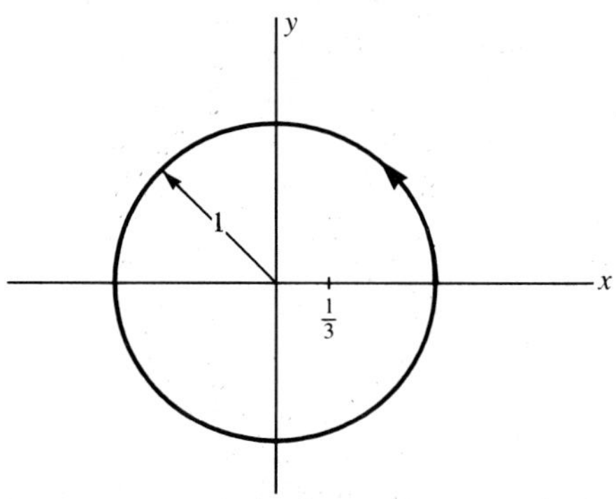

Figure 16.1

If the interval of integration for the real integral is not from 0 to 2π, the integrand may sometimes be modified before applying the substitution.

Example 16.2 Evaluate

$$\int_0^{\pi} \frac{d\theta}{5 - 3\cos\theta}$$

Solution Under the substitution $\phi = 2\pi - \theta$, the integral of this integrand over the interval $(\pi, 2\pi)$ is seen to be identical with the integral over $(0, \pi)$. We have, therefore,

$$\int_0^{\pi} \frac{d\theta}{5 - 3\cos\theta} = \frac{1}{2}\int_0^{2\pi} \frac{d\theta}{5 - 3\cos\theta} = \frac{\pi}{4}$$ □

Exercises for Sections 16.1 and 16.2

Evaluate the following real integrals.

1. $\displaystyle\int_0^{2\pi} \frac{dx}{3 + \cos x}$

2. $\displaystyle\int_0^{2\pi} \frac{dx}{a + \cos x}, \quad a > 1$

3. $\displaystyle\int_0^{2\pi} \frac{dx}{1 + a\cos x}, \quad 0 < a < 1$

4. $\displaystyle\int_0^{2\pi} \frac{dx}{\frac{5}{4} + \sin x}$

5. $\displaystyle\int_0^{2\pi} \frac{\sin^2 x}{5 - 4\cos x}\,dx$

6. $\displaystyle\int_0^{2\pi} \frac{dx}{5 - 4\cos 2x}$

7. $\displaystyle\int_0^{\pi/2} \frac{dx}{1 + \sin^2 x}$

8. $\displaystyle\int_0^{\pi/2} \frac{dx}{a + \sin^2 x}, \quad a > 0$

9. $\displaystyle\int_0^{\pi} \frac{dx}{a + \cos^2 x}, \quad a > 1$

10. $\displaystyle\int_0^{2\pi} \frac{dx}{1 - 2a\sin x + a^2}, \quad 0 < a < 1$

11. $\displaystyle\int_0^{2\pi} \frac{dx}{1 - 2a\cos x + a^2}, \quad 0 < a < 1$

12. $\displaystyle\int_0^{2\pi} \frac{dx}{(1 + a\cos x)^2}, \quad 0 < a < 1$

13. $\displaystyle\int_0^{2\pi} \frac{dx}{(a + \sin x)^2}, \quad a > 1$

14. $\displaystyle\int_0^{\pi} \frac{dx}{a^2\sin^2 x + b^2\cos^2 x}, \quad a, b > 0$

15. $\displaystyle\int_0^{\pi} \tan(x + ia)\,dx, \quad a > 0$

16.3 Rational Functions

For these considerations, we need the following well-known definitions and theorem concerning improper real integrals.

Definition 16.1 The improper integral $\int_{-\infty}^{\infty} f(x)\,dx$ converges in the ordinary sense to a number L (called its *value*) if

$$\lim_{B\to\infty}\int_0^B f(x)\,dx + \lim_{A\to\infty}\int_{-A}^0 f(x)\,dx = L$$

Definition 16.2 The improper integral $\int_{-\infty}^{\infty} f(x)\,dx$ converges in the Cauchy sense to a number L_c (called the *Cauchy principal value*, C.P.V.) if

$$\lim_{A\to\infty}\int_{-A}^{A} f(x)\,dx = L_c$$

Note that for ordinary convergence, the limiting processes at the upper and lower limit are independent of one another whereas in finding the C.P.V. they are dependent.

Theorem 16.1 If an improper real integral $\int_{-\infty}^{\infty} f(x)\,dx$ converges in the ordinary sense, it converges in the Cauchy sense and, moreover, the value and the C.P.V. are the same. ■

In some improper real integrals, negative contributions on one side of the origin cancel out positive ones on the other side so that the integral converges in the Cauchy sense but not in the ordinary sense.

Example 16.3 Discuss $\int_{-\infty}^{\infty} x\,dx$.

Solution

$$\begin{aligned}\int_{-\infty}^{\infty} x\,dx &= \lim_{B\to\infty}\int_0^B x\,dx + \lim_{A\to\infty}\int_{-A}^0 x\,dx\\ &= \lim_{B\to\infty}\frac{B^2}{2} - \lim_{A\to\infty}\frac{A^2}{2}\end{aligned}$$

which doesn't exist.

On the other hand,

$$\text{C.P.V.}\int_{-\infty}^{\infty} x\,dx = \lim_{A\to\infty}\int_{-A}^{A} x\,dx = \lim_{A\to\infty}\left(\frac{A^2}{2}-\frac{A^2}{2}\right) = 0 \qquad \square$$

In cases where the integrand does not change sign, of course, the Cauchy and ordinary approaches must be equivalent.

In this section we show how the Cauchy principal value of improper real integrals with rational integrands may be determined. In virtually all our examples and exercises, since the integrals will actually converge in the usual sense, the C.P.V. obtained will also be the value in the ordinary sense.

In the following general outline of the technique, we assume that $f(x)$ is a real function of the order of $1/x^p$ for some integer $p > 1$, that is, $\lim_{x\to\infty} x^p f(x) =$ a constant. For rational functions this means that the polynomial in the denominator has degree at least *two* greater than the degree of the polynomial in the numerator.

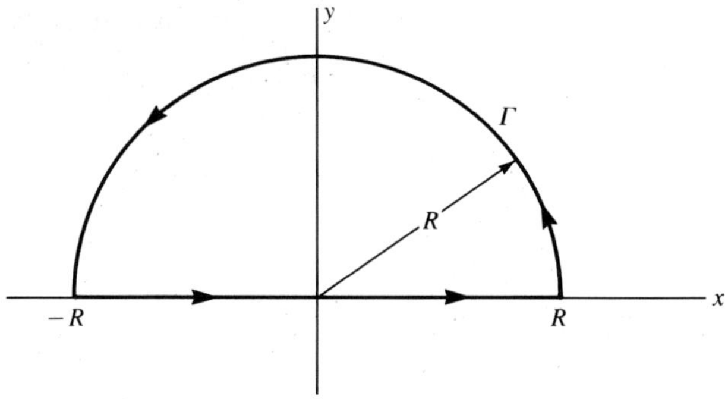

Figure 16.2 *Semicircular contour*

To find C.P.V. $\int_{-\infty}^{\infty} f(x)\,dx$ by *contour integration*, we specify

1. the function of a complex variable which serves as the integrand, and
2. the path or "contour" of integration.

In the case where the real integrand is a rational function of x we simply change x to z and use a path of integration which bounds a semicircular region in the upper half-plane. Then

$$\oint_C f(z)\,dz = \int_{-R}^{R} f(x)\,dx + \int_\Gamma f(z)\,dz$$

where Γ is the semicircular path shown in Figure 16.2. Hence

$$\int_{-R}^{R} f(x)\,dx = \oint_C f(z)\,dz - \int_\Gamma f(z)\,dz$$

and therefore

$$\text{C.P.V.} \int_{-\infty}^{\infty} f(x)\,dx = 2\pi i \sum \operatorname{Res} f(z) - \lim_{R\to\infty} \int_\Gamma f(z)\,dz$$

where the summation is over the points in the *upper* half-plane that are singularities of $f(z)$. In the last integral on the right, represent Γ by $z = Re^{it}$. Then

$$\int_\Gamma f(z)\,dz = i \int_0^{\pi} f(Re^{it})\,Re^{it}\,dt$$

Therefore

$$\left|\int_\Gamma f(z)\,dz\right| \leqq \int_0^{\pi} |f(Re^{it})|\,R\,|e^{it}\,dt|$$

If we assume $f(x)$ to be of the order of $1/x^p (p > 1)$, it follows that for large z, $|f(z)| < M/R^p$ and hence

$$\left|\int_\Gamma f(z)\,dz\right| \leqq \frac{M}{R^p} R \int_0^{\pi} |e^{it}\,dt| = \frac{M}{R^p} R\pi = \frac{M\pi}{R^{p-1}}$$

which approaches 0 as R becomes unbounded. Therefore

$$\text{C.P.V.} \int_{-\infty}^{\infty} f(x)\,dx = 2\pi i \sum \operatorname{Res} f(z)$$

Example 16.4 Evaluate

$$\int_0^{\infty} \frac{dx}{1 + x^4}$$

Solution

$$\int_0^{\infty} \frac{dx}{1 + x^4} = \frac{1}{2}\text{C.P.V.} \int_{-\infty}^{\infty} \frac{dx}{1 + x^4}$$
$$= \pi i \sum \operatorname{Res}\left(\frac{1}{1 + z^4}\right)$$

The sum is taken over the simple poles in the upper half-plane at $z_1 = e^{\pi i/4}$ and $z_2 = e^{3\pi i/4}$. Since

$$\operatorname*{Res}_{z=z_j}\left(\frac{1}{1 + z^4}\right) = \frac{1}{4z_j^3}$$

Then

$$\int_0^{\infty} \frac{dx}{1 + x^4} = \pi i\left(\frac{1}{4e^{3\pi i/4}} + \frac{1}{4e^{\pi i/4}}\right)$$
$$= \frac{\pi i}{4}(-e^{\pi i/4} + e^{-\pi i/4})$$
$$= \frac{\pi}{2}\sin\frac{\pi}{4} = \frac{\pi}{2\sqrt{2}}$$

□

16.4 Exponential and Power Functions

The (semi-) circular contour is no longer appropriate for real integrals wherein the integrand consists solely of exponential functions. In such situations it is often important to take advantage of the periodicity of these functions in the complex plane.

Example 16.5 Evaluate

$$\int_{-\infty}^{\infty} \frac{e^{mx}}{1 + e^x}\,dx, \qquad 0 < m < 1$$

Solution Choose $e^{mz}/(1 + e^z)$ as the function of a complex variable to integrate and let C be the contour shown in Figure 16.3.

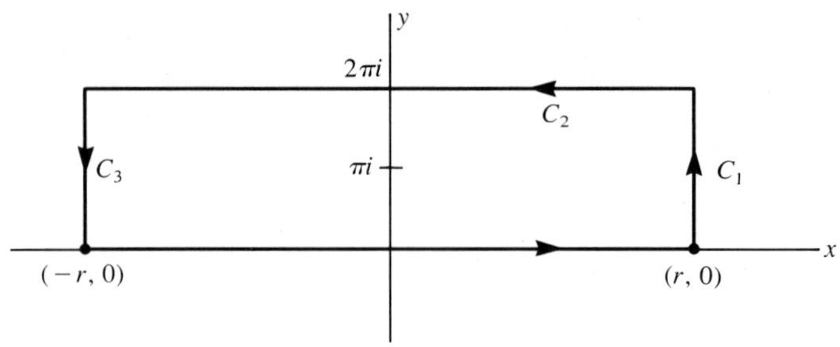

Figure 16.3

The function $e^{mz}/(1+e^z)$ has only the one pole at $z=\pi i$ within the contour. Since

$$\operatorname*{Res}_{z=\pi i}\frac{e^{mz}}{1+e^z}=\frac{e^{m\pi i}}{e^{\pi i}}=-e^{m\pi i}$$

we have by the residue theorem

$$\oint_C \frac{e^{mz}}{1+e^z}\,dz=-2\pi i e^{m\pi i}$$

Further,

$$\oint_C \frac{e^{mz}}{1+e^z}\,dz=\int_{C_1}\frac{e^{mz}}{1+e^z}\,dz+\int_{C_2}\frac{e^{mz}}{1+e^z}\,dz+\int_{C_3}\frac{e^{mz}}{1+e^z}\,dz+\int_{-r}^{r}\frac{e^{mx}}{1+e^x}\,dx$$

On C_1, $z=r+iy$, so that

$$\left|\int_{C_1}\frac{e^{mz}}{1+e^z}\,dz\right|\leqq\int_{C_1}\frac{|e^{mz}|}{|1+e^z|}|dz|\leqq|e^{mr}|\int_{C_1}\frac{|e^{myi}||dz|}{|1-|e^r||e^{iy}||}$$
$$\leqq\frac{e^{mr}2\pi}{e^r-1}=\frac{2\pi e^{(m-1)r}}{1-e^{-r}}$$

Since $m<1$, this becomes arbitrarily small as r becomes unbounded.

A similar argument shows that

$$\lim_{r\to\infty}\left|\int_{C_3}\frac{e^{mz}}{1+e^z}\,dz\right|=0$$

On C_2, $z=x+2\pi i$. Therefore

$$\int_{C_2}\frac{e^{mz}}{1+e^z}\,dz=\int_{r}^{-r}\frac{e^{mx}e^{2m\pi i}}{1+e^xe^{2\pi i}}\,dx=-e^{2m\pi i}\int_{-r}^{r}\frac{e^{mx}}{1+e^x}\,dx$$

and thus

$$\lim_{r\to\infty}(1-e^{2m\pi i})\int_{-r}^{r}\frac{e^{mx}}{1+e^x}\,dx=\lim_{r\to\infty}\oint_C\frac{e^{mz}}{1+e^z}\,dz=-2\pi i e^{m\pi i}$$

from which we obtain

$$\int_{-\infty}^{\infty} \frac{e^{mx}}{1+e^x}\,dx = \text{C.P.V.} \int_{-\infty}^{\infty} \frac{e^{mx}}{1+e^x}\,dx = \frac{2\pi i e^{m\pi i}}{e^{2m\pi i}-1} = \frac{\pi}{\sin m\pi} \qquad \square$$

Whenever the integrand of the real integral under consideration includes fractional powers, care must be taken regarding the domain of definition of the complex functions which result when we change x to z.

Example 16.6 Evaluate

$$\int_0^{\infty} \frac{x^{p-1}}{1+x}\,dx, \qquad 0 < p < 1$$

Solution It is possible to change variables in the integral of the previous example to obtain the one given here. We give a different method of evaluation, however, which enables us to exhibit the kind of contour which is needed when branch functions are present. (See Figure 16.4.)

We note that z^{p-1} is a power function and agree for the purpose of maintaining single-valuedness to restrict z such that $0 \le \arg z < 2\pi$. Then since $z^{p-1}/(1+z)$ has a pole at $z = -1$ and nowhere else,

$$\oint_C \frac{z^{p-1}}{1+z}\,dz = 2\pi i \operatorname*{Res}_{z=-1} \frac{z^{p-1}}{1+z} = 2\pi i (e^{\pi i})^{p-1} = -2\pi i e^{\pi i p}$$

where C is the composite contour shown in Figure 16.4 in which C_U and C_L are on "opposite sides" of a cut along the positive x-axis.

On the other hand,

$$\int_C \frac{z^{p-1}}{1+z}\,dz = \left(\int_{C_U} + \int_{\Gamma} + \int_{C_L} + \int_{\gamma}\right) \frac{z^{p-1}}{1+z}\,dz$$

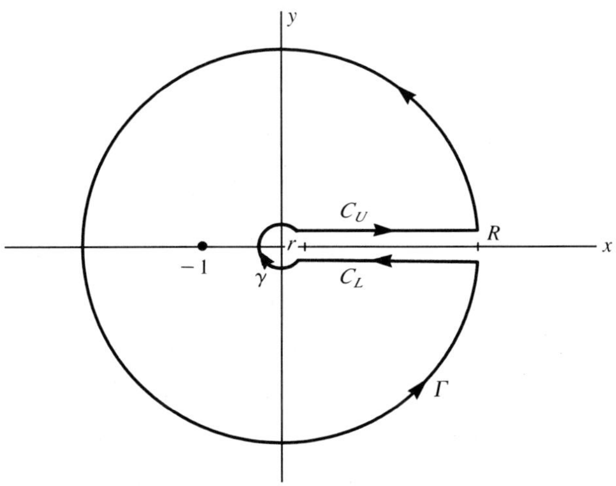

Figure 16.4

On γ and Γ the circular paths may be represented by $z = re^{it}$ and $z = Re^{it}$, respectively. With the given limitations on the parameter p, it is then relatively easy to show that the integrals over γ and Γ can be made arbitrarily small as $r \to 0$ and $R \to \infty$.

Moreover, in view of our restriction on $\arg z$, on C_U we have $z^{p-1} = x^{p-1}$ while on C_L,

$$\begin{aligned} z^{p-1} &= e^{(p-1)\ln z} \\ &= e^{(p-1)\log_e x} e^{(p-1)2\pi i} \\ &= x^{p-1} e^{2\pi i p} \end{aligned}$$

Hence

$$\begin{aligned} \oint_C \frac{z^{p-1}}{1+z} dz &= \lim_{\substack{r\to 0 \\ R\to\infty}} \left(\int_{C_U} + \int_{C_L} \right) \frac{z^{p-1}}{1+z} dz \\ &= \lim_{\substack{r\to 0 \\ R\to\infty}} \left[\int_r^R \frac{x^{p-1}}{1+x} dx + e^{2\pi i p} \int_R^r \frac{x^{p-1}}{1+x} dx \right] \end{aligned}$$

and therefore

$$(1 - e^{2\pi i p}) \int_0^\infty \frac{x^{p-1}}{1+x} dx = -2\pi i e^{\pi i p}$$

from which

$$\int_0^\infty \frac{x^{p-1}}{1+x} dx = \frac{\pi}{\sin \pi p} \qquad \square$$

Exercises for Sections 16.3 and 16.4

Find the value of the following improper real integrals.

1. $\displaystyle\int_{-\infty}^{\infty} \frac{dx}{1+x^4}$

2. $\displaystyle\int_0^{\infty} \frac{x^2}{1+x^6} dx$

3. $\displaystyle\int_{-\infty}^{\infty} \frac{x^5}{1+x^{10}} dx$

4. $\displaystyle\int_{-\infty}^{\infty} \frac{dx}{x^2+2x+2}$

5. $\displaystyle\int_{-\infty}^{\infty} \frac{x^2}{(1+x^2)^2} dx$

6. $\displaystyle\int_{-\infty}^{\infty} \frac{x}{(x^2-2x+2)^2} dx$

7. $\displaystyle\int_{-\infty}^{\infty} \frac{x^2}{x^4+6x^2+13} dx$

8. $\displaystyle\int_{-\infty}^{\infty} \frac{dx}{(x^2+1)^2(x^2+9)}$

9. $\displaystyle\int_0^{\infty} \frac{x^6}{(1+x^4)^2} dx$

10. $\displaystyle\int_{-\infty}^{\infty} \frac{x^4}{(x^2+1)^2(x^2+4)} dx$

11. $\displaystyle\int_{-\infty}^{\infty} \frac{dx}{(x^2+9)^2}$

12. $\displaystyle\int_{-\infty}^{\infty} \frac{x^2}{(x^2+1)^3} dx$

13. $\displaystyle\int_{-\infty}^{\infty} \frac{\cos mx}{e^x+e^{-x}} dx, \quad m > 0$

14. $\displaystyle\int_0^{\infty} \frac{x^{p-1}}{1+x^2} dx, \quad 0 < p < 2$

15. Do Example 16.6 by changing variables in the integral of Example 16.5.

Occasionally the reasoning and estimates of the last two sections can be applied not to evaluate real integrals but rather to sum infinite series of the form

$$\sum_{n=-\infty}^{\infty} a_n$$

We illustrate this technique by considering

$$I = \oint_{C_N} \frac{1}{z^2} \pi \cot \pi z \, dz$$

where C_N is the square with vertices $\pm(N + \frac{1}{2}) \pm i(N + \frac{1}{2})$ with N a large integer. On the one hand, the integrals along each of the sides of the square tend to zero as $N \to \infty$. On the other hand, by the residue theorem

$$\begin{aligned} I &= 2\pi i \sum_{n=-N}^{N} \operatorname*{Res}_{z=n} \frac{\pi \cot \pi z}{z^2} \\ &= 2\pi i \left(-\frac{1}{3}\pi^2 + 2 \sum_{n=1}^{N} \frac{1}{n^2} \right) \end{aligned}$$

Taking the limit as N becomes unboundedly large, we deduce that

$$\sum_{n=1}^{\infty} \frac{1}{n^2} = \frac{\pi^2}{6}$$

(Compare Exercise 8, p. 248.)

16. Integrate $\pi/(z^2 \sin \pi z)$ around the square C_N and show that

$$\sum_{n=1}^{\infty} \frac{(-1)^{n+1}}{n^2} = \frac{\pi^2}{12}$$

17. Let $f(z)$ be analytic in the entire complex z-plane except for poles at the points $b_1, b_2, \ldots, b_k$, none of which is an integer. Integrate $\pi f(z) \cot \pi z$ around the square C_n to deduce that if $|zf(z)| \to 0$ as $z \to \infty$,

$$\sum_{n=-\infty}^{\infty} f(n) = -\sum_{i=1}^{k} \operatorname*{Res}_{z=b_i} [\pi \cot \pi z \, f(z)]$$

18. In analogy with Exercise 17, show that under the same hypotheses on $f(z)$,

$$\sum_{n=-\infty}^{\infty} (-1)^n f(n) = -\sum_{i=1}^{k} \operatorname*{Res}_{z=b_i} \left[\frac{\pi}{\sin \pi z} f(z) \right]$$

Apply the results of Exercises 17 and 18 to determine the explicit values of the following sums:

19. $\displaystyle\sum_{n=-\infty}^{\infty} \frac{(-1)^n}{n^2 + a^2}, \quad a \neq 0$

20. $\displaystyle\sum_{n=-\infty}^{\infty} \frac{1}{(n - a)^2}, \quad a$ nonintegral

16.5 Fourier Integrals

The Fourier series representation of a function satisfying, say, the Dirichlet conditions (see Theorem 7.7) is useful for applications involving periodic functions or functions defined only on a finite interval. Frequently, however, the functions of interest are nonperiodic and defined for all values of the independent variable. In these situations the Fourier integral (transform) is often employed, as suggested earlier in Chapters 7 and 8.

The defining relationships for the Fourier integral representation are

$$f(x) = \frac{1}{2\pi}\int_{-\infty}^{\infty} F(\omega)e^{i\omega x}\,d\omega$$

where

$$F(\omega) = \int_{-\infty}^{\infty} f(x)e^{-i\omega x}\,dx$$

The Fourier integral expression for $f(x)$ can be obtained by considering a periodic function $f_p(x)$ that approaches f as p increases, and taking the limit, as p approaches infinity, of the complex exponential Fourier series. Indeed, if we take the expressions for a periodic function defined over $(-L, L)$ given earlier in Section 7.5, namely,

$$f(x) = \sum_{n=-\infty}^{\infty} c_n e^{in\pi x/L}$$

where

$$c_n = \frac{1}{2L}\int_{-L}^{L} f(x)e^{-in\pi x/L}\,dx$$

and make the associations $\omega = n\pi/L$ and $F(\omega) = 2Lc_n$, the infinite summation can be rewritten as

$$\frac{1}{2\pi}\sum_{n=-\infty}^{\infty} F(\omega)e^{i\omega x}\Delta\omega$$

with

$$F(\omega) = \int_{-L}^{L} f(x)e^{-i\omega x}\,dx$$

where $\Delta\omega = \pi/L$ and the values of the function $F(\omega)$ are calculated at the points $n\,\Delta\omega$. The new expression for the summation suggests the definition of a definite integral; and as $L \to \infty$ we formally obtain the relationships as stated.

In view of this derivation, the Fourier integral $F(\omega)$ can be thought of as measuring the intensity of f in the frequency interval $d\omega$ just as the coefficients in a Fourier series measure intensity or amplitude at a specific frequency. Moreover, the striking resemblance between the expression for $F(\omega)$ and a Laplace transform is not accidental as we will see in Section 16.7. We do not go far astray, therefore, by thinking of a Fourier integral as a sort of two-sided Laplace transform.

As an alternative expression for the Fourier integral representations, we also have

$$f(x) = \frac{1}{\pi}\int_0^\infty A(\omega)\cos\omega x\,d\omega + \frac{1}{\pi}\int_0^\infty B(\omega)\sin\omega x\,d\omega$$

where

$$A(\omega) = \int_{-\infty}^{\infty} f(x)\cos\omega x\,dx$$

$$B(\omega) = \int_{-\infty}^{\infty} f(x)\sin\omega x\,dx$$

In either representation, the integrals involved appear to be merely variants of those already discussed in this chapter. Indeed, it turns out for example, if the function $f(x)$ is rational, the determination of $F(\omega)$ or of $A(\omega)$ and $B(\omega)$ can be effected by a slight modification of the procedure outlined earlier in Section 16.3.

We assume that $f(x)$ is a real function of the order of $1/x^p$ for $p > 1$ and consider $\int_{-\infty}^{\infty} f(x)e^{i\omega x}\,dx$. (This will give us $\overline{F(\omega)}$.) Just as in Section 16.3, we let $x = z$ and use as the path of integration the semicircular contour depicted in Figure 16.2. Then

$$\int_{-R}^{R} f(x)e^{i\omega x}\,dx = \oint_C f(z)e^{i\omega z}\,dz - \int_\Gamma f(z)e^{i\omega z}\,dz$$

Since $|e^{i\omega z}| = e^{-wy} \leq 1$ (note that $y \geq 0$ in the upper half-plane and the expression for $f(x)$ in terms of $A(\omega)$ and $B(\omega)$ involves $\omega \geq 0$), the proof that

$$\lim_{R\to\infty}\left|\int_\Gamma f(z)e^{i\omega z}\,dz\right| = 0$$

is essentially the same as in the previous case and therefore,

$$\text{C.P.V.}\int_{-\infty}^{\infty} f(x)e^{i\omega x}\,dx = 2\pi i\sum \text{Res}\,[f(z)e^{i\omega z}]$$

As before, with the given hypothesis the improper integral actually converges in the ordinary sense, and $A(\omega)$ and $B(\omega)$ can be obtained by taking real and imaginary parts:

$$A(\omega) = \int_{-\infty}^{\infty} f(x)\cos\omega x\,dx = \text{Re}(2\pi i\sum \text{Res}[f(z)e^{i\omega z}])$$

$$B(\omega) = \int_{-\infty}^{\infty} f(x)\sin\omega x\,dx = \text{Im}(2\pi i\sum \text{Res}[f(z)e^{i\omega z}])$$

In applying these formulas it is well to remind ourselves that the summation is taken just over the poles in the *upper* half-plane.

Example 16.7 Find

$$\int_{-\infty}^{\infty} \frac{\cos\omega x}{x^2+4}\,dx, \qquad \omega \geq 0$$

Solution

$$\begin{aligned}\int_{-\infty}^{\infty} \frac{\cos \omega x}{x^2+4}\,dx &= \text{Re}\left(2\pi i \sum \text{Res}\frac{e^{i\omega z}}{z^2+4}\right)\\ &= \text{Re}\left(2\pi i \frac{e^{i\omega(2i)}}{2(2i)}\right)\\ &= \frac{\pi e^{-2\omega}}{2}\end{aligned}$$

□

We can use the result of the previous example in the following illustrative application.

Example 16.8 Determine an integral representation for the temperature in a thin bar of infinite length whose sides are well insulated if the bar has the initial temperature $u(x,0) = 1/(x^2+4)$.

Solution As we recall from Section 8.9, the mathematical model for this problem is given by the heat equation

$$u_{xx} = ku_t$$

and the auxiliary condition

$$u(x,0) = \frac{1}{x^2+4}$$

The analog of the series representation discussed earlier is

$$u(x,t) = \frac{1}{2\pi}\int_{-\infty}^{\infty} u(x,t;\lambda)\,d\lambda$$

where

$$u(x,t;\lambda) = F(\lambda)\exp\left(i\lambda x - \frac{\lambda^2 t}{k}\right)$$

with $F(\lambda)$ given by the Fourier integral of the initial temperature distribution

$$F(\lambda) = \int_{-\infty}^{\infty} u(x,0)e^{-i\lambda x}\,dx$$

Our particular auxiliary condition leads to

$$\begin{aligned}F(\lambda) &= \int_{-\infty}^{\infty} \frac{1}{x^2+4}e^{-i\lambda x}\,dx\\ &= \int_{-\infty}^{\infty} \frac{\cos \lambda x}{x^2+4}\,dx \qquad \text{(since } \sin \lambda x \text{ is odd)}\\ &= \frac{\pi}{2}e^{-2|\lambda|}\end{aligned}$$

by Example 16.7. Therefore

$$\begin{aligned} u(x,t) &= \frac{1}{4}\int_{-\infty}^{\infty} e^{-2|\lambda|}\exp\left(i\lambda x - \frac{\lambda^2 t}{k}\right)d\lambda \\ &= \frac{1}{2}\int_0^{\infty} e^{-2\lambda}\cos\lambda x\, e^{-\lambda^2 t/k}\,d\lambda \end{aligned}$$

(again because $\sin\lambda x$ and $\cos\lambda x$ are odd and even, respectively). □

The above example is indicative of the use of the Fourier integral in the solution of one-dimensional heat problems. A similar approach can be applied with equal ease to the solution of boundary valuc problems for Laplace's equation and initial value problems for the wave equation in infinite domains. We treat a representative case in the next example.

Example 16.9 Find the bounded, harmonic function $u(x, y)$ in the semi-infinite strip $0 < x < a$, $y > 0$, which satisfies the auxiliary conditions $u(x,0) = 0$, $u(0, y) = 0$, and $u(a, y) = 2y/(y^4 + 4)$.

Solution Laplace's equation

$$u_{xx} + u_{yy} = 0 \qquad 0 < x < a,\ y > 0$$

is to be solved with the given boundary conditions. Here then the analog of the series representations used previously in Section 8.5 is

$$u(x,y) = \frac{1}{\pi}\int_0^{\infty} u(x,y;\lambda)\,d\lambda$$

where

$$u(x,y;\lambda) = F(\lambda)\sinh\lambda x\sin\lambda y$$

with $F(\lambda)$ given by the Fourier integral (in alternative form)

$$F(\lambda)\sinh\lambda a = 2\int_0^{\infty} u(a,y)\sin\lambda y\,dy$$

For our specific boundary condition $u(a, y)$ we have

$$\begin{aligned} F(\lambda)\sinh\lambda a &= 2\int_0^{\infty}\frac{2y}{y^4+4}\sin\lambda y\,dy \\ &= \int_{-\infty}^{\infty}\frac{2y}{y^4+4}\sin\lambda y\,dy \\ &= \operatorname{Im}\left(2\pi i \operatorname*{Res}_{z=1+i}\frac{2ze^{i\lambda z}}{z^4+4} + 2\pi i \operatorname*{Res}_{z=-1+i}\frac{2ze^{i\lambda z}}{z^4+4}\right) \\ &= \pi e^{-\lambda}\sin\lambda \end{aligned}$$

It follows that

$$u(x, y) = \int_0^\infty e^{-\lambda}\left(\frac{\sin \lambda}{\sinh \lambda a}\right) \sinh \lambda x \sin \lambda y \, d\lambda$$

is the desired integral representation for our harmonic solution. □

16.6 Fourier Integrals with Singular Integrands

If the integrand in a Fourier integral has a pole on the x-axis, the situation can often be handled by *indenting* the contour to circumvent this point. Specifically, suppose we wish to evaluate

$$\int_{-\infty}^{\infty} f(x)e^{i\omega x}\,dx \qquad \text{for } \omega > 0$$

As a bit of variation, assume that f is a rational function with a denominator exactly *one* degree higher than the numerator, and with a simple pole at the origin. (See Figure 16.5.) Then,

$$\oint_C f(z)e^{i\omega z}\,dz = \int_{-R}^{-r} f(x)e^{i\omega x}\,dx + \int_\gamma f(z)e^{i\omega z}\,dz + \int_r^R f(x)e^{i\omega x}\,dx + \int_\Gamma f(z)e^{i\omega z}\,dz$$

Since the pole at the origin is simple, the integrand has a Laurent series with center at 0 of the form

$$f(z)e^{i\omega z} = \frac{\underset{z=0}{\operatorname{Res}}\,[f(z)e^{i\omega z}]}{z} + g(z)$$

where $g(z)$ is analytic at $z = 0$. On γ, the path may be represented by $z(t) = re^{it}$:

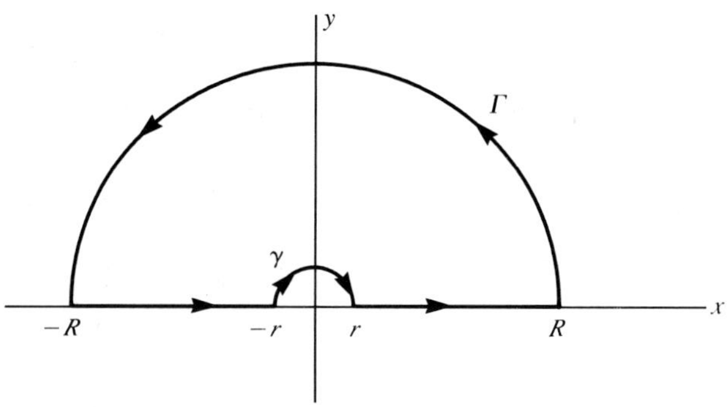

Figure 16.5 *Indented contour*

$$\int_\gamma f(z)e^{i\omega z}\,dz = \int_\gamma \frac{\operatorname*{Res}_{z=0}[f(z)e^{i\omega z}]}{z}\,dz + \int_\gamma g(z)\,dz$$

$$= \operatorname*{Res}_{z=0}[f(z)e^{i\omega z}]\int_\pi^0 i\,dt + \int_\pi^0 g(re^{it})(ire^{it})\,dt$$

Since $g(z)$ is analytic at $z = 0$, it is bounded near the origin and hence the last of these integrals can be made as small as desired by making r small enough. Thus

$$\lim_{r\to 0}\int_\gamma f(z)e^{i\omega z}\,dz = -\pi i \operatorname*{Res}_{z=0}[f(z)e^{i\omega z}]$$

On Γ, the path can be written $z(t) = Re^{it}$:

$$\int_\Gamma e^{i\omega z}f(z)\,dz = \int_0^\pi f(Re^{it})e^{i\omega Re^{it}}iRe^{it}\,dt$$

$$|e^{i\omega Re^{it}}| = |e^{i\omega R(\cos t + i\sin t)}| = e^{-\omega R\sin t}$$

Therefore

$$\left|\int_\Gamma e^{i\omega z}f(z)\,dz\right| = \left|\int_0^\pi f(Re^{it})e^{i\omega R(\cos t+i\sin t)}iRe^{it}\,dt\right|$$

$$\leqq \int_0^\pi e^{-\omega R\sin t}\varepsilon R\,dt$$

where ε can be made *arbitrarily* small for sufficiently large R since $f(z)$ is a rational function in which the numerator is dominated by the denominator. Using the fact that $\sin t$ assumes the same values on 0 to $\pi/2$ as on $\pi/2$ to π, this last integral is equal to $\int_0^{\pi/2} 2e^{-\omega R\sin t}\varepsilon R\,dt$. Owing to the fundamental trigonometric inequality $\sin t \geq 2t/\pi$ for $0 \leq t \leq \pi/2$, it then follows that

$$\left|\int_\Gamma e^{i\omega z}f(z)\,dz\right| \leqq 2\varepsilon R\int_0^{\pi/2} e^{-2R\omega t/\pi}\,dt$$

$$= 2\varepsilon R\left(\frac{-\pi}{2R\omega}\right)e^{-2R\omega/\pi}\Big|_0^{\pi/2}$$

$$= \frac{\pi\varepsilon}{\omega}(1 - e^{-R\omega}) < \frac{\pi\varepsilon}{\omega}$$

As a consequence, the integral over Γ tends to zero as $R \to \infty$. The end result of our analysis then is the formula

$$\text{C.P.V.}\int_{-\infty}^{\infty} f(x)e^{i\omega x}\,dx = \pi i \operatorname*{Res}_{z=0}[f(z)e^{i\omega z}] + 2\pi i \sum_{y>0}\operatorname{Res}[f(z)e^{i\omega z}] \qquad \square$$

The classic application of this type is given in the following example.

Example 16.10 Evaluate

$$\int_{-\infty}^{\infty}\frac{\sin x}{x}\,dx$$

Solution We integrate e^{iz}/z over the semicircular contour of Figure 16.5 indented at the origin. The residue of the integrand at 0 is 1. Since the integrand has no other poles,

$$\text{C.P.V.} \int_{-\infty}^{\infty} \frac{e^{ix}}{x}\,dx = \pi i$$

which means that

$$\int_{-\infty}^{\infty} \frac{\sin x}{x}\,dx = \pi$$

(Owing to the alternating sign character of the integrand, the improper integral can be shown to converge in the ordinary sense using the alternating series test applied to the series $\sum_{n=N}^{\infty} \int_{n\pi}^{(n+1)\pi} \frac{\sin x}{x}\,dx$.) □

Exercises for Sections 16.5 and 16.6

Evaluate the following improper integrals:

1. $\int_{-\infty}^{\infty} \frac{\sin x}{x^2+1}\,dx$

2. $\int_{-\infty}^{\infty} \frac{\cos 4x}{x-2i}\,dx$ $\left(\textit{Hint: } \frac{1}{x-2i} = \frac{x+2i}{x^2+4}\right)$

3. $\int_{-\infty}^{\infty} \frac{\cos 3x}{x^4+4}\,dx$

4. $\int_{-\infty}^{\infty} \frac{e^{-ix}}{x^2+9}\,dx$

5. $\int_{0}^{\infty} \frac{x \sin \omega x}{x^2+a^2}, \quad \omega, a > 0$

6. $\int_{-\infty}^{\infty} \frac{x \sin x}{x^2-2x+10}\,dx$

7. $\int_{-\infty}^{\infty} \frac{x \sin 2x}{x^4+4}\,dx$

8. $\int_{0}^{\infty} \frac{\cos 2x}{(x^2+1)^2}\,dx$

9. $\int_{0}^{\infty} \frac{x^3 \sin 2x}{(x^2+1)^2}\,dx$

10. $\int_{0}^{\infty} \frac{x^3 \sin \omega x}{x^4+4}\,dx, \quad \omega > 0$

11. $\int_{-\infty}^{\infty} \frac{\sin^2 \omega x}{x^2+1}\,dx, \quad \omega > 0$

12. $\int_{-\infty}^{\infty} \frac{\cos x}{(x^2+1)(x^2+4)}\,dx$

13. Determine an integral representation for the temperature in a thin bar of infinite length if the initial temperature is given by $u(x,0) = x^2/(x^4+4)$.

14. Find the bounded, harmonic function $u(x, y)$ in the semi-infinite strip $0 < x < a$, $y > 0$ that satisfies the auxiliary conditions $u_y(x,0) = 0$, $u(0, y) = 0$, and $u(a, y) = y^2/(y^2+1)^2$.

Recall that the voltage (current) on a balanced ($LG = RC$) transmission line can be written as

$$v(x,t) = e^{-Rt/L}V(x,t)$$

where V satisfies the one-dimensional wave equation

$$V_{xx} = LC\,V_{tt}$$

(see p. 293.) If the line is infinite in extent, a Fourier transform procedure analogous to that used in Example 16.8 can be applied in order to derive an integral representation for the particular solution which satisfies given initial conditions. The result is

$$V(x,t) = \frac{1}{2\pi}\int_{-\infty}^{\infty} A(\lambda)e^{i\lambda x}\cos\lambda\sqrt{LC}t\,d\lambda + \frac{1}{2\pi}\int_{-\infty}^{\infty} B(\lambda)e^{i\lambda x}\sin\lambda\sqrt{LC}t\,d\lambda$$

with

$$A(\lambda) = \int_{-\infty}^{\infty} V(x,0)e^{-i\lambda x}\,dx$$

and

$$\lambda\sqrt{LC}\,B(\lambda) = \int_{-\infty}^{\infty} V_t(x,0)e^{-i\lambda x}\,dx$$

15. Determine an integral representation for the voltage on a balanced transmission line of infinite extent if the initial conditions are $v(x,0) = 0$, $v_t(x,0) = 1/(x^4+4)$.

16. Find an integral representation for the voltage on an infinitely long balanced transmission line under the initial conditions $v(x,0) = 3/(x^2+4)$, $v_t(x,0) = 0$. (Be careful!)

17. Carry out an analysis similar to the one before Exercise 15 in order to derive the form of the integral representation for the voltage (current) on a semi-infinite ($x > 0$) balanced transmission line in the situation where the auxiliary conditions are $v(0,t) = f(t)$, $v_x(0,t) = g(t)$.

As discussed previously (see p. 294), applications involving spherically symmetric solutions of the three-dimensional wave equation can be approached in a manner which mirrors the transmission line analysis. We find

$$u(\rho,t) = \frac{1}{\rho}v(\rho,t)$$

where

$$v(\rho,t) = \frac{1}{2\pi}\int_{-\infty}^{\infty} C(\lambda)e^{i\lambda ct}\cos\lambda(\rho-\rho_0)\,d\lambda + \frac{1}{2\pi}\int_{-\infty}^{\infty} D(\lambda)e^{i\lambda ct}\sin\lambda(\rho-\rho_0)\,d\lambda$$

Here we have written the expressions in a form appropriate for the imposition of auxiliary conditions on the surface of the sphere $\rho = \rho_0$.

18. Verify that the integral representation for spherically symmetric solutions of the three-dimensional wave equation has the above form and derive formulas for the functions $C(\lambda)$ and $D(\lambda)$ in terms of known data $v(\rho_0,t)$, $v_\rho(\rho_0,t)$.

19. Determine an integral representation for the spherically symmetric solution of the three-dimensional wave equation in the case where $v(\rho_0,t) = 0$, $v_\rho(\rho_0,t) = \dfrac{t^2}{16t^4+1}$.

Verify the following integral formulas:

20. $\displaystyle\int_0^{\infty} \frac{1-\cos x}{x^2}\,dx = \frac{\pi}{2}$

21. $\displaystyle\int_{-\infty}^{\infty} \frac{\sin^2\omega x}{x^2}\,dx = \pi|\omega|$
(Compare with Exercise 19, p. 318.)

22. C.P.V. $\displaystyle\int_{-\infty}^{\infty} \frac{x\,dx}{x^3+1} = \frac{\pi}{\sqrt{3}}$

23. C.P.V. $\displaystyle\int_{-\infty}^{\infty} \frac{e^{-2ix}}{1-x}\,dx = \pi i e^{-2i}$

24. C.P.V. $\displaystyle\int_{-\infty}^{\infty} \frac{\cos \omega x}{1-x^2}\,dx = \pi \sin \omega, \quad \omega > 0$

(The contour will need to be indented at two points.)

25. C.P.V. $\displaystyle\int_{-\infty}^{\infty} \frac{e^{ix}}{(x-1)(x-2)}\,dx = \pi i(e^{2i} - e^{i})$

26. $\displaystyle\int_{0}^{\infty} \frac{\sin x \cos ax}{x}\,dx = \begin{cases} 0, & a > 1 \\ \dfrac{\pi}{4}, & a = 1 \\ \dfrac{\pi}{2}, & 0 \le a < 1 \end{cases}$

27. $\displaystyle\int_{-\infty}^{\infty} \frac{\sin \omega x}{x(x^2+a^2)}\,dx = \frac{\pi}{a^2}(1 - e^{-a\omega}), \quad a > 0, \omega > 0$

16.7 Inversion of Laplace Transforms

In Chapter 5, and briefly again in Chapter 8, the Laplace transformation was seen to be a magnificent operational tool for the solution of various applied problems. In each case, however, the final step in the analysis was the determination of an inverse transform. We are now ready to discuss a comprehensive approach to this inversion process which not only contains the basic results presented earlier but also enables us to handle systematically the general situation.

We first recall the Fourier integral relations given in Section 16.5:

$$G(\omega) = \int_{-\infty}^{\infty} g(x)e^{-i\omega x}\,dx$$

$$g(x) = \frac{1}{2\pi}\int_{-\infty}^{\infty} G(\omega)e^{i\omega x}\,d\omega$$

For a sufficiently well-behaved function $g(x)$, we can compute the function $G(\omega)$, called the *Fourier transform* of g, from the first of these expressions. The second relation then allows us to reverse or invert the process. Indeed, it can be shown that what one relation "does" the other "undoes." Thus, if the first expression defines the Fourier transform of g, the second expression defines the *inverse Fourier transform* of G, and the two relations together constitute the Fourier transform pair.

Perhaps surprisingly, the comparable relationships for the Laplace transform initially considered in Chapter 5 can be derived from these Fourier integrals. We replace x by t, ω by $-is + i\sigma$ (so that $s = \sigma + i\omega$, where σ is fixed), $e^{\sigma x}g(x)$ by $f(t)$, and $G(\omega)$ by $F(s)$ and agree that $f(t)$ is causal, that is, vanishes for $t < 0$. Upon substitution, the first expression above becomes

$$F(s) = \int_{0}^{\infty} e^{-st}f(t)\,dt$$

which we recognize as the Laplace transform of Definition 5.1. In similar fashion the second expression gives rise to

$$f(t) = \mathscr{L}^{-1}(F) = \frac{1}{2\pi i}\int_{\sigma - i\infty}^{\sigma + i\infty} F(s)e^{st}\,ds$$

We call this the *general inversion formula for the Laplace transform.*

Since this latter relation involves a line integral in the complex s-plane, it is clear why this expression was not discussed earlier. On the other hand, with the knowledge of contour integration that we now possess, we can carry out the inversion process in many practical cases using the residue calculus and a modest variation of techniques already considered in this chapter. We illustrate with a number of examples, some old, some new.

Example 16.11 Find the inverse Laplace transform of

$$F(s) = \frac{1}{s^2} \qquad \operatorname{Re} s > 0$$

Solution We consider a contour integral of e^{st}/s^2 along a path of integration which bounds the semicircular region in the s-plane depicted in Figure 16.6. Then

$$\oint_C \frac{e^{st}}{s^2}\,ds = \int_{\sigma - iR}^{\sigma + iR} \frac{e^{st}}{s^2}\,ds + \int_\Gamma \frac{e^{st}}{s^2}\,ds$$

and thus, by virtue of the residue theorem,

$$f(t) = \mathscr{L}^{-1}(F) = \sum \operatorname{Res}\left(\frac{e^{st}}{s^2}\right) - \frac{1}{2\pi i}\lim_{R\to\infty}\int_\Gamma \frac{e^{st}}{s^2}\,ds$$

For the integration along the semicircular arc we represent Γ by $s = \sigma + iRe^{i\phi}$. It follows that

$$\int_\Gamma \frac{e^{st}}{s^2}\,ds = -\int_0^\pi \frac{e^{\sigma t}\exp(itRe^{i\phi})Re^{i\phi}\,d\phi}{(\sigma + iRe^{i\phi})^2}$$

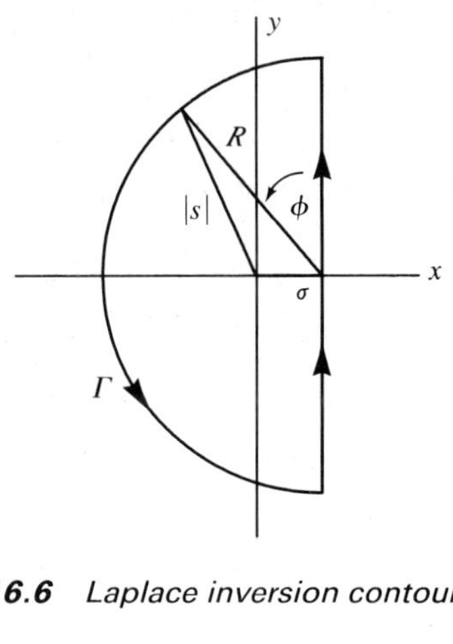

Figure 16.6 Laplace inversion contour for $t > 0$

and hence, for large R and nonnegative t,

$$\left|\int_\Gamma \frac{e^{st}}{s^2}\,ds\right| \le \frac{Re^{\sigma t}}{(R-\sigma)^2}\int_0^\pi e^{-tR\sin\phi}\,d\phi \le \frac{\pi\, Re^{\sigma t}}{(R-\sigma)^2}$$

For each t (and fixed σ), this last expression approaches 0 as R becomes unbounded, and therefore we are able to conclude that

$$f(t) = \sum \operatorname{Res}\left(\frac{e^{st}}{s^2}\right) = t, \qquad t \ge 0$$

(Recall Example 5.1.) □

Example 16.12 Determine $\mathscr{L}^{-1}\left(\dfrac{b}{s^2+b^2}\right)$, for $\operatorname{Re} s > 0$.

Solution We again employ the contour C shown in Figure 16.6 and calculate

$$\oint_C \frac{b}{s^2+b^2}e^{st}\,ds = \left(\int_{\sigma-iR}^{\sigma+iR} + \int_\Gamma\right)\frac{b}{s^2+b^2}e^{st}\,ds$$

Since the integral along the arc Γ again approaches zero as $R \to \infty$ ($t \ge 0$), we find

$$\begin{aligned} f(t) &= \frac{1}{2\pi i}\int_{\sigma-i\infty}^{\sigma+i\infty}\frac{b}{s^2+b^2}e^{st}\,ds \\ &= \operatorname*{Res}_{s=ib}\frac{b}{s^2+b^2}e^{st} + \operatorname*{Res}_{s=-ib}\frac{b}{s^2+b^2}e^{st} \\ &= \sin bt, \qquad t \ge 0 \end{aligned}$$

(See Table 5.1.) □

Example 16.13 Invert $F(s) = \dfrac{1}{s-a}$, $\operatorname{Re} s > a$.

Solution In this case we choose $\sigma > a$ in the general inversion formula. Our now familiar procedure using the contour C of Figure 16.6 then leads to

$$\mathscr{L}^{-1}\left(\frac{1}{s-a}\right) = \operatorname*{Res}_{s=a}\frac{e^{st}}{s-a} - \frac{1}{2\pi i}\lim_{R\to\infty}\int_\Gamma \frac{e^{st}}{s-a}\,ds$$

Here, however, a little more care (as in Section 16.6) must be taken to show that the contribution from the integration along Γ approaches zero as R becomes unbounded. The final result is

$$f(t) = e^{at}, \qquad t > 0$$

as expected (recall Example 5.2). □

The approach used in the above examples is typical of what is needed in the general inversion process with rational $F(s)$. The straight-line contour ($\sigma - i\infty$ to $\sigma + i\infty$) always lies to the *right* of the singularities of the function whose

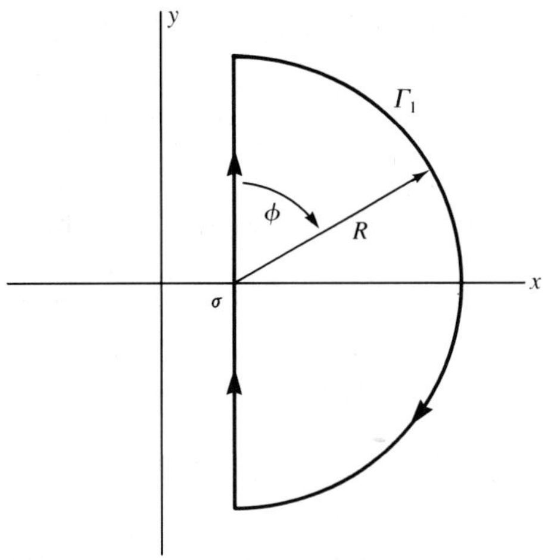

Figure 16.7 *Laplace inversion contour for $t < 0$*

inverse Laplace transform we are determining. Closing the contour with a large semicircle to the left as shown in Figure 16.6 then allows us to relate the desired function $f(t)$ for $t > 0$ to the residues of $F(s)e^{st}$ via the residue theorem. We remark that if the contour is closed off with a large semicircle to the right as in Figure 16.7, it may be deduced that $f(t) = 0$ for $t < 0$. This last observation is important in the next example.

Example 16.14 Find the inverse Laplace transform of

$$F(s) = \frac{e^{-as}}{s^2 - 5s + 6}, \qquad \text{Re}\, s > 3, \text{ with } a \text{ real}$$

Solution With $\sigma > 3$, using the contours depicted in Figures 16.6 and 16.7, we calculate

$$f(t) = \begin{cases} \sum \text{Res} \dfrac{e^{(t-a)s}}{s^2 - 5s + 6}, & t > a \\ 0, & t < a \end{cases}$$

Hence in the notation of Section 5.8, $f(t) = u_a(t)[e^{3(t-a)} - e^{2(t-a)}]$. □

As in Example 16.6, a different contour is needed when branch functions are involved.

Example 16.15 Show that $\mathscr{L}^{-1}(\sqrt{\pi/s}) = 1/\sqrt{t}, \qquad t > 0.$

Solution We consider a contour integral of $e^{st}\sqrt{\pi/s}$ along the *indented* contour depicted in Figure 16.8. If we restrict s by $-\pi < \arg s < \pi$, the integrand is analytic inside the composite contour and

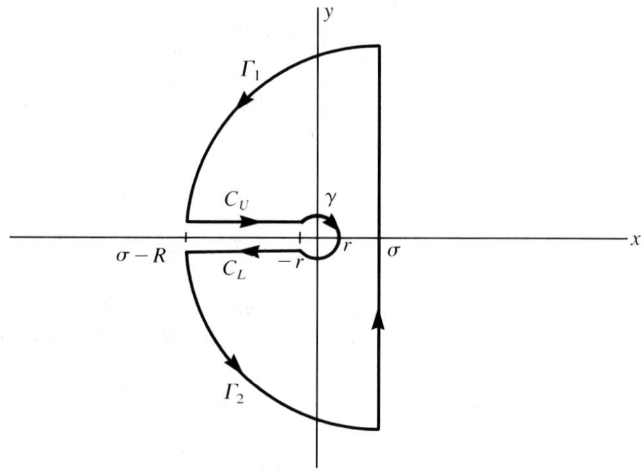

Figure 16.8

$$\left(\int_{\sigma-iR}^{\sigma+iR} + \int_{\Gamma_1} + \int_{C_U} + \int_{\gamma} + \int_{C_L} + \int_{\Gamma_2}\right)\sqrt{\frac{\pi}{s}}\,e^{st}\,ds = 0$$

On γ and Γ we introduce the parametrizations $s = re^{i\phi}$ and $s = \sigma + iRe^{i\phi}$, respectively. Using arguments similar to those employed in Example 16.6 and Section 16.6, we then can show that the integrals over these circular paths can be made arbitrarily small as $r \to 0$ and $R \to \infty$ ($t > 0$). It remains to note that on C_U, $s = |x|e^{i\pi}$ and $\sqrt{s} = \sqrt{|x|}i$, while on C_L, $s = |x|e^{-i\pi}$ and $\sqrt{s} = -\sqrt{|x|}i$. Hence as $R \to \infty$

$$\int_{\sigma-iR}^{\sigma+iR} \sqrt{\frac{\pi}{s}}\,e^{st}\,ds - \int_{\sigma-R}^{0} i\sqrt{\frac{\pi}{|x|}}\,e^{xt}\,dx - \int_{0}^{\sigma-R}\left(-i\sqrt{\frac{\pi}{|x|}}\,e^{xt}\right)dx = 0$$

and

$$\int_{\sigma-iR}^{\sigma+iR} \sqrt{\frac{\pi}{s}}\,e^{st}\,ds = 2i\int_{0}^{R-\sigma} \sqrt{\frac{\pi}{x}}\,e^{-xt}\,dx$$

from which it follows that

$$\begin{aligned} f(t) &= \frac{1}{\pi}\int_0^\infty \sqrt{\frac{\pi}{x}}\,e^{-xt}\,dx \\ &= \frac{2}{\sqrt{\pi}}\int_0^\infty e^{-ty^2}\,dy = \frac{1}{\sqrt{t}}, \qquad t > 0 \end{aligned}$$

□

Exercises for Section 16.7

Determine the inverse Laplace transforms of the following functions using the residue calculus.

1. $\dfrac{s}{s^2 + b^2}$, $\operatorname{Re} s > 0$

2. $\dfrac{b}{s^2 - b^2}$, $\operatorname{Re} s > |b|$

3. $\dfrac{1}{s(s^2+9)}$, $\operatorname{Re} s > 0$

4. $\dfrac{s^2}{(s+2)^3}$, $\operatorname{Re} s > -2$

5. $\dfrac{s+2}{(s+1)^2(s+3)}$, $\operatorname{Re} s > -1$

6. $\dfrac{e^{-3s}}{s^2}$, $\operatorname{Re} s > 0$

7. $\dfrac{e^{-4s}-e^{-s}}{s^3}$, $\operatorname{Re} s > 0$

8. $\dfrac{e^{-s}}{s^2+4}$, $\operatorname{Re} s > 0$

9. $\dfrac{e^{-s}}{(s-1)(s-2)}$, $\operatorname{Re} s > 2$

10. $\dfrac{se^{-2s}}{s^2+16}$, $\operatorname{Re} s > 0$

Verify the following Laplace transform pairs directly from the general inversion formula. (Restrict s to $-\pi < \arg s < \pi$ and employ the contour shown in Figure 16.8.)

11. $\mathscr{L}^{-1}\left(\dfrac{\Gamma(\nu)}{s^\nu}\right) = t^{\nu-1}, \quad 0 < \nu < 1$

12. $$\mathscr{L}^{-1}\left(\sqrt{\frac{\pi}{s}}e^{-2k\sqrt{s}}\right) = \frac{2}{\sqrt{\pi}}\int_0^\infty e^{-ty^2}\cos 2ky\,dy = \frac{1}{\sqrt{t}}e^{-k^2/t}$$

13. $$\mathscr{L}^{-1}(\sqrt{\pi}e^{-2k\sqrt{s}}) = \frac{2}{\sqrt{\pi}}\int_0^\infty ye^{-ty^2}\sin 2ky\,dy = (k/t^{3/2})e^{-k^2/t}$$

14. $$\mathscr{L}^{-1}\left(\frac{\sqrt{\pi}}{s}e^{-2k\sqrt{s}}\right) = \sqrt{\pi} - \frac{2}{\sqrt{\pi}}\int_0^\infty \frac{e^{-ty^2}\sin 2ky}{y}\,dy = \sqrt{\pi}[1-\operatorname{erf}(k/\sqrt{t})]$$

(Note that the integral over the inner circular path γ has a nonzero contribution in this case.)

The Laplace transform can be used to advantage in the solution of a variety of simple boundary and initial value problems. (Recall the earlier application in Example 8.22, p. 304.) As another representative example, consider the one-dimensional heat flow problem

$$u_{xx} = u_t, \qquad 0 < x < 1,\ t > 0$$
$$u(0,t) = 0 = u(1,t)$$
$$u(x,0) = 2\sin 3\pi x$$

If we take the Laplace transform with respect to the t variable, considering x for the moment as a parameter, the given partial differential equation and auxiliary conditions give rise to the following ordinary differential equation and boundary conditions for $U(x,s)$, the transformed temperature:

$$U_{xx}(x,s) - sU(x,s) = -2\sin 3\pi x$$
$$U(0,s) = 0 = U(1,s)$$

This latter problem may be readily solved to obtain

$$U(x,s) = \frac{2}{s+9\pi^2}\sin 3\pi x$$

which, when inverted, yields

$$u(x,t) = 2e^{-9\pi^2 t}\sin 3\pi x$$

15. Use a Laplace transform approach to determine the solution of the heat flow problem

$$\begin{aligned} u_{xx} &= u_t, & 0 < x < 2, t > 0 \\ u(0,t) &= 0, & u(2,t) = 6 \\ u(x,0) &= 3x \end{aligned}$$

(*Hint:* Note that $U(2,s) = 6/s$.)

16. Solve the following heat flow problem using a Laplace transform procedure.

$$\begin{aligned} u_{xx} &= u_t, & x > 0, t > 0 \\ u(0,t) &= 2t \\ u(x,0) &= x^2 \end{aligned}$$

(This problem is somewhat artificial in that it leads to an unbounded solution.)

17. Determine the solution of the following heat flow problem via a Laplace transform approach.

$$\begin{aligned} u_{xx} &= u_t, & x > 0, t > 0 \\ u(0,t) &= u_0, & u_0 \text{ constant} \\ u(x,0) &= 0 \end{aligned}$$

(*Hint:* From boundedness considerations, use only the term with $e^{-x\sqrt{s}}$ and apply the result of Exercise 14.)

18. Verify that if the constant u_0 is replaced by $f(t)$ in Exercise 17, the solution can be represented as the convolution

$$u(x,t) = \frac{x}{2\sqrt{\pi}} \int_0^t \frac{f(t-\tau)e^{-x^2/4\tau}}{\tau^{3/2}}\, d\tau$$

(*Hint:* Employ the result of Theorem 5.9 and Exercise 13.)

19. Make the appropriate modifications in the above Laplace transform procedure so as to solve the finite string vibration problem

$$\begin{aligned} u_{xx} &= \frac{1}{c^2} u_{tt}, & 0 < x < 1, t > 0 \\ u(0,t) &= 0 = u(1,t) \\ u(x,0) &= 0 \\ u_t(x,0) &= 5 \sin \pi x \end{aligned}$$

20. Solve the following string vibration problem using a Laplace transform approach

$$\begin{aligned} u_{xx} &= \frac{1}{c^2} u_{tt}, & x > 0, t > 0 \\ u(0,t) &= u_0, & u_0 \text{ constant} \\ u(x,0) &= 0 \\ u_t(x,0) &= 0 \end{aligned}$$

(*Hint:* From boundedness considerations, use only the term with $e^{-xs/c}$.)

21. Verify that if the constant u_0 is replaced by $f(t)$ in Exercise 20, the solution can be expressed in terms of the unit step function as

$$u(x,t) = f\left(t - \frac{x}{c}\right)u_{x/c}(t)$$

(*Hint:* Employ the second shifting theorem, p. 178.)

16.8 Looking Ahead

This final chapter has touched on a number of areas of application of the residue calculus. Much more can be said, however. For example, in addition to numerous worked examples of real and complex integrals, the problem book [11] (see also [9]) has a chapter on summation of infinite series which complements the several exercises on p. 595. Also see [1] for a variety of interesting examples as well as the essence of Watson's transformation which uses the residue calculus to "accelerate the convergence" of infinite series. Further discussion of contour integration in the vicinity of branch points appears in [14] and [15]. Applications to difference equations and sampled-data systems of the operational procedure called the Z-transform, based upon Laurent and residue series notions, are given in [4] and [8]. Finally, the extraordinary utility of other transform techniques in applied problem solving, and the role of the calculus of residues therein, is more fully considered in [1], [2], [10], and [17].

In many practical situations integral expressions arise which depend upon a parameter. In such cases the interest may not be so much in an exact evaluation of the integral as in a reasonably accurate approximation to its value for, say, large values of the parameter in question. Brief introductions to the asymptotic evaluation of integrals appear in [6] and [16]; extended discussion is contained in [1], [7], and [13]. In these and other references we see how considerations of the residue calculus, extended to finite Laplace and Fourier transforms (and their generalizations), lead naturally to Watson's lemma, the method of stationary phase, and the steepest descent procedure. Application of these asymptotic ideas specifically to entire functions is the subject of [5] (see also [3] and [12]).

Intermediate and More-Advanced References

[1] Carrier, G. F., et al. 1966. *Functions of a Complex Variable*. New York: McGraw-Hill.

[2] Churchill, R. V. 1971. *Operational Mathematics*. 3d ed. New York: McGraw-Hill.

[3] Cochran, J. A. 1982. *Applied Mathematics: Principles, Techniques, and Applications*. Belmont, Calif.: Wadsworth.

[4] Doetsch, G. 1971. *Guide to the Application of Laplace and Z-Transforms*. 2d ed. New York: Van Nostrand-Reinhold.

[5] Evgrafov, M. A. 1962. *Asymptotic Estimates and Entire Functions*. New York: Gordon and Breach.

[6] Evgrafov, M. A. 1978. *Analytic Functions*. New York: Dover.

[7] Fuchs, B. A., and V. I. Levin. 1961. *Functions of a Complex Variable and Some of Their Applications*. London: Pergamon Press.

[8] Kaplan, W. 1962. *Operational Methods for Linear Systems*. Reading, Mass.: Addison-Wesley.

[9] Keckic, J. D., and D. S. Mitrinovic. 1984. *The Cauchy Method of Residues*. Holland Reidel.

[10] Kyrala, A. 1972. *Applied Functions of a Complex Variable*. New York: John Wiley.

[11] Mitrinovic, D. S. 1966. *Calculus of Residues*. Groningen: P. Noordhoff.

[12] Olver, F. W. J. 1974. *Asymptotics and Special Functions*. New York: Academic Press.

[13] Polya, G., and G. Latta. 1974. *Complex Variables*. New York: John Wiley.

[14] Saff, E. B., and A. D. Snider. 1976. *Fundamentals of Complex Analysis for Mathematics, Science and Engineering*. Englewood Cliffs, N. J.: Prentice-Hall.

[15] Silverman, R. A. 1974. *Complex Analysis with Applications*. Englewood Cliffs, N. J.: Prentice-Hall.

[16] Volkovysky, L., et al. 1977. *Problems in the Theory of Functions of a Complex Variable*. 2d ed. Moscow: Mir Publishers.

[17] Weinberger, H. F. 1965. *A First Course in Partial Differential Equations with Complex Variables and Transform Methods*. Waltham, Mass.: Blaisdell.

Answers to Exercises

Chapter 1

Exercises 1.1–1.3, p. 4

1. **(a)** $(1, 0, 7)$ **(b)** $(-3, 2, 1)$ **(c)** $(6, -3, 9)$ **(d)** $\sqrt{18}$ **3.** **(a)** $8\mathbf{i} + \mathbf{j} - \mathbf{k}$ **(b)** $-3\mathbf{i} - 3\mathbf{j} + 10\mathbf{k}$ **(c)** $-5\mathbf{i} + 2\mathbf{j} - 9\mathbf{k}$ **5.** 3 units **7.** $\mathbf{A} - \mathbf{B} + \mathbf{C} = \mathbf{0}$

9. $\mathbf{A} = \frac{1}{2}(\mathbf{S} + \mathbf{D})$; $\mathbf{B} = \frac{1}{2}(\mathbf{S} - \mathbf{D})$ **11.** $\mathbf{C} = \mathbf{A} + \dfrac{m}{m+n}(\mathbf{B} - \mathbf{A})$

Exercises 1.4–1.5, p. 13

1. **(a)** $\sqrt{29}$ **(b)** 7 **(c)** $\operatorname{Cos}^{-1}\dfrac{7}{\sqrt{29}\sqrt{11}}$ **(d)** $(-1/\sqrt{11}, 3/\sqrt{11}, 1/\sqrt{11})$

3. **(a)** $\{(1/\sqrt{10}, 3/\sqrt{10}), (3/\sqrt{10}, -1/\sqrt{10})\}$ **(b)** $\{(\frac{1}{3}, \frac{2}{3}, \frac{2}{3}), (-\frac{2}{3}, \frac{2}{3}, -\frac{1}{3}), (\frac{2}{3}, \frac{1}{3}, -\frac{2}{3})\}$

5. **(a)** 1 unit of work **(b)** 15 units of work **7.** Sphere with **A** as one of its diameters

9. $2x + y - z = 6$ **13.** **(b)** Use the quadratic formula on $\lambda^2|\mathbf{B}|^2 + 2\lambda\mathbf{A}\cdot\mathbf{B} + |\mathbf{A}|^2$

15. **(a)** $1/\sqrt{3}$ **(b)** $1/\sqrt{5}$ **(c)** $1/\sqrt{2}$ **(d)** 1 **(e)** $1/\sqrt{2}$ **19.** Only if it is the zero vector

21. $a_1 = 1/\sqrt{2}$, $b_1 = \sqrt{\frac{3}{2}}$, $b_2 = 0$, $c_1 = 3\sqrt{10}/4$, $c_2 = 0$, $c_3 = -\sqrt{10}/4$ **23.** $(\frac{\pi^3}{3} - \frac{3\pi}{2})^{1/2}$

25. **(a)** $2\sqrt{2}$ **(b)** 0

Exercises 1.6, p. 17

1. $\mathbf{A} \times \mathbf{B} = (-1, -7, -5)$; $\mathbf{C} \times (\mathbf{A} \times \mathbf{B}) = (4, 18, -26)$; $\mathbf{C}\cdot(\mathbf{A} \times \mathbf{B}) = -28$

3. $\mathbf{A} = (3, 0)$, $\mathbf{B} = (1, 2)$, $|\mathbf{A} \times \mathbf{B}| = |6\mathbf{k}| = 6$ **5.** $\mathbf{M} = -4\mathbf{i} - \mathbf{j} + 7\mathbf{k}$

7. $\mathbf{v} = \Omega\left(\dfrac{1}{\sqrt{3}}, \dfrac{1}{\sqrt{3}}, \dfrac{1}{\sqrt{3}}\right) \times (1, 2, 2) = -\dfrac{\Omega}{\sqrt{3}}\mathbf{j} + \dfrac{\Omega}{\sqrt{3}}\mathbf{k}$ **9.** -10

Exercises 1.7, p. 24

1. $\begin{pmatrix} 1 & -1 \\ 7 & 2 \end{pmatrix}$ **3.** $\begin{pmatrix} -11 & 4 \\ 2 & 8 \end{pmatrix}, \begin{pmatrix} 6 & 7 \\ 6 & -9 \end{pmatrix}$ **5.** **(a)** -3 **(b)** 19 **(c)** 2 **(d)** 10

9. **(c)** No. Why? **11.** $b = d = 0$; $c = d = 0$; $b = c = 0, d = 1$ **17.** $\mathbf{A}^T = \begin{pmatrix} 2 & 3 \\ 1 & 0 \\ 3 & 1 \end{pmatrix}$

Exercises 1.8, p. 31

7. **(a)** $\frac{1}{\sqrt{2}}(\mathbf{i}+\mathbf{j}), \frac{1}{\sqrt{2}}(\mathbf{i}-\mathbf{j}), \mathbf{k}$ **(b)** $\frac{1}{\sqrt{14}}(3\mathbf{i}+2\mathbf{j}+\mathbf{k}), \frac{1}{\sqrt{27}}(\mathbf{i}+\mathbf{j}-5\mathbf{k}), \frac{1}{\sqrt{378}}(11\mathbf{i}-16\mathbf{j}-\mathbf{k})$
9. Dimension is 1 **11.** $a+bx+cx^2 = \frac{1}{2}(a+3b-c)(x-1)+c(x^2+1)+\frac{1}{2}(c-a-b)(x-3)$
17. Investigate $c_1x + c_2|x|$ on the two spaces. **23.** Yes

Exercises 1.9, p. 38

1. $x = 11, y = -4, z = 3$ **3.** No solution **5.** No solution **7.** $x_1 = 1, x_2 = 2, x_3 = 3, x_4 = 4$
9. $x_1 = x_2 = x_3 = 0$ only **17.** $\lambda = \frac{2}{3}, \mathbf{X} = (\frac{1}{2}); \lambda = \frac{1}{2}, \mathbf{X} = (\frac{1}{3})$

19. $\lambda = 1, \mathbf{X} = \begin{pmatrix} 0 \\ 1 \\ -1 \end{pmatrix}; \lambda = \frac{1}{2}(7+3\sqrt{5}), \mathbf{X} = \begin{pmatrix} 10 \\ 5+3\sqrt{5} \\ 10 \end{pmatrix}; \lambda = \frac{1}{2}(7-3\sqrt{5}), \mathbf{X} = \begin{pmatrix} 10 \\ 5-3\sqrt{5} \\ 10 \end{pmatrix}$

21. $\lambda = 1, \mathbf{X} = \begin{pmatrix} 1 \\ -1 \\ 0 \end{pmatrix}; \lambda = 1, \mathbf{X} = \begin{pmatrix} 1 \\ 0 \\ -1 \end{pmatrix}$

Chapter 2

Exercises 2.1–2.3, p. 49

5. $(\sin x)y = c$ **7.** $y = -\ln(c - e^x)$ **9.** $a \sec\frac{y+c}{a} = x$ **11.** $y\ln[c(1-x)] = 1$ **13.** $i = ce^{-Rt/L}$

15. $y = \frac{1}{3}x^3 + 2$ **17.** $y = 3e^{1+\cos x}$ **19.** $y = \ln\left(\frac{3-e^{-x^2}}{2}\right)$ **21.** $y = 1$

23. **(a)** 38° **(b)** $t \doteq 41.91$ min **25.** 78.6 grams **27.** $y = \pm(c-2x)^{-1/2}$ **29.** $y^2 = 2x + c$

31. $k = \frac{1}{450}$; $C = 45$ mol/L **33.** **(a)** $v = \frac{mg}{k} + \left(v_0 - \frac{mg}{k}\right)e^{-kt/m}$ **(b)** $\lim_{t\to\infty} v(t) = \frac{mg}{k}$

35. $v(t) = 8\sqrt{5}\left\{\frac{1-e^{-8t/\sqrt{5}}}{1+e^{-8t/\sqrt{5}}}\right\}$ **37.** **(a)** 10^6 **(b)** 52.9 months

39. **(a)** $y(x) = ce^{-(a-x)^2/2b}$
(b) a = mean; value of x about which distribution is symmetric. b = variance; controls width of distribution and hence area under the curve for various intervals about the mean.

Exercises 2.4–2.5, p. 59

1. $\ln y - \frac{x^2}{2y^2} = c$ **3.** $x^2(x^2 - 2y^2) = c$ **5.** $\ln(xy^3 + y^4) - \frac{x}{y} = c$ **7.** $\cos(y/x) - \ln x = c$
9. $(x + y) = c(x^2 + 4y^2)$
11. $(\sqrt{2}-1)\ln\left(\sqrt{2}+1-\frac{y-13}{x-5}\right) - (\sqrt{2}+1)\ln\left(\sqrt{2}-1+\frac{y-13}{x-5}\right) - 2\ln(x-5) = c$
13. $x^2 + 2xy - y^2 = c$ **15.** Exact, $3x^2y + y^3 = c$ **17.** Not exact **19.** Exact, $ye^x - x^2 = c$
21. Exact, $3x^2 + y^2x - y^3 = c$ **23.** Not exact **25.** $\frac{\partial f}{\partial y} = 0 = \frac{\partial g}{\partial x}$ **27.** $a = 2d$ and $c = 2b$
29. $N(x, y) = xe^{xy} - 6xy^2 + g(y)$ **31.** $x^3 - 6x^2y + y^3 = c$ **33.** $y + \sqrt{x^2 + y^2} = 3$
35. $\sin^2 x - x^2y^2 + y^2 = 4$ **37.** $x^2 + \sin xy = 3$ **39.** **(a)** There are no such values. **(b)** $b = -1/a$
41. $x^2 + y^2 = ay$

Exercises 2.6, p. 66

1. $y = ce^{-2x/3} - x/2 + \frac{3}{4}$ **3.** $y = ce^{4x} + x^2/4 - x/8 - \frac{1}{32}$ **5.** $y = (c + x)e^{-2x}$

7. $y = x^2(c + 3x)e^{-x}$ **9.** $y = \dfrac{1}{x}(c + \sin x) - \cos x$ **11.** $y = x$ **13.** $y = 2e^{-\sin x}$ **15.** $y = -\ln x$

17. $y = \cos x(\sin x - 1)$ **19.** $y = \begin{cases} \frac{1}{2} + \frac{3}{2}e^{-x^2}, & 0 \le x < 2 \\ \frac{1}{2}(e^4 + 3)e^{-x^2}, & x > 2 \end{cases}$ **21. (a)** $i = \dfrac{e^{-t} - e^{-Rt/L}}{R - L}$

(b) $i = \dfrac{R\sin t - L\cos t + Le^{-Rt/L}}{R^2 + L^2}$ **(c)** $i = \dfrac{L}{R^2}(e^{-Rt/L} - 1) + \dfrac{1}{R}t$

23. $i = \begin{cases} 2 - 2\left(1 - \dfrac{t}{5}\right)^{10}, & 0 \le t \le 5 \\ 2, & t \ge 5 \end{cases}$ **25.** $i = \begin{cases} 1, & 0 \le t \le 1 \\ -1 + 2e^{1-t}, & 1 \le t \le 2 \\ (2e - e^2)e^{-t}, & t \ge 2 \end{cases}$ **27.** $\dfrac{1}{1 + e}$

29. (a) $y = M(1 - e^{-At})$ **(b)** $\lim_{t\to\infty} y(t) = M$ **31.** $x = ce^y - y^2 - 2y - 2$

33. $x = \left(\dfrac{1}{2} - \dfrac{1}{2y} + \dfrac{1}{4y^2}\right)e^y + \dfrac{c}{y^2}e^{-y}$ **35.** $y = \dfrac{e^x}{\sqrt{x^2 + c}}$

Exercises 2.7, p. 74

7. First line: $y = -2x - 2$; second line: $y = -6x + 6$; $y(4) = -18$

9. First line: $y = x + 1$; second line: $y = 4x - 5$; third line: $y = 9x - 20$; $y(4) = 16$ **11.** $y = ce^x$

13. $y = e^{x^2/2}\left(c - \int_0^x e^{-t^2/2}\,dt\right)$ **15.** $xy = c$ **17.** $y = -2e^x$; $y(4) \doteq -109.2$

19. $y = \frac{5}{3} + x^3/3$; $y(4) = 23$

Exercises 2.8, p. 78

1. $k = \dfrac{\ln 2}{15} \doteq 0.0462$ **3.** $ay + dx - \ln(y^b x^c) = \text{constant}$ **5.** $y(60) = 100 - 90e^{-60/25} \doteq 91.8$ lb

7. $y^2 = a(b + t)^2 - 1/a$ **9.** $y = a$ or $y = ae^{bt}$ **11.** $y = \dfrac{1}{a}\cosh(at + b) + c$ **13.** $y = a\,\mathrm{Tan}^{-1}\,at + b$

15. $y = (t + a)\ln(t + a) - t + b$ **17.** $y = \mathrm{Sin}^{-1}(ae^x) + b$

Chapter 3

Exercises 3.2, p. 86

1. $y = c_1e^x + c_2e^{2x} + c_3e^{-4x}$ **3.** $y = c_1e^{-2x} + c_2e^{2x}$ **5.** $y = c_1 + c_2e^x + c_3e^{7x}$

7. $y = c_1\cos\dfrac{x}{3} + c_2\sin\dfrac{x}{3}$ **9.** $y = c_1\cos\dfrac{x}{2} + c_2\sin\dfrac{x}{2} + c_3\cos 3x + c_4\sin 3x$

11. $y = c_1e^{-x} + c_2\cos 2x + c_3\sin 2x$ **13.** $y = c_1e^{(1+\sqrt{2})x} + c_2e^{(1-\sqrt{2})x} + c_3e^{(-1+\sqrt{2})x} + c_4e^{(-1-\sqrt{2})x}$

15. $y = c_1x^3 + c_2x^{-1}$ **17.** $y = c_1 + c_2x + c_3x^{-1}$ **23.** $y = x + c_1 + c_2e^x$

25. $y = c_1e^{2x} + c_2e^{-x} - 3x + \frac{3}{2} - 2xe^{-x}$ **29. (b)** $y(t) = y_0\cos\sqrt{\dfrac{k}{m}}\,t + v_0\sqrt{\dfrac{m}{k}}\sin\sqrt{\dfrac{k}{m}}\,t$

Exercises 3.3, p. 89

1. $y = c_1e^{2x} + c_2e^x$ **3.** $y = c_1x^2 + c_2x^2\ln x$ **5.** $y = c_1\dfrac{\sin x}{\sqrt{x}} + c_2\dfrac{\cos x}{\sqrt{x}}$

7. $y = c_1\cos x + c_2\sin x + x\sin x + \cos x\ln\cos x$ **9.** $y = c_1\cos x + c_2\sin x + \frac{1}{2}\csc x$

11. $y = c_1e^{2x} + c_2(2x + 5) - e^x$ **13.** $y = c_1x^2 + c_2\sin x$

15. $y = c_1\sin x + c_2(\cos x - \sec x\sin^2 x) + \frac{1}{2}\cos x$ **19.** $y = c_1x + c_2x^2 + c_3x^3$

Exercises 3.4, p. 94

1. $\{\sin 3x, \cos 3x\}$ **3.** $\{e^{-x}, e^{-4x}\}$ **5.** $\left\{e^{-x}, e^{x/2}\sin\frac{\sqrt{3}}{2}x, e^{x/2}\cos\frac{\sqrt{3}}{2}x\right\}$ **7.** $\{1, x, e^{-4x}, xe^{-4x}\}$

9. $\{e^{2x}, e^{-x}, xe^{-x}\}$ **11.** $\{e^{-x}, \cos 2x, \sin 2x\}$ **13.** $\{\cos x, \sin x, \cos 2x, \sin 2x, x\cos 2x, x\sin 2x\}$

15. $y = c_1e^{-2x} + c_2e^x + c_3xe^x + c_4x^2e^x$

17. $y = c_1e^{2x} + c_2e^{-2x} + c_3e^x\cos\sqrt{3}x + c_4e^x\sin\sqrt{3}x + c_5e^{-x}\cos\sqrt{3}x + c_6e^{-x}\sin\sqrt{3}x$ **19.** $y = e^x$

21. $y = 9\sin\frac{x}{3}$ **23.** $y = e^{2x}(\cos\sqrt{3}x - \sqrt{3}\sin\sqrt{3}x)$ **25.** $y = 2\cos\sqrt{2}x - xe^{2x}$

27. $y = -(2x + x^2)e^{2x}$ **29. (a)** $y = (1 - x)e^{2x}$ **(b)** $y = \frac{1}{\varepsilon}(\varepsilon + 1 - e^{\varepsilon x})e^{2x}$ **31.** $y = 2e^{-x}$

33. $u = c_1e^{kx} + c_2e^{-kx} + c_3\cos kx + c_4\sin kx$ **35.** $P = P_0 + A\cos\sqrt{k}(t + B)$; period $= 2\pi/\sqrt{k}$

Exercises 3.5, p. 99

1. $y = c_1e^{2x} + c_2xe^{2x} + e^x$ **3.** $y = c_1e^{2x} + c_2xe^{2x} + \frac{1}{2}x^2e^{2x}$

5. $y = c_1e^{2x} + c_2xe^{2x} + \frac{1}{6}x^3e^{2x} + \frac{1}{2}x^2e^{2x}$ **7.** $y = c_1\cos x + c_2\sin x - \frac{1}{3}\sin 2x$

9. $y = c_1 + c_2e^{-4x} - \frac{1}{20}(2\cos 2x + \sin 2x)$ **11.** $y = c_1e^x\cos 2x + c_2e^x\sin 2x + \frac{1}{4}xe^x\sin 2x$

13. $y = c_1e^{-x} + c_2e^x + \frac{x}{2}\sinh x$ **15.** $y = c_1e^x + c_2e^{2x} - xe^x + xe^{2x} + \frac{1}{6}e^{-x}$ **17.** $y = c_1 + c_2e^x - 3x$

19. $y = c_1\cos 2x + c_2\sin 2x + \frac{3}{4}$ **21.** $y = c_1e^{-x} + c_2xe^{-x} + c_3e^{2x} + \frac{1}{100}(7\cos 2x - \sin 2x)$

23. $y = c_1e^{-2x} + c_2e^{-x}\cos 2x + c_3e^{-x}\sin 2x - \frac{1}{24}e^x$

25. $y = c_1e^x + c_2e^{-x}\cos x + c_3e^{-x}\sin x - \frac{1}{2}(x^2 + 1) - \frac{1}{5}(3\cos 2x + 4\sin 2x)$

27. $y = 1 - e^{-3x} + 2x - 3x^2$ **29.** $y = \frac{1}{2}(x\cos 3x - \sin 3x)$ **31.** $y = x - 3$

33. $y = e^{-x}(3\cos 2x + x\sin 2x)$ **35.** $y = 5x + 1 + \frac{1}{6}(x^3 - 3x^2)e^x$

37. $A\cos x + B\sin x + Cx\cos x + Dx\sin x + (Ex^4 + Fx^3 + Gx^2 + Hx)e^{3x}$

39. $Ax\cos x + Bx\sin x + Cx^6 + Dx^5 + Ex^4 + Fx^3 + Gx^2$

41. (a) $u = \frac{\sin x}{EI}$ **(b)** $u = \frac{1}{360EI}(x^6 - 3x^5)$

Exercises 3.6, p. 104

1. $y = c_1e^{-x} + c_2e^{2x} + \frac{1}{3}xe^{2x}$ **3.** $y = c_1\cos 2x + c_2\sin 2x - \frac{1}{4}\cos 2x\ln(\sec 2x + \tan 2x)$

5. $y = c_1\cos 2x + c_2\sin 2x + \frac{1}{12}(\sin^4 2x + 3\cos^2 2x - \cos^4 2x)$ **7.** $y = c_1e^{2x} + c_2xe^{2x} + xe^{2x}\ln x$

9. $y = c_1e^{-x} + c_2xe^{-x} + \frac{1}{4}x^2e^{-x}(2\ln x - 3)$ **15.** $y = c_1x + c_2x\ln x$

17. $y = c_1x + c_2x^2 - 4x\ln x + \frac{1}{10}[\sin(\ln x) + 3\cos(\ln x)]$ **19.** $y = x^2(2 - 3\ln x)$

21. $y = 4x^{-2} - x + 2x\ln x$ **23.** $y = -2\ln x\cos(\ln x)$

25. $y = c_1 + c_2\cos x + c_3\sin x + \ln(\sec x) - \sin x\ln(\sec x + \tan x)$

27. $y = c_1(x + 1) + c_2x^2 - 2x + x^2\ln x$ **29.** $y = c_1x + c_2\left(1 + \frac{1}{x}\right) + x\ln x - 2$

Exercises 3.7–3.8, p. 115

1. $i_t = \frac{225}{656}e^{-9t} - \frac{1025}{656}e^{-t}$; $i_{ss} = \frac{50}{41}\cos t + \frac{125}{82}\sin t$

3. $i_t = e^{-5t/18}\left(-\frac{35}{37}\cos\frac{\sqrt{551}}{18}t - \frac{625}{37\sqrt{551}}\sin\frac{\sqrt{551}}{18}t\right)$; $i_{ss} = \frac{35}{37}\cos t + \frac{25}{37}\sin t$

7. $y = \frac{1}{8}\sin 2t - \frac{1}{4}t\cos 2t$ **9.** At the positive values of $\tan^{-1}(-\alpha)$; $e^{2\pi\alpha}$ **11.** $\sqrt{10}$ kcycles

13. $y = 5\sin 0.95t - \frac{4.75}{1.05}\sin 1.05t$

15. $u_t = e^{-at/2}\left[\frac{\alpha b - \beta a}{b}\cos\sqrt{b - a^2/4}\,t + \frac{\alpha ab + \beta(2b - a^2)}{2b\sqrt{b - a^2/4}}\sin\sqrt{b - a^2/4}\,t\right]; u_{ss} = a\beta/b$

17. (a) Overdamped: $u = Ne^{-at/2}\left[\cosh\sqrt{a^2/4 - b}\,t + \frac{a}{2\sqrt{a^2/4 - b}}\sinh\sqrt{a^2/4 - b}\,t\right]$

(b) Critically damped: $u = Ne^{-at/2}\left[1 + \frac{a}{2}t\right]$

(c) Underdamped: $u = Ne^{-at/2}\left[\cos\sqrt{b - a^2/4}\,t + \frac{a}{2\sqrt{b - a^2/4}}\sin\sqrt{b - a^2/4}\,t\right]$

All solutions decay exponentially to zero as $t \to \infty$.

19. Roots of characteristic equation are $m = -\alpha \pm \sqrt{\alpha^2 - \beta^2}$ and α is positive.

21. $g_{ss}(t) = \begin{cases} F/\beta^2, & 0 \le t \le 1 \\ \frac{F}{\sqrt{\beta^2 - \alpha^2}} e^{-\alpha(t-1)}[\sqrt{\beta^2 - \alpha^2}\cos\sqrt{\beta^2 - \alpha^2}\,(t-1) + \alpha\sin\sqrt{\beta^2 - \alpha^2}\,(t-1)], & t > 1 \end{cases}$

Chapter 4

Exercises 4.2, p. 122

1. $\mathbf{X} = c_1\begin{pmatrix}2\\1\end{pmatrix}e^{3t} + c_2\begin{pmatrix}-2\\1\end{pmatrix}e^{-t}$ 3. $\mathbf{X} = c_1\begin{pmatrix}1\\1\end{pmatrix}e^{-5t} + c_2\begin{pmatrix}-1\\1\end{pmatrix}e^{t}$

5. $\mathbf{X} = c_1\begin{pmatrix}2\\3\\1\end{pmatrix}e^{2t} + c_2\begin{pmatrix}1\\0\\0\end{pmatrix}e^{t} + c_3\begin{pmatrix}1\\0\\2\end{pmatrix}e^{-t}$ 7. $\mathbf{X} = c_1\begin{pmatrix}1\\0\\0\end{pmatrix}e^{t} + c_2\begin{pmatrix}-5\\-2\\2\end{pmatrix}e^{-t} + c_3\begin{pmatrix}-2\\0\\3\end{pmatrix}e^{-2t}$

9. $\mathbf{A} = \begin{pmatrix}3 & 2\\-3 & -4\end{pmatrix}; \mathbf{X} = c_1\begin{pmatrix}-2\\1\end{pmatrix}e^{2t} + c_2\begin{pmatrix}-1\\3\end{pmatrix}e^{-3t}$

11. $\mathbf{A} = \begin{pmatrix}-3 & -2\\-3 & -1\end{pmatrix}; \mathbf{X} = c_1\begin{pmatrix}-2\\1+\sqrt{7}\end{pmatrix}e^{(-2+\sqrt{7})t} + c_2\begin{pmatrix}-2\\1-\sqrt{7}\end{pmatrix}e^{(-2-\sqrt{7})t}$ 13. $\mathbf{X} = \begin{pmatrix}3e^{4t} - 3e^{-t}\\2e^{4t} + 3e^{-t}\end{pmatrix}$

17. $\mathbf{C} = a_1\begin{pmatrix}1\\1\end{pmatrix} + a_2\begin{pmatrix}-V_2\\V_1\end{pmatrix}e^{-k((1/V_1)+(1/V_2))t}$

Exercises 4.3, p. 127

1. $\mathbf{X} = c_1\begin{pmatrix}-4\cos 2t\\\cos 2t + \sin 2t\end{pmatrix}e^{4t} + c_2\begin{pmatrix}-4\sin 2t\\\sin 2t - \cos 2t\end{pmatrix}e^{4t}$

3. $\mathbf{X} = c_1\begin{pmatrix}3\\1\end{pmatrix}e^{-3t} + c_2\left[\begin{pmatrix}\frac{7}{2}\\1\end{pmatrix} + t\begin{pmatrix}3\\1\end{pmatrix}\right]e^{-3t}$

5. $\mathbf{X} = c_1\begin{pmatrix}0\\1\\0\end{pmatrix}e^{t} + c_2\begin{pmatrix}-\cos t\\\frac{1}{2}(\cos t - \sin t)\\\cos t + \sin t\end{pmatrix} + c_3\begin{pmatrix}-\sin t\\\frac{1}{2}(\cos t + \sin t)\\\sin t - \cos t\end{pmatrix}$

7. $\mathbf{X} = c_1\begin{pmatrix}-1\\2\\2\end{pmatrix}e^{-3t} + c_2\begin{pmatrix}2\\1\\0\end{pmatrix}e^{6t} + c_3\begin{pmatrix}2\\0\\1\end{pmatrix}e^{6t}$

9. $\mathbf{A} = \begin{pmatrix}-3/2 & -5/2\\5/2 & 3/2\end{pmatrix}; \mathbf{X} = c_1\begin{pmatrix}-5\cos 2t\\3\cos 2t - 4\sin 2t\end{pmatrix} + c_2\begin{pmatrix}-5\sin 2t\\4\cos 2t + 3\sin 2t\end{pmatrix}$

11. $\mathbf{A} = \begin{pmatrix} -3 & -4 \\ 1 & 1 \end{pmatrix}$; $\mathbf{X} = c_1\begin{pmatrix} -2 \\ 1 \end{pmatrix}e^{-t} + c_2\left[\begin{pmatrix} -1 \\ 1 \end{pmatrix} + t\begin{pmatrix} -2 \\ 1 \end{pmatrix}\right]e^{-t}$ **13.** $\mathbf{X} = \begin{pmatrix} 1 \\ 3 \\ 1 \end{pmatrix}e^{t} - \begin{pmatrix} 1 \\ 2 \\ 2 \end{pmatrix}te^{t}$

15. $\mathbf{X} = c_1\begin{pmatrix} \cos \alpha t \\ -\alpha \sin \alpha t \\ -\alpha^2 \cos \alpha t \\ \alpha^3 \sin \alpha t \end{pmatrix} + c_2\begin{pmatrix} \sin \alpha t \\ \alpha \cos \alpha t \\ -\alpha^2 \sin \alpha t \\ -\alpha^3 \cos \alpha t \end{pmatrix} + c_3\begin{pmatrix} \cos \omega t \\ -\omega \sin \omega t \\ -\omega^2 \cos \omega t \\ \omega^3 \sin \omega t \end{pmatrix} + c_4\begin{pmatrix} \sin \omega t \\ \omega \cos \omega t \\ -\omega^2 \sin \omega t \\ -\omega^3 \cos \omega t \end{pmatrix}$

$\theta = c_1 \cos \alpha t + c_2 \sin \alpha t + c_3 \cos \omega t + c_4 \sin \omega t$ $\psi = -c_1 \cos \alpha t - c_2 \sin \alpha t + c_3 \cos \omega t + c_4 \sin \omega t$

Exercises 4.4, p. 135

1. $\mathbf{X} = c_1\begin{pmatrix} 1 \\ 3 \end{pmatrix}e^{5t} + c_2\begin{pmatrix} 1 \\ 1 \end{pmatrix}e^{3t} + \begin{pmatrix} \frac{1}{2} \\ \frac{1}{2} \end{pmatrix}e^{5t}$

3. $\mathbf{X} = c_1\begin{pmatrix} -2 \sin 2t \\ -\cos 2t \end{pmatrix} + c_2\begin{pmatrix} 2 \cos 2t \\ -\sin 2t \end{pmatrix} + \begin{pmatrix} 2 \sin 2t \cos^3 t - 3 \cos 2t \sin t + 2 \cos 2t \sin^3 t \\ \cos 2t \cos^3 t + \frac{3}{2} \sin t \sin 2t - \sin^3 t \sin 2t \end{pmatrix}$

Skillful use of trigonometric identities shows that the last term above is equal to $\begin{pmatrix} \sin t \\ \cos t \end{pmatrix}$. (See Exercise 9.)

5. $\mathbf{X} = c_1\begin{pmatrix} -1 \\ 0 \\ 1 \end{pmatrix} + c_2\begin{pmatrix} -1 \\ 2 \\ 0 \end{pmatrix}e^{-t} + c_3\begin{pmatrix} -1 \\ 0 \\ 2 \end{pmatrix}e^{-t} + \begin{pmatrix} 2 \\ 0 \\ -2 \end{pmatrix}t + \begin{pmatrix} -1 \\ 0 \\ 2 \end{pmatrix} + \begin{pmatrix} -2 \\ 0 \\ 2 \end{pmatrix}e^{-t} + \begin{pmatrix} -2 \\ 0 \\ 4 \end{pmatrix}te^{-t}$

13. $\mathbf{X} = \begin{pmatrix} -1 \\ -1 \end{pmatrix}e^{t} + \begin{pmatrix} 1 \\ 2 \end{pmatrix}e^{2t} + \begin{pmatrix} 0 \\ -1 \end{pmatrix}e^{3t}$ **15.** $\mathbf{X} = \begin{pmatrix} 1 \\ 1 \\ 1 \\ 1 \end{pmatrix}e^{t} + \begin{pmatrix} 0 \\ -1 \\ 0 \\ 0 \end{pmatrix}t^2 + \begin{pmatrix} 0 \\ 0 \\ -2 \\ 0 \end{pmatrix}t + \begin{pmatrix} 0 \\ 0 \\ 0 \\ -2 \end{pmatrix}$

19. $\mathbf{C} = \begin{pmatrix} c_1 \\ c_2 \end{pmatrix} = b_1\begin{pmatrix} 1 - \alpha/k \\ 1 \end{pmatrix}e^{-\alpha t} + b_2\begin{pmatrix} 1 - \beta/k \\ 1 \end{pmatrix}e^{-\beta t} + \begin{pmatrix} a \\ a \end{pmatrix}$ where $\alpha, \beta = \dfrac{2k + h \pm \sqrt{4k^2 + h^2}}{2}$

Exercises 4.5, p. 142

1. $\mathbf{X}' = \begin{pmatrix} 0 & 1 & 0 \\ 0 & 0 & 1 \\ -1 & -\frac{2}{3} & \frac{7}{3} \end{pmatrix}\mathbf{X} + \begin{pmatrix} 0 \\ 0 \\ \frac{t^2}{3} \end{pmatrix}$

3. **(a)** $\mathbf{X} = \begin{pmatrix} x \\ y \end{pmatrix} = c_1\begin{pmatrix} 1 \\ 1 \end{pmatrix}\cos 9t + c_2\begin{pmatrix} 1 \\ 1 \end{pmatrix}\sin 9t + c_3\begin{pmatrix} 1 \\ -1 \end{pmatrix}\cos 11t + c_4\begin{pmatrix} 1 \\ -1 \end{pmatrix}\sin 11t$

(b) $\mathbf{X} = \begin{pmatrix} 1 \\ 1 \end{pmatrix}\cos 9t + \begin{pmatrix} 1 \\ -1 \end{pmatrix}\cos 11t$ **5.** **(a)** $m_1 x'' = -k_1 x + k_2(y - x) + f(t)$; $m_2 y'' = -k_2(y - x)$

(b) $\begin{pmatrix} x \\ y \\ u \\ v \end{pmatrix}' = \mathbf{X}' = \begin{pmatrix} 0 & 0 & 1 & 0 \\ 0 & 0 & 0 & 1 \\ -\dfrac{k_1 + k_2}{m_1} & \dfrac{k_2}{m_1} & 0 & 0 \\ \dfrac{k_2}{m_2} & -\dfrac{k_2}{m_2} & 0 & 0 \end{pmatrix}\mathbf{X} + \begin{pmatrix} 0 \\ 0 \\ \dfrac{f(t)}{m_1} \\ 0 \end{pmatrix}$

7. 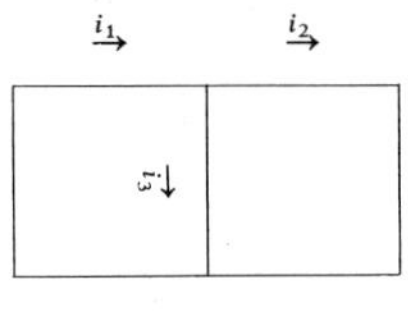

$$\begin{pmatrix} i_1 \\ i_2 \end{pmatrix} = \tfrac{15}{17}\begin{pmatrix} -3 \\ 2 \end{pmatrix} e^{-t/5} - \tfrac{10}{17}\begin{pmatrix} 4 \\ 3 \end{pmatrix} e^{-4t/3} + \begin{pmatrix} 5 \\ 0 \end{pmatrix}; \quad i_3 = i_1 - i_2$$

9. **(a)** $\begin{pmatrix} x \\ u \end{pmatrix}' = \begin{pmatrix} 0 & 1 \\ 0 & 0 \end{pmatrix}\begin{pmatrix} x \\ u \end{pmatrix} \quad \begin{pmatrix} y \\ v \end{pmatrix}' = \begin{pmatrix} 0 & 1 \\ 0 & 0 \end{pmatrix}\begin{pmatrix} y \\ v \end{pmatrix} + \begin{pmatrix} 0 \\ -g \end{pmatrix}$

(b) $\begin{pmatrix} x \\ u \\ y \\ v \end{pmatrix}' = \mathbf{X}' = \begin{pmatrix} 0 & 1 & 0 & 0 \\ 0 & 0 & 0 & 0 \\ 0 & 0 & 0 & 1 \\ 0 & 0 & 0 & 0 \end{pmatrix}\mathbf{X} + \begin{pmatrix} 0 \\ 0 \\ 0 \\ -g \end{pmatrix}$

11. $\begin{pmatrix} x \\ u \\ y \\ v \end{pmatrix} = \begin{pmatrix} x_0 + tv_0 \cos\alpha \\ v_0 \cos\alpha \\ y_0 + tv_0 \sin\alpha - \frac{1}{2}gt^2 \\ v_0 \sin\alpha - gt \end{pmatrix}$ **13.** $u = c_1 + c_2 x + c_3 x^2 + c_4 x^3 + \dfrac{fx^4}{24EI}$

15. $\begin{pmatrix} e^{2x} & e^{-2x} & \sin 2x & \cos 2x \\ 2e^{2x} & -2e^{-2x} & 2\cos 2x & -2\sin 2x \\ 4e^{2x} & 4e^{-2x} & -4\sin 2x & -4\cos 2x \\ 8e^{2x} & -8e^{-2x} & -8\cos 2x & 8\sin 2x \end{pmatrix}$ **19.** $\begin{pmatrix} 1 & -e^{-3bt} & -e^{(-2a-b)t} \\ 1 & -e^{-3bt} & e^{(-2a-b)t} \\ 1 & 2e^{-3bt} & 0 \end{pmatrix}$

Chapter 5

Exercises 5.2–5.3, p. 153

9. $\dfrac{(s+1)}{s^2}(e^{-s} - e^{-2s})$ **11.** $\dfrac{e^{-5(s+1)}}{s+1}$ **13.** $\dfrac{1}{s-2} - \dfrac{2}{s} + \dfrac{1}{s+2}$ **15.** $\dfrac{2}{s(s^2-4)}$ **17.** $\dfrac{s-2}{(s-2)^2+1}$

19. $\dfrac{(e^{-as} - e^{-bs})}{(b-a)s}$ **21.** e^{-as}

Exercises 5.4, p. 160

1. $\dfrac{s}{s^2+4}$ **3.** $\dfrac{\Gamma(3/2)}{(s-1)^{3/2}}$ **5.** $\dfrac{6}{(s-4)^4} + \dfrac{4}{(s-4)^3} + \dfrac{1}{(s-4)^2} = \dfrac{s^2 - 4s + 6}{(s-4)^4}$

7. $\dfrac{1}{2}\left[\dfrac{s}{s^2+1} - \dfrac{s}{s^2+9}\right] = \dfrac{4s}{(s^2+1)(s^2+9)}$ **9.** $\dfrac{6}{(s+3)^4}$ **11.** $\dfrac{1 + 2e^{-4(s-2)}}{s-2}$ **13.** $\dfrac{e^{-7(s+4)}}{s+4}$

15. $\dfrac{1}{s^2+1}$ **17.** $\dfrac{2s}{(s^2+1)^2} - \dfrac{\sin 1}{s^2}$ **19.** $\dfrac{1}{s^3} - \dfrac{2}{s}$ **21.** $\dfrac{6s^2 + 8s + 10}{s^2(s^2+2s+5)^2}$ **23.** $\dfrac{8(s+5)}{s[(s+5)^2+16]^2}$

25. $\dfrac{1}{s(s-1)^3} - \dfrac{1}{s(s+1)^3}$ **31.** $\dfrac{-5}{s^2+3s-4}$ **33.** $\dfrac{s}{(s^2+4)^2} - \dfrac{s+1}{s^2+4}$

35. $\dfrac{3+2s^2}{s^3+s} + \dfrac{s-2}{(s^3+s)(s^2-4s+5)}$ **37.** $\dfrac{Cs}{(RCs+1)(s+1)}$ **41.** $\dfrac{2}{9!}t^9 e^{-2t}$ **43.** $\frac{1}{2} - \frac{1}{2}e^{-2t}$

45. $-\frac{2}{3} + \frac{5}{3}e^{-3t}$ **47.** $1 - \cos t$ **49.** $3e^{-2t}\sin t$ **51.** $-(\sin t)/t$ **53.** $-e^{-t}/t - e^{t}/t + 2/t$

Exercises 5.5, p. 164

3. $\dfrac{2}{s(1 + e^{-Ps/2})}$

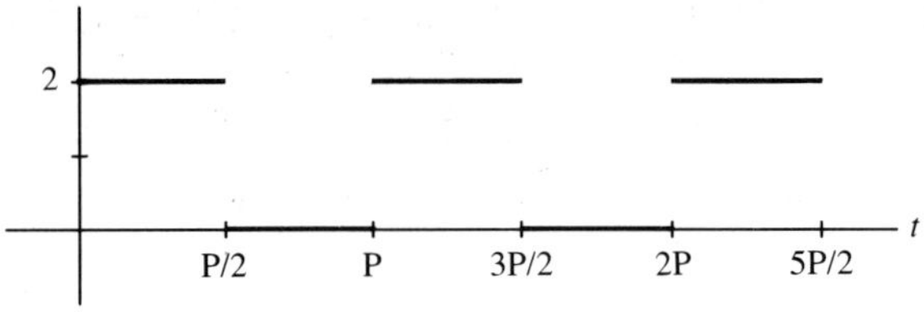

5. $\dfrac{1 - e^{-s} - e^{-2s} + e^{-3s}}{s(1 - e^{-3s})}$

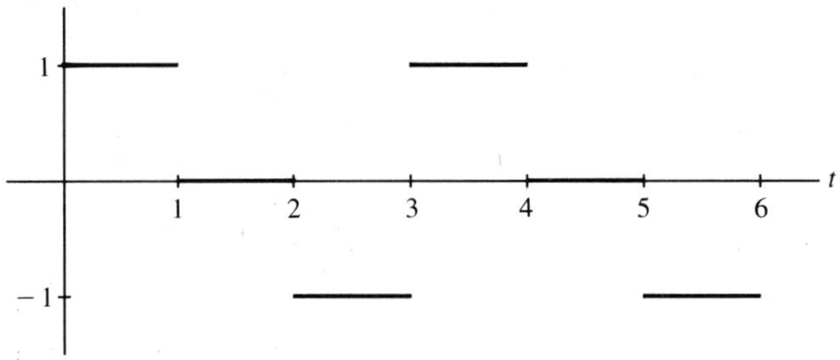

7. $\dfrac{1}{(1 + s^2)(1 - e^{-\pi s})}$

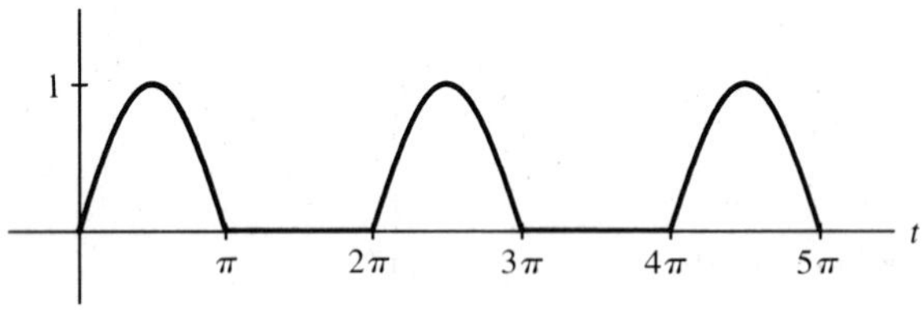

9. $\dfrac{s + 2be^{-\pi s/2b} - se^{-\pi s/b}}{(s^2 + b^2)(1 - e^{-\pi s/b})} = \dfrac{s + b \operatorname{csch} \dfrac{\pi s}{2b}}{b^2 + s^2}$

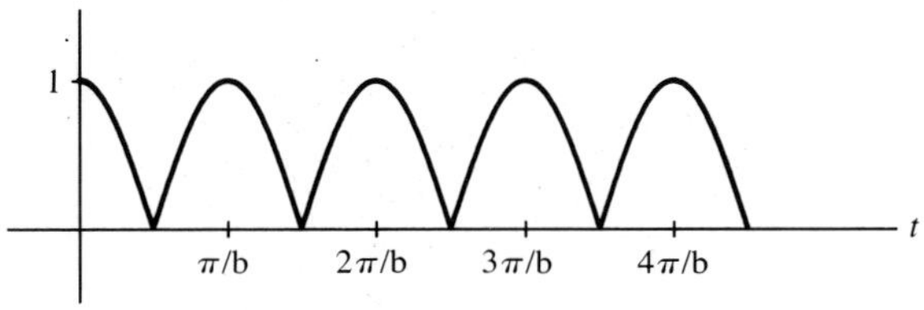

11. $\dfrac{1 - e^{-s} - e^{-2s} + e^{-3s}}{s^2(1 - e^{-3s})}$

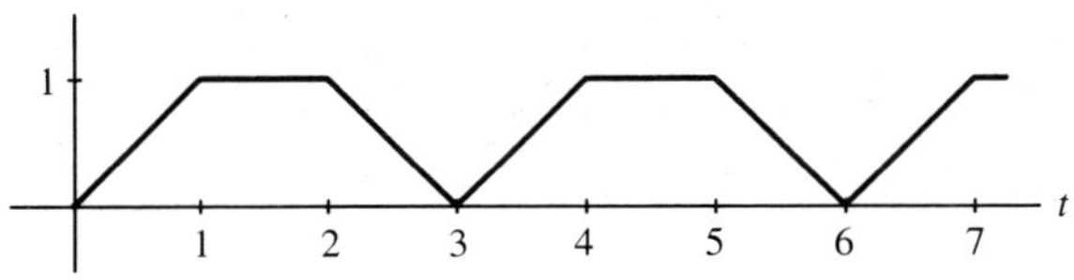

Exercises 5.6, p. 170

1. $\dfrac{t^6 e^{2t}}{144}$ **3.** $e^{-2t} \cos t$ **5.** $\dfrac{1 - \cos \omega t}{\omega}$ **7.** $e^{-t} - e^{-2t}$ **9.** te^{-2t}

11. $e^{-2t} \cosh \sqrt{2}t - \sqrt{2}e^{-2t} \sinh \sqrt{2}t$ **13.** $e^{-at} \cos bt - \dfrac{a}{b} e^{-at} \sin bt$ **15.** $\frac{1}{2}e^{-t} - 4e^{-2t} + \frac{9}{2}e^{-3t}$

17. $-\frac{1}{9}e^t + \frac{1}{3}te^t + \frac{1}{9}e^{-2t}$ **19.** $e^t(3 + 2t) - 2e^{2t}$ **21.** $4t - \sin t$ **23.** $\frac{t}{2} \sin 2t$ **25.** $(\sin t)/t$

27. $\displaystyle\int_0^t \frac{\sin u}{u}\, du$ **29.** $\frac{1}{4}[\sin 2t + |\sin 2t|]$

Exercises 5.7, p. 174

1. $y = \frac{2}{3}e^t - \frac{2}{3}e^{-2t}$ **3.** $y = \frac{13}{12}\cosh 4t - \frac{1}{24}\sinh 4t - \frac{1}{12}e^{-2t}$ **5.** $y = -\frac{17}{64}e^{-3t} + \frac{17}{64}e^t + (\frac{1}{8}t^2 - \frac{1}{16}t)e^t$
7. $y = e^t$ **9.** $y = \frac{33}{20}e^{2t} - \frac{52}{20}e^t + \frac{3}{20}\sin 2t - \frac{1}{20}\cos 2t$ **11.** $y = \cos at$
13. $y = -\frac{19}{8}\cosh t + \frac{39}{8}\sinh t + \frac{7}{3} + \frac{1}{24}e^{3t}$ **15.** $y = e^t + e^{-2t} + 2te^{-2t}$ **17.** $i = \frac{1}{25}e^{-200t} - \frac{1}{25}e^{-800t}$
19. $i = \frac{22}{25}e^{-2t} - \frac{12}{5}te^{-2t} + \frac{3}{25}\cos t + \frac{4}{25}\sin t$ **21.** $\theta = \theta_0 \cos \omega t = \theta_0 \cos \sqrt{\frac{g}{l}}\,t$ **23.** $y_1(t) = 0,\ y_2(t) = 1$
25. $y_1 = 3\sin t + t^2,\ y_2 = 3\cos t - t^3$
27. **(a)** $c_1 = \dfrac{a+b+c}{3} + \left(\dfrac{2a-b-c}{3}\right)e^{-3t};\ c_2 = \dfrac{a+b+c}{3} + \left(\dfrac{2b-a-c}{3}\right)e^{-3t};$

$c_3 = \dfrac{a+b+c}{3} + \left(\dfrac{2c-a-b}{3}\right)e^{-3t}$

Exercises 5.8, p. 181

1.

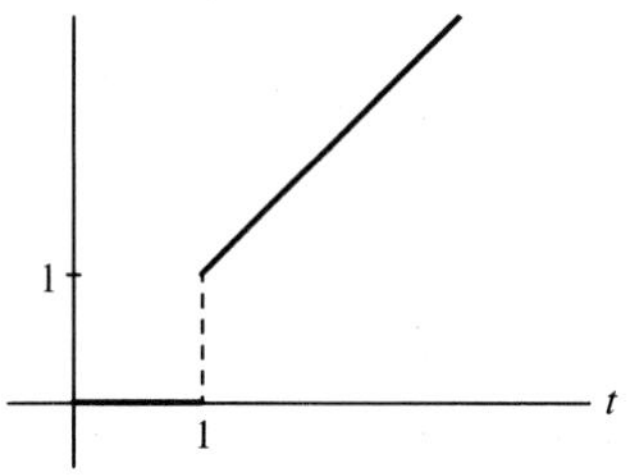

3.

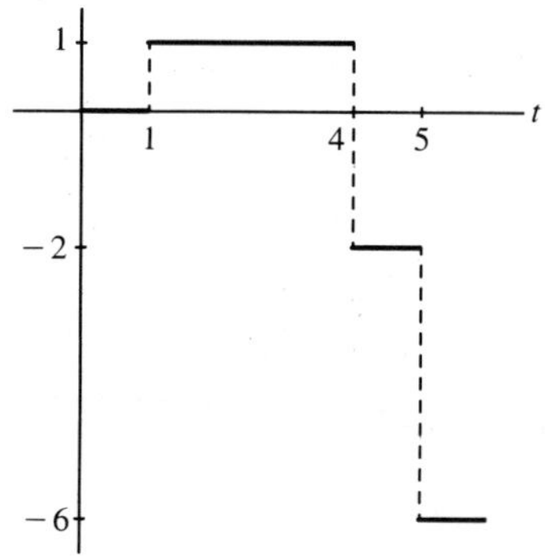

5.

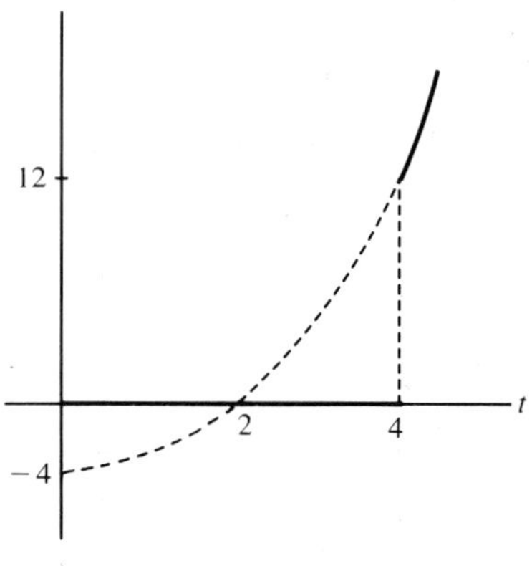

7.

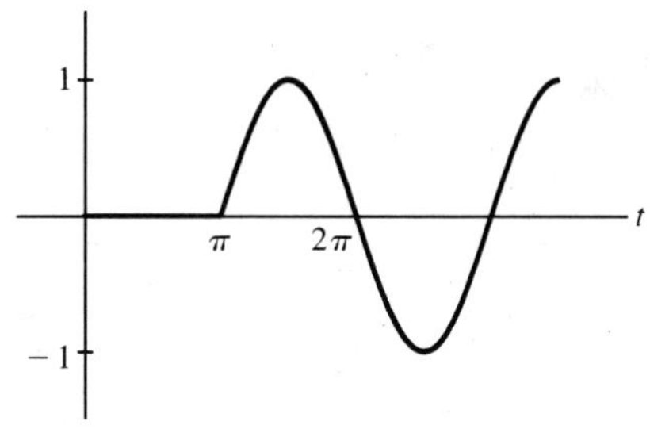

9.

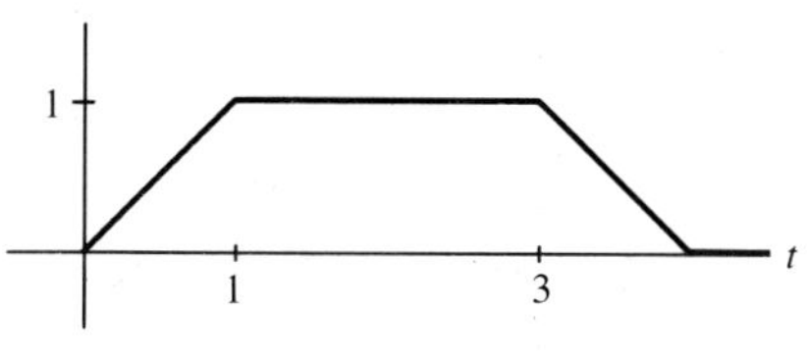

11.

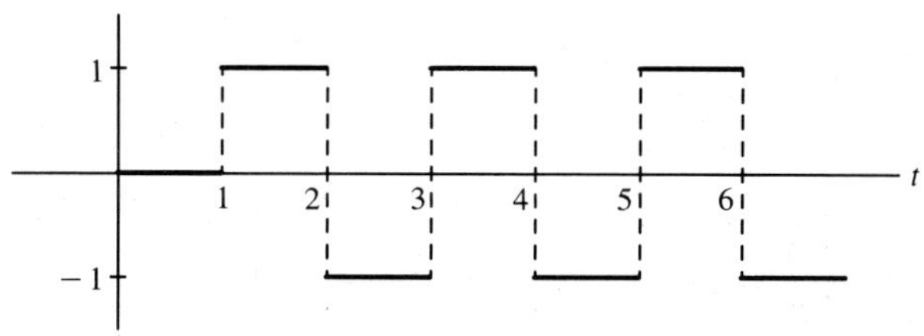

13. $\dfrac{3}{s}+\dfrac{e^{-2s}}{s^2}$ **15.** $\dfrac{3}{s^2+9}(1+e^{-\pi s})$ **17.** $\dfrac{1}{s^2}(1-e^{-s}-e^{-3s}+e^{-4s})$

19. $\dfrac{1}{s^2}-\dfrac{1}{s}(e^{-s}+e^{-2s}+e^{-3s}+\cdots)=\dfrac{1}{s^2}-\dfrac{e^{-s}}{s(1-e^{-s})}$ **21.** $\dfrac{1}{s^2}-2\dfrac{e^{-s}}{s^2}+\dfrac{e^{-3s}}{s^2}+\dfrac{e^{-4s}}{s}$

23. $\frac{1}{2}(t-4)^2u_4-\frac{1}{2}(t-1)^2u_1$ **25.** $\frac{1}{3}u_3\sinh 3(t-3)$ **27.** $[e^{2(t-1)}-e^{(t-1)}]u_1$

29. $\frac{1}{2}e^{-(t-1)}(t-1)^2u_1$ **31.** $\frac{1}{2}[e^{3(t-1)}+e^{(t-1)}]u_1$ **33.** $u_1e^{(t-1)}\sin(t-1)$ **35.** $\dfrac{e^{-2s}}{s^2}+\dfrac{2e^{-2s}}{s}$

37. $-e^{-\pi s}[s/(s^2+1)]$ **39.** $\dfrac{e^{3-3s}}{s-1}$ **41.** $e^{-(1/2)\pi s}[s/(s^2+1)]$ **43.** $i=(V_0/R)(1-e^{-(R/L)t})$

45. $i=(L/R^2)[e^{-(R/L)t}-1]+t/R+(1/R)[u_1(e^{-(R/L)(t-1)}-1)+u_2(e^{-(R/L)(t-2)}-1)+u_3(e^{-(R/L)(t-3)}-1)+\cdots]$

47. $\theta(t)=H(t)-u_1H(t-1)$ where $H(t)\equiv\dfrac{f_0l}{\omega g}\left[\omega-e^{-(\mu/2m)t}\left(\dfrac{\mu}{2m}\sin\omega t+\omega\cos\omega t\right)\right]$ with $\omega^2=\dfrac{g}{l}-\left(\dfrac{\mu}{2m}\right)^2$

49. $y(t)=f(t)+2u_1f(t-1)+4u_2f(t-2)+8u_3f(t-3)+\cdots$

Exercises 5.9, p. 188

1. $\dfrac{2}{s^2(s^2+4)}$ **3.** $y=1-\cos t$ **5.** $y=-\frac{1}{2}e^{-4t}+\frac{1}{2}e^{-2t}$ **7.** $y=\frac{1}{8}-\frac{1}{8}e^{-2t}\cos 2t-\frac{1}{8}e^{-2t}\sin 2t$

9. $y=t+t^3/6$ **11.** $y=-e^{-2t}+e^{-t}$ **13.** $y=\frac{1}{6}e^{2t}-\frac{1}{6}e^{-t}\cos\sqrt{3}t+(\sqrt{3}/6)e^{-t}\sin\sqrt{3}t$

15. $y=t+\frac{1}{2}t^2$ **17.** $y=\int_0^t f(t-u)ue^{-bu}\,du$ **21.** **(a)** $\psi(t)=-(2/\omega^2)e^{-t}+(1+1/\omega^2)te^{-t}$

23. $r(t)=P_0t+2P_0$

Chapter 6

Exercises 6.1–6.3, p. 197

1. $y=1+x+x^2+\frac{1}{3}x^3+\frac{1}{12}x^4+\cdots=2e^x-x-1$ **3.** $y=e^x$

5. $y=1+x-\frac{7}{2}x^2+\frac{25}{6}x^3-\frac{79}{24}x^4+\cdots=2e^{-x}-e^{-3x}$

7. $y=2-2(x-1)+(x-1)^2-\frac{1}{12}(x-1)^4+\frac{1}{20}(x-1)^5+\cdots$

9. $y=1+(x-1)+\frac{1}{6}(x-1)^3-\frac{1}{24}(x-1)^4+\frac{1}{40}(x-1)^5+\cdots$

11. $y=1+\frac{1}{2}x^2+\frac{1}{6}x^3+\frac{1}{24}x^4+\frac{1}{30}x^5+\cdots$

13. $y=a_0(1-\frac{1}{2}x^2+\frac{1}{24}x^4+\cdots)+a_1(x-\frac{1}{6}x^3+\frac{1}{120}x^5-\cdots)$ **15.** $y=a_0+a_1(x-1)-\frac{3}{2}a_1(x-1)^2+\cdots$

17. $y=(1+2x^2+3x^4+\cdots+(n+1)x^{2n}+\cdots)+\left(x+\dfrac{5}{3}x^3+\dfrac{7}{3}x^5+\cdots+\dfrac{2n+3}{3}x^{2n+1}+\cdots\right)$

19. $y=x-\dfrac{1}{3}x^3+\left(1-\dfrac{3}{2}x^2+\dfrac{1}{8}x^4+\cdots+\dfrac{3}{2^nn!(2n-1)(2n-3)}x^{2n}+\cdots\right)$

23. $y=(x-1)+\frac{1}{6}(x-1)^3+\frac{1}{12}(x-1)^4+\frac{1}{120}(x-1)^5+\frac{1}{120}(x-1)^6+\cdots$

Exercises 6.4, p. 205

1. $y=a_0\displaystyle\sum_{n=0}^{\infty}\frac{(-1)^nx^n}{n!}$ **3.** $y=\displaystyle\sum_{n=0}^{\infty}\frac{(-1)^nx^n}{n!}(a_0+x)$

5. $y=\displaystyle\sum_{n=0}^{\infty}\frac{x^{2n}}{(2n)!}\left[a_0+\frac{1}{2}(-1)^{n+1}\right]+\sum_{n=0}^{\infty}\frac{x^{2n+1}}{(2n+1)!}\left[a_0+\frac{1}{2}(-1)^{n+1}\right]$ **7.** $y=a_0\displaystyle\sum_{n=0}^{\infty}\frac{(-1)^n}{3^nn!}x^{3n}$

9. $y=\displaystyle\sum_{n=0}^{\infty}a_{3n}x^{3n}+\sum_{n=0}^{\infty}a_{3n+1}x^{3n+1}$ where

$$a_{3n}=\frac{(-1)^n(3n-2)(3n-5)\ldots7\cdot4\cdot1}{(3n)!}a_0$$

$$a_{3n+1}=\frac{(-1)^n(3n-1)(3n-4)\ldots8\cdot5\cdot2}{(3n+1)!}a_1$$

11. $y = \sum_{n=0}^{\infty} a_{5n}x^{5n} + \sum_{n=0}^{\infty} a_{5n+1}x^{5n+1}$ where

$$a_{5n} = \frac{(-1)^n a_0}{5n(5n-1)(5n-5)(5n-6)\ldots 10\cdot 9\cdot 5\cdot 4}$$

$$a_{5n+1} = \frac{(-1)^n a_1}{(5n+1)(5n)(5n-4)(5n-5)\ldots 11\cdot 10\cdot 6\cdot 5}$$

13. $y = \sum_{n=0}^{\infty} \frac{x^{2n}}{(2n)!}\left[a_0(-1)^n + \frac{1}{2}\right] + \sum_{n=0}^{\infty} \frac{x^{2n+1}}{(2n+1)!}\left[a_1(-1)^n + \frac{1}{2}\right]$

15. $y = y_1 + \sum_{n=1}^{\infty} \frac{2^{2n-2}[(n-1)!]^2}{(2n+1)!}x^{2n+1}$ where

$$y_1 = a_1 x + a_0\left[1 + \frac{1}{2}x^2 + \sum_{n=2}^{\infty} \frac{(2n-3)!}{2^{2n-2}n!(n-2)!(2n-1)}x^{2n}\right]$$

17. $y = a_0\left(1 + 2x^2 + \frac{1}{3}x^4\right) + a_1 \sum_{n=0}^{\infty} \frac{3(-1)^n}{2^n n!(2n+1)(2n-1)(2n-3)}x^{2n+1}$

19. $y = a_1 x + a_0\left[1 + \frac{3}{2}x^2 + \sum_{n=2}^{\infty} \frac{(-1)^{n+1}(2n+1)(2n-3)!}{2^{2n-2}n!(n-2)!}x^{2n}\right]$

21. $y = a_0[1 - \frac{1}{6}(x-2)^3 + \frac{1}{180}(x-2)^6 + \cdots] + a_1[(x-2) - \frac{1}{12}(x-2)^4 + \frac{1}{504}(x-2)^7 + \cdots]$

23. $y = a_0 + a_1(x-1) + \frac{1}{2}(1 - 3a_0 - a_1)(x-1)^2 + \cdots$ **27.** $y = a_1(x - \frac{4}{3}x^3) + a_0(1 - \frac{9}{2}x^2 + \frac{15}{8}x^4 + \cdots)$

29. $p = 0, y = c_1 + c_2 \operatorname{Cos}^{-1} x; p = 1, y = c_1 x + c_2\sqrt{1-x^2}; p = 2, y = c_1(2x^2 - 1) + c_2 x\sqrt{1-x^2}$

Exercises 6.5–6.6, p. 210

1. R.S.P. at 0 **3.** R.S.P. at 0; I.S.P. at 1 **5.** I.S.P. at 2 **7.** $y = a_0 x^{1/2} + b_0 x^{1/2} \sum_{n=1}^{\infty} \frac{x^n}{2^{n-1}n!}$

9. $y = a_0(1 + x)$ **11.** $y = a_0 \sum_{n=0}^{\infty} \frac{(-1)^n x^{2n+2}}{2^{2n-1}n!(n+2)!}$ **13.** $y = a_0 \sum_{n=0}^{\infty} \frac{(-1)^n x^{2n}}{2^{2n}(n!)^2}$ **15.** $y = a_0 x^{1/2} \sum_{n=0}^{\infty} \frac{x^n}{2^n n!}$

17. $y = x^{-1} \sum_{n=0}^{\infty} a_n x^n + x^{1/3} \sum_{n=0}^{\infty} b_n x^n$, where a_0 is arbitrary, $a_1 = a_0$, and $a_n = \frac{(n-2)a_{n-1} - 2a_{n-2}}{n(3n-4)}, n \geq 2$;
b_0 is arbitrary, $b_1 = \frac{1}{21}b_0$, and $b_n = -\frac{(3n-2)b_{n-1} - 6b_{n-2}}{3n(3n+4)}, n \geq 2$

19. $y = x^{-1} \sum_{n=0}^{\infty} a_n x^n + x^{1/3} \sum_{n=0}^{\infty} b_n x^n$, where a_0 is arbitrary, $a_1 = -a_0$, and $a_n = \frac{a_{n-1} - (n-3)(n-4)a_{n-2}}{n(3n-4)}, n \geq 2$;
b_0 is arbitrary, $b_1 = \frac{1}{7}b_0$, and $b_n = \frac{9b_{n-1} - (3n-8)(3n-5)b_{n-2}}{9n(3n+4)}, n \geq 2$

21. $y = \left(x + \frac{1}{2}\right) \sum_{n=0}^{\infty} a_n \left(x + \frac{1}{2}\right)^n$, where a_0 is arbitrary and $a_n = -\frac{2}{n(n+1)}a_{n-1}, n \geq 1$

23. $y = 1 + 9(x-1) + 12(x-1)^2 + 4(x-1)^3$ **25.** $y = \sum_{n=0}^{\infty} \frac{\Gamma(n-p)}{\Gamma(-p)(n!)^2}x^n$

Exercises 6.7, p. 216

1. $y_1 = \frac{1}{x}\sum_{n=0}^{\infty} \frac{(2x)^n}{(n!)^2}; y_2 = y_1 \ln x - 4 - 3x - \frac{22}{27}x^2 - \cdots$

3. $y_1 = 1 + \frac{1}{2}x + \frac{1}{12}x^2; y_2 = x^5 \sum_{n=0}^{\infty} \frac{60x^n}{n!(n+5)(n+4)(n+3)}$ **5.** $y_1 = \frac{1}{x} - \frac{1}{2}; y_2 = x^2 \sum_{n=0}^{\infty} \frac{(-1)^n 6(n+1)}{(n+3)!}x^n$

7. $y_1 = 2(x-1); y_2 = y_1 \ln(x-1) + 1 - 3(x-1) + \sum_{n=2}^{\infty} \frac{(-1)^{n+1}(n+1)}{(n-1)}(x-1)^n$

9. $y_1 = 1 - 3x + \frac{3}{2}x^2 - \frac{1}{6}x^3$; $y_2 = y_1 \ln x + 7x - \frac{23}{4}x^2 + \frac{11}{12}x^3 + x^4 \sum_{n=0}^{\infty} \frac{(-1)^{n+1} 6n!}{[(n+4)!]^2} x^n$

11. $y_1 = 1 + 6(x-1) + \frac{15}{2}(x-1)^2 + \frac{5}{2}(x-1)^3$; $y_2 = y_1 \ln(x-1) + \frac{11}{3} + \frac{19}{2}(x-1) + \frac{17}{8}(x-1)^2 + \cdots$

Chapter 7

Exercises 7.2–7.3, p. 224

1. $\left\{\sqrt{\frac{2}{L}} \sin \frac{n\pi x}{L}\right\}_{n=1}^{\infty}$ **11. (a)** No **(b)** No **(c)** No **(d)** Yes **(e)** Yes; 4 must divide L.

Exercises 7.4, p. 231

1. $0, \quad \frac{\pi^2}{3} + 4\sum_{n=1}^{\infty} \frac{(-1)^n}{n^2} \cos nx$ **3.** $\frac{1}{2}(1 + \cos 2x)$ **5.** $\frac{2}{\pi} + \frac{4}{\pi}\sum_{n=1}^{\infty} \frac{(-1)^{n+1}}{4n^2 - 1} \cos 2nx$

7. $\frac{1}{3} + \frac{4}{\pi^2}\sum_{n=1}^{\infty} \frac{(-1)^n}{n^2} \cos n\pi x$ **9.** $\frac{1}{2} + \frac{2}{\pi}\sum_{n=0}^{\infty} \frac{1}{2n+1} \sin(2n+1)\frac{\pi x}{2}$

11. $u(x) = \frac{B_n \sin n\pi x/L}{EI(n\pi/L)^4 - P(n\pi/L)^2} = \frac{B_n \sin n\pi x/L}{(n\pi/L)^2(P_n - P)}$

13. $u(x) = -\frac{12}{\pi^3}\sum_{n=1}^{\infty} \frac{\sin(2n+1)\pi x}{(2n+1)^3(P_{2n+1} - P)}$ where $P_{2n+1} = (2n+1)^2\pi^2 EI$

Exercises 7.5, p. 235

1. $\frac{3}{2} + \frac{2}{\pi}\sum_{n=0}^{\infty} \frac{1}{(2n+1)} \cos\left[(2n+1)x - \frac{\pi}{2}\right]$ **3.** $\frac{2}{\pi} + \frac{4}{\pi}\sum_{n=1}^{\infty} \frac{\cos(2nx - \pi)}{4n^2 - 1}$ **5.** $\frac{4}{\pi}\sum_{n=0}^{\infty} \frac{\cos[(2n+1)x + \pi/2]}{2n+1}$

7. $\frac{8}{\pi}\sum_{n=1}^{\infty} (-1)^{n+1} \frac{n}{4n^2 - 1} \cos\left(2nx - \frac{\pi}{2}\right)$ **9.** $\frac{3}{2} - \frac{i}{\pi}\sum_{n=-\infty}^{\infty} \frac{1}{(2n+1)} e^{i(2n+1)x}$ **11.** $-\frac{2}{\pi}\sum_{n=-\infty}^{\infty} \frac{1}{4n^2 - 1} e^{i2nx}$

13. $\frac{2i}{\pi}\sum_{n=-\infty}^{\infty} \frac{e^{i(2n+1)x}}{2n+1}$ **15.** $\frac{4i}{\pi}\sum_{n=-\infty}^{\infty} (-1)^n \frac{n}{4n^2 - 1} e^{i2nx}$

19. $i(t) = A_1 \cos(\omega t - \delta_1) + A_2 \cos(2\omega t - \delta_2) + A_3 \cos(3\omega t - \delta_3) + \cdots$

where $A_1 = \omega C B_1 / \sqrt{(\omega^2 LC - 1)^2 + \omega^2 R^2 C^2}$, $\delta_1 = \text{Tan}^{-1} \frac{\omega RC}{1 - \omega^2 LC}$

$A_2 = 2\omega C B_2 / \sqrt{(4\omega^2 LC - 1)^2 + 4\omega^2 R^2 C^2}$, $\delta_2 = \text{Tan}^{-1} \frac{2\omega RC}{1 - 4\omega^2 LC}$

$A_3 = 3\omega C B_3 / \sqrt{(9\omega^2 LC - 1)^2 + 9\omega^2 R^2 C^2}$, $\delta_3 = \text{Tan}^{-1} \frac{3\omega RC}{1 - 9\omega^2 LC}$

Exercises 7.6–7.7, p. 240

1. (a), (g), (h), (l), (n) are odd; (b), (d), (e), (f), (k), (m), are even; (c), (i), (j) are neither odd nor even.

3. 0 **5.** $\frac{4}{\pi}\sum_{n=0}^{\infty} \frac{\sin(2n+1)x}{2n+1}$; $\frac{2}{\pi}\sum_{n=0}^{\infty} \frac{\sin(2n+1)x}{2n+1}$ **7.** $5\cos 2x - 3 + \frac{8}{\pi}\sum_{n=1}^{\infty} (-1)^{n+1} \frac{n}{4n^2 - 1} \sin nx$

9. (a) $\cos x$ **(b)** $\cos x$ **(c)** $\frac{2}{\pi} + \frac{4}{\pi}\sum_{n=1}^{\infty} \frac{(-1)^{n+1}}{4n^2 - 1} \cos 2nx$ **11.** $\frac{\pi}{2} + \frac{4}{\pi}\sum_{n=0}^{\infty} \frac{\cos(2n+1)x}{(2n+1)^2}$

13. $\frac{8}{\pi^2}\sum_{n=0}^{\infty} \frac{1}{(2n+1)^2} \cos(2n+1)\pi x$ **15.** $\frac{2}{\pi}\sum_{n=1}^{\infty} \frac{(-1)^{n+1}}{n} \sin \frac{n\pi x}{2} + \frac{4}{\pi^2}\sum_{n=0}^{\infty} \frac{(-1)^n}{(2n+1)^2} \sin(2n+1)\frac{\pi x}{2}$

Exercises 7.8, p. 247

1.

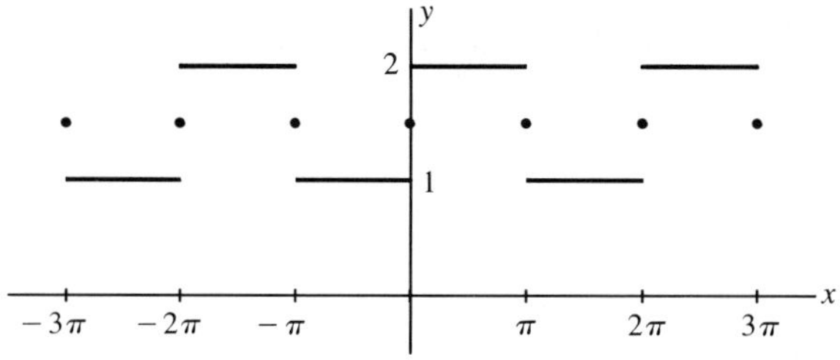

3.

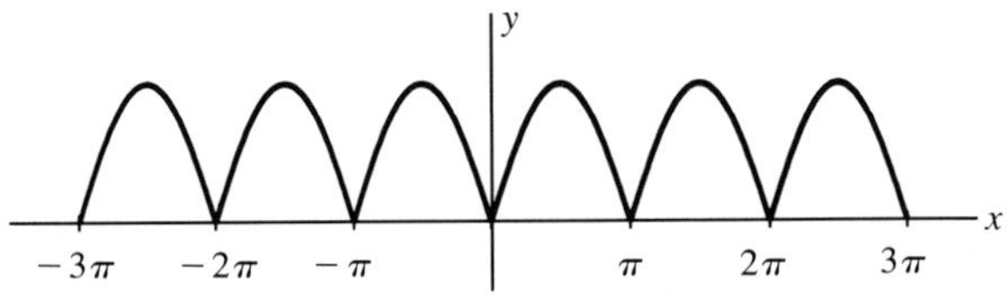

5.

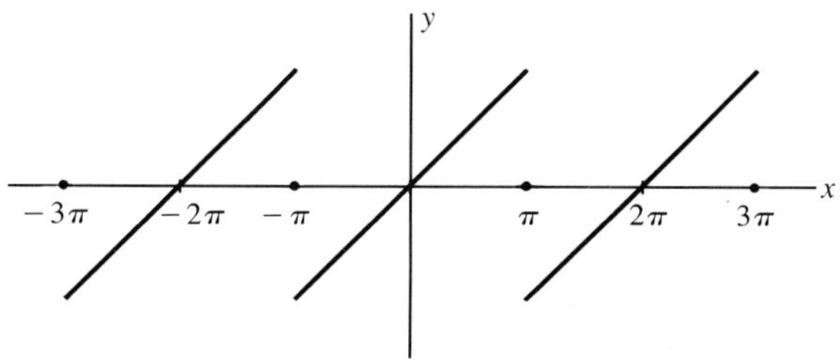

7.

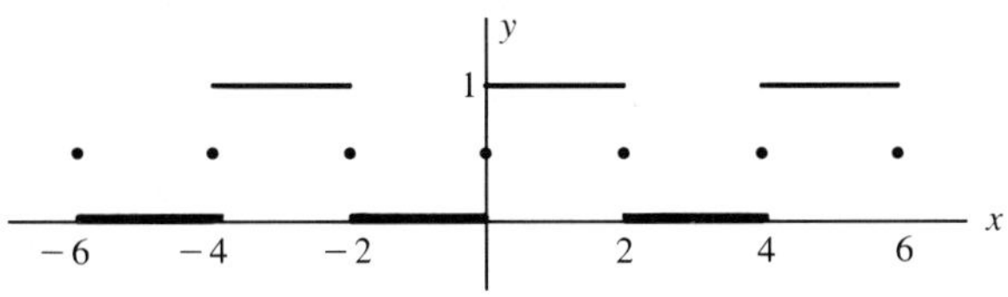

13. $\dfrac{2\pi^2}{3} - 1 + 8\sum_{n=1}^{\infty} \dfrac{(-1)^n}{n^2}\cos nx$ **15.** $\dfrac{7\pi}{4} + 3\sum_{n=1}^{\infty} \dfrac{(-1)^{n+1}}{n}\sin nx + \dfrac{2}{\pi}\sum_{n=0}^{\infty} \dfrac{\cos(2n+1)x}{(2n+1)^2}$

17. One example is $f(x) = 3x^5 - 10x^3 + 7x$.

Exercises 7.9–7.10, p. 255

1. **(a)** $f_o(x) = \begin{cases} -x-1, & -2 < x < -1 \\ x+1, & -1 < x < 0 \end{cases}$ $f_e(x) = \begin{cases} x+1, & -2 < x < -1 \\ -x-1, & -1 < x < 0 \end{cases}$

(b) $f_o(x) = -x^2 + x,\ -2 < x < 0;\quad f_e(x) = x^2 - x,\ -2 < x < 0$

(c) $f_o(x) = \sin\dfrac{\pi x}{2},\ -2 < x < 0;\quad f_e(x) = -\sin\dfrac{\pi x}{2},\ -2 < x < 0$

(d) $f_o(x) = -\cos\dfrac{\pi x}{2},\ -2 < x < 0;\quad f_e(x) = \cos\dfrac{\pi x}{2},\ -2 < x < 0$

(e) $f_o(x) = -e^{-x},\ -2 < x < 0;\quad f_e(x) = e^{-x},\ -2 < x < 0$

3. **(a)** Fourier sine series: $\frac{8}{\pi}\sum_{n=1}^{\infty}\frac{n}{4n^2-1}\sin 2nx$

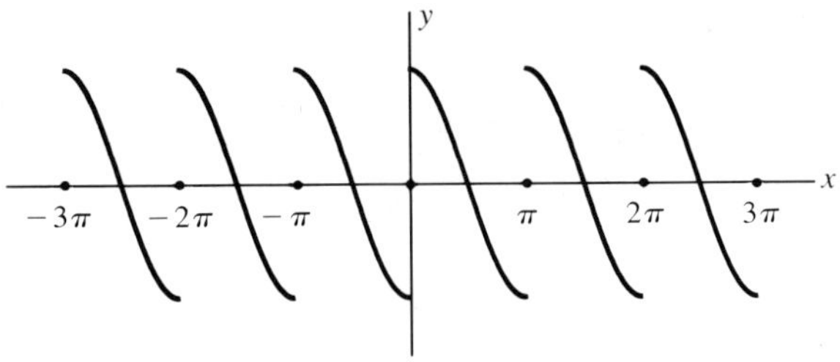

(b) Fourier cosine series: $\cos x$

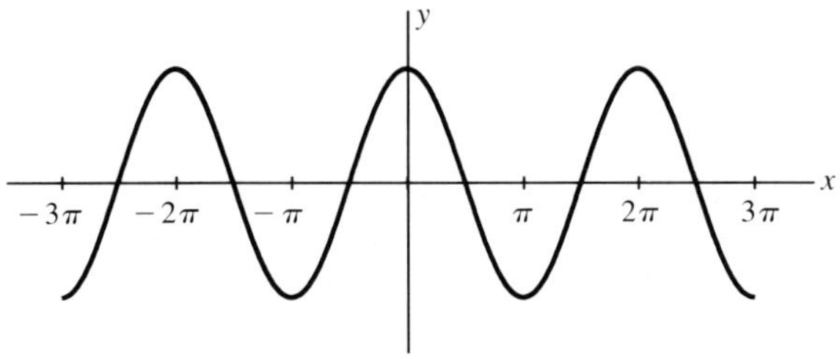

5. $\frac{8}{\pi}\sum_{n=1}^{\infty}\frac{n(-1)^{n+1}}{4n^2-1}\sin nx;\quad \frac{2}{\pi}-\frac{4}{\pi}\sum_{n=1}^{\infty}\frac{\cos nx}{4n^2-1}$

7. $\frac{96}{\pi^3}\sum_{n=1}^{\infty}\frac{(-1)^n}{n^3}\sin\frac{n\pi x}{2};\quad -2+\frac{16}{\pi^2}\sum_{n=1}^{\infty}\left[\frac{1+2(-1)^n}{n^2}\right]\cos\frac{n\pi x}{2}+\frac{192}{\pi^4}\sum_{n=0}^{\infty}\frac{1}{(2n+1)^4}\cos(2n+1)\frac{\pi x}{2}$

9. $\frac{4}{\pi}\sum_{n=0}^{\infty}\frac{\sin(2n+1)\pi x}{(2n+1)};\quad 1$ **11.** **(a)** $(1/n)^2$ **(b)** $(1/n)$ **(c)** $(1/n)^2$

13. One example is $3x^5-40x^3+112x$.

Exercises 7.11, p. 258

1. $F(t)=2\pi^2+24\sum_{n=1}^{\infty}\frac{(-1)^n}{n^2}\cos nt;\ y(t)=A_0+A_1\cos(t-\delta_1)+A_2\cos(2t-\delta_2)+\cdots$, where $A_0=\pi^2/10$, $A_1=12\sqrt{2}/\sqrt{113}$, $\delta_1=\operatorname{Tan}^{-1}\frac{1}{15}\pm\pi$, $A_2=3$, $\delta_2=\pi/2$

3. **(a)** $\omega_r \doteq 3.123$, $\omega_0 \doteq 3.127$. Note the significantly larger amplitude of the term with $\omega=3$ in the result of Example 7.25.

(b) $\omega_r \doteq 1.995$, $\omega_0 \doteq 1.997$. Note the larger amplitude of the term with $\omega=2$ in the result for Exercise 1.

5. $v'(t)=\frac{4}{\pi}\sum_{n=0}^{\infty}\frac{\sin(2n+1)t}{(2n+1)};\ i(t)=A_1\cos(t-\delta_1)+A_3\cos(3t-\delta_3)+A_5\cos(5t-\delta_5)+\cdots$, where $A_1=4/(\sqrt{17}\pi)$, $\delta_1=\pi-\operatorname{Tan}^{-1}\frac{1}{4}$, $A_3=4/(3\sqrt{673}\pi)$, $\delta_3=\operatorname{Tan}^{-1}\frac{23}{12}\pm\pi$, $A_5=4/(5\sqrt{5441}\pi)$, $\delta_5=\operatorname{Tan}^{-1}\frac{71}{20}\pm\pi$

7. $f(t)=\sin\omega t;\ g(t)=A\cos(\omega t-\delta)$, where $A=1/\sqrt{(\omega^2-0.0004)^2+0.0025\omega^2}$, $\delta=\operatorname{Tan}^{-1}\left(\frac{\omega^2-0.0004}{0.05\omega}\right)\pm\pi$

9. If $f(t)=\sin\omega t$, then $g(t)=A\cos(\omega t-\delta)$ with $\delta=\operatorname{Tan}^{-1}\left(\frac{\omega^2-\beta^2}{2\alpha\omega}\right)\pm\pi$. For $\beta^2<\omega^2$, as α increases from 0 to ∞, δ decreases (monotonically) from $3\pi/2$, say, to π. For $\beta^2>\omega^2$, as α increases from 0 to ∞, δ increases (monotonically) from $\pi/2$, say, to π.

Exercises 7.12, p. 262

1. $\lambda=n^2\pi^2/L^2$, $y=\cos(n\pi x/L)$ with $n=0, 1, 2, \ldots$ **3.** $y''+\lambda y=0$, $y(0)=y(2L)$, $y'(0)=y'(2L)$

5. $\alpha=-1$ **7.** $\lambda=n^2\pi^2$, $y=\cos(n\pi\ln x)$ with $n=0, 1, 2, \ldots$ **11.** $(e^{-x^2}y')'+2ne^{-x^2}y=0;\ r(x)=e^{-x^2}$

13. Bessel's equation in self-adjoint form is $(xy')'+\left(\lambda x-\frac{n^2}{x}\right)y=0$ with $r(x)=x$.

Chapter 8

Exercises 8.1–8.3, p. 273

3. **(a)** One possibility: $yu_x + xu_y = 0$ **(b)** $yu_x - xu_y = 2xy$

9. **(a)** $u = f(x)$ **(b)** $u = f(y)\cos x + g(y)\sin x$ **(c)** $u = f(y) + g(y)\cos x + h(y)\sin x$

(d) $u = \frac{1}{2}x^2 y + f(x) + xg(y) + h(y)$ **11.** $u = -\sinh\frac{\pi}{L}(L - x)\sin\frac{\pi y}{L}$ **13.** $u = 3\sin 10\pi x \cos 10\pi t$

15. One possibility: $u = 1/(1 + y - x)$

Exercises 8.4, p. 279

1. **(a)** $u = ce^{(1-2b)x+by}$ **(b)** $u = ce^{\pm\sqrt{b^2-b}\,x+by}$ **3.** **(a)** $u = ce^{(1-2b)x+by}$ **(b)** $u = ce^{ax+(1\pm\sqrt{1+4a^2})y/2}$

5. **(a)** $u = f(x + \frac{1}{2}y) + g(x + \frac{1}{3}y)$ **(b)** $u = f(x) - g(x + \frac{1}{4}y)$ **9.** $u = ce^{-a/x - y^3/a}$

11. $u = (c_1 x + c_2 + c_3 \cosh ax + c_4 \sinh ax)y^{-a^2}$ or $u = (c_1 x + c_2 + c_3 \cos ax + c_4 \sin ax)y^{a^2}$ or $u = c_1 x^3 + c_2 x^2 + c_3 x + c_4$ **13.** $u = e^{a\,\mathrm{Tan}^{-1} y}(c_1 + c_2 e^{ax})$ **15.** $u = ce^{x^2/2 + y^2/2 + a(x-y)}$

19. **(a)** $u(x, y) = (c_1 e^{\beta x} + c_2 xe^{\beta x} + c_3 e^{-\beta x} + c_4 xe^{-\beta x})\sin \beta y$

(b) $u(x, y) = \cos \alpha x\,(c_1 e^{\alpha y} + c_2 ye^{\alpha y} + c_3 e^{-\alpha y} + c_4 ye^{-\alpha y})$

Exercises 8.5, p. 284

3. $u(x, y) = \sum_{n=1}^{\infty} C_n \sinh\frac{n\pi}{K}(L - x)\sin\frac{n\pi y}{K}$, where $C_n \sinh\frac{n\pi L}{K} = \frac{2}{K}\int_0^K g(y)\sin\frac{n\pi y}{K}\,dy$

5. $u(x, y) = D_0 + \sum_{n=1}^{\infty} D_n \cosh\frac{n\pi}{K}(L - x)\cos\frac{n\pi y}{K}$, where

$D_0 = \frac{1}{K}\int_0^K g(y)\,dy$ and for $n \geq 1$, $D_n \cosh\frac{n\pi L}{K} = \frac{2}{K}\int_0^K g(y)\cos\frac{n\pi y}{K}\,dy$.

7. $u(x, y) = -\sum_{n=1}^{\infty} \frac{2}{n\pi \sinh 3n\pi}\sinh[n\pi(3 - x)]\sin n\pi y$ **9.** $u(x, y) = \sum_{n=1}^{\infty} \frac{2(-1)^{n+1}}{n^2\pi^2 \sinh 2n\pi}\cosh n\pi x \sin n\pi y$

11. $u(r, \theta) = r\cos\theta$ **13.** $u(r, \theta) = \frac{10}{\pi}\sum_{n=0}^{\infty} \frac{(-1)^n}{(2n + 1)} r^{2n+1}\cos(2n + 1)\theta + \frac{10}{\pi}\sum_{n=0}^{\infty} \frac{r^{4n+2}}{(2n + 1)}\sin(4n + 2)\theta$

15. $u(r, \theta) = D_0 + \sum_{n=1}^{\infty} r^n(C_n \sin n\theta + D_n \cos n\theta)$, where D_0 is arbitrary and for $n \geq 1$,

$nC_n = \frac{1}{\pi}\int_0^{2\pi} g(\theta)\sin n\theta\,d\theta$, $nD_n = \frac{1}{\pi}\int_0^{2\pi} g(\theta)\cos n\theta\,d\theta$ **17.** $u(r, \theta) = -2\sum_{n=1}^{\infty} \frac{r^n}{n^2}\sin n\theta$ **21.** $u(r, \theta) = r\cos\theta$

Exercises 8.6–8.7, p. 292

3. There are six combinations of values of x, t to consider:

(i) $u(x, t) = 0$ for $x < -1 - ct$, **(ii)** $u(x, t) = (x + 1)/c + t$ for $-1 - ct < x < |1 - ct|$,

(iii) $u(x, t) = 2/c$ for $|x| < ct - 1$, **(iv)** $u(x, t) = 2t$ for $|x| < 1 - ct$,

(v) $u(x, t) = (1 - x)/c + t$ for $|1 - ct| < x < 1 + ct$, **(vi)** $u(x, t) = 0$ for $x > 1 + ct$

5. $u(x, t) = \frac{1}{2c}[\mathrm{Tan}^{-1}(x + ct)^2 - \mathrm{Tan}^{-1}(x - ct)^2]$ for $0 \leq t < x/c$

$u(x, t) = \frac{1}{2c}[\mathrm{Tan}^{-1}(ct + x)^2 - \mathrm{Tan}^{-1}(ct - x)^2]$ for $t > x/c$

9. $i(x, t) = e^{-Rt/L}[f(x + t/\sqrt{LC}) + g(x - t/\sqrt{LC})]$

13. $u(\rho, t) = t/\rho$ for $0 \leq t < (\rho - 1)/c$, $u(\rho, t) = (1/\rho)[u_0 + (\rho - 1)/c]$ for $t > (\rho - 1)/c$

15.

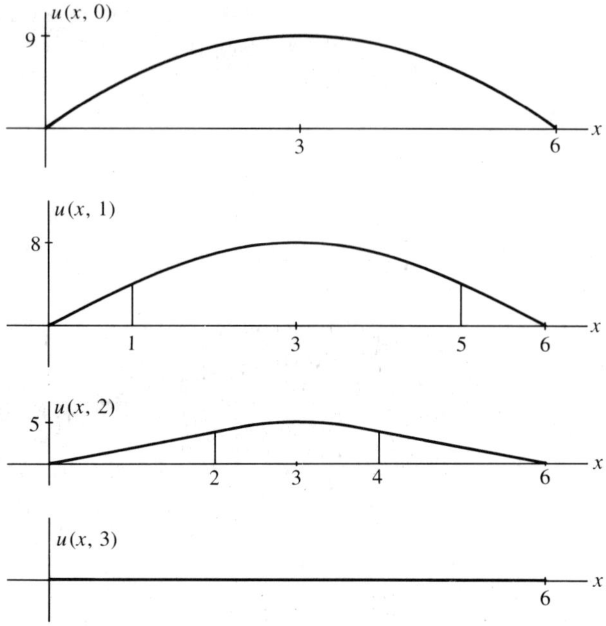

19. $u(x,1) = \begin{cases} 4x, & 0 < x < 1 \\ -x^2 + 6x - 1, & 1 < x < 5 \\ -4x + 24, & 5 < x < 6 \end{cases}$ $\quad u(x,2) = \begin{cases} 2x, & 0 < x < 2 \\ -x^2 + 6x - 4, & 2 < x < 4 \\ -2x + 12, & 4 < x < 6 \end{cases}$

$u(x,3) = 0,\ u(x,4) = -u(x,2),\ u(x,5) = -u(x,1)$

Exercises 8.8, p. 299

1. $u(x,t) = \sin \pi x \cos \pi c t$ **3.** $u(x,t) = \sin 4\pi x \cos 4\pi c t - \sin 100\pi x \cos 100\pi c t$

5. $u(x,t) = \dfrac{12}{\pi^2} \sum_{n=0}^{\infty} \dfrac{1}{(2n+1)^2} \sin\dfrac{(2n+1)\pi}{3} \sin\dfrac{(2n+1)\pi x}{3} \cos\dfrac{(2n+1)\pi c t}{3}$

9. $u(x,t) = \dfrac{64}{c\pi^4} \sum_{n=0}^{\infty} \dfrac{1}{(2n+1)^4} \sin\dfrac{(2n+1)\pi x}{2} \sin\dfrac{(2n+1)\pi c t}{2}$ **11.** $u(x,y,t) = \sin \pi x \sin \pi y \cos \sqrt{2}\pi c t$

13. $u(x,y,t) = \dfrac{16}{\pi^4} \sum_{m=0}^{\infty} \sum_{n=0}^{\infty} \dfrac{(-1)^{m+n}}{(2m+1)^2(2n+1)^2} \sin(2m+1)\pi x \sin(2n+1)\pi y \cos\sqrt{(2m+1)^2 + (2n+1)^2}\,\pi c t$

17. $u(x,y,t) = -\dfrac{64}{c\pi^3} \sum_{m=0}^{\infty} \sum_{n=0}^{\infty} \dfrac{1}{(2m+1)[(2m+1)^2 - 4](2n+1)[(2n+1)^2 - 4]}$

$\cdot \dfrac{\sin(2m+1)\pi x \sin(2n+1)\pi y \sin\sqrt{(2m+1)^2 + (2n+1)^2}\,\pi c t}{\sqrt{(2m+1)^2 + (2n+1)^2}}$

Exercises 8.9–8.10, p. 307

1. $u(x,t) = \dfrac{8}{\pi^2} \sum_{n=0}^{\infty} \dfrac{(-1)^n}{(2n+1)^2} \sin\dfrac{(2n+1)\pi x}{2} e^{-(2n+1)^2\pi^2 t/(4k)}$ **3.** $u_{ss} = 10 + 80x$

7. $u(x,t) = 1/2 + x\sqrt{k/4t} - (1/3)x^3(k/4t)^{3/2} + \cdots$

11. $u(r,\theta) = (Ar^{\alpha} + Br^{-\alpha})(C\cos\alpha\theta + D\sin\alpha\theta),\quad \alpha \neq 0$ $\quad u(r,\theta) = (A + B\ln r)(C + D\theta),\quad \alpha = 0$

13. $u(r,t) = [AJ_0(\alpha r) + BY_0(\alpha r)](C\cos\alpha c t + D\sin\alpha c t),\quad \alpha \neq 0$ $\quad u(r,t) = (A + B\ln r)(C + Dt),\quad \alpha = 0$

15. $u(r,\theta,z) = [AJ_n(\alpha r) + BY_n(\alpha r)](Ce^{\alpha z} + De^{-\alpha z})\sin n\theta,\quad \alpha \neq 0$

$u(r,\theta,z) = (Ar^n + Br^{-n})(C + Dz)\sin n\theta,\quad \alpha = 0$

Chapter 9

Exercises 9.1–9.2, p. 316

1. $-\Gamma(\frac{4}{3})$ **3.** $\Gamma(\frac{7}{4})$ **5.** $\Gamma(a+1)/(\ln a)^{a+1}$

Exercises 9.3–9.4, p. 324

3. $\frac{1}{2}B(\frac{5}{2},\frac{1}{2}) = 3\pi/16 \doteq 0.589$ **5.** $\frac{1}{4}B(\frac{1}{4},\frac{3}{4}) = \pi\sqrt{2}/4$ **7.** $B(\frac{1}{2},\frac{1}{2}) = \pi$ **9.** $\frac{1}{2}B(3,4) = \frac{1}{120}$
11. $\frac{1}{2}B(\frac{3}{4},\frac{5}{4}) = \pi\sqrt{2}/8$ **13.** $4B(\frac{1}{2},\frac{1}{4})$ **17.** $\pm 1/\sqrt{2}$ **23.** $H_n(x) = \sqrt{\pi}(-1)^n e^{x^2}(d^{n+1}\Phi(\sqrt{2}x)/dx^{n+1})$

Exercises 9.5–9.8, p. 331

1. $y(x) = \sum_{k=0}^{\infty} a_k(x-1)^k;\ a_{k+1} = \dfrac{n(n+1)-k(k+1)}{(k+1)^2}a_k \quad (k \geq 0)$
3. $P_4(x) = \frac{1}{8}(35x^4 - 30x^2 + 3),\ P_5(x) = \frac{1}{8}(63x^5 - 70x^3 + 15x),\ P_6(x) = \frac{1}{16}(231x^6 - 315x^4 + 105x^2 - 5)$
13. $\dfrac{2a}{5}P_3 + \dfrac{2b}{3}P_2 + \dfrac{5c+3a}{5}P_1 + \dfrac{3d+b}{3}P_0$ **15.** For $n > m$, let $f(x) = P_m(x)$ **17.** $c_0 = \frac{1}{4}, c_1 = \frac{1}{2}, c_2 = \frac{5}{16}$
19. $u = \rho^2 \cos 2\theta\, P_1(\cos\phi) - 2\rho^2\cos^2\theta\, P_2(\cos\phi)$

Exercises 9.9–9.10, p. 337

5. $y = c_1 J_{1/2}(x) + c_2 J_{-1/2}(x)$ or $y = c_1 J_{1/2}(x) + c_2 Y_{1/2}(x)$ **7.** $y = c_1 J_{1/2}(x/2) + c_2 J_{-1/2}(x/2)$
9. $J_{1/2}(x) = \sqrt{\dfrac{2}{\pi x}}\left(x - \dfrac{x^3}{6} + \dfrac{x^5}{120} - \cdots\right);\ J_{-1/2}(x) = \sqrt{\dfrac{2}{\pi x}}\left(1 - \dfrac{x^2}{2} + \dfrac{x^4}{24} - \cdots\right)$
11. $J_{1/2}(x/2) = \dfrac{1}{\sqrt{\pi x}}\left(x - \dfrac{x^3}{24} + \dfrac{x^5}{1920} - \cdots\right)$
23. $u(r,\theta,z) = \displaystyle\int_0^1 g(t)\,dt + 2\sum_{n=1}^{\infty}\left(\int_0^1 g(t)\cos n\pi t\,dt\right)\frac{I_0(n\pi r)}{I_0(n\pi)}\cos n\pi z$

Exercises 9.11–9.13, p. 344

11. $-xJ_0(x) + \int J_0(x)\,dx + c$ **13.** $x^2 J_1(x) + xJ_0(x) - \int J_0(x)\,dx + c$
15. $J_0(0.1) \approx 1, J_0(0.2) \approx 1, J_0(0.3) \approx 1$
$J_1(0.1) \approx 0.05, J_1(0.2) \approx 0.1, J_1(0.3) \approx 0.15$
$J_0(6) \approx 0.157, J_0(10) \approx -0.247, J_0(14) \approx 0.170$
$J_1(6) \approx -0.286, J_1(10) \approx 0.0526, J_1(14) \approx 0.129$

Exercises 9.14–9.15, p. 351

1. $f(x) \sim \sum_{j=1}^{\infty} A_j J_1(s_{1j}x)$ where the $3s_{1j}$ are the zeros of
$J_1(x)$ and $A_j = \dfrac{2}{9s_{1j}^2 J_2^2(3s_{1j})}\left[-3s_{1j}J_0(3s_{1j}) + \displaystyle\int_0^{3s_{1j}} J_0(x)\,dx\right]$
3. $f(x) \sim \sum_{j=1}^{\infty} A_j J_0(s_{0j}x)$ where the s_{0j} are the zeros of $J_0(x)$ and $A_j = \dfrac{4J_2(s_{0j})}{s_{0j}^2 J_1^2(s_{0j})} = \dfrac{8}{s_{0j}^3 J_1(s_{0j})}$
7. $u(r,t) = \sum_{m=1}^{\infty} a_m \cos k_m ct J_0(k_m r)$ where the $k_m R$ are the zeros of $J_0(x)$ and $a_m = \dfrac{2}{k_m^3 R^3 J_1^2(k_m R)}\displaystyle\int_0^{k_m R} J_0(x)\,dx$
9. $u(r,z) = \sum_{m=1}^{\infty} a_m J_0(s_{0m}r)e^{-s_{0m}z}$ where the $s_{0m}R$ are the zeros of $J_0(x)$ and $a_m = \dfrac{2}{R^2 J_1^2(s_{0m}R)}\displaystyle\int_0^R rf(r)J_0(s_{0m}r)\,dr$
11. $u(r,z) = \sum_{m=1}^{\infty} a_m J_0(s_{0m}r)\cosh(s_{0m}z)$ where the $s_{0m}R$ are the zeros of $J_0(x)$ and

$$\cosh(s_{0m})a_m = \frac{2}{R^2 J_1^2(s_{0m}R)} \int_0^R rg(r)J_0(s_{0m}r)\,dr$$

13. $u(r, \theta, t) = \sum_{m=1}^{\infty} a_m J_0(\beta_m r)e^{-\beta_m^2 t/k}$ where the β_m are the zeros of $J_0(x)$ and

$$a_m = \frac{2}{\beta_m J_1(\beta_m)}\left[1 - \frac{2J_2(\beta_m)}{\beta_m J_1(\beta_m)}\right] = \frac{2(\beta_m^2 - 4)}{\beta_m^3 J_1(\beta_m)}$$

15. $u(r, z, t) = (AJ_0(\alpha r) + BY_0(\alpha r))(C\cos\beta z + D\sin\beta z)(E\cos\sqrt{\alpha^2+\beta^2}\,ct + F\sin\sqrt{\alpha^2+\beta^2}\,ct)$

Chapter 10

Exercises 10.2, p. 360

3. $\mathbf{F}(u, v) = \cos u \sin v\mathbf{i} + \sin u \sin v\mathbf{j} + \cos v\mathbf{k}$. The function has period 2π in the sense that $\mathbf{F}(u + k2\pi, v + n2\pi) \equiv \mathbf{F}(u, v)$ for all integers k, n.

5. $\mathbf{F}(t) = \frac{t^2}{2}\mathbf{i} + \frac{t^3}{3}\mathbf{j} + \frac{t^4}{4}\mathbf{k} + \mathbf{c}$, where $\mathbf{c}$ is a constant vector. **9.** 2 **15.** $\mathbf{r}(t) = \mathbf{a}\cos\omega t + (\mathbf{b}/\omega)\sin\omega t$

Exercises 10.3–10.4, p. 368

1. **(a)**

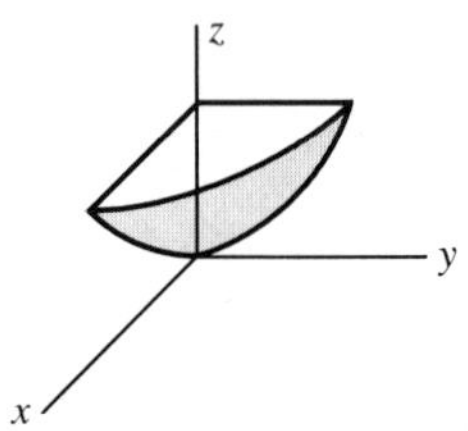

(b)

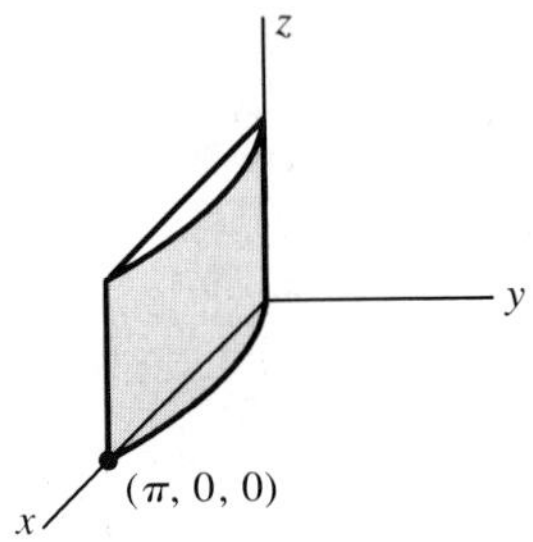

(c)

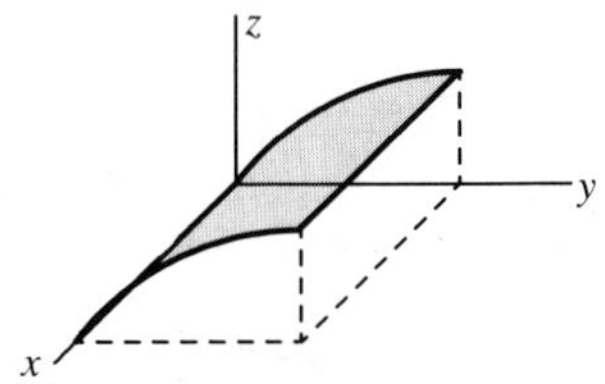

3. **(a)**

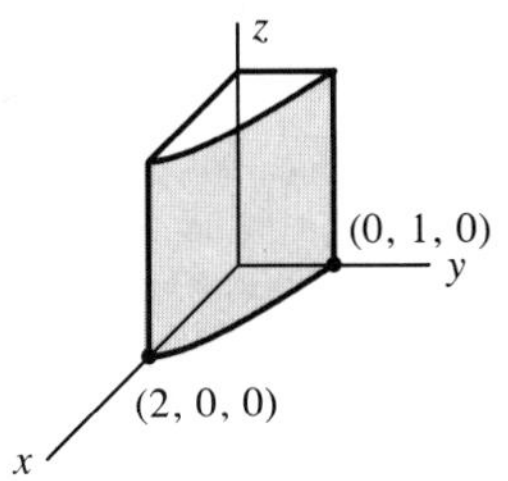

(b)

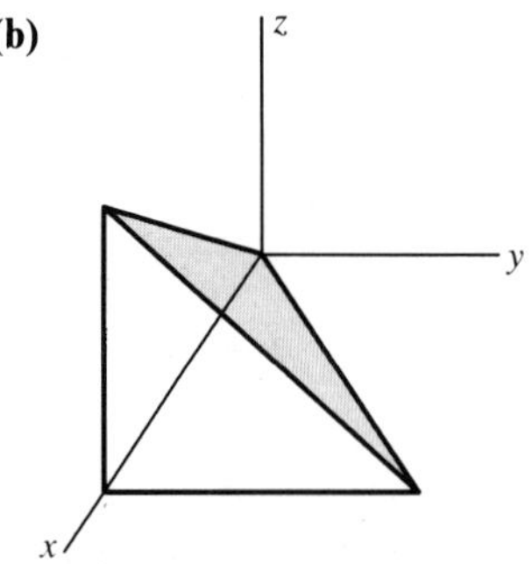

(c)

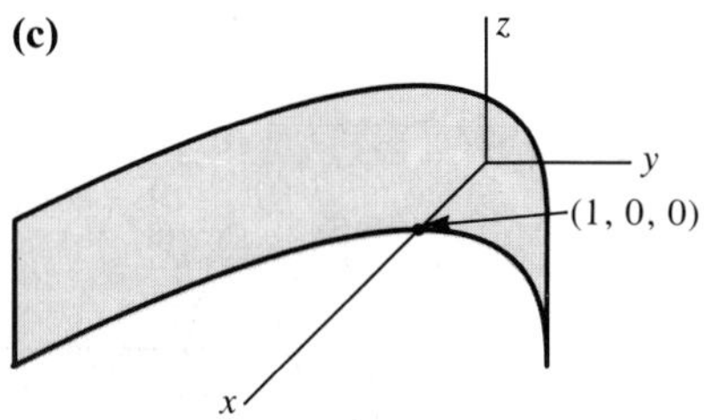

5. (a)

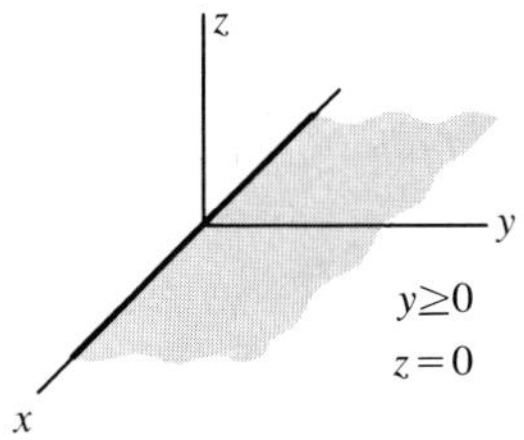
z
y
x
y≥0
z=0

(b)

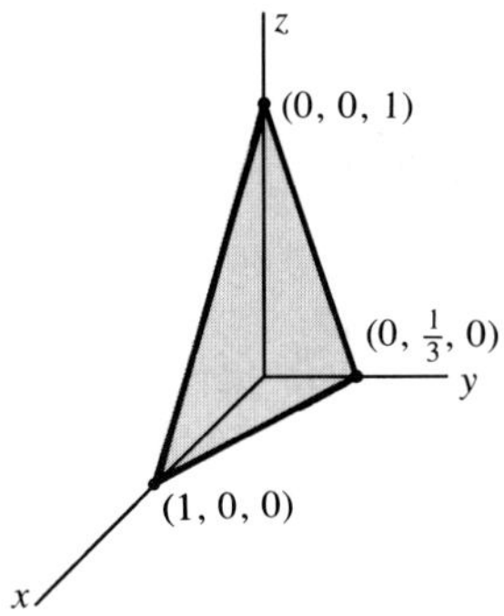
z
(0, 0, 1)
(0, 1/3, 0)
y
(1, 0, 0)
x

7. (a)

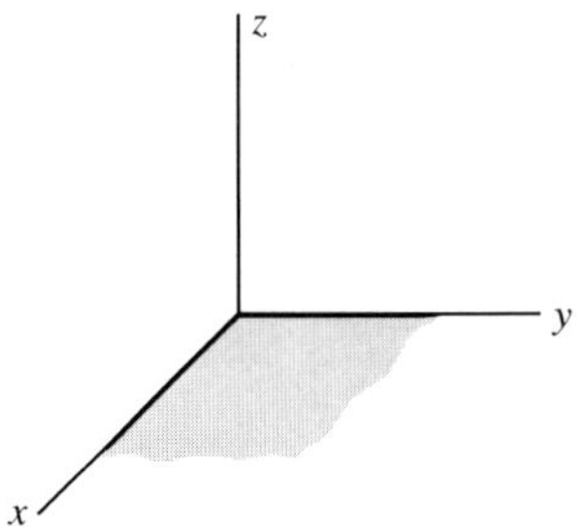
z
y
x

(b)

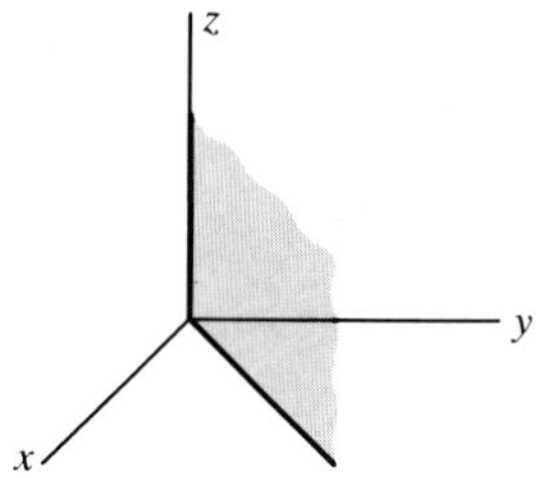
z
y
x

9. (a)

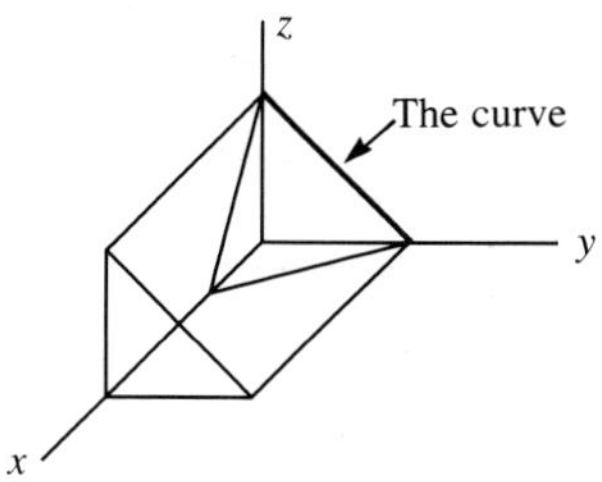
z
The curve
y
x

(b)

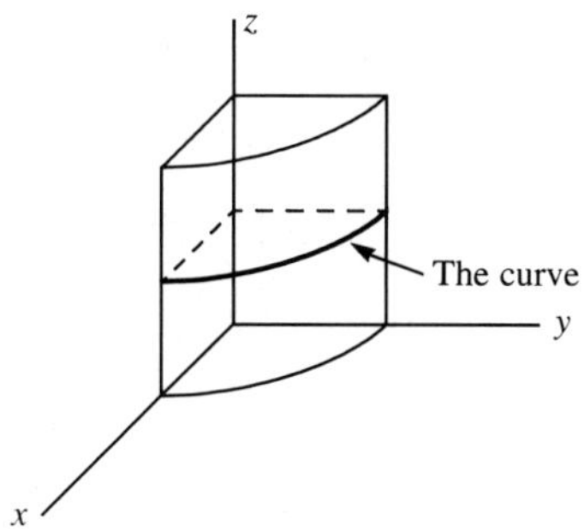
z
The curve
y
x

(c)

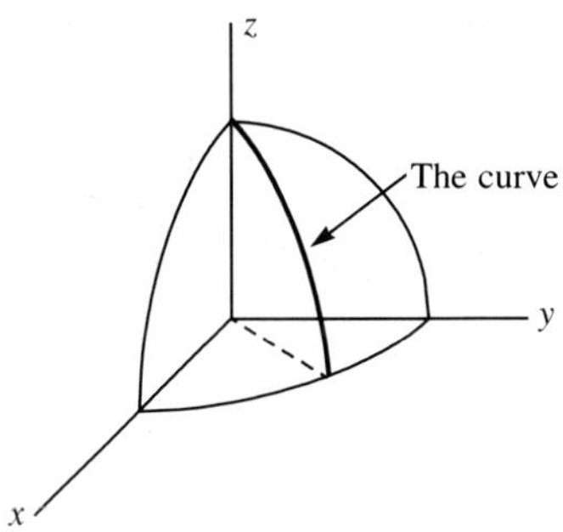
z
The curve
y
x

11. (a)

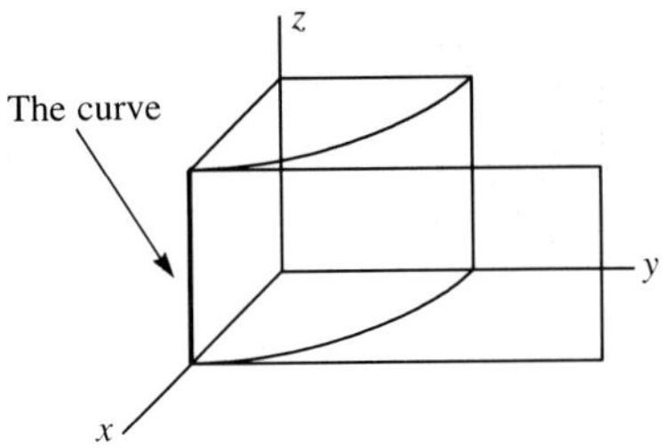

(b)

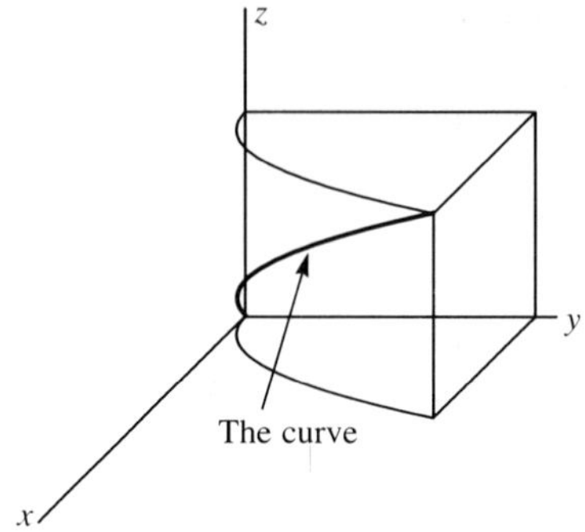

(c)

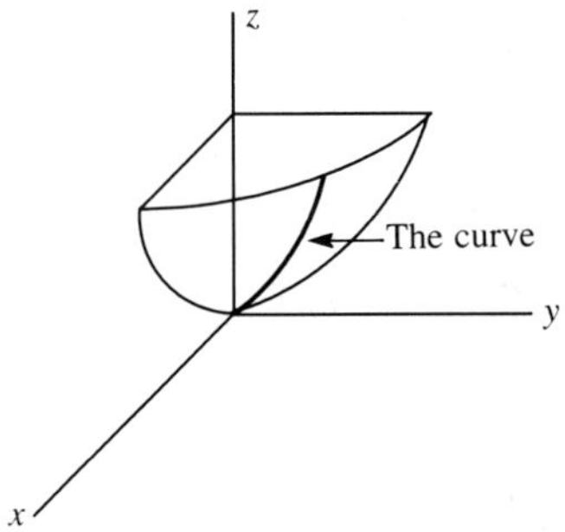

13. (a)

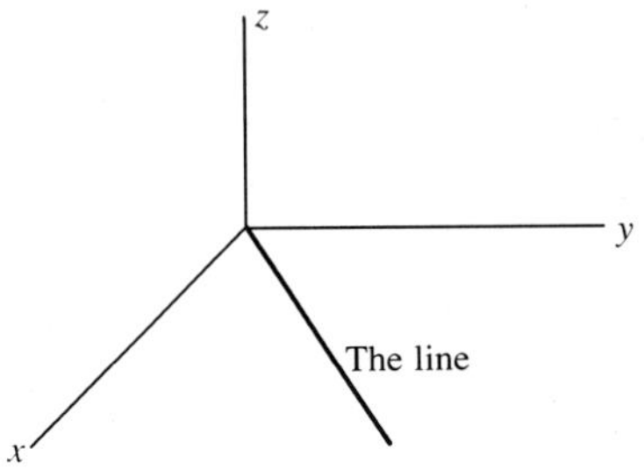

(b)

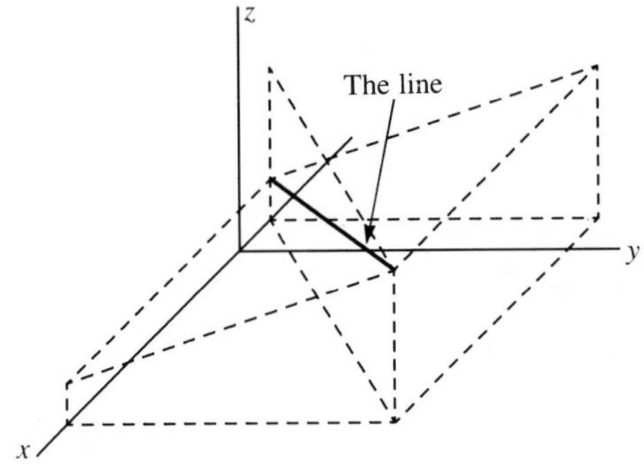

15. 1(a) $x = \sqrt{3}u\cos v,\ y = u\sin v,\ z = 3u^2$ **(b)** $x = v,\ y = \sin v,\ z = u$ **(c)** $x = u,\ y = v^2,\ z = v$
2(a) $x = u,\ y = u\sin v,\ z = u\cos v$ **(b)** $x = u,\ y = 2,\ z = v$ **(c)** $x = u,\ y = 2 - u,\ z = v$
3(a) $x = 2\cos v,\ y = \sin v,\ z = u$ **(b)** $x = u,\ y = v,\ z = u - v$ **(c)** $x = \cosh v,\ y = \sinh v,\ z = u$
4(a) Two surfaces: $x = u,\ y = v,\ z = 2$; and $x = u,\ y = v,\ z = -2$ **(b)** $x = u\cos v,\ y = u\sin v,\ z = u^2$
(c) $x = u\cos v,\ y = u\sin v,\ z = 4 - u^2$ **17. 12(a)** $x^2 + y^2 = 1,\ z = 2$ **(b)** $y = z,\ x = y^2$
13(a) $z = 0,\ x = y,\ x \geq 0$ **(b)** $x - 2y + 1 = 0,\ z - 3y - 1 = 0$ **14(a)** $x = 2,\ y = 3z$
(b) $y = 2x,\ z = 0$ **19.** $\mathbf{R}(t) = \mathbf{i} - (\mathbf{i} - 2\mathbf{j})t$ **21.** By a constant vector in the direction of the line
23. $\mathbf{R}(t) = (2\cos t)\mathbf{i} + (2 + \sqrt{2}\sin t)\mathbf{j} + (2 + \sqrt{2}\sin t)\mathbf{k}$
25. $\mathbf{R}(u, v) = (2\cos u\cos v)\mathbf{i} + 2(1 + \sin u\cos v)\mathbf{j} + (3 + 2\sin v)\mathbf{k}$

Exercises 10.5–10.6, p. 374

1. $\mathbf{R}(s) = (4\mathbf{i} + 2\mathbf{j} + 8\mathbf{k}) + (4\mathbf{i} + \mathbf{j} + 12\mathbf{k})s$ **3.** $\mathbf{T} = \frac{1}{\sqrt{2}}(\mathbf{i} + \mathbf{k}), \quad \mathbf{N} = -\mathbf{j}$

5. at $t = 2$: $\mathbf{v} = 4\mathbf{k}$, $\mathbf{a} = 2\pi^2\mathbf{i} - \pi^2\mathbf{j} + 2\mathbf{k}$, speed $= 4$ $\mathbf{a}_T = 2\mathbf{k}$, $\mathbf{a}_N = 2\pi^2\mathbf{i} - \pi^2\mathbf{j}$

at $t = \frac{1}{4}$: $\mathbf{v} = \pi\mathbf{i} - \frac{\pi}{\sqrt{2}}\mathbf{j} + \frac{1}{2}\mathbf{k}$, $\mathbf{a} = -\frac{\pi^2}{\sqrt{2}}\mathbf{j} + 2\mathbf{k}$ speed $= (\pi^2 + \frac{1}{2}\pi^2 + \frac{1}{4})^{1/2}$ $\mathbf{a}_T = \frac{(\pi^3 + 2)}{6\pi^2 + 1}(2\pi\mathbf{i} - \pi\sqrt{2}\mathbf{j} + \mathbf{k})$

$$\mathbf{a}_N = -\frac{2\pi(\pi^3 + 2)}{6\pi^2 + 1}\mathbf{i} + \left(\frac{\pi^4\sqrt{2} + 2\pi\sqrt{2}}{6\pi^2 + 1} - \frac{\pi^2}{\sqrt{2}}\right)\mathbf{j} + \left(2 - \frac{\pi^3 + 2}{6\pi^2 + 1}\right)\mathbf{k}$$

9. $\mathbf{r}(t) = \frac{t^3}{3}\mathbf{i} + \frac{t^2}{2}\mathbf{j} - \frac{t^4}{4}\mathbf{k} + \mathbf{c}$ where $\mathbf{c}$ is a constant vector

Exercises 10.7–10.8, p. 382

1. **(a)**

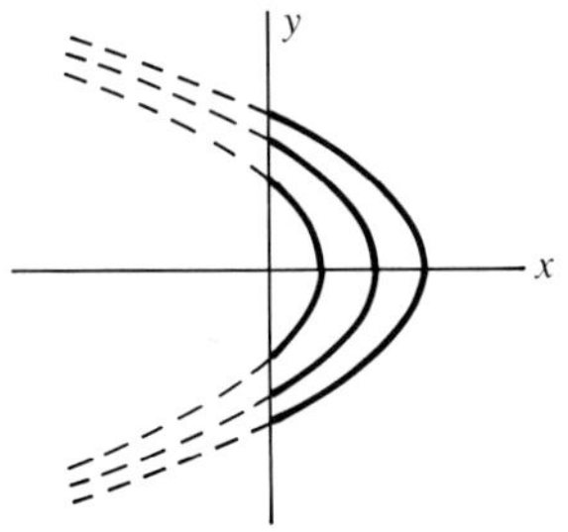

(b)

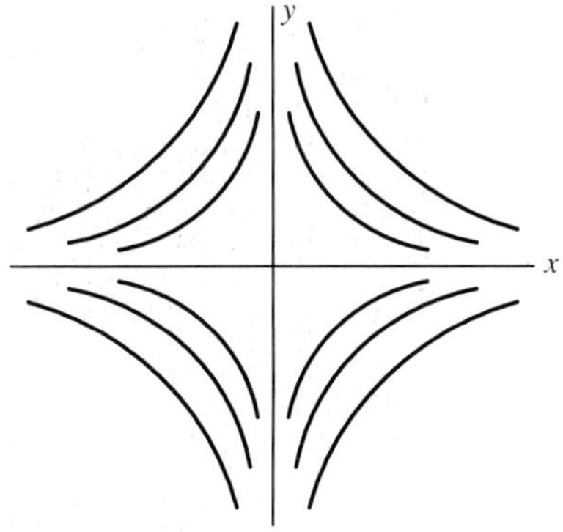

(c) All curves

3. **(a)**

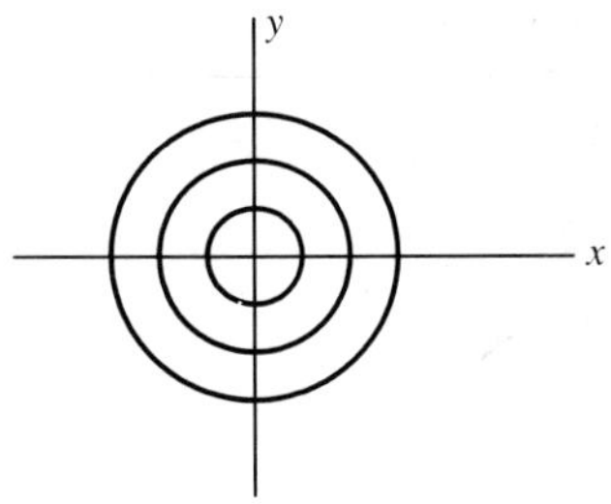

(b) Same as **(a)**. **(c)** Same as **(a)**.

5. **(a)**

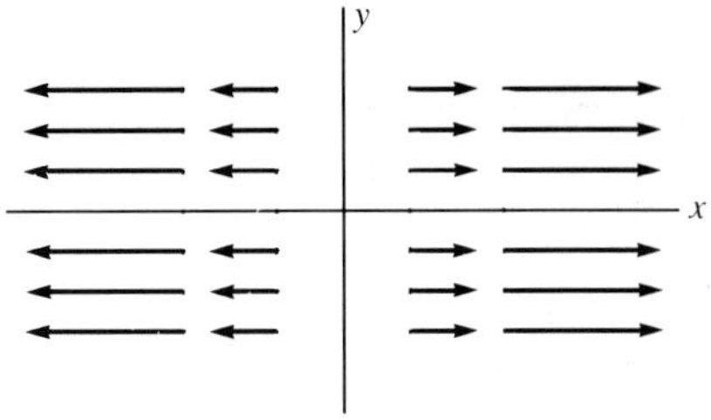

(b)

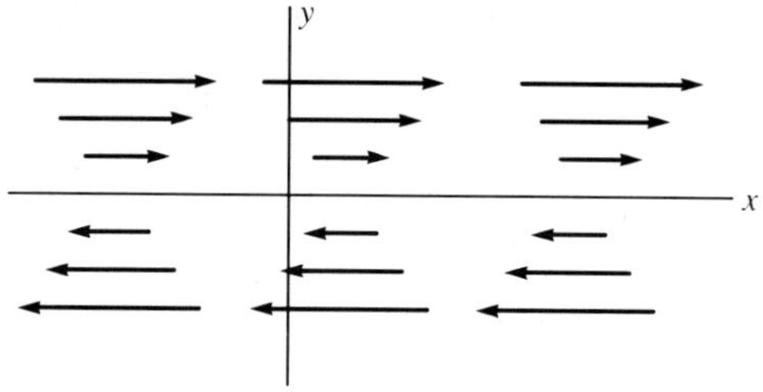

(c)

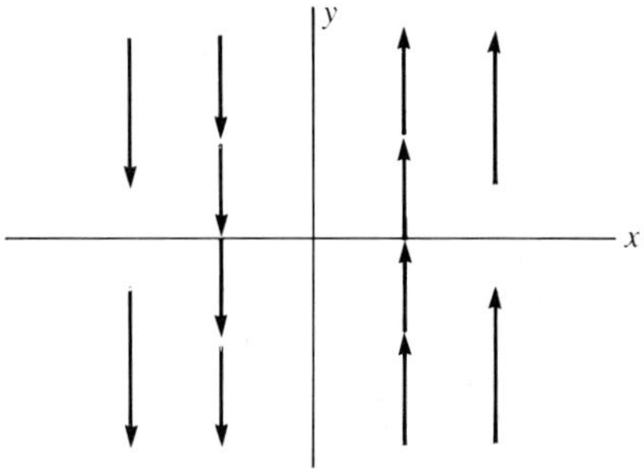

7. **(a)**

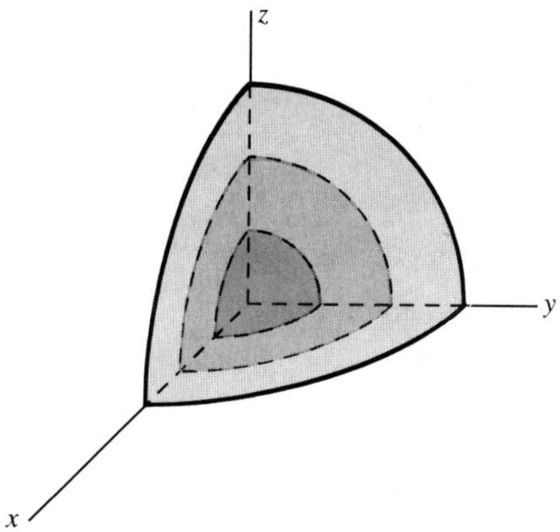

(b) Same as **(a)**.

(c)

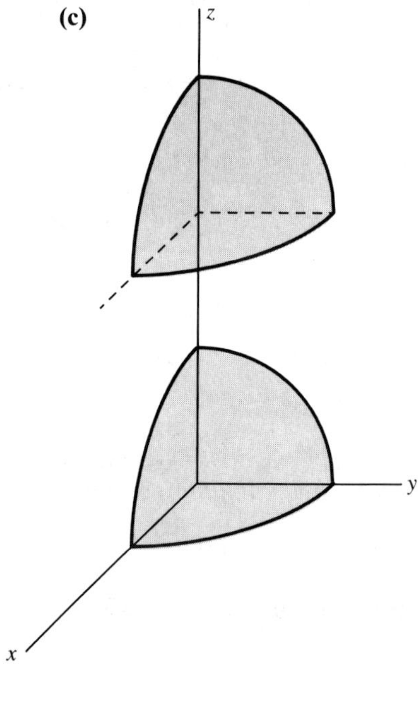

9. **(a)** **(b)**

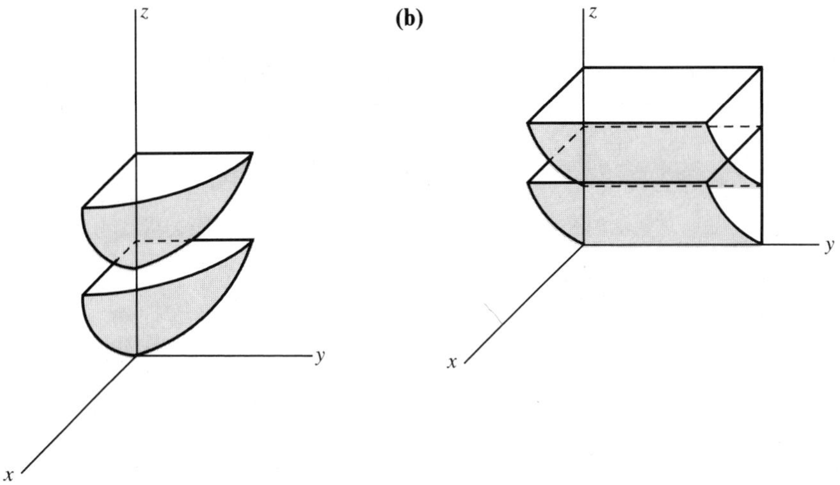

11. **(a)** $y = ce^{2\sqrt{x}}$ **(b)** $y = c$ **(c)** $y = \log_e(c \sin x)$
13. $|\mathbf{r}_1|, |\mathbf{r}_2|$ and the angle between $\mathbf{r}_1$ and $\mathbf{r}_2$ are unchanging with time

Exercises 10.9, p. 388

1. **(a)** $yz\mathbf{i} + xz\mathbf{j} + xy\mathbf{k}$ **(b)** $2x\mathbf{i} + 2y\mathbf{j} + 2z\mathbf{k}$
(c) $(\cos y - z \sin x)\mathbf{i} + (\sin z - x \sin y)\mathbf{j} + (y \cos z + \cos x)\mathbf{k}$ **3.** **(a)** $(x\mathbf{i} + y\mathbf{j} + z\mathbf{k})/(x^2 + y^2 + z^2)$
(b) $[\mathbf{k} - (xz\mathbf{i} + yz\mathbf{j})/(x^2 + y^2)](x^2 + y^2 - z^2)^{-1/2}$ **(c)** $4x\mathbf{i} + 6y\mathbf{j} + 2z\mathbf{k}$
5. **(a)** -2 **(b)** 6, in direction $2\mathbf{i} + 2\mathbf{j} + \mathbf{k}$ **(c)** $(y + z)\mathbf{i} + (x + z)\mathbf{j} + (x + y)\mathbf{k}$
7. **(a)** $f(x, y, z) = x^2yz + \cos(yz) - \frac{1}{4}\cos 2x^2$ **(b)** $f(x, y, z) = x^2y + y^2z + z^2x$ **9.** **(a)** $x\mathbf{i} + y\mathbf{j} + z\mathbf{k}$
(b) $yz\mathbf{i} + xz\mathbf{j} + xy\mathbf{k}$ **(c)** $(\sin z - \sin x)\mathbf{i} + \cos y\mathbf{j} + x \cos z\mathbf{k}$ **11.** $\mathbf{E} = p\dfrac{3xz\mathbf{i} + 3yz\mathbf{j} + (2z^2 - x^2 - y^2)\mathbf{k}}{(x^2 + y^2 + z^2)^{5/2}}$
17. **(a)** $e^{\rho \cos \phi}(\cos \phi\, \mathbf{u}_\rho - \sin \phi\, \mathbf{u}_\phi)$ **(b)** $\cos \theta (\sin \phi\, \mathbf{u}_\rho + \cos \phi\, \mathbf{u}_\phi) - \sin \theta\, \mathbf{u}_\phi$
(c) $\rho[\sin^2 \phi \sin 2\theta + \sin 2\phi (\sin \theta + \cos \theta)]\mathbf{u}_\rho + \rho[\frac{1}{2} \sin 2\phi \sin 2\theta + \cos 2\phi (\sin \theta + \cos \theta)]\mathbf{u}_\phi$
$+ \rho[\sin \phi \cos 2\theta + \cos \phi (\cos \theta - \sin \theta)]\mathbf{u}_\theta$
19. $f(\rho, \phi, \theta) = \text{const.}\ p\dfrac{\cos \phi}{\rho^2}$

Exercises 10.10, p. 394

1. **(a)** $yz + xz + xy$ **(b)** $6x + 6z$ **(c)** 0 **3.** **(a)** 12 **(b)** 0 **(c)** 0 **7.** $38/\sqrt{14}$
19. **(a)** 0 **(b)** 0 **(c)** 0

Exercises 10.11, p. 399

1. **(a)** $f(x)\mathbf{i}$ **(b)** $f(y)\mathbf{i}$ **3.** **(a)** 0 **(b)** 0 **(c)** 0
5. No, a field may be both irrotational and solenoidal.
11. Unidirectional fields must vary in the direction of the field to be nonsolenoidal. They must vary crosswise to the field to be nonirrotational. **19.** **(a)** 0 **(b)** $\rho \sin \phi\, \mathbf{u}_\theta$ **(c)** 0

Chapter 11

Exercises 11.2, p. 409

3. 219 **5.** 153 **7.** $\pi; 2\sqrt{2}$ **9.** $\frac{17}{2}$ **11.** 24

Exercises 11.3–11.4, p. 416

3. 10 **5.** $\frac{61}{30}$ **7.** 0 **9.** -6π **11.** 2π **13.** $\sqrt{\pi/2} - 7$ **15.** 0 **17.** $3\sin 8 - \sin 1$
19. **(a)** 0 **(b)** 4 **(c)** 310 **21.** **(a)** Path independent for all finite x, y, z
(b) Path independent in any domain in which $y + z \neq 0$ **(c)** Not path independent
23. $\frac{\pi}{6}\ln 2$ **25.** $\frac{1}{2}F(x_1^2 + y_1^2) - \frac{1}{2}F(x_0^2 + y_0^2)$ where $F(z)$ is an antiderivative of $f(z)$
27. **(a)** 0 **(b)** $2\pi R^2$

Exercises 11.5–11.6, p. 427

3. 12π **5.** 4 **7.** 108 **9.** 132 **11.** 2 **13.** $8\pi/3$ **15.** $4\pi c$ **17.** $4\pi^2 ab$
19. $(5^{3/2} - 1)/12$ **21.** $19\sqrt{2}/6$ **23.** 1 unit **25.** $\frac{1}{2}L$ units **27.** 0 **29.** 0

Exercises 11.7, p. 436

3. $\pi a^2/4$ **5.** 0 **7.** 0 **9.** $-\frac{4}{3}$ **13.** **(a)** $3\pi/2$ **(b)** 0
15. $\mathbf{F}$ and $\mathbf{V}$ are orthogonal; $\nabla \cdot \mathbf{V} = \mathbf{k} \cdot (\nabla \times \mathbf{F})$ **17.** 0 **19.** $\oint_C \frac{1}{2}(-y\mathbf{i} + x\mathbf{j}) \cdot d\mathbf{r}$; 2
21. **(a)** $243\pi/2$ **(b)** $525\pi/2$ **23.** **(a)** 0 **(b)** Solenoidal **25.** **(a)** 0 **(b)** 0

Exercises 11.8, p. 443

3. πa^2 **5.** $2\pi a^2$ **7.** $-9\pi/\sqrt{2}$ **9.** -1
11. $\iint_S (-2x)\mathbf{j} \cdot \mathbf{N}\, dS$ where S is the interior of the ellipse $x^2 + y^2 = 1$ in the plane $z = y$; 0
23. πab

Exercises 11.9, p. 449

1. $\bar{z} = 1$ **3.** $\frac{64}{9}$ **5.** **(a)** 0 **(b)** 0 **7.** $\frac{8}{15}$ **11.** $\iiint_V (1 + 2xy + x^2)\, dV$; $7\pi/12$
13. $\frac{128}{15}$ **19.** 1

Exercises 11.10, p. 452

1. 4 **3.** $\pi/2$ **5.** $-\frac{1}{3}$ **7.** $\pm 4\pi/\sqrt{2}$ **9.** $\sqrt{2\pi}$ **11.** 20,951/12 **13.** -12π **15.** 1
17. $\pi\sqrt{2}/16$ **19.** $16\pi/3$ **21.** $\frac{7}{4}$ **23.** $64\pi/3$ **25.** $\frac{1}{2}$ **27.** 8π **29.** 0 **31.** 8π

Chapter 12

Exercises 12.2–12.3, p. 461

1. **(a)** $32 + 47i$ **(b)** $-i$ **(c)** $5 - 5\sqrt{3}i$ **(d)** $8i$ **(e)** $8i$ **(f)** $3 + 4i$ **3.** **(a)** -8
(b) $-1 + i$ **(c)** $1 - \frac{1}{2}\sqrt{3} - i(1 + \frac{1}{2}\sqrt{3})$ **5.** **(a)** $2e^{-i(\pi/3 + 2n\pi)}$ **(b)** $2e^{i(\pi/2 + 2n\pi)}$ **(c)** $5e^{i2n\pi}$
7. Rotation by 90° counterclockwise **9.** All 1 **17.** **(a)** Closed, bounded
(b) Open, bounded, connected **(c)** Closed, unbounded, connected

19. **(a)**

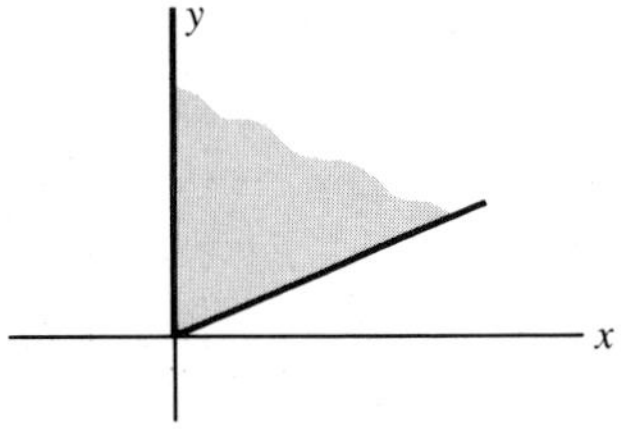

(b)

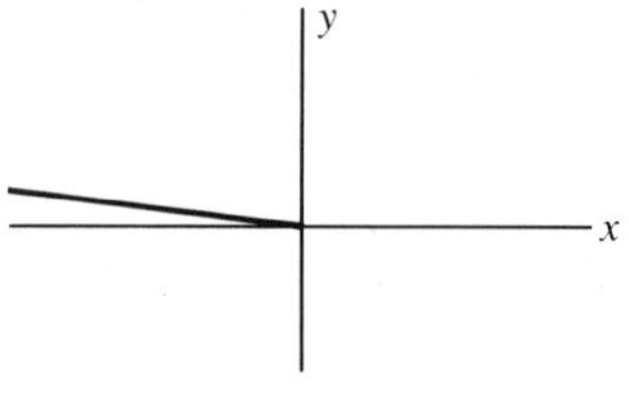

(c)

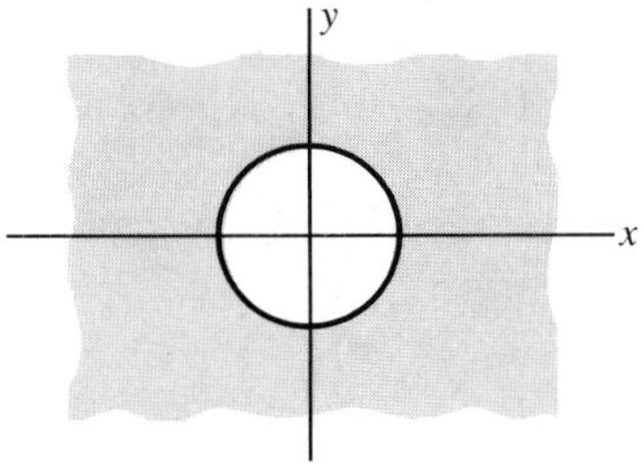

(d)

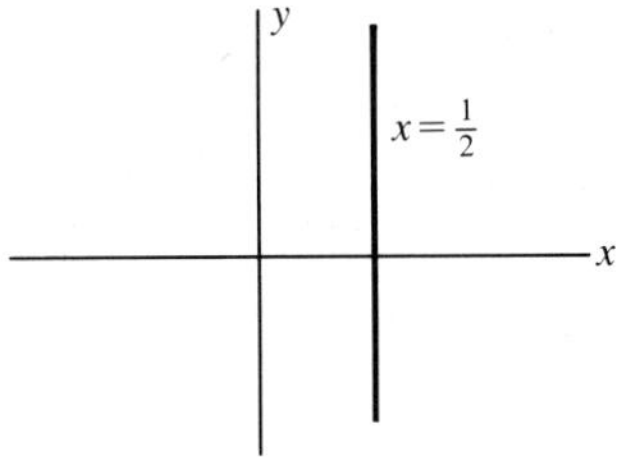

(e)

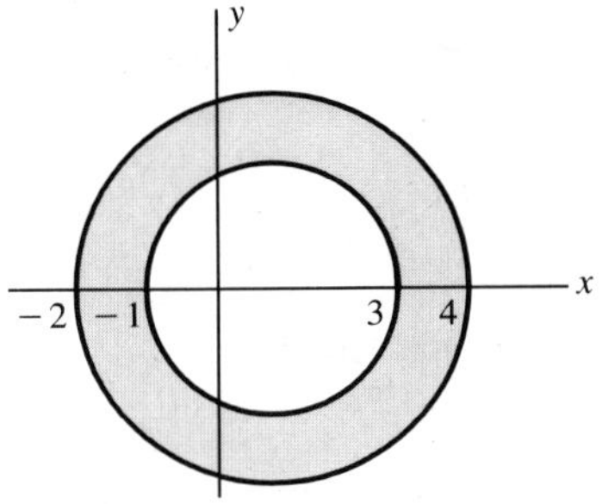

21. The straight-line segment joining z_1 and z_2

Exercises 12.4, p. 466

1. **(a)** $x^2 - y^2 - x + 1 + i(2xy - y)$ **(b)** $\dfrac{1-x}{(x-1)^2+y^2} + i\dfrac{y}{(x-1)^2+y^2}$

(c) $\dfrac{x^2+y^2-1}{x^2+(y+1)^2} - i\dfrac{2x}{x^2+(y+1)^2}$ **(d)** $\dfrac{x^2-y^2}{(x^2-y^2)^2+(2xy+1)^2} - i\dfrac{2xy+1}{(x^2+y^2)^2+(2xy+1)^2}$

3. **(a)** $|w| < 1, 0 < \operatorname{Arg} w < \pi$ **(b)** $(u-\frac{1}{2})^2 + v^2 < \frac{1}{4}$

5. **(a)** First quadrant portion of the annulus $1 \le u^2 + v^2 \le e^{\pi}$ **(b)** $|w| > \frac{1}{2}$ **(c)** $u < \log_e 3, \quad 0 < v < \dfrac{\pi}{3}$

7. **(a)** The circle $(u-\frac{1}{2})^2 + (v-\frac{1}{2})^2 = \frac{1}{2}$ **(b)** The line $v + u = 0$

9. Let $z = re^{i\theta}$ with $r \le 1$ and $-\pi < \theta < 0$

11. The point of intersection in the $z =$ plane is $\sqrt{3}$ and $\tan\psi_z = 1 - \frac{1}{2}\sqrt{3}$

The image curves are $u^2 + v^2 = \dfrac{\sqrt{3}}{3}u + v$ and $u^2 + v^2 = \dfrac{\sqrt{3}}{3}(u+v)$

The point of intersection in the $w =$ plane is $\dfrac{\sqrt{3}}{3}$ and $\tan\psi_w = 1 - \frac{1}{2}\sqrt{3}$

13. $x^3 - 3xy^2 + 3x^2 - 3y^2 + i(3x^2y - y^3 + 6xy)$ **15.** **(a)** $x^2 - y^2 + x + i(y - 2xy)$

(b) $x^2 + y^2 + 5x + i5y$ **(c)** $\dfrac{x^3+x^2+x+1-3xy^2-y^2}{(x^2-y^2+1)^2+4x^2y^2} + i\dfrac{3x^2y-y^3+2xy+y}{(x^2-y^2+1)^2+4x^2y^2}$ **17.** The y-axis

19. $|a| = |c|$ **21.** $\psi(x,y) = \dfrac{U_0 y}{x^2+y^2}(x^2+y^2-a^2)$

Exercises 12.5, p. 474

5. **(a)** $\cos x \cosh y - i \sin x \sinh y$ **(b)** $1/z$ **(c)** $-\sin x \cosh y - i \cos x \sinh y$

9. **(a)** $e^x \cos y + ie^x \sin y$ **(b)** z^3 **(c)** $-iz$ **13.** **(a)** $v = -y/(x^2+y^2)$ **(b)** $v = \cos x \sinh y$

(c) $v = \frac{1}{2}(y^2 - x^2) + y - x$ **15.** $a(x^2 - y^2) + bxy$

Exercises 12.6, p. 479

1. **(a)** $(1+i)/\sqrt{2}, (-1+i)/\sqrt{2}, (-1-i)/\sqrt{2}, (1-i)/\sqrt{2}$ **(b)** $\sqrt{3}+i, -\sqrt{3}+i, -2i$
(c) $\sqrt{3}+i, 2i, -\sqrt{3}+i, -\sqrt{3}-i, -2i, \sqrt{3}-i$ **(d)** $\sqrt{2}(-1+i), \sqrt{2}(1-i)$
3. $2^{1/10}[\cos(\pi/20)+i\sin(\pi/20)]$, $2^{1/10}[\cos(9\pi/20)+i\sin(9\pi/20)]$, $2^{1/10}[\cos(17\pi/20)+i\sin(17\pi/20)]$, $2^{1/10}[\cos(5\pi/4)+i\sin(5\pi/4)]$, $2^{1/10}[\cos(33\pi/20)+i\sin(33\pi/20)]$
5. **(a)** $z=\pm(1+i)/\sqrt{2}$ **(b)** $z=\pm 2^{1/4}[\cos(\pi/8)-i\sin(\pi/8)]$
(c) $z=\pm 2^{-1/4}[\cos(\pi/8)-i\sin(\pi/8)]$ **(d)** $z=\pm(1-i)/\sqrt{2}$
11. $f_1(z)=|z|^{1/5}[\cos(\theta/5)+i\sin(\theta/5)]$, $-3\pi/2<\theta<\pi/2$, and so on. This yields a 5-sheeted Riemann surface.

Exercises 12.7, p. 483

3. **(a)** $\frac{1}{2e}(1+\sqrt{3}i)$ **(b)** $-\frac{1}{2}(\sqrt{3}+i)$ **(c)** $-\frac{e^4}{2}(1+\sqrt{3}i)$
5. **(a)** $z=\log_e 2+i(2n+1)\pi$ **(b)** No solution **(c)** $z=i(4n+1)\pi/2$ **(d)** $z=2n\pi i$
7. $x>0$ **11.** $u^2+v^2=e^{(2/c)\,\mathrm{Tan}^{-1}(v/u)}$ **15.** $T(x,y)=10e^x\sin y$

Exercises 12.8–12.10, p. 490

1. **(a)** $\cosh\pi$ **(b)** $-i\sinh 2\pi$ **(c)** $i\tanh 2\pi$

5. Typical image

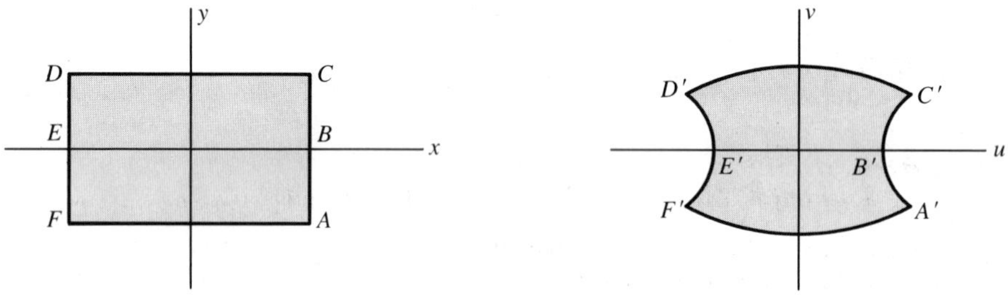

7. The line $y=0$ and the lines $x=(2n+1)\pi/2$ **13.** **(a)** $z=e^{-1+2n\pi i}$ **(b)** $z=e^{i(1+2n\pi)}$ **(c)** $e^{2n\pi i}$
15. **(a)** $e^{(1+8n)\pi/4}[\cos(\log_e\sqrt{2})-i\sin(\log_e\sqrt{2})]$ **(b)** $ie^{(1+4n)\pi/2}$ **(c)** $e^{-2n\pi}$
(d) $\cos(2\sqrt{3}n\pi)+i\sin(2\sqrt{3}n\pi)$ **17.** $w=\sin^{-1}z=-i\ln(iz+\sqrt{1-z^2})$
19. $\sinh z=\frac{1}{2}(e^z-e^{-z})=\sinh x\cos y+i\cosh x\sin y$ $\cosh z=\frac{1}{2}(e^z+e^{-z})=\cosh x\cos y+i\sinh x\sin y$
21. $\psi(x,y)=-\mathrm{Im}(\cos z)=\sin x\sinh y$

Chapter 13

Exercises 13.1–13.3, p. 497

1. **(a)** $u=\dfrac{t}{t^4-t^2+1}, v=\dfrac{1-t^2}{t^4-t^2+1}$ or $(u^2+v^2)^2-v(u^2+v^2)-u^2=0$
(b) $(u+\frac{1}{24})^2+v^2=(\frac{5}{24})^2$ **(c)** $\{w \mid |w-\frac{1}{2}|<\frac{1}{2}\}\cap\{w \mid \mathrm{Im}\,w<0\}$
3. **(a)** No rotation **(b)** Rotation by π **(c)** No rotation
5. **(a)** $v=$ constant, $u=$ constant **(b)** Positive imaginary axis **(c)** Negative real axis
(d) $-\pi/3<\mathrm{Arg}\,w<\pi/6$; $1<|w|<16$
7. **(a)** $|w|=|z|^2$, $\arg w=2\arg z$; conformal everywhere except $z=0$
(b) $|w|=|z|^3$, $\arg w=3\arg z$; conformal everywhere except $z=0$
9. **(a)** $z=\frac{1}{2}$ **(b)** $z=0$ **(c)** $z=0, \pm1, \frac{1}{2}(n\pi\pm\sqrt{n^2\pi^2-4})$ **13.** $(x^2+y^2)^2-2(x^2-y^2)+\frac{8}{9}=0$
15. **(a)** $z(t)=t+i(1-2t)/3$ **(b)** $z(t)=2\cos t+i\sin t$ **(c)** $z(t)=(1+2\cos t)+i(2\sin t-1)$
(d) $z(t)=\cosh t+i\sinh t$ **17.** **(a)** $z(t)=(t+2)+4i(t+1)$ **(b)** $z(t)=(t+1)+i(t+3)e/3$
(c) $z(t)=t+2i$ **19.** $z=1$ and $z=-1$ are critical points of order 1; angles are doubled at these points.

Exercises 13.4, p. 504

1. **(a)** Rotation by $\pi/4$, dilation by $\sqrt{2}$ **(b)** Inversion-reflection plus rotation by $\pi/2$
(c) Translation by -1 followed by inversion-reflection
3. **(a)** $w = -z + 1 + i$ **(b)** $w = (2z+1)/(z-1)$ **(c)** $w = z$
5. If $c \neq 0$, $z = \frac{1}{2c}(a - d \pm \sqrt{(a-d)^2 + 4bc})$; if $c = 0$, $z = b/(d-a)$
7. **(a)** $w = az + b$ or, if $c \neq 0$, $w = \dfrac{az+b}{cz+d}$ with $(a-d)^2 + 4bc = 0$
(b) No such bilinear transformation exists. **9.** **(a)** $w = (az+1)/(z+a)$ with $a \neq \pm 1$
(b) $w = (az-1)/(z+a)$ with $a \neq \pm i$ **11.** $w_1 = 2z$, $w_2 = w_1 + 1$, $w_3 = 1/w_2$, $w_4 = -\frac{7}{2}w_3$, $w = \frac{5}{2} + w_4$
15. The exterior of the circle $|w - \frac{1}{2}| = \frac{1}{2}$ in the first quadrant **17.** $\left\{w \,\middle|\, \left|w + \dfrac{i}{2}\right| > \dfrac{1}{2}\right\} \cap \left\{\operatorname{Im} w > -1\right\}$

Exercises 13.5, p. 507

1. **(a)** The half-plane $v > u$ **(b)** The right half-plane $u > 0$ **(c)** The lower half-plane $v < 0$
5. $w = i\dfrac{z(i-1)+2}{z(1-i)+2i}$; $|w| < 1$ **7.** If $z = x$, then $|w| = \left|\dfrac{x-\lambda}{x-\bar{\lambda}}\right| = 1$; $w(\lambda) = 0$
9. $w = \left(\dfrac{z-i}{z+i}\right)^4$ **11.** **(a)** The upper half-plane **(b)** The lower half-plane
13. $\phi(x, y) = \operatorname{Im} w = \operatorname{Im} \lambda \dfrac{1 - x^2 - y^2}{1 + x^2 + y^2 - 2x\cos\alpha - 2y\sin\alpha}$ **15.** $\operatorname{Im} \dfrac{\bar{\lambda}(x-\gamma) - \lambda e^{i\alpha}(x - \bar{\gamma})}{(x-\gamma) - e^{i\alpha}(x-\bar{\gamma})} = 0$; $w(\gamma) = \lambda$

Exercises 13.6, p. 512

1. $x^2 - y^2 = c$, $xy = d/2$
11. The entire unit disk, and hence the image points A', B', C', D', E', are all rotated counterclockwise by α.
13. The image is the exterior of the unit circle.

Exercises 13.7, p. 517

1. $V(x, y) = \dfrac{100}{\pi}\operatorname{Arg}\left(\dfrac{z-a}{z+a}\right) = \dfrac{100}{\pi}\operatorname{Cos}^{-1}\dfrac{x^2 + y^2 - a^2}{\sqrt{(x^2+y^2-a^2)^2 + 4a^2y^2}}$

3. $T(x, y) = 100\left(1 - \dfrac{2}{\pi}\operatorname{Arg} z\right) = 100\left(1 - \dfrac{2}{\pi}\operatorname{Tan}^{-1}\dfrac{y}{x}\right)$

5. $T(x, y) = \dfrac{100}{\pi}\operatorname{Arg}\left(-i\dfrac{z+1}{z-1}\right) = \dfrac{100}{\pi}\operatorname{Cos}^{-1}\dfrac{-2y}{\sqrt{(1-x^2-y^2)^2 + 4y^2}}$

7. $T(x, y) = \dfrac{100}{\pi}\operatorname{Arg}\dfrac{1+z_1}{1-z_1} = \dfrac{100}{\pi}\operatorname{Cos}^{-1}\dfrac{1 - x_1^2 - y_1^2}{\sqrt{(1-x_1^2-y_1^2)^2 + 4y_1^2}}$ where $x_1 = \frac{1}{2}x\left(1 + \dfrac{1}{x^2+y^2}\right)$,
$y_1 = \frac{1}{2}y\left(1 - \dfrac{1}{x^2+y^2}\right)$

9. **(a)** $\psi(x, y) = U_0(x^2+y^2)^{\beta/2}\sin\left(\beta\operatorname{Cos}^{-1}\dfrac{x}{\sqrt{x^2+y^2}}\right)$

(b) $$\psi(x, y) = -U_0(x^2+y^2)^{\beta/(4\beta-2)} \begin{cases} \sin\left[\dfrac{\beta}{2\beta-1}\operatorname{Cos}^{-1}\dfrac{x}{\sqrt{x^2+y^2}}\right] & y < 0 \\ \sin\left[\dfrac{\beta}{2\beta-1}\left(\operatorname{Cos}^{-1}\dfrac{x}{\sqrt{x^2+y^2}} - \dfrac{\pi}{\beta}\right)\right] & y > 0 \end{cases}$$

11. $\psi(x, y) = U_0 \operatorname{Im} a\left(\dfrac{z}{a} + \dfrac{a}{z}\right) = U_0 y\left(1 - \dfrac{a^2}{x^2+y^2}\right)$

Chapter 14

Exercises 14.1–14.2, p. 527

1. **(a)** $S_n = a(z_1 - a) + z_1(z_2 - z_1) + \cdots + z_{n-1}(b - z_{n-1})$
(b) $S_n^* = z_1(z_1 - a) + z_2(z_2 - z_1) + \cdots + b(b - z_{n-1})$ **(c)** $\frac{1}{2}(S_n + S_n^*) = \frac{1}{2}(b^2 - a^2)$

3. **(a)** $6 + 5i$ **(b)** $\sqrt{61}$ **5.** $\frac{1}{2}\pi e$ **7.** $\dfrac{\pi}{4} + \text{Tan}^{-1}\dfrac{1}{2} + \dfrac{1}{2}\log_e 50$ **9.** **(a)** $49\pi i$ **(b)** $-2\pi i$

11. **(a)** $1 + \frac{1}{3}i$ **(b)** 1 **13.** **(a)** 2 **(b)** $\pi^2 e^{2\pi}$ **15.** **(a)** $(1/\pi)(\cosh 3\pi - 1)$ **(b)** $(2/\pi)\cosh \pi$

17. 0 **19.** $2\pi i$ **25.** 15 **27.** $\dfrac{2}{3}$ **29.** $\sqrt{949}$ **31.** $\sqrt{10}[\sqrt{5} + \frac{1}{2}\log_e(2 + \sqrt{5})]$

Exercises 14.3, p. 533

1. $1 - \cos 1 \cosh 1 + i \sin 1 \sinh 1$ **3.** $\frac{2}{3}$ **5.** πi **7.** 0 **9.** $\frac{2}{3}(1 + i)$ **11.** $\frac{2}{3}(-1 + i)$
13. **(a)** $\frac{2}{3}\sinh 6\pi(\sin 3 - i\cos 3)$ **(b)** $\frac{63}{4} - 6i$ **(c)** $\frac{1}{2}\tan^{-1}(i/2) = (i/4)\ln 3$ **15.** 0
17. $-2\pi i$ **19.** **(a)** $f(z) = -aiz^2, \Gamma(C) = 0$ **(b)** $f(z) = z, \Gamma(C) = 0$
21. $f(z) = 1 - a^2/z^2; \Gamma(C) = 0$

Exercises 14.4–14.5, p. 540

3. **(a)** 0 **(b)** $-4\pi i$ **(c)** 0 **(d)** $-4\pi i$ **5.** **(a)** $4\pi i$ **(b)** $2\pi i$ **(c)** 0 **(d)** 0
7. 0 **9.** 0 **11.** **(a)** Everywhere in the finite z-plane.
(b) The right half-plane $x > 0$; the left half-plane $x < 0$. **(c)** Such domains do not exist.
(d) Any simply connected domain not containing the origin $z = 0$. **13.** (a), (b), (e), (f)

Exercises 14.6–14.7, p. 546

1. $2\pi i$ **3.** **(a)** 0 **(b)** $2\pi i$ **(c)** $\pi(-\sin\frac{1}{2} + i\cos\frac{1}{2})$ **(d)** 0 **(e)** $2\pi i$
5. **(a)** 0 **(b)** -2π **(c)** 2π **(d)** -2π **(e)** 0 **7.** **(a)** 0 **(b)** $-\pi i$
(c) $14\pi i$ **(d)** $\frac{1}{3}\pi i$ **(e)** $-\frac{1}{3}\pi i$ **9.** **(a)** $14\pi i$ **(b)** $2\pi(-6 + 7i)$ **(c)** 0
11. **(a)** $\pi/8$ **(b)** 0
13. By the principle of path deformation, C may be chosen to be the circle $|z - z_0| = 1$. Then use the parametrization $z = z_0 + e^{it}, 0 \le t \le 2\pi$ **21.** $10\pi i$
23. Consider the difference between two harmonic functions with the same boundary values.

Chapter 15

Exercises 15.1–15.2, p. 554

1. Diverges **3.** Converges to 0 **5.** Diverges **7.** Converges to $\frac{2}{3}$ **9.** Converges to $\frac{2}{3}$
11. Converges **13.** Diverges **15.** Converges **17.** Converges **19.** Converges

23. **(a)** $S_n = 2 - 2^{-n}$ **(b)** $S_n = \dfrac{a - a^{-n}}{a - 1} < \dfrac{a}{a - 1}$ **25.** No **27.** Yes **29.** Yes **31.** No

Exercises 15.3–15.4, p. 559

1. $\text{Im}\, z = 0$ **3.** $|z + \frac{15}{8}| > \frac{5}{8}$ **5.** **(a)** $|z| < 1$ **(b)** $|z| < \frac{1}{3}$ **7.** $|z - a| < 1$ **9.** $|z| < 2$

11. No, since $|-3 + i| > |2 - 2i|$ **13.** Yes, since $\left|-3 + i - \dfrac{i}{3}\right| = \left|2 - 2i - \dfrac{i}{3}\right|$ **15.** $w(z) = e^z$

17. $|z| < 1$; $f'(z) = \sum_{n=1}^{\infty} (-1)^n 2nz^{2n-1}$; $F(z) = \sum_{n=0}^{\infty} (-1)^n \dfrac{z^{2n+1}}{2n + 1} + c$

19. $|z| < 3; f'(z) = 3 \sum_{n=1}^{\infty} n^4 \frac{z^{3n-1}}{27^n}; F(z) = \sum_{n=1}^{\infty} \frac{n^3 z^{3n+1}}{(3n+1)27^n} + c$

21. $z = 1; f'(z) = \sum_{n=1}^{\infty} n^{n+1}(z-1)^{n-1}; F(z) = \sum_{n=1}^{\infty} \frac{n^n}{n+1}(z-1)^{n+1} + c$

23. **(b)** $|z-1| < 2;$ $\left(\text{Recall } \Gamma(x)\Gamma(1-x) = \frac{\pi}{\sin \pi x}.\right)$

Exercises 15.5, p. 566

1. $\sum_{n=0}^{\infty} (-1)^n \frac{z^{2n}}{(2n)!}$; all finite z **3.** $\sum_{n=0}^{\infty} (-1)^n \frac{z^n}{n!}$; all finite z **5.** $\sum_{n=0}^{\infty} \frac{(z-i)^n}{(1-i)^{n+1}}; |z-i| < \sqrt{2}$

7. $\sum_{n=0}^{\infty} (-1)^n (n+1)(z-1)^n; |z-1| < 1$ **9.** $\sum_{n=0}^{\infty} (-1)^n \frac{z^{4n+2}}{(2n+1)!}$; all finite z

11. $(z-1)^2 + 2(z-1) + 1$; all finite z **13.** $(z+1) - 2$; all finite z

15. $(z-1)^2 - 6(z-1) - 5$; all finite z **17.** $-\frac{1}{4}\sum_{n=0}^{\infty}\left[1 + \frac{(-1)^n}{3^{n+1}}\right](z-1)^n; |z-1| < 1$

19. $\frac{1}{6}\sum_{n=0}^{\infty} (-1)^n (n+3)(n+2)(n+1)z^n; |z| < 1$ **21.** $-\sum_{n=0}^{\infty} z^{4n}; |z| < 1$ **23.** $\sum_{n=0}^{\infty} (-1)^n \frac{z^{4n+1}}{4n+1}; |z| < 1$

25. $e(1 - z + \frac{3}{2}z^2 - \frac{13}{6}z^3 + \cdots)$ **27.** $1 + z + \frac{1}{2}z^2 - \frac{1}{8}z^4 + \cdots$ **29.** $-z - \frac{3}{2}z^2 - \frac{4}{3}z^3 - z^4 + \cdots$

31. $\sum_{n=0}^{\infty} (-1)^n \frac{(z-1)^{n+1}}{n+1}; |z-1| < 1$ **33.** $-\sum_{n=1}^{\infty} (-1)^n \frac{z^n}{nn!}$; all finite z

35. $\sum_{n=0}^{\infty} (-1)^n \frac{z^{4n+3}}{(4n+3)(2n+1)!}$; all finite z

Exercises 15.6, p. 572

1. $\sum_{n=0}^{\infty} \frac{z^{3-n}}{n!}; |z| > 0$ **3.** $-\sum_{n=0}^{\infty} z^{4n}; |z| < 1$ $\sum_{n=0}^{\infty} z^{-4n-4}; |z| > 1$

5. $\frac{1}{2}\sum_{n=0}^{\infty} (-1)^n (n+2)(n+1)(z-1)^n; |z-1| < 1$ $\frac{1}{2}\sum_{n=0}^{\infty} (-1)^n (n+2)(n+1)(z-1)^{-n-3}; |z-1| > 1$

7. $\sum_{n=0}^{\infty} (-1)^n z^{2n-2}; 0 < |z| < 1$ $\sum_{n=0}^{\infty} (-1)^n z^{-2n-4}; |z| > 1$

9. $-\sum_{n=0}^{\infty}\left(\frac{z+1}{2}\right)^n + \frac{1}{z+1}; 0 < |z+1| < 2$ $\sum_{n=0}^{\infty}\left(\frac{2}{z+1}\right)^{n+1} + \frac{1}{z+1}; |z+1| > 2$

11. $\frac{1}{8}\sum_{n=0}^{\infty} \frac{(-1)^n (2z)^{3-2n}}{(2n)!}; |z| > 0$

13. $-2\sum_{n=0}^{\infty} (z-1)^n - \frac{1}{z-1}; 0 < |z-1| < 1$ $2\sum_{n=0}^{\infty} (z-1)^{-n-1} - \frac{1}{z-1}; |z-1| > 1$

15. $\frac{1}{z} + \frac{z}{6} + \frac{7z^3}{360}$ **17.** $\frac{1}{z^2} + \frac{1}{z} - \frac{1}{2}$ **19.** **(a)** $-\sum_{n=0}^{\infty} z^{2n}; |z| < 1$ $\sum_{n=0}^{\infty} z^{-2n-2}; |z| > 1$

(b) $\frac{1}{4}\sum_{n=0}^{\infty} (-1)^n \left(\frac{z-1}{2}\right)^{n-1}; 0 < |z-1| < 2$ $\frac{1}{4}\sum_{n=0}^{\infty} (-1)^n \left(\frac{2}{z-1}\right)^{n+2}; |z-1| > 2$

(c) $-\frac{1}{4}\sum_{n=0}^{\infty}\left(\frac{z+1}{2}\right)^{n-1}; 0 < |z+1| < 2$ $\frac{1}{4}\sum_{n=0}^{\infty}\left(\frac{2}{z+1}\right)^{n+2}; |z+1| > 2$

21. $\frac{i}{z} + \frac{1}{i+1}\sum_{n=0}^{\infty} [i^{-n} + (-1)^n] z^n; 0 < |z| < 1$ $\frac{i}{z} + \frac{1}{i+1}\sum_{n=0}^{\infty} [-i^{n+1} + (-1)^n] z^{-n-1}; |z| > 1$

23. $\sum_{n=0}^{\infty}(-1)^n\left[\frac{2^{-n-1}}{1+i}+i-\frac{i(1-i)^{-n-1}}{i+1}\right](z-1)^n;\ |z-1|<1$

$\sum_{n=0}^{\infty}(-1)^n\left\{\left[\frac{2^{-n-1}}{1+i}-\frac{i(1-i)^{-n-1}}{i+1}\right](z-1)^n+i(z-1)^{-n-1}\right\};\ 1<|z-1|<\sqrt{2}$

$\sum_{n=0}^{\infty}(-1)^n\left\{\frac{2^{-n-1}}{1+i}(z-1)^n+\left[i-\frac{i(1-i)^n}{1+i}\right](z-1)^{-n-1}\right\};\ \sqrt{2}<|z-1|<2$

$\sum_{n=0}^{\infty}(-1)^n\left[i-\frac{i(1-i)^n}{1+i}+\frac{2^n}{1+i}\right](z-1)^{-n-1};\ |z-1|>2$

25. **(a)** $|z|>1$ **27.** Note that $\ln\left(1+\frac{1}{z}\right)\equiv\int_0^{1/z}\frac{d\zeta}{1+\zeta}$ **29.** Use antidifferentiation

Exercises 15.7, p. 575

1. Simple zeros at ± 1 and $\pm i$ **3.** Simple zeros at $\pm n\pi$
5. Double zeros at $\pm n\pi$, double poles at $\pm(2n+1)\pi/2$ **7.** Simple zeros at $\pm 2n\pi i$
9. Simple zero at 1, simple poles at $\pm i$ **11.** Simple zeros at $\pm i$ **13.** Essential singularity at 0
15. Simple zeros at $\pm i$, simple pole at 0 **17.** Removable singularity at 0, simple zeros at $\pm n\pi$ $(n \neq 0)$
19. Simple pole at 1, double pole at 2, triple pole at -1; simple zeros at 0, -2, and three other points
21. Simple zero at -1, simple pole at 1, double pole at 0
23. Simple zeros at $\pm(2n+1)\pi/2$, simple poles at $\pm n\pi$
25. Simple zeros at $\pm i(2n+1)\pi/2$, simple poles at $(\pm 4n-1)\pi/4$

Exercises 15.8, p. 579

1. e at $z=1$ **3.** -1 at $z=(2n+1)\pi/2$ **5.** $(-1)^{n+1}$ at $z=(2n+1)\pi/2$
7. $(-1\mp i)\sqrt{2}/8$ at $z=e^{\pm\pi i/4}$; $(1\mp i)\sqrt{2}/8$ at $z=e^{\pm 3\pi i/4}$ **9.** $\frac{3}{2}$ at $z=0$, $-\frac{5}{2}$ at $z=-2$, 1 at $z=-1$
11. $-\frac{1}{2}$ at $z=0$ **13.** 0 at $z=1$, 0 at $z=-1$ **15.** -2 at $z=0$, $\frac{11}{2}e^{-1}$ at $z=-1$
17. $\pi i/2$ **19.** 0 **21.** 0 **23.** $\sqrt{e}\pi i/2$ **25.** $-2\pi i$ **27.** 0
29. Branch points at $z=\pm\frac{1}{2}e^{i\alpha}$, no other singularities; $\Gamma(C)=0$
31. Residue at $z=-\frac{5}{6}$ is $Ai/(2\pi)$; $\Gamma(C)=-A$

Exercises 15.9, p. 582

1. $-2\pi i$ **3.** $\pi i(1-e^{-2})$ **5.** $2\pi i$ **7.** $4\pi i$ **9.** $-4\pi i$ **11.** $-\pi i$ **13.** $-20\pi i/(3e)$
15. 0 **17.** **(a)** $8\pi i$ **(b)** πi **(c)** πi **(d)** Does not exist **19.** **(a)** $-8\pi i$
(b) $208\pi i$ **(c)** $208\pi i$ **(d)** $208\pi i$ **(e)** $216\pi i$ **21.** 0 **23.** 0
25. Rewrite $f(z)$ as $(z-z_0)g(z)$ where $g(z_0)\neq 0$.

Chapter 16

Exercises 16.1–16.2, p. 588

1. $\pi/\sqrt{2}$ **3.** $2\pi/\sqrt{1-a^2}$ **5.** $\pi/4$ **7.** $\sqrt{2}\pi/4$ **9.** $\pi/\sqrt{a^2+a}$ **11.** $2\pi/(1-a^2)$
13. $2\pi a/(a^2-1)^{3/2}$ **15.** πi

Exercises 16.3–16.4, p. 594

1. $\pi/\sqrt{2}$ **3.** 0 **5.** $\pi/2$ **7.** $(\pi/2)[(\sqrt{13}-3)/2]^{1/2}$ **9.** $3\pi\sqrt{2}/16$ **11.** $\pi/54$
13. $\pi/[2\cosh(\pi m/2)]$ **15.** Change x into e^x. **19.** $\pi/(a\sinh \pi a)$

Exercises 16.5–16.6, p. 602

1. 0 **3.** $\frac{1}{4}\pi e^{-3}(\cos 3 + \sin 3)$ **5.** $\frac{1}{2}\pi e^{-\omega a}$ **7.** $\frac{1}{2}\pi e^{-2}\sin 2$ **9.** 0 **11.** $\frac{1}{2}\pi(1 - e^{-2\omega})$

13. $F(\lambda) = \frac{1}{2}\pi e^{-|\lambda|}(\cos\lambda - \sin|\lambda|)$ $u(x,t) = \frac{1}{2}\int_0^\infty e^{-\lambda}(\cos\lambda - \sin\lambda)\cos\lambda x\, e^{-\lambda^2 t/k}\,d\lambda$

15. $\lambda\sqrt{LC}\,B(\lambda) = \dfrac{\pi}{4}e^{-|\lambda|}(\cos\lambda + \sin|\lambda|),\ A(\lambda) = 0$ $v(x,t) = \dfrac{1}{4}e^{-Rt/L}\displaystyle\int_0^\infty e^{-\lambda}(\sin\lambda + \cos\lambda)\cos\lambda x\frac{\sin\lambda\sqrt{LC}t}{\lambda\sqrt{LC}}\,d\lambda$

17. $v(x,t) = \dfrac{1}{2\pi}\displaystyle\int_{-\infty}^{\infty} A(\lambda)\cos\lambda x\, e^{i\lambda\sqrt{LC}t}\,d\lambda + \dfrac{1}{2\pi}\int_{-\infty}^{\infty} B(\lambda)\sin\lambda x\, e^{i\lambda\sqrt{LC}t}\,d\lambda$ with

$A(\lambda) = \displaystyle\int_{-\infty}^{\infty} f(t)e^{-i\lambda\sqrt{LC}t}\,dt$ and $\lambda B(\lambda) = \displaystyle\int_{-\infty}^{\infty} g(t)e^{-i\lambda\sqrt{LC}t}\,dt$

19. $C(\lambda) = 0,\ \lambda D(\lambda) = \dfrac{\pi}{8\sqrt{2}}e^{-\lambda c/(2\sqrt{2})}\left(\cos\dfrac{\lambda c}{2\sqrt{2}} - \sin\dfrac{\lambda c}{2\sqrt{2}}\right)$ $(\lambda \geq 0)$

$$u(\rho,t) = \frac{1}{8\rho\sqrt{2}}\int_0^\infty e^{-\lambda c/(2\sqrt{2})}\left(\cos\frac{\lambda c}{2\sqrt{2}} - \sin\frac{\lambda c}{2\sqrt{2}}\right)\cos\lambda ct\frac{\sin\lambda(\rho - \rho_0)}{\lambda}\,d\lambda$$

Exercises 16.7, p. 608

1. $\cos bt$ **3.** $\frac{1}{9}(1 - \cos 3t)$ **5.** $e^{-t}(\frac{1}{4} + \frac{1}{2}t) - \frac{1}{4}e^{-3t}$ **7.** $\frac{1}{2}(t-4)^2 u_4 - \frac{1}{2}(t-1)^2 u_1$

9. $(e^{2(t-1)} - e^{(t-1)})u_1$ **15.** $u(x,t) = 3x$ **17.** $u(x,t) = u_0\left[1 - \operatorname{erf}\left(\dfrac{x}{2\sqrt{t}}\right)\right]$

19. $u(x,t) = \dfrac{5}{c\pi}\sin\pi x\sin\pi ct$

Subject Index

Note: The letter "e" after a page number indicates that the indexed item will be found in an exercise set; the letter "n" after a number refers to a footnote.

Symbol Index